D0085078

THE DNA STORY

THE DNA STORY
A DOCUMENTARY HISTORY OF GENE CLONING

JAMES D. WATSON
COLD SPRING HARBOR LABORATORY

JOHN TOOZE
EUROPEAN MOLECULAR BIOLOGY ORGANIZATION

W. H. FREEMAN AND COMPANY
SAN FRANCISCO

We dedicate this book to Francis Crick.

ACKNOWLEDGEMENTS

We thank all those who have given us permission to reproduce their articles, commentaries, and correspondence, and the many among them who have encouraged us during the preparation of this book. We are particularly grateful to Waclaw Szybalski and Norton Zinder for many suggestions concerning the selection of documents, and for giving us access to the ones in their files. It is a pleasure to thank Susan Gensel and the staff of the library of Cold Spring Harbor Laboratory for their hospitality, help, patience, and fortitude. Karen Herrmann helped in many ways with the preparation of the Scientific Background; we also thank John Fiddes, Doug Hanahan, and Dave Kurtz for their criticism of that section, and Tom Broker for providing micrographs.

The design of the cover and interior of the book was created by Fred Weiss, who spent many hours patiently educating us in the art of layout and design by transforming a collection of ill-matched printed matter into a consistent and attractive form. We are indebted to him for this and for guiding it through to a bound book. We have been fortunate to have the confidence of our long-time friend Neil Patterson, who, as President of W. H. Freeman and Company, now makes it possible for the DNA story to be visualized in a form that begins, we hope, to approximate reality.

The person to whom we owe our deepest gratitude is Sharon Queally, who has acted as our assistant at all stages and in so many ways. More than any of us she has lived this book, being throughout so cheerfully conscientious and dependable as to command our sincerest admiration.

Library of Congress Cataloging in Publication Data

Watson, James D., 1928-
The DNA story

Bibliography: p.
Includes index.
1. Recombinant DNA—Research—History—Addresses, essays, lectures. 2. Cloning—Research—History—Addresses, essays, lectures. I. Tooze, John. II. Title.
QH442.W37 574.87'3282 81-3299
ISBN 0-7167-1292-X AACR2

Printed in the United States of America

9 8 7 6 5 4 3 2 1

Contents

F. CRICK

Prologue

It is only the rare scientific revolution that immediately excites the public. The elucidation of the double-helical structure of DNA in 1953, for example, created almost no stir outside the then small band of scientists who were waiting for the discovery to be made. To be sure, within those narrow circles there was intense excitement, for we could now tackle genetics at the molecular level, and there would soon emerge many key insights about gene structure and function. But we did not foresee any immediate practical consequences for the world about us and no reason why the man on the street, as opposed to the young student, should know that we existed, much less try to understand the increasingly complicated facts about genes that were to be worked out over the next twenty years.

Certainly each new powerful discovery in molecular genetics has brought its attendant intellectual satisfaction. But this was not everyone's game, and the fact that the genetic code is based on a four-nucleotide (letter) alphabet had to seem to most lawyers as irrelevant to their well-being as the meaning of torts to most molecular biologists. As all busy people know, we should not waste time with abstract concepts that we shall never in our daily lives use.

In 1973, however, Herbert Boyer, Stanley Cohen, and their collaborators devised simple test-tube procedures to produce "recombinant DNA molecules" derived from two different parental DNA molecules (for example, from mouse DNA and *E. coli* bacterial DNA) and the picture changed dramatically. For with the development of recombinant DNA, genetic engineering—at least for microorganisms—was now a very practical proposition. Although previously there had been numerous speculative scenarios for altering our genetic compositions, not one had a ring of plausibility, and for all practical purposes they were indistinguishable from science fiction, a genre upon which neither of us had ever wasted any time. But with the new recombinant DNA tricks the genetic engineering of microorganisms, and later of higher plants and animals, would help to shape the world of the future. Without doubt molecular geneticists now had the power to alter life on a scale never before thought possible by serious scientists.

***FRANCIS CRICK**, in 1956. This photograph was taken shortly after the elucidation of the double helix in 1953 by James Watson and Francis Crick. (Photo by Francis DiGennaro.)*

Our first reaction was one of pure joy. Now genetics would not grind to a halt when we exhausted the potential of current experimental techniques. Instead, with recombinant DNA we now might understand, if not rearrange, the chromosomes of the cells in higher plants and animals. Each such cell contains roughly 1000 times more DNA than the simple bacterial cells on which virtually all molecular geneticists had focused after the discovery of the double helix. And until recombinant DNA came along even the smallest of bacteria had much more DNA than we could realistically explore.

Then increasingly during 1973 we began to ask whether in the process of possibly discovering the power of "unlimited good" we might simultaneously be setting the stage for discovering the power of "unlimited bad." Might some of the new genetic combinations that we would create in the test-tube rise up like the genie from Aladdin's lamp and multiply without control, eventually replacing preexisting plants and animals, if not man himself? If we assume that evolution can generate harmful variants, shouldn't we worry that creating novel combinations of DNA might have consequences orders of magnitude worse than such natural disasters as the lethal swine flu epidemic of 1918? But from the start we knew that no one had concrete facts by which to gauge these scenarios of possible doom. So perhaps we best proceed in the fashion of the past 500 years of Western civilization, striking ahead and only pulling back if we find the savages not of normal size but of the King Kong variety, against which we have no chance.

But this was not the mood of the early 1970s when many academics thought that science was already out of control, and the emergence of the first Earth Day (1970) was widely greeted by intellectuals as a watershed marking the moment when we stopped moving ahead willy-nilly without rigorously thinking through the consequences. This feeling was widely shared, and it was by the combined efforts of Nixon's Republican administration and the Democratic Congress that, in 1972, the U.S. Environmental Protection Agency (EPA) came into existence. From then on, Americans were to be protected by Environmental Impact Statements that would characterize every new project potentially having a significant effect on our environment, be it local or national.

Thus for us to move full steam ahead with recombinant DNA without considering deeply what havoc it might wreak obviously seemed socially irresponsible. Then most molecular biologists who, outraged by the Vietnamese war, reflexively rejected virtually anything proposed by the inherently conservative (be it bigger missiles, lower taxes, or no more asinine governmental regulations). So disregarding the unforeseen consequences of crying wolf, by April 1974 we did not think recombinant DNA could be kept under the rug, and we had to ask ourselves seriously whether or not our genie could safely be released into the outer world.

Although some fringe groups (such as Science for the People) thought this was a matter to be debated and decided by all and sundry, it was never the intention of those who might be called the Molecular Biology Establishment to take this issue to the general public to decide. The matter was not only too technical but in a way also too fuzzy for responsibility to be easily shared with outsiders. We did not want our experiments to be blocked by overconfident lawyers, much less by self-appointed bioethicists with no inherent knowledge of, or interest in, our work. Their decisions could only be arbitrary. Given that there were no definite facts on which to base "danger" signals, we might find ourselves at the mercy of Luddites who do not want to take the chance of any form of change. And as we learned the hard way over the next several years, our paucity of data could be used to buttress either the go-ahead signal for recombinant DNA or the decision to put off its use until it was proven safe.

Recombinant DNA as a social issue thus started as a dialogue between scientists. Even though we saw no harm and realized as inevitable that some newspapers would take note of our deliberations, not even the most pessimistic (experienced?) of those initially involved raised the possibility that we would have to bring in the common man once we proposed (in July 1974) temporarily stopping certain experiments (the Moratorium). Yet if we were worried that our laboratory-made bacteria might cause epidemics of cancer, what right did we have not to ask others whether they wanted to be at risk? It would not be sufficient for us as scientists to decide ten months later at the Asilomar Conference Center that we saw no reason to put off most recombinant DNA research so long as we followed the so-called guidelines. In a democratic society, must not a matter of such potential consequences also be thought through by others?

This has in fact happened—and with such a vengeance that it has left many of us temporarily in the state of shock that accompanies not knowing whether it is oneself who is stark raving mad or whether the problem is only that all others have gone out of their minds. Already the recombinant DNA debate has taken its place as a major event in defining the never cozy relationship between the practitioners of science and the populace around them, and by now over ten books have appeared that try to put the matter in perspective. But descriptions of others' moods and motives often reflect more the view of the author than of the living characters they deal with. So we believe it to be still valuable to collect for unhurried public gaze as many of the key documents and speeches as can be encompassed within a book of publishable length.

As we collected together the vast numbers of documents, articles, cartoons, speeches, and letters that might have a claim to inclusion, and then as we began to distill them, it became obvious that we had a severe space problem on our hands. Our publisher was exceptionally generous, but many things had to be left out; the choice was not always easy to make, even though three documents we wished to include are missing because their authors refused us permission to reproduce them. It became clear also that the documents would require two sorts of introduction: one simply to place them into the historical perspective of the recombinant DNA controversy; the other to make intelligible to the layperson some of the more technical documents and, even more importantly, to try to explain the scientific significance of recombinant DNA and its roots in modern biology. To provide readers with a thumbnail sketch of major historical events of the DNA story, we have included a "Recombinant DNA Dateline" following the Prologue.

Writing the short introductions to each chapter of documents in order to guide the reader from one section to another was a relatively simple matter because we decided not to write our own evaluations of the various positions, which we felt should stand on their own intrinsic merits. We personally believe that recombinant DNA research is best left virtually unregulated, but we know that others saw—and may still see—the matter as less clearcut. We have striven, therefore, to present the opinions in as impartial a way as possible.

The scientific introduction, or more appropriately the scientific background, began to assume unenvisaged proportions as we tried to outline simply the techniques of recombining DNA in the test-tube, how cloned genes can be identified, and what we have already learned by using the techniques. As this section developed in size and scope and ultimately became the final portion of the book, we and our publisher came to realize that the essence of what we were saying might best be conveyed to those not interested in scientific details in a pictorial manner.

At this point the help of George Kelvin was sought. In sessions with him we explained our problem. Could the essence of the DNA story be encapsulated in four or five large full-color illustrations to precede the sections containing the documents and the hardcore scientific background for the old or new addict? His artistry fulfilled all our hopes and expectations, even to the extent of symbolizing the commercial promise of recombinant DNA. And so our book became a three-part documentary of gene cloning: a technicolor preview to a black and white drama in two acts, beginning with calls for a worldwide moratorium on recombinant DNA research and culminating almost eight years later in a worldwide boom industry based on DNA.

Recombinant DNA Dateline

- *1871* Discovery of DNA in the sperm of trout from the Rhine River.
- *1943* DNA proved to be a genetic molecule capable of altering the heredity of bacteria.
- *1953* Postulation of a complementary double-helical structure for DNA.
- *1956* Genetic experiments support the hypothesis that the genetic messages of DNA are conveyed by its sequence of base pairs.
- *1958* Proof that DNA replication involves separating the complementary strands of the double helix.
- *1958* Isolation of first enzyme (DNA polymerase I) that makes DNA in a test-tube.
- *1959* Discovery of an enzyme (RNA polymerase) that makes RNA chains on the surface of single-stranded DNA.
- *1960* Discovery of messenger RNA and demonstration that it carries the information that orders amino acids in proteins.
- *1961* Use of a synthetic messenger RNA molecule (poly-U) to work out the first letters of the genetic code.
- *1965* Appreciation that genes conveying antibiotic resistance in bacteria are often carried on small supernumerary chromosomes called plasmids.
- *1966* Establishment of the complete genetic code.
- *1967* Isolation of the enzyme DNA ligase that can join DNA chains together.
- *1970* Isolation of the first enzyme (the restriction enzyme) that cuts DNA molecules at specific sites.
- *1972* Use of the joining enzyme DNA ligase to link together DNA fragments created by restriction enzymes. The first recombinant DNA molecules generated at Stanford University.
- *1973* Foreign DNA fragments inserted into plasmid DNA to create chimeric plasmids. Finding that they can be functionally reinserted into the bacterium *E. coli*. Potential now exists for cloning bacteria of any gene.

Photo by Walther Goebel, courtesy The Rockefeller University Archives.

ca. 1930s **OSWALD AVERY**, *pictured in his lab where DNA was found to be a genetic molecule.*

Photo courtesy Cold Spring Harbor Laboratory Library Archives.

1947 **EDWARD TATUM** *(left) and* **JOSHUA LEDERBERG** *(center), who first sexually crossed mutant bacteria, at the 1947 Cold Spring Harbor Symposium on Nucleic Acids, together with* **SOL SPIEGELMAN** *(right), who later pioneered nucleic-acid hybridization methods.*

Photo courtesy Cold Spring Harbor Laboratory Library Archives.

1953 **SALVADOR LURIA** *(standing) and* **MAX DELBRÜCK** *(seated), in front of the Cold Spring Harbor Laboratory where they had done many of their early experiments on the genetics of bacteria and bacterial viruses.*

Photo by Jill Hershey, courtesy Cold Spring Harbor Laboratory Library Archives.

1955 **AL HERSHEY** *(foreground), who showed that DNA is the genetic component of phages, sailing with* **JIM WATSON** *at Cold Spring Harbor.*

Photo courtesy Cold Spring Harbor Laboratory Library Archives.

ca. 1950s **FRANK STAHL** *(left), who with Matthew Meselson proved that the two strands of a DNA molecule separate when it replicates, photographed during a meeting at Cold Spring Harbor with* **MAX DELBRÜCK.**

Photo courtesy of the Archives, California Institute of Technology.

1958 **MATTHEW MESELSON** *next to the analytical ultracentrifuge used in his classic experiments with Frank Stahl.*

1973 First public concern that recombinant DNA procedures might generate potentially dangerous, novel microorganisms.

1974 Call for a worldwide moratorium on certain classes of recombinant DNA experiments.

1975 Government report in the United Kingdom calls for special laboratory precautions for recombinant DNA research.

1975 International meeting at Asilomar, California, urges adoption of guidelines regulating recombinant DNA experimentation. Call for the development of safe bacteria and plasmids that cannot escape from the laboratory.

1976 Release of the first guidelines by the National Institutes of Health; prohibition of many categories of recombinant DNA experimentation. Rising public concern that the guidelines might not be effective. *The New York Times Magazine* article urges prohibiting the awarding of the Nobel Prize for recombinant DNA research.

1977 Formation of the first genetic engineering company (Genentech), specifically founded to use recombinant DNA methods to make medically important drugs.

1977 Creation of the first recombinant DNA molecules containing mammalian DNA, and the discovery of split genes.

1977 Development of procedures for the rapid sequencing of long sections of DNA molecules.

1978 The Nobel Prize in Medicine is awarded for the discovery and use of restriction enzymes.

1978 Production of the first human hormone somatostatin by using recombinant DNA.

1979 General relaxation of the NIH guidelines allows viral DNAs to be studied by using recombinant DNA procedures.

1980 Construction work begins on the first industrial plant designed to make insulin by using recombinant DNA procedures.

1980 The Nobel Prize in Chemistry is awarded dually for the cloning of the first recombinant DNA molecules and the development of powerful methods for sequencing DNA.

1981 Offer to the general public of stock in the first recombinant DNA company (Genentech). Valuation by Wall Street in excess of 200 million dollars.

Photo courtesy Stanford University.

ca. 1960s ARTHUR KORNBERG, *the master pioneer of the enzymology of DNA replication, photographed soon after he moved his lab to Stanford.*

Photo courtesy Sydney Brenner.

1962 SYDNEY BRENNER, *soon after his incisive experiments with Francis Crick showing that amino acids are coded by groups of three nucleotides.*

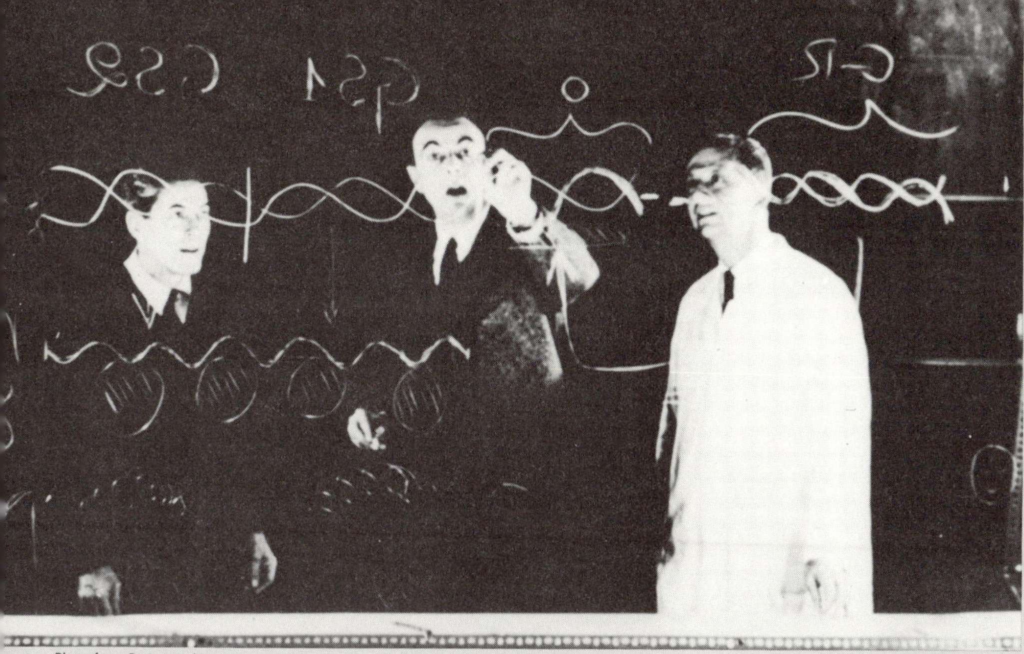

Photo from *Paris-Match*.

1966 ***FRANÇOIS JACOB (left) and JACQUES MONOD (center), the proposers of the operon model of bacterial genes, photographed at the Institute Pasteur with ANDRÉ LWOFF (right) soon after the announcement of their Nobel Prize.***

Photo courtesy Cold Spring Harbor Laboratory Library Archives.

1966 ***GOBIND KHORANA (left), the first to synthesize chemically a gene, photographed at the 1966 Cold Spring Harbor Symposium on the Genetic Code, together with FRANCIS CRICK (center), and MARIANNE GRUNBERG-MANAGO (right), who was the first to isolate an enzyme capable of making ribonucleic-acid molecules.***

Photo by D.N. Leff/NSF-Mosaic.

1980 ***WALTER GILBERT, the isolator with Benno Müller-Hill of the first repressor, standing next to a large DNA molecule in his Harvard University lab.***

Photo by Arthur Foster from The Observer Ltd.

1980 ***FRED SANGER, the first person to determine the amino-acid sequence of a protein, who later devised equally important rapid methods for analyzing the sequences of RNA and DNA, photographed after the announcement of his second Nobel Prize.***

Visualizing DNA

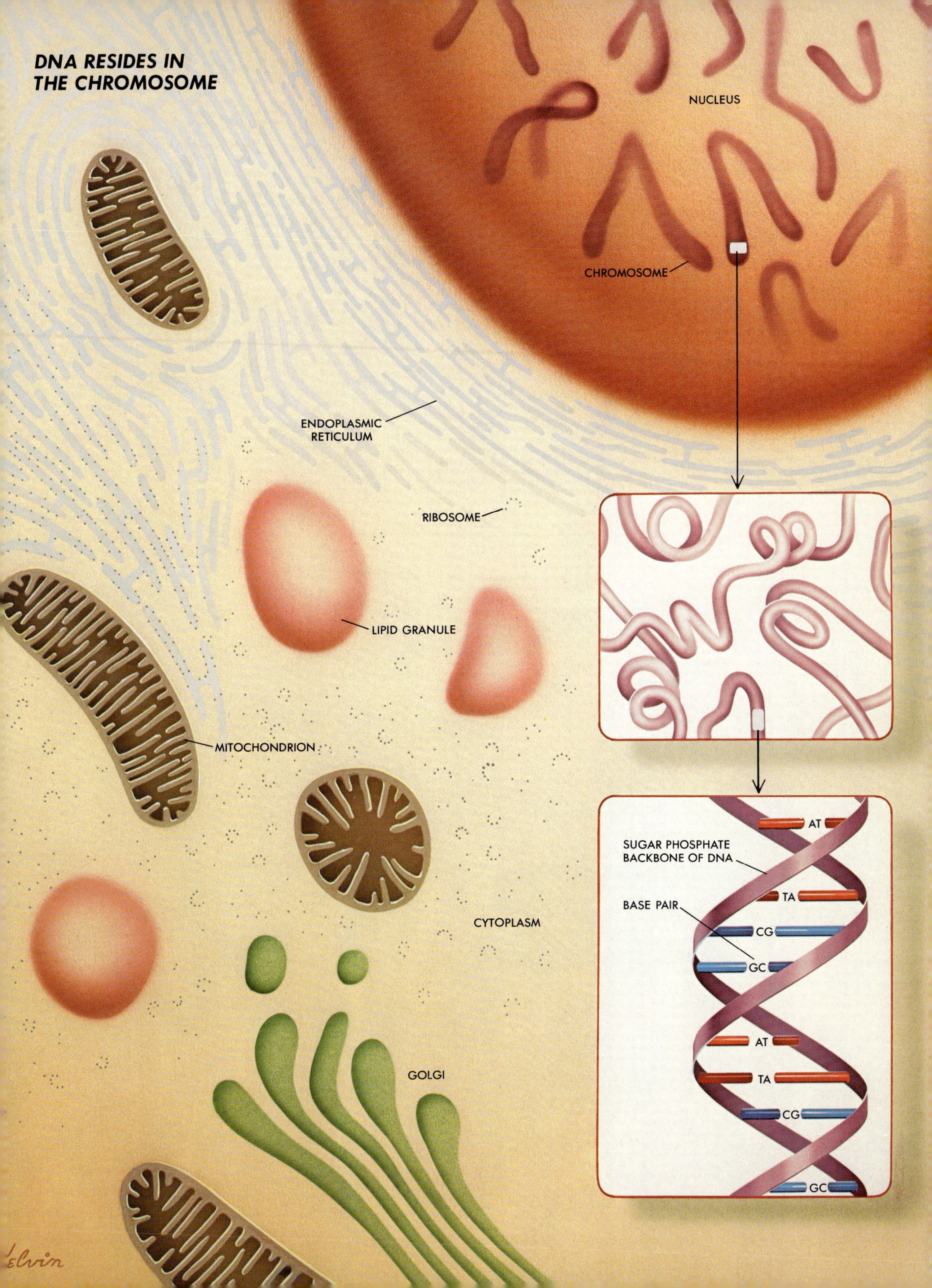
DNA RESIDES IN THE CHROMOSOME
NUCLEUS
CHROMOSOME
ENDOPLASMIC RETICULUM
RIBOSOME
LIPID GRANULE
MITOCHONDRION
CYTOPLASM
GOLGI
SUGAR PHOSPHATE BACKBONE OF DNA
BASE PAIR
AT
TA
CG
GC
AT
TA
CG
GC
Elvin

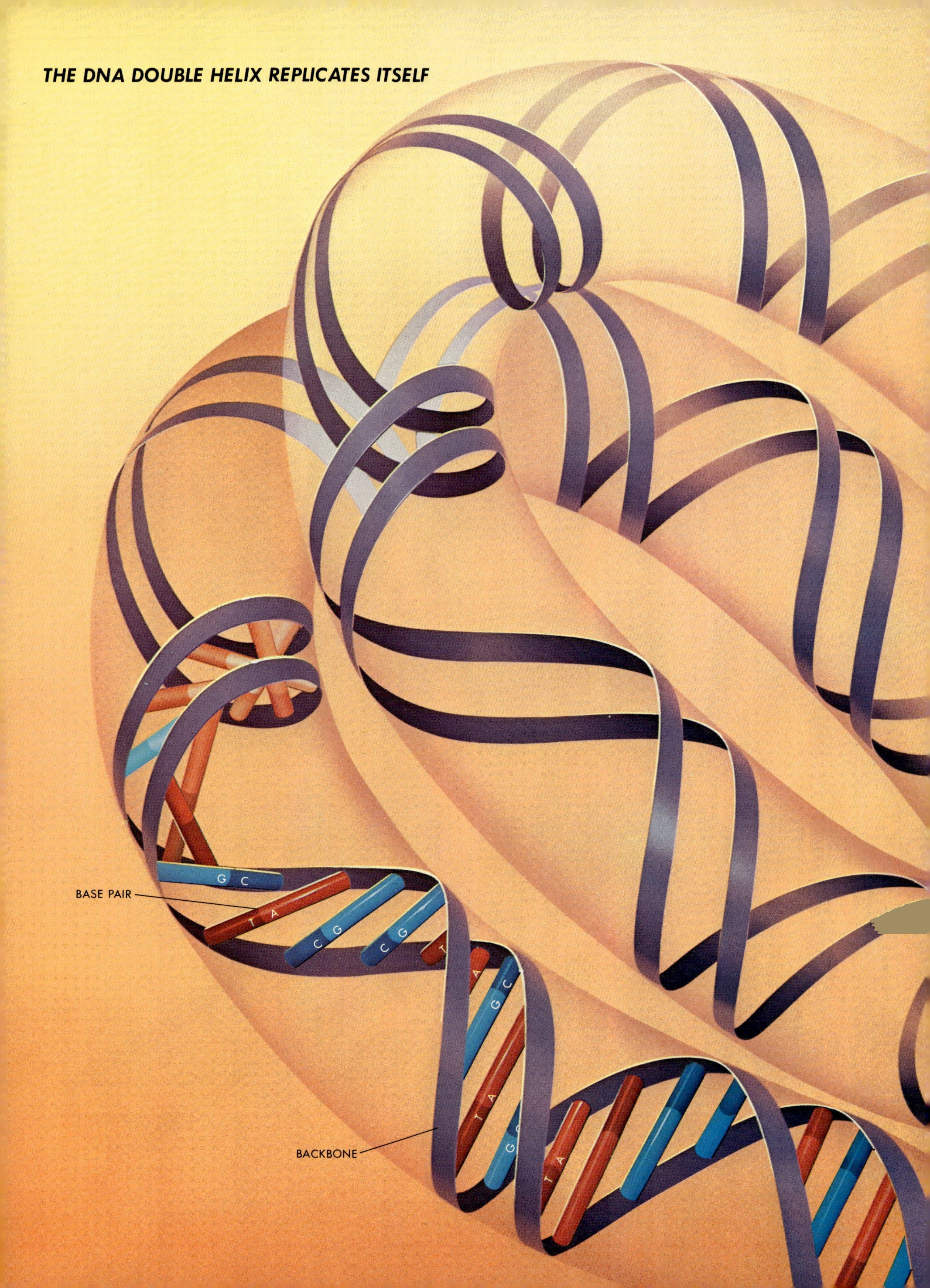
THE DNA DOUBLE HELIX REPLICATES ITSELF
BASE PAIR
BACKBONE

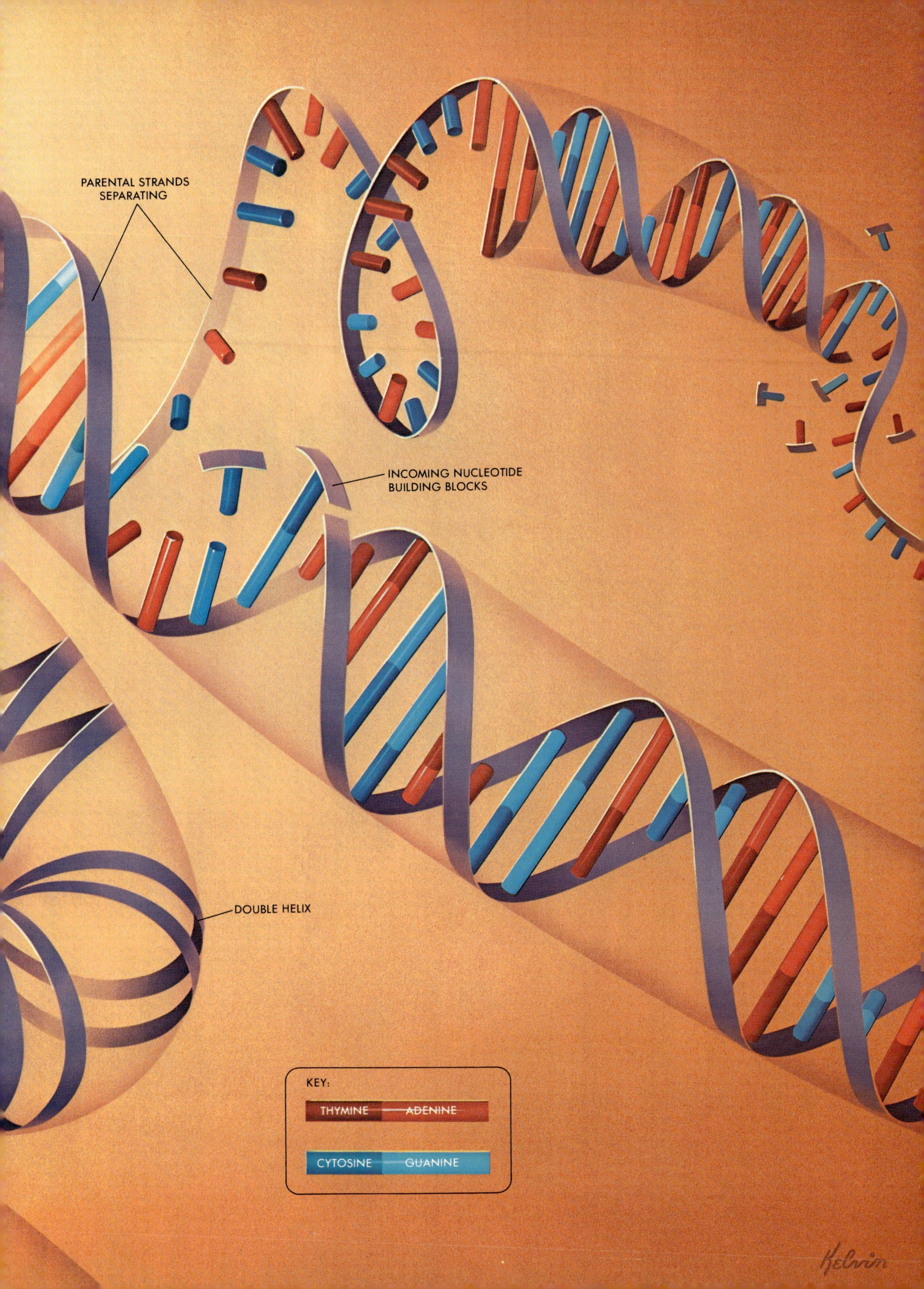
PARENTAL STRANDS SEPARATING
INCOMING NUCLEOTIDE BUILDING BLOCKS
DOUBLE HELIX
KEY:
THYMINE
ADENINE
CYTOSINE
GUANINE
Kelvin

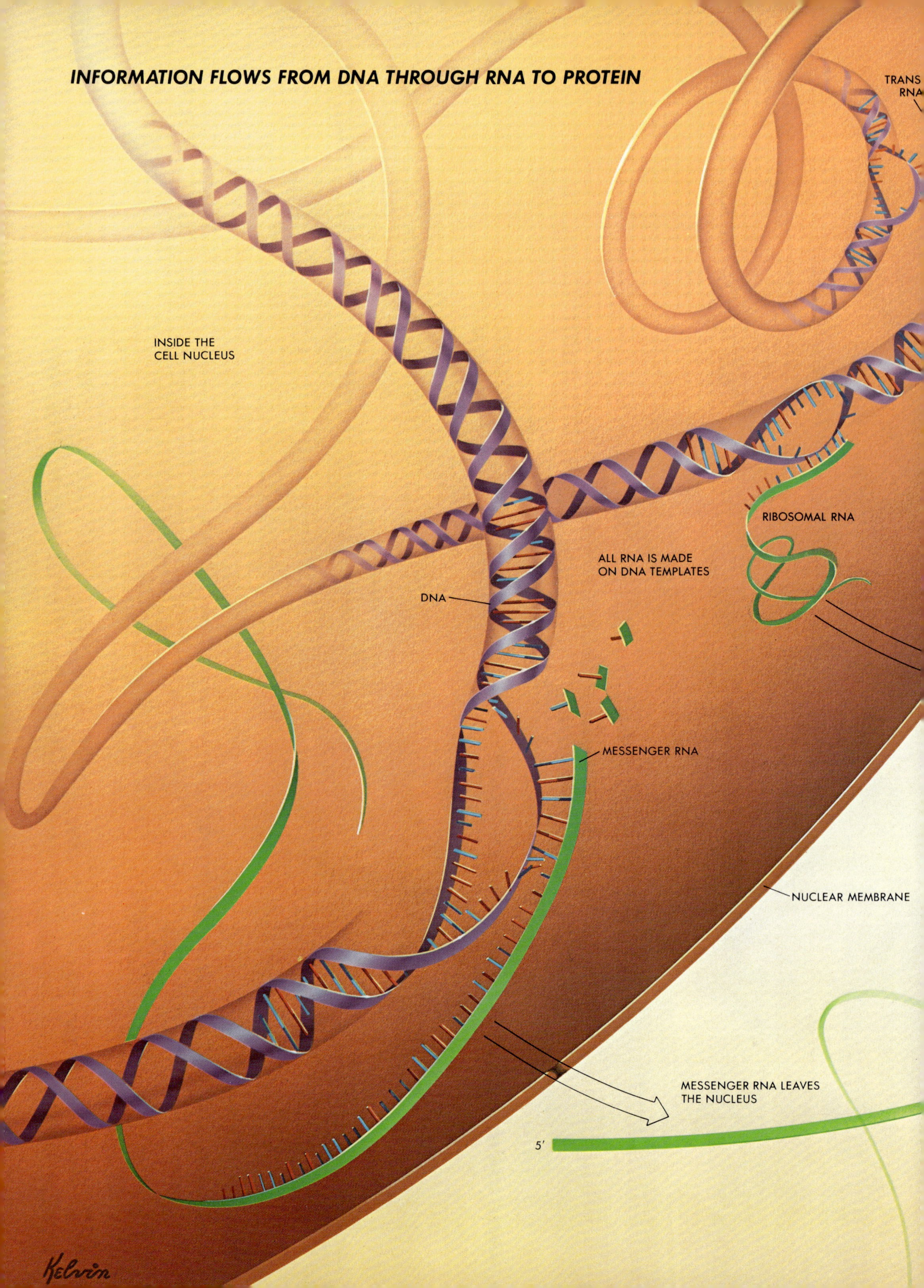

INFORMATION FLOWS FROM DNA THROUGH RNA TO PROTEIN
TRANS
RNA
INSIDE THE
CELL NUCLEUS
RIBOSOMAL RNA
ALL RNA IS MADE
ON DNA TEMPLATES
DNA
MESSENGER RNA
NUCLEAR MEMBRANE
MESSENGER RNA LEAVES
THE NUCLEUS
5′
Kelvin

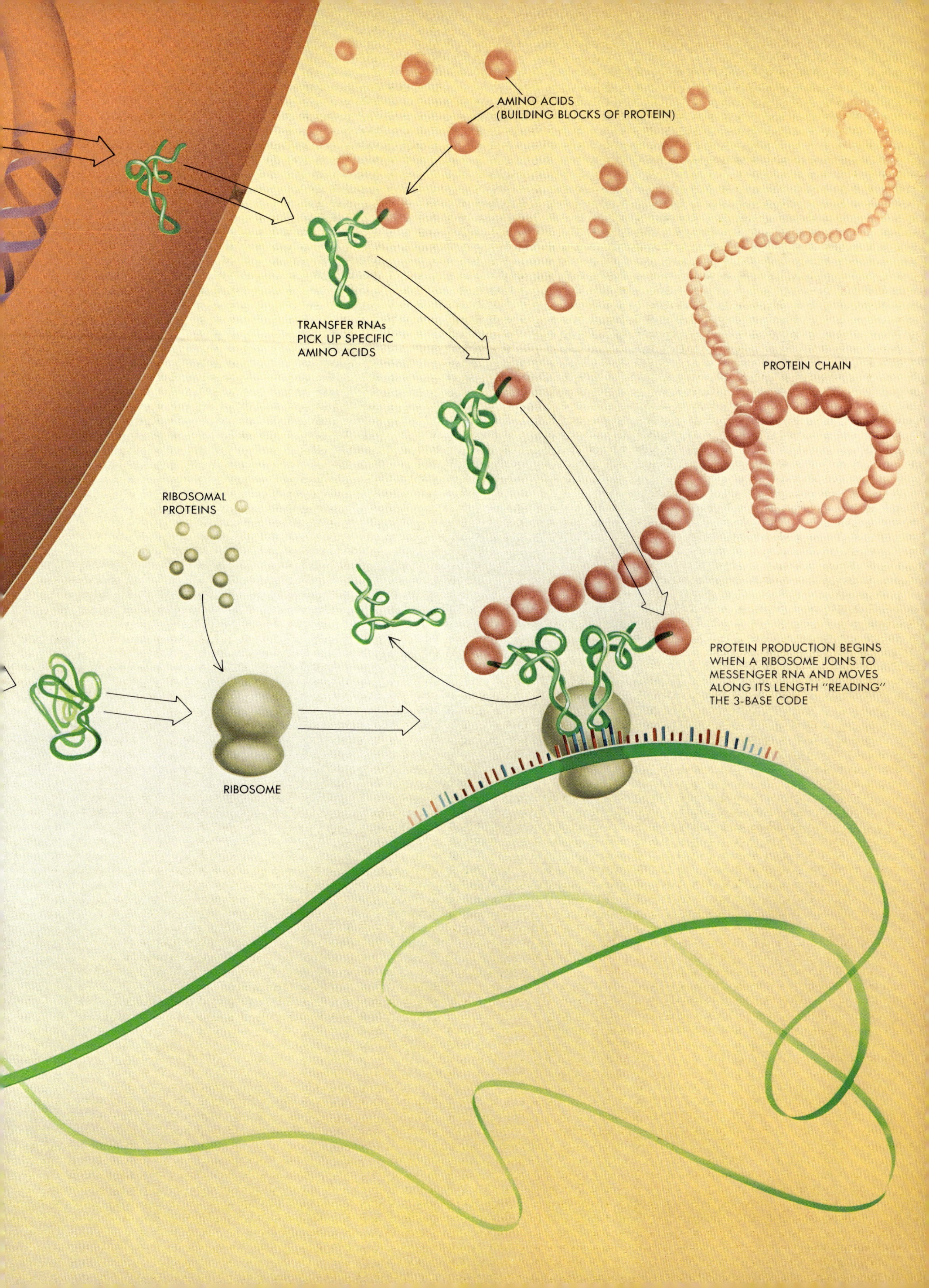

AMINO ACIDS
(BUILDING BLOCKS OF PROTEIN)
TRANSFER RNAs
PICK UP SPECIFIC
AMINO ACIDS
PROTEIN CHAIN
RIBOSOMAL
PROTEINS
PROTEIN PRODUCTION BEGINS
WHEN A RIBOSOME JOINS TO
MESSENGER RNA AND MOVES
ALONG ITS LENGTH "READING"
THE 3-BASE CODE
RIBOSOME

THE CLONING AND EXPRESSION OF AN INTERFERON GENE IN THE BACTERIUM E. COLI

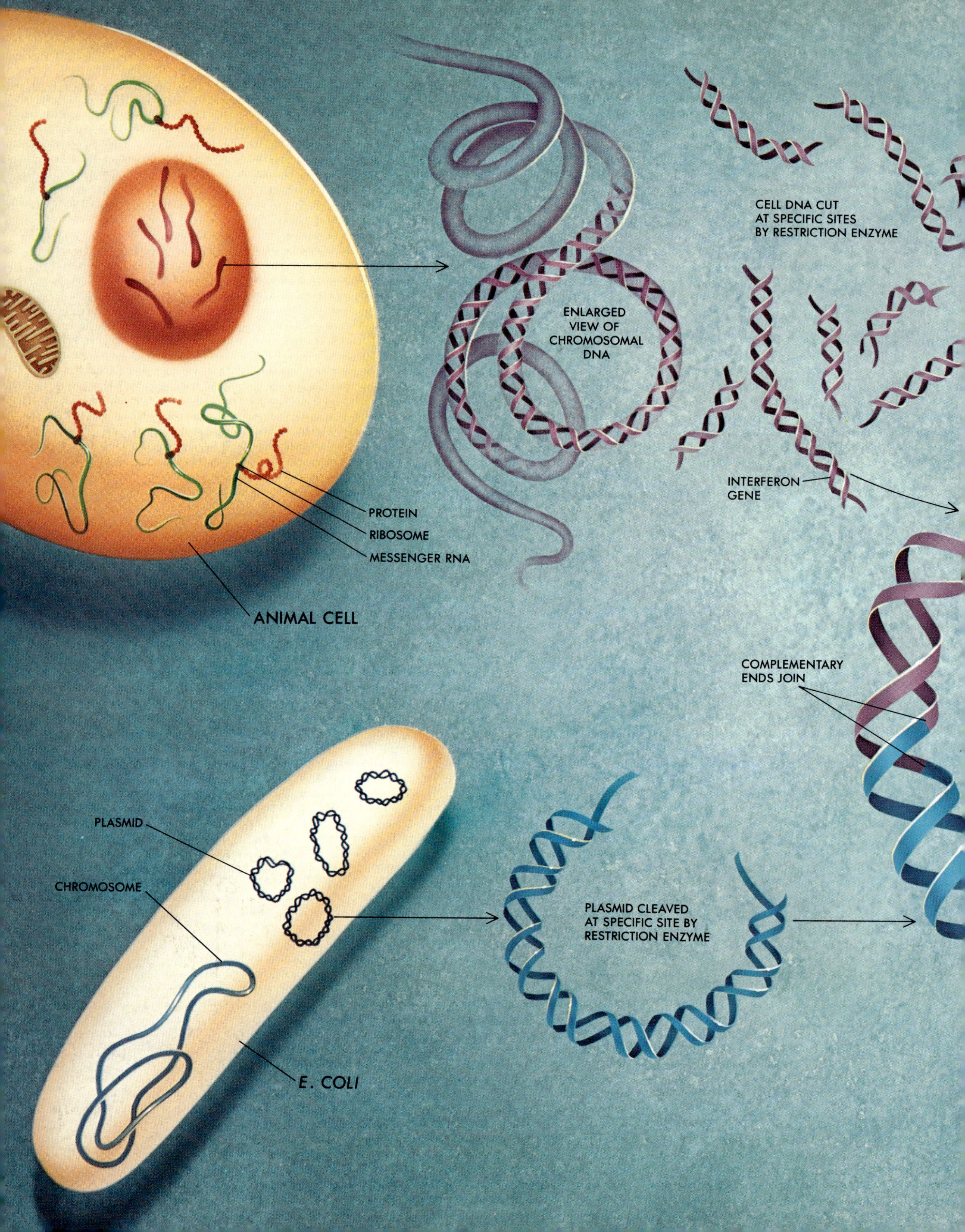

ENLARGED VIEW OF CHROMOSOMAL DNA

CELL DNA CUT AT SPECIFIC SITES BY RESTRICTION ENZYME

INTERFERON GENE

PROTEIN

RIBOSOME

MESSENGER RNA

ANIMAL CELL

COMPLEMENTARY ENDS JOIN

PLASMID

CHROMOSOME

PLASMID CLEAVED AT SPECIFIC SITE BY RESTRICTION ENZYME

E. COLI

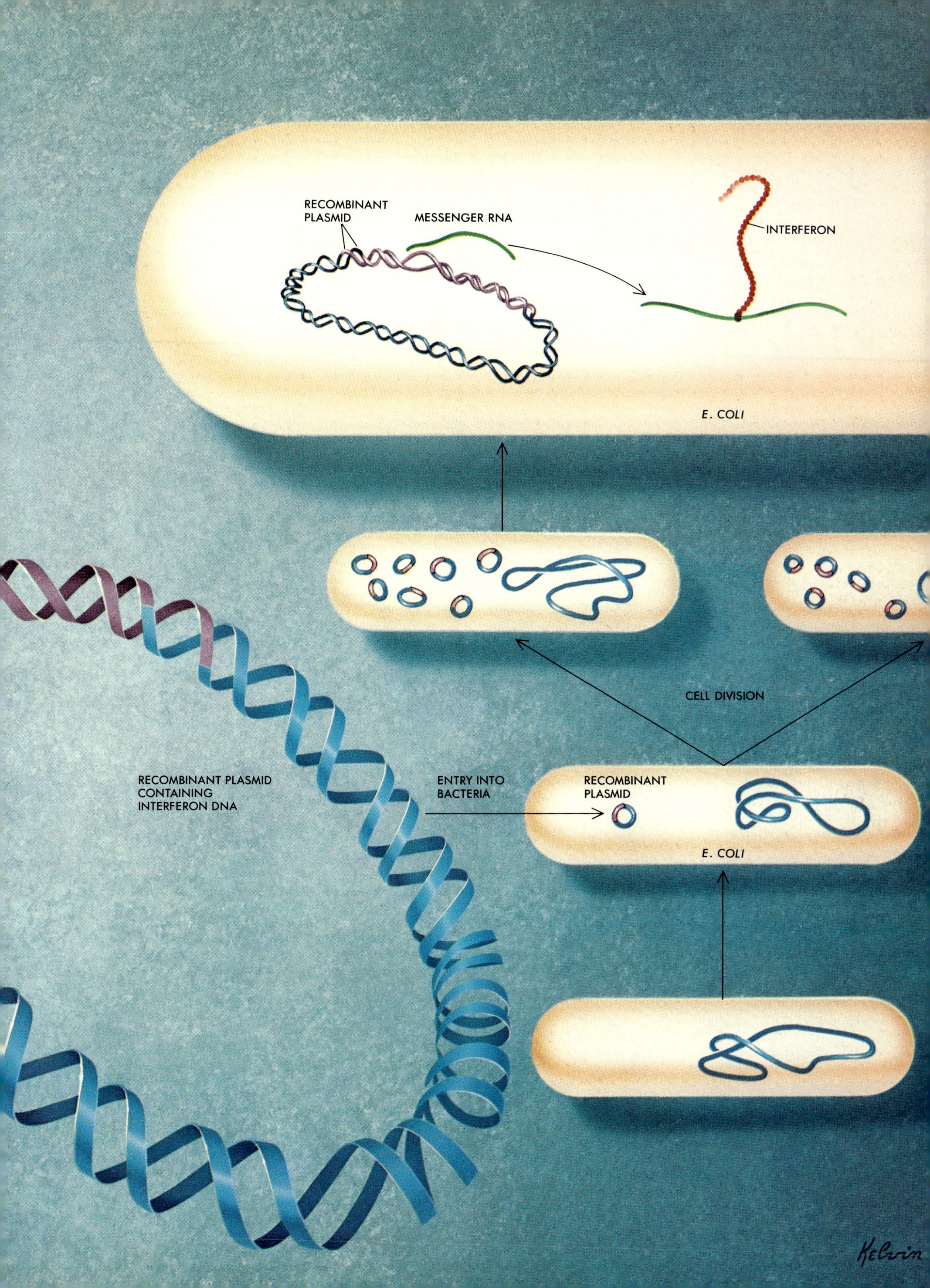

RECOMBINANT
PLASMID
MESSENGER RNA
INTERFERON
E. COLI
CELL DIVISION
RECOMBINANT PLASMID
CONTAINING
INTERFERON DNA
ENTRY INTO
BACTERIA
RECOMBINANT
PLASMID
E. COLI
Kelvin

PRODUCTION OF PHARMACEUTICALS

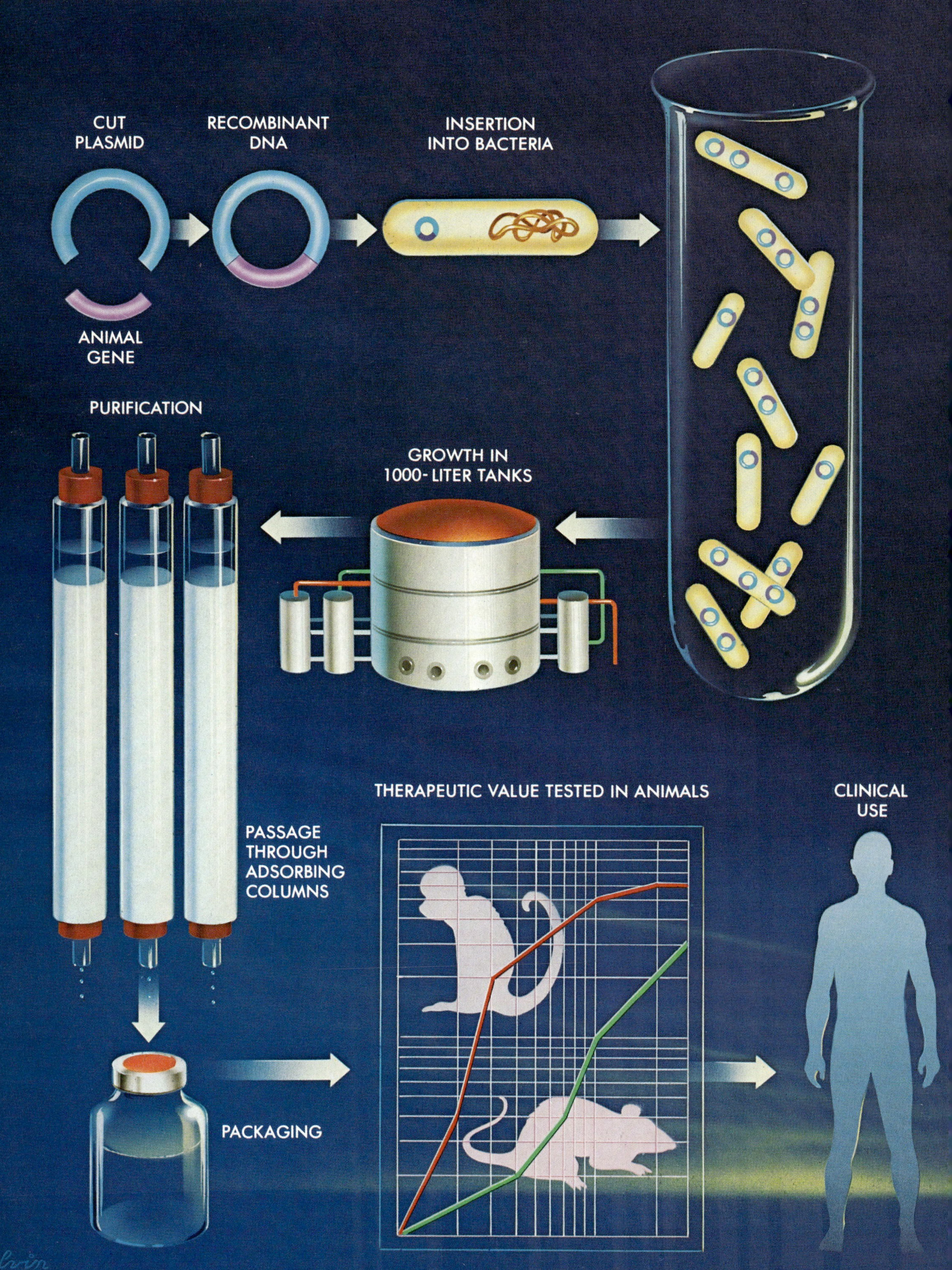

THE DNA STORY

"We are writing to you, on behalf of a number of scientists, to communicate a matter of deep concern."

—Singer and Soll Letter to Handler

(See Doc. 1.1)

THE ROAD TO ASILOMAR

The problem all historians face is knowing where to begin. The decision is inevitably arbitrary, since every important historical turning point has its antecedents. In the case of the recombinant DNA debate, one could say it began in June 1973 during the Gordon Conference on Nucleic Acids, a private meeting in New Hampshire of some 130 molecular biologists. Discussions at the meeting led its cochairpersons Dr. Maxine Singer (National Institutes of Health) and Dr. Dieter Soll (Yale) to write letters first to the President of the U.S. National Academy of Sciences (Doc. 1.1) requesting that an Academy committee be set up to investigate the possible consequences of the recombinant DNA technique, and second to the weekly journal *Science* (Doc. 1.2) to announce publicly the concern felt by some molecular biologists that recombinant DNA experiments might entail novel biohazards.

But in reality this debate began earlier. Events in the United States within the community of molecular biologists between 1971 and early 1973 made the Gordon Conference discussion and its aftermath almost an historical inevitability. During these years increasing numbers of molecular biologists began to switch from studying *E. coli* and its bacteriophages to work with animal cells in tissue culture and animal viruses. In particular they became interested in tumor viruses, which cause tumors either in their natural animal hosts or under special laboratory conditions in species that they do not infect in the wild. Most molecular biologists had not been trained as medical microbiologists and were not familiar with the disciplined laboratory practices used by those who study serious pathogens. The increase in the numbers of laboratories handling animal viruses led to the feeling that there was a growing need to consider potential health hazards. Apprehension increased in the summer of 1971 when Janet Mertz, then a young colleague of Paul Berg's at Stanford University, described at a meeting at the Cold Spring Harbor Laboratory, Long Island, an experiment in which it was intended to introduce the chromosome of a tumor virus called simian virus (SV40) into a laboratory strain of the human intestinal bacterium *E. coli*. Berg had for some time been thinking about using SV40 as a vehicle for introducing foreign genes into animal cells, but the prospect of a bacterium—especially one

Photo courtesy Tom Broker, Cold Spring Harbor Laboratory.

related to bacteria naturally occurring in the human intestine—being engineered to carry the DNA of a virus capable, under experimental conditions, of inducing tumors in rodents alarmed some molecular biologists. Robert Pollack of the Cold Spring Harbor Laboratory phoned Paul Berg about his concern, and the proposed experiments with SV40 were postponed.

By January 1973 this unease led to the organization at Asilomar, California, of a meeting to discuss the biohazards of work with animal viruses and in particular with tumor viruses. The proceedings of this meeting, which had no press coverage, were published by the Cold Spring Harbor Laboratory as a book entitled *Biohazards in Biological Research*. The book was virtually the only written evidence of the first Asilomar meeting.

A few quotations from an article published in *Science* (Doc. 1.3) in November 1973 convey the atmosphere of uncertainty that prevailed both before and after this first Asilomar Conference:

> I'm afraid that the National Cancer Institute avoids facing up to its moral, if not legal, responsibility by declaring almost all of the viruses we work with as unlikely to be of sufficient long-term danger to require first-grade safety equipment giving absolute control.
>
> —J. D. Watson

> I have no doubt that if you gave enough of some of these agents to a susceptible person he would get a tumor. It's entirely a guess as to risk but my guess is that it is considerably less dangerous than smoking two packs of cigarettes a day.
>
> —George J. Todaro

> We're in a pre-Hiroshima situation. It would be a real disaster if one of the agents now being handled in research should in fact be a real human cancer agent.
>
> —Robert Pollack

> The Berg experiment scares the pants off a lot of people, including him.
>
> —Wallace Rowe

Given such edginess, which in some minds amounted to actual alarm, it is not so surprising that announcements at the Gordon Conference in June 1973 of a new, easy method for recombining DNAs of quite unrelated species—the restriction-enzyme method described in Document 1.4 and the Scientific Background—led to the unscheduled debate and votes on biohazards. The Singer and Soll letters (Doc. 1.1 and Doc. 1.2) voice this uncertainty and concern. What the brief letter to *Science* does not say is that the discussion on the conjectural hazards of recombinant DNA took place hurriedly on the last day of the Conference when several participants had already departed, and that two votes were taken. In the first vote, which passed with an overwhelming majority, 78 out of the some 90 participants still at the meeting were in favor of sending a letter expressing concern to the National Academy of Sciences. The second vote, which passed with a very much narrower majority (48 to 42), was in favor of publishing the letter of concern much more widely. In short, these 48 molecular biologists took upon themselves the responsibility of announcing publicly, without further private discussions within appropriate professional bodies in the United States and elsewhere, their unease about the direction in which molecular biology was moving. Whether or not a vote on this issue, in which a larger proportion of the world's molecular biologists and medical microbiologists had been enfranchised, would have led to the same result is anyone's guess. In fact, the letter to *Science* was not seized upon by the daily press, although it was reported and given a wider circulation as a result of Edward Ziff's (Rockefeller University) article in *New Scientist* (Doc. 1.4), a British weekly journal.

The experiments described at the Gordon Conference continued and the letter of concern notwithstanding, press releases from laboratories pursuing recombinant DNA research were written up in the daily newspapers as a marvellous achievement of modern science rather than as dangerous, irresponsible tinkering with genes, which might jeopardize the health and future of the biosphere. However, the July 1974 publication in both *Science* and the British weekly journal *Nature* of the so-called Moratorium or Berg letter (Doc. 1.5) by the blue-ribbon Committee on Recombinant DNA Molecules, set up by the National Academy of Sciences in response to the letter to its President, brought the issue to the headlines of the U.S. daily press (Doc. 1.6).

With the press and other media now alerted, at least some industrial companies began to take notice. Biolabs—one of the producers of the restriction enzymes used in recombinant DNA experiments—offered to collaborate in any moratorium and help persuade similar companies to do likewise (Doc. 1.7). Discussion within U.S. universities concerning possible new safety procedures also began in earnest (Doc. 1.8). Waclaw Szybalski (University of Wisconsin) first suggested the use of biological containment (Doc. 1.8). Meanwhile Paul Berg, together with three of the group who had signed the Moratorium letter, and Sydney Brenner from Cambridge, England, began preparing for the second Asilomar Conference (Asilomar II), which was to be international (Doc. 1.9).

Outside the United States it was the Moratorium letter that caused the issue to be taken seriously for the first time (Doc. 1.10 and Doc. 1.11), and in Britain the government became involved. Why was it that in Britain, rather than in the United States or elsewhere, a government first directly intervened? The reasons may be manifold, but the fact that in 1973 there had been a fatal accident with smallpox virus at the London School of Hygiene and Tropical Medicine was surely one of them. The Working Party on the Experimental Manipulation of the Genetic Composition of Micro-organisms, set up in the United Kingdom in the summer of 1974 under the chairmanship of Lord Ashby, promptly set to work. By January 1975, just a few weeks before the Asilomar Conference, its report had been published by the British government. The report's conclusions (Doc. 1.12), which many felt were sober, common-sensical, and measured, did not meet everyone's expectations. The editor of *New Scientist* in London, for one, felt it was "Not Good Enough" (Doc. 1.13).

Document 1.1

July 17, 1973

Singer and Soll Letter to Handler

Yale University Box 1937 Yale Station, New Haven, Connecticut 06520

DEPARTMENT OF MOLECULAR BIOPHYSICS
AND BIOCHEMISTRY

July 17, 1973

Dr. Philip Handler, President
National Academy of Sciences
2101 Constitution Avenue
Washington, DC 20418

Dear Doctor Handler:

We are writing to you, on behalf of a number of scientists, to communicate a matter of deep concern. Several of the scientific reports presented at this year's Gordon Research Conference on Nucleic Acids (June 11-15, 1973, New Hampton, New Hampshire) indicated that we presently have the technical ability to join together, covalently, DNA molecules from diverse sources. Scientific developments over the past two years make it both reasonable and convenient to generate overlapping sequence homologies at the termini of different DNA molecules. The sequence homologies can then be used to combine the molecules by Watson-Crick hydrogen bonding. Application of existing methods permits subsequent covalent linkage of such molecules. This technique could be used, for example, to combine DNA from animal viruses with bacterial DNA, or DNAs of different viral origin might be so joined. In this way new kinds of hybrid plasmids or viruses, with biological activity of unpredictable nature, may eventually be created. These experiments offer exciting and interesting potential both for advancing knowledge of fundamental biological processes and for alleviation of human health problems.

Certain such hybrid molecules may prove hazardous to laboratory workers and to the public. Although no hazard has yet been established, prudence suggests that the potential hazard be seriously considered.

A majority of those attending the Conference voted to communicate their concern in this matter to you and to the President of the Institute of Medicine (to whom this letter is also being sent). The conferees suggested that the Academies establish a study committee to consider this problem and to recommend specific actions or guidelines should that seem appropriate. Related problems such as the risks involved in current large-scale preparation of animal viruses might also be considered.

A list of participants in the Conference is attached for your interest.

Sincerely yours,
Maxine Singer and Dieter Söll (MS)

Maxine Singer
Dieter Soll
Co-Chairmen of the 1973 Gordon
Research Conference on Nucleic Acids

Maxine Singer
National Institutes of Health
Room 9N-119, Building 10
Bethesda, MD 20014

Dieter Soll
Associate Professor of Molecular Biophysics
Yale University
New Haven, CT 06520

Enclosure

Document 1.2

September 21, 1973
Science (**181**:1114)

LETTERS

Guidelines for DNA Hybrid Molecules

Those in attendance at the 1973 Gordon Conference on Nucleic Acids voted to send the following letter to Philip Handler, president of the National Academy of Sciences, and to John R. Hogness, president of the National Institute of Medicine. A majority also desired to publicize the letter more widely.

We are writing to you, on behalf of a number of scientists, to communicate a matter of deep concern. Several of the scientific reports presented at this year's Gordon Research Conference on Nucleic Acids (June 11–15, 1973, New Hampton, New Hampshire) indicated that we presently have the technical ability to join together, covalently, DNA molecules from diverse sources. Scientific developments over the past two years make it both reasonable and convenient to generate overlapping sequence homologies at the termini of different DNA molecules. The sequence homologies can then be used to combine the molecules by Watson-Crick hydrogen bonding. Application of existing methods permits subsequent covalent linkage of such molecules. This technique could be used, for example, to combine DNA from animal viruses with bacterial DNA, or DNA's of different viral origin might be so joined. In this way new kinds of hybrid plasmids or viruses, with biological activity of unpredictable nature, may eventually be created. These experiments offer exciting and interesting potential both for advancing knowledge of fundamental biological processes and for alleviation of human health problems.

Certain such hybrid molecules may prove hazardous to laboratory workers and to the public. Although no hazard has yet been established, prudence suggests that the potential hazard be seriously considered.

A majority of those attending the Conference voted to communicate their concern in this matter to you and to the President of the Institute of Medicine (to whom this letter is also being sent). The conferees suggested that the Academies establish a study committee to consider this problem and to recommend specific actions or guidelines, should that seem appropriate. Related problems such as the risks involved in current large-scale preparation of animal viruses might also be considered.

MAXINE SINGER

Room 9N-119, Building 10, National Institutes of Health, Bethesda, Maryland 20014

DIETER SOLL

Department of Molecular Biophysics and Biochemistry, Yale University, New Haven, Connecticut 06520

Microbiology: Hazardous Profession Faces New Uncertainties

Since the turn of the century, some 3500 cases of laboratory-acquired infections have been reported, more than 150 of which resulted in death. Although with this accident rate it may still make more sense to be a microbiologist than a steeplejack, the profession is not entirely without risk. The risks are, if anything, increasing as more people take up work with viruses, including viruses suspected of causing cancer in man. Besides the risk to scientists themselves, there are also dangers posed by the new kinds of virus that can now be created in the laboratory and which, if they escaped, might constitute a threat to public health.

The degree to which people have become infected with the agents they work with depends on the care they take and the nature of the agent, but even under the most stringent safety conditions that can be devised, such as those at the former biological warfare laboratories at Fort Detrick, Maryland, infections do occur. During the quarter-century that the Fort Detrick laboratory was in operation, there were 423 cases of infection and three deaths. Since the cost of building even a moderate-sized laboratory to the same standards of safety is about $125,000, most civilian laboratories have to make do with less. One experienced virologist reckons that, when working with agents which infect man, about 5 percent of the laboratory staff may become infected each year. "Every microbiologist has inhaled or absorbed significant amounts of any organism he has worked with," says A. Wedum, former safety director at Fort Detrick.

Bacteria were once the most common cause of laboratory infections, a role that has now been taken over by viruses. According to Wallace Rowe of the National Institute of Allergy and Infectious Diseases (NIAID), the hazards to laboratory workers are probably on the increase. One reason is that many of the people now coming into virology are, for example, biochemists who do not have the safety instincts of the trained microbiologist and tend to regard viruses simply as another chemical reagent. Another is the trend to use viruses in more and more highly concentrated forms. Infection depends on the dose of virus to which a person is exposed, and solutions now in common laboratory use contain 100 to 1000 times more virus than they did a few years ago. A third kind of hazard is the creation of hybrid or otherwise new viruses, which pose unknown risks both to scientists who work with them and the population at large.

According to Wedum, about a quarter of all laboratory infections can be traced to accidents, such as self-inoculation with a syringe. For the rest, a precise cause is usually hard to find, but inhalation is often the reason. Many common laboratory operations, such as blending, sonicating, or simple spillage, can lead to the formation of an aerosol containing viral particles.

Probably the most dangerous single source of viruses is monkeys, in which occur a number of agents fatal to man. There have been 20 suspected cases of human infection with herpesvirus B, with only three possible survivors. Another monkey agent, Marburg virus, infected 31 laboratory workers and others in an outbreak in Germany in 1967, resulting in seven deaths.

These are known dangers for which there are known precautions. Harder to assess is the degree of risk involved in dealing with tumor viruses. So far, the only known death from cancer caused by a laboratory accident is that of a French medical student, Henri Dadon, who in 1926 pricked his hand with a syringe containing fluid from a cancer patient; a nodule developed on his palm, and he died a year later from metastasized tumors. A different risk is presented by the numerous animal tumor viruses that have since been discovered and are now under intensive study. On the theoretical assumption that a species is able to combat its own tumor viruses quite well but may be more susceptible to those of other species, the animal tumor viruses may be just as hazardous to work with as a natural human tumor agent which, if it exists, is unlikely to be highly virulent. "I have no doubt that if you gave enough of some of these agents to a susceptible person he would get a tumor," says George J. Todaro of the National Cancer Institute. At the same time, Todaro does not rate the danger of working with animal tumor agents as very high—"It's entirely a guess as to risk, but my guess is that it is considerably less dangerous than smoking two packs of cigarettes a day." What if the guess is wrong? "We're in a pre-Hiroshima situation," says Robert Pollack of the Cold Spring Harbor Laboratory: "It would be a real disaster if one of the agents now being handled in research should in fact be a real human cancer agent."

James D. Watson, also of Cold Spring Harbor, is another who takes a serious view of the possibilities. "I'm afraid that the NCI avoids facing up to its moral, if not legal, responsibility by declaring almost all of the viruses we work with as unlikely to be of sufficient long-term danger to require [first grade safety equipment giving absolute control]," Watson said at a recent conference on laboratory biohazards.* W. Emmett Barkley, an NCI official concerned with safety, responds that, in the feeling of the NCI, the hazard of working with tumor viruses is not such as to require absolute control. In any case, 90 percent of safety comes from good technique on the part of the investigator, only 10 percent from the equipment and facilities, Barkley says.

Even more intractable than tumor viruses are the theoretical hazards posed

* *Biohazards in Biological Research*, proceedings of a conference held at the Asilomar Conference Center, California, January 1973 (Cold Spring Harbor Laboratory, Cold Spring Harbor, N.Y., 1973).

by the creation in the laboratory of viruses that do not exist in nature. An empirical reason for believing such viruses would not be dangerous is that many millions of dollars were invested at Fort Detrick in trying to improve upon the lethality of viruses harmful to man, but, according to Wedum, they "never had much luck." Yet what Fort Detrick could not accomplish by design, others may achieve by accident. There is concern in virological circles with at least three kinds of study now in progress. One is the attempt to devise a better influenza vaccine by means of hybrid flu viruses. The danger here is that the ability to genetically manipulate flu viruses could lead to a new combination that might escape from the laboratory, by infecting an employee, say, and spread to the population at large. "This could recreate the conditions for an influenza pandemic like that of 1918," says Rowe of NIAID.

Tumorigenic Monkey Virus

Another kind of virus is a combination of the DNA of SV40, a monkey virus that causes tumors in lower animals, and certain bacterial genes. The hybrid DNA molecule was synthesized by Paul Berg and colleagues at the Stanford University School of Medicine for the purpose of studying how the bacterial genes work. One of the possible experiments with the hybrid virus calls for it to be made to infect *E. coli*, a bacterium that is a common inhabitant of the human gut. There is considerable concern that if an SV40-infected *E. coli* should escape from Berg's laboratory it might become established in the population at large, which would then forevermore be exposed to SV40 genes, the effects of which in man are unknown. "The Berg experiment scares the pants off a lot of people, including him," says Rowe. According to Todaro, the Berg experiment "is one of those which I think just shouldn't be done." Berg says he cannot prove the experiment to be absolutely safe and has decided not to do it for the time being.

The general issue of hybrid DNA molecules so concerned a group of scientists at the Gordon Research Conference this June that they sent a letter to the president of the National Academy of Sciences stating that such hybrid molecules "may prove hazardous to laboratory workers and the public" and suggesting that a committee be set up to study the problem (see *Science*, 21 September, p. 1114).

A third kind of virus that causes concern is the group of hybrids that occur between SV40 and some of the human adenoviruses. The hybrids were first created in the course of manufacturing adenovirus vaccine, which is produced in monkey cells. (The vaccine was received by military conscripts, but no follow-up study has been done to observe the effects, if any.) Certain of the adenovirus/SV40 viruses, known as non-defective hybrids, also happen to be a useful research tool for mapping the genes of SV40. Should the viruses escape, the risk they pose is that, like pure adenovirus, they could become established in the tonsils of young children and become part of the human experience for generations to come. Although some believe that concern about SV40 is a tempest in a teapot, at the same time it is impossible to know for certain what SV40 may do in man.

Because of these concerns, the virologist who developed the adenovirus-SV40 hybrids, Andrew M. Lewis of the NIAID, was uneasy about distributing the virus. It is now official NIAID policy that those wanting samples must sign a memorandum of understanding, in which they agree to take certain safety measures and not to pass the virus on to anyone who does not promise to do likewise.

Such a policy, moderate though it may seem, in fact cuts across the scientific ethic that materials should be freely exchanged with any fellow scientist who wants them—so much so, in fact, that Lewis has been suspected by some of raising public safety as a ploy to keep the viruses to himself. Lewis feels that an informal moral commitment is preferable to legal regulations.

"If the public feels the scientific community is acting irresponsibly, there will be an immediate reaction and the freedom of research will be curtailed. If we don't exercise due caution we are heading for trouble," Lewis believes.

It is probably true to say that virologists and other microbiologists are more safety conscious now than they used to be, but safety practices vary widely from one laboratory to another. The history of safety regulations in almost every field of activity is that it takes a disaster to arouse effective concern. Virology, maybe, will prove an exception.

—Nicholas Wade

Document 1.4
October 25, 1973
New Scientist (**60**: 274)

Benefits and hazards of manipulating DNA

Molecular biologists at a recent Gordon conference in America suggested that a committee be set up to assess the possible health hazards presented by new techniques involving manipulating strands of DNA. Here, one of the participants (and a supporter of the suggestion), describes the basis of their concern

Dr Edward Ziff
is a visiting American scientist at the MRC Laboratory of Molecular Biology, Cambridge, England

Meeting in New Hampshire in June at the Gordon Conference on Nucleic Acids, a group of 130 scientists by majority vote instructed the co-chairmen of the conference, Maxine Singer of the US National Institutes of Health and Dieter Söll of Yale University, to draft a letter of concern about the possible future implications of their research. This letter, directed to the presidents of the US National Academy of Sciences and the National Institute of Medicine, and reprinted in the 21 September issue of Science states that ". . . we presently have the technical ability to join together, covalently, DNA molecules from diverse sources." Noting that this ability makes possible the preparation of various combinations of bacterial and viral DNA, the letter continues, "Certain such hybrid molecules may prove hazardous to laboratory workers and to the public. Although no hazard has yet been established, prudence suggests that the potential hazard be seriously considered."

What new developments in molecular biology enable the formation of hybrid combinations of DNA, what is the interest in the preparation of such hybrids, and how could they constitute a possible health hazard? To examine these questions it is helpful to consider the discovery of a class of enzymes termed "restriction enzymes".

'Restriction enzymes'

Certain bacteria, notably strains of Haemophilus and Escherichia coli, contain enzymes distinguished by their ability to cleave DNA at specific nucleotide sequences. These enzymes, restriction enzymes, provide a mechanism for defending the bacteria against foreign DNA, such as DNA injected by bacteriophage. While the bacterium protects its own DNA from the restriction enzyme by methylating particular residues within the potential cleavage sequences, unprotected DNA is cut into fragments, rendering it nonfunctional. Hamilton Smith and Thomas Kelly working at Johns Hopkins University Medical School in Baltimore, purified one restriction enzyme from the bacterium Haemophilus influenzae (abbreviated "Hin" enzyme), and identified the sequence of six nucleotide base pairs which it recognises as a cleavage site. A second restriction enzyme isolated from E. coli and named RI has been analysed in the laboratories of Herbert Boyer and Howard Goodman at the University of California Medical Center in San Francisco. They found that the nucleotide sequence recognised by RI is eight base pairs long.

Because the cleavage sequences of restriction enzymes are specific, these enzymes reduce DNA to fragments which are themselves specific. Daniel Nathans and Kathleen Dana, also working at Johns Hopkins, have studied the action of Hin enzymes upon the DNA chromosome of the animal virus SV40. This chromosome, which is a double stranded and circular piece of DNA approximately 5000 base pairs long, is cut into 11 unique fragments by Hin enzyme. The action of the RI enzyme upon SV40 DNA has been examined by workers in the Paul Berg's laboratory at Stanford, and by Hajo Delius and Carel Mulder of the Cold Spring Harbor Laboratory in New York. In the case of RI they found that SV40 DNA is cleaved only once, converting it from a circular form to a unique linear molecule. Restriction enzymes therefore provide a convenient tool for cutting simple chromosomes at specific points.

People soon recognised that the action of RI differs from that of Hin not only in the nucleotide sequence which it recognises, but also in the type of cleavage which it causes. Whereas the cleavages described by Kelly and Smith for Hin are at precisely opposing points on the two strands of the DNA molecules, the cuts introduced by RI are staggered

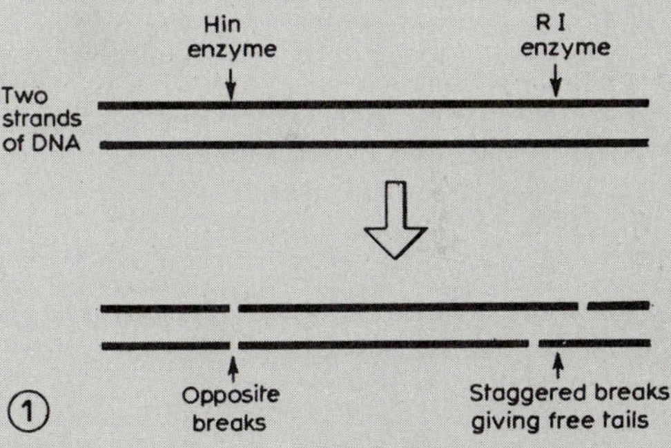

(Figure 1). That is, the cut in one strand is displaced from the cut in the other by four nucleotide residues. Because of this displacement, RI fragments contain a long body of double stranded DNA, with single stranded "tails" four residues long at each end. These tails have nucleotide sequences which are complementary to another, and as the Stanford workers show, the tails can reassociate by forming of Watson-Crick base pairs. The ends of the SV40 linear molecule can therefore rejoin to regenerate a circular molecule. Because the RI enzyme is capable of cutting within only one particular sequence of nucleotides, the tails of all RI fragments are the same. Thus it is possible to bring together RI fragments of DNA from diverse sources, and through Watson-Crick base pairing of their tails, join them together.

These facts at once suggest a means for splicing together the genetic material of different organisms. Two DNAs could be treated with the RI restriction enzyme, and the fragment products separated from one

This article first appeared in *New Scientist*, London, the weekly review of science and technology.

another. After mixing together of any two fragments under the proper conditions, a tail of one fragment could bind to a tail of the other. The DNA strands could be covalently joined through the action of DNA ligase, an enzyme whose function is the repair of breaks in DNA molecules. Methods for the generalisation of these reactions to other fragments of DNA have been proposed by the Stanford workers. And it is likely that a wide spectrum of new restriction enzymes will soon be described, increasing the number and types of DNA fragments which one could produce and combine.

What sorts of DNA hybrids might one wish to prepare? The chromosomes of small bacteriophages and animal viruses are the smallest functional DNA molecules, and therefore represent logical starting materials. One might place a segment of animal virus DNA bearing a viral gene into the DNA of a bacteriophage. By infecting bacteria with such a hybrid, the expression of the viral gene might be studied in the context of the bacterium, which is very much more simple than that of the normal animal cell host. Conversely, a bacterial gene (or even one from a higher organism) might be inserted into an animal virus DNA. Because certain animal viruses such as SV40 and polyoma are capable of integrating their DNA into the host cell's chromosome, such a hybrid might carry the new gene into the host cell's gene pool. In this manner, faulty genes might be replaced by functional ones, and a basis for curing genetic diseases might be established.

Finally, it has been suggested that one might splice a viral DNA into a bacterial plasmid (a piece of DNA which can replicate itself from generation to generation within a bacterium, independently from the bacterial chromosome). Once the virus was so installed, its genetic structure and protein gene products might be readily analysed, simplifying an understanding of viral function in animal cells. It is likely that many varied and fundamental biological research problems could be fruitfully approached through the use of DNA hybrids.

Why such concern?

What then is the basis of concern about the production of DNA hybrids? Such molecules constitute new combinations of genetic material, and thus their biological properties remain untested, either in the laboratory or in nature. Many sorts of DNA hybrids are likely to prove entirely innocuous. However, certain cases have aroused concern. Some viral DNAs are capable of transforming cells in tissue culture to a malignant state. While most of these viruses are not known to cause disease in man, experimental procedures which could increase the chance of human exposure, or provide new avenues of transmission, should be treated with caution. In particular it has been suggested that the linkage of human DNA, or the DNA of a related species, to an oncogenic viral DNA (one capable of causing malignancies) could prove hazardous because such linkage could conceivably allow the hybrid to enter human chromosome. Perhaps more important is the concern over the linkage of oncogenic DNA to a DNA which could permit its replication in a bacterium associated with man. A person harbouring such a composite organism could suffer an increased risk of contracting cancer. The performance of experiments requiring the formation of such hybrids under laboratory conditions of isolation would be a safeguard for reducing potential dangers.

I should emphasise that so far no danger has been established for the preparation of hybrid DNA molecules. It remains possible that an animal virus DNA cannot be maintained in a bacterium. And humans may have sufficient mechanisms for protecting themselves against these foreign DNAs. These possibilities, however, remain untested. Until they are, a conscientious scientific community will give especially careful attention to the means of its research as well as its goals, and take the full precautions necessary for ensuring public safety.

Cancer-causing viruses

A second area of research to which the Gordon conference letter calls attention is the large scale production and isolation of potential cancer causing viruses. The link between certain viruses and cancer in laboratory and domestic animals has led to an upsurge in the study of the molecular biology of these viruses, and the search for their human equivalents. Of particular interest at the moment are a class of RNA viruses which contain reverse transcriptase, an enzyme which makes a DNA copy of the virus's RNA chromosome. To study this enzyme, or other protein or nucleic acid components, viruses often must be cultured on a large scale, and then purified in concentrated form. The handling of concentrated viruses can present special problems of dissemination, such as the formation of aerosols (micro-droplets of airborne virus). No virus has yet been conclusively demonstrated to cause cancer in man. The effects of exposure to a carcinogenic factor, however, may not be evident until years afterwards. Thus a limited history of safe handling of a virus may not be sufficient to regard it as free of risk. The massive search which is in progress for viruses which are responsible for human cancers may well uncover agents which do present a clear hazard. Therefore, as research directions change, new safety guidelines will be required, and adequate protective facilities provided.

The Gordon conference letter proposed that a committee be formed to consider the health implications of these new areas of biological investigation. Such a committee, which the National Academy of Sciences has now agreed to set up, will be hampered by the lack of concrete evidence which distinguishes harmless cases from potentially hazardous ones. And dissenting scientific opinions would certainly be voiced. But until such time as reasonable guidelines are established, vigorous discussion and debate within the scientific community will continue to provide the greatest safeguard of the public good.

Document 1.5

July 26, 1974
Science (**185**:303)

LETTERS

Potential Biohazards of Recombinant DNA Molecules

Recent advances in techniques for the isolation and rejoining of segments of DNA now permit construction of biologically active recombinant DNA molecules in vitro. For example, DNA restriction endonucleases, which generate DNA fragments containing cohesive ends especially suitable for rejoining, have been used to create new types of biologically functional bacterial plasmids carrying antibiotic resistance markers (*1*) and to link *Xenopus laevis* ribosomal DNA to DNA from a bacterial plasmid. This latter recombinant plasmid has been shown to replicate stably in *Escherichia coli* where it synthesizes RNA that is complementary to *X. laevis* ribosomal DNA (*2*). Similarly, segments of *Drosophila* chromosomal DNA have been incorporated into both plasmid and bacteriophage DNA's to yield hybrid molecules that can infect and replicate in *E. coli* (*3*).

Several groups of scientists are now planning to use this technology to create recombinant DNA's from a variety of other viral, animal, and bacterial sources. Although such experiments are likely to facilitate the solution of important theoretical and practical biological problems, they would also result in the creation of novel types of infectious DNA elements whose biological properties cannot be completely predicted in advance.

There is serious concern that some of these artificial recombinant DNA molecules could prove biologically hazardous. One potential hazard in current experiments derives from the need to use a bacterium like *E. coli* to clone the recombinant DNA molecules and to amplify their number. Strains of *E. coli* commonly reside in the human intestinal tract, and they are capable of exchanging genetic information with other types of bacteria, some of which are pathogenic to man. Thus, new DNA elements introduced into *E. coli* might possibly become widely disseminated among human, bacterial, plant, or animal populations with unpredictable effects.

Concern for these emerging capabilities was raised by scientists attending the 1973 Gordon Research Conference on Nucleic Acids (*4*), who requested that the National Academy of Sciences give consideration to these matters. The undersigned members of a committee, acting on behalf of and with the endorsement of the Assembly of Life Sciences of the National Research Council on this matter, propose the following recommendations.

First, and most important, that until the potential hazards of such recombinant DNA molecules have been better evaluated or until adequate methods are developed for preventing their spread, scientists throughout the world join with the members of this committee in voluntarily deferring the following types of experiments.

► *Type 1*: Construction of new, autonomously replicating bacterial plasmids that might result in the introduction of genetic determinants for antibiotic resistance or bacterial toxin formation into bacterial strains that do not at present carry such determinants; or construction of new bacterial plasmids containing combinations of resistance to clinically useful antibiotics unless plasmids containing such combinations of antibiotic resistance determinants already exist in nature.

► *Type 2*: Linkage of all or segments of the DNA's from oncogenic or other animal viruses to autonomously replicating DNA elements such as bacterial plasmids or other viral DNA's. Such recombinant DNA molecules might be more easily disseminated to bacterial populations in humans and other species, and thus possibly increase the incidence of cancer or other diseases.

Second, plans to link fragments of animal DNA's to bacterial plasmid DNA or bacteriophage DNA should be carefully weighed in light of the fact that many types of animal cell DNA's contain sequences common to RNA tumor viruses. Since joining of any foreign DNA to a DNA replication system creates new recombinant DNA molecules whose biological properties cannot be predicted with certainty, such experiments should not be undertaken lightly.

Third, the director of the National Institutes of Health is requested to give immediate consideration to establishing an advisory committee charged with (i) overseeing an experimental program to evaluate the potential biological and ecological hazards of the above types of recombinant DNA molecules; (ii) developing procedures which will minimize the spread of such molecules within human and other populations; and (iii) devising guidelines to be followed by investigators working with potentially hazardous recombinant DNA molecules.

Fourth, an international meeting of involved scientists from all over the world should be convened early in the coming year to review scientific progress in this area and to further discuss appropriate ways to deal with the potential biohazards of recombinant DNA molecules.

The above recommendations are made with the realization (i) that our concern is based on judgments of potential rather than demonstrated risk since there are few available experimental data on the hazards of such DNA molecules and (ii) that adherence to our major recommendations will entail postponement or possibly abandonment of certain types of scientifically worthwhile experiments. Moreover, we are aware of many theoretical and practical difficulties involved in evaluating the human hazards of such recombinant DNA molecules. Nonetheless, our concern for the possible unfortunate consequences of indiscriminate application of these techniques motivates us to urge all scientists working in this area to join us in agreeing not to initiate experiments of types 1 and 2 above until attempts have been made to evaluate the hazards and some resolution of the outstanding questions has been achieved.

PAUL BERG, *Chairman*
DAVID BALTIMORE
HERBERT W. BOYER
STANLEY N. COHEN
RONALD W. DAVIS
DAVID S. HOGNESS
DANIEL NATHANS
RICHARD ROBLIN
JAMES D. WATSON
SHERMAN WEISSMAN
NORTON D. ZINDER

Committee on Recombinant DNA Molecules Assembly of Life Sciences, National Research Council, National Academy of Sciences, Washington, D.C. 20418

Document 1.6

July 18, 1974

Newspaper Headlines and Articles

The Washington Post, July 18, 1974

NAS Panel Warns of 'Hazards'

Halt in Genetic Work Urged

By Stuart Auerbach
Washington Post Staff Writer

A committee of the National Academy of Science fears that the new science of transferring genetic material from animals to bacteria could increase the incidence of cancer and could create drug-resistant strains of mutant germs that would cause new diseases.

In a report to be released today, the committee calls on scientists to postpone or abandon research in this field because of its "potential hazards," even though the research could benefit mankind.

The committee includes the researchers who developed during the past year new techniques that allow them to transfer bits of deoxyribonucleic acid (DNA) from animals or bacteria that contains genetic material into genetic material from bacteria.

This is the first time that scientists actively working in a field have called for voluntary restraints in their research, said Dr. Paul Berg of Stanford University, the chairman of the national academy's committee.

"Although such experiments are likely to facilitate the solution of important theoretical and practical biological problems, they would also result in the creation of novel types of infectious DNA elements whose biological properties cannot be completely predicted in advance," the committee reports in a Science magazine issue to be published Friday.

"Many expert investigators have ex-

See DNA, Col. 4

New York Times

POSSIBLE DANGER HALTS GENE TESTS

The Washington Post, July 18, 1974

AN INDEPENDENT NEWSPAPE

The Scientific Conscienc

EVER SINCE HIROSHIMA, scientists have been concerned that probing the secrets of nature without caution and moral restraint might open a Pandora box of ills, if not disasters, that could do terrible damage. But who is to be the keeper of the keys? Who could be presumed to possess the wisdom to decide what secrets might be unlocked and what had better not be tampered with? Nature knows neither good nor evil. The same discovery or intervention that might do harm might also bring new blessings. Scientific research and experimentation is surely not a matter for the police to control. The best we can hope for is that the collective conscience of scientists themselves asserts itself to weigh the risks in each specific instance.

NEW YORK, THURSDAY, JULY 18, 1974

Genetic Tests Renounced Over Possible Hazards

Document 1.7

August 2, 1974
Memo from Berg

OFFICE MEMORANDUM • STANFORD UNIVERSITY • OFFICE MEMORANDUM

DATE: August 2, 1974

TO : Signators of "Biohazard" Letter
Maxine Singer

FROM : Paul Berg

SUBJECT:

What do you think of this? Any advice to offer? Should this be discussed at the February meeting?

I can't see how banning commercial production of these enzymes will thwart the construction of unwanted molecules. On the other hand, I know of an instance where a person with weird ambitions was probably stopped by the unavailability of the enzymes and his unwillingness or inability to make them. But good labs will not be limited in what they do by the lack of commercially available enzymes!

PB:af
Enc.

STANFORD UNIVERSITY • OFFICE MEMORANDUM • STANFORD UNIVERSITY • OFFICE MEMORANDUM

July 29, 1974
Comb Letter to Berg

BIOLABS 283 CABOT STREET, BEVERLY, MASSACHUSETTS 01915 (617) 927-5054

July 29, 1974

Dr. Paul Berg
Committee on Recombinant DNA Molecules
Dept. of Biochemistry
Stanford University School of Medicine
Stanford, Calif. 94305

Dear Dr. Berg:

This letter is to request guidance by the Committee on Recombinant DNA Molecules as to the suitability of the commercial production of restriction endonucleases. Does your committee feel that introducing these enzymes to the scientific community at large is beneficial and outweighs their possible use for the type of experiments your committee recommended against?

Many commercial suppliers of biologicals are preparing to market these enzymes in the near future. Hin II & III has already been prepared by this laboratory and is marketed through a large supplier.

I strongly endorse the goals of your recent letter in Science but I'm bothered by the social responsibility of commercial laboratories in distributing these enzymes to a world-wide market. I'm sure that most, if not all, of us in this business are not knowledgeable enough to project any long-term effects of their distribution. Thus, I urge your committee to take on this difficult judgement.

If you decide a moratorium is advisable, this laboratory will stop production and try to induce others to follow. It's important that a decision be made early before large investments in production and marketing are made.

You may argue that this is not a decision scientists should make and I would remind you of DDT and napalm--a few among many products that have not been used wisely.

Sincerely yours,

Don Comb

Donald G. Comb, PhD
Director

Document 1.8

October 25, 1974
Szybalski Letter to Reznikoff

MEDICAL CENTER

McARDLE LABORATORY
FOR CANCER RESEARCH
UNIVERSITY OF WISCONSIN • MADISON, WISCONSIN 53706

October 25, 1974

Dr. W. Reznikoff
Department of Biochemistry

Dear Bill:

This letter was requested in your memo of October 14, 1974 regarding safety factors in chimeric DNA research. Let me begin with the following statement.

When deciding whether to institute a new type of safety regulation one should first weigh the advantages versus the disadvantages of such a proposal. The advantages are very hypothetical, since they are based on the wrong presumptions that (1) "bad" genes could be transmitted with reasonable efficiency from bacteria or bacteriophages to humans and their beasts, (2) this is very different from what already happens spontaneously in nature, (3) many scientists working in this field are very sloppy or malicious, and (4) you could effectively control the malicious or sloppy scientists. The disadvantages are very real and the include: (1) creating a new bureaucracy, which would interfere with research, (2) diverting funds from research to this bureaucracy, (3) discouraging research in this very important area, (4) delay or elimination of discoveries leading to new kinds of treatments of many diseases, including cancer, and thus to alleviation of human miseries, (5) general delay in scientific progress. I would like to conclude that the real disadvantages outweigh very strongly the hypothetical dangers and thus, the research on chimeric DNA should not be discouraged in any way, especially by restrictive regulations and bureaucratic controls. Let me now answer your four questions.

(a) The real dangers might involve the resistance transfer factors and some oncogenic viruses; these should be handled as any infectious material. Other "dangers" are comparable with the dangers of using any new mutants of new bacterial species of unknown origin. Perhaps they also carry "bad" genes.

(b) The scientific merits of this type of research are enormous and certainly it should be encouraged if the experiments are well and cleverly designed. For instance, creating DNA chimeras between E. coli and antigen-producing genes of tumor viruses could produce immunity against oncogenic viruses. The opportunities of studying isolated mammalian operons carried by bacteriophages cannot be overemphasized.

(c) The standard bacteriological procedures are quite adequate for most of the research. In the case of some resistance transfer factors and oncogenic viruses the techniques used with infectious material should be applied.

(d) The Biological Safety Committee should prepare literature informing the scientist working with chimera DNAs about the necessity to follow the procedures mentioned in point (c). In applications pertaining to the work with chimeras, including resistance transfer factors and pathogenic viruses, a statement should be made that procedures applying to infectious microbiological materials will be used, including proper sterilization of all materials.

Let me conclude by stating that even good intentions could easily lead to creation of a self-sustaining and even growing bureaucratic monster which would discourage and delay very important research. After all, the roads to hell are paved with good intentions.

Sincerely yours,

Waclaw Szybalski
Professor of Oncology

McARDLE LABORATORY
FOR CANCER RESEARCH
MEDICAL CENTER UNIVERSITY OF WISCONSIN • MADISON, WISCONSIN 53706

October 8, 1974

Dr. Paul Berg
Dept. of Biochemistry
Stanford University Medical School
Stanford, CA 94305

Dear Paul:

This is to follow up my recent letter of October 4, hopefully before you will have a chance to answer it. Specifically, I would like to draw your attention to a point which I have never seen brought up in the discussion of potential dangers of the recombinant DNA chimeras, namely the development of safe, "biocontained" DNA vector which cannot escape from the laboratory. I have in mind special "self-destructing" mutants of bacteriophage λ, which cannot establish a symbiotic state with the E. coli host since they (1) cannot form a prophage state (att-int-xis deleted, cI deleted, vir) (2) cannot form a plasmid (cl7, nin5) (3) cannot readily recombine (Red$^-$ in Rec$^-$ host, deleted homologies) and (4) are very virulent and thus kill the host. I have many ideas along these lines and I would be glad to contribute them to any of your future deliberations on this subject. I strongly believe that clever genetic design is much superior to any costly and basically inefficient solutions based on "brute force" mechanical containment systems.

With best wishes,

Sincerely yours,

Waclaw Szybalski
Professor of Oncology

WS:kt

Document 1.9

December 4, 1974
Berg Invitation to Watson

STANFORD UNIVERSITY MEDICAL CENTER
STANFORD, CALIFORNIA 94305

DEPARTMENT OF BIOCHEMISTRY

PAUL BERG
Jack, Lulu and Sam Willson
Professor of Biochemistry

December 4, 1974

Dr. James Watson
Professor of Biology
Harvard University
Cambridge, Massachusetts 02138

Dear Dr. Watson:

A meeting on Recombinant DNA Molecules will be held in Pacific Grove, California (at the Asilomar Conference Center) during February 24-27, 1975. The Conference is being sponsored by the United States National Academy of Sciences with funds provided by the National Institutes of Health and the National Science Foundation. The purpose of the meeting is to review the progress, opportunities, potential dangers and possible remedies associated with the construction, and introduction of new recombinant DNA molecules into living cells [see Science 185: 303 (1974)].

The program will include discussions of a) the present and possible future methodologies for constructing, propagating and amplifying recombinant nucleic acid molecules; b) the molecular biology and natural history of autonomously replicating plasmids and their microbiol, plant and animal hosts; c) the molecular biology of free and cryptic oncogenic virus genomes including an examination of the rationale and potential risks of introducing such genetic infection into microbiol plasmids; d) the methodology, scientific and practical benefits, and the risks of linking segments of eukaryote genomes to autonomously replicating microbiol plasmids; e) approaches to assess, and to minimize or eliminate any serious biohazards stemming from this line of research.

On behalf of the Organizing Committee I should like to invite you to be a participant at the Conference. Your experience, knowledge and interest in these problems would be valuable in attempts to assess the potential benefits and risks of this line of investigation.

Though the Conference grant is sufficient to provide the entire room and food expense for each of the participants, there are insufficient funds to reimburse the travel expenses of any but the session chairmen, the speakers and panelists. If you wish to attend the Conference but would be unable to do so for lack of travel funds, please let me know as soon as possible the amount that you would need.

Dr. James Watson
December 4, 1974
Page 2

Inasmuch as attendance is being limited to about 150 people and there are many possible participants, we must insist that this invitation not be transferred to someone else without our knowledge. Because of a shortage of time your prompt reply to this invitation would be deeply appreciated. If you agree to attend, you will receive a future mailing describing the Asilomar Conference Center and specific instructions and transportation schedules for reaching the facility.

Sincerely yours,

Paul Berg

Paul Berg
for the Committee on Recombinant DNA Molecules

December 20, 1974

Dr. Paul Berg
Department of Biochemistry
Stanford University Medical Center
Stanford, California 94305

Dear Paul:

I'm coming to the Asilomar meeting.

Yours sincerely,

J. D. Watson
Director

JDW/mh

EMBO

EUROPEAN MOLECULAR BIOLOGY ORGANIZATION

From the Secretary-General

MRC LABORATORY OF MOLECULAR BIOLOGY
HILLS ROAD, CAMBRIDGE CB2 2QH, ENGLAND

Telephone: Cambridge 48011

19 July 1974

Professor Paul Berg,
Stanford University School of Medicine,
Department of Biochemistry,
Stanford,
Calif. 94305, USA

Dear Paul,

I am writing in my capacity as Secretary-General of EMBO to say that we have received the statement of your committee on 'Potential Biohazards of recombinant DNA molecules'. The Council of EMBO has not yet had an opportunity to meet to discuss this matter, but I have talked about it to our Chairman, Niels Jerne, and some other individual members of Council. We all share your concern, and would very much appreciate being kept in touch with your further activities.

Specifically we note your fourth recommendation that a meeting be convened in Spring 1975 to review the progress in this area and to discuss appropriate ways to deal with the potential biohazards. On behalf of Niels Jerne I am writing to say that EMBO would be very glad to be associated with you in this enterprise, and would be ready to collaborate with you in organizing such a meeting, if you felt this to be appropriate. I should be very grateful for your reactions to this suggestion.

I should add that both the EMBO Council and the Scientific Advisory Committee of the European Molecular Biology Laboratory have been approached on the same subject by a group of European biologists including Charles Weissmann. Further, you may like to know that Sydney Brenner will probably be involved with a British group organized at the official level to discuss the same problems.

With kind regards,

Yours sincerely,

John

John C. Kendrew

Document 1.10

July 26, 1974
Nature (**250:**278)

What British scientists say . . .

Molecular dirty tricks ban

THE critics of molecular biology are fond of pointing out the scarcity of practical benefits. It is a sad paradox that the very developments which could ultimately have immense value for production of useful, but rare, molecules, should result in an urgent cry of 'halt' from a committee of leading scientists involved, supported by no less than the United States National Academy of Sciences. The amplification of selected genes and their products by synthetic recombination with freely replicating DNA of bacterial plasmids, could, for example, revolutionise the commercial production of substances like insulin, or pituitary hormones, while the bulk synthesis of a transforming gene protein from an oncogenic virus might allow the design of specific antagonists. As more genes are identified on cleavage products of DNA from animal cells and viruses, so the potential for practical application will increase, to say nothing of the basic knowledge gained.

Yet few will deny the wisdom of the appeal for a moratorium on certain classes of experiment until the implications are more clearly understood. It is not just the risk of unpredicted and explosive self replication of some sinister DNA sequences which causes concern, it is the possibility that by using *Escherichia coli*, the workhorse of molecular biology, the explosion might occur in someones gut and then be transmitted like antibiotic resistance, throughout the world. No doubt a good many dirty tricks have been attempted and discarded by nature in the course of evolution, but the disquiet arises from the utterly novel associations of genetic material which are now possible. The potential benefits should, therefore, be delayed, not for ever, but until consequences can be assessed, and preliminary experiments carried out under conditions of maximum secruity.

Most of the technology involved in all this has been developed in the United States and it is encouraging that the very leaders in the field have taken the initiative and have been supported by the acadamy. It is now to be hoped that academics and learned societies in other countries will add their weight, and that international organisations such as the European Molecular Biology Organisation will lend support. Granting agencies, and even editors of scientific journals, will also need to consider their policies in the face of wide support for a moratorium. For many it will be a test of self denial and social responsibility in the face of strong intellectual temptation.

Michael Stoker, ICRF

"*Walter, explain DNA just once more and I promise I won't ask you again.*"

Document 1.11
August 28, 1974
Medical Tribune (**15**: 3)

Europeans Divided on E. Coli Manipulation

Medical Tribune World Service

GENEVA—Investigators in Europe are reacting with concern to an appeal by a group of prominent U.S. biologists to halt certain types of genetic experiments.

The council of the 290-member European Molecular Biologists Organization (EMBO) is already planning a meeting to consider whether the appeal, published in *Science* and *Nature*, should be followed, Prof. Niels Jerne, EMBO president, told MEDICAL TRIBUNE in a telephone interview.

European scientists are taking positions along a broad spectrum of views on the controversy, which relates to genetic manipulation of Escherichia coli.

Wide Range of Views

Views range from support for a clear-cut ban on the two specific types of experiments described in the appeal, to opposition to any restrictions.

The U.S. appeal was launched by a group that includes Dr. Paul Berg, of Stanford University School of Medicine; Dr. David Baltimore, of the Massachusetts Institute of Technology; and Dr. James Watson, the Nobelist who now directs the Cold Spring Harbor laboratory of quantitative biology on Long Island.

The U.S. scientists warned that work planned by several research teams, including signatories to the appeal, would create new types of infectious DNA elements whose biological properties cannot be predicted in advance.

They have asked scientists throughout the world to join them in voluntarily deferring experiments which would involve:

- Implanting various types of drug resistance in bacteria now lacking it.
- Attaching oncogenic or other viruses to the plasmids of bacteria, or to the DNA of other viruses.

Support for the idea of warning the scientific community of the possible risks was voiced by Professor Werner Arber, of Basle University, Switzerland.

Because the experiments are simple to do, Dr. Arber fears "headline seekers" may rush to be first to recombine a possibly oncogenic virus with a fragment of a bacterial strain in a race for fame. "This is only one very specific application which comes to mind because the money flows easily in cancer research," he commented.

Dr. Arber says it is urgent to define the guidelines for controls in this area of research. Control is not complicated, he points out. "There are laboratories using pathogenic material all over the world. The problem is that people working with the new protein enzymes that can cut up DNA are biochemists. They have no idea of the danger of a virus or of pathogenic bacteria and do not work in sterilized conditions. Such work should be done only in a microbiological laboratory where all protective measures are taken."

Responsible scientists acquainted with the field should get together and discuss precisely what action should be taken, in the view of Prof. John H. Suback-Sharp of the institute of virology at the University of Glasgow, Scotland. "A case has been made by serious scientists which persuades me it is reasonable for this problem to have the consideration of an expert committee," he told MEDICAL TRIBUNE.

The area of experimentation has wide significance and is likely to attract "all sorts of people." Given the money to take adequate precautions, the experiments could well be done, "but I'm not at all sure that anyone who wishes to do them would necessarily do them under safe conditions."

It is better to stop this experimentation right now, in the view of Dr. Georges Cohen, molecular biologist at the Pasteur Institute in Paris. As for the argument of advancing knowledge: "Things have been learned from producing atomic bombs, also. This is a matter that transcends science and not all experiments should be done. Scientists should have some ethics, too."

In agreement with Dr. Cohen, is Prof. Hans Zadhau of the Institute for Physiological Chemistry in Munich, West Germany. "We in Europe are just as aware as those in the U.S. of the risks involved. Personally, I have the feeling certain experiments should not be done, I am not doing them, and I would prefer that nobody else does them either."

Dr. Jerne's Position

EMBO President Dr. Jerne, who is director of the Basle institute for immunology, Switzerland, takes a more neutral position. It could have been considered dangerous to culture polio virus in large amounts, Dr. Jerne points out. "Still, it was done to make the vaccine."

The new techniques offer advantages for the study of mammalian genes, Dr. Jerne went on. One of the questions not so far off in his laboratory is the genetic basis for antibody formation in vertebrates. "If part of the genetic material during this antibody formation in vertebrates could be studied in bacteria, this would offer advantages and be entirely harmless. There are no safety problems involved here."

"There are inherent dangers in much of our work. We must get some knowledgeable people together to determine whether we should limit this technique or not—and that is what we intend to do in EMBO. Simply calling a halt to the research does not necessarily solve any problems at this point."

Dr. Peter Hofschneider of the Max-Planck Institute for Biochemistry in Munich, West Germany, agrees. "All research can be dangerous—so are knives," says Dr. Hofschneider. "If you can avoid escape from the laboratory of the E. coli with new plasmids, then it is not dangerous."

In the coming 10 to 20 years it may be possible to produce important biological materials for use in genetic therapy of inborn diseases. "Rather than stop the experiments, it is better to take the proper precautions in well-equipped laboratories," he says.

The potential for medical progress is far too great to call a halt to such research, according to Prof. Martin Billeter of the Institute of Molecular Biology at Zurich University, Switzerland. With proper controls, he feels, the danger involved is not going to be great. There is now the possibility of replicating all sorts of DNA as plasmids in E. coli, Dr. Billeter said.

Document 1.12

January 1975

Ashby Working Party Report (HMSO Cmnd. 5880), Excerpt

REPORT OF THE WORKING PARTY ON THE EXPERIMENTAL MANIPULATION OF THE GENETIC COMPOSITION OF MICRO-ORGANISMS

6.8 Finally, we would put on record our impression of the general concern among scientists themselves over this matter. It was a group of research workers using these techniques who called for a pause in the work to give time for the risks and benefits to be assessed. It can be no more than a pause, because the techniques open up exciting prospects both for science and for its applications to society, and the evidence we have received indicates that the potential hazards can be kept under control. The tradition of prompt and open publication of results is likely to acquaint all workers in the field with the progress of their peers. This can help in dissuading people from conducting irresponsible or unnecessarily hazardous experiments.

VII. RECOMMENDATIONS

7.1 We have not been asked to make recommendations, and the gist of our report is summarised in sections I and VI. However, our assessment would not be valid without certain recommendations, which we set out below.

7.2 We recommend that, subject to rigorous safeguards based on those set out in para. 6.5, these techniques should continue to be used because of the great benefits to which they may lead (3.1–3.4).

7.3 We recommend that all those who work with these techniques should have training in the handling of pathogens and that laboratories where the work is done should have the basic equipment for the containment of ordinary pathogens (6.5 (*a*) and (*b*)).

7.4 We recommend that in laboratories using these techniques someone should be designated as a biological safety officer and that some sort of central advisory service should be established (6.5 (*d*)).

7.5 We recommend that continued epidemiological monitoring of workers in the laboratories where these techniques are used should become a routine practice (4.4, 5.2, 5.10, 6.5 (*h*)).

7.6 We recommend that large-scale work with any potential hazard involving quantities of 10 litres or more, should not be done except in specially equipped laboratories (6.5 (*f*)).

7.7 We recommend that workers using these techniques should seek to reinforce the safeguards of traditional containment procedures by genetical devices to " disarm " the organisms they use for experiments and by conducting other kinds of research on possible hazards of the techniques (5.12, 5.13).

7.8 We recommend that further investigations should be sponsored forthwith to improve our knowledge of the viability in the human intestine of laboratory strains of *E. coli*, especially K12, and their capacity to transfer plasmids to resident strains (4.5).

Presented to Parliament by the Secretary of State for Education and Science by Command of Her Majesty January 1975

LONDON

HER MAJESTY'S STATIONERY OFFICE

26p net

Cmnd. 5880

Document 1.13
January 23, 1975
New Scientist (**65**:86)

Comment

Not good enough

The report by Lord Ashby's working party on the genetic manipulation of microbes (Cmnd 5880, HMSO, 26p), published on Tuesday, is a document of historic significance. The working party was set up last year by the Advisory Board for the Research Councils, to assess the "potential benefits and potential hazards" of techniques that allow the genetic composition of microorganisms to be manipulated in the laboratory. The immediate stimulus was a call by a US National Academy of Sciences committee, chaired by Professor Paul Berg, for a temporary halt to two types of experiments. These were the artificial introduction into bacteria of genes conferring resistance to antibiotics or causing the formation of toxins, and secondly the introduction of virus genes into bacteria. The committee also urged that experiments involving the incorporation of animal genes into bacteria "should not be undertaken lightly" (Nature, vol 250, p 175 and New Scientist, vol 60, p 274). Central to the committee's anxiety was the possibility that strains of Escherichia coli containing artificially-created combinations of DNA could become disseminated outside of the laboratory and cause unpredictable dangers in man, in other animals, and possibly in plants. E. coli, the bacterium most commonly used in research of this nature, also lives in abundance in the human gut. In theory, therefore, the novel and potentially dangerous strains could establish themselves in the intestine and/or transfer their deleterious genes to its normal inhabitants.

Lord Ashby's report is notable, therefore, in examining a potential hazard of unique seriousness—and one which has been brought into the arena of public debate by a group of scientists anxious about the repercussions of their own work. It is also unprecedented in the clarity and simplicity of its prose. The committee felt, quite rightly, that the subject should be made accessible to people not familiar with the increasingly opaque jargon of microbial genetics. One cannot but applaud the motives of both the researchers and the Ashby panel in stimulating public awareness and discussion in this way.

The working party was clearly impressed by the benefits likely to come from manipulating microbial genes. These include the incorporation of nitrogen-fixing genes into crop plants that do not usually fix nitrogen, extension of the climatic range of crops by similar techniques, and the use of animal genes to turn microbes into miniature factories for making products such as hormones. The research could also reveal much about gene action and possibly about the aetiology of cancer.

Turning to the hazards, much of the report underlines the argument of Professor E. S. Anderson (Nature, vol 250, p 279) that the risks could be greatly reduced if molecular biologists doing experiments of this sort were to adopt the precautions that are routine in medical microbiology laboratories handling pathogens. Such techniques are far from universal at present, and the working party rightly advises that they should be used wherever microbial genetics research is in progress. They also recommend appropriate training of personnel, the creation of biological safety officers in such laboratories, and the restriction of investigations with large cultures to specially equipped laboratories.

After urging these long-overdue reforms, however, the report falls down badly. It describes for the first time some recent and very important experiments by Professor Anderson showing (unexpectedly) that K12, the favoured laboratory strain of E. coli, can indeed survive for a few days in the human gut, while suggesting that the likelihood of accidental genetic transfer to the resident flora is "very low". The report also draws attention to the possibility of developing strains of E. coli in which genetic manipulation might be conducted but which *cannot* survive in the gut. Yet while suggesting that "this search for genetic safeguards should be pursued", the report concludes that genetic manipulation should continue to be used. No embargo. Not even a voluntary pause while such practical safeguards are developed.

The Ashby group specifically points out that it was not concerned with wider ethical questions about the use of the new techniques. Even so, it is surely unsatisfactory for the working party to have reported for the first time the persistence (even temporarily) of E. coli K12 in the gut; to suggest model systems that might be developed to obviate all risk; and yet to conclude that not even a brief moratorium is needed. That it should have reached this disquieting conclusion, just a month before an international conference in California is to consider the whole question, is even more regrettable.

Bernard Dixon

This article first appeared in *New Scientist*, London, the weekly review of science and technology.

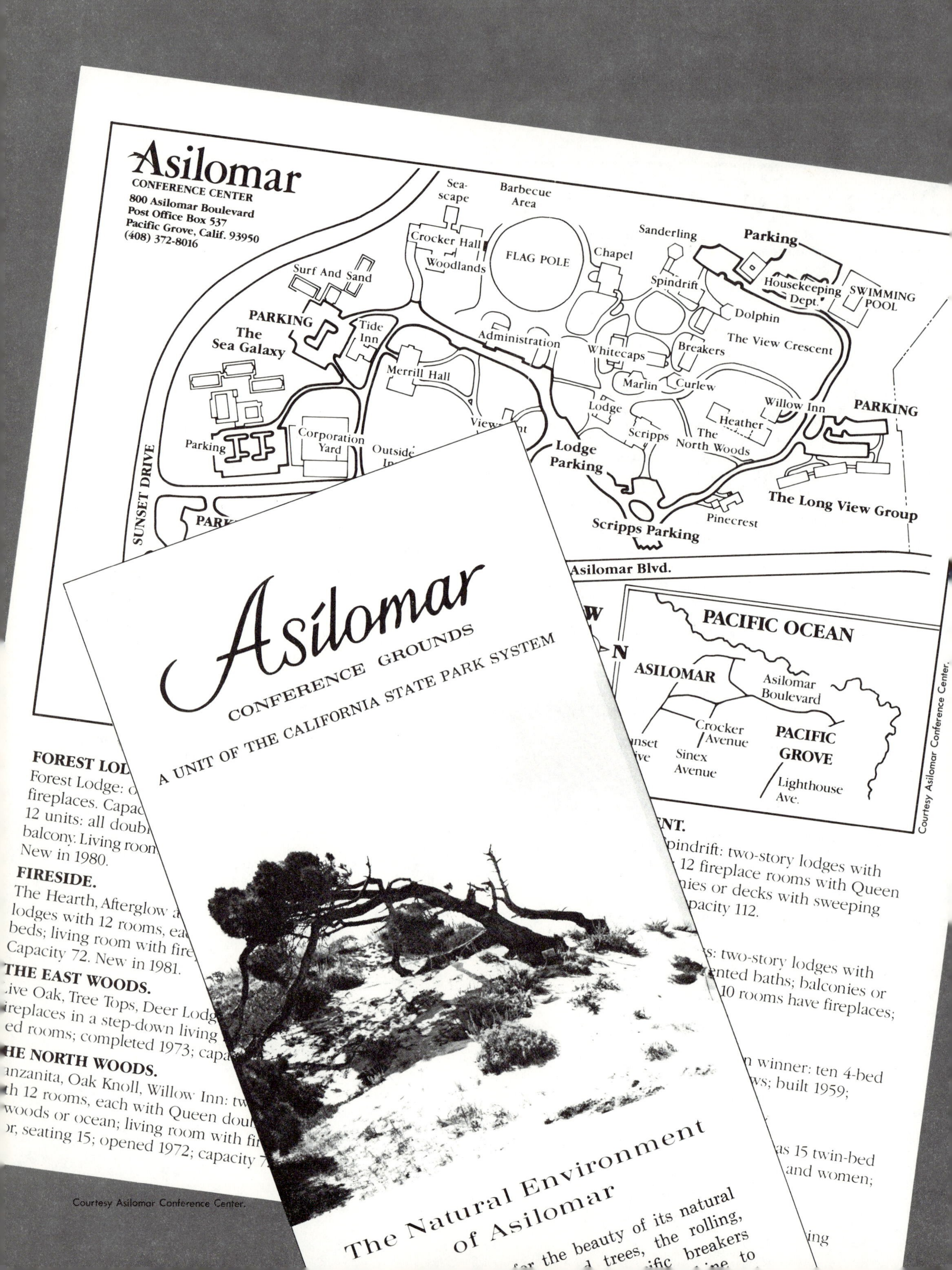

Courtesy Asilomar Conference Center.

THE ASILOMAR CONFERENCE

It would be extremely difficult to improve upon Michael Rogers' masterly description of the atmosphere at the Asilomar Conference in February 1975, and we shall not try. (See Doc. 2.1.) Rogers' book *Biohazard* (Knopf, 1977), describing the early part of the recombinant DNA debate, is also enjoyable reading.

The significant events at the Conference took place during the last morning, which was a confused (some would say chaotic) affair. After a crowded three-day program (Doc. 2.2), members of the organizing committee spent the final night drafting guidelines that included a scale of special laboratory safety procedures to match a scale of conjectural hazards imagined to be associated with specific classes of recombinant DNA experiments. Paul Berg, Chairman of the Organizing Committee, was clearly set upon ensuring that a majority of the participants voted in favor of his committee's draft before the Conference dispersed. Repeatedly, however, precisely what was being put to the vote as the guidelines were discussed, item by item, became totally obscure as suggestions and amendments from the floor threatened to overwhelm the chair. Moreover, a small minority—notably Stanley Cohen (Stanford), Joshua Lederberg (Stanford), and James Watson (Cold Spring Harbor Laboratory)—expressed their opposition to the whole proceeding. Watson, for example, had proposed that the moratorium be lifted and no special precautions or guidelines be self-imposed. With a different audience and in a different atmosphere, this view might well have found more favor (Doc. 2.3), but the majority at Asilomar either declared, or failed vocally to deny, that special guidelines were necessary; they at least agreed that there were valid reasons for concern, that there might be significant hazards.

The confusion and opposition notwithstanding, the requisite majorities were marshalled for each section of the organizing committee's report, although it was obvious to everyone that such hastily conceived and written proposals required much polishing before they could form a report fit for transmission to the National Academy and for publication. In any event the delay was of several months, and the Report of the Asilomar Conference eventually appeared in *Science* on June 6, 1975 (Doc. 2.4).

Despite the confusion of the last session, many participants left

Asilomar as exhilarated as they were exhausted. As one reflective participant, DeWitt Stetten (NIH), who was to play a major role in subsequent events as the first Chairman of the NIH Recombinant DNA Advisory Committee, has shrewdly recognized in retrospect, Asilomar had many of the elements of a revivalist meeting (see Chapter 12, Doc. 12.6). The participants had been praised by lawyers who addressed the meeting, for their social responsibility and the exemplary way in which they had publicized their concern about the safety of their research as soon as the first element of doubt entered their minds, and before any real hazards, as opposed to conjectural ones, had arisen. They had voted to impose upon themselves special safety precautions. Having cleared their consciences, they now felt able to return to their laboratories and carry on, or begin those exciting experiments, which since the publication of the Moratorium letter had seemed of dubious propriety. Having demonstrated their integrity, they naively believed that they would now be free of outside intervention, supervision, and bureaucracy.

At the time the additional safety precautions seemed a price worth paying. The safety practices ranged from the most elementary, such as not pipetting by mouth and not eating or smoking in the laboratory, to the installation of sophisticated containment cabinets to protect against aerosols, in specially ventilated rooms. In addition the use of biologically disabled host-vector systems for the recombinant DNAs was agreed. Indeed the idea of biological containment seemed to be one of the most important and novel results of the Conference, although the concept had been raised previously by Waclaw Szybalski (Doc. 1.8). It should be possible to render the host bacteria and vectors used for recombinant DNA research so weak as to be incapable of survival outside special environments created in the laboratory. Fear of an Andromeda strain being generated and wreaking havoc among humans or another natural population could then be completely discounted. Everyone at Asilomar felt confident that such biologically disabled and therefore containable bacteria and vectors could be produced by conventional genetic manipulation in a few weeks or at most months, and Sydney Brenner led a discussion on their genetic design. Subsequent events were, however, to prove otherwise.

The daily press recognized the results of the Conference as a vote both to continue the research and to require tighter safety precautions to contain the associated biohazards (Doc. 2.5). At that stage the U.S. environmentalist groups (such as Friends of the Earth) that would subsequently align themselves with the small minority of biologists who wished to restrict severely if not entirely prohibit recombinant DNA research had not yet taken up the issue and therefore played no part at Asilomar. Science for the People, a group of new-left scientists in Cambridge and Boston, which included several molecular biologists, was also not represented. The molecular biologists in this organization had won much publicity in 1969 by holding a press conference to denounce their own research as a step toward the eventual manipulation of human genes, a capability that they believed would inevitably give rise to evil eugenics policies. The experiments precipitating this self-denunciation were the isolation of a genetically characterized fragment of DNA, in close to purity, from the bacterium *E. coli* by a combination of physical and conventional genetic methods. The recombinant DNA technique provided a simpler, more general method for achieving this same end—the purification of genes in large amounts. Not surprisingly, therefore, the open letter sent by Science for the People to the participants at the Asilomar conference reiterated the group's familiar denunciation of such research, as well as the structure and priorities of U.S. biological research in general. They called for a continuation of the moratorium (Doc. 2.6).

Document 2.1

June 19, 1975
Rolling Stone (**189**:36)

THE PANDORA'S BOX CONGRESS

By Michael Rogers

140 Scientists Ask: Now that We Can Rewrite the Genetic Code, What Are We Going To Say?

The event was sufficiently historic that not until it was nearly over did anyone have time to think of taking a group portrait. And by then the official photographer had already departed and thus the International Conference on Recombinant DNA Molecules—a diverse mix of 140 scientists who manipulate the most fundamental life processes in laboratories from Moscow to Memphis—will remain pictorially unrepresented in the history of modern science.

But their activities, almost certainly, will not: The conference—four intense 12-hour days of deliberation on the ethics of genetic manipulation—should survive, in texts yet to be written, as both landmark and watershed in the evolution of social conscience in the scientific community.

And perhaps in the evolution of humanity itself. "Nature," as a middle-European microbiologist told me late one night, "does not need to be legislated. But playing God does."

Abruptly, in a matter of months, the young science of molecular biology had happened upon the first real tools of genetic engineering: the ability to create, in the test tube, creatures never before seen on this planet. Thus far the technique was crude and extremely limited—but even so, the molecular biologists had clearly reached the edge of an experimental precipice that may ultimately prove equal to that faced by nuclear physicists in the years prior to the atomic bomb.

But in the last hectic hours before the conference slid, with the momentum of a base runner, into its conclusion, it would briefly appear that even those scientists with fingers on the most intimate genetic self-regulation processes of nature were themselves incapable of any similar scientific self-regulation.

The Arrival

"I can't room with you, man!" a middle-aged Scottish delegate in wool sports jacket exclaims to his newly assigned roommate at the Asilomar registration desk as, through broad picture windows, a crystalline orange Pacific sunset fills the sky above the white beach 200 yards to the west. The Scot, it appears, has just learned that his roommate, a young American, specializes in the lower mammals.

"I'm an invertebrate man, myself," the Scot explains, mock serious.

"Well," says the young American tentatively, somewhat taken aback. "Well," he admits slowly, "I'm an insomniac."

"Great!" says the Scot. "So am I. We can rewrite the proposal at three in the morning!" And then off they go, baggage laden, into the dusk and the van that carries attendees to the small redwood dormitories scattered throughout the wooded grounds.

The molecular biologists descend upon California's Monterey Peninsula on very nearly the same day as do the monarch butterflies. The Asilomar Conference Center, three hours south of San Francisco, is a scatter of rustic dormitories and spacious meeting halls, hidden in a seaside forest of redwood and Monterey pine, just outside the tiny town of Pacific Grove. Traditionally, each spring immense flocks of migrating monarchs, numbering in the millions, briefly cover the trees here in thick sheets of orange and black—an event of no small significance to the merchants of Pacific Grove.

This spring, along with the butterflies, come the biologists, arriving from everywhere and fueled by $100,000 put up jointly by two prestigious American organizations: the National Institutes of Health and the National Science Foundation. The conference was organized by the National Academy of Science.

They arrive on a bright blue Sunday afternoon—the finest February weather that the Monterey Peninsula has to offer—shuttled in from the local airport, often still clad in overcoats donned (hours earlier) in Cambridge or Kraków. Prominent on the Asilomar registration desk is a stack of mimeographed maps that detail the route one need follow to view the migrating monarchs but no one, this crisp Sunday, seems terribly interested in Lepidoptera.

"Yes!" a plump and bulkily sweatered New Yorker exclaims to a Japanese colleague. "We tried that; first you mutagenize the cell, then you cut. . . ." Sandwiched between pool and ping-pong tables, researchers meet for the first time in months, and even in the middle of an overwaxed linoleum floor, their discussions suggest both the vitality of small boys with new chemistry sets and the electricity of back yard gossip. The excitement is unmistakable. Clearly these people think they are *on* to something.

And the fact is, they are. And that, moreover, is precisely why they are at Asilomar.

It was, after all, only about a century ago that the Austrian monk Gregor Mendel, browsing in his monastery sweet-pea patch, first described the phenomenon of genetic inheritance. And far more recently that human beings identified the minute chemical container — deoxyribonucleic acid: the DNA molecule—in which that genetic information was actually stored.

That container — an intricate, lengthy, ladder-shaped organic molecule—holds the most widely understood language on the planet. The design of every living organism—from the paramecium in the mud puddle to Albert Einstein at Princeton—was at one point described in variations of precisely the same sinuous patterns. The most understood—and also the most difficult to translate. A single strand of human DNA, microscopically small, contains at least the information of a library of 1000 volumes.

The chemical keys to that library have been hard to find. To translate one volume, even harder. And to write one's own book—impossible.

Until recently. "Science," as a British biologist observed at Asilomar, "has built-in pauses; some last 100 years. But the thing about recombinant DNA engineering is that it's suddenly made many things very easy that were once very difficult." Recombinant DNA engineering is the reason for Asilomar: the discovery of the first rudiments of grammar for that previously unspeakable genetic tongue.

The ancient Greeks believed in a mythologic being called a Chimera—a female monster composed of pieces of two or more animals. Molecular biologists now believe in DNA molecules that they call precisely the same thing. Moreover, they make them themselves. Recombinant DNA engineering uses certain newly discovered enzymes to disassemble the long DNA molecule in so orderly a fashion that the loose bits of genetic coding may then be rejoined, grammatically, into coherent sentences. And such a sentence may well describe—and create—the mutual offspring of two altogether different creatures incapable of mating in nature. "To join duck DNA," as the same British biologist was fond of saying, "with orange DNA."

If one knows the grammar, one can begin to make up new sentences. Dial-a-baby, then? Or better, dial-a-monster? Not, by a long shot, yet. The brand-new techniques work, thus far, only with bacteria and viruses—organisms so small that human beings really only notice them when they make us ill.

But there, precisely, lies the rub—and also fit meat for countless science fiction scenarios.

A Science Fiction Scenario

For starters, we'll cast a young molecular biologist who looks like Woody Allen, tends to shuffle a bit and mumbles, say, in the vaguely sullen fashion of a Cambridge-educated Stanley Kowalski. He should be exceedingly bright but not always terribly careful about laboratory hygiene.

He is most likely interested in cancer because cancer is where the money is just now, but he might just as well be curious about, say, the way bacteria learn to resist antibiotics.

The molecular biologist has several million laboratory helpers: a colleague or two, a couple of graduate students, a handful of technicians and an immense colony of bacteria called *Escherichia coli*—the last of whom work only for room and board.

Andrew A. Stern/NAS

Senior Soviet academician W.A. Englehardt.

Those particular bacteria were recruited from a single human gut more than half a century before—and since then they have existed, almost exclusively, within the laboratory. A whole host of their relatives still reside quite happily in literally millions of human intestines — but since its isolation, this laboratory strain of K12 *E. coli*, as it is known to its friends, has had a far less placid existence.

In our scenario, the K12 *E. coli* is about to serve as something of a factory hand—an experimental workhorse — in a procedure called plasmid engineering.

A plasmid is a tiny circular bit of genetic information—DNA—that floats around inside the sack-like cell of a simple bacterium. Plasmids can affect the bacterium in some fairly significant ways; they are, for example, responsible for the increasing number of bacteria that are now learning resistance to old-line antibiotics like penicillin and tetracycline.

And here it is that our scenario really begins, for what our molecular biologist has set out to do is to produce plasmids of a variety never before seen in nature and then pump those novel plasmids into the patient old *E. coli*, whereupon *E. coli* will not only begin to obey the plasmid information but to reproduce it as well.

From here on, the scenario almost writes itself. The researcher, say, manages to isolate what he suspects to be a bit of the genetic information that causes tumor growth—an ability known technically as "oncogenicity." By grafting that bit of DNA onto a plasmid and introducing that modified plasmid into *E. coli*, he might just determine whether he has in fact identified those spurious genetic orders that cause normal cells to lose their biochemical minds.

If he has, and he follows up carefully, it could mean the Big Time; a major break in cancer research and—who knows?—the limelight, prizes, prestige, a funded research chair: all of the not inconsiderable gravy that can accrue to the very good or very lucky medical researcher in this country.

But this isn't a Cinderella scenario. So he gets sloppy; just once. Perhaps he has only recently turned to molecular biology from, say, chemistry and while he does his best, he still hasn't fully comprehended that his glassware now contains something far different from lifeless arrangements of molecules. Or perhaps a laboratory assistant is at fault, borrowing a page from an old story told about Fort Detrick—for years the highest-security center of American biologic warfare research—where an enlisted man, not so fresh from a weekend pass, once failed to seal securely a high-speed centrifuge and thus managed to spray the entire laboratory with a fine aerosol mist of concentrated and monstrously contagious plague.

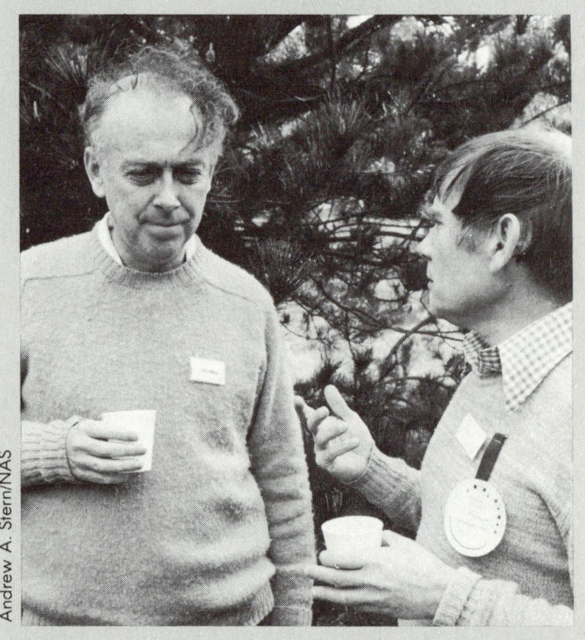
Andrew A. Stern/NAS

Nobel laureate J.D. Watson and Sydney Brenner confer amid the cypress.

At any rate, the long imprisoned *E. coli*, laden with a brand-new bit of biological ability, suddenly finds itself liberated; floating in a minute droplet on a technician's finger, then onto a tuna-fish sandwich and thence into a luckless human gut. Or, in a culture not quite completely killed, down some stainless steel laboratory sink and thus into a sewer system teeming with billions of close relatives.

And now what? Nothing to this point is excessively speculative: It was, after all, only two years ago that smallpox virus managed to escape from an experimental laboratory in London, killing two women in its characteristically swift—and incurable—fashion. So the really speculative part is yet to come: Precisely what could our artificially mutated *E. coli* do with its sudden freedom?

An epidemic cancer that spreads through the sewer system? A once conquered disease—like bubonic plague—now, abruptly, again incurable? Or a brand-new disease, sudden and mysterious, that has never before appeared in human beings?

At this point, there's no certain answer to the question. There is, simply, no further information on which to proceed—and there's no information precisely because the question deals with organisms that have never before existed on the planet. But the concern that brought 140 molecular biologists to Asilomar is clear: Human beings have once again happened upon the ability to threaten themselves with a blight that might someday prove to be the biological equivalent of nuclear radiation leakage.

But gamma rays, of course, cannot reproduce.

The Letter

"Eight months ago," says conference organizer Paul Berg late one afternoon at Asilomar, "the telephone calls were coming into our laboratory daily: 'Send us pSC101 [a variety of plasmid used for recombinant engineering].' 'What do you want to do?' we'd ask. And we'd get a description of some kind of horror experiment and you'd ask the person whether in fact he'd *thought* about it and you found that he really hadn't thought about it at all. And that's not to call down criticism on anybody because two years earlier I had been in the same position."

Thus it was, in July of 1974, that the letter first appeared, published simultaneously in the slick pages of three major scientific journals — the first major self-regulation appeal to the scientific community since the early Forties, when physicists agreed to deny German access to nuclear data. The letter, signed by 11 leaders in the field, filled only a single page—headed "Potential Biohazards of Recombinant DNA Molecules"—but that single page said enough to send a major tremor through the entire international fraternity of molecular biologists. In blunt language, the statement called for a global moratorium on certain experiments in the brand-new field of recombinant DNA engineering—specifically those which threatened to introduce antibiotic resistance, oncogenicity or the other poisonous qualities known generally as "pathogenicity" into microorganisms that presently do not possess these abilities.

But wait a minute. One would think we already have all of the dangerous microbes we need and bac-

Andrew A. Stern/NAS

Molecular biologists on chapel steps.

teria seem to be learning antibiotic resistance quickly enough on their own. So who the hell would *want* to do these experiments in the first place?

Lots of people, it develops—and for some fairly good technical reasons. Antibiotic resistance, for example, just happens to be a very convenient genetic trait to transplant. Once the previously nonresistant bacteria have been modified through recombinant engineering, one can find out whether the genetic transplant has taken simply by dosing the modified bugs with antibiotic. If the bacteria die, the experiment is a failure. If they live, the experiment is a success and one likely has learned something fairly valuable about molecular biology. And at the same time created a brand-new strain of antibiotic-resistant bacteria. And equally, how will one ever figure out how cancer works without using the same techniques to take a cancer virus apart?

So reaction to the letter was both swift and vocal —from the decision of the Medical Research Council in Great Britain to make the suspect experiments virtually illegal to the opinion expressed by one microbiologist at an international conference in Switzerland: "You should not hamper basic science. You cannot slow down research."

"Such experiments," the letter warned, "should not be undertaken lightly." And that was in fact part of the problem: Nobody in the field was taking them *lightly*. Molecular biology had been in the experimental horse latitudes too long and, as another attendee at the Swiss conference observed: "This new approach is likely to revolutionize not only our knowledge of gene and chromosome organization but possibly of genetic diseases, perhaps cancer. The potential benefits are so great that this sort of research is gaining uncontrollable momentum. . . ."

Potential benefits. In a crasser perspective this seemed, as well, to be one of those periods of scientific inquiry when the prizes and the plums hang from a somewhat lower bough on the tree of knowledge. One of the signatories of the letter—Nobel laureate James Watson, who, with Francis Crick, first deciphered the structure of DNA two decades ago—himself describes that peculiar brand of breathless scientific competition in his book *The Double Helix*.

So feelings were high and—as some of the signers of the letter were themselves pioneers in the field—there were even dark mutterings about the moratorium as "intellectual lockout." But even so the moratorium was almost universally observed during the eight months between the publication of the letter and the first sunny Sunday at Asilomar. And by that time there was no desire to talk about the few violators. Instead the questions were of the immediate future: Under what conditions may we proceed with these experiments?

And far more urgently: When can we start?

First Session

Monday morning, the full moon still bright over the blue-black Pacific, the breakfast bell tolls in the center of the compound and soon the molecular biologists begin to file through the dawn light and into the redwood chapel that will serve as center for the next four days.

Inside, the chapel is dim and gloomy, with theater-type seats, exposed beams and an elevated stage that, even stripped of ecclesiastical accoutrements, is still unmistakably reminiscent of an altar. Debating the ethics of human interference with the mechanics of evolution in a church at the edge of the immense saline test tube where it all started: Rarely does one find one's metaphors so cheap—or so apt. "Here we are," a young scientist from the East Coast will tell me later that night over beers, "sitting in a chapel, next to the ocean, huddled around a forbidden tree, trying to create some new Commandments—and there's no goddamn Moses in sight."

The caliber of the conference is such that, should some wrathful hand wipe out this chapel and the scientists within, it would likely set back the progress of molecular biology a decade or so. And to add cheap irony to cheap metaphor—it is immediately clear, this morning, that of this vanguard now fiddling with the most basic mechanics of reproduction, no more than six are female.

And no real candidates for Moses, either, among the jet-lagged congregation. Paul Berg, the Stanford biologist who headed the moratorium group, just doesn't look enough like Charlton Heston. This morning he appears the quintessential young California academic: tanned, intense, athletic, in studiedly casual sport clothes and a suitably collegiate sweater donned against the early morning Monterey chill. He might as easily be dressed for sailing or an early round of golf, but here he is, standing onstage in front of 140 international colleagues, expressing once again a concern that some privately consider obsessive: "What is new," he says with flat certainty, "is that recombinant DNA can now be made from organisms not usually joined by mating—and hence can give rise to DNA molecules not previously seen in nature."

"If we come out of here split and unhappy," says another conference organizer—a young, successful molecular biologist named David Baltimore, clad in trim beard and elaborately embroidered Levi jacket —"then we will have failed the mission before us."

The first session rolls on well past the lunch bell, much of it fetchingly anal. A major question, vociferously argued, is just how likely it is that these *E. coli* K12 bacteria, so long laboratory pampered, will survive in the human gut, should they escape their test tubes. A series of British researchers demonstrate a consistent penchant for mixing K12 cultures into half pints of milk, swallowing same and then monitoring their subsequent stool for evidence of bacterial survival. The topic offers some opportunity for drollery ("A nice, quiet, boring person," someone describes a chart of stool flora, "as far as his colon is concerned"), but by the end of the session, the implications of K12 ingestion seem far from resolved.

But by the end of the same session, another implication seems all too clearly defined: pure and unadulterated paranoia.

The Press, the Public and Paranoia

"These proceedings," announces David Baltimore at the opening of the first session, "will be taped—for the archives and for review, not for release. And anyone who does not want to be taped may ask and the machine will be turned off."

Immediately someone rises in the audience. "But what about the *press*?"

There is a brief silence in the still somnolent audience. What *about* the press, with those nasty Sony cassette machines perched stage left, right beside the official Academy of Science sound equipment?

After some deft reassurance by the NAS press officer, a vote is taken. The press is permitted, with many abstentions, its recording equipment. But it is not, by any means, yet permitted any real welcome.

And that's no surprise to the press. By now the vibes are unmistakable—and have been so since first application for permission to attend Asilomar. Press attendance was not actively encouraged by anyone involved and in the case, say, of a reporter from Rolling Stone, it took some persistence even to find out whom to *ask*. A writer from Washington told the conference organizers straight out: "A secret international meeting of molecular biologists to discuss biohazards? If the press isn't allowed, I'll guarantee you nightmare stories." Or as a journalist from Southern California said: "The scientists loved the press when we got Nixon. But when we start hanging around their own back yard, they get very nervous."

Nervous indeed. One young researcher from Stanford—a fellow sufficiently in the vanguard to have had a variety of plasmid named after him—is so compulsively press shy that even when the official photographer approaches him at the chapel entrance, he retreats, face covered, like a newly busted big-time mobster hiding behind his fedora on the steps of a precinct house.

It seemed almost a paradigm of the unsatisfactory relationship between the press and science. Paranoid behavior is guaranteed to engender the journalistic suspicion that something is up. And the more attention the press paid, the more paranoid the attendees became—and not entirely without reason. A suitably hysteric story about the antics of an international cabal of biologists devoted to some blackly humorous campaign of creating new cancer viruses might be just the thing to stampede Joe Public.

And the press was not always altogether reassuring; after four days of intense sessions, some individuals were still asking questions that suggested they had passed the previous days locked in a very dark closet. As some cornered scientist explained for the fifth time a fairly fundamental concept of cell biology, the question in his eyes was clear: How the *hell* does this befuddled individual with the notebook think he's going to explain the subtleties of plasmid engineering?

Or perhaps even worse were the questions clearly designed to elicit the quotable lead for some mythic housewife to digest over morning coffee: "Dr. X, would you say that this technique of plasmid engineering is the most important advance in science since the invention of the mammal?" Or: "Dr. Y [who has already expressed the utter impossibility of answering this question at least half a dozen times], in how many years will we have a genetic cure for diabetes?"

Welcome or not, however, the press was there, hunkered down in the front row, Sonys turning, and there was really nothing to be done about it. By the end of the sessions, it was clear that press presence caused the conference attendees both some discomfort and some extra efforts toward public caution.

And that, finally, seemed fairly healthy.

Happy Talk

"This is what we know how to do," one Eastern microbiologist notes plaintively midway in the proceedings. "This is what we're used to doing. I mean, we all get together, we want to know what everybody else is doing."

Indeed. During the first two days of sessions, it becomes immediately clear that the conference attendees would really rather talk about almost anything than the issue at hand. "Molecular biology of

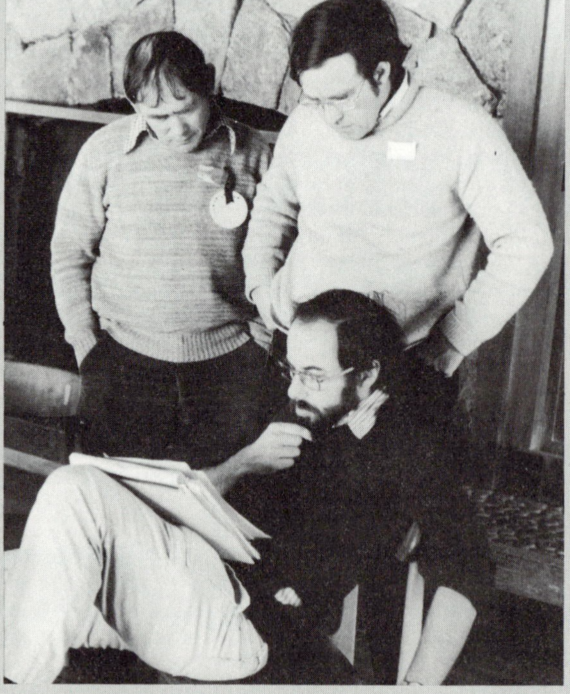

Andrew A. Stern/NAS

"You will," said one lawyer, "be a hard act to follow."

bacterial conjugation and conjugative mobilization of plasmid and other DNAs," say, or "molecular cloning of DNA as a tool for the study of plasmid and eukaryotic biology," each illustrated with slides that appear either like children's plastic building blocks or the tracings of snails in debris-scattered rain puddles.

The talk is exceedingly technical and wanders over a spectrum of topics—some including information sufficiently original that afterward researchers queue up at Asilomar's two pay telephones to relay the word back to their laboratories.

It all goes very smoothly—this is, after all, what these people know how to do—and the presentations offer an interesting inside view of the potentials that the popular press has so long trumpeted for genetic engineering.

At one point, for example, a tan young Southern Californian clambers onstage holding a three-foot-tall weed, freshly harvested from a Monterey roadside, as a colleague passes out similar plants to each row of molecular biologists. It is a legume, says the plant specialist, and if one shakes the dirt off, it's possible to see the tiny nodules on the root structure that "fix" nitrogen—adapt it to forms useful to living organisms—directly from the soil. Not a bad trick, since most world food crops are not legumes and thus require doses of artificial nitrogen fertilizer—created, in turn, from a commodity somewhat scarcer than dirt: petroleum.

Clearly, if one were able to isolate the gene that teaches nitrogen fixation and then ally that, say, with a *wheat* plant—then one might just have one hell of a food crop. One in five human beings already harbors a strain of nitrogen-fixing bacteria in their gut; the same bacterium is almost universal among certain New Guinea tribesmen who eat ten pounds of sweet potatoes a day. It might not, the plant specialist suggests, be such a difficult thing to teach a plant. . . .

Andrew A. Stern/NAS

Serious talk in the Asilomar cafeteria: Only the corned beef was a failure.

It is obviously a far more imminent prospect for genetic engineering than dial-a-baby. So also is the possibility of recruiting some remarkably cheap labor in the pharmaceutical industry. One might, say, isolate the gene that codes for insulin production, pump that gene into a colony of *E. coli* and then sit back as the colony, suitably well fed, begins to pump out pure insulin. The prospect is sufficiently realistic that the conference includes representatives from the research arms of drug manufacturers Merck, Roche, G. D. Searle—and even General Electric.

A Primitive Tribe of Beach-Dwelling Molecular Biologists

But the state of the art, entrancing as it is, isn't really the issue at hand. Yet each time the real issue arises—as, say, when some advisory group, previously appointed, introduces "Proposed Guidelines for Plasmid-Cell DNA Recombinant Experiments" —the proceedings rapidly develop the appearance of some obscure primitive tribe eons ago, accidentally stumbling by trial and error onto the secret of parliamentary procedure.

Odd, one might think, that a roomful of the leading minds on the leading edge of science can't agree on how to run a meeting. But there is little evidence of any concerted drive toward organization here. Sometimes the derailments are benign: An irrepressible gentleman from Switzerland monopolizes a microphone for a baffling ten-minute dissertation on scientific ethics, by the end of which it is difficult to decide whether the researcher's English or his thought processes are the more twisted. Or some well-intentioned American drones for an equivalent period about his role in the licensing of an obscure vaccine some years earlier, which, it grows quickly clear, has very little to do with anything. Or the entire discussion lurches off into a swamp of technical detail: "Well, why can I use *Xenopus* DNA under low risk conditions when Josh Lederberg has to use *Bacillus subtilis* under moderate risk?"

But sometimes it's not clear precisely *what* is at stake. One afternoon a paunchy and influential Englishman, who each morning in the Asilomar dining hall salts his cornflakes, rises to quote a sentence from one of the working papers.

" 'For our purposes,' " he recites, " 'pathogenicity and virulence are defined similarly as "the capacity to cause disease." ' " He pauses briefly and then announces in precisely cultured intonation: "This must rank as the greatest oversimplification of all time."

The panel of seven who drafted the report, arrayed behind microphones onstage, looked slightly hurt. Their report is, after all, 35 single-spaced pages, and getting scientists to write 35 single-spaced pages of anything is no small accomplishment. The Englishman, however, proceeds then to sharply question the qualifications of the whole bunch of them.

Finally the panel leader cuts him off, thanks him for the "input" and asks for a written critique. "This is, after all," he apologizes, "a rather terse document."

The Englishman continues to stand. "You could have fooled me," he observes acidly.

Alterations will be made, the panel leader assures him; after all, this working document was assembled in only six days.

"And why couldn't you do it in six days?" the Englishman wants to know. "After all, the Lord created the world in only seven."

The Russians

"These Russians," an elderly microbiologist from Wisconsin says one night at dinner, "they just send over the old guys from the academy [the politically powerful Academy of Sciences of the U.S.S.R.] who don't know *anything*. You ask them something and they hedge around—and it's not that they're hiding things . . . they just don't *know* in the first place."

The Russian presence is in fact somewhat puzzling. The Soviets number five: two oldsters, in black suits and narrow ties, both with thick, healthy shocks of nearly pure white hair, along with two younger scientists—"The Rolling Stones? Iss a jass band, no?" —and a charming, dapper, San Francisco vice-consul, performing chaperon service.

The oldest is 88 and the talk at Asilomar has him the highest of honchos in Soviet biology. Talk has it also that, due to the massive doctrinal detour posed under Stalin by Trofim Lysenko's curious evolutionary notions, Russian molecular biology is not yet precisely state of the art. And, in fact, the eldest Russian's primary contribution to the conference seems to be his stolid, front-row use of a small pocket camera to photograph the charts and slides projected onstage. His diligence provides more than a bit of amusement, however: For unfathomable reasons, he persists in using a bright miniature flashgun each time he photographs the screen and the consensus is that he may have taken home a bit less than expected.

The oldest Russian's English seems baffling as well; our few conversations stall quickly, as I admit in my college Russian that "I do not speak very much Russian," to which the aged academician smiles, nods and pronounces somewhat ambiguously: "Berry goodt!" What, I wonder, does he expect to get out of these presentations, all in the most technical of English? Until one day I hear the oldster speaking to a science reporter from the East, an attractive young brunette who has fetchingly left the top two buttons of her thin cotton blouse undone. For her the old Russian's English flows like the Volga.

At the final session on Wednesday night, there is a handout at the door: a xerox of a telegram, transmitted through an intermediary in Toronto and signed by a young Russian named Alexander Goldfarb. Goldfarb has a fairly thorny request for the conference:

. . . TO DISCUSS AND EXPRESS YOUR OPINION ON THE POSSIBILITY OF USING A RESEARCH ON ENZYME RNA POLYMERASE FROM ESCHERICHIA COLI PARTICULARLY BY ITS MODIFICATION INDUCED BY P-EVEN BACTERIOPHAGES FOR WARFARE.

HOWEVER CRAZY THIS REQUEST MAY SEEM TO YOU, THE FACT IS THAT LAST SPRING MY APPLICATION FOR A VISA TO EMIGRATE TO ISRAEL WAS TURNED DOWN BY THE SOVIET AUTHORITIES JUST BECAUSE THIS RESEARCH WHICH I WAS DOING IN THE BIOLOGY DEPARTMENT OF THE KURCHATOV INSTITUTE OF ATOMIC ENERGY, MOSCOW, WAS "CONSIDERED IMPORTANT FOR THE STATE SECURITY OF THE U.S.S.R."

Goldfarb's request is something of a tall order. Plasmid engineering, in suitably amoral hands, might produce microbes to make superanthrax or concentrated botulinus toxin look like German measles. There seems enough hazard already in pure and simple carelessness, and at the outset of the conference it has been agreed that the issue of new horizons in biologic warfare will not even be raised; for the moment, it is first things first.

And hence Goldfarb's request requires no action and in fact elicits little comment. Except for a reporter who, after the evening session, follows the two elderly Russians out into the cool night air.

Immediately one of the younger Soviets slides in to cover the situation. "Of course," he says in flawless English. "We knew this might come; we had been briefed."

What will happen to Goldfarb's appeal?

The young Russian shakes his head. It is standard, he explains, that anyone associated with security work is required to wait three to five years, after leaving his job, before leaving the country. It is as simple as that, he says, and there is nothing that can be done.

So the Soviet Union considers molecular biology as militarily significant?

It is as simple as that, the Russian repeats, and there is nothing than can be done.

At this point the older academician, who has been observing this exchange with glacier-blue eyes, suddenly shakes his head. "We knew," he says shortly, "we knew he was writing these letters. He has been writing these letters to everyone. He started out writing these letters to the heads of state and now he is writing them to the porters at the door." He snorts, shakes his head again and then the three Russians move into the Monterey night.

"Jesus," says a young American scientist, hair touching his shoulders, who has listened to the last portion of our discussion. "Goldfarb's fucked," he says. "I bet his next assignment is cryobiology."

Cryobiology?

"You know," he nods. "In Siberia."

Disarming the Bug

"What I would like to do," says Sydney Brenner late one afternoon, "and what certainly seems incumbent upon me, is to erect the highest barriers possible between my laboratory, where the work is performed, and the people outside."

Brenner, a compact Englishman in his 40s, with bushy eyebrows, gleaming eyes and nonstop animation that blend to an impression midway between leprechaun and gnome, soon emerges as the single most forceful presence at Asilomar. Repeatedly, when the sessions wander off into a technical morass that threatens to engulf larger considerations, Brenner rises to redirect deftly.

"Does anyone in the audience believe," he asks,

in one such redirection, "that this work—prokaryotes at least—can be done with absolutely no hazard?" There is no immediate response. "This is not a conference," Brenner goes on, "to decide what's to be done in America next week. If anyone thinks so, then this conference has not served its purpose.

"In some countries," he says, "this would be done by the government, and once guidelines were set and you broke them there would be no question of peer censure—the police would simply come out and arrest you."

This is an opportunity, Brenner concludes, for scientists to show that they can regulate themselves —"to reject the attitude that we'll go along and pretend there's a biohazard and hope we can arrive at a compromise that won't affect my own small area, and I can get my tenure and grants and be appointed to the National Academy and all the other things that scientists seem to be interested in."

Brenner—a leader in the field—also takes charge of a series of afternoon sessions devoted to a task called "disarming the bug." The sessions are lively

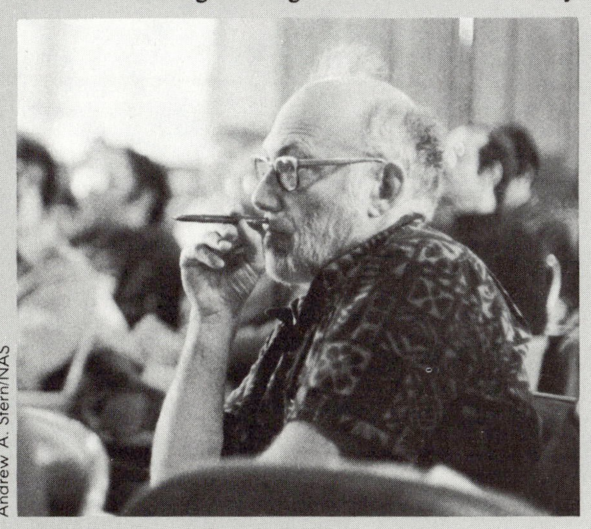

Andrew A. Stern/NAS

Nobel laureate Joshua Lederberg.

and well attended and they represent a curious tangent of mankind's new involvement in the processes of evolution. These people are trying to create novel organisms that are by design incapable of living in the real world.

"Self-destructing vectors" is one phrase for the new bugs. A "vector," in biologic terms, is a mode of transmission; "ecologically disabled organisms" is another. What these will be, ultimately, are bacteria or bacterial viruses—tools in plasmid engineering—that will be unable to live outside the laboratory. Should they manage an escape, even into sewer or stomach, they—and their novel genetic content—will die without reproducing.

Brenner has great faith in the notion of self-destructing vectors; such faith, in fact, that the first of the proposed strains, dubbed "Mark One," has been renamed by cynics "Mach One"—in honor of the speed with which Brenner believes it can be produced.

Even so, Brenner has his doubts about the course that the biohazard question will run in the United States. He is, after all, from a country that has already had an exceedingly expensive biohazard accident—and one, moreover, where laboratory technicians are sufficiently organized that they can shut down a laboratory when a biohazard question remains unsolved.

"The competitive nature of the institutions themselves can affect the situation," he says one afternoon on the chapel steps. "Sooner or later, at some place like MIT or Stanford, some laboratory assistant may well contract something like leukemia—and he will sue the place for everything they've got."

At the time, it seems a fairly pessimistic scenario —but by the time Asilomar lurches to its conclusion, it grows clear that Brenner has described an issue that ultimately proves thoroughly telling.

Chosen Sons of Alfred Nobel

"Nobel laureates can't believe in their own scientific fallibility," says a young molecular biologist one day at lunch. "I've seen lots of them and it's common to the phenotype."

"If you're a Nobel laureate in this country," agrees a plant biologist, "then there's nobody who can touch you."

Perhaps so. While William Shockley might disagree, it's clear that the two American Nobel laureates at Asilomar—Joshua Lederberg from the West and James Watson from the East—exert powerful presences during the conference proceedings. And not always in a terribly popular fashion.

Watson and Lederberg seem almost perfect opposites. Lederberg is a large, bearded, well-nourished man, given to wearing loose sport shirts, brightly patterned in Hawaiian style. He has the healthy look of the senior California academic who spends weekends in hot baths at Esalen or working in a manicured garden. Watson, on the other hand, seems almost to cultivate the persona of absent-minded professor: tall, pale, thin, shirt collar turned up, wispy brown hair tugged so constantly that it stands out from his head in total disarray. He speaks with a regular punctuation of grimaces and, in the midst of any given sentence, his gaze can wander off into space; a consummate 2000-yard stare.

"If we can't communicate the tentativeness of this document," Lederberg says early one afternoon in the chapel, "then we are in trouble." There is, he suggests ominously, "a graver likelihood of this paper crystallizing into legislation than some of us would like to think."

At this point, one major approach to the problem has been to classify experiments in six numbered categories of risk—from those sufficiently safe that they can be done with only standard precautions, through class-IV, which require a fairly complex and costly set of containment procedures, to class-VI, which at this time, goes the recommendation, simply should not be attempted. Lederberg fears that such detailed restrictions might be taken too literally—and inflexibly—by some well-intentioned legislative body and thus thoroughly garrote future research. An alternative suggestion has been to create three less specific risk categories: high, moderate and low.

But shouldn't, someone asks, we benefit from all the experience we already have?

Watson, slumped low in the middle of the audience, mutters to his neighbor: "But there *is* no experience."

"I have to emphasize," says the onstage panel leader, "that there was a great deal of consensus among the members of our panel."

"So is there in the State Department!" Watson exclaims quietly. He sits for a moment, then whispers to his seat mate: "These people have made up guidelines that don't apply to their own experiments."

"Stand up and say it," his companion urges softly. "You can say it; I can't."

Finally, prompted, Watson rises to ask the question: Why, according to the panel, is this particular form of DNA considered safer than another?

The chairman frowns. "It wouldn't be fair for me to answer that question," he says and turns to the panel. "Anybody like to defend *Xenopus*?"

Nobody really wants to and finally Watson sits, shaking his head. "He refused to answer the question," Watson announces softly to anyone within range.

Paul Berg stands to get the session back on course. "We have to make a decision," he says. "Can we measure the risks numerically?"

Watson, sotto voce, explodes: "We can't even *measure* the fucking risks!"

From here on, the discussion begins to fragment. A long-haired researcher from Alabama suggests, aptly, that "anything that comes out of this meeting should self-destruct in 12 months." Someone from Stanford wonders what will happen when local committees have to assess biohazards: "If we can't agree on the danger of experiments here, imagine the situation of a local university committee!"

"Legislation," says one experimenter, "is inevitable. I can't believe that we'll be allowed to continue to control ourselves. But something that could set back the progress of science even more than legislation is if, in a few years, there's a sudden epidemic around Stanford, say, or Cold Spring Harbor."

Finally, just before the dinner bell rings, an English researcher rises to suggest that the problem here is sufficiently complex that those in attendance should go home, brood a bit and then offer suggestions in writing.

"So you don't believe we can arrive at a statement by Thursday noon?" Berg asks him.

The Englishman pauses, shakes his head. "I don't know," he says finally. "I don't know."

"But we are the people," says a young American, "who are supposed to *know* about this and we can't go home from here and decide nothing."

"But we haven't," says Lederberg, "been told what the vector will be. Unless Berg and the organizers tell us the utilization of this document, I'm going to be very hesitant about making any recommendations."

Berg rises again. "If our recommendations," he says, "look self-serving, we will run the risk of having standards imposed. We must start high and work down. We can't say that 150 scientists spent four days at Asilomar and all of them agreed that there was a hazard—and they still couldn't come up with a single suggestion. That's telling the government to do it for us."

Andrew A. Stern/NAS

Conference organizer Paul Berg.

At this, Watson, inspired, is up like a shot: "We can tell them they couldn't do it either!"

A Bit of Little-Known Recent History

Tuesday afternoon at Asilomar the early spring air is crisp and the sky a cloudless Kodachrome blue. Inside the chapel, however, the curtains are drawn, the air is still and heavy and only an occasional shaft of sunlight manages to penetrate, striking a balding head here, a graying one there. The texture and color in the rustic chapel this afternoon are early Rembrandt; the content of the program is pure and simple Modern Dilemma.

The final speaker this afternoon—Andrew Lewis of the National Institutes of Health—has the distinction of being the only member of a working group who felt it necessary to submit a minority opinion. ". . . Given the limited amount of information available at this time," he wrote in that dissension, "I believe that the risks associated with the widespread, semicontained use of this procedure exceed the rewards from the information to be obtained."

Lewis—in his 30s, conservatively dressed, unmistakably serious—also offers a singular perspective on the problem. He is the first person in this country to be burdened with the distribution of a brand-new, laboratory-created—and potentially hazardous—variety of cancer virus.

Adenovirus 2 is a member of a common family of

viruses found, usually fairly harmlessly, in human beings. Simian Virus 40 is a virus found in the kidneys of certain Asiatic monkeys. SV40, however, has also been shown to cause tumors in newborn laboratory animals and, moreover, to cause similar cancerous changes in human tissue in test tubes. In 1969, the isolation of an accidental hybrid between adenovirus 2 and SV40 was reported—apparently combining genetic material from both viruses and, moreover, capable of independent reproduction in both human and monkey cells.

The new hybrid virus represented an altogether unknown hazard to human beings—and also an exceedingly interesting subject for cancer research work. And thus it was that Lewis found himself responsible for distributing a virus strain of unknown pathogenicity to other research laboratories.

It wasn't as if SV40—for a monkey virus—hadn't already had enough to do with human beings. SV40 was not discovered until the early Sixties—and by then a considerable amount of polio vaccine had already been grown in monkey kidney cultures ripe with SV40, which in some cases survived to inhabit the vaccine. Thus, from ten to 30 million Americans presently between 15 and 35 years of age received, along with their brand-new polio vaccine, a dose of live SV40.

Well. More than a bit of discreet medical surveillance has been directed, by now, at known SV40 recipients—and thus far there has been no evidence of any mass malignant onset, which, considering the numbers involved, could make thalidomide appear small potatoes in the history of self-inflicted human suffering. A handful of recent studies, however, has suggested the presence of SV40 in association with some thoroughly unpleasant human tumors and neurologic disorders. And so the scrutiny continues. If medical science has learned anything about viruses thus far, it is that they are tricky.

Andrew Lewis agreed to distribute seed stocks of the first SV40 hybrids, along with a letter describing "reasonable precautions" and requesting that the recipient laboratory distribute none of the virus on its own. But by then, four more hybrid viruses had been located—all equally suspect—and these Lewis refused to send out.

"The question we faced," Lewis says this afternoon at Asilomar, "was whether one individual had the right to decide to distribute potentially hazardous laboratory-created recombinants." The reaction from the research community was immediate: threats of congressional action or administrative pressure from NIH—even the suggestion of a group letter to *Science*. And, on the other hand, concerned scientists warned that if Lewis went ahead with the distribution, they would file for a federal environmental impact statement.

"I felt," says Lewis, "that voluntary compliance by interested investigators was the most satisfactory method," and so he decided to require a formal document from each laboratory that requested the viruses, stipulating that the researchers assumed full moral and legal responsibility for the viral agents.

Fifteen months later at Asilomar, Lewis is no longer so certain about volun- [*Continued*] tary compliance. "Several major laboratories," he says, "have thus far not supported the Memorandum of Understanding and Agreement. In addition, our original request to restrict the distribution of the first hybrid appears to have been ignored by one or more of these same laboratories."

Lewis's opinion, clearly, is not popular. The audience is cold and midway in his presentation, Lewis begins to lean intently over the podium, grasping his pointer, straight up, like a spear carrier in an Italian opera. The source of the hostility is no mystery: This conference is, after all, *about* self-regulation — and regardless of Lewis's unfortunate experience, one bad apple need not spoil the barrel.

Or does that depend on the apples?

THE LAWYERS: REALITY THERAPY

By Wednesday night—with only a single morning session remaining—the situation at Asilomar seems as unsettled as the Monterey weather. The mild blue skies have begun to turn and by now a massive bank of thick gray fog lies a mile or so off the coast, sending in low, damp clouds both morning and evening. The conference meals have started to deteriorate as well and this evening's corned beef and cabbage barely achieves summer camp standards. Worst of all, however, is the possibility that the conference, enmeshed in bickering, may not actually be able to arrive at a group statement. But clearly, with three days of talk about biohazards thoroughly soaked up by the press, it is too late to stop now. The question remaining is exactly what to do about it.

Part of the impetus, as it develops, is about to come.

The Wednesday night program looks fairly innocuous: presentations by lawyers regarding ethics and legal liability. Lawyers, of course, are supposed to have some knack for public speaking, as opposed to—as anyone who has attended a scientific conference can testify—men of science. The evening promises, at least, a nice diversion.

And so, at first, it seems. The first speaker, dapper, pleasant, goateed — the husband, in fact, of the only female conference organizer — spends a mild-mannered quarter-hour eloquently dissecting a dictum familiar to all medical students—first of all, do no harm—and concludes with a fairly abstract three-part analysis of risk versus benefit that covers, in thorough generalities, just about exactly the major issues of the past three days.

So far, so good. Analysis like this isn't going to help

Andrew A. Stern/NAS

Conference organizers drafting final statement.

anyone decide between numerical hazard rating and high/moderate/low, but it's nice to hear just how important the responsibility *is*. And just how complicated the problem.

The second speaker—a professor of international law—promises to be equally entertaining. He approaches the podium in sports coat and open-collar shirt, appearing not much older than 22. His demeanor, however, is distinctly confident as he starts with a nice joke: A scientist and a lawyer are arguing about which of theirs is the older profession. The argument goes back and forth, from Pericles to Hippocrates to Maimonides to Hammurabi, until it reaches all the way back to God.

God, the scientist states, must have been a scientist, to have brought order out of chaos.

Yes, the lawyer responds. But where do you think the chaos came from?

The joke turns out to be less joke than promise, as the young lawyer proceeds into a merciless "outsider's analysis" that, within minutes, has jaws dropping all over the chapel. Much of the conference, he suggests, has been irrelevant to the central issue.

The audience is suddenly very quiet.

Many of the specific arguments, he goes on, have been equally inapplicable. "Academic freedom," he points out, "does not include the freedom to do physical harm." And "prior restraint"—a notion advanced the previous day by a Nobel laureate—makes perfect sense when it involves restraint from doing physical damage.

"This group," the young lawyer suggests flatly, "is not competent to assign overall risk."

What? But that's the *point* —most here, likely, would consider themselves *uniquely* competent. Who else could do it?

"It is the right of the public," continues the speaker, "to act through the legislature and to make erroneous decisions."

Jesus. Now that's a hell of a reassuring thing to hear from a lawyer and it's clear, in the still air of the redwood chapel, that the audience is growing just a bit discomfited.

And it's only worse when the lawyer suggests a hypothetical situation wherein Congress might insert its grubby political fingers into the delicate process; "Congress" manages to draw a low but audible groan. Legislation, however, might not be all bad, he explains: The law might provide, say, for liability in cases of biohazard accidents.

But that, by now, seems faint comfort. "Legal institutions," the young lawyer intones in civics-class fashion, "are a part of your world, whether you like it or not." And with what the conference attendees are into, he concludes, it's time to involve those institutions.

Well. There is a low buzz of conversation as the next speaker is introduced; all of this really isn't that pleasant, compared to the technical papers. Clearly, one group of slightly arrogant professionals are here being dressed down by another group of at least equally arrogant professionals. And the lawyers probably make more money, too.

But the best is yet to come. The final speaker is short, middle-aged, fairly nondescript in mismatched suit and tie and thick glasses. One would have noticed him earlier only because of his ceaseless squirming during the technical sessions—the scientists did not squirm; they either paid attention and took notes or went to sleep. This speaker appears something of a manic milquetoast but by the time he is behind the podium and has completed a sentence or so, his dry sharp delivery makes it clear that this is the lawyer who slices one mercilessly to very tiny ribbons in the witness box.

Which is, coincidentally, precisely his topic: "Conventional aspects of the law," as he puts it, "and how they may sneak up on you—in the form, say, of a multimillion-dollar lawsuit."

Hmmmm. Now, abruptly, the audience is *very* quiet. The subject of legal responsibility has really not yet been dealt with in the nitty-gritty terms of "Who, exactly, gets sued when something goes dreadfully wrong?"

This lawyer aims to explain and, having himself just squirmed through three days of abstruse technical jargon, he takes some relish in trotting out his own — torts, liability, proximate cause, OSHA—and illustrating just how finely—if not fairly—the wheels of law can grind.

Professional negligence, the lawyer suggests, is a failing that finds juries exceedingly unsympathetic. Take the case of the ophthalmologist in Oregon who lost a malpractice suit on his failure to perform a glaucoma test on a young and asymptomatic patient in whom the chances in glaucoma were one in 25,000. Judges, he points out, are experts only in law—and juries, quite intentionally, are experts in nothing at all.

By now, the gathered molecular biologists are conspicuously attentive.

Oh, it's not totally hopeless, the lawyer reassures. If, say, a burglar were to break into one's lab, steal a vial of deadly virus and then strew the contents all over Brooklyn, then maybe—just maybe—the intervention of a third party might get one off the hook. Or if one's work is for national security purposes, then one is probably also fairly immune from prosecution.

But, of course, there are already laws under which recombinant engineering might be controlled: OSHA, for example—the Occupational Safety and Health Act—could conceivably be invoked to protect laboratory workers. According to OSHA, the lawyer explains, "the work place must be free of hazard. Not relatively free," he says. "The statute says *free*." And the person who sets those standards is — the Secretary of Labor.

Jesus God. All of this, clearly, is the most violent intrusion of the real world into these proceedings thus far. Some goon from the Department of Labor, waltzing into one's lab for a surprise inspection, on the outcome of which might hang a $10K fine. Or the notion of one's own laboratory technician, bizarrely diseased and setting out for revenge on the basis of a bloodless legal principle called "deepest pockets." While there has been no lack of real and humane concern among the attendees these past three days, there has been something about this brief legal seminar that has brought home, rather forcefully, just how *unpleasant* things could get.

Within minutes, however, the molecular biologists have rallied to the defense, the more vocal bravely citing legal precedents—fetal experimentation, medical research on prisoners—as argumentatively as possible. It is, however, as effective as if one of the lawyers had earlier risen to question a certain enzymatic manipulation of a lambda bacteriophage.

Finally, Nobel laureate Lederberg stands and, with some eloquence, presents an intricate analogy involving the risks and responsibilities of accidentally bringing home a deadly African virus.

"That argument," says one of the lawyers, "with all due respect, is almost entirely beside the point. If we are remiss about our international travel regulations, we should move to correct that situation, rather than taking it as reason for being equally remiss about our approach to the biohazard question."

Lederberg returns to his seat, big tan arms folded across his chest like a wounded Buddha. The next question is more along the lines of, *precisely*, who is likely to get sued? And by the end of the evening, one of the lawyers is actually advising the conference to look into the possibilities of extended personal liability insurance. "At least," he says, "then you won't have *quite* so much to worry about."

Final Session

Thursday morning provides the grayest sky thus far at Asilomar. "At noon," Paul Berg announces at 9:00 a.m. inside the dim chapel, "I would like to terminate this meeting and I hope that by noon we can reach a point where such termination is possible."

Berg and his organizing committee have done their best toward that goal: up all night revising, condensing, rewriting the working papers and discussion of the past three days into what is now a freshly xeroxed five-page handout titled "Provisional Statement of the Conference Proceedings."

The statement, clearly, is a compromise: The six-category classification of risk has now, uniformly, been condensed to low/moderate/high. And there is no flat proscription of the experiments that some have earlier called unreasonably hazardous.

But it is still, in context, a strong document: If adopted, many researchers will have to go home and spend thousands of dollars on laboratory containment — no small sacrifice in these days of tight funding —to do the experiments they could have done for nothing eight months earlier.

Behind the scientists are the pressure of the lawyers' dark predictions and the hovering presence of the press; before them are the plain realities of research funding. Who, after all, really wants to drop $40,000 on a negative-pressure laboratory equipped with laminar-flow hoods unless it's absolutely necessary?

And so the discussion begins, lurching along, sidetracking, backtracking, and before long, the sense of it is so tangled that weeks later it survives best as a collection of disparate and anonymous exchanges.

"But the input of 150 people has been ignored!"

"If we wanted to ignore the input of this group, we'd all be asleep right now."

"But are you talking about a vote? How will a consensus be arrived at?"

"No. No votes will be taken. We will arrive at consensus through discussion."

"But we *have* to have some kind of vote. Maybe we could ask the press to go away."

"I would like to see something added to this paragraph to say explicitly that there are experiments we can imagine that are too dangerous to perform at present."

"May I respond to that? Based on the split at this meeting as to whether that is a philosophical or a practical question, we decided to put such experiments into the high-risk category. That is by no means a license, since those facilities are both extremely cumbersome and not widely available."

"We have no power but moral censure if someone goes off to Uganda tomorrow and puts General Amin into plasmids. And I don't want to carry the can for *any* of you. I don't want a situation where people can say, well, I did it because the organizing committee said I could. In the end, it is your individual judgment. Certain experiments should have to cross higher barriers of judgment — and others should cross even higher."

"It will save me a great deal of trouble later to get your judgment on a specific case: Would it require a moderate-risk facility, comparable to that used for oncoviruses, to do the following experiment: the transmittal of pSC101 into *Bacillus subtilis*?"

"I'd have to say that it would."

(*silence*)

"I think we're not going to try out all the scenarios here. I think we'll have to move on."

"We modified your decision by drawing a line at cold-blooded versus warmblooded, rather than mammal versus nonmammal, because the avian viruses are known to grow in and transform human cells."

"It's also fair to point out that there are tumor viruses in frogs."

"Might I ask your committee how they propose to deal with the question of type-VI experiments? It has been put to me by people in the U.K. that they can see no possible current experiment involving, say, smallpox DNA. It's not that they couldn't *design* such experiments—it's just that the combination of benefits and risks, at the moment, put these experiments into class-VI."

"I thought we'd agreed to indicate a split of opinion as to whether certain experiments should be ruled not permissible. At this time."

"Could we perhaps, for my benefit as much as anyone else's, test the feeling? Perhaps there *isn't* a split."

"Okay, let me see if I can put the question simply. One view says that there are experiments that should be performed only in the highest containment facilities available today. The other view is that there is a class of experiments that should not be done at all, with present containment methods."

(*a show of hands*)

"Well, it's clear what the sentiment is. The predominant view supports the latter. If we have time, perhaps we can come back to this section.

"Please make your comments brief."

"Is low risk in quotations one thing and low risk without quotations another?"

"I must say, personally, that I really feel, having worked on the plasmid document, that it's been prostituted. And that's all I have to say."

"This refers to someone who goes into a shotgun experiment with the warmblooded beasts and recombines with a safe vector; they may then be reassigned to low risk."

"I think we should remove the phrase 'pharmacologically active agents' and just talk about 'toxic substances.' Insulin could be considered a pharmacologically active agent and some agency might take that very seriously."

"By putting these experiments into high risk, requiring a facility like Fort Detrick— which is probably filled up with some total waste of time —we seem to be discouraging them entirely."

"We're coming up on twelve o'clock."

"It seems to me that there are invertebrates—take mosquitoes, for example — that contain diseases that are dangerous to human beings and if people go willy-nilly joining mosquito DNA with *E. coli*, I'd hate to think I'd get malaria from walking around on the street."

(*front row, sotto voce*): "Huh?"

"How could you get malaria . . .?"

"How could you . . ."

"You're right, taking field-caught mosquitoes and extracting DNA and trying to clone it . . ."

"I just ask to consider this . . ."

(*front row, sotto voce*): "Can they explain . . .?"

"*Say* it."

(*loudly*): "Can You Explain, please, how a protozoan parasite could result from an *E. coli* and thus cause malaria?"

"We will, ah, take the comment under advisement."

(*front row, sotto voce*): "It's bloody nonsense."

"It's clear that the sentiment is overwhelmingly in favor."

"I don't think that's clear at all."

"Well, as far as I could tell, that would be overwhelming. I think we'll have to go on to item . . ."

"*May I have ten seconds*?"

"Ten seconds."

"Under the pressure of time, very complex issues are being railroaded through. If you're willing to say, in your preliminary document, that that characterizes the nature of the consensus here, then I could go along."

"Okay. We'll move on, then, to the next section."

"Could you ask how many people abstained at including a high-risk rating for insulin?"

"How many people did not vote?"

(*show of hands*)

"A small number."

"Let's have the show of hands again. Please! There's no point in abstaining. What are your sentiments?"

(*low mumbling*)

"What are we voting on now?"

Suddenly Sydney Brenner interrupts: "Please," he says slowly, sounding like a sleep-deprived leprechaun. "Could I ask whether, in paragraph six, if the phrase 'this document represents our best assessment of the potential biohazards,' we might change that word 'best' to 'first'?"

For the first time in two hours there is laughter. And then the noon bell tolls for the first time and, shortly, the final vote is taken. The working document, murkily amended, is adopted, with only a few dissenters.

The noon bell tolls again, and as the scientists stir for the early afternoon dash to Monterey airport, one of the lawyers briefly takes the podium. "In many ways," he says, "it was a moving experience for anyone outside the scientific community to watch this group grapple with a very difficult problem. It's been nice being here," he says; "you will be a hard act to follow."

But the act, of course, has only just begun.

Conclusion: Human Sex as Moderate-Risk Experiment

"There's only one way to control scientists in institutions," a young researcher noted midway at Asilomar: "Take away their money."

Less than 24 hours after the tentative consensus at Asilomar, a select group of scientists and administrators sit at a polished wooden conference table in an ornate San Francisco hotel meeting room. They represent the National Institutes of Health and their job, this bright morning, is to transform the general conclusions of Asilomar into formal guidelines that involve funding restrictions, local biohazard committees, the variegated details of specific enforcement.

This is, of course, what many conference attendees feared most. "Already," a microbiologist admitted one night by the dormitory fireplace, "we spend two months a year applying for grants; now, we'll spend another month filling out more forms. And the forms don't protect anybody—they just take more time."

Even so, the molecular biologists at Asilomar were the first modern researchers to assume voluntarily some measure of social responsibility for their work.

Perhaps it is that the life sciences have come later to their crisis of responsibility. "Physics is checkers," a biologist once told me; "biology is chess." Even the terrible notion of biologic warfare has never really been a particularly efficient way to commit violence. It's simply too uncontrollable.

In a sense, then, the biologic sciences have only now suffered their first real loss of innocence: out of Mendel's monastery garden, so to speak, and into J. D. Watson's stainless steel laboratory.

And the box is only just opened. An antique science fiction story comes to mind that describes a military plague that drops upon a city, propagates briefly, infects all within and then — bacterial generations later — destroys itself, just as the conquering forces march in unscathed.

Pure science fiction. Yet only a few months ago, the top molecular biologists on the planet met to discuss, precisely — but benignly — how one might artificially create microorganisms programmed to self-destruct.

What is not science fiction, however, is that when artificial control of the evolutionary process—this "creation of novel biotypes never seen before in nature" — comes more firmly into our grasp, it will represent as profound an expulsion from the Garden as human intelligence has thus far managed. We will finally— on a molecular level — have cut ourselves loose from the dictates of primal nature. And at that point we will need all the foresight and self-control that we can muster.

And perhaps that was the final, foggy significance of Asilomar: a promise that the scientists who deal with the most fundamental of life stuff will not sequester themselves beneath Chicago stadiums or within blockhouses in the New Mexico desert — that their work, at least as significant as the most subtle of subnuclear manipulations, will be done with care and public scrutiny.

The territory, God knows, is uncharted. Not until late in the afternoon in that gilt San Francisco hotel room — deep in the midst of guidelines for experimental DNA recombination—does one middle-aged woman, staring intently at her working paper, abruptly begin to laugh.

The other panel members look up, puzzled.

"Do you realize," she says finally, "what this *means*?"

No one seems quite certain.

"This means," the woman announces brightly, "that we have just made human sex a moderate-risk experiment."

There is brief silence, then tentative laughter around the table.

"But that's only," someone else notes a moment later, "in the *laboratory*."

Revisions ensue immediately.

Document 2.2

February 24–27, 1975
Asilomar Conference Program

P R O G R A M

INTERNATIONAL CONFERENCE ON RECOMBINANT DNA MOLECULES

Asilomar Conference Center, Pacific Grove, California February 24-27, 1975

Sponsored by the U.S. National Academy of Sciences with financial support from the National Cancer Institute of the National Institutes of Health, and the National Science Foundation.

TIME	SUBJECT	SPEAKER
MONDAY—*FEBRUARY 24, 1975*		
8:30	Introduction: History, Aim of Meeting, Outline of Task	D. Baltimore *(MIT—USA)*
8:50	Assessment of present technology, joining methods and vectors	P. Berg *(Stanford—USA)*
9:10	*Ecology of Plasmids and Enteric Organisms*	
	Ubiquity of organisms, natural history, plasmid ecology and introduction to normal bowel flora	S. Falkow *(U. Washington—USA)*
	General problem of Salmonella infection in man and domestic animals and the use and misuse of antibiotics in medicine and agriculture, enhancement of R plasmid emergence	E. Anderson *(Colindale—UK)*
	Plasmids which contribute to virulence or colonization of enteric species	H. Williams-Smith *(Houghton Poultry Station—UK)*
10:00	Coffee Break	
	Gene transfer *in vivo* of both enteric organisms and *Staphylococcus*	M. Richmond *(Bristol—UK)*
	Panel discussion on ecology and potential dissemination of *E. Coli* strains	S. Falkow *(Chairman)*
12:00	Lunch	
1:00	A Public Policy Perspective	H. Green *(The George Washington Univ. National Law Ctr.)*
4:30	Molecular biology of bacterial plasmids	R. Novick *(Pub. Health Res. Inst.—USA)*
	Molecular biology of bacterial conjugation and conjugative mobilization of plasmid and other DNAs	R. Curtiss *(U. Alabama—USA)*
	Phenotypic expression of plasmids in bacteria	J. Davies *(U. Wisconsin—USA)*
	Some features of DNA uptake by bacteria	M. Fox *(MIT—USA)*
6:00	Dinner	
7:30	Genetics and biochemistry of DNA restriction	H. Boyer *(U. Calif., San Francisco—USA)*
	Molecular cloning of DNA as a tool for the study of plasmid and eukaryotic biology	S. Cohen *(Stanford Univ.—USA)*
	Development of bacteriophage systems as cloning vehicles	K. Murray *(U. Edinburgh—UK)*
TUESDAY—*FEBRUARY 25, 1975*		
8:30	Panel discussion on experimental guidelines for novel recombinant prokaryotes	R. Novick (Chairman) *(Pub. Health. Res. Inst.—USA)*
12:00	Lunch	
4:30	*Plasmid-viral DNA Recombinants*	
	Design, construction and value of new kinds of molecules using SV40 (deletion mutants, transducing viruses, plasmid-viral gene recombinants)	D. Nathans *(Johns Hopkins Univ.—USA)*
	"Natural history of viruses"; especially, are viral genes attached to plasmids a special case?	J. Sambrook *(Cold Spring Harbor Lab—USA)*
	Adeno-SV40 hybrid DNAs as a case history; including: a) NIH distribution policy, b) conditions required for use in lab, c) results of risk assessment experiments.	A. Lewis *(NIH—USA)*
6:00	Dinner	
7:30	Comparison of SV40 and human papova viruses: pathogenicity, (cross reactivity and oncogenicity)	D. Walker *(U. Wisconsin—USA)*
	Infectivity and transforming capacity of SV40 + SV40 DNA; use of recombinant DNAs to infect bacteria vs. mammalian cells	D. Jackson *(U. Michigan—USA)*
	Isolation and study of Rous sarcoma proviral DNA presence of "virogenes" in "normal" cells	M. Bishop *(U. Calif., San Francisco—USA)*
	Possible experiments with herpesviruses DNA fragments	B. Roizman *(U. Chicago—USA)*
	Panel discussion on guidelines for virus-plasmid recombinants	A. Shatkin (Chairman)

WEDNESDAY—*FEBRUARY 26, 1975*

8:30	*Plasmid-cell DNA Recombinants*	
	What has already been done in linking cell and plasmid DNAs	D. Hogness *(Stanford—USA)* D. Brown *(Carnegie Inst.—USA)* D. Carroll *(Carnegie Inst.—USA)*
10:00	Coffee Break	
	Major problems which can be solved by making cell DNA-plasmid-cell DNA recombinant	S. Brenner *(MRC—UK)*, P. Day *(Conn. Ag. Exp. Station—USA)*, R. Valentine *(U. Calif., San Diego—USA)*
12:00	Lunch	
4:30	Panel discussion of guidelines for plasmid-cell DNA recombinant experiments.	D. Brown (Chairman)
6:00	Dinner	
7:30	Brief analysis of responsibility of research scientists and of risk balancing. Will touch on ethical issues insofar as responsibility is concerned, and focus on the several components involved in balancing the risks and benefits of creating recombinant DNA molecules	D. Singer *(Fried, Frank, Harris, Shriver and Kampelman, Washington, D.C.; Hastings Inst. of Society, Ethics and the Life Sciences—USA)*
	Legal liability of investigators and institutions in the event of proximate or remote injury arising out of work with recombinant DNAs. Applicability of Occupational Health and Safety Act. Informed consent, insurance and indemnity.	R. Dworkin *(Indiana Univ.—USA)*
	Institutionalization of current and future experimental guidelines, role of NIH Advisory Committee, international co-ordination of policies of different nations.	A. Capron *(U. Penn. Law Sch.—USA)*

THURSDAY—*FEBRUARY 27, 1975*

8:30	Discussion and adoption of conference statement
12:00	Ajournment

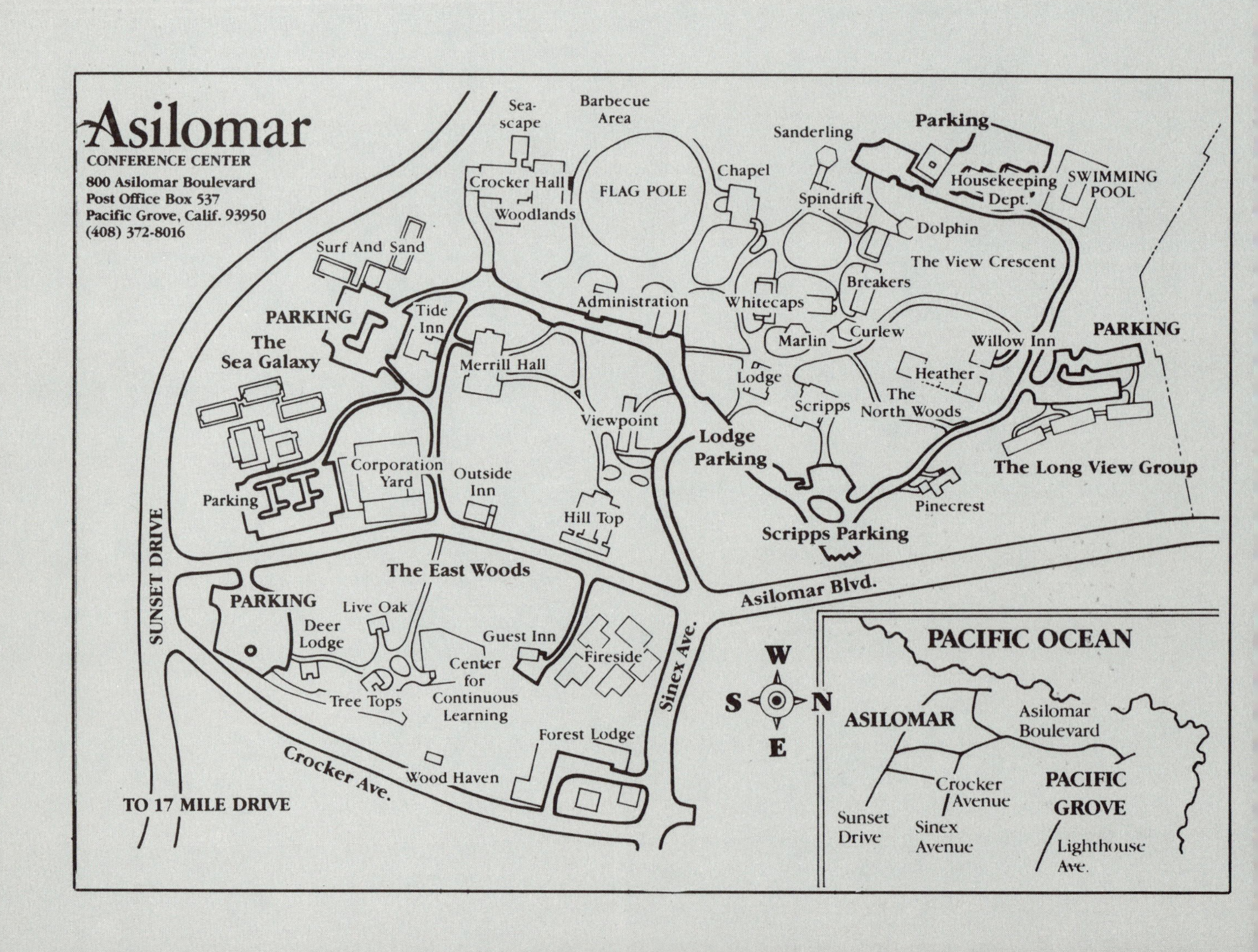

Document 2.3

March 20, 1975

Rogers Letter to Watson

THE UNIVERSITY OF TENNESSEE
CENTER FOR THE HEALTH SCIENCES
MEMPHIS, TENNESSEE 38163

DEPARTMENT OF BIOCHEMISTRY
PHONE 528-6150

March 20, 1975

Dr. James Watson
Cold Spring Harbor
Biological Laboratories
Cold Spring Harbor
Long Island, New York

Dear Jim:

I missed the last two days on the Conference at Asilomar because of a previous commitment in New York.

What I read in Science about your statements I agreed with completely. If such rules were to be applied to Medicine the Hospitals would have to be closed. Dewitt Stetten told me that what these people needed was a course in Medical Microbiology.

It was very good to see and talk with you.

Sincerely yours,

Stanfield Rogers, M.D.

sr/mj

Document 2.4

June 6, 1975
Science (**188**:991)

Asilomar Conference on Recombinant DNA Molecules*

Paul Berg, David Baltimore, Sydney Brenner,
Richard O. Roblin III, Maxine F. Singer

I. Introduction and General Conclusions

This meeting was organized to review scientific progress in research on recombinant DNA molecules and to discuss appropriate ways to deal with the potential biohazards of this work. Impressive scientific achievements have already been made in this field, and these techniques have a remarkable potential for furthering our understanding of fundamental biochemical processes in pro- and eukaryotic cells. The use of recombinant DNA methodology promises to revolutionize the practice of molecular biology. Although there has as yet been no practical application of the new techniques, there is every reason to believe that they will have significant practical utility in the future.

Of particular concern to the participants at the meeting was the issue of whether the pause in certain aspects of research in this area, called for by the Committee on Recombinant DNA Molecules of the National Academy of Sciences in the letter published in July 1974 (*1*), should end, and, if so, how the scientific work could be undertaken with minimal risks to workers in laboratories, to the public at large, and to the animal and plant species sharing our ecosystems.

The new techniques, which permit combination of genetic information from very different organisms, place us in an area of biology with many unknowns. Even in the present, more limited conduct of research in this field, the evaluation of potential biohazards has proved to be extremely difficult. It is this ignorance that has compelled us to conclude that it would be wise to exercise considerable caution in performing this research. Nevertheless, the participants at the Conference agreed that most of the work on construction of recombinant DNA molecules should proceed, provided that appropriate safeguards, principally biological and physical barriers adequate to contain the newly created organisms, are employed. Moreover, the standards of protection should be greater at the beginning and modified as improvements in the methodology occur and assessments of the risks change. Furthermore, it was agreed that there are certain experiments in which the potential risks are of such a serious nature that they ought not to be done with presently available containment facilities. In the longer term serious problems may arise in the large-scale application of this methodology in industry, medicine, and agriculture. But it was also recognized that future research and experience may show that many of the potential biohazards are less serious and/or less probable than we now suspect.

II. Principles Guiding the Recommendations and Conclusions

Although our assessments of the risks involved with each of the various lines of research on recombinant DNA molecules may differ, few, if any, believe that this methodology is free from any risk. Reasonable principles for dealing with these potential risks are: (i) that containment be made an essential consideration in the experimental design and (ii) that the effectiveness of the containment should match, as closely as possible, the estimated risk. Consequently, whatever scale of risks is agreed upon, there should be a commensurate scale of containment. Estimating the risks will be difficult and intuitive at first, but this will improve as we acquire additional knowledge; at each stage we shall have to match the potential risk with an appropriate level of containment. Experiments requiring large-scale operations would seem to be riskier than equivalent experiments done on a small scale and therefore require more stringent containment procedures. The use of cloning vehicles or vectors (plasmids, phages) and bacterial hosts with a restricted capacity to multiply outside of the laboratory would reduce the potential biohazard of a particular experiment. Thus, the ways in which potential biohazards and different levels of containment are matched may vary from time to time, particularly as the containment technology is improved. The means for assessing and balancing risks with appropriate levels of containment will need to be reexamined from time to time. Hopefully, through formal and informal channels of information within and between nations of the world, the way in which potential biohazards and levels of containment are matched would be consistent.

Containment of potentially biohazardous agents can be achieved in several ways. The most significant contribution to limiting the spread of the recombinant DNA's is the use of biological barriers. These barriers are of two types: (i) fastidious bacterial hosts unable to survive in natural environments and (ii) nontransmissible and equally fastidious vectors (plasmids, bacteriophages, or other viruses) able to grow only in specified hosts. Physical containment, exemplified by the use of suitable hoods or, where applicable, limited access or negative pressure laboratories, provides an additional factor of safety. Particularly important is strict adherence to good microbiological practices which, to a large measure, can limit the escape of organisms from the experimental situation and thereby increase the safety of the operation. Consequently, education and training of all personnel involved in the experiments is essential to the effectiveness of all containment measures. In practice, these different means of containment will complement one another and documented substantial improvements in the ability to restrict the growth of bacterial hosts and vectors could permit modifications of the complementary physical containment requirements.

*Summary statement of the report submitted to the Assembly of Life Sciences of the National Academy of Sciences and approved by its Executive Committee on 20 May 1975.

Stringent physical containment and rigorous laboratory procedures can reduce but not eliminate the possibility of spreading potentially hazardous agents. Therefore, investigators relying upon "disarmed" hosts and vectors for additional safety must rigorously test the effectiveness of these agents before accepting their validity as biological barriers.

III. Recommendations for Matching Types of Containment with Types of Experiments

No classification of experiments as to risk and no set of containment procedures can anticipate all situations. Given our present uncertainties about the hazards, the parameters proposed here are broadly conceived and meant to provide provisional guidelines for investigators and agencies concerned with research on recombinant DNA's. However, each investigator bears a responsibility for determining whether, in his particular case, special circumstances warrant a higher level of containment than is suggested here.

A. Types of Containment

1) *Minimal risk*. This type of containment is intended for experiments in which the biohazards may be accurately assessed and are expected to be minimal. Such containment can be achieved by following the operating procedures recommended for clinical microbiological laboratories. Essential features of such facilities are no drinking, eating, or smoking in the laboratory, wearing laboratory coats in the work area, the use of cotton-plugged pipettes or preferably mechanical pipetting devices, and prompt disinfection of contaminated materials.

2) *Low risk*. This level of containment is appropriate for experiments which generate novel biotypes but where the available information indicates that the recombinant DNA cannot alter appreciably the ecological behavior of the recipient species, increase significantly its pathogenicity, or prevent effective treatment of any resulting infections. The key features of this containment (in addition to the minimal procedures mentioned above) are a prohibition of mouth pipetting, access limited to laboratory personnel, and the use of biological safety cabinets for procedures likely to produce aerosols (for example, blending and sonication). Though existing vectors may be used in conjunction with low-risk procedures, safer vectors and hosts should be adopted as they become available.

3) *Moderate risk*. Such containment facilities are intended for experiments in which there is a probability of generating an agent with a significant potential for pathogenicity or ecological disruption. The principal features of this level of containment, in addition to those of the two preceding classes, are that transfer operations should be carried out in biological safety cabinets (for example, laminar flow hoods), gloves should be worn during the handling of infectious materials, vacuum lines must be protected by filters, and negative pressure should be maintained in the limited access laboratories. Moreover, experiments posing a moderate risk must be done only with vectors and hosts that have an appreciably impaired capacity to multiply outside of the laboratory.

4) *High risk*. This level of containment is intended for experiments in which the potential for ecological disruption or pathogenicity of the modified organism could be severe and thereby pose a serious biohazard to laboratory personnel or the public. The main features of this type of facility, which was designed to contain highly infectious microbiological agents, are its isolation from other areas by air locks, a negative pressure environment, a requirement for clothing changes and showers for entering personnel, and laboratories fitted with treatment systems to inactivate or remove biological agents that may be contaminants in exhaust air and liquid and solid wastes. All persons occupying these areas should wear protective laboratory clothing and shower at each exit from the containment facility. The handling of agents should be confined to biological safety cabinets in which the exhaust air is incinerated or passed through Hepa filters. High-risk containment includes, in addition to the physical and procedural features described above, the use of rigorously tested vectors and hosts whose growth can be confined to the laboratory.

B. Types of Experiments

Accurate estimates of the risks associated with different types of experiments are difficult to obtain because of our ignorance of the probability that the anticipated dangers will manifest themselves. Nevertheless, experiments involving the construction and propagation of recombinant DNA molecules using DNA's from (i) prokaryotes, bacteriophages, and other plasmids; (ii) animal viruses; and (iii) eukaryotes have been characterized as minimal, low, moderate, and high risks to guide investigators in their choice of the appropriate containment. These designations should be viewed as interim assignments which will need to be revised upward or downward in the light of future experience.

The recombinant DNA molecules themselves, as distinct from cells carrying them, may be infectious to bacteria or higher organisms. DNA preparations from these experiments, particularly in large quantities, should be chemically inactivated before disposal.

1) *Prokaryotes, bacteriophages, and bacterial plasmids*. Where the construction of recombinant DNA molecules and their propagation involve prokaryotic agents that are known to exchange genetic information naturally, the experiments can be performed in minimal-risk containment facilities. Where such experiments pose a potential hazard, more stringent containment may be warranted.

Experiments involving the creation and propagation of recombinant DNA molecules from DNA's of species that ordinarily do not exchange genetic information generate novel biotypes. Because such experiments may pose biohazards greater than those associated with the original organisms, they should be performed, at least, in low-risk containment facilities. If the experiments involve either pathogenic organisms or genetic determinants that may increase the pathogenicity of the recipient species, or if the transferred DNA can confer upon the recipient organisms new metabolic activities not native to these species and thereby modify its

The authors are members of the Organizing Committee for the International Conference on Recombinant DNA Molecules, Assembly of Life Sciences, National Academy of Sciences–National Research Council, Washington, D.C. 20418. Dr. Berg is chairman of the committee and he is professor of biochemistry, Department of Biochemistry, Stanford University Medical Center, Stanford, California. Dr. Baltimore is American Cancer Society Professor of Microbiology, Center for Cancer Research, Massachusetts Institute of Technology, Cambridge, Massachusetts. Dr. Brenner is a member of the Scientific Staff of the Medical Research Council of the United Kingdom, Cambridge, England. Dr. Roblin is professor of microbiology and molecular genetics, Harvard Medical School, and assistant bacteriologist, Infectious Disease Unit, Massachusetts General Hospital, Boston, Massachusetts. Dr. Singer is head of the Nucleic Acid Enzymology Section, Laboratory of Biochemistry, National Cancer Institute, National Institutes of Health, Bethesda, Maryland.

relationship with the environment, then moderate- or high-risk containment should be used.

Experiments extending the range of resistance of established human pathogens to therapeutically useful antibiotics or disinfectants should be undertaken only under moderate- or high-risk containment, depending upon the virulence of the organism involved.

2) *Animal viruses.* Experiments involving linkage of viral genomes or genome segments to prokaryotic vectors and their propagation in prokaryotic cells should be performed only with vector-host systems having demonstrably restricted growth capabilities outside the laboratory and with moderate-risk containment facilities. Rigorously purified and characterized segments of nononcogenic viral genomes or of the demonstrably nontransforming regions of oncogenic viral DNA's can be attached to presently existing vectors and propagated in moderate-risk containment facilities; as safer vector-host systems become available such experiments may be performed in low-risk facilities.

Experiments designed to introduce or propagate DNA from nonviral or other low-risk agents in animal cells should use only low-risk animal DNA's as vectors (for example, viral or mitochondrial), and manipulations should be confined to moderate-risk containment facilities.

3) *Eukaryotes.* The risks associated with joining random fragments of eukaryote DNA to prokaryotic DNA vectors and the propagation of these recombinant DNA's in prokaryotic hosts are the most difficult to assess.

A priori, the DNA from warm-blooded vertebrates is more likely to contain cryptic viral genomes potentially pathogenic for man than is the DNA from other eukaryotes. Consequently, attempts to clone segments of DNA from such animals and particularly primate genomes should be performed only with vector-host systems having demonstrably restricted growth capabilities outside the laboratory and in a moderate-risk containment facility. Until cloned segments of warm blood vertebrate DNA are completely characterized, they should continue to be maintained in the most restricted vector-host system in moderate-risk containment laboratories; when such cloned segments are characterized, the may be propagated as suggested above for purified segments of virus genomes.

Unless the organism makes a product known to be dangerous (for example, a toxin or virus), recombinant DNA's from cold-blooded vertebrates and all other lower eukaryotes can be constructed and propagated with the safest vector-host system available in low-risk containment facilities.

Purified DNA from any source that performs known functions and can be judged to be nontoxic may be cloned with currently available vectors in low-risk containment facilities. (Toxic here includes potentially oncogenic products or substances that might perturb normal metabolism if produced in an animal or plant by a resident microorganism.)

4) *Experiments to be deferred.* There are feasible experiments which present such serious dangers that their performance should not be undertaken at this time with the currently available vector-host systems and the presently available containment capability. These include the cloning of recombinant DNA's derived from highly pathogenic organisms (that is, Class III, IV, and V etiologic agents as classified by the U.S. Department of Health, Education and Welfare), DNA containing toxin genes, and large-scale experiments (more than 10 liters of culture) using recombinant DNA's that are able to make products potentially harmful to man, animals, or plants.

IV. Implementation

In many countries steps are already being taken by national bodies to formulate codes of practice for the conduct of experiments with known or potential biohazard (*2*). Until these are established, we urge individual scientists to use the proposals in this document as a guide. In addition, there are some recommendations which could be immediately and directly implemented by the scientific community.

A. Development of Safer Vectors and Hosts

An important and encouraging accomplishment of the meeting was the realization that special bacteria and vectors, which have a restricted capacity to multiply outside the laboratory, can be constructed genetically, and that the use of these organisms could enhance the safety of recombinant DNA experiments by many orders of magnitude. Experiments along these lines are presently in progress and, in the near future, variants of λ bacteriophage, nontransmissible plasmids, and special strains of *Escherichia coli* will become available. All of these vectors could reduce the potential biohazards by very large factors and improve the methodology as well. Other vector-host systems, particularly modified strains of *Bacillus subtilis* and their relevant bacteriophages and plasmids, may also be useful for particular purposes. Quite possibly safe and suitable vectors may be found for eukaryotic hosts such as yeast and readily cultured plant and animal cells. There is likely to be a continuous development in this area, and the participants at the meeting agreed that improved vector-host systems which reduce the biohazards of recombinant DNA research will be made freely available to all interested investigators.

B. Laboratory Procedures

It is the clear responsibility of the principal investigator to inform the staff of the laboratory of the potential hazards of such experiments, before they are initiated. Free and open discussion is necessary so that each individual participating in the experiment fully understands the nature of the experiment and any risk that might be involved. All workers must be properly trained in the containment procedures that are designed to control the hazard, including emergency actions in the event of a hazard. It is also recommended that appropriate health surveillance of all personnel, including serological monitoring, be conducted periodically.

C. Education and Reassessment

Research in this area will develop very quickly, and the methods will be applied to many different biological problems. At any given time it is impossible to foresee the entire range of all potential experiments and make judgments on them. Therefore, it is essential to undertake a continuing reassessment of the problems in the light of new scientific knowledge. This could be achieved by a series of annual workshops and meetings, some of which should be at the international level. There should also be courses to train individuals in the relevant methods, since it is likely that the work will be taken up by laboratories which may not have had extensive experience in this area. High priority should also be given to research that could improve and evaluate the containment effectiveness of new and existing vector-host systems.

V. New Knowledge

This document represents our first assessment of the potential biohazards in the light of current knowledge. However, little is known about the survival of laboratory strains of bacteria and bacteriophages in different ecological niches in the outside world. Even less is known about whether recombinant DNA molecules will enhance or depress the survival of their vectors and hosts in nature. These questions are fundamental to the testing of any new organism that may be constructed. Research in this area needs to be undertaken and should be given high priority. In general, however, molecular biologists who may construct DNA recombinant molecules do not undertake these experiments and it will be necessary to facilitate collaborative research between them and groups skilled in the study of bacterial infection or ecological microbiology. Work should also be undertaken which would enable us to monitor the escape or dissemination of cloning vehicles and their hosts.

Nothing is known about the potential infectivity in higher organisms of phages or bacteria containing segments of eukaryotic DNA, and very little is known about the infectivity of the DNA molecules themselves. Genetic transformation of bacteria does occur in animals, suggesting that recombinant DNA molecules can retain their biological potency in this environment. There are many questions in this area, the answers to which are essential for our assessment of the biohazards of experiments with recombinant DNA molecules. It will be necessary to ensure that this work will be planned and carried out; and it will be particularly important to have this information before large-scale applications of the use of recombinant DNA molecules are attempted.

References and Notes

1. P. Berg *et al.*, *Science* **185**, 303 (1974).
2. Advisory Board for the Research Councils, "Report of the Working Party on the Experimental Manipulation of the Genetic Composition of Micro-Organisms, Presented to Parliament by the Secretary of State for Education and Science by Command of Her Majesty, January 1975" (Her Majesty's Stationery Office, London, 1975); National Institutes of Health Recombinant DNA Molecule Program Advisory Committee, Bethesda, Maryland.
3. The work of the committee was assisted by the National Academy of Sciences National Research Council Staff: Artemis P. Simopoulos, Executive Secretary, Division of Medical Sciences, Assembly of Life Sciences; Elena O. Nightingale, Resident Fellow, Division of Medical Sciences, Assembly of Life Sciences. Supported by the National Institutes of Health (contract NO1-OD-52103) and the National Science Foundation (grant GBMS75-05293). Requests for reprints should be addressed to: Division of Medical Sciences, Assembly of Life Sciences, National Academy of Sciences, 2101 Constitution Avenue, NW, Washington, D.C. 20418.

A P-4 Laboratory of the NIH.

Photo courtesy of NIAID.

Document 13.8

February – March 1975

Newspaper Headlines and Articles

Washington Star Telex Quotes

San Francisco Chronicle

The Largest Daily Circulation in Northern California

113th Year No. 65 ★★★★ 777-1111 20 CENTS

'Fail-Safe' Plan On Gene Research

By David Perlman
Science Correspondent

NEW YORK, FRIDAY, FEBRUARY 28, 1975

World Biologists Tighten Rules On 'Genetic Engineering' Work

By VICTOR K. McELHENY
Special to The New York Times

Los Angeles Times, Friday, February 28, 1975

Scientists to Resume Genetic Research

BY GEORGE ALEXANDER

From "New Biology — 2nd Genesis or Pandora's Box?" by Judith Randal, *Washington Star*, March 11, 1975

SAN FRANCISCO—"We desperately wanted to have this thing go forward. But at the same time we had reached a point where we would suddenly see that there was a potential hazard to others and the environment, as well as ourselves."

The speaker was Dr. David Baltimore, Professor of Microbiology of the Massachusetts Institute of Technology's Center for Cancer Research.

Sydney Brenner, of the Medical Research Council of the United Kingdom, said the odds against escape of potentially dangerous viruses are enormous. But he added, "The essential thing is that these things are self-replicating. It would be like a crash on a motor way leading to other crashes."

Paul Berg, a professor of biochemistry at the Stanford University Medical Center, said, "We decided to set our standards rather high so that the containment requirements could be reduced later where it showed the hazard was not as great as believed."

A great deal of confidence, for example, was placed in what Dr. Sydney Brenner of Britain's Medical Research Council called "Messiah Genes." These genes, he said, can confer on bacteria traits—such as a dependence on ultraviolet light, some nutrient or an extreme of temperature—that would rob them of the ability to reproduce except under laboratory conditions. Because each experimental strain of bacteria can be fitted with many such dependencies, the chances that any one of them would be able to multiply if they escaped into the world at large could be made as small as one in a trillion trillion.

10 THE WALL STREET JOURNAL Friday, Feb. 26, 1975

Gene-Grafting Research to Resume Soon Under Stri Laboratory-Safety Measure

By JERRY E. BISHOP
Staff Reporter of THE WALL STREET JOURNAL

PACIFIC GROVE, Calif.—Biologists are preparing to resume a series of experiments that they voluntarily halted several months ago for fear they might accidentally create unnatural organisms that could be hazardous to humans.

Many of the experiments, however, will be done in laboratories using safety and containment measures heretofore used only when scientists were dealing with the most dangerous of known germs and viruses. And some of the experiments will be deferred until new safety precautions and countermeasures can be developed.

The experiments involve creation of an entirely new combination of genes and the grafting of these genes to common bacteria. This gene-grafting technology has been perfected only in the last couple of years. It has created considerable excitement among scientists as the technique potentially is of enormous benefit. It might be used, for example, to uncover the genes that turn a normal cell cancerous, or to create food plants that make their own fertilizer.

For the past 18 months, however, a debate has been raging among biologists over whether there is some unknown hazard to such gene grafting. This concern reached a peak when it became apparent, for example, that it would be possible to take the genes of a virus that infects animals, notably one that causes cancer in hamsters, and graft them to bacteria that normally flourish harmlessly in the gut of animals, including humans.

Call for World-Wide Delay

As a result of such concern, a committee of U.S. biologists last summer issued an unprecedented call for such experiments to be delayed world-wide until an international conference of scientists could be convened to determine what the possible hazards of the experiments might

Document 2.6

February 1975

Open Letter from Science for the People to the Asilomar Conference

OPEN LETTER TO THE ASILOMAR CONFERENCE ON HAZARDS OF RECOMBINANT DNA

In recent years we have witnessed the rapid development of unusually potent biological technologies. This conference has been called to consider the public and occupational hazards inherent in one of these technologies, the linking together of DNA molecules across natural species barriers. Regardless of how slight the *current hazards* of introducing man made hybrid microorganisms into the environment appear at the moment, the hazards of such activities are unknown and possibly great. We have seen how technologies which appeared completely beneficial at the time of their introduction have become intentionally or accidentally destructive of human life and the environment. Molecular biologists are in a position to benefit from the lessons of our technological present and not contribute to the inventory of tragic results already caused by, for example, radium, asbestos, thalidomide, vinyl chloride and dieldrin.

There are even broader social issues that must be considered. The growing preoccupation with technologies involving genetic manipulation, and parallel developments such as cell fusion and in vitro fertilization, all point to the application of these techniques for human genetic manipulation. Technology and scientific development, even when labelled biomedical, is not intrinsically socially beneficial. Specifically, technologies pointing to the modification of human genetic material must be examined with the greatest care to understand why they are being so eagerly developed, and for precisely whose benefit.

Decisions at this crossroad of biological research must not be made without public participation.

There is little evidence that the technologies being discussed at this meeting arise from social or medical needs of large segments of the population. Rather, they represent specialized interests including those of the scientific community itself. The consequences are that experiments that happen to be conceived, get done, regardless of whether or not they should be done. The public rationale for these rapid developments in genetic engineering generally involves positing hope for individuals suffering from rare genetic diseases. In fact, considerable risk may be taken by clinicians eager to apply advanced knowledge to effect new cures. However, the search for such dramatic cures often diverts attention from the massive health needs of the population as a whole and the need to prevent the epidemics of our time, such as environmental and industrial carcinogenesis, malnutrition and coronary heart disease.

The dangers inherent in the new technologies mandate some regulation of their development. We do not believe that the molecular biology community, which is actively engaged in the development of these techniques, is capable of wisely regulating this development alone. This is like asking the tobacco industry to limit the manufacture of cigarettes. Although we could imagine a scientific community in which the spirit of social cooperation would be sufficiently developed so as to require no external regulation, this is not our case today. We have all had personal experience of the competitive and professional pressures which remove caution, prudence and a larger concern for social benefits, from the path of hazardous experiments. Scientific careers are not built solely on a concern for public health, for the well being of the underprivileged, or right action.

Since the risks and danger of these technologies are borne by the society at large, and not just scientists, the general public must be directly involved in the decision making process. Yet we see even in the structure of this conference that a scientific elite is here alone trying to determine the direction that such regulation should take. The presence of scientists from specialized government agencies is an important input in this discussion, but not a sufficient one.

The moratorium was called until attempts have been made to evaluate the hazards and some resolution of the outstanding questions has been achieved. These conditions still have not been met. From scientists should come the initiative to open and create areas of participation in order to achieve a resolution to these questions in a socially balanced manner. This has not been done and should be done as soon as possible. So far the lead has come from people who are beginning to organize their priorities in science and technology in order to define and implement options that will affect present and future societies, very differently than those conceived by scientific experts alone.

In our efforts to listen to people outside the scientific establishment, we offer five proposals:

1. Involve those most immediately at risk—technicians, students, custodial staff, etc., in collective decision making on safety policy for the laboratory.
2. Integrate into the curriculum of biology and medical courses the social implications of present and future biomedical research.
3. Require social and environmental impact statements on the means and goals of biological research projects.
4. Continue examination of these matters at public sessions of scientific meetings.
5. Expand participation in the advisory committee of the National Institutes of Health requested by the moratorium. The N.I.H. could be the structure through which the involvement of non-scientists in decision making could be implemented.

Since the original call for the moratorium, 7 months ago, the outstanding questions posed by it have become even more relevant, given the number of university laboratories and commercial interests that are preparing to work with artificial recombinant DNA molecules and their applications. Clearly at least the minimal prudence of the moratorium should be continued until the above proposals are put into meaningful effect.

Genetic Engineering Group of Science for the People:

Jon Beckwith, *Harvard Medical School*
Luigi Gorini, *Harvard Medical School*
Fred Ausubel, *Harvard University*
Paolo Strigini, *Boston University*

Kostia Bergman, *M.I.T.*
Kaaren Janssen, *M.I.T.*
Jonathan King, *M.I.T.*
Ethan Signer, *M.I.T.*
Annamaria Torriani, *M.I.T.*

Courtesy of Avoine.

OPPOSING VIEWS

The two articles comprising this chapter (Doc. 3.1 and Doc. 3.2) are both written by distinguished molecular biologists. They illustrate the opposing views that developed before and after the Asilomar Conference about the wisdom of society's placing constraints on the research. Prior to the recombinant DNA debate, Robert Sinsheimer (Caltech), well known for his research on the replication of bacteriophage ϕX174, had written articles about the long-term prospects of utilizing DNA research to benefit mankind—for example, in the cure of human genetic disabilities. Joshua Lederberg was among the first to recognize that molecular biology would in the long run have a great, favorable impact on human life and had often stated as much in his column in the *Washington Post*.

How could two eminent molecular biologists fully familiar with all the technicalities of recombinant DNA research reach opposing views? The reason, as both documents make clear, is that the hazard of this research is entirely conjectural. It has not been measured and may not exist. Whether or not it is reasonable to advocate placing restriction on an activity that has not been proven to be significantly hazardous, and that therefore may not be, is a social and philosophical rather than a scientific question. Molecular biologists inevitably held different opinions on that issue and on the related question of whether or not the proposed restrictions upon the freedom of scientific inquiry posed different but equally severe hazards for society.

Document 3.1
October 16, 1975
New Scientist (**68**:148)

Troubled dawn for genetic engineering

Biologists have found themselves equipped with molecular tools which confer on them profound powers of good or ill. After a pause to assess this new potential and the magnitude of the dangers involved, researchers are now moving forward more or less cautiously. Here we present a perspective on these important developments

Dr Robert Sinsheimer
is chairman of the division of biology, California Institute of Technology, Pasadena

The essence of engineering is design and, thus, the essence of genetic engineering, as distinct from applied genetics, is the introduction of human design into the formulation of new genes and new genetic combinations. These methods thus supplement the older methods which rely upon the intelligent selection and perpetuation of those chance genetic combinations which arise in the natural breeding process.

The possibility of genetic engineering derives from major advances in DNA technology—in the means of synthesising, analysing, transposing and generally manipulating the basic genetic substance of life. Three major advances have all neatly combined to permit this striking accomplishment: these are, 1, the discovery of means for the cleavage of DNA at highly specific sites; 2, the development of simple and generally applicable methods for the joining of DNA molecules; and 3, the discovery of effective techniques for the introduction of DNA into previously refractory organisms.

The art of DNA cleavage and degradation languished in a crude and unsatisfactory state until the discovery and more recent application of enzymes known as restriction endonucleases. These enzymes protect the host cells against invasion by foreign genomes by specifically severing the intruding DNA strands. For the purposes of genetic engineering, restriction enzymes provide a reservoir of means to cleave DNA molecules reproducibly at a limited number of sites by recognising specific tracts of DNA ranging for four to eight nucleotides in length. These sites may be deliberately varied by the choice of the restriction enzyme.

The enzymes cut both strands of the DNA double helix, and the break may be at the same base pair or staggered by several bases (Figure 1). In the latter case the two fragments of DNA are each left with a terminal unpaired strand—a so-called cohesive or "sticky" end. This is particularly valuable in joining together two pieces of DNA end to end.

The number of susceptible tracts in a DNA obviously depends on the particular DNA and the particular enzyme. In some important instances there is only one such tract. For instance, the restriction enzyme coded by the *E. coli* drug resistance transfer factor I—Eco RI—cleaves the DNA of the simian virus 40 at only one site. Similarly it cleaves the circular DNA of the plasmid PSC 101 at only one site. The DNA of bacteriophage lambda is, however, severed at five sites. It is possible to produce mutants of lambda with progressively fewer sites, until lambda strains are now available with just one or two sites.

For some purposes more numerous cleavage sites are useful. In a number of laboratories, including my own, the ØX virus RF can be cut at up to 13 sites using selected restriction enzymes. Because these enzymes yield overlapping fragments, a physical map of the DNA can be formed and correlated with the viral genetic map.

Restriction enzymes thus permit us to obtain specific fragments of DNA. For genetic engineering one would like to be able to rejoin such fragments in arbitrary ways. Two general methods have been developed to achieve this, both of which depend on the "sticky end" principle in which complementary single strand ends combine (Figures 1 and 2). Restriction enzymes which inflict staggered cuts automatically produce "sticky ends" in the DNA chain severed. Alternatively, a combination of enzymic and chemical manipulation can create a "stick end".

Modified plasmids in E. coli

By these means, then, any arbitrarily selected piece of DNA from any source can be inserted into the DNA of an appropriately chosen plasmid or virus. The new combination must then, for most purposes, be reintroduced into an appropriate host cell. This was achieved just a few years ago when, Stanley Cohen, at Stanford, discovered that plasmid DNA could be reintroduced, albeit with low efficiency, into appropriately treated *E. coli* cells and that these could then subsequently grow and propagate the plasmid. Foreign genes can therefore be introduced into *E. coli* plasmids which can be propagated indefinitely in ordinary bacterial cultures. As one instance, the ribosomal RNA genes of *Xenopus laevis* (the African clawed toad) have been introduced into an *E. coli* plasmid and propagated for over 100 cell generations. And these genes are transcribed in their new host (Figure 2).

A similar result can, in principle, be achieved with the bacteriophage lambda. A foreign gene can be inserted into lambda DNA; spheroplasts or treated cells infected with this DNA will yield virus which can then be used to infect normal cells. By clever manipulation a recombinant DNA can be obtained which can subsequently be integrated into the host chromosome and propagated thereafter with the host.

To what purposes may these novel genetic combinations be put? One can conceive of a variety of benign purposes. Unfortunately one can also conceive of malign purposes, and of

This article first appeared in *New Scientist*, London, the weekly review of science and technology.

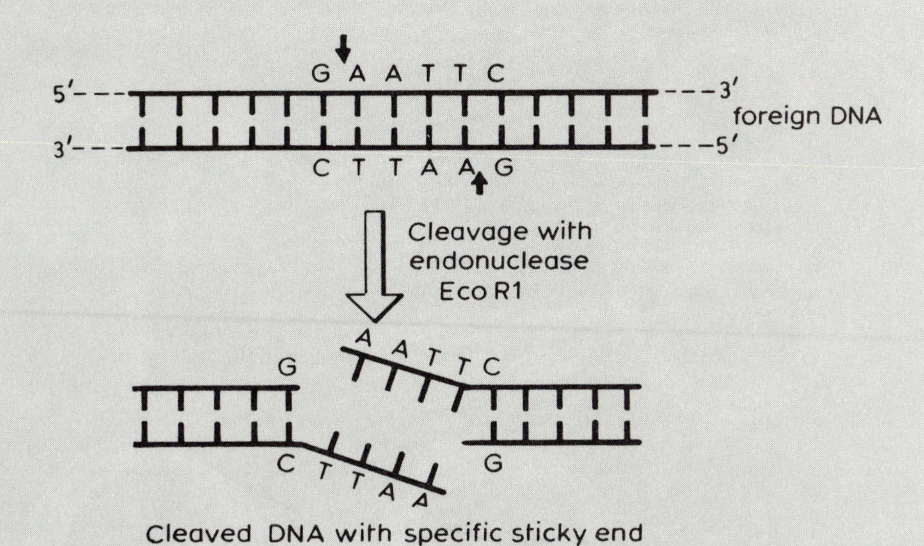

Figure 1 Endonucleases can cleave double stranded DNA at one point, or staggered as shown in the diagram. The staggered cut produces sticky ends which can join with sections of DNA severed by the same enzyme

major, if unintended, hazards.

The first purposes that come to mind are of a purely scientific character. The structure and organisation of the eukaryotic (higher organism) genome is currently being studied intensively. This research has been grossly impeded by the complexity of these genomes and the lack of means to isolate particular portions in adequate quantities for experimental analysis. The insertion of fragments of eukaryotic DNA into plasmids, followed by cloning (cellular multiplication), permits one to grow cultures of any size containing just one particular fragment. At present the choice of fragments to be inserted cannot in general be precisely defined, although some prior selection can be introduced. However, ingenious methods are being devised to permit subsequent selection of those bacterial clones carrying fragments of particular interest.

Clones of bacteria bearing, say, histone genes, ribosomal RNA genes, genes from individual bands of *Drosophila* DNA, DNA of a certain degree of repetition in the sea urchin genome, and so forth, are currently being investigated. There are numerous questions to ask and numerous matters of interest concerning the transcription and translation of such genes in the bacterial host: for instance, the rates at which they may mutate, and the use of such cloned genes as probes of the eukaryotic genome.

It is very probable that in time the appropriate genes can be introduced into bacteria to convert them into biochemical factories for producing complex substances of medical importance: for example, insulin (for which a shortage seems imminent), growth hormone, specific antibodies, and clotting factor VIII which is defective in hemophiliacs. Even if these specific genes cannot be isolated from the appropriate organisms, the chances of synthesising them from scratch are now significant.

Other more grandiose applications of microbial genetic engineering can be envisaged. The transfer of genes for nitrogen fixation into presently inept species might have very significant agricultural applications. Appropriate design might permit appreciable modifications of the normal bacterial flora of the human mouth with a significant impact upon the incidence of dental caries. Even major industrial processes might be carried out by appropriately planned microorganisms.

However, we must remember that we are creating here novel, self-propagating organisms. And with that reminder, another darker side appears on this scene of brilliant scientific enterprise. For instance, for scientific purposes there is great interest in the insertion of particular regions of viral DNA into plasmids—particularly, portions of oncogenic (cancer-inducing) viral DNA—so as to be able to obtain such portions and their gene products in quantity and subsequently to study the effects of these substances on their normal host cells. Abruptly we come to the potential hazard of research in this field, in fact the specific hazard which inspired the widely known "moratorium" proposed last year by a committee of the US National Academy, chaired by Paul Berg.

This moratorium and its related issues deserves very considerable discussion. Briefly, it became apparent to the scientists involved —at almost the last hour when all of the techniques were really at hand—that they were about to create novel forms of self-propagating organisms—derivatives of strains known to be normal components of the human intestinal flora—with almost completely unknown potential for biological havoc. Could an *Escherichia coli* strain carrying all or part of an oncogenic virus become resident in the human intestine? Could it thereby become a possible source of malignancy? Could such a strain spread throughout a human population? What would be the consequence if even

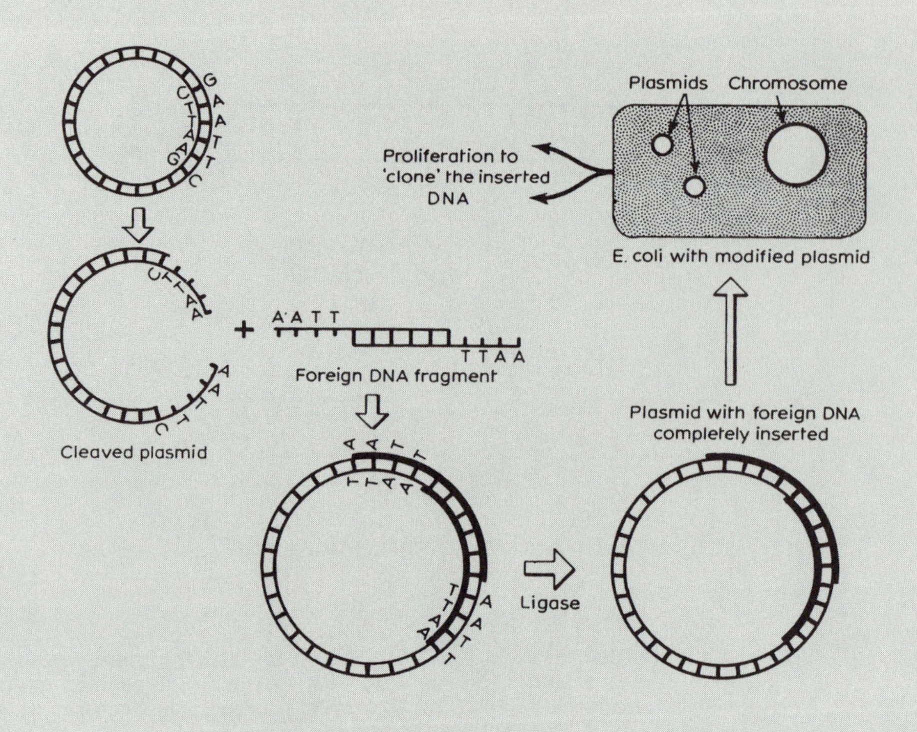

Figure 2
Cloning a gene: a nick is made in the circular DNA of a plasmid; the require DNA sequence, excised with the same restriction enzyme, is inserted into the gap; the DNA chains are repaired by a ligase enzyme; the plasmids are reintroduced into E. coli; *when the coli culture multiplies, the plasmids, and the foreign genes with them, are multiplied (cloned) too*

an insulin-secreting strain became an intestinal resident? Not to mention the more malign or just plain stupid scenarios such as those which depict the insertion of the gene for botulinus toxin into *Escherichia coli.*

Unknown probabilities

Unfortunately the answers to these questions in terms of probabilities that some of these strains could persist in the intestines, the probabilities that the modified plasmids might be transferred to other strains, better adapted to intestinal life, the probabilities that the genome of an oncogenic virus could escape, could be taken up, could transform a host cell, are all largely unknown.

Following the call for a moratorium a conference was held at Asilomar at the end of last February to assess these problems. While it proved possible to rank various types of proposed experiments with respect to potential hazard, for the reasons already stated it proved impossible to establish, on any secure basis, the absolute magnitude of hazard. Various distinguished scientists differed very widely, but sincerely, in their estimates. Historical experience indicated that simple reliance upon the physical containment of these new organisms could not be completely effective.

In the end a broad, but not universal, consensus was reached which recommended that the seemingly more dangerous experiments be deferred until means of "biological containment" could be developed to supplement physical containment. By biological containment is meant the crippling of all vehicles—cells or viruses—intended to carry the recombinant genomes through the insertion of a variety of genetic defects so as to reduce very greatly the likelihood that the organisms could survive outside of a protective, carefully supplemented laboratory culture.

This seems a sensible and responsible compromise. However, several of the less prominent aspects of the Asilomar conference also deserve much thought. The lens of Asilomar was focused sharply upon the potential biological and medical hazard of this new research, but other issues drifted in and out of the field of discussion. There was, for instance, no specific consideration of the wisdom of diverting appreciable research funds and talent to this field, in lieu of others. An indirect discussion of this question was perhaps implicit in the description of the significance and scientific potential of research in this field presented by those who were impatient of any delay.

Indeed the eagerness of the researchers to get on with the work in this field was most evident. To a scientist this was exhilarating. Obviously these new techniques open many previously closed doorways leading to the potential resolution of long-standing and important problems. I think also there is a certain romance in this joining together of DNA molecules that diverged billions of years ago and have pursued separate paths through all of these millenia. Personally I feel confident one could easily justify this new research direction. But a sociologist of science might see other under-currents in this impetuous eagerness, and the bright scientific promise should not blind us to the realities of other concerns.

Nor was there any sustained discussion at Asilomar of ancillary issues such as the absolute right of free inquiry claimed quite vigorously by some of the participants. Here, I think, we have come to recognise that there are limits to the practice of any human activity. To impose any limit upon freedom of inquiry is especially bitter for the scientist whose life is one of inquiry; but science has become too potent. It is no longer enough to wave the flag of Galileo.

Rights are not found in nature. Rights are conferred within a human society and for each there is expected a corresponding responsibility. Inevitably at some boundaries different rights come into conflict and the exercise of a right should not destroy the society that conferred it. We recognise this in other fields. Freedom of the press is a right but it is subject to restraints, such as libel and obscenity and, perhaps more dubiously, national security. The right to experiment on human beings is obviously constrained. Similarly, would we wish to claim the right of individual scientists to be free to create novel self-perpetuating organisms likely to spread about the planet in an uncontrollable manner for better or worse? I think not.

This does not mean we cannot advance our science or that we must doubt its ultimate beneficence. It simply means that we must be able to look at what we do in a mature way.

There was, at Asilomar, no explicit consideration of the potential broader social or ethical implications of initiating this line of research—of its role, as a possible prelude to longer-range, broader-scale genetic engineering of the flora and fauna of the planet, including, ultimately, man. It is not yet clear how these techniques may be applied to higher organisms but we should not underestimate scientific ingenuity. Indeed the oncogenic viruses may provide a key; and mitochondria may serve as analogues for plasmids.

Controlled evolution?

How far will we want to develop genetic engineering? Do we want to assume the basic responsibility for life on this planet—to develop new living forms for our own purpose? Shall we take into our own hands our own future evolution? These are profound issues which involve science but also transcend science. They deserve our most serious and continuing thought. I can here mention only a very few of the more salient considerations.

Clearly the advent of genetic engineering, even merely in the microbial world, brings new responsibilities to accompany the new potentials. It is always thus when we introduce the element of human design. The distant, yet much discussed application of genetic engineering to mankind would place this equation at the centre of all future human history. It would in the end make human design responsible for human nature. It is a responsibility to give pause, especially if one recognises that the prerequisite for responsibility is the ability to forecast, to make reliable estimates of the consequence.

This article is based on a lecture given recently to the Genetics Society of America

Can we really forecast the consequence for mankind, for human society, of any major change in the human gene pool? The more I have reflected on this the more I have come to doubt it. I do not refer here to the alleviation of individual genetic defects—or, if you will, to the occasional introduction of a genetic clone—but more broadly to the genetic redefinition of man. Our social structures have evolved so as to be more or less well adapted to the array of talents and personalities emergent by chance from the existing gene pool and developed through our cultural agencies. In our social endeavours we have, biologically, remained cradled in that web of evolutionary nature which bore us and which has undoubtedly provided a most valuable safety net as we have in our fumbling way created and tried out varied cultural forms.

To introduce a sudden major discontinuity in the human gene pool might well create a major mismatch between our social order and our individual capacities. Even a minor perturbation such as a marked change in the sex ratio from its present near equality could shake our social structures—or consider the impact of a major change in the human life span. Can we really predict the results of such a perturbation? And if we cannot foresee the consequence, do we go ahead?

It is difficult for a scientist to conceive that there are certain matters best left unknown, at least for a time. But science is the major organ of inquiry for a society—and perhaps a society, like an organism, must follow a developmental programme in which the genetic information is revealed in an orderly sequence.

The dawn of genetic engineering is troubled. In part this is the spirit of the time —the very idea of progress through science is in question. People seriously wonder if through our cleverness we may not blunder into worse dilemmas than we seek to solve. They are concerned not only for the vagrant lethal virus or the escaped mutant deadly microbe, but also for the awful potential that we might inadvertently so arm the anarchic in our society as to shatter its bonds or conversely so arm the tyrannical in our society as to forever imprison liberty.

It is grievous that the elan of science must be tempered, that the glowing conviction that knowledge is good and that man can with knowledge lift himself out of hapless impotence must now be shaded with doubt and caution. But in this we join a long tradition. The fetters that are part of the human condition are not so easily struck.

We confront again, the enduring paradox of emergence. We are each a unit, each alone. Yet, bonded together, we are so much more. As individuals men will have always to accept their genetic constraints, but as a species we can transcend our inheritance and mould it to our purpose—if we can trust ourselves with such powers. As geneticists we can continue to evolve possibilities and take the long view.

Document 3.2

November 1975
Prism (**3**:33)

DNA Research: Uncertain Peril and Certain Promise

Joshua Lederberg
Stanford University & Center for Advanced Study in the Behavioral Sciences
Stanford Calif., 94305

Nucleic acids were first isolated from cells and recognized as distinctive substances over 100 years ago. However, their crucial role in biology was long obscured for two reasons:

1) a methodological one—the lack of a direct biological assay for the specificity of a nucleic acid; and

2) a doctrinal one—the fallacy of chemical simplicity that characterized the organic chemists model of nucleic acids as homogeneous polymers devoid of informational variety.

As a result, although nucleic acids were known quite early to be systematically associated with the chromosomes (indeed these are named from the colorability of nucleic acids with the basic dyes used in cytology) there seemed to be no way to relate them to the vital functions of the cell, much less to put them to any utilitarian application in medicine, in agriculture, or in industry.

The scientific revolution of modern biology may be dated to 1944 when O.T. Avery, Colin MacLeod and Maclyn McCarty working at the Rockefeller Institute announced that the "transforming agent" of the pneumococcus consisted of DNA. This agent had already been speculatively compared with genes transferable from one bacterial strain to another. The announcement quickly provoked a high level of critical discussion, as befit a claim of such fundamental import. It was crucial to be certain that DNA itself, rather than a trace contamination with an active protein, was the active agent. Furthermore, too little was known of bacterial genetics to be sure that an isolated example of the modification of a bacterial outer coat really did reflect a gene transfer. However, so many new experiments were prompted by these new claims that, by 1953, the detailed atomic structure of DNA had been embodied in the now familiar double-helix model of Watson and Crick; and a whole new discipline of microbial (and viral) genetics had grown up in an area that a few years before had been widely thought to be wholly disconnected from Mendelian-Darwinian biology.

This article originally appeared in *Prism*, November 1975, in slightly different form.

The unprecedented acceleration of fundamental chemical biology since 1945 can be related to the intellectual stimulation of these findings, and the analytical tools that accompany them, as well as to the policy of continued federal support of scientific research in the era since World War II. We may also note that medical microbiology was just then past its most heroic era, the identification of all of the principle agents of infectious disease, the very empirical success of which may have tended to hinder closer cooperation with academic biology.

Although our scientific understanding of the cell has been completely transformed in the last 30 years, we have yet to see practical applications of molecular biology of nearly comparable importance. There is very little indeed in the practice of contemporary medicine (even of clinical genetics) that depends on the knowledge that DNA has a bihelical structure, which is an appropriate shibboleth for the underlying doctrine of this new field. This has not eroded our faith that basic approaches to the understanding of viruses, of the neoplastic and of the aging cell, of the mechanism of the immune response, or of aberrant chromosomes, will be instrumental to far-reaching changes in medical technology. The human benefit from these will match the theoretical impact that DNA study has already demonstrated for cell biology.

These expectations for a possibly long-delayed future fruit have been exalted within the last year or so by some new findings that give us much greater technical capacity for the laboratory manipulation of microbial DNA. These new methods of "DNA-splicing" have already opened up many lines of experimental investigation of the structure of eukaryotic [higher life forms] chromosomes. For it is now possible to fragment an animal or human cell's DNA into perhaps a million segments, and transfer a single one to a bacterial host wherein it can be studied in a microcosm, or whence large quantities of a specific DNA can be produced for more elaborate analysis than will ever be possible with the enormously complex original source material.

This technique of gene implantation can also be used to transfer the genetic information for a given product from one species of cell to another; and this is the direction that, in my own view, leads to an early chance for a technology of untold importance for diagnostic and therapeutic medicine: the ready production of an unlimited variety of human proteins. Analogous applications may be foreseen in fermentation processes for cheaply manufacturing essential nutrients, and in the improvement of microbes for the production of antibiotics and of special industrial chemicals. It has also been foreseen by many workers in the field that such technical potency may also be associated with public hazards, and that we then face a Promethean dilemma.

Public policy decision in this field can only lead to social benefit if it is well informed, equally about the potential risks as well as benefits of further work on DNA-splicing. Indeed, if any substantial risk can be identified, there is no question of the need for ethical and operational standards; the only question must be whether their actual form and implementation result in a true net advantage to the public health and safety. Too often, the superficially easy way to cope with such a problem is to invoke a formal regulatory statute, ignoring how well the actual bureaucratic implementation or policing of the rules meets the intended balance of risks and benefits. Before elaborating on the policy issues, it may therefore be incumbent to outline some further detail on 1) the present state of DNA-splicing technology, 2) some promising applications, and 3) the sources of risk in further work in this field.

1. How we splice DNA in the laboratory.

DNA recombination as the be-all and end-all of sexual reproduction is of course one of the major happenings of the natural world of life. Among higher forms, exchange of DNA is almost always limited to members of the same or closely related species and is generally associated with sexual differentiation, that is males and females. Bacteria and viruses exhibit some analogous barriers; but also many exceptions: which may reflect the fragility of the idea of "species" when applied to these forms. For example, the entire group of "enteric bacteria" including also such forms as Shigella, Escherichia coli, Proteus, and Serratia can readily be shown to exchange genetic fragments without special interventions—we have to assume that this also occurs in nature, and that it would be difficult to aggravate the risks already latent in that natural occurrence.

An especially interesting and important level of genetic organization in bacteria is the plasmid: a bit of circular DNA that behaves like an extra small chromosome, and one that seems to survive in nature by virtue of its easy transmissibility from one bacterial strain to another. Many different kinds of plasmids are known; medically, the most prominent today are those which confer transmissible antibiotic resistance on human pathogens, notably staphylococci and some enteric pathogens like Shigella. These plasmids are part of the evolutionary history of their host organisms: the spread of antibiotic-resistance plasmids is the most formidable ecological response that pathogenic bacteria have yet discovered to our widespread use of antibiotics. Other plasmids are undoubtedly involved in altering the pathogenicity and host-specificity of various bacteria; therefore in simple self-defense of our species it is urgently important that we learn what we can about them.

Plasmids have now achieved special prominence also for a technical reason, that they are especially convenient vehicles for DNA splicing and for the transmission of DNA segments from one species to another, especially in conjunction with another elegant tool: the R- [for restriction] enzyme. The R-nucleases are widely distributed among cell types; they may be an important mechanism by which a cell fends off any "foreign" DNA while protecting its own. For historical reasons, this particular mode of defense has been called "restriction", and now that we know the responsible enzymes, they have inherited a name that may be confusing to bystanders. The important attribute of an R-nuclease is that it recognizes a special sequence of bases in double-stranded DNA, depending on the particular enzyme, and that it will then cut that DNA leaving a proruded "sticky" end. The "stickiness" arises from the fact that the same sequence can be recognized on other protrusions, and that the Watson-Crick rules for uniting separated DNA strands will then apply.

For example, Dr. Stanley N. Cohen has used an R-enzyme to simplify a naturally occurring plasmid to the point where it consisted of a small circle of DNA, embracing the minimum genetic information needed to be able to replicate, plus a single R-enzyme recognition site. This artificial plasmid, pSC-101, has been an important tool for DNA-splicing research. When it is exposed to R-enzyme, the circle is cut into a single open length with sticky ends. Another enzyme, ligase, can be used to rejoin such ends; However, it is also possible to insert other sticky-ended pieces of DNA from divers sources into the plasmid, then close it up with ligase. This is then the key to the convenient design and construction of new DNA molecules which can then be transferred to a bacterial host—an experiment which is today's counterpart of Avery's report of 30 years ago.

The new DNA does not have to be of the same bacterial species; for example, Dr. Cohen and his collaborators have already reported on the effective transfer of DNA from a toad, Xenopus, into E. coli with evidence of the production of toad-like ribosomal nucleic acids by the modified bacteria.

Besides these plasmids, bacterial viruses are being used in a similar fashion; less elegantly, segments of DNA from intact bacteria may also be used both for insertions and as the acceptors. All of these experiments so far depend on the innate (and poorly understood) ability of bacterial cells to incorporate DNA furnished from without. There have been many published claims of similar phenomena with plant and animal cell acceptors, but none to date have the reproducibility and general acceptance as undoubted fact that pertains to the bacterial work. There is no immediate substance to the idea that these techniques are applicable to the "genetic engineering of human beings". (In the long run, the possibility of such technical capabilities cannot be denied in principle, no more than we can disprove the possibility of a peaceful world, or of a global morale capable of the wisest disposition of our existing powers for good and evil.)

The special power of these enzymological techniques is that they depend on the basic chemical structure of DNA (recall the shibboleth!) rather than on the biological adaptations of particular microbial species. Laboratory manipulations may then synthesize constructs that occur rarely, if ever, in the natural world. It is difficult, however, to assess just what can or cannot occur in the entire immense mass of the earth's biosphere. R-enzymes, mixed DNA and acceptor bacteria surely occur with some frequency in natural habitats. The likelihood of a prevalent natural transmission of plasmids among "unrelated" forms is also enhanced by recent findings of agents with extraordinarily broad host range. From a practical standpoint, there can be no doubting the power of DNA-splicing as a means of acquiring specific gene sequences and studying and applying them. Possible regulatory strictures against promoting the transfer of plasmids among "species that do not normally do so" may be mooted by the behavior of the newly found agents—unless, of course, we are prevented from making just such discoveries by the restriction of research.

DNA-splicing is however but the most powerful of a range of techniques and processes whose end-effect is the bringing together of more or less natural assemblages of DNA information. Indeed, it may prove to be less powerful than older methods—of sexual crossing, of transduction with bacteriophage, of DNA transformation—for special constructions that involve larger complexes than the segments yielded by the R-enzymes. These methods in turn are an extrapolation of the artificial breeding of domestic animals and plants upon which civilization was founded. In any event, the most efficient application of DNA-splicing requires intimate knowledge of the genetic structure of both the donor and the acceptor strains, for which breeding methods are important if not indispensable.

Perhaps the most important single conclusion is that this technology is just in its infancy, but has already made great leaps; and that it is simple enough that it can be practised in any laboratory that can handle pure bacterial cultures. Just this simplicity, which makes for great convenience and rapidity of experimental advance, has been a source of concern about the proliferation of the methods in the hands of people with less than mature professional and ethical judgment and with deficiencies in the skills entailed in containing bacterial cultures in the laboratory.

2. The promise: biosynthetic human proteins

DNA segmentation and splicing is certain to play a vital role in the further domestication of microbes for such roles as antibiotic development and production of high quality protein supplements. However, the unique strength of this procedure is to allow the large scale production of gene products of a less easily domesticated species: the human. In particular, human proteins already play a substantial role in medicine but one which is hindered by scarce supplies.

The most attractive options that are visible are the human antibody globulins. This preference has a theoretical basis. Of all the body proteins, the immune globulins have evolved to be variable within and between individuals in the fulfillment of a vital function. Compared to the rare genetic defects in other proteins (like hemophilia), failures or errors in production of antibody globulin are quite prevalent, and are known to play a major role in (1) defense against infectious diseases, (2) autoimmune and allergic disease, and (3 perhaps also in cancer.

The most comprehensive role of biosynthetic proteins would be in passive immunization against infectious diseases. Animal antisera were once used but had to be abandoned because of the anti-animal antibody that they provoked in man. Priority targets for passive globulin therapy are those diseases where either technical or social factors may lead to gaps in protection by active immunization. They include influenza, hepatitis, smallpox, encephalitis virus, rubella, herpes, rabies and perhaps also trypanosomes, malaria, schistosomes, tuberculosis and leprosy and many others.

I believe there is reason for special urgency to develop a backup capability of passive immunization to prevent a global catastrophe that may result from our becoming too complacent about active immunization against diseases like smallpox and polio, and the technical inadequacy of vaccines like rubella and hepatitis. Our general posture of defense against viral pandemic is a feeble one. We have no assurance that the next influenza epidemic will not be slightly more virulent and cost a million lives for lack of a ready defense.

A broader need applies to polyvalent prophylaxis for infants. The principal medical argument for breast-feeding is the provision of colostrum and of a continuing supply of maternal mixed globulins in the milk. There would be a huge and valid market for polyvalent gamma globulin supplements to infant dietaries both in industrialized and in poorer countries. An analagous veterinary demand speaks to further efficiency in food production.

Specific antibodies are, of course, already very widely applicable as diagnostic reagents of high specificity and selectivity. Blocking antibodies may also be expected to play a useful role in protecting transplanted tissues and organs from more aggressive immunological attack by the new host. Conversely, tissue-specific ligating antibodies, although not necessarily themselves carrying cytotoxic capability, may also be expected to be useful in enhancing the cell specific toxicity of cancer controlling compounds. Cell specific reagents will also be invaluable for diagnostic purposes and for the specific separation of human cell types to be used further for either diagnostic or therapeutic applications.

Anti-sperm immunity is also being very seriously proposed as an approach to durable male contraception. I have been quite uneasy about such proposals that involve the vaccination of men against their own sperm for fear of unwonted side-effects and also on account of probable difficulties in reversibility. Passive antibody directed against sperm flagella is demonstrably able to interfere with fertilization simply by the immobilization of the sperm and should have a minimum of other side-effects. Such immunizations would be reversible by the spontaneous decay of passive immunity over periods of from 3 to 6 months. Comparable possibilities exist for the immunization of women against sperm.

Besides the specific antibody globulins, a number of important but less specific proteins (complement, properidin) play an important part in defense against infection. Fibrinolysis (plasmin) and urokinase (plasminogen-activator) represent a group of enzymes that are experimentally promising for the control of embolism. Many protein hormones are equally scarce for clinical trials. The list could be extended substantially! The most important products are perhaps those that are invisible by present methodology.

Microbial biosynthesis may well be supplemented by organic synthesis by the elaboration of proteins in human and hybrid somatic cell culture and by cell free ribosomal synthesis with contrived m-RNA. Each of these options has its own peculiar difficulties and hazards and the whole field will be most rapidly advanced by using the best available methods for a given problem.

3. The risks: will dangerous organisms escape?

At the present time, perhaps a half-dozen bacterial species are well enough understood to be prime vehicles for laboratory study of DNA-splicing. For safety and convenience, investigators have preferred not to use pathogenic forms wherever feasible. Significant concern arises from the possibility that the introduction of new genetic information may (inadvertently) generate a new pathogen for man, or its analogue, a source of ecological disruption at some other point in the biosphere. The most likely, but not necessarily the only, source of such genes for pathogenicity are precisely the organisms that most urgently need further study—the subtle and insidious killers that are not now amenable to medical treatment and prevention. These include slow virus infections that may be involved in a wide range of chronic diseases and cancer, and more familiar viruses like herpes for which satisfactory vaccines are not now available.

Public discussion of advances in DNA-splicing indeed have been almost totally focussed on the possible hazards of escape of new forms of microorganisms, rather than on their utilitarian merits. The most urgent source of concern has been for the prospect of introducing potential cancer-causing DNA into common bacteria. While it is recognized how speculative this hazard is, the general territory is so poorly understood that no one can argue against the need for cautious laboratory procedures. Having gone so far, a number of workers—particularly those not previously experienced or trained in medical microbiology—have confessed having given almost no thought in the past to problems of microbial safety; and some of these are now among the most zealous in demanding tight regulation of further research. As part of this process, there has been a sincere, almost frantic, effort to seek out the most remote conceivable hazards.

Viewed as a rather public soul-searching and self-education, these discussions are invaluable. The main danger is that tentative questions will be incorporated by some political imperative into ironclad regulations that will be with us long after anyone has forgotten why they were instituted. One can after all raise similar questions about the widest range of human activities: should it be lawful to keep domestic cats now that they are under suspicion of harboring toxoplasmosis, and possibly leukemia as well? The same kinds of questions that are asked of microbiology could be lodged against plant breeding: what positive assurance can there be against the next artificial pollination producing the weed that will ruin the wheat crop a decade from now? Closer to home, should we forbid international travel, given the certain knowledge that our quarantine procedures are quite unable to hinder the importation of exotic diseases?

For each of these cases, and many more, the apparently innocuous doctrine: "As long as there is any risk, don't do it!" can only lead to a loss in human welfare. We must instead make every feasible effort to assess both the risks and the benefits of a given course of action—only then are we in a position to weigh the optimal balance. This in no way may deny the rights of individuals to make voluntary decisions about their exposure to risk, even if for public benefit. But individuals can hardly make the best policy about their own future, including their expectations for what medicine will offer for the infirmities of their own later years, without expert assessment.

These assessments are difficult, problematical, and controversial but a committee of the National Academy of Sciences has made some headway in trying to classify different categories of hazard. Where that hazard is reasonably predictable (in the current atmosphere, now so highly sensitized) laboratory containment precautions akin to those appropriate for known pathogens have been recommended, and these will doubtless be comprehendingly complied with. This applies, for example, to experiments involving the recombination of known tumor virus DNA with bacterial plasmids.

For more conjectural hazards—like the introduction of antibiotic resistance into common, non-pathogenic species—the requirements for high security laboratories may be an inordinate burden (who, in fact will pay for them?) in relation to the prospective gains. The best strategy here seems to be the development of safe vectors: plasmids and bacteria engineered to have little chance of survival outside the laboratory. In fact, in the long run this is a safer procedure than relying upon uncertain human compliance with fixed rules and regulations.

Remaining controversies in this field center upon rather complicated analyses of the remotest kinds of risks. Given some additional time, and a decision to postpone a rigorous framework of external regulation, most research institutions will be able to work out their own reasonable plans based on the national guidelines. A premature crystallization of such regulation will not only frustrate the achievement of pragmatically useful goals, but will hinder the research needed to make more refined assessments. Those who regard themselves as guardians of the public safety must count not only the speculative hazards of these marginal situations, but also the costs to the public health of impeding their investigation.

This partly voluntaristic approach will not satisfy a demand for absolute assurance that no foolish experiment is ever attempted. But the history of human institutions should suffice to show that NO system of sanctions can have such a perfect outcome. The human species is constantly and inevitably attended by contaminating and parasitic microbes—the person suffering from an enteric infection who fails to wash his hands, or the influenza victim who insists on going to work is behaving unethically, and to the peril of his fellows. But we would scarcely invoke serious regulatory sanctions in preference to public education except where there is an unusual public risk, and some evidence that an enforced quarantine was likely to yield a positive gain.

While a newly awakened scientific conscience is the main source of public discussion of the hazards of microbiological research, some of it also appears to be directed to the frustration of its benefits. For some such critics, any measure that might prolong lives (especially of our own citizens) is merely worsening the population problem and the excessive consumption of limited world resources. Others are sure that any technological advance in a free-enterprise society will inevitably benefit only the already powerful and wealthy. Still others harbor deep-seated resentments against the professional and economic dominance of the medical profession. These motifs do not alter by one whit the actual substance of concerns about the hazards of research; we must understand their prevalence to foresee the ways in which the issue may be exploited for reasons quite far from scientific conscience.

Senator Kennedy has remarked that society must give its informed consent to technological innovation. The power of the purse is enough to enforce that doctrine, nor can there be any quarrel with it on ethical grounds. The relevant information surely includes the hazards of saying no to prospects of significant medical advances. The particular field of DNA-splicing research, far from being an idle scientific toy, or the basis of expensive and specialized aid to a few lives, promises some of the most pervasive benefits for the public health since the discovery and promulgation of the antibiotics.

"THAT'S THE PROBLEM – EVEN THOUGH IT'S *RE*COMBINANT, WE CAN'T MAKE IT *DE*COMBINANT."

DRAFTING THE NIH GUIDELINES

The Asilomar Conference triggered the establishment of recombinant DNA committees in many countries at a private or semigovernmental rather than governmental level, usually within the framework of academies or research councils. In Britain, government publication of the Ashby Report (Doc. 1.12) was not immediately followed by further governmental measures. Indeed, seven months elapsed before a second governmental Working Party was set up to draft a code of laboratory practice (see Chapter 11). With the exception of Britain, most countries decided to wait for further events in the United States rather than take separate initiatives.

In the United States, the naivete of those Asilomar participants who believed that they would be allowed to continue their work without further interference was made immediately apparent. On April 22, 1975, Senator Edward Kennedy as Chairman of the Subcommittee on Health of the Committee on Labor and Public Welfare of the U.S. Senate held a hearing on the relationship of a free society to its scientific community. With two of the four expert witnesses, scientists involved in recombinant DNA experiments (Stanley Cohen of Stanford and Donald Brown of the Carnegie Institution), the hearing used the recombinant DNA issue as an example of the problem of determining the legitimate role of the public and their legislative representatives in deciding the direction of scientific research and the utilization of scientific knowledge. Whatever else it may or may not have achieved, the hearing certainly signaled the beginning of the U.S. legislature's interest in the issue.

And although few at Asilomar may have realized it, on October 7, 1974, a few months *before* the Conference, the Director of the National Institutes of Health (NIH) formed a committee with the tortuous title the Recombinant DNA Molecule Program Advisory Committee (RAC for short). Although the mandate of the RAC was to advise "concerning a program (a) for the evaluation of potential biological and ecological hazards of DNA recombinations of various types, (b) for developing procedures which will minimize the spread of such molecules within human and other populations, and (c) for devising guidelines to be followed by investigators working

Biohazard symbol construction and photo by Kenneth Kneitel.

with potentially hazardous recombinants" after Asilomar the RAC became almost entirely preoccupied with its third task: devising guidelines. During the spring and summer of 1975, the RAC began to draft a comprehensive and detailed set of guidelines, taking over where the Asilomar Conference had left off. The task proved difficult, as Nicholas Wade's article reports (Doc. 4.1).

At the incentive of David Hogness (Stanford), Chairman of the Drafting Subcommittee, the RAC attempted to produce encyclopaedic guidelines in which all possible classes of recombinant DNA experiments were included. Faithfully following the Asilomar Report, the RAC attempted to categorize the experiments according to its estimate of the magnitude of the conjectural hazards they might entail. This analysis resulted in the scientifically unsubstantiated conclusion that the closer the phylogenetic and evolutionary relationship between humans and the species whose DNA was being used, the greater the conjectural hazard. The DNA of known natural pathogenic microorganisms was also assessed by the Committee to be particularly hazardous.

The RAC then tried to match its hierarchy of conjectural hazards with a graded series of containment precautions. Two types of containment were available: physical containment of the conventional sort already used in medical microbiology laboratories handling real pathogens, and biological containment, which had been vigorously promoted at the Asilomar Conference.

Four categories of physical containment (P1 through P4) were arbitrarily defined. P4 corresponded to the highest level of containment required for handling the most lethal microorganisms and viruses such as anthrax bacilli, smallpox viruses, lassa fever viruses, and so on. P1 essentially comprised elementary measures of laboratory hygiene.

The RAC then established necessarily arbitrary criteria for biological containment. A host-vector system based on a wild-type *E. coli* isolated from an animal or person was considered to have zero containment. The use of *E. coli* K12 as a host was considered to offer the first grade of containment (EK1), since this standard laboratory strain is far less robust than the wild types and cannot colonize the intestines of people and animals. The RAC then defined further increased levels of biological containment and corresponding test criteria to be met by combinations of any derivative of bacteriophage or plasmid vectors* and *E. coli* K12 host cells in order to qualify as biologically contained EK2 or EK3 host-vector systems. In practice, it was only in April 1976 that the first host-vector system sufficiently disabled to meet the EK2 definition was officially approved, and we still await—and shall probably now never have—an EK3 host-vector system meeting the most exacting standards.

Finally, the RAC attempted to match its scale of increasing conjectural biohazard with a scale of increasing overall containment achieved by combinations of physical and biological containment. It was not until December 1975 that a consensus was reached, after months of haggling (Doc. 4.2). Molecular biologists developed a new jargon, speaking of their experiments as P2 + EK2 or P4 + EK1, and so on.

The guidelines proposed by the RAC in December 1975 were very stringent indeed, as reported by the daily press (Doc. 4.3). For example, to study by recombinant DNA techniques random fragments of the DNA of an adult primate required P4 + EK2 or P3 + EK3 containment. Neither EK2 nor EK3 host-vector systems that had been officially approved existed. P4 physical-containment laboratories existed only in special centers designed initially by military authorities for biological warfare research, or by medical authorities for handling exotic lethal pathogens. They were not available to molecular biologists and had not been approved by the RAC and the NIH as meeting the P4 critieria.

One immediate consequence of these guidelines would be a halt to recombinant DNA experiments with DNA from warm-blooded animals and from viruses. The enormous opportunities that the new method offered for studying human and mammalian genes and viruses, thereby contributing to our understanding and prevention of diseases such as cancer and virus infections, were to be sacrificed, as were opportunities for developing bacteria as new sources for the production from inserted genes of human proteins such

*Definitions of these and other terms can be found in the Scientific Background section at the end of the book.

as interferon, insulin, and growth hormones. The only experiments with eukaryotes that could be continued without serious delays were with the DNAs of amphibians, fish, and invertebrates.

Notwithstanding this stringency, commentators noted that the NIH was not proposing to back the guidelines by any special legal measures. The NIH and other U.S. governmental agencies could insist that institutions receiving federal funds follow the guidelines, but private laboratories (such as those of industrial companies) were not legally obliged to follow them. In January 1976 *New Scientist* raised the question, which was soon to be heard in several quarters, of possible direct legislation of recombinant DNA research (Doc. 4.4).

Inevitably those who wished to see recombinant DNA research progress quickly found the proposed guidelines much too restrictive; those who wished a virtual halt to the research found them too lax. Another six months passed—the "last look before the leap," in Nicholas Wade's words (Doc. 4.5)—before the Director of the NIH, with the concurrence of the Secretary of Health, Education and Welfare, issued the guidelines on June 23, 1976. Document 4.6 presents parts of the introduction to the guidelines by the Director of the NIH. Instead of reproducing from the *Federal Register* the full text of the guidelines, which makes turgid reading, we include as Document 4.7 Colin Norman's digest of them from *Nature*. The State Department's help was enlisted to broadcast the news to the world (Doc. 4.8).

Almost a year and a half had elapsed since the Asilomar Conference, during which time an increasing number of investigators had begun recombinant DNA experiments, following the Asilomar guidelines. Several of them found themselves with completed or partially completed experiments that were no longer permissible under the new guidelines. As the article from *Science and Government Report* (Doc. 4.9) and the correspondence in Document 4.10 show, the RAC ordered the destruction of recombinant DNA molecules, the fruits of over a year's work, because the researchers could not comply with the new regulations.

Document 4.1

November 21, 1975
Science (**190**:767)

Recombinant DNA: NIH Group Stirs Storm by Drafting Laxer Rules

A flurry of objections has been touched off in the biological research community by a National Institutes of Health committee's first attempt to draft the terms under which work may proceed on a potentially revolutionary but currently embargoed technique of genetic manipulation. Some think the proposed rules will impede research by their unnecessary strictness, but the majority of objections, including a petition signed by 50 biologists, hold that the NIH committee has set safety standards considerably laxer than those agreed upon by the international conference held at Asilomar to discuss how the new technique should be controlled (*Science*, 14 March 1975).

The technique, in brief, involves the use of recently discovered enzymes to rearrange the genetic material of living organisms in novel combinations which may never before have occurred in nature. The reason for the embargo is that the recombinant DNA molecules, as they are called, might escape from the laboratory with consequences which cannot be foretold but which, at the worst imagining, include the generation of novel and uncontrollable epidemics.

The technique will doubtless procure several Nobel prizes for those skilled and lucky enough to bring home first fruits, not to say many practical benefits in medicine, industry, and agriculture. There is considerable impatience in many laboratories for the NIH to complete its guidelines so that work can begin. The NIH committee* is working in a charged atmosphere in which suspicion is rampant and in which everyone has heard rumors that embargoed experiments have been clandestinely performed at certain laboratories. (*Science* could confirm none of these rumors; Paul Berg of Stanford University, the guiding spirit of the Asilomar conference, says he has seen no published experiment which contravenes the principles adopted there.)

The root cause of the objections attracted by the committee's first draft should probably be sought not in any lack of goodwill on the part of the committee—although some charges of conflict of interest are being voiced—but rather in the extraordinary difficulty of translating the general principles laid down at Asilomar into practical guidelines that everyone can live with.

Because of the way it has reacted to the criticism, however, the NIH committee has woven itself into a procedural tangle which may not be resolved without some internal friction. Essentially what has happened is that a subcommittee under David S. Hogness of Stanford University drafted a set of guidelines which were substantially weakened during a July meeting at Woods Hole attended by 8 of the committee's 12 members. The weakened, Woods Hole version attracted serious criticism on that account from Berg and from the signatories of a petition organized by Richard Goldstein of the Harvard Medical School and Harrison Echols of the University of California at Berkeley.

In response to the criticisms, committee chairman Dewitt Stetten, NIH Deputy Director for Science, asked Elizabeth Kutter of Evergreen State College in Olympia, Washington, to form a new subcommittee and propose alternative guidelines. The Kutter guidelines are due to be completed and sent to committee members this week, and will probably be more stringent than either the original Hogness guidelines or the Woods Hole version. According to committee secretary William J. Gartland, even the Hogness version does not adequately reflect the tone of caution implicit in the Asilomar conference's recommendations. The committee will discuss the Kutter guidelines at its meeting in San Diego early next month, but discussion may be complicated because of the irritation felt by some members at the way the Woods Hole version has apparently been abandoned.

What the Woods Hole guidelines essentially do is to set up categories of physical and biological containment and assign combinations of the two to the various types of recombinant DNA experiments at present envisaged. There are four levels of physical containment, named P1 to P4 in ascending order of strictness, and three levels of biological containment, designated EK1 to EK3 because it is assumed that most experiments will take place in the common laboratory strain K12 of the human gut bacterium *Escherichia coli*.

At the risk of some slight distortion, these safety levels may be summarized as follows:

P1: Use standard microbiological techniques.

P2: Same as P1 but hang a Keep Out notice on the door while the experiment is in progress.

P3: Same as P2 but put the lab under negative air pressure, or if you can't manage that, at least use negative pressure cabinets.

P4: Same as for handling really dangerous agents—air locks, negative pressure, change clothes and shower, and so forth.

The biological containment levels are intended to satisfy the Asilomar requirement that the organisms used in possibly hazardous experiments be rendered demonstrably incapable of surviving outside the laboratory environment. Again at the risk of simplification, the Woods Hole guidelines stipulate the following three levels:

EK1: Just use the standard laboratory version of *E. coli* K12 as the host for your recombinant DNA molecule, and use *E. coli*'s standard virus or plasmids (independently replicating bacterial chromosomes) as the vector (that is, as the means of getting your molecule inside the bug).

EK2: Use strains of *E. coli* genetically altered so as to be *in theory* 10^6 times less likely to escape successfully from the lab than standard *E. coli* K12.

EK3: Same as EK2, except that someone has gotten around to actually con-

*Members of the committee, known as the Recombinant DNA Molecule Program Advisory Committee, are as follows: DeWitt Stetten, NIH (chairman); Edward A. Adelberg, Yale; Ernest H. Y. Chu, University of Michigan; Roy Curtiss, University of Alabama; James E. Darnell, Rockefeller University; Stanley Falkow, University of Washington, Seattle; Donald R. Helinski, University of California, San Diego; David S. Hogness, Stanford University; John W. Littlefield, Johns Hopkins Hospital; Wallace P. Rowe, NIH; Jane K. Setlow, Brookhaven National Laboratory; Waclaw Szybalski, University of Wisconsin; Charles A. Thomas, Harvard Medical School; Elizabeth M. Kutter, Evergreen State College; John Spizizen, Scripps.

firming by empirical test that the disarmed bug is indeed 10^6 times less likely to make a successful breakout.

One of the most unpredictably hazardous class of experiments made possible by the new technique is the so-called "shotgun" experiment, in which the whole DNA of an organism is chopped enzymatically into segments a few genes or so long, and inserted into bacteria for cloning. The most dangerous of all possible shotgun experiments, some people believe, is that involving the genetic complement of the organisms closest to man, such as other primates. Under the Woods Hole guidelines, shotgun experiments with all mammalian genomes could be carried out under conditions of P3 physical containment and EK2 biological containment. For warm-blooded animals other than mammals you could loose off your shotgun under conditions of P3 and EK1 or, if you didn't like that, with P2 and EK2 safety requirements instead. (The Asilomar guidelines specify the equivalent of P3 and EK3.) For cold-blooded and other lower animals, unless they are known pathogens, Woods Hole asks you to use P2 and EK1 (that is, completely standard laboratory techniques and materials, except for the Keep Out notice).

The original Hogness subcommittee guidelines are understood to be essentially the same as the Woods Hole version, except that various experiments were assigned stricter containment levels, and the "disarmed" *E. coli* cited in EK2 and EK3 were required to be 10^8, not 10^6, times less likely to survive outside the laboratory. Another change, according to Waclaw Szybalski of the University of Wisconsin, was to make things easier for people accustomed to using plasmids as their vectors rather than phages (bacterial viruses). Says Szybalski, in whose laboratory a "disarmed" phage has been constructed: "There are more people on the committee who have an investment in plasmids than in phages, which happen to be safer vectors, so the committee is a little opinion slanted toward making adjustment to plasmid standards."

The most influential critic of the Woods Hole guidelines is Paul Berg, chairman of the National Academy of Sciences committee that first called for the moratorium and the convening of the Asilomar conference 16 months ago (*Science*, 26 July 1974). The nature of Berg's criticisms is fairly widely known but the following account is unauthenticated because Berg believes it premature to discuss his views. Berg is said to believe that the Woods Hole version is much weaker than the Hogness draft which he regards as perfectly acceptable; that methods of physical containment are overrated because they are all vulnerable to human error; that even the P3 level of containment would reduce but not eliminate human exposure within and outside the laboratory; and that the only sensible way to avoid this leakiness is to enfeeble the host and vector organisms so greatly that even mistakes can't create potential tragedies.

Quite similar objections are raised by the 50 petitioners. They are concerned in general that the Woods Hole draft "appears to lower substantially the safety standards set and accepted by the scientific community as represented at the meeting at Asilomar." Specifically, they urge that the most hazardous experiments be postponed until experimental determination has been made of their risks. Second, the petitioners express concern at the P3 and EK2 conditions recommended for mammalian shotgun experiments; such experiments should take place only under P4 conditions until biological barriers of proved efficacy (EK3) are available.

Third, the petition raises questions about the composition of the NIH committee. There should be more members who have no direct interest in shotgun experiments, say the petitioners; there should be representation of the public at large; and the committee should span a broader range of scientific disciplines.

Goldstein, one of the organizers of the petition, is now preparing with two others a radical critique of the Woods Hole guidelines in which he argues that use of standard *E. coli* as a host organism is unsafe "in any size, shape or form." The NIH committee envisages *E. coli* as the organism of choice because, through its having become a standard laboratory workhorse, more is known about its behavior than about that of any other bacterium. But the fact that *E. coli* infects man, argues Goldstein, that it easily becomes airborne and lodges in the throat, makes it a "reckless" choice and "ecologically unsuitable" as host to recombinant DNA molecules of potential hazard. (The committee's microbiology expert, Stanley Falkow of the University of Washington, Seattle, considers the K12 strain of *E. coli* to be enfeebled to the point of being relatively harmless, but even he believes that "we are ignorant in large part of the ecology of *E. coli* and of its plasmids and its phages.")

Because of the infectivity of *E. coli*, Goldstein says, the physical levels of containment recommended in the Woods Hole guidelines are "practically meaningless" except for the highest level, P4. *Escherichia coli* should be used as a host only for comparatively safe experiments until a new bacterial host is developed which cannot infect man, Goldstein contends.

Another critique of the Woods Hole guidelines is being prepared by the Genetics and Society group of Scientists and Engineers for Social and Political Action. A member of the SESPA group, Jonathan King of the Massachusetts Institute of Technology says that the function of the NIH committee, as presently constituted, "is to protect geneticists, not the public." Hogness, chairman of the subcommittee that wrote the original guidelines, is an active worker in the recombinant DNA field, which King likens to "having the chairman of General Motors write the specifications for safety belts."

Szybalski, a member of the Hogness subcommittee, agrees that there was a potential conflict of interest but defends Hogness by saying he acted with impartiality: "Hogness did an admirable job and tried to be fair, but he is very vulnerable to that criticism; I admire him for doing the job well and for his courage in taking it on," Szybalski says. Hogness rejects the charge of conflict of interest, saying that in the area he is working in, shotgun experiments with *Drosophila*, there is no disagreement he knows of on what the appropriate safety precautions should be.

The Hogness subcommittee has now been disbanded and the initiative at present seems to rest with the Kutter group. Who is Elizabeth Kutter? She became a member of the committee only after the July meeting, which she attended as an observer. Her name was proposed to the committee by Szybalski, who had met her at a conference in Canada. She was co-opted partly in response to the committee's desire to have a layperson, or at least a semilayperson, as well as some one from a small college, among their ranks. Kutter in fact has a Ph.D. in biophysics and works with phages.

At a meeting in May Kutter suggested to the committee that she hold a session on constructing safer phages at the Cold

Spring Harbor phage conference in August. Before the conference she expressed her concern about the Woods Hole guidelines to Goldstein and King in Boston. Her session at the phage meeting turned into a general criticism of the guidelines, and it was from this session that the Goldstein-Echols petition was set in motion.

Panicky Reaction

In response to this and maybe other criticisms, the NIH asked Kutter to draw up new guidelines, an action which has caused some distress among committee members. Szybalski regards it as a "terribly panicky reaction to criticism." All that was necessary was to revert to the Hogness draft, he says, since it was in fact the changes in the Hogness draft that the critics were objecting to. Committee chairman Stetten comments that in retrospect, "It is possible we did not react as judiciously as we might have, but there was an emotional and significant wave of criticism in some quarters against the Woods Hole draft."

Kutter's task has been made more difficult because of a report emanating from Goldstein (who says Kutter told him so) that the Woods Hole guidelines have been scrapped. "That made me climb the wall," says one committee member. According to Stetten, the NIH committee has not set the Woods Hole draft aside, but rather is "looking at it again." Kutter, however, is using the Hogness draft, not the Woods Hole version, as her basic text; the Woods Hole version, she says, "is not being put into effect."

The Kutter subcommittee, which met early this month, consists of herself, Falkow, and Joe Sambrook of Cold Spring Harbor. Sambrook is not a member of the committee but represented a subgroup on animal viruses working under Wallace Rowe of NIH. Kutter is taking input from a large number of sources, including Hogness, Berg, Joshua Lederberg of the Stanford University Medical Center, and the various letters received by the committee. Her goal, she says, is "to get together all the dissenting ideas and come up with compromises."

Several committee members, Stetten included, are anxious to prevent the committee becoming polarized into opposing camps. Given the paucity of data on which to make a decision, and the conflicting pressures on the committee, it is not surprising that there should be a range of views. "We are being asked to set guidelines based upon hazards based upon accidents which have not yet happened. Even Lloyds of London is unwilling to write insurance on accidents for which there are no actuarial data," says Stetten.

Despite the darkness in which the committee is working, pressures are mounting for it to take a leap anyway. "If you keep everybody waiting, there is going to be stuff done on Saturday night," says committee member Jane K. Setlow of Brookhaven National Laboratory. "Many people I know have invested in P3 containment facilities and are being held up for lack of guidelines," notes Hogness. Stetten intends to produce a set of guidelines at next month's meeting. But the committee is in the unenviable position that however hard it tries, it is unlikely to make everyone happy.—NICHOLAS WADE.

POINT OF VIEW

The Right to Free Inquiry

What degree of restriction on the recombinant DNA technique can reasonably be accepted without infringing the right to free inquiry? A suggestion that no such absolute right exists has been put forward by Robert Sinsheimer of Caltech. At the Asilomar conference, he noted in a recent lecture to the Genetics Society of America, "there was no sustained discussion of ancillary issues such as the absolute right of free enquiry claimed quite vigorously by some of the participants. . . . To impose any limit upon freedom of inquiry is especially bitter for the scientist whose life is one of inquiry; but science has become too potent. It is no longer enough to wave the flag of Galileo.

"Rights are not found in nature. Rights are conferred within a human society and for each there is expected a corresponding responsibility. . . . Would we wish to claim the right of individual scientists to be free to create novel self-perpetuating organisms likely to spread about the planet in an uncontrollable manner for better or worse? I think not. "This does not mean we cannot advance our science or that we must doubt its ultimate beneficence. It simply means that we must be able to look as what we do in a mature way. . . .

"It is difficult for a scientist to conceive that there are certain matters best left unknown, at least for a time. But science is the major organ of inquiry for a society—and perhaps a society, like an organism, must follow a developmental program in which the genetic information is revealed in an orderly sequence."

Document 4.2
December 19, 1975
Science (**190**:1175)

Recombinant DNA: NIH Sets Strict Rules to Launch New Technology

La Jolla, Calif. The signal to proceed with slow motion was given here on 5 December to a new technology whose ultimate benefits and potential risks may prove comparable in extent to those of harnessing the atom. Guidelines drawn up during a tensely argued 2-day meeting of a National Institutes of Health (NIH) committee will allow researchers to experiment with a new technique of genetic manipulation which because of its potential hazards has been under almost complete embargo for the last 18 months.

The technique involves the use of recently discovered enzymes to cut and splice the hereditary material of living organisms with unprecedented and possibly undreamed-of precision. A DNA segment carrying one or more genes can be excised from a chromosome and tacked onto another segment which may come from a quite different organism. The ability to construct recombinant DNA molecules, as they are known, is of both heuristic and practical significance. It offers in principle the means of obtaining a complete set of the genetic plans of any organism, including man. Biologists are already describing the technique in terms such as "revolutionary" and "one of the most significant advances of 20th century biology."

The practical applications so far envisaged range from equipping crop plants with nitrogen-fixing genes to make nitrogen fertilizer unnecessary, to the construction of microorganisms capable of synthesizing some of the products now obtained from oil. The recombinant DNA technique offers man power over nature in a more fundamental way than that of any other technology, because it is the power to intervene in evolution, to design and create combinations of genes in ways radically different from the slow reshufflings by which new organisms are created in nature.

Despite its promise, the new technique has been voluntarily forsworn by the scientific community because of theoretical hazards which most biologists consider to be extremely remote. The hazards stem from the fact that the properties of many recombinant DNA's likely to be constructed cannot always be predicted and may be deleterious. Should the addition of new genes confer a selective advantage on a virus or bacterium harmful to man or other forms of life, the outcome could be a catastrophe of possibly epidemic proportions.

Such horror scenarios, however incredible, are made more conceivable by the circumstance that the standard laboratory microorganism which will serve as the host for many recombinant DNA's is *Escherichia coli*, a common inhabitant of the human gut and throat. Laboratory workers often get infected by the organisms they handle, and through this means, if not by direct escape, a recombinant-containing bacterium might become established in the population at large. What cannot yet be excluded is the possibility that whatever genes have been built into the recombinant might be switched into action and interfere with the metabolism of those infected by the escaped bacterium.

This risk attaches in particular to one of the technique's most immediate uses, the so-called "shotgun" experiment, in which the total DNA of an organism is cut into segments and inserted into bacteria so that each segment may be grown in bacterial clones. Several of these segments are likely to contain harmful genes, such as those specifying any toxins the organism may produce, or the cryptic tumor viruses postulated to exist in certain animals' genomes.

That such harmful molecules should be able to wreak damage even if they were to escape from the laboratory appears highly improbable to many who have studied the question. It is this kind of consideration, however, which has occasioned a research moratorium that is unique in the history of science and has set in motion the train of events that culminated in the elaborate guidelines laid down at the La Jolla meeting. The moratorium was called for in July 1974 by a committee headed by Paul Berg of Stanford University. An international group of scientists who met this February at Asilomar, California, voted in principle to lift the moratorium provided that certain general safety principles were met. It was left to national committees in each country to devise specific guidelines, pending which the moratorium has effectively remained in force. The NIH committee appointed to this task drafted a set of guidelines that were judged too lax by many critics including Berg, a group of 50 biologists who signed a petition of protest, and several of the committee's own members (*Science*, 21 November 1975).

The problem stated by the NIH committee here was one of considerable delicacy. On the one hand, it was faced with mounting impatience among biological researchers to set rules that would allow research to begin. Had the committee postponed decision once again, or set rules that were indeed too restrictive, there are signs that the moratorium would have been flouted, and that the ubiquitous rumors of Saturday-night experiments would have rapidly turned out to be true.

On the other hand, the rules had to be sufficiently tight to convince outsiders, particularly in Congress, that the scientific community was doing a reasonably disinterested job of self-regulation. That task is the harder because of the committee's obvious vested interest. Of its 15 voting members , all but the chairman are active biological researchers who may one day wish to use the technique, and at least three members (Edward A. Adelberg, David S. Hogness, and Charles A. Thomas) are personally involved in recombinant DNA experiments of the limited type permitted by the Asilomar conference.

That the committee rose at least someway above its vested interest is shown by the fact that the guidelines set here are more severe than many members believe are necessary. That position was not reached easily. By the end of the first day, the committee had drawn up rules almost as loose as the draft version which provoked the initial outcry. European countries, in one foreign delegate's opinion, would probably not have found such rules acceptable. But the next day, through some mysterious alchemy, the committee changed its collective mind and rewrote the rules more strictly.

The tightness of the final version, which was accepted unanimously, probably owes much to the presence of three members of the group that organized the Asilomar conference, Paul Berg, Sydney Brenner of the Laboratory of Molecular Biology in Cambridge, England, and Maxine Singer of the NIH. Another factor that probably made tight guidelines easier to write was the apparently imminent availability of means of biological containment. This idea, one of the key principles laid down at Asilomar, calls for the use in recombinant DNA experiments of genetically enfeebled viruses and bacteria which cannot survive outside the laboratory. Despite assiduous attempts, committee member Roy Curtiss of the University of Alabama had been unable to construct a disarmed strain of *E. coli* at the time the committee drew up its first draft. A few weeks ago he succeeded, which means that the requirement for an experiment to use biologically safe *E. coli* is no longer tantamount to an embargo.

Much of the debate focused on where to place particular classes of experiments on the two-valued scale the committee had devised earlier. In brief, the scale consists of four levels of physical containment, designated P1 to P4, and three of biological, labeled EK1 to EK3 after the *E. coli* K-12 strain commonly used in laboratories. P1 consists of standard microbiological practice, P2 requires a few extra precautions, such as not creating aerosols, and P3 means putting the whole laboratory under negative air pressure. The highest category, P4, involves techniques such as airlocks, protective clothing, and showering on exit, which are used in handling the most dangerous known pathogens. Some believe that the stringency necessary to operate a P4 facility is incompatible with a university atmosphere.

The lowest level of biological containment, EK1, requires the experimenter simply to use the standard K-12 strain of *E. coli*, which most—but not all—microbiologists believe is unable to colonize the normal human bowel. EK2, as now defined, stipulates the use of K-12 strains genetically altered so that on average only one bacterium in 100 million would be expected to survive in the environment outside the laboratory (the earlier draft had this safety factor set at 1 million). EK3 is an EK2 system (that is, the bacterium and associated viruses used to introduce recombinant molecules into it) for which the postulated safety factor has been proved by test feeding the bacteria to animals.

Discussion about what levels to assign to various classes of experiments was clearly influenced by particular cases that people had in mind. At one point Berg, as spokesman for a group setting rules on animal virus vectors, announced the highly detailed rules shown in the summary table and apologized for doing so at so late a stage. Whereupon someone remarked that the rules "show what beautiful progress Dan Nathans' experiments are making." (Nathans has been working in the area of animal virus recombinants.) "I don't think we can tailor the guidelines to suit the progress of an investigator," Berg replied, "to tell us this is to keep Dan Nathans in business, well—I'd like to slow him down" (laughter).

Another instance where argument was evidently guided by a particular experiment in progress was the debate that raged back and forth about how to classify shotgun experiments with the genomes of cold-blooded vertebrates. The Asilomar guidelines said that these could go ahead in conditions equivalent to P2 plus EK1, and an experiment using recombinants from the frog genome has already been started by Donald D. Brown of the Carnegie Institution of Washington. Half a dozen other researchers (none of them on the committee) are also said to be interested in the field.

Debate on this class of experiments was opened by Hogness (Stanford University) whose recombinant DNA experiments, also permitted under Asilomar guidelines, involved the *Drosophila* fruit fly. Hogness pressed for details from those who believed there was a hazard in shotgun experiments with cold-blooded vertebrates. He was answered by Brenner, who observed that "the essence of a shotgun experiment is that it explores a very large sample of the genome. That issue is the same whether we use *Bacillus subtilis*, *Drosophila*, or humans. So the production of that hazard is uniform." While *Bacillus subtilis* presented

NIH Committee Guidelines

Summary of NIH committee guidelines for containing experiments with recombinant DNA. Each class of experiment has been assigned both a physical level of containment, designated P1 to P4 in increasing order of severity, and a biological level, designated EK1 to EK3. See text for a description of the levels. Table is not the authorized committee version and is subject to error and revision.

A. Shotgun experiments with *Escherichia coli* (use of recombinants to introduce undefined segments of an organism's genome into *E. coli*, classified by type of organism)
- (i) Eukaryotic DNA recombinants
  - Nonembryonic primate: P3 + EK3 or P4 + EK2
  - Embryonic primate: P3 + EK2
  - Other mammals: P3 + EK2
  - Birds: P3 + EK2
  - Cold-blooded vertebrates: P2 + EK2
  - Invertebrates and lower plants (ferns to algae): P2 + EK1
  - Higher plants: P2 + EK2, but P2 + EK1 if cells are taken from embryonic or germline tissue
  - Higher plants that produce pathogenic or toxic agents: P3 + EK2
  - Purification: If a cloned recombinant DNA can be made 99 percent pure on a weight-for-weight basis, the P value of containment may be reduced by one level
- (ii) Prokaryotic DNA recombinants
  - Prokaryotes that naturally exchange genes with *E. coli*:
    - Class 1 agents (as classified by the Center for Disease Control), such as enterobacteria: P1 + EK1
    - Class 2 agents, such as *Salmonella typhi*: P2 + EK2
    - Class 3 and higher: Experiments banned
  - Prokaryotes that do not naturally exchange genes with *E. coli*
    - Nonpathogens: P2 + EK1
    - Pathogens: P3 + EK2 if of low pathogenicity; P3 + EK3 or P4 + EK2 if of moderate pathogenicity

B. Use of recombinants to insert genes from viruses, eukaryotic plasmids, and organelles into *E. coli*
- Animal viruses: P4 + EK2 or P3 + EK3
- Plant viruses: P3 + EK1 or P2 + EK2
- Eukaryotic plasmids or organelles: As for the shotgun categories, unless the recombinant DNA has been rendered 99 percent pure, in which case either the P or the EK value may be reduced by one level

C. Use of animal virus vectors
- Defective polyoma virus + class 1 virus or nonpathogens: P3
- Defective polyoma virus + class 2 viruses: P4, but if the polyoma host range has not been changed and the virus segment can be proved harmless, then P3
- Defective SV40 + class 1 virus or nonpathogen: P4
- Defective SV40 + nonpathogenic and purified DNA, whether prokaryotic or eukaryotic: P3
- Defective SV40 or defective polyoma (lacking late genes) + prokaryotic or eukaryotic DNA: P3, as long as no virus particles are produced by infected cells.

little cause for concern, Brenner implied, "I would worry just a little about insects. I think that the rationale [for treating one organism differently from another] ought to be spelled out, because to people from the outside this thing looks like the settling of all sorts of different bargains. That may sound obnoxious but that is how it looks."

A vote was taken and the committee agreed by 9 votes to 4 to keep the class at P2 plus EK1. But the next morning, John W. Littlefield (Johns Hopkins Hospital) reopened the issue and proposed that shotgun experiments with cold-blooded vertebrates be upgraded to P2 plus EK2. The motion passed by 7 to 6, whereupon Hogness successfully proposed an experiment using embryonic tissue of these animals that could still be conducted in P2 plus EK1.

The next twist in the debate was a bid by Charles A. Thomas (Harvard Medical School) to create a grandfather clause for experiments already initiated under the Asilomar guidelines. When that lost, Thomas proposed that the actual clones of recombinant DNA already constructed could continue to be used. He was rebuked by a consultant to the committee, Peter Day of the Connecticut Agricultural Experiment Station, who said that "the whole question of a grandfather clause is germane to the credibility of the committee because it is quite clear that it concerns vested interests." The committee proceeded to follow a suggestion of Brenner's that it simply require those with clones constructed under Asilomar guidelines to consult the committee about their future use.

Other changes made by the committee to its earlier draft included the abolition of a loophole in the definition to P3, the upgrading of experiments with animal viruses, and more rigorous definition of the conditions under which purification of recombinant DNA may allow containment levels to be downgraded.

Implementation of the guidelines will proceed by having the NIH committee certify EK2 and EK3 systems when they become available. For physical containment, an investigator's laboratory must be certified both by his local biohazard committee and by the NIH peer review committee to which he applies for a grant. The granting agency must also receive proof of purity when a researcher wishes to downgrade the containment level of an experiment. According to the NIH's physical containment expert Emmett Barkley, the safety cabinets required for P2 conditions cost $5000 each; to convert a P2 facility to P3 can cost up to $50,000; and rather than trying to convert an old laboratory to P4, it would be cheaper to build one from scratch at a cost of about $200,000.

Stricter than Asilomar Guidelines

The rules that the committee has now produced are demonstrably stricter than the Asilomar guidelines, even though nothing has happened since then to make the speculated risks seem any more likely. At least within the scientific community, the NIH committee's guidelines are likely to be favorably received. James E. Darnell, for example, a committee member who considers the levels stricter than necessary to protect either scientists or the public, also believes that they will not constitute a serious impediment to research.

It seems likely that European countries will adopt the same general levels of containment as those hammered out at La Jolla, thus preventing a potentially embarrassing split in the world's scientific community. The Europeans have not yet written detailed guidelines and have, for the most part, been waiting to see what would happen in the United States.

Berg considers that the new guidelines satisfy all the objections he voiced to the earlier draft, and that they are "a faithful translation of the spirit of Asilomar." They are tough on him personally, requiring that he abandon a whole series of experiments involving recombinants made with the monkey virus SV40. He will lose 5 or 6 months, he estimated, in switching to the mouse polyoma virus. It was an experiment with an SV40 recombinant that aroused Berg's first scruples about the technique some 3 years ago.

It is too early to judge how the guidelines will appear to those outside the scientific community, but the committee is likely to receive some criticism on the grounds of vested interest and lack of public representation. The guidelines may look like a document of "byzantine complexity," as one observer termed it, tailored to fit particular experiments that are already on the drawing boards. It contains such apparent inconsistencies as that shotgun experiments with higher plants (which no one at present plans to do) are rated more hazardous than those with many types of animal genomes.

Yet those who argued in favor of lower containment levels were reflecting not just a personal bias but a widely held view that the hazards are being overemphasized. As evidence that it rose above its own interests the committee can point to the fact that its final guidelines are more stringent than those of Asilomar.

There are at present only a few positions in which the committee is outflanked by more conservative critics. Four scientists calling themselves the Boston Area Recombinant DNA Group has argued that *E. coli*, because of its ability to infect man, is an unsuitable host for recombinant DNA experiments and should be phased out of use within 2 years. Committee member Wallace P. Rowe (NIH) also feels strongly that *E. coli* is the wrong host but thinks people would not wait for a new host to be developed. Rowe also headed a group which recommended much higher containment levels for all shotgun experiments on the grounds that the expected hazard does not vary with the species. He, however, accepts the La Jolla guidelines.

The committee did not quite come to grips with a point raised by Brenner, that as the containment levels for an experiment are lowered, the number of laboratories attempting it will proliferate; moreover, an experiment that may be safely performed at Stanford may not be contained so well in less skilled establishments. Physical containment levels up to P3 are vulnerable to human error: of the 5000 laboratory-acquired infections in the last 30 years, one-third occurred in laboratories with special containment facilities. Even in the P4 conditions of the Army's biological warfare laboratories at Fort Detrick, there were 423 cases of infection and 3 deaths over some 25 years. Argument can thus be made about the P level assigned to any experiment, but the committee's levels are as strict as most.

Congressmen tempted to write legislation on the subject might pause to consider whether they would really do a better job. On the basis of an as yet purely speculative hazard, scientists have for 18 months held off from the use of the new technique, an act of self-denial unique in their own and perhaps most other professions. If the experiments now to be conducted make the hazards seem any more tangible, the same sense of responsibility will presumably continue to be manifested.

—NICHOLAS WADE

Document 4.3

December 9, 1975

New York Times

THE NEW YORK TIMES, TUESDAY, DECEMBER 9, 1975

Gene Experiments Panel Urges Stiffer Guidelines

By VICTOR K. McELHENY

After months of controversy a committee of scientists advising the United States Government has completed a set of guidelines for lessening possible risks from newly developed techniques of manipulating genetic chemicals that could foreshadow what has been called "genetic engineering."

The guidelines are expected to be issued to the scientific community and the public in about a month, according to Dr. DeWitt Stetten Jr., chairman of the committee, which held a quarterly meeting in La Jolla, Calif., Thursday and Friday.

As in previous versions drafted at an international scientific conference last February and by the Government advisory committee in July, the guidelines detail increasingly severe standards of "containment," achieved with laboratory equipment and procedures, for experiments that are graded according to their potential for harming the health of laboratory workers or the public.

Such graded levels of containment are traditionally used in research on disease-bearing organisms but have been unfamiliar to so-called molecular biologists up to now.

The experiments involve the use of enzyme catalysts to rearrange genes, chemical units of heredity, and move them from the cells of complex organisms, such as a mammal, into simple, rapidly multiplying bacteria in hopes that the genes can be studied with greater precision.

The bacteria used are laboratory variants of a type commonly found in the human intestine, and scientists have been concerned that altered bacteria might infect humans or cause cancer in an inadvertent sort of "genetic engineering."

The term genetic engineering is used to describe the possible future use of modern biological techniques to insert genes, such as the gene specifying human growth hormone, into animals or people lacking them. Such medical uses of modern genetics are widely regarded as remote.

Dr. Stetten is deputy director of the National Institutes of Health, the Government's leading agency for supporting biological and medical research.

In a telephone interview Dr. Stetten said that the committee had endorsed an experiment, to be conducted soon under maximum security conditions to test some of the potential hazards of using the new techniques in working with the genetic material of viruses known to cause cancer in mice.

The experiment was devised by the noted molecular biologist, Dr. Sydney Brenner of Cambridge, England, and two American scientists, Dr. Stetten said.

He added that the committee had also determined that further safety in some genetic chemistry experiments would result from using virtually virus-free cells from human or monkey fetuses, when available.

This sort of "biological containment" for the experiments, Dr. Stetten said, would be in addition to the protection provided by so-called "safe bugs," specially selected bacteria or viruses that would not be able to multiply in human or animal bodies if they happened to escape from laboratory equipment.

The concept of such safe vehicles for the studies was detailed last February at an international conference on potential risks and benefits of genetic engineering, held at the Asilomar grounds near Monterey, Calif., after a voluntary seven-month moratorium on some aspects of the research.

The ground rules reached in La Jolla replaced guidelines drafted July 2 and 3 in Stanford, Calif., by a subcommittee headed by Dr. David S. Hogness of Stanford University Medical School, and revised by the N.I.H. committee at Woods Hole, Mass., July 18 and 19.

The revisions were attacked by some scientists as too restrictive and by others as not restrictive enough.

Fears on Safety Raised

On one side, critics said the guidelines would unnecessarily delay work that could benefit both medicine and agriculture. On the other, critics said the guidelines were drafted by a group oriented to promoting the research and did not protect the public sufficiently from potential health hazards.

The Woods Hole version was attacked in a letter of Aug. 27 signed by Dr. Richard Goldstein of Harvard Medical School, Dr. Harrison Echols of the University of California at Berkeley, and 46 other scientists attending a course at the Cold Spring Harbor Laboratory on Long Island.

The scientists said the Woods Hole version "appears to lower substantially the safety standards set and accepted by the scientific community as represented at the meeting at Asilomar."

In response, the N.I.H. committee set up a new drafting group under Dr. Elizabeth Kutter of Evergreen State College in Olympia, Wash., which presented its conclusions at the La Jolla meeting.

Dr. Stetten said the result at La Jolla was "probably closer to the Hogness original report in its attitude" than to later drafts, "but amplified with a lot more information."

He said that the committee would continue to review research results and guidelines of subsequent quarterly meetings.

The La Jolla guidelines were described as "strict" by Dr. James E. Darnell of Rockefeller University, a committee member, interviewed by telephone.

At the same time, Dr. Darnell said, many of the potential dangers that gained wide attention in the last year were "only barely imaginable." The new document, he said, represented "the spirit of Asilomar."

Much of the work since the Asilomar meeting, Dr. Stetten said, "has increased scientists' confidence that risks can be controlled."

Document 4.4
January 29, 1976
New Scientist (**69**:218)

Comment

Recombinant DNA—rules without enforcement?

Suddenly, we seem to have a plethora of committees dealing with safety guidelines for scientists working with recombinant DNA—that assembled by joining together DNA from different sources, such as bacteria and animals. In addition to the National Institutes of Health's committee in the US (*New Scientist*, vol 68, p 618), the UK working party under Professor R. E. O. Williams, and similar bodies in other countries, the European Molecular Biology Organisation has now announced the establishment of an "Advisory Standing Committee on Recombinant DNA". Unlike most of the other bodies, the EMBO group will be purely advisory and will have no regulatory or legislative functions.

The committee's main function, it seems, will be in dealing with enquiries on scientific and technical matters from both individual scientists and governments and other organisations. In answering individual queries, the committee will emphasise that it is in no sense a licensing authority and that it is the responsibility of scientists to ensure that their experiments conform to any national and international standard in force in the country where they work. In addition to a very welcome emphasis on training programmes (many of those who have drifted into molecular biology in recent years have never received the sort of instruction and training that is commonplace for medical microbiologists), the committee believes that one of its chief roles will be in advising research workers about strains of bacteria, plasmids and bacteriophages best suited to for particular experiments.

It is, of course, in agreeing which hypothetical dangers which can be adequately contained by physical stratagens, such as scrupulous aseptic technique and rigorous structural safeguards in the laboratory, and those which require more fundamental biological "fail-safe" measures, that the central dilemma lies. There has been much talk recently about the fabrication of "self-destruct" bacteria which would be entirely safe as model systems because they would be unable to survive other than in the artificial cultural conditions of the laboratory. At the same time, there is no doubt that experiments involving a lower level of risk could be conducted without danger simply by insisting on standard pathology laboratory techniques. It will be the task of Professor Williams's working party, now beginning to take evidence from scientists, scientific bodies, and even trade unions in Britain, to assess the relative necessity of biological as against physical containment for various categories of experiments.

But there is another problem, this time administrative and political rather than technical. The National Institutes of Health committee, which recently laid down a very precise list of guidelines for recombinant DNA studies (*New Scientist*, vol 68, p 682), is in a uniquely strong position to enforce regulations of this sort. The NIH dispenses money for research, and can therefore ensure through its grant awarding machinery and other channels that the rules laid down by its committee are strictly adhered to. Yet it was intended that the La Jolla guidelines, whose origin can be traced back to a genuinely international gathering at Pacific Grove, California a year ago (*New Scientist*, vol 65, p 546), would also serve as a framework for recombinant DNA research in other parts of the world. The question, then, is one of enforcement—emphasised by the EMBO committee's disavowal of any intention of trying directly to regulate work in this field.

Similarly in Britain, the terms of reference of the Williams working party include an instruction to draft a central code of practice for experiments on the genetic manipulation of microorganisms. It is reasonable to assume that any such code could be enforced throughout the universities and the research councils. But what of those teams and individuals supported by industry, by the foundations, or by any of the other multifarious bodies involved in financing research. The industrial situation is perhaps the most worrying. At least one company, ICI, already has a substantial stake in "genetic engineering". Coincidentally, the news last June, that the company had completed its high security facility at the Corporate Laboratory at Runcorn came just one day after publication of a cautionary statement after the Pacific Grove conference (*New Scientist*, vol 66, p 594). As we noted at the time, it would be absurd to suggest that ICI is proceeding with its plans unmindful of the warnings from so many responsible scientists in recent months. It would, however, be just as imprudent not to be disturbed by the establishment of a commercial base for work which is still hotly controversial and upon which specific safety guidelines have yet to be agreed.

Sir John Kendrew has called for an international commission to monitor potential dangers behind contemporary research trends in biology. For some of this work to be carried out beneath a cloak of military or commercial security, he believes, would be dangerous. Precisely that situation is now developing. And despite the plethora of committees, we still have no reassurance that any commission or other authoritative international body is doing, or can do, anything about it. Sooner or later, legislation may be the only answer.

Bernard Dixon

"Run for the hills—the recombinant DNA has escaped!"

This article first appeared in New Scientist, London, the weekly review of science and technology.

Document 4.5

April 16, 1976
Science (**192**:236)

NEWS AND COMMENT

Recombinant DNA: The Last Look Before the Leap

The tortuous and possibly historic debate on whether to proceed with research on recombinant DNA is now nearing the end of its first round, with a clear victory in sight for those who wish research to go ahead under stiff but not grossly inconvenient safety conditions.

This is the course that is favored by probably a vast majority of biological researchers. Yet it is worth noting the strong dissent of two scientists who are as eminent as any of the contributors to the debate, and who in addition have no personal interest in using the technique. Robert Sinsheimer, chairman of the biology division at Caltech, believes that all research should be confined to one site, such as the former biological warfare laboratories at Fort Detrick. Erwin Chargaff of Columbia University would like to see the research prohibited altogether to allow a two-year period of "cooling off" and reflection.

These views occur in written comments solicited by National Institutes of Health director Donald S. Fredrickson. On the basis of the comments, and of the record of a public hearing on the issue (*Science*, 27 February 1976), Fredrickson has proposed some minor emendations to the present draft guidelines on recombinant DNA research prepared by an NIH committee. At a two-day meeting held on the NIH campus last week, the same committee considered and rejected most of them.

Whether or not Fredrickson accepts the committee's advice, the guidelines that he will issue within the next few weeks will not differ greatly from the present draft.

It is perhaps a pity that Sinsheimer's views were not discussed by the NIH committee last week because, though not widely held, they are by no means negligible. Moreover, Sinsheimer seems to have a broader sense of perspective than others about the place of the new technique both in history and in evolution. His critique of the guidelines is premised on a fundamental and so far unrefuted theorem, that there is a barrier to genetic exchange between the two great classes of living things, the prokaryotes and the eukaryotes. (Prokaryotes are primitive cells, such as bacteria and blue-green algae, which lack a nuclear membrane; eukaryotes, the cells of all higher organisms, have a quite different and more sophisticated organization.)

Many of the proposed experiments with recombinant DNA involve inserting segments of eukaryotic DNA into prokaryotic cells, and the whole thrust of the guidelines has been to rank these experiments in a graded series of risks based on the nature of the eukaryotic DNA segment. Sinsheimer, however, believes that the risk lies not in the particular DNA being inserted, but in the very fact of putting eukaryotic genes into prokaryotes. If he is right, the elaborate edifice of rules constructed by the NIH committee is built on a foundation of sand.

Sinsheimer's argument, as expressed in two letters sent to Fredrickson in February, goes as follows.

Though prokaryotes and eukaryotes interact intensely with each other as organisms, they are not known to interact with any frequency at the genetic level. One evident reason for this lack of genetic intercourse is that, though they use the same genetic code, they have different control elements, different genetic signals for governing how the code is to be put into operation. The great danger of putting any piece of eukaryotic DNA into a prokaryote is that it may endow prokaryotes with the eukaryote control signals, a sort of betrayal of state secrets at the molecular level. Even if this occasionally happens by accident in nature, Sinsheimer says, numerous experiments of the type envisaged can only increase the risk.

What might be the consequences of breaching the natural barrier between prokaryotes and eukaryotes? One is that the prokaryotic viruses, particularly the lysogenic species, could acquire the capacity to infect eukaryotes. A bacterial virus carrying the gene for a restriction enzyme, for example, could wreak havoc inside a eukaryotic cell. Another possibility is that bacteria might acquire the capacity to serve as reservoirs for some of the common eukaryotic viruses. "One need not continue to spin out potential horror stories," Sinsheimer says. "The point is that we will be perturbing, in a major way, an extremely intricate ecological interaction which we understand only dimly."

Because of these risks, Sinsheimer would like to see all recombinant DNA work performed at one site in the country, such as Fort Detrick. Meanwhile a major program should be launched to find a more suitable host for recombinant DNA molecules than *Escherichia coli*, the present candidate of choice. With recombinant DNA experiments being performed in hundreds of laboratories about the United States, organisms "will inevitably escape—and enter into the various ecological niches inhabited by *E. coli*." It would be better to employ as host a bacterium that only grows in special environments (such as the thermophiles that live in hot springs), or else to incorporate DNA into an animal virus, such as cowpox, against which we already have a viable defense in the form of vaccination.

"Obviously," Sinsheimer concludes, "neither I nor anyone else can say that if the present committee guidelines are adopted, disaster will ensue. I will say, though, that in my judgment, if the guidelines are adopted and nothing untoward happens, we will owe this success far more to good fortune than to human wisdom."

Sinsheimer's barrier theorem arouses strong disagreement in many other biologists, though the strength of the reaction generally relates to fear that his views will be used to impede research rather than to any obvious flaw in the argument. Those who disagree point out that the barrier to genetic exchange may only seem to exist because of our ignorance about the flow of genes between species. For example, it now seems possible that viruses may play an important evolutionary role by transferring genes between species and allowing one species to sample the genetic progress being made by others. Nevertheless, such mechanisms are not yet known to operate across the presumptive prokaryote-eukaryote barrier.

Another argument raised against Sinsheimer's hypothesis is that the barrier, if it indeed exists, may have come into being as the accidental by-product of some other process and have no inherent purpose in itself. In other words it is contingent, not a specific mechanism designed for evolutionary reasons to keep eukaryotes and prokaryotes in genetic apartheid.

Nevertheless, Sinsheimer's theorem does not appear at present to be literally refutable, even though many disagree with it. "It's evolutionary speculation. I don't believe it for a minute," says David Botstein of MIT. If Sinsheimer cannot be directly refuted, probably the best of the indirect arguments raised against his position is the view that further experiments will answer the questions he raises. "What will deal with Sinsheimer is experiments, on the basis of which we will know what to be careful about," says Botstein. For similar reasons, Matthew Meselson of Harvard says that the work itself will reduce the hazards.

Another strongly held belief of those who oppose Sinsheimer's position is that recombinant DNA research is urgently required as insurance against forthcoming catastrophes. David Hogness of Stanford, for example, points to the need to feed increasing populations. Meselson sees the technique as the key to reducing man's total vulnerability to viruses. "Most species that have ever existed are gone now, and they seem to vanish randomly in time," says Meselson. "We don't know why, but there is nothing like a virus infection to give that kind of statistics. You could argue that knowing how viruses spread, which this research will tell us, should be of the highest priority."

This is not a trivial consideration, but neither is Sinsheimer's view that scientists simply may not have the right to create novel organisms likely to spread about the planet in an uncontrollable manner for better or worse. There has been no explicit consideration, he said in a recent lecture to the Genetics Society of America, "of the potential broader social or ethical implications of initiating this line of research—of its role, as a possible prelude to longer-range, broader-scale genetic engineering of the fauna and flora of the planet, including, ultimately, man. . . . Do we want to assume the basic responsibility for life on this planet—to develop new living forms for our own purpose? Shall we take into our hands our own future evolution?"

At its meeting last week the NIH Recombinant DNA Molecule Program Advisory Committee addressed itself not to these questions but, at Fredrickson's behest, to changing jobs and tittles in its draft. Fredrickson's position, which in effect is to endorse the present guidelines, may indeed be merited, but it also happens to fall within the limits of two quite cogent political constraints. The first is the attitude taken by European countries toward the present guidelines. In a maneuver of some finesse, the European Molecular Biology Organization (EMBO) won itself almost a veto power over Fredrickson's decision by making known that it would only go along with the NIH guidelines if they became no stricter. Thus if European-American unity were to be preserved, a generally desirable objective, Fredrickson could make few substantive changes in the guidelines. As he observed at last week's meeting, "Without a certain measure of conformity, the whole exercise would be futile."

A second constraint is the possibility of the guidelines being ignored altogether if made unacceptably rigorous. Were the Sinsheimer suggestion to be adopted, says an NIH staff member who helped analyze the public's comments for Fredrickson's decision, "the research would go on under other conditions anyway, so that wouldn't be an effective stance for NIH to take even if we agreed with it."

[Most of Fredrickson's proposed emendations to the guidelines concern raising the safety levels required for particular kinds of experiments. (The safety levels are designated, in increasing severity, P1 to P4 for physical methods of containment, and EK1 to EK3 for biological methods.) His strongest suggestion, unless there "are compelling arguments to the contrary," is to raise the so-called shotgun experiment with the genomes of cold-blooded vertebrates to the level of P3 + EK2. The committee voted to keep it at P2 + EK2. Fredrickson also suggested raising the level of the shotgun experiment with other cold-blooded animals (including insects) to P2 + EK2. In a carefully worded paragraph, the committee essentially agreed to this if the species carries a known toxin, but said that laboratory grown animals such as *Drosophila* should stay at P2 + EK1. (Higher plants, by contrast, whose genomes are not obviously more threatening than those of insects, stay at P2 + EK2). Fredrickson was also turned down on a suggestion to raise work with the SV40 virus from P3 to P4 conditions.]

Since the NIH committee has passed final word on its guidelines, now may be as good a time as any to comment on the balance of forces within it. The two most dominant members have in general been David Hogness of Stanford University and Charles Thomas of Harvard, both of whom have forcefully argued the case against stricter levels of containment. Both, as it happens, are personally interested in doing recombinant DNA experiments, a circumstance which has led to suggestions of a conflict of interest. However that may be, they represent a legitimate and widely held point of view that they would doubtless have argued anyway.

Hogness and Thomas have in fact put their case so effectively that other members felt the issue was being railroaded. Several even turned for help to a group of young Cambridge scientists who, calling themselves the Boston Area Recombinant DNA Group,* produced a cogent position paper in favor of tighter guidelines. It has in fact been largely in response to outside pressures, such as that exerted by the Boston Area Recombinant Group and others, that the guidelines have been increased in stringency.

The NIH committee's hardest working member has undoubtedly been Roy Curtiss of the University of Alabama. He and 8 colleagues have worked overtime for about a year to develop the enfeebled strain of *Escherichia coli* which the guidelines require to be used for many categories of recombinant DNA experiments. Since safety measures for some reason lack glamor, Curtiss and his team may not get the credit they deserve, but it is only through his voluntary efforts that the bacterium will be available just when it is needed. (The committee approved for use an enfeebled bacterial virus developed by Philip Leder and others at NIH. It is expected to certify Curtiss's *E. coli* imminently).

The NIH committee has clearly succeeded in producing a reasonable and scientifically acceptable set of guidelines that will probably be adopted or closely copied throughout the world. Yet Sinsheimer's arguments have raised awkward questions which nobody yet seems able to directly answer. So the present plan is to go ahead anyway and let them be answered by events. That is maybe what has to be done, but it would look better if Sinsheimer's Cassandra-like fears could be proved imaginary first.—NICHOLAS WADE

*The group consists of Richard Goldstein, Paul Primakoff, Margaret Duncan, and Hiroshi Inouye, all of the Harvard Medical School, and Cristian Orrego of Brandeis University. The group is not affiliated with Science for the People, as was erroneously stated in *Science* (27 February).

Document 4.6

July 7, 1976

Introduction to the NIH Guidelines, *Federal Register,* Excerpt

DEPARTMENT OF HEALTH, EDUCATION, AND WELFARE

National Institutes of Health

RECOMBINANT DNA RESEARCH

Guidelines

On Wednesday, June 23, 1976, the Director, National Institutes of Health, with the concurrence of the Secretary of Health, Education, and Welfare, and the Assistant Secretary for Health, issued guidelines that will govern the conduct of NIH supported research on recombinant DNA molecules. The NIH is also undertaking an environmental impact assessment of these guidelines for recombinant DNA research in accordance with the National Environmental Policy Act of 1969.

The NIH Guidelines establish carefully controlled conditions for the conduct of experiments involving the production of such molecules and their insertion into organisms such as bacteria. These Guidelines replace the recommendations contained in the 1975 *Summary Statement of the Asilomar Conference on Recombinant DNA Molecules.* The latter would have permited research under less strict conditions than the NIH Guidelines.

The chronology leading to the present Guidelines is described in detail in the NIH Director's decision document that follows. In summary, scientists engaged in this research called, in 1974, for a moratorium on certain kinds of experiments until an international meeting could be convened to consider the potential hazards of recombinant DNA molecules. They also called upon the NIH to establish a committee to provide advice on recombinant DNA technology.

The international meeting was held at the Asilomar Conference Center, Pacific Grove, California, in February 1975. The consensus of this meeting was that certain experiments should not be done at the present time, but that most of the work on construction of recombinant DNA molecules should proceed with appropriate physical and biological barriers. The Asilomar Conference report also made interim assignments of the potential risks associated with different types of experiments. The NIH then assumed responsibility for translating the broadly based Asilomar recommendations into detailed guidelines for research.

The decision by the NIH Director on these Guidelines was reached after extensive scientific and public airing of the issues during the sixteen months which have elapsed since the Asilomar Conference. The issues were discussed at public meetings of the Recombinant DNA Molecule Program Advisory Committee (Recombinant Advisory Committee) and the Advisory Committee to the NIH Director. The Recombinant Advisory Committee extensively debated three different versions of the Guidelines during this period.

The Advisory Committee to the NIH Director, augmented with consultants representing law, ethics, consumer affairs and the environment, was asked to advise as to whether the proposed Guidelines balanced responsibilty to protect the public with the potential benefits through the pursuit of new knowledge. The many different points of view expressed at this meeting were taken into consideration in the decision.

The NIH recognizes a special obligation to disseminate information on these guidelines as widely as possible. Accordingly, the Guidelines will be sent to all of the approximately 25,000 NIH grantees and contractors. Major professional societies which represent scientists working in this area will also be asked to endorse the Guidelines. The Guidelines will be sent to medical and scientific journals and editors of these journals will be asked to request that investigators include a description of the physical and biological containment procedures used in any recombinant research they report on. International health and scientific organizations will also receive copies of the guidelines for their review.

Filing of an environmental impact statement will provide opportunity for the scientific community, Federal, State and local agencies and the general public to address the potential benefits and hazards of this research area. In order for there to be further opportunity for public comment and consideration, these guidelines are being offered for general comment in the FEDERAL REGISTER. It must be clearly understood by the reader that the material that follows is *not* proposed rulemaking in the technical sense, but is a document on which early public comment and participation is invited.

Please address any comments on these draft policies and procedures to the Director, National Institutes of Health, 9000 Rockville Pike, Bethesda, Maryland 20014. All comments should be received by November 1, 1976.

Additional copies of this notice are available from the Acting Director, Office of Recombinant DNA Activities, National Institute of General Medical Sciences, National Institutes of Health, 9000 Rockville Pike, Bethesda, Maryland 20014.

DONALD S. FREDRICKSON,
Director,
NIH National Institutes of Health.

JUNE 25, 1976.

Document 4.7

July 1, 1976
Nature (**262**: 2)

Genetic manipulation: guidelines issued

The NIH ground rules for genetic manipulation experiments may not mark the end of an unprecedented debate within the scientific community. **Colin Norman** reports from Washington

AFTER two years of controversy and uncertainty, the National Institutes of Health (NIH) last week issued a complex set of guidelines governing the use of a powerful new technique for manipulating genes in living organisms. The guidelines establish safety rules for experiments which may revolutionize biology, but which also provide man with unprecedented ability to alter the characteristics of living things. They are, however, far from being the final word on whether, and under what circumstances, such research should be allowed to go ahead, for they are already being overtaken by events in some places.

On July 7, for example, the City Council in Cambridge, Mass., will vote on a resolution, proposed by the Mayor, which would ban for two years all such experiments at Harvard and MIT. And months of bitter debate at the University of Michigan have resulted in the adoption of regulations there which are more strict than those issued by NIH last week. The focus of the debate is clearly shifting from Washington into the university communities where the research will take place.

Nevertheless, the NIH guidelines will provide important ground rules for genetic manipulation experiments in many institutions, and their impact will extend far beyond the borders of the United States, for they are likely to influence the establishment of guidelines in many other countries. They have been developed by an extraordinary process of self-regulation by the scientific community.

The process began in 1973, when scientists familiar with the nascent technique began to worry about potential hazards associated with its use. Their concerns led a committee of the National Academy of Sciences to issue a public statement in July 1974, urging scientists around the world to defer two types of experiments until the hazards have been defined. The moratorium lasted until last February, when it was partly replaced by general guidelines recommended by an international group of geneticists which met at Asilomar, California. Then an NIH advisory committee, consisting of scientists, took centre stage. It laboured hard during most of last year trying to cast the Asilomar guidelines into more specific rules, completing its task by hammering out a set of complex proposals last December. The proposals went to NIH Director Donald S. Fredrickson, who called a public meeting to discuss them, solicited the views of numerous scientists and non-scientists, and asked the advisory committee to reconsider some of its suggestions. The guidelines issued last week represent Fredrickson's distillation of the conflicting advice presented to him.

They differ a little in detail, but not in philosophy, from the recommendations of the advisory committee. They will allow most planned experiments to go ahead, albeit under strict safety controls, outlawing only a handful of the more hazardous types of experiments. They are, however, stricter than the Asilomar guidelines, a fact which Fredrickson suggested means that "the research will go forward in a manner responsive and appropriate to hazards that may be realised in the future". And Dr DeWitt Stetten, Deputy NIH Director for Science and chairman of the advisory committee, argued last week that "the issuance of the guidelines is in no sense an opening of the floodgates, rather it is a closing of the leaks (in the Asilomar guidelines)".

Be that as it may, the guidelines are not even NIH's final say on the matter. Their publication falls under the terms of the National Environmental Policy Act, which means that NIH must prepare an assessment of the potential impact of the research on the environment. A draft assessment should be ready by September, and since it will be opened up for public comments before being cast in final form, another round of discussion of the risks and benefits of the research is in store. In the meantime, the guidelines will take effect, governing NIH's support of the research.

New dimension

Why have the guidelines taken so long to produce, and caused so much strife? The answer is that the research offers a spectacular mix of potential benefits and possible hazards, and it opens up an entirely new dimension in biology. The experiments consist, in short, of snipping genes from the DNA of any organism and splicing them into the DNA of another, perhaps entirely unrelated, organism. The resulting molecule—a recombinant DNA molecule—is copied (cloned) each time the new host reproduces, producing large quantities of the transplanted genes. The technique offers a powerful tool for probing the working of genes and their arrangement in complex organisms and, more distantly, it may offer a means of constructing special micro-organisms for a variety of medical, commercial and industrial uses.

But the worry is that the technique allows biologists to breach genetic barriers between species which have evolved over thousands of years. In short, it allows biologists to by-pass the processes of evolution. More specifically, foreign genes inserted into an organism may cause that organism to behave in a dangerous, and perhaps unpredictable, manner. The guidelines thus seek to ensure that micro-organisms bearing transplanted genes are contained in the laboratory.

They spell out our levels of physical containment to be used in such experiments, designated P1 to P4, ranging from use of standard microbiological techniques (P1) to the use of specially equipped facilities akin to biological warfare laboratories (P4). And, as a second line of defence, they spell out three levels of biological containment, EK1 to EK3, which must be used for experiments involving the insertion of genes into a strain of the common gut bacterium *E. coli*—the organism which will be used for most experiments. The levels are as follows:

- EK1—use of standard *E. coli* K12, a laboratory strain of *E. coli* which has been used for genetic experiments for decades. Foreign genes are inserted into the bacterium by splicing them into a plasmid (a ring of bacterial DNA which reproduces independently from the bacterium's chromosomes) and reintroducing the recombinant into the bacterium, or by splicing them into the DNA of a bacteriophage which then infects *E. coli* K12.

- EK2—the use of specially mutated strains of *E. coli* or bacteriophage which, according to laboratory tests, are virtually incapable of surviving outside the laboratory so that the recombinant DNA will have less than 1 in 10^{-8} chances of surviving in the natural environment.

- EK3—the same as EK2 except that the survivability has been tested in animals, plants and other environments.

The guidelines assign specific physical and biological safety levels to various types of experiments on the basis of their potential hazards (see box). They also spell out safety rules for experiments involving recombinants formed by splicing genes into the DNA of animal viruses, which are then grown in cell cultures to provide multiple copies of the foreign genes. The final guidelines alter some of the containment levels proposed by the advisory committee, but no substantial alterations have been made.

Cautious approach

How did Fredrickson arrive at his final decision when confronted with such a wealth of conflicting advice? His first, and most fundamental, consideration was simply whether the research should be allowed to go ahead at all, in view of the potential hazards. The majority of the commentators on the guidelines recommended that it should, but he was advised to take an extremely cautious approach by two of the most eminent of the commentators, neither of whom intends to conduct experiments with recombinant DNA.

Dr Robert Sinsheimer, chairman of the Department of Biology at California Institute of Technology, argued in a letter to Fredrickson that the experiments present a serious hazard if they breach the genetic barrier between higher organisms (eukaryotes) and lower organisms (prokaryotes). "One need not continue to spin out potential horror stories", he wrote, "the point is that we will be perturbing, in a major way, an extremely intricate ecological interaction which we understand only dimly". As for the guidelines themselves, Sinsheimer stated: "I cannot believe that under these proposed guidelines the organism can be contained. If the work is going on in a hundred laboratories about the United States, performed by technicians, graduate students, etc., the organism will inevitably escape—and will enter into the various ecological niches known to be inhabited by *E. coli.*" He therefore proposed that all work with recombinant DNA should be performed under maximum containment conditions at a single institution in the United States, and that there should be an intensive effort to seek a microorganism more suitable than *E. coli* for the work.

The other eminent, though more flamboyant, critic of the guidelines, Erwin Chargaff of Columbia University, recommended that all work on recombinant DNA should be halted for at least two years to allow time for the hazards to be assessed. Chargaff asked, in a letter published in *Science*, "Have we the right to counteract, irreversibly, the evolutionary wisdom of millions of years, in order to satisfy the ambition and the curiosity of a few scientists?"

The advisory committee did not consider Sinsheimer's or Chargaff's pro-

Guidelines in detail

The guidelines define four levels of physical containment, designated, in order of increasing stringency, P1 to P4, and three levels of biological containment, EK1 to EK3, and assign experiments to them on the basis of potential risk. The following is a summary of containment levels specified for various sources of DNA.

a. Shotgun experiments using *E. coli* as the host	
Non-embryonic primate tissue	P3+EK3 or P4+EK2
Embryonic primate tissue or germ line cells	P3+EK2
Other mammals	P3+EK2
Birds	P3+EK2
Cold blooded vertebrates, non-embryonic	P2+EK2
embryonic or germ line	P2+EK1
If vertebrate produces a toxin	P3+EK2
Other cold blooded animals and lower eukaryotes	P2+EK1
If Class 2 pathogen*, produces a toxin, or carries a pathogen	P3+EK2
Plants	P2+EK1
Prokaryotes that exchange genes with *E. coli*	
Class 1 agents (non-pathogens)	P1+EK1
Low risk pathogens (for example, enterobacteria)	P2+EK1
Moderate risk pathogens (for example, *S. typhi*)	P2+EK2
Higher risk pathogens	banned
Prokaryotes that do not exchange genes with *E. coli*	
Class 1 agents	P2+EK2 or P3+EK1
Class 2 agents (moderate risk pathogens)	P3+EK2
Higher pathogens	banned

In all above cases, if DNA is at least 99% pure before cloning and contains no harmful genes, either physical or biological containment levels can be reduced one step.

b. Cloning plasmid, bacteriophage and other virus genes in *E. coli*	
Animal viruses	P4+EK2 or P3+EK3
If clones free from harmful regions	P3+EK2
Plant viruses	P3+EK1 or P2+EK2
99% pure organelle DNA, Primates	P3+EK1 or P2+EK2
other eukaryotes	P2+EK1
Impure organelle DNA: shotgun conditions apply.	
Plasmid or phage DNA from hosts that exchange genes with *E. coli*	
If plasmid or phage genome does not contain harmful genes or if DNA segment 99% pure and characterised	P1+EK1
Otherwise, shotgun conditions apply.	
Plasmids and phage from hosts which do not exchange genes with *E. coli*	
Shotgun conditions apply, unless minimal risk that recombinant will increase pathogenicity or ecological potential of the host, then	P2+EK2 or P3+EK1

NB. cDNAs synthesised *in vitro* from cellular or viral RNAs are included in above categories.

c. Animal virus vectors	
Defective polyoma virus+DNA from non-pathogen	P3
Defective polyoma virus+DNA from Class 2 agent	P4
If cloned recombinant contains no harmful genes and host range of polyoma unaltered, reduce to	P3
Defective SV40+DNA from non-pathogens	P4
If inserted DNA is 99% pure segment of prokaryotic DNA lacking toxigenic genes, or a segment of eukaryotic DNA whose function has been established and which has previously been cloned in a prokaryotic host-vector system, and if infectivity of SV40 in human cells unaltered	P3
Defective SV40 lacking substantial section of the late region+DNA from non-pathogens, if no helper used and no virus particles produced	P3
Defective SV40+DNA from non-pathogen can be used to transform established lines of non-permissive cells under P3 provided no infectious particles produced. Rescue of SV40 from such cells requires	P4

d. Plant host-vector systems

P2 conditions can be approximated by insect-free greenhouses, sterilization of plant, pots, soil and runoff water, and use of standard microbiological practice.

P3 conditions require use of growth chambers under negative pressure and routine fumigation for insect control.

Otherwise, similar conditions to those prescribed for animal systems apply.

*Classes for pathogenic agents as defined by the Center for Disease Control.

posals during its public meetings, but Fredrickson alluded to them in a lengthy statement published along with the guidelines. Recognising that the breaching of genetic barriers might pose a hazard, Fredrickson nevertheless argued that the research can be controlled so that it is carried out safely. Noting that "the international scientific community . . . has indicated a desire to proceed with research in a conservative manner", and that "most of the considerable public commentary on the subject, while urging caution, has also favoured proceeding", Fredrickson pointed out that there is, in any case, no way to prohibit the research throughout the world. "There is," he added, "no reason to attempt it."

Having decided that there should be no flat proscription on the research, Fredrickson turned to some of the chief concerns raised by the critics of the guidelines. The most prominent concern arises from the fact that most work with recombinant DNA will take place with the *E. coli* K12 bacterium. Since *E. coli* is a common inhabitant of the human gut, many observers have considered it a dangerous choice for the research. In particular, a group of scientists from the Boston area urged that a different host for transplanted genes be developed, and that the use of *E. coli* be phased out as swiftly as possible.

Fredrickson argues, however, that since the bacterium has been used for decades as the geneticists' workhorse, there is extensive knowledge of its behaviour and, moreover, there is good evidence that it is unlikely to survive for long in the environment in competition with wild strains of *E. coli*. In other words, *E. coli* K12 itself provides a level of biological containment. "I believe that because of this experience, *E. coli* K12 will provide a host-vector system that is safer than other systems", Fredrickson argued, and he declined to set a limit on when it should be phased out of the research.

Another area of concern with the proposed guidelines centred on how they should be implemented, and that section has been extensively revised. The guidelines strictly apply only to research supported by NIH, and investigators must comply with them before they can receive a grant. Some commentators suggested that principal investigators be required to obtain informed consent from all laboratory personnel before proceeding with recombinant DNA experiments, but Fredrickson opted instead for a requirement that the investigators simply inform all laboratory workers of the real and potential hazards associated with the experiments.

The guidelines also require that each institution where recombinant DNA experiments will be conducted should establish a biohazards committee, to ensure that facilities meet specified requirements, training is adequate and so on. The advisory committee had firmly recommended that such committees should not be responsible for determining the containment conditions for specific experiments, but Fredrickson deleted that prohibition, leaving the matter up to individual institutions.

A final point concerning implementation which is clearly worrying many people is that the guidelines do not apply to industry. Early last month, however, Fredrickson briefed officials from various industries on the guidelines and received from them expressions of support for their intent and their general provisions. Some officials expressed reservations about specific items, however, such as the provision prohibiting large-scale experiments with recombinant DNA, and the Pharmaceutical Manufacturers' Association has decided to convene a committee to review whether the guidelines are applicable to the drug industry.

Effect of guidelines

Now that these guidelines have been issued, how do they affect academic scientists conducting, or hoping to conduct, recombinant DNA experiments? First, they specify that many experiments should use EK2 or EK3 biological containment, and until crippled micro-organisms which meet those criteria are available, such experiments should not be conducted. The advisory committee which drafted the guidelines is responsible for certifying whether crippled strains meet the criteria.

A strain of *E. coli* produced by Roy Curtiss of the University of Alabama (called χ1776) has been proposed as an EK2 system with two specific plasmids (pSC101 and Col El-*kan*). Curtiss, who has built many crippling mutations into the bacteria in an effort which took the best part of a year, has provided the committee with reams of data on its survivability. Last month, a subcommittee voted unanimously that the strain meets the EK2 specifications, and the full committee is expected to approve it in the next few weeks. That would pave the way for many experiments.

But a confusing situation has developed with respect to two other candidate EK2 strains. At its last meeting in April, the committee approved a strain of bacteriophage lambda, developed by Philip Leder of NIH as an EK2 strain. But a subcommittee which met last month to consider a second candidate strain, developed by Fred Blattner at the University of Wisconsin, was divided on a matter which affects both strains. In short, two subcommittee members refused to endorse the use of either phage as an EK2 system unless they are used in conjunction with crippled bacteria. The full committee will take up that thorny issue when it next meets in September. Curtiss's *E. coli* strain, incidentally, is resistant to infection by phage λ.

Many of the objections and reservations which critics have levelled at the guidelines are now beginning to crop up in local debates in and around the universities where the research will be conducted. In that regard, the situation brewing in Cambridge, Mass., may be an indication of things to come. It began quietly a few months ago, when some researchers at Harvard proposed that a laboratory, meeting P3 requirements, be established in the Harvard Biology Labs. The facility, which would cost some $350,000, would have involved converting some existing laboratory space, and the intent was to undertake a variety of work in it, including some recombinant DNA experiments.

The proposal met with opposition within Harvard, however, largely because the biology labs are old, infested with cockroaches and with a species of ant which has so far evaded attempts at eradication. In short, critics suggested that the building is unsuitable for a P3 facility. Their criticisms broadened into an assault on the proposal to conduct recombinant DNA experiments at Harvard, however, and the dispute spilled over into the city when the matter was reported at length in Boston's weekly newspaper.

Approval of the laboratory by the Harvard authorities was all but assured, since the concept had been endorsed by a biohazards committee, Harvard's Committee on Research Policy, and the Dean of Arts and Sciences. But, when he read about the dispute in the newspapers, Cambridge city Mayor Alfred Velucci stepped in. He called a council meeting on June 23—ironically the day the NIH guidelines were issued—to discuss the matter, and a string of scientists testified about the potential hazards and benefits of recombinant DNA research. By all accounts, the atmosphere was highly charged, and there was considerable heated discussion.

The matter has now gone well beyond the issue of whether the laboratory should be built, however, for Velucci has introduced a resolution which would prohibit all recombinant DNA experiments in Cambridge—even those deemed to have minimal risk—for two years. The council will vote on the resolution on July 7, and the outcome will be closely watched around the country. □

Document 4.8

June 23, 1976

Department of State Telegram to Countries Believed to Have Substantial Biomedical Research

Department of State **TELEGRAM**

UNCLASSIFIED 7766

DRAFTED BY HEW/NIH:JRQUINN
APPROVED BY OES/APT/BMP:WJWALSH, III
NEA/EX:EGABINGTON(INFO)
ARA/EX:GECHAFIN(INFO)
EA/EX:JEFFREY CUNNINGHAM(INFO)
EUR/EX:CEREDMAN(INFO)
AF/EX:TFORD(INFO)
DHEW/OIH-JRKING

--------------------- 040862

R 222057Z JUN 76
FM SECSTATE WASHDC
TO AMEMBASSY ANKARA
AMEMBASSY ATHENS
AMEMBASSY BELGRADE
AMEMBASSY BEIRUT BY POUCH
AMEMBASSY BERLIN
AMEMBASSY BERN
AMEMBASSY BOGOTA
AMEMBASSY BONN
AMEMBASSY BRASILIA
AMEMBASSY BRUSSELS
AMEMBASSY BUDAPEST
AMEMBASSY BUENOS AIRES
AMEMBASSY CAIRO
AMEMBASSY CANBERRA
AMEMBASSY CARACAS
AMEMBASSY COPENHAGEN
AMEMBASSY DUBLIN
USMISSION GENEVA
AMEMBASSY THE HAGUE
AMEMBASSY HELSINKI
AMEMBASSY LISBON
AMEMBASSY MADRID
AMEMBASSY MONTEVIDEO
AMEMBASSY MOSCOW
AMEMBASSY OSLO
AMEMBASSY OTTAWA
AMEMBASSY PARIS
USLO PEKING
AMEMBASSY PRAGUE
AMEMBASSY PRETORIA
AMEMBASSY ROME
AMCONSUL RIO DE JANEIRO
AMEMBASSY SAN JOSE
AMEMBASSY SANTIAGO
AMEMBASSY SOFIA
AMEMBASSY STOCKHOLM
AMEMBASSY TAIPEI
AMEMBASSY TEL AVIV
AMEMBASSY TOKYO
AMEMBASSY VIENNA
AMEMBASSY WELLINGTON
USMISSION USUN NEW YORK
USMISSION OECD PARIS
AMCONSUL CAPE TOWN

UNCLAS STATE 154658

SUBJECT: NIH GUIDELINES ON DNA/RNA(GENETIC ENGINEERING ETC.) RECOMBINANT RESEARCH

1. NATIONAL INSTITUTES OF HEALTH PLANS ANNOUNCE GUIDELINES ON JUNE 23, 1976, CONCERNING RECOMBINANT DEOXYRIBO NUCLEIC ACID(DNA) RESEARCH DEVELOPED AS A RESULT OF OPEN DISCUSSIONS WITH U.S. SCIENTISTS.

2. IN VIEW OF SERIOUS IMPORTANCE OF SUBJECT TO WORLD'S SCIENTISTS AND NEED FOR WORLD PUBLICITY AND COOPERATION ON PROBLEM, DEPARTMENT REQUESTS ADDRESSEE MISSIONS ALERT APPROPRIATE OFFICIALS OF HOST GOVERNMENT'S AGENCY MOST CONCERNED WITH SCIENCE AND/OR SUPPORT OF BIOMEDICAL RESEARCH AND OFFER TO PROVIDE COPY OF GUIDELINES SOONEST. COMMENTS ON GUIDELINES MAY BE ADDRESSED TO DIRECTOR, NATIONAL INSTITUTES OF HEALTH, BETHESDA, MARYLAND 20014.

3. TWO COPIES OF THE GUIDELINES WILL BE AIRPOUCHED SEPARATELY TO POST SHORTLY AFTER RELEASE DATE OF JUNE 23. APPRECIATE DELIVERY OF GUIDELINES ASAP.

4. NIH ALSO PLANS SEND COPY GUIDELINES DIRECTLY TO ALL GRANTEES AND CONTRACTORS, DOMESTIC AND FOREIGN.

5. PLEASE NOTIFY NIH (ATTN: DR. J.R. QUINN) OF PERSON IN HOST GOVERNMENT ADVISED OF ABOVE MESSAGE AND TO WHOM GUIDELINES DELIVERED. ROBINSON

Document 4.9

March 15, 1977

Science & Government Report (**7**:1)

SCIENCE & GOVERNMENT REPORT

The Independent Bulletin of Science Policy

Vol. VII, No. 5 P.O. Box 6226A, Washington, D.C. 20015 March 15, 1977

DNA Rules Force Dumping of Research Project

In what is believed to be the first major action to enforce federal safety guidelines governing recombinant DNA research, a University of California geneticist has been forced to destroy the best part of two years' work because his experiments were performed under conditions proscribed by the final version of the guidelines.

The incident is complicated by the fact that the research was carried out during a period when the safety guidelines were being developed and the ground rules were changing. Early versions of the guidelines didn't rule out the experiments, and even the final version is a bit ambiguous in one key area. Nevertheless, the incident provides an interesting example of the way the guidelines are working, though it raises a number of questions.

The work was carried out by John Carbon, a researcher at the University of California at Santa Barbara. It consisted, essentially, of slicing up the entire DNA contained in yeast cells with a special enzyme, and splicing the DNA fragments onto a circular ring of DNA known as a plasmid. Carbon then inserted the plasmids containing the "foreign" yeast genes into bacteria known as *E. Coli,* and grew large colonies of the bacteria containing copies (clones) of the transplanted yeast genes.

None of that would ordinarily be ruled out by the guidelines, but the problem was that Carbon used a strain of *E. Coli* which was capable of transferring plasmids from one bacterium to another (a so-called conjugative system). That property was, in fact, an essential feature of Carbon's experiments.

The danger in using such a system, however, is that the "foreign" genes may be transferred into bacteria capable of surviving easily in the environment. In other words, the system lowers the barrier against escape of recombinant DNA molecules from the laboratory.

Shortly after the final version of the guidelines was published last June by the National Institutes of Health, Carbon wrote to two members of the NIH Advisory Committee on Recombinant DNA (which played a key role in drafting the guidelines) to clarify whether his experiments would be outlawed by the guidelines.

The committee itself decided at a meeting in September last year that, although the guidelines were not explicit, they rule out the use of conjugative systems in recombinant DNA research. After some debate, the committee decided to recommend that Carbon should destroy the bacterial colonies into which he had transplanted the foreign genes. Carbon says he complied with the recommendation last fall.

During the past two years, Carbon had built up a bank containing some 50,000 clones of yeast genes, and he had also begun to build up a bank containing 10,000 clones of genes from the fruit fly *Drosophila.* Though he was able to extract some of the plasmids containing specific genes and transfer them to bacterial strains from which they would not be easily transmitted to other bacteria (non-conjugative systems), most of the clones were simply put into an autoclave and destroyed.

One irony in all of this is that Carbon had demonstrated that some of the transplanted yeast genes were functioning in the bacteria, a demonstration which has caused considerable interest because it opens up the possibility of studying what makes genes switch on and off. It also makes some of the potential dangers associated with recombinant DNA research more plausible, however, since bacteria containing transplanted genes pose no hazards unless the transplanted genes are capable of functioning in some manner.

If Carbon had used a non-conjugative system to transplant yeast genes, his work would have been considered low risk under the NIH guidelines, because the yeast used in the experiments is not known to contain "harmful" genes.

Asked last week when he first had doubts about whether or not his work fell outside the NIH guidelines, Carbon said that "the realization dawned on me very slowly." He said that he began to wonder about the system last Spring, and was challenged on the matter at a recombinant DNA symposium held at MIT last June, shortly before the NIH guidelines were published. He wrote to the NIH advisory committee soon after the guidelines were published.

Carbon published the first results of his experiments in the February 1977 issue of the *Proceedings of the National Academy of Sciences,* and accompanied the report with a statement pointing out that although his system "provides a valuable tool," he has discontinued its use because it is prohibited by the NIH guidelines.

Asked last week for his opinion of the matter, Carbon said, "We are willing to go along with the guidelines", but he added, "we are not happy about it".

A number of other researchers queried last week said that although the guidelines are indeed fuzzy on the use of conjugative systems, it has always been generally assumed that such systems would be proscribed since they lower the safety barrier in recombinant DNA research.

It took nearly two years for Carbon to realize that the research would probably fall outside the scope of the guidelines, but in the future, the level of consciousness of the scientific community will probably have been raised to such a point that inadvertant violations of the guidelines would be difficult to explain. —CN

Document 4.10

July 22, 1976

Carbon Letter to Gartland

UNIVERSITY OF CALIFORNIA, SANTA BARBARA

BERKELEY · DAVIS · IRVINE · LOS ANGELES · RIVERSIDE · SAN DIEGO · SAN FRANCISCO  SANTA BARBARA · SANTA CRUZ

DEPARTMENT OF BIOLOGICAL SCIENCES SANTA BARBARA, CALIFORNIA 93106

Dr. William Gartland
Office of Recombinant DNA Activities
National Institute of Health
Bethesda, Maryland 20014

July 22, 1976

Dear Dr. Gartland:

One important aspect of current work on recombinant DNA, the use of F-mediated transfer for the detection of specific hybrid plasmids, is not covered or discussed in the current NIH guidelines. Roy Curtiss suggested that I inform you of the issue, so that it could be discussed at the September meeting of the Advisory Committee.

As you know, cloning vectors such as col El which are normally non-transmissible by direct conjugation are readily transferred to certain other bacterial cells in the presence of a conjugative plasmid such as the fertility factor F. We have made use of this property to develop an efficient and rapid screening procedure for the detection, in collections of F^+ transformant colonies, of specific hybrid plasmids capable of complementing particular E. coli auxotrophic mutations. Using this method we have isolated hybrid col El plasmids containing yeast DNA segments which can complement various E. coli mutations. This observation is particularly important, since it is an indication that eukaryotic DNA, in this case from yeast, can be expressed meaningfully in the bacterial cell. I am enclosing a preprint of a paper describing the method, along with a copy of a recent news article which summarizes the scientific importance of the observations.

The problem is that the current guidelines do not treat the issue of recombinant DNA plasmids or phage in a conjugative or transferable system. I believe that the method is an important contribution to recombinant DNA technology, and I am convinced that it will eventually be used in many laboratories. Thus it is important that the guidelines be modified to address the issue as soon as possible.

One suggested approach for work done in the presence of transferable factors would be to require the use of physical containment one level higher than normally required for a given type of DNA. For example, the cloning of DNA normally requiring P2-EK1 conditions would become P3-EK1 in the presence of a conjugative plasmid. An alternative proposal, suggested by Roy Curtiss, would be the use of a tra I mutant of F as the conjugative factor, which would prevent the transfer of the F plasmid to other bacterial recipients, but would still allow the initial transfer of the hybrid plasmid.

I would appreciate it if this issue could be considered with respect to how this type of work should be conducted in the future. We are naturally anxious to carry out our research under the safest possible conditions, and seek guidance from your office and the Advisory Committee.

Sincerely yours,

John A. Carbon, Ph.D.
Professor of Biochemistry

JAC:ag
Encl.

October 22, 1976
Gartland Letter to Carbon

DEPARTMENT OF HEALTH, EDUCATION, AND WELFARE
PUBLIC HEALTH SERVICE
NATIONAL INSTITUTES OF HEALTH
BETHESDA, MARYLAND 20014

October 22, 1976

Dr. John A. Carbon
Department of Biological Sciences
University of California
Santa Barbara, CA 93106

Dear Dr. Carbon:

In response to your letters of July 22, 1976 to me and July 26, 1976 to Dr. Brian W. Kimes, the NIH Recombinant DNA Molecule Program Advisory Committee, at its September meeting, considered the issues raised by you concerning the cloning of foreign DNA on ColE1-derived vectors into a cell that possessed the derepressed F plasmid. In its first action, the Committee reaffirmed its belief and intent that hosts utilized in EK1 systems should not contain a conjugative plasmid. Since the present wording in the Guidelines is not totally explicit on this point, the Committee unanimously voted to make the statement explicit so as to preclude use of hosts that contained either conjugative plasmids or known transducing prophages. The changed wording adopted by the Committee and to be inserted in future versions of the Guidelines is as follows:

> The E. coli K-12 hosts should not contain wild-type conjugative plasmids whether autonomous or integrated and preferably should be non-lysogenic for potential transducing phages.

In further consideration of the specific cases you mentioned in your correspondence, the Committee did not agree that using these clones containing Drosophila or yeast inserts under P3 containment conditions was a sufficient compensation for the increased probabilities of transmission of cloned DNA by the organisms you have constructed. The basis for this decision is as follows. First, F is a depressed conjugative plasmid and such derepressed plasmids in nature are rare, if they occur at all. Thus, since derepressed plasmids allow donor cells to transmit DNA at 1000 times higher frequencies than repressed plasmids, it was considered that this factor alone would increase the probability that recombinant DNA cloned on the ColE1-derived vectors would be transmitted to other organisms likely to be encountered in nature at one million times higher frequency than would be the case if one had cloned the DNA segments in an E. coli host cell lacking a conjugative plasmid.

Second, if one considers other natural barriers for the acquisition of a conjugative plasmid by a host cell and its subsequent transmission to other likely recipients in nature (e.g., density effects, cell surface differences, restriction, etc.), one can estimate that the probability of transmission of cloned DNA by a host cell already possessing a conjugative plasmid to another cell encountered in nature would be increased by another 100- to 1000-fold as a minimum. Indeed, studies on acquisition of repressed conjugative plasmids *in vivo* based on feeding experiments in rodents, chickens, calves and humans indicate that this overall possibility for receipt of a conjugative plasmid is at least 10^{-6} if not 10^{-8}. Third, the Committee also noted the efficiency with which pCR1 chimeric plasmids were mobilized by F and considered that such a high average frequency of mobilization would not be encountered in nature using R plasmids frequently encountered in enteric microorganisms. Based on all of these considerations the Committee thus felt that your use of F^+ or Hfr cells for introduction of chimeric plasmids with inserts of yeast and/or Drosophila DNA was not sufficiently contained to justify permission for continued use of these clones, even under P3 conditions. The Committee felt that if these clones were to be used that the minimum physical containment that might be acceptable would be the P4 type of facility. It was therefore recommended that you have a reasonable amount of time to recover those cloned segments that are of particular interest and then destroy the remaining cultures unless P4 facilities are available for use.

The above decision raises other questions concerning the use of these clones by individuals to whom you may have sent them. It was felt that you should notify these individuals of the decision of the Committee but it was also suggested that you indicate to the NIH Office of Recombinant DNA Activities the names of these individuals so that NIH could directly inform them of its decision.

In a related discussion, the Committee discussed the possibility of cloning foreign DNA into a strain that did possess a conjugative plasmid to see if a safe means to accomplish this might be developed. The Committee believed that the use of an F plasmid containing a *traI* mutation as suggested by Dr. Curtiss was little different than cloning into an Hfr cell and therefore deemed this an insufficient safeguard. Because of the important potential of the system you have developed for screening functional attributes of cloned DNA segments by using a system with a conjugative plasmid and the acknowledged advantages of minimizing repetitious shotgun experiments, the questions of how to develop a sufficiently contained system was referred to a subcommittee chaired by Dr. E. A. Adelberg and whose membership includes Drs. Roy Curtiss and Waclaw Szybalski. This subcommittee has been charged with suggested revisions in the sections of the Guidelines pertaining to biological containment and their proposed suggested revisions will be considered at the January meeting of the Recombinant DNA Molecule Program Advisory Committee.

Sincerely yours,

William J. Gartland, Jr.

William J. Gartland, Jr., Ph.D.
Director
Office of Recombinant DNA Activities
National Institute of General Medical Sciences

November 3, 1976
Carbon Letter to Davidson

UNIVERSITY OF CALIFORNIA, SANTA BARBARA

BERKELEY · DAVIS · IRVINE · LOS ANGELES · RIVERSIDE · SAN DIEGO · SAN FRANCISCO

SANTA BARBARA · SANTA CRUZ

DEPARTMENT OF BIOLOGICAL SCIENCES

SANTA BARBARA, CALIFORNIA 93106

November 3, 1976

Dr. Norman Davidson
Division of Chemistry
California Institute of Technology
Pasadena, California

Dear Norman:

As you will note from the enclosed letter, recently forwarded to me from the NIH, the cloning of eukaryotic DNA on non-conjugative plasmid vectors in the presence of conjugative plasmids such as F is specifically prohibited by the NIH guidelines. We have been directed to destroy all bacterial clones that fall in the above category.

As directed by the enclosed letter from Dr. Gartland, I am requesting that you destroy all bacterial clones that I sent to you or that were brought to Cal Tech by Dr. Chris Ilgen, and that contain hybrid Col El-Drosophila DNA plasmids in F^+ cells. You apparently will have "a reasonable amount of time" to isolate plasmid DNA from these clones before destroying them.

Sincerely yours,

John Carbon

John Carbon, Ph.D.
Professor of Biochemistry

cc: Dr. William J. Gartland

November 3, 1976
Carbon Letter to Gartland

UNIVERSITY OF CALIFORNIA, SANTA BARBARA

BERKELEY · DAVIS · IRVINE · LOS ANGELES · RIVERSIDE · SAN DIEGO · SAN FRANCISCO

SANTA BARBARA · SANTA CRUZ

DEPARTMENT OF BIOLOGICAL SCIENCES

SANTA BARBARA, CALIFORNIA 93106

November 3, 1976

Dr. William J. Gartland, Jr.
Office of Recombinant DNA Activities
National Institutes of Health
Bethesda, Maryland

Dear Dr. Gartland:

We have your letter of October 22, 1976, concerning the cloning of foreign DNA on non-conjugative plasmid vectors in the presence of a derepressed F plasmid.

We will of course comply with the revised guidelines, and we will destroy the clones in question as you request. These clones have only been given to two other investigators, Dr. Norman Davidson at Cal Tech and Dr. Christine Ilgen at Princeton. I have written to these individuals requesting that they destroy the clones and have sent them copies of your letter.

I would be interested to learn the outcome of the deliberations of the sub-committee chaired by Dr. Adelberg, since I feel that the methods we have developed and used are quite valuable.

Sincerely yours,

John Carbon, Ph.D.
Professor of Biochemistry

November 29, 1976
Carbon Letter to Hogness

UNIVERSITY OF CALIFORNIA, SANTA BARBARA

BERKELEY · DAVIS · IRVINE · LOS ANGELES · RIVERSIDE · SAN DIEGO · SAN FRANCISCO

SANTA BARBARA · SANTA CRUZ

DEPARTMENT OF BIOLOGICAL SCIENCES

SANTA BARBARA, CALIFORNIA 93106

November 29, 1976

Dr. David Hogness
Biochemistry Department
Stanford University
Stanford, California 94305

Dear Dave:

I am enclosing the Gartland Letter along with copies of our Cell paper and the PNAS preprint. We will hold off destroying the F^+ Col El hybrid yeast clones until after the January meeting, just in case something can be done about changing the original committee decision.

In addition to the hisB and leuB complementations described in the PNAS paper, we now have isolated Col El-hybrid yeast DNA plasmids capable of complementing trpAB deletions and argH mutations (non-reverting). The cloned DNA segments have been shown to be yeast DNA by reassociation kinetics; reassociation of labeled single stranded plasmid DNA is driven by excess unlabeled yeast DNA, but not by E. coli DNA.

We have attempted complementation of 15 different genes (argH, galK, hisB, hisC, hisD, ilvA, ilvC, ilvE, leuB, metE, pyrB, thyA, trpA, trpE, hisF). Of these, four were definitely complemented, and two (metE and hisF) are possibilities (not checked out yet). Thus, the "succcess rate" is at least 25%, but maybe as high as 40%. With cloned E. coli DNA, the success rate is about 80-85%, using the same methods.

We do not know if the yeast enzyme synthesized in the E. coli cell is identical to the enzyme isolated from yeast extracts. However, in the case of the leuB enzyme (leu2 in yeast), β-isopropylmalate dehydrogenase, the activity synthesized by the hybrid plasmid strain is extremely cold-sensitive, as is the wild-type yeast enzyme. Both activities are nearly completely inactivated after only 20 minutes at 0°, while the wild-type E. coli enzyme is stable in the cold. The pYetrp plasmid complements both trpA and trpB deletions, even though it was originally selected by trpA complementation. This is in line with the idea that both activities are on a single polypeptide in yeast.

I hope that this gives you most of the information that you need. Please let me know if I can be of further assistance.

Best regards,

John Carbon

February 2, 1977
Davidson Letter to Gartland

CALIFORNIA INSTITUTE OF TECHNOLOGY

PASADENA, CALIFORNIA 91125

DIVISION OF CHEMISTRY AND CHEMICAL ENGINEERING
THE CHEMICAL LABORATORIES

Feb. 2, 1977

Dr. William Gartland
Director, Office of
Recombinant DNA Activities
National Institute of General Medical
Sciences
Bethesda, Maryland 20014

Dear Dr. Gartland:

This is in response to the enclosed letter from Professor John Carbon. I wish to inform you that, in November 1976, we destroyed all the bacterial clones of F^+ cells containing Col El-Drosophila DNA plasmids referred to in Dr. Carbon's letter. The plasmids have all been transferred to F^- cells for future work.

Sincerely yours,

Norman Davidson

ND:eg

Encl. 1

cc: John Carbon
R.L. Sinsheimer

Photo by Rick Stafford, Stock, Boston.

THE CAMBRIDGE, MASSACHUSETTS EPISODE

By the early summer of 1976 it was evident that the NIH had no intention of recommending that recombinant DNA research be prohibited, but its guidelines, very significantly stricter than those of Asilomar, would *de facto* prevent many experiments that required levels of containment not then available. A small minority of molecular biologists and biologists, however, remained convinced that recombinant DNA work should be either postponed or better prohibited. Among these was Erwin Chargaff (Columbia), a biochemist whose contributions in the late 1940s and early 1950s to our understanding of the chemistry of DNA were used by Watson and Crick in the discovery of the double helical structure of DNA. In an influential letter published in *Science* on June 4, 1976 (vol. 192, pp. 938–940), Chargaff called for a cessation of recombinant DNA research. In a letter in the same issue of *Science*, Francine Robinson Simring, Chairman of the Committee for Genetics of the Friends of the Earth, an environmentalist organization, called for caution and careful assessment of the risks and impact of recombinant DNA research. This letter (Doc. 5.1) reflected the growing alignment of environmentalist groups in the United States with those who opposed the rapid development of the techniques and believed that the assessments of possible risks had been inadequate.

Unfortunately, Dr. Chargaff has denied us permission to reprint his letter to *Science*. On October 21, 1976, a few months after its publication, however, he testified at a public hearing called by the Attorney General of the State of New York, Louis J. Lefkowitz, where he repeated several of the points raised in the letter. Although the transcript of this public testimony is of course less eloquent and succinct than the letter, which we would have preferred to use, we reproduce it as Document 5.2 so that readers may judge for themselves.

The misgivings of the general public, encouraged by the media, that the guidelines were not adequate arose not only because molecular biologists had played a major and potentially self-serving role in their formulation, but also because of specific criticisms of the guidelines by some scientists and public-interest groups. The extreme view still held by some biologists and biochemists that all recombinant DNA research should be prohibited in-

Photo by JoAnn Verburg.

creased public uncertainty and unease. Matters came to a head in Cambridge, Massachusetts, when the City Council intervened.

Although we include as Documents 5.3, 5.4, and 5.5 several reports of the events in Cambridge, some additional background information is needed. Early in 1976 the Molecular Biology Department at Harvard University proposed to convert several rooms in the Biological Laboratories into a well-equipped P3 laboratory. They met with resistance from a number of colleagues in the Biology Department, and the dispute quickly broadened to involve other members of the university, which in late May held a campus forum. Mark Ptashne (Harvard) was the prime mover of the case in favor of the containment laboratory, while Richard Goldstein (Harvard Medical School, and now a member of the expanded RAC; see Chapter 12) argued against it. He based part of his case on the fact that some wild strains of *E. coli* are pathogenic. Bernard Fields (Harvard Medical School) and David Botstein (MIT) both argued that this information was irrelevant to experiments that would be done using laboratory *E. coli* K12.

The May meeting was attended by City Councilwoman Barbara Ackerman, a long-time friend of the University, who announced that the City Council was closely following events. After the forum, Mayor Vellucci, whose relations with Harvard had often been strained and acrimonious, was encouraged to raise the issue in the City Council by George Wald, Harvard biologist and Nobel Laureate, and his wife and fellow biologist Ruth Hubbard, who both vociferously opposed the P3 laboratory at Harvard. The Mayor also received much advice and encouragement from Richard Goldstein, Jonathan King (MIT), Jon Beckwith (Harvard Medical School), and other members of Science for the People. (Representative articles by these scientists are reprinted in several other chapters of this book.)

Debates in the City Council followed on June 23 and July 7, 1976. Afterwards the Cambridge City Council declared a three-month moratorium on recombinant DNA research in the city and therefore at Harvard and MIT. Moreover, in August 1976 the City Council set up its own local committee, the Cambridge Experimentation Review Board, to decide what safety and health measures would be required of anyone working with recombinant

DNA in the city. Cambridge would set an example to the nation; local government would act, if central government was impotent and failed in its responsibility to the public (Doc. 5.6).

In the correspondence columns of *Science*, Freeman Dyson (Princeton) rebutted Chargaff's arguments for a ban on the research, and Bernard Davis (Harvard) argued that the risks of recombinant DNA did not warrant public anxiety. However, Ruth Hubbard felt otherwise (Doc. 5.7). Inevitably the events in Cambridge made headlines, although Dyson's polite rebuttal of Chargaff's stance did not attract the attention of the editors of the mass media.

After the Cambridge City Council's decision, opponents of recombinant DNA research were further encouraged by a long article in *The New York Times Magazine* of August 22, 1976 (Doc. 5.8) by Liebe Cavalieri (Sloan Kettering Institute), another biologist sharing Chargaff's sentiment, who called not for a mere three-month moratorium in Cambridge but a worldwide moratorium of indefinite length pending further assessment of the risks. George Wald, with the lustre of a Nobel Prize and now an ally of Mayor Vellucci on this issue, added more prophesies of what might happen if molecular biologists were given a free rein (Doc. 5.9). Diagonally across the country from Cambridge, the city government of San Diego, following the Cambridge lead, set up its own Subcommittee on Recombinant DNA.

Individual private citizens, their imaginations running wild, began to lobby individual scientists and politicians (Doc. 5.10). The scientists, at least, unaccustomed to so much direct contact with the public, found it irksome that so much of their time was being taken up in this way (Doc. 5.11). Meanwhile the Cambridge Experimentation Review Board (set up in August) held its hearings and drafted its report and recommendations, which were announced in January 1977. Anyone doing recombinant DNA research in Cambridge would have to comply not only with the NIH guidelines but also with more requirements and restrictions specified by this Board (Doc. 5.12).

The Cambridge episode placed recombinant DNA squarely in the arena of local politics, and for some months there was the serious prospect of larger and smaller units of government across the United States vying with one another in local democracy or populist fervor, as they drew up local ordinances and regulations to superimpose on those of the NIH (Doc. 5.13). At this time some molecular biologists, although basically opposed to any legislation of recombinant DNA research, began to envision the passage of a federal bill, which would at least preempt a series of local and unpredictable initiatives, as the lesser of two evils.

In retrospect it is perhaps surprising that so few cities attempted to emulate Cambridge by adopting the additional provisions specified by the Cambridge Review Board. Indeed in contrast to the situation in Cambridge, some local authorities—for example, the Dane County Board of Supervisors in Wisconsin (a county-elected government unit that includes the town of Madison in which the University of Wisconsin is located)—passed a resolution in support of recombinant DNA research conducted within the NIH guidelines (Doc. 5.14).

An event at Princeton University following the Cambridge decisions is an example of what might have happened, but fortunately did not, all over the country. Princeton University authorities informed the city officials that no recombinant DNA work was going on in the university. The university's Biohazard Committee, chaired by ecologist Robert May, was then informed that Abraham Worcel, a molecular biologist at the university, had obtained from David Hogness (Stanford) some recombinant DNA molecules containing pieces of the genome of the fruitfly *Drosophila*. Even though Worcel had not made these molecules himself, the Biohazard Committee asked him to destroy them. Worcel could not bring himself to commit such a wasteful, almost medieval act and instead returned them to Stanford.

Document 5.1

June 4, 1976
Science (**192:** 940)

On the Dangers of Genetic Meddling

The recombinant DNA research controversy is permeated by the assumptions that (i) the work will go ahead; (ii) benefits outweigh the risks; (iii) we can act now and learn later; and (iv) any given problem has a solution. We therefore had better take those steps necessary to ensure that the April meeting of the Recombinant DNA Molecule Program Advisory Committee at the National Institutes of Health (NIH) is *not* the "last look before the leap" (News and Comment, 16 Apr., p. 236).

Several serious questions must be addressed.

On what basis were the scientists on that committee chosen so that a mutually reinforcing group was able to vote down almost every safety suggestion requested by NIH director Donald S. Frederickson? Why was there no committee discussion about or reference to the myriad reports, statements, letters, and varied data submitted to the committee from throughout the United States by eminent scientists stressing the necessity for (i) more stringent control measures; (ii) centralized P4 facilities; (iii) rejection of *Escherichia coli* as host; or (iv) postponement of recombinant DNA research? When and why was it decided that the work will go ahead merely pending guideline ratification?

As recently as this past February, after NIH's public hearings on recombinant DNA research, David L. Bazelon, Chief Judge of the District of Columbia Court of Appeals, advised that Frederickson, in assessing the varied testimonies covering a spectrum ranging from laboratory safety procedures to ramifications of interference with evolution, should set forth in great detail the reason for each step he takes or does not take. Yet at the April meeting, the advisory committee reviewed the details of laboratory containment facilities and procedures as if the public hearings had never taken place. That the benefits outweigh the risks of recombinant DNA research was taken as a matter of course, not a matter of discussion. This "act now and learn later" approach gave rise to a vote for the use of an "enfeebled" strain of bacteriophage (lambda), predicated in the proposed guidelines on its use with "enfeebled" *E. coli* bacteria. Yet we cannot predict whether, within the human organism, either host or vector, or both, will not later revert to greater strength.

We should seriously question whether these DNA committee meetings are window dressing for those scientists, many currently involved in recombinant DNA research, who are committed to pushing this research ahead with as little impediment as possible. Producing guidelines serves not only as a sop to Cerberus but distracts from the basic alternatives of (i) postponement of research and (ii) open, unbiased discussions of benefits and risks. We should be wary of self-imposed guidelines which experimenters may cite as a defense in lawsuits for punitive monetary damages. It is none too soon to consider federal legislation that would prevent limiting the liability of the experimenter, laboratory, institution, manufacturer, distributor, and direct agent in the case of disease, injury, or death resulting from recombinant DNA research.

There are striking parallels between the recombinant DNA and the nuclear energy controversies. Thirty years ago, when nuclear energy development was initiated, the problems of waste transport and disposal, sabotage, weapons proliferation, and low-level radiation were either not foreseen or not deemed worthy of consideration.

Proponents of nuclear energy defined the problems and proposed their own solutions. Questionable data were classified and talk centered on design criteria, reactor safety, and regulation. The unquantifiable problems—the genetic risk to future generations, human fallibility, the vulnerability of centralized electric generation, acts of malevolence, the threat to civil liberties by massive security measures, and the economic investment and subsidies required—were not addressed.

In like manner, proponents of DNA research have set up the question of laboratory containment as the pivotal problem, for which their guidelines will be the solution. What scientist would claim that complete laboratory containment is possible and that accident due to human fallibility and technical failures will not inevitably occur?

It is therefore essential that open discussion include the entire range of problems in the field of genetic engineering and take into account the biohazards of accidental release of uncontrollable new organisms, the implications of interference with evolution, reduction of diversity in the gene pool, the imposition of complex medical decisions on individuals and society, and the inherent fallibility (not to mention corruptibility) of inspection, enforcement, and regulatory bodies.

We have the unique opportunity now, *before* the intellectual and economic investment in the development of recombinant DNA research grows much greater, to assess the benefits and risks. Such assessment should include acknowledgement that not all problems necessarily have solutions and that problems will arise that cannot possibly be foreseen. The vast number of human and technical variables precludes adequate anticipation of the problems of new technologies. At a recent recombinant DNA conference, NIH deputy director for science DeWitt Stetten, Jr., warned that ". . . the real hazard is the one no one around this table has dreamed of yet, and this you cannot specify against."

Francine Robinson Simring
Committee for Genetics,
Friends of the Earth,
72 Jane Street, New York 10014

Document 5.2

October 21, 1976
Chargaff, New York State Testimony,
Public Hearing on Recombinant DNA Research

PUBLIC HEARING
RECOMBINANT DNA RESEARCH

BEFORE:

LOUIS J. LEFKOWITZ, Attorney General of the State of New York

PHILIP WEINBERG, Assistant Attorney General of the State of New York, in charge of the Environmental Protection Bureau.

DEBORAH FEINBERG, Environmental Scientist, New York State Department of Law, Environmental Protection Bureau.

RICHARD BERGER, Assistant Attorney General of the State of New York.

Held on October 21, 1976 at the Office of the Attorney General of the State of New York, Two World Trade Center, New York, New York 10047.

MR. BERGER: Dr. Erwin Chargaff, Columbia University, School of Physicians and Surgeons.

DR. CHARGAFF: Well, you see, I have been going to scientific meetings for fifty years and I have developed an allergy against handing in papers. I have notes. And so I have to apologize for the random manner of an old ex-Columbia professor. I am forty years a professor of biochemistry at Columbia University until my retirement and I suppose the reason I am here is I wrote a letter to Science Magazine of June 4, 1976 which in a way might serve as a text of my statement. Perhaps you can get a copy of it. In any event, the scientific expert has the great interest in keeping this discussion on a high scientific level. The public, on the other hand, has the even greater interest, and in my opinion the overwhelming interest, in bringing the discussion down to its own level, which may be higher or lower as you look at it. In fact, when the expert speaks to the public, he sometimes sounds like a magician and very often sounds like an idiot. In my statement, in fact, I might say one other thing. I started my career as a chemical assistant in the Department of Bacteriology at the University of Berlin. I am a Ph.D. I am a chemist by training, but I was surrounded at that time by real bacteriologists. It seems even at that time, that was in the early 30's, the precautions that we observed in the handling of pathogenic materials, I shudder to think that our graduate students and our post-doc's as I see them now, even equipped with double copies of the Guidelines, are let loose on experiments that they cannot even really foresee the outcome. They are not trained. They are not trained as were the bacteriologists of the old days, and they are certainly now—immunologists, bacteriologists of a very wide kind—these people are not trained for it and the biologists in general do not go through the training that these people require. So, I would say, it is absurd really to discuss the possible dangers of organisms unless we make sure that the personnel that is supposed to handle them are really up to the highest standards. This is one objection that I have to the Guidelines. They do not really enforce the expertise that ought to be required. But do they enforce anything? When I got the Guidelines, I looked at them, and I had the paranoic feeling that many of the experiments that were almost proscribed had been going on already for the last two or three years because there's no one to look at it. Everyone who worked in a research laboratory knows that even the boss has a hard time supervising what goes on, so let alone the public law of an outside agency. In my letter to Science I first of all objected to the use of E. coli as the host in these experiments. This has been very often discussed. Many people come to me and castigate me that I was slowing down progress, I was anti-science because after all two or three scientific generations had spent all their lives on investigating E. coli and now I was telling them they should go and look for another organism that did not live in our stomachs. Well, I don't know whether such organisms could be found, or would be as usable as E. coli in genetic experimentation, but I'm almost sure that it could be found if one searched for it. One could think of a host of microorganisms of which so little is known that we can't even say there are no phages, no plasmids and so on. And the moratorium that I proposed of two or three years would be very well spent on supporting, on publicly supporting, the research on alternative organisms. I grant that there is some importance in genetic experimentation, and I have been too long a scientist to say certain kinds of experiments should be forbidden by the authorities. But they should be controlled and I would not leave it to the scientists themselves to control themselves to ensure safe operations. Who will watch the watchers? So, the first proposal then involves the search for another organism, or other organisms that do not invade or infect human beings or animals and that could be used for recombinant DNA research instead of E. coli. My second recommendation was that the experiments should all be done in facilities such as Fort Detrick, as was mentioned earlier by speakers before me. It stands to reason that if you have a hundred laboratories of different quality and different moral and scientific standards, all hacking away, then something is bound to happen. They are not all eating it. One facility, even if it is Fort Detrick, and I am no great admirer of monopoly research, even Fort Detrick has the minimal requirements for well supervised containment procedures. Even so, something could happen eventually. Fort Detrick incidentally has had laboratory infections at the time its biological warfare was openly going on. So even that would not prevent the worst, but it would minimize it. I also raised some other questions. When the Legionnaires' Disease was discovered in Philadelphia, and I was in Europe at that time, I got three letters, two scientists and one crackpot, who asked me whether I thought that this mysterious disease could be due to some recombinant DNA organisms that had escaped. My answer was, I don't think they are that good yet in Philadelphia. But it shows that we are putting a cloud of uncertainty, and I spoke about it in my letter, over the future generation by telling them everything is okay because we watch out for you, while at the same time, we don't know the first thing about it. Because we know really very little about the biochemistry of genetics, we know very little about the structure of DNA, but we go ahead as if this were a completely innocuous thing, but it is not. I just don't know. I don't know whether there will be infections. I don't know whether there could be epidemics, but that possibility we raise in the public's mind and on the part of many countries. This question is itself enough for me to advocate the most strict controls. Because we could not even say comfortably, I don't know what an organism containing the insulin gene, whether mine, or a rabbit, or a dog insulin gene, could do when it gets into the gut or into these areas. I just don't know what the supposedly isolated gene will carry with it when it is transferred to a plasmid or what it will do. I just don't know. We have no real animal or other experimentation yet going on. We are jumping into the middle and assuring everybody it's okay. I feel it is not. I think we are taking a good step in considering how at least these minimal guidelines, which as I say, I don't consider sufficient, how these minimal guidelines could be enforced. I do however think about this also in my letter that it will be a terrible job to look after it and that we should contemplate the creation of an authority similar to the original atomic energy authority which will license the work and supervise it and be responsible for it. In addition, I would advise all laboratories in which the work is going on to take out very high amounts of liability. For this I see coming the Legionnaires' Disease was only the first beginning of something which eventually will be found to be particularly frequent in Palo Alto or in Cambridge and I would advise therefore the presidents of these institutions to insure themselves very heavily. Thank you.

MR. WEINBERG: One question, Dr.Chargaff, could you elaborate a little bit on the kinds of training that were employed when you were at the University of Berlin — whether or not something like that could be used here?

DR. CHARGAFF: Well, the people were handling pathogens. First of all, M.D.'s trained in the old way — very heavy in bacteriology. If they had any function, they also had a state license — a license might help. I really don't know how this could be adopted to our present standards, but as I say the students and the post-doc's and others who call themselves molecular biologists could not have passed these exams in my time.

MS. FEINBERG: Dr. Chargaff, you suggested a two- or three-year moratorium on all this research. During that period, what do you suggest should go on; any particular studies?

DR. CHARGAFF: Yes; I say that, in fact, I would be very happy if NIH, or whomever, would support selected research on the other hosts, not E. coli. I could think of marine organisms — of simple cryophilic or thermophilic organisms. In other words, I'm not convinced that the weakened, enfeebled strain is really disabled since genetic exchange in the gut or in other circumstances has often been found and I don't suppose it's an acceptable argument in this connection that everything that could happen has already happened. Our time has brought about the breaking of the atomic nucleus which certainly did not happen in historical times before. We are the first really to get inside the parameter of life. That is the reason of this scientific conference, and I think we are very well able to do damage to ourselves and especially to the following generation.

MS. FEINBERG: Thank you.

MR. WEINBERG: Let's take a recess of two or three minutes, and then we'll resume.

(Whereupon, at this time, a short recess was taken.)

Document 5.3

July 1, 1976
New Scientist (**71**:14)

Threat to US genetic engineering

Local resistance to genetic engineering in Cambridge Massachusetts may halt research there for two years. Other parts of the country may soon follow suit

Graham Chedd An extraordinary confrontation took place last Wednesday in the council chambers of the city of Cambridge, Massachusetts. At times unnervingly like an inquisition, it was the first major public clash between scientists planning genetic engineering experiments with the new recombinant DNA technology and representatives of the community in which those experiments will be performed. By coincidence, the debate—in the form of a public hearing before the Cambridge City Council—was held on the day the long-awaited federal guidelines governing recombinant DNA research were issued. The immediate cause of the town-gown confrontation was a new laboratory planned at Harvard University. But the hearing also provided the first real test of the acceptability of the new federal guidelines to the man in the street.

For most scientists, the result can only have been disheartening and, regrettably, confirmation of their prejudice that the man in the street simply doesn't—and maybe cannot—understand the issues. For biologists at Harvard and MIT, both within the City boundaries, the consequences could be calamitous. In the middle of the triangle, Cambridge mayor Alfred Vellucci announced his intention of introducing before the council on 7 July a resolution insisting "that no experimentation involving recombinant DNA should be done within the City of Cambridge for at least two years." And the precedent established by the council is likely to extend far beyond its area of jurisdiction. Vellucci last Sunday set off to Milwaukee for the US Conference of Mayors armed with a resolution urging other cities to demand public hearings on the issue of recombinant DNA research.

At the heart of the Cambridge affair is Harvard's insensitivity to the city council's dignity—and arguably to its own public responsibilities—and Vellucci's fine sense of outrage. Helping to ignite the issue is the fact that Cambridge is not only a major focus of research involving recombinant DNA but also the centre of passionate and articulate opposition to the research from among the science community itself.

The affair began with Harvard's planning to construct on the fourth floor of the biology building on Divinity Avenue in Cambridge a so-called P3 facility. This is a laboratory with the standard of containment required by the new federal guidelines for many recombinant DNA experiments. (The guidelines match types of experiment, according to their conjectured hazard to public health, with

"... there is no significant risk involved in experiments authorised to be done ... in P3 laboratories" ... **Mark Ptashne**

Pictures by Noeleen Chedd

"We caught Harvard just in time"... **Mayor Alfred Vellucci**

four levels of containment: P1 and P2, which are minimal and can both be achieved in ordinary laboratories; P3; and P4, a containment level found only in specialised facilities like those at Fort Detrick, Maryland, the ex-germ warfare establishment). A P3 facility involves limited access of personnel, a negative air pressure, safety hoods, provision for solid and liquid waste sterilisation, and strict operating procedures. The measures are designed to minimise the risk of an experimental organism escaping from the laboratory.

Harvard wants the laboratory principally for work on animal tumour viruses, where the major problem is contamination of animal cell cultures from outside. But it also intends that some recombinant DNA research shall be performed there. Under the terms of the new federal guidelines, this research can now go forward once local biohazard committees are satisfied that the containment facilities match those laid down for specific experiments. The construction of the facility has been approved by the university biohazards committee, chaired by Daniel Branton. Chairman of the committee responsible for the laboratory is Mark Ptashne. An NIH grant has been obtained for work.

While there was some dissent about the facility from within the university—notably from biology professor Ruth Hubbard who shares the fourth floor with the lab—discussion had been confined exclusively within the Harvard faculty. Then on 8 June the Boston *Phoenix*, an ex-underground weekly newspaper, published a detailed and critical article about Harvard's plans, drawing attention particularly to the university's failure to consult with the local community. Mayor Vellucci was by his own account outraged to learn from the newspaper that Harvard was planning to build a laboratory where new and nameless horrors were to be perpetrated in the names of science and to the potential detriment of the health of his constituents. Last week's public hearing, held in a council chamber packed literally to the ceiling and overflowing into the halls, was his response.

Vellucci was in his element; the scientists appearing before him were not. The consequences were sometimes farcical, sometimes painful and on occasion—such as when Vellucci demanded that everyone present in favour of the new laboratory stand up and identify themselves—positively frightening.

Mark Ptashne presented the main case for the facility's being built. He argued that no experiments would be performed there that had not previously been approved by the biohazards committee; and that experiments designated for P3 containment presented no risks to the health of the local population. He also argued against the implicit assumption that the single act of building the facility was going to open the floodgates to unprecedented and uncontrollable new experiments. Recombinant experiments allowed in P1 and P2 facilities would continue while P4 experiments—to his personal relief—are not planned.

But Ptashne was trapped in the apparent logical paradox of the whole safety issue. If it was true, the councillors wanted to know, that the experiments posed no significant health threat, then why was it necessary to construct a special facility? To build in an extra margin of safety, Ptashne replied. But you just said the experiments were safe, came the response. Yes, said Ptashne, but there is a *hypothetical* risk. In that case, why don't you do the experiments in a P4 facility out in the desert, and not in a laboratory that you yourself cannot guarantee is escape-proof in the middle of one of the most densely settled areas of the country? The councillors wanted things in black and white and Ptashne was stuck trying to explain shades of grey.

He was supported valiantly by Maxine Singer, who had come up from Washington for the meeting to explain the federal guidelines. After some bullying from Vellucci about whether she was or wasn't "on the Harvard team", Singer explained that she was among the first to express doubts from within the scientific community about the safety of recombinant DNA research; thus she had been a critic of the guidelines as they evolved through their various drafts, but that she was now satisfied that they "provide a high degree of confidence that hazardous agents won't escape." She added that she felt "very comfortable with experiments to be carried out in P3 facilities."

"Everyone agrees that these facilities will not contain *E. coli* . . ." . . . **Ruth Hubbard**

But the councillors' doubts remained. How could they be expected to make up their minds on these issues when scientists couldn't decide among themselves? The most persistent and telling criticism came from councillors who argued that Harvard University—and indeed, scientists as a group—did not have the right to take decisions about matter that I may expose the public to danger, however hypothetical, without first consulting them. Maxine Singer and later David Baltimore argued in vain that the whole decision process leading up to the new research guidelines had been conducted openly and with an unprecedented awareness of social responsibility.

Whether the public's role in formulating the guidelines has been real or cosmetic is the real issue at the core of both the specific Harvard case and the wider debate. Harvard's Daniel Branton claimed, for example, that his biohazards committee had attempted to contact Cambridge public health commissioner about the proposed P3 facility, but Vellucci effortlessly tied him in knots, making it apparent that Branton had not tried as hard as he might have. And the lack of a genuine attempt to involve the public has always been the major thrust of the criticism of recombinant DNA research from Science for the People. Jonathan King of MIT presented their case, arguing that it was right for the debate at last to be taking place on the floor of the council chamber because the real issues "are political rather than scientific." The last person he would trust to protect his health, he argued, "is someone who doesn't believe there *is* a threat to my health." In a democracy, only the community should have the right to decide whether or not a potentially hazardous area of research is worth the potential benefits, not the scientists themselves. King was pleased with the council's initiative: "It empowers people in other parts of the country to realise there is something they can do." He concluded on a note of slightly embarrassed rhetoric. Genetic engineering is "tampering at the most profound biological level . . . I think it's sacrilegious."

Ruth Hubbard was also among the numerous opponents to the Harvard plan. Her objections focused more on the scientific arguments, in particular the use of *E. coli* on the host organism for the experiments. She stressed the ubiquity of *E. coli,* the fact that it is bound to be carried out of the laboratory on people's skin and clothing, and argued that *E. coli* can get all over the place before we know its out, and we cannot call it back."

But the major issue remained that of public participation and Harvard's perceived arrogance in planning the P3 facility without informing Vellucci or the council. Dean Howard Hiatt of Harvard's School of Public Health made the flat admission that "we in the medical and scientific community have been very remiss in not involving the public more." Harvard's specific failure to do so has now blown up in its face—and the fall-out may well involve MIT too, at the other end of Cambridge and much more deeply into recombinant DNA research. Should Vellucci's resolution to bar all such research in Cambridge for two years succeed next week, it will devastate the biology, biochemistry and cancer research activities of both institutions. The blanket warning of the resolution would prevent experiments which pose no conceivable risk, and doubtless send many more biologists along the route already adopted by Harvard researcher Tom Maniatis, who for the last year has been working in the P3 facility at Cold Spring Harbor out on Long Island.

But it's conceivable that there soon will be nowhere for the putative gene engineers to turn. Vellucci is carrying the banner of community involvement to 2000 mayors in Milwaukee this week, and already such public interest groups as Friends of the Earth have begun to take up the issue. While the debate so far has taken place almost exclusively at the federal level—resulting in last Wednesday's guidelines—it is likely now, with Cambridge's initiative, to resume at the level of the local community. □

". . . a high degree of confidence that hazardous agents won't escape." . . . **Maxine Singer**

Document 5.4

July 9, 1976

Newspaper Headlines and Article

Policing genetic research

THE WASHINGTON POST *Friday, July 9, 1976*

City Blocks DNA Research

By Edward Schumacher
Special to The Washington Post

CAMBRIDGE, Mass., July 8—The Cambridge City Council early this morning voted a three-month moratorium on potentially dangerous genetic research at Harvard University and Massachusetts Institute of Technology that federal research officials fear will set a precedent of community control over such work.

The nine-member council voted 5 to 3 with one abstention to establish a review committee of scientists and citizens to recommend by the end of the moratorium a city policy on the "recombinant DNA" research. The city has the legal power to ban the experiments by declaring them a public health hazard.

Mayor Alfred E. Vellucci, who is also head of the city council, said today, "Cambridge has six square miles and we're boss here. They're going to do what we tell them."

The moratorium will have little effect on the university for the time being. Harvard does not plan to have the requisite laboratory until next spring. MIT has a lab, but it has not been certified yet under new federal guidelines on such research.

The experiments involve combining the DNA-deoxyribonucleic acid, which carries an organism's genetic information—of two types of organisms, commonly a warm-blooded animal and strain of E-coli bacteria.

The experiment creates a new organism. The possibility exists that it will be an unknown one and that its properties will be unpredictable, scientists on all sides of the issue agree.

The fear is that new diseases with unknown cures will be created and spread.

Proponents of the research say that it offers the basic scientific understanding of cell reproduction that could lead to cures for cancer and other diseases, as well as to the production of organic things such as insulin and self-fertilizing plants.

William J. Garland, head of genetic research at the National Institutes of Health, attended the five-hour hearing, which was packed with several hundred local residents, students, scientists and two Nobel laureates. He later said the council's action may be "obstructive" by starting a wave of non-uniform regulations across the country.

He said the recombinant research is expected to escalate rapidly. A voluntary national moratorium on the research had been in effect since 1974 until two weeks ago when NIH issued long-awaited guidelines on the safety issue.

Mayor Vellucci, a burly, rough-hewn man, has built a political reputation among Cambridge's largely blue collar electorate of stalking the two wealthy, private universities.

On this issue, however, he formed an alliance with many of the liberal intellectuals opposing the genetic research. These include outspoken Nobel laureate George Wald and his wife, Ruth Hubbard, as well as many of the university students Vellucci has ridiculed in the past.

"It's nice to know the city can expose an issue like this and have all the Nobel scientists come to us" Vellucci said.

Attention has centered on Harvard, where recently the administration approved plans for a recombinant DNA lab in the biology building after months of debate among students and professors. One biology professor opposing the work has since demanded her office be moved farther away from the lab.

Harvard geneticist William Petrie said at the hearings that the lab should be moved to a desert area. "If it blows up, only a few persons will be hurt," he said.

But Matthew S. Meselson, chairman of the department of biochemistry and molecular biology and a supporter of the research, said that if he thought the lab were dangerous, "I would not subject myself to it . . . The work is too important to be stopped.

"For a civilization whose whole development is based on understanding ourselves and our world . . . this technology is so profound and of such enormous scientific importance for all of us."

Controls in gene research

Dealing with DNA

NEWS AND COMMENT

Recombinant DNA: Cambridge City Council Votes Moratorium

Cambridge, Massachusetts. The City Council chamber here was packed to overflowing late on the night of 7 July as the Council of seven men and two women, used to dealing with taxes and street closings and similar civic matters, tried to grapple with one of the most perplexing problems in contemporary biology—the safety of certain types of research involving recombinant DNA. Before them was the question of whether to allow investigators at Harvard and the Massachusetts Institute of Technology (MIT) to proceed with controversial and potentially dangerous experimentation or whether to ask them to hold off awhile until the councillors could better understand what is at stake.

By a vote of 5 to 3, with one abstention, the Council asked the researchers to hold off when it declared a 3-month, "good faith" moratorium on the work. In addition, the Council voted to establish a permanent body—the Cambridge Laboratory Experimentation Review Board—of scientists and citizens to investigate recombinant DNA (and, in the future, other types of research) and report back with a recommendation about allowing it to take place in Cambridge.

With those two votes, the councillors of this largely working-class town of 100,000 citizens took what are thought to be unprecedented steps to involve themselves in decision-making regarding biological research. There has been a good deal of discussion recently about so-called "public participation" in science (*Science*, 30 April), much of it led by Senator Edward M. Kennedy (D–Mass.) and other members of the Senate health subcommittee of which he is chairman. But there has been precious little said about just what public participation means. Ironically, it is the citizens of Kennedy's own state that are providing one of the first concrete examples of public participation. And they are calling on him to hold hearings to provide a national forum for debate. (He well may).

But opinion is divided over whether the City Council's example is one to be admired or deplored. There are some who hope that the Cambridge precedent of local involvement will be followed by similar action in other communities, and there are others who think it is a disaster.

For nearly 2 years, biologists voluntarily have been observing a moratorium on certain types of research with recombinant DNA, while looking to Washington and the National Institutes of Health (NIH) for guidelines on how safely to proceed with potentially hazardous experiments that involve combining in the laboratory the genes of organisms that do not combine in nature. On 23 June, the NIH guidelines finally came out (*Science*, 16 July) and, throughout most of the country, scientists began making preparations to get on with work they have held in abeyance since 1974.

But here in Cambridge, recombinant DNA researchers at Harvard and MIT were making preparations of a different sort. They were preparing a defense of their proposed research to put before the City Council that was holding a public hearing that night to find out what recombinant DNA is all about. Alfred E. Vellucci, the city's flamboyant mayor, had been saying that with recombinant DNA "those people in white coats" could build a Frankenstein or turn loose upon the populace a deadly organism like the fictional Andromeda strain. There was talk of a 2-year moratorium on controversial types of recombinant DNA research within the Cambridge city limits, and many scientists, who had already waited 2 years, saw their most exciting projects slipping out of their hands. As one city councillor said later on, "The Harvard and MIT people thought that, because Washington had said it was OK to go ahead, that was that. They were flabbergasted to discover that Al Vellucci could have a noose around their neck in just a few days' time. Here's a guy ranting and raving about monsters and germs in the sewers and they have to stop what they want to do because of him. They just didn't understand."

At that first City Council hearing, which lasted until 1 a.m. and was described as a "circus" by those who were there, the nine councillors, who had never before even heard of recombinant DNA, listened to testimony from those who spoke of its potential benefits to mankind and those who dwelled on its potential hazards. It was 2 weeks later that the Council held its second hearing and voted in favor of the moratorium.

The precipitating factor in the present situation was a split within the Harvard biology faculty over the renovation of one floor of the biology laboratories, but many observers believe that the issue of recombinant DNA would have come before the public sooner or later in any case because of the strong opposition to it from the Science for the People group, which is active at both Harvard and MIT.

Briefly stated, the NIH guidelines distinguish four classes of recombinant DNA research, designated from P1, which is safe enough to conduct in any open laboratory, to P4, which is to be conducted only under conditions of strict physical containment, such as those prevailing at National Cancer Institute facilities at Fort Detrick in Frederick, Maryland. Harvard wants to build a "moderate" containment or P3 laboratory by renovating existing space on campus. Some biologists, informally led by Matthew Meselson, favor this plan. Others, led by Nobel laureate George Wald and his wife, Harvard biologist Ruth Hubbard, are opposed. Three sets of circumstances apparently came together over this issue to bring it to public attention. One of the members of the City Council attended a hearing that the Harvard faculty held on the subject. So did a reporter for the Boston *Phoenix*, who wrote up the internal debate for that "alternate" paper. And Wald went to see Mayor Vellucci, whom he persuaded that the potential threat of P3 recombinant DNA experiments to the public health is a very real one.

From there, the course to City Council hearings was simple, especially since

Vellucci for years has gotten a lot of political mileage out of attacking Harvard. He is well known around town, for example, for his periodic rhetoric about turning Harvard Yard into a parking lot. With something as esoteric as recombinant DNA, Vellucci had an ideal opportunity to go after Harvard (which has far poorer relations with the city than MIT) while protecting innocent women and children from the menaces of science. He was also able to jump on a favorite theme, that Harvard never communicates with City Hall. "All of these plans for research were going on and I had to read about it in the *Phoenix*," Vellucci fumed, choosing to leave out reference to Wald's visit to him.

Arguments for and against doing certain types of recombinant DNA experimentation were pretty much the same ones that were heard nationally as scores of scientists contributed their thoughts to the debate about the NIH guidelines. Those in favor claim the risks are minuscule but the potential rewards are great—the cure of cancer and the production of new kinds of organisms to eat up oil spills being frequently mentioned. On the other side, it is said that to dangle the cure of cancer before the public is to make an empty promise and that bugs that eat spilled oil will eat oil from other sources as well. According to those who were present at the first hearing, the City Council listened to it all but did not really come alive until the matter of the Cambridge city health commissioner came up.

Responding to the mayor's taunts about Harvard not involving the city in its research plans, one university scientist declared in prepared testimony that the health commissioner had been invited to attend meetings of the Harvard committee on the regulation of hazardous biological agents. It was a grievous mistake, for, as one observer told *Science*, "The members of the City Council didn't know a thing about DNA but there was one thing they did know and that is that Cambridge doesn't have a health commissioner. Hasn't had one for 19 months, and it's something of a sore point with them."

But now the mayor has promised to find a health commissioner posthaste because whoever fills that long-empty position already has a central role to play in the current DNA contretemps. It is the health commissioner who has the authority of last resort in this matter—the power to ban the research by declaring it a health hazard. (The reason the City Council issued only a "good faith" moratorium is that it lacks legal authority to decree anything more forceful.) And it is the health commissioner who is likely to be chairman of the Laboratory Experimentation Review Board that must recommend a course of action to the City Council. It is easy to see why recombinant DNA research proponents feel discouraged about having their fate in the hands of a nonexistent board, but there it is.

In all of this, the city councillors say, the most important issues are political, in part because it is nearly impossible to grapple with the scientific ones. During the weeks between the two City Council hearings, every councillor was lobbied by scientists hoping to convince them that the work is safe and a moratorium not necessary. But they found it hard to know what was true in the face of mountains of conflicting statements from scientists themselves. Councillor Leonard J. Russell told *Science* that listening to the scientific debate made him feel "fuzzy" because "every time I think I understand an argument, someone pokes holes in it." Councillor Saundra Graham tried to help but missed the point when she moved to change the 3-month moratorium to a 6-month one, so that the scientists themselves could resolve their differences. But they cannot, of course, and that is why the political process is going to help them.

Councillor David E. Clem, a city planner by training, put it this way: "I tried to understand the science, but I decided I couldn't make a legitimate assessment of the risk. When I realized I couldn't decide to vote for or against a moratorium on scientific grounds, I shifted to the political." In the end, Clem, who voted for the moratorium, was influenced by his concern for public participation and the need for scientists to educate the public, which he called "cumbersome but necessary," and by his fears that NIH is not the right agency to assume responsibility for monitoring work on recombinant DNA.

The issue comes down to this: Can an agency that promulgates research as its primary mission also effectively regulate that research? Clem is among those who think the answer is "no." He recalls what happened to the Atomic Energy Commission when it tried to do two jobs. What is needed, Clem maintains, is a separate, federal regulatory body to oversee recombinant DNA research not just in universities but in industry as well. He is urging the City Council to petition Congress on this point and believes that, short of federal regulation, NIH should at the very least provide funds to enable local communities to monitor for themselves research at local institutions.

The members of the City Council are adamant in saying that they do not want to stop work on recombinant DNA in its tracks, and, on the whole, most of them say they are more persuaded by its proponents than by its detractors. But the fact that federal guidelines have been written is not, in itself, enough to satisfy them. As one of the mayor's aides said, "We looked at the process by which they arrived at those guidelines and found it was anything but placid. We were not reassured." And so Cambridge is going to go through at least part of that process itself, redundant though it may be, until the local community is satisfied that all is well. Clem put it aptly when he said, "Science is just going to have to learn to bear with it."

—Barbara J. Culliton

Cambridge DNA Politics and Wald's Response

Al Vellucci was unhappy. The Cambridge Experimentation Research Board (CERB), created by the City Council last July to investigate proposed DNA research at Harvard and MIT, was going to present its report to the press and public in two days, and none of the councillors had seen it. Vellucci knew that CERB had unanimously recommended to the city's acting health commissioner, Francis Comunale, that experiments with recombinant deoxyribonucleic acid (DNA) at the P3 level be allowed. (P3 experiments require separation from public areas, controlled access corridors and air locks, and biological safety cabinents.)

At Monday's City Council meeting, Vellucci told Daniel Hayes,the CERB chairperson (and a former mayor and city councillor),"Our fight isn't with you, it's with the city manager and Dr. Comunale . . . but I'll never call on you, I'll make sure the manager never calls on you to serve on a committee anymore." Hayes had just told the councillors they didn't need to see the report before Wednesday, and he didn't think it would be a good idea for his committee to be questioned about the report by the City Council in front of the press and public on Wednesday night. Barbara Ackermann helped bring about a compromise; Hayes promised to bring the rest of the CERB to a Wednesday night meeting and present the report to the public.

Everybody came Wednesday including Channels 2, 4, 5 and 7, CBS News, reporters from English science magazines, and hundreds of Cambridge residents. There were enough television lights to make the Council Chambers look like Fenway Park on a Friday night.

The CERB report allows recombinant DNA research which adheres to guidelines formulated by the National Institute of Health (NIH), and expanded by CERB. It recommends manuals of procedures to be followed during research, a Biohazards Committee to monitor research sites, systems capable of screening and containing strains of E. coli (EK) 12 (a strain that is of no-or minimal risk, according to NIH), and regular monitoring of laboratory workers to make sure none of the bacilli have escaped into their intestines, causing disease.

Nobel laureate George Wald Eric A. Roth

Nobel laureate George Wald, the Harvard biologist who was instrumental in bringing the DNA issue to the City Council and who is an opponent of recombinant DNA research, called the CERB report "sober, sophisticated and thoughtful." Wald still would like a national moratorium on DNA research.

"The really biggest problem," he told us, "presented a choice between letting this kind of experiment be done in crowded cities or universities or segregating it into one or a few national laboratories. Clearly they came out on the side of allowing this research to go on in universities, in the crowded city of Cambridge, so that was the big decision. Actually, it's hardly mentioned as a decision, so that one hardly realizes that the committee had a choice of alternatives, and took one of them. . . . It's a very serious matter. The claim that these strains of E. coli don't survive in the human gut, and hence are supposed to be safe, doesn't cover at all what happens when one has transformed these E. coli by putting in recombinant DNA from other organisms."

Document 5.7

July 2, 1976 — *Science* (**193**:6)

August 6, 1976 — *Science* (**193**:442)

Costs and Benefits of Recombinant DNA Research

The discussions in *Science* of the recombinant DNA problem, beginning with Singer and Soll (Letters, 21 Sept. 1973, p. 1114) and Berg *et al.* (Letters, 26 July 1974, p. 303) and continuing with Chargaff (Letters, 4 June, p. 938) and Simring (Letters, 4 June, p. 940), present the issue as if it were a balance between the costs to society of possible disastrous infections and the benefits to a few biologists of pursuing their professional careers. Chargaff asks, "Have we the right to counteract, irreversibly, the evolutionary wisdom of millions of years, in order to satisfy the ambition and the curiosity of a few scientists?" If this were really the question at issue, the inevitable answer would be negative. But in fact the technology of recombinant DNA offers potential public benefits which are at least as significant as the dangers. The public costs of saying no to further development may in the end be far greater than the costs of saying yes. Unfortunately, our legal and political institutions were designed to count the costs of saying yes to unsound technological ventures, and have no established procedures for counting the costs of saying no.

Biologists who defend their work on the ground that it may be of benefit to humanity come under suspicion of serving their own interests. Biologists feel comfortable saying that they do their work for fun or for a living; they feel uncomfortable posing as saviors of humanity. Therefore I find it appropriate, as a physicist having no personal stake in recombinant DNA, to make certain claims which the biologists are inhibited about making for themselves. I claim that the exploitation of recombinant DNA techniques may lead to an understanding, and conceivably to a cure, of cancer. It may lead to the creation of improved food plants which could save hundreds of millions of people from imminent starvation. It may lead to the creation of energy crops which offer benign alternatives to nuclear fission and fossil fuels. These claims are of course impossible to substantiate. There is no way to estimate numerically the probability that these things will happen. I can only say that in my nonexpert opinion, and in spite of Chargaff's eloquent derision, these possible benefits of recombinant DNA research are more likely to materialize than any of the most extreme dangers. I do not deny or belittle the dangers. I say only, let us not leave the starving millions of humanity out of account when we balance the dangers against the benefits. It is perhaps not irresponsible, but rather an act of enlightened courage, to expose ourselves to an unknown risk of disastrous epidemics in order to give ourselves a change of lifting some hundreds of millions of our fellow humans out of the degradation of poverty.

Finally there is the warning of DeWitt Stetten, Jr., quoted by Simring, "the real hazard is the one no one around this table has dreamed of yet, and this you cannot specify against." This is true. But it is equally true that the real benefit to humanity from recombinant DNA will probably be the one no one has dreamed of. Our ignorance lies equally on both arms of the balance. All that we can say with certainty is that prodigious changes in the conditions of human life must come within the next century if civilization is to survive. The exploitation of recombinant DNA is only one of these changes, and perhaps not the most dangerous nor the least hopeful.

FREEMAN J. DYSON
Institute for Advanced Study, Princeton, New Jersey 08540

Evolution, Epidemiology, and Recombinant DNA

In attempting to assess the hazards of incorporating eukaryotic DNA into bacteria it is not enough simply to set up hypothetical scenarios: we must also try to judge critically the underlying assumptions. The first assumption is that these experiments will breach an ancient barrier between eukaryotes and prokaryotes and will thereby produce a radically novel class of organisms.

Principles from evolution and bacterial ecology offer our best guides for judgment. Bacteria in nature have long been exposed to DNA from lysed mammalian cells—for example, in the gut and in decomposing corpses. *Escherichia coli* can take up DNA after damage to the cell envelope, and one would expect random phenotypic variation to produce such damage occasionally (perhaps at frequencies of 10^{-5} to 10^{-10}). Homologous DNA is efficiently incorporated after entry, because its potential pairing with long regions of host cell DNA facilitates enzymatic crossover. Indeed, genetic recombination between bacteria (transformation) has even been observed in the human host. Incorporation of nonhomologous DNA is much less efficient but nevertheless can occur, presumably by transient pairing between adventitious short regions of complementarity. For example, deletions based on such "illegitimate recombination" occur at frequencies of about 10^{-9}.

With such low frequencies of both entry and incorporation, one could not expect to demonstrate natural hybridization between bacteria and man. Nevertheless, its scale almost certainly compensates for its inefficiency. Every person's gut is a huge chemostat, and the total population excretes about 10^{22} bacteria per day. Hence over the past 10^6 years human-bacterial hybrids are exceedingly likely to have already appeared and been tested in the crucible of natural selection. If so, experimental DNA recombination will not be yielding a totally novel class of organisms.

A second assumption is that some of the recombinant strains are likely to spread and cause epidemics. Evolutionary principles are again pertinent. Nature selects for genetic balance: the contribution of a gene to Darwinian fitness

September 3, 1976
Science (**193**:834)

depends on the rest of the genome. In bacteria, specifically, the introduction of a substantial block of foreign DNA would almost always lower the growth rate. With the short generation time of bacteria such a difference would lead to rapid outgrowth by competitors (unless the introduced genes promoted adaptation to alterations in the environment, such as the wide use of an antibiotic).

This argument is reinforced by a large body of epidemiological and experimental evidence. To cause communicable disease a potentially pathogenic organism must be able to survive in nature, in competition with other strains. It must also be able to be transmitted to a host, reach a susceptible tissue, and express its toxic potentialities there. Much current anxiety seems to be based on unawareness that microbial pathogenicity and communicability are complex and depend on a balanced genome. *Escherichia coli* carrying a gene for diphtheria toxin would be poorly suited to cause a diphtheria epidemic.

While bacteria carrying mammalian genes are thus unlikely to menace the public health, the risk of laboratory infection is much larger, since a heavy infecting dose of even a poorly communicable organism can cause disease in an individual. But this danger resembles that encountered with known pathogens, and it can be minimized by similar means. Perhaps the most valuable outcome of the current debate would be the requirement that those working on recombinant DNA be trained and supervised like medical bacteriologists.

I conclude that the risks in research on recombinant DNA require reasonable precautions but do not warrant public anxiety: A greater danger may be that the presumed analogy to nuclear weapons will lead to demands for virtually absolute freedom from risk. Yet the analogy to our mastery over infectious diseases is more apt. And if this field had faced similar demands, from its start, we might still be losing one-quarter of our children to communicable diseases. Is the balance of risk and benefit in research on recombinant DNA so much more unfavorable?

Bernard D. Davis
Bacterial Physiology Unit, Harvard Medical School, Boston, Massachusetts

---

Recombinant DNA: Unknown Risks

Bernard D. Davis's letter (6 Aug., p. 442) makes interesting reading in a week in which 25 Legionnaires have died and more than 100 are still ill—some critically—of a disease of unknown origin, unusual virulence, and unprecedented epidemiology. If scientists do not conclude from this juxtaposition that it would be good to face up to ignorance in areas in which we have no experience, instead of engaging in facile speculation, I hope the public will.

I am not claiming that the mystery disease is a result of genetic manipulation, since obviously no one knows its cause. But I wish to point out that to pretend to know more than we do about causes and prevention of disease can only discredit science and scientists. (At this writing, infectious and toxic agents have in turn been ruled out as causes of "Legionnaire's Disease" and today's newspaper talks about Fort Detrick and possible unknown varieties of infectious agents.)

A further point: *if* a recombinant (and perhaps short-lived) coliform organism ever were to produce an outbreak of an epidemic, it might well be nearly impossible to identify or to culture as the cause in the presence of all the other, normal strains of *Escherichia coli* that grow in us.

Davis suggests that medical history shows such risks must be taken and implies that the high child mortality rate of a century ago was reduced through medical intervention. This is not true. Almost nine-tenths of the decline in the combined death rate from scarlet fever, whooping cough, diphtheria, and measles in children under age 15 occurred before the introduction of specific therapies or vaccinations; and similarly with tuberculosis, cholera, typhoid, and most other infectious diseases. The most probable reasons for these reductions were improvements in nutrition and public health measures—better housing, clean water, and so forth. The specific medical measures of the last three to four decades only clipped the tail off the asymptotic curve. This is not to underrate the importance of every life saved. Furthermore, those risks were taken to cure known diseases, not to create new ones.

Ruth Hubbard
Biological Laboratories,
Harvard University,
Cambridge, Massachusetts 02138

Document 5.8

October 21, 1976

Cavalieri, New York State Testimony,
Public Hearing on Recombinant DNA Research

New Strains of Life—or Death

We had hoped to reprint here an article with the above title written by Liebe Cavalieri and published in *The New York Times Magazine* of August 22, 1976. We regret that Dr. Cavalieri felt unable to grant permission.

In August 1976 Dr. Cavalieri was by no means alone in these opinions, although few had opportunities such as his to expound them in so widely read and prestigious a publication as *The New York Times Magazine*. On October 21, 1976, Dr. Cavalieri testified at a public hearing on recombinant DNA research before Louis J. Lefkowitz, Attorney General of the State of New York. His testimony, less emotive than his article, was as follows:

PUBLIC HEARING
RECOMBINANT DNA RESEARCH

BEFORE:

LOUIS J. LEFKOWITZ, Attorney General of the State of New York.

PHILIP WEINBERG, Assistant Attorney General of the State of New York, in charge of the Environmental Protection Bureau.

DEBORAH FEINBERG, Environmental Scientist, New York State Department of Law, Environmental Protection Bureau.

RICHARD BERGER, Assistant Attorney General of the State of New York.

Held on October 21, 1976 at the Office of the Attorney General of the State of New York, Two World Trade Center, New York, New York 10047.

L.J. Lefkowitz after giving an introduction said "Well, we will have our first witness, Mr. Berger, if you're ready."

MR. BERGER: Dr. Liebe Cavalieri from the Sloan-Kettering Institute.

DR. CAVALIERI: I'm afraid I have a respiratory infection and I hope my voice holds out. Because of that, in the interest of time, I think I will read my statement and as slowly as possible.

I want to discuss several points which have been touched upon at various committee meetings held by the NIH in connection with the Guidelines, but which I feel have not been thoroughly reviewed. These are: First, the benefit/risk ratio of recombinant DNA research; two, what I call the hazards of success as opposed to the inadvertent dangers of the research; three, the moral responsibility of scientists and others to future generations; and finally, suggestions for a possible course of action.

First, regarding the benefit to risk ratio. In all of the discussions which have taken place in the last year and a half, it has been agreed among scientists that there are risks involved in this technology. The guidelines have been formulated as an attempt to minimize these risks. Without going into too much detail, the major immediate risk is the escape into the biosphere of bacteria or viruses in which foreign DNA has been inserted in the form of recombinant DNA. The inserted DNA may alter, in a variety of unpredictable ways, the effect of the bacterium or virus on other organisms which it infects, be they plant, animal or man. I would like to quote from the draft of the Environmental Impact Statement of the NIH, issued in August of this year.

"In the case of a cancer-producing DNA fragment, evidence of harmful effects might not be apparent for many years. The agent might be so widespread as to make control difficult or impossible. Subsequent cessation of experiments would not stop the diffusion of the hazardous agent. It is possible that means will not be available to prevent or stop untoward events."

The benefits expected from recombinant DNA research are divided into two broad categories: First, an increased understanding of basic biological processes; and secondly, practical applications for medicine, agriculture and industry. No one can quarrel with the first of these, but the same statement can be made about so many other areas of research that it is not a compelling argument for any single approach, as I will discuss later. As for the second category, according to the Environmental Impact Statement, these applications are only speculative since the basic mechanisms have yet to be proven. The facile synthesis of drugs and other therapeutic agents or industrial products by recombinant microorganisms is indeed a possibility, but it is fraught with danger should the laboratory-constructed organisms escape to some site—say for instance a human being—where their products could be highly deleterious. It would be foolish to assume that other, safer methods of producing the needed materials cannot be found. In agriculture, it has been proposed that the world's food supply might be increased by recombinant DNA techniques. In actual fact, this technique has little or nothing to offer that cannot be more conveniently accomplished by more traditional plant breeding methods. The only exception is the hypothetical possibility of endowing plants with the ability to utilize nitrogen from the air, a property now confined to bacteria. But even if this were to prove feasible, it would not solve the fundamental problem that faces the earth—that of bringing its food supply and its population into a state of equilibrium. One can ask whether it is worthwhile to invite these potential dangers, in order to attain the possibility of a temporary gain.

I don't mean to say that there are no occasions when a benefit/risk analysis may not validly opt to accept serious dangers. The swine flu vaccination program may be an example: In a massive program designed to protect millions from the threat of illness and death, unforeseen complications such as the 10 recent deaths are to be expected. But the case of recombinant DNA technology is altogether different. The risks, however they may be minimized, are worldwide and terrifying in their scope, while the benefits are speculative and in fact their value or uniqueness may be questioned. A careful and dispassionate evaluation of the facts is called for, and has not yet taken place.

Why not? Primarily because an important element has been omitted from the deliberations concerning the level of risk that may be considered tolerable in pursuing the aims of the research. The evaluation of benefits and risks and the guiding rules of the game have been made by scientists, with only token input by non-scientists. Is it not the public to whom the results, good or bad, will accrue? It is presumptuous of the scientific community to assume that its interests coincide with those of society as a whole. Furthermore, in acting on this assumption they have pre-empted a public function, that of decision-making where its own fate is concerned. Scientists obviously must play a role in this process, but not the only role.

The next point I wish to discuss deals with the hazards of success. Let us imagine for the moment that scientists will be in fact successful in producing plants which will yield more food and faster; and in producing variously designed microorganisms and so forth. This will inevitably result in a change in the balance of our natural environment, an intricate balance which has taken millions of years to establish. Are scientists so pretentious as to think they can foresee all of the variables and control all of the interactions necessary to do this successfully? I hope not. Because of the unplumbed complexities of nature, success in specific genetic manipulations carries with it the hazard of devastating nature and defacing the earth. These matters require much more careful contemplation than they have been given.

If recombinant DNA technology is permitted to proceed, its application to genetic engineering will most certainly follow. Experiments along these lines are already on the drawing boards. The manipulation of animal and eventually human embryos may one day be a reality, all in the name of improving the human situation. Not only the environment of future generations, but their very being may be determined before we have the knowledge and the wisdom to know what we are doing. From an ethical point of view, do we want to accept this heavy moral responsibility?

It can be argued that the Wright brothers who invented the airplane, affected future men because supersonic aircraft pollute the atmosphere with noise and noxious gases. It can be argued that the inventors of the internal combustion engine affected future man because about 50,000 people die each year in this country in automobile accidents. These counter-arguments are not germane to the present issue. Those earlier days were not filled with sophisticated technology, and we may not indict those men who could not have predicted the consequences of their inventions. But as a modern scientist, I know that DNA, as my colleagues do know, that DNA is a life substance, that DNA is a synonym for genes, and that genetics determine the nature of life. I would be fooling myself and you if I said that I cannot know what will happen. I cannot predict in detail, but I can say with certainty that because of the complexity of nature and the limitations of man's knowledge, a headlong rush into the unexamined territory of genetic manipulation is sure to result in catastrophe.

The proponents of recombinant DNA research tend to give the impression that there are no other paths to the knowledge they seek. This is not true. There are always more avenues to biological and medical knowledge than our limited resources will permit us to explore, and most of these appear to be free of the serious hazards that are inherent in the very process of research on recombinant DNA.

The issue can be viewed as a problem of priorities. We have at our disposal, a potential for biological research which can help to solve some of the pressing problems of humanity. It is up to society to see to it that the research effort is directed toward the solution of major problems and not to the production of new problems. This means that society at large, and not the scientists whose vision is limited by their specialization, must evaluate major research directions and choose among them. This need has been apparent for some time, but is particularly pressing in the case of recombinant DNA research, not only because its drawbacks are so clearly palpable, but because it is the beginning of a new technology, one with vast implications for life on earth, but one in which we have as yet no stake. We can still choose. If the process of evaluation and choice takes time, biological research need not stand still, but can press ahead in the many directions that are not so heavily charged with unexplored and unpredictable implications.

What can we do to make it possible to reach a socially responsible decision? The first thing, I think, that we should do is to call a halt to this research in order to provide time for a broad based study of all of its implications. Representatives from all segments of society should be involved in this effort. I urge that New York State impose a moratorium on recombinant DNA research involving the crossing of genetic barriers, and I hope that New York State will take the initiative in creating a commission to study the question. Such an idea has already been proposed at the national level. Let me read a statement concerning a proposal made by Senator Mondale, according to one of his aides. Incidentally, this is not a plug for the Carter-Mondale ticket. It was made about six or seven years ago.

It says it would establish a two year study commission with 15 members appointed by the President from a broad variety of disciplines. The commission would study the ethical, social and legal implications of advances in bio-medical research and technology. It would make full use of relevant studies conducted by other public and private groups. After two years it would report its findings and conclusions to the President and to the Congress. Its final report would include such recommendations for actions by public and private bodies and individuals as it deems advisable.

I believe that a commission or other deliberative body of this sort is urgently needed both nationally and internationally. A real positive effort by New York State would in my opinion have a strongly catalytic effect in bringing about a resolution to this serious problem.

MS. FEINBERG: I have one question. I'll make it brief, because we're running out of time. You mentioned the hazards of recombinant DNA research, but you didn't mention whether or not NIH guidelines provide safety measures to greatly reduce these hazards.

DR. CAVALIERI: I think they do minimize the hazards.

MS. FEINBERG: You think they do?

DR. CAVALIERI: Yes

MS. FEINBERG: Do you have any problem with these guidelines or are they adequate as they stand?

DR. CAVALIERI: I think they evade the issue. That's my personal opinion. I think they evade the issue. I think that the research should be not continued, at least in cases where one is dealing with the crossing of genetic barriers. I'm talking about a moratorium and I think it should continue. I don't think the guidelines do not minimize risks. They certainly do, and they have been thought out very well on that technical basis. I think it's the wrong basis to address the issue. It's too narrow a basis.

MS. FEINBERG: Thank you.

MR. WEINBERG: Your position is in effect that no guidelines would be adequate because the research itself is so fraught with danger that it is not clearly a question of containment procedures, but of the research itself.

DR. CAVALIERI: If pressed, I would say that some research could continue under very restrictive conditions and in restrictive areas and with some specific problems in hand. I object, I think, mostly to having everyone doing everything which is essentially the way it is and the Guidelines will not stop that, and incidentally, the question of industry where Guidelines will not apply, is terrifying. We have no control over what they can do.

MR. BERGER: Thank you, Dr. Cavalieri.

Document 5.9

September 1976
The Sciences (**16**:6)

The Case Against Genetic Engineering

by George Wald

During hearings before the Cambridge, Massachusetts City Council, Harvard biologist George Wald — among others — testified in opposition to performing genetic recombination research at Harvard University. Proponents of the experiments included Harvard scientists Matthew Meselson and Mark Ptashne and MIT Nobel prize-winner David Baltimore. Despite the fact that the NIH had issued its voluntary Guidelines days earlier, permitting such research to go on under special laboratory conditions, on July 7 the city council voted a three-month recombinant DNA research moratorium to study the issue further. In this article, Nobel laureate George Wald outlines his objections to continuing genetic recombinant DNA research at Harvard, even under the restrictions imposed by the NIH Guidelines.

Photograph copyright 1977 Eric A. Roth.

Recombinant DNA technology faces our society with problems unprecedented not only in the history of science, but of life on the Earth. It places in human hands the capacity to redesign living organisms, the products of some three billion years of evolution.

Such intervention must not be confused with previous intrusions upon the natural order of living organisms: animal and plant breeding, for example; or the artificial induction of mutations, as with X-rays. All such earlier procedures worked within single or closely related species. The nub of the new technology is to move genes back and forth, not only across species lines, but across any boundaries that now divide living organisms, particularly the most fundamental such boundary, that which divides prokaryotes (bacteria and bluegreeen algae) from eukaryotes (those cells with a distinct nucleus in higher plants and animals). The results will be essentially new organisms, self-perpetuating and hence permanent. Once created, they cannot be recalled.

This is the transcendent issue, so basic, so vast in its implications and possible consequences, that no one is as yet ready to deal with it. We can't deal with it until we know a lot more; and to learn those things we would have to venture out into this no-man's land. It is nothing like making new transuranic elements. New elements only add to the simple series of integral atomic numbers that underlie the Periodic System. Their numbers are limited and their properties highly predictable. Not so new organisms. They can be as boundless and unpredictable as life itself.

Up to now living organisms have evolved very slowly, and new forms have had plenty of time to settle in. It has taken from four to 20 million years for a single mutation, for example the change of one amino acid in the sequence of hemoglobin or cytochrome c, to establish itself as the species norm. Now whole proteins will be transposed overnight into wholly new associations, with consequences no one can foretell, either for the host organisms or their neighbors.

Technologically Redesigning Living Organisms

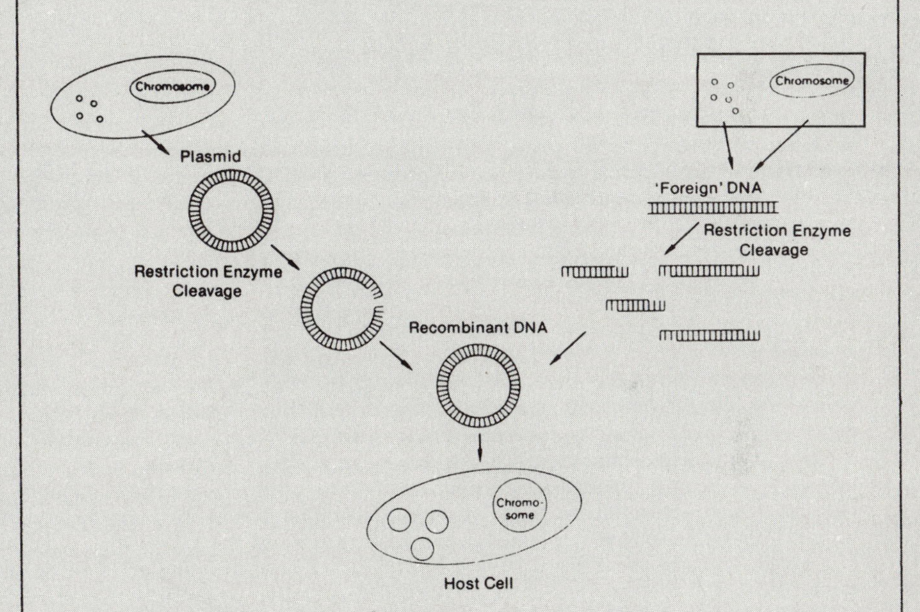

Recombinant DNA technology was launched in 1973 and 1974, largely through researches carried out in the laboratories of Stanley Cohen at Stanford University and Herbert Boyer at the University of California in San Francisco. A rapidly growing number of available restriction enzymes can be used to cut short specific segments of DNA usually containing several genes out of the chromosomes of any type of cell. These segments are then spliced with the help of the same and other enzymes, ligases, into viruses or the naturally-occurring small circular extrachromosomal particles of DNA called plasmids. The plasmids can then be taken up by bacteria or animal or plant cells in which they reproduce, either in phase with the host cell or sometimes independently and many times faster. On occasion, the new genetic material fuses with the host chromosomes and reacts thereafter as a normal component of the host's genetic apparatus. In effect, such cells that have received foreign genes are new organisms, permanent hybrids of the host cells and whatever organism donated the transplanted genes. Their properties and capacities may differ profoundly from either host or donor.

—*George Wald*

It is all too big, and is happening too fast. So this, the central problem, remains almost unconsidered. It presents probably the largest ethical problem that science has ever had to face. Our morality up to now has been to go ahead without restriction to learn all that we can about nature. Restructuring nature was not part of the bargain; nor was telling scientists not to venture further in certain directions. That comes hard. With some relief, most biologists turn away from so vast and uncomfortable an issue and take refuge in the still knotty but infinitely easier technical questions: not *whether* to proceed, but *how*. For going ahead in this direction may be not only unwise but dangerous. Potentially, it could breed new animal and plant diseases, new sources of cancer, novel epidemics.

We must never forget that the first intimation of these potential hazards came from workers in this field. All honor to them. Faced with unique problems, as they alone then realized, they did unprecedented things. They brought about a voluntary moratorium on certain, more clearly dangerous kinds of experiments. And now, after three years of debate, consultation and negotiation, the National Institutes of Health issued its Guidelines on June 23.

The Nobel Letters

During the hearings a number of leading biologists wrote letters to Cambridge Mayor Alfred E. Vellucci defending genetic recombinant research at Harvard. Below are excerpts from letters by three Nobel laureates.

In my view these Guidelines are far more stringent than is reasonably necessary for the protection of public health. In every case where reasonable doubt could be entertained, it has been resolved in a way that imposes the most serious and conservative protective requirements. Most of the risks in question are purely conjectural and no substantive basis can be found for the dire prediction that the public health could be endangered by recombinant organisms. Nevertheless, the Guidelines in their present form have accepted every such speculation as if it were accepted reality. In summary, even the most cautious view of the NIH Guidelines should give citizens ample assurance that they go far beyond what is necessary to protect their health.

Elsewhere I have commented that the very act of setting up such elaborate precautions would frighten people because they go so far beyond what we do in other spheres of life. This seems to have happened in the present case—it is the very security precautions having been doubled and redoubled that has generated an unjustified fear. On the other side of the coin, I take the opportunity to indicate that research in this area has the potential for the most extraordinary contributions to medical advance and I would hope that Cambridge, Massachusetts would be proud to be the seat of major accomplishments in this direction.

Joshua Lederberg
Stanford University Medical School

In terms of our present knowledge, I feel that there are no real accidental dangers involved in research on animal virus and vertebrate cell DNAs under the NIH Guidelines. The specific dangers that have been suggested involve combinations of events that are either known not to occur or occur only at very low probabilities. Therefore, the likelihood of the occurrence of any specific danger is so low that it can be considered zero. In fact, I consider that the Guidelines are probably too restrictive in terms of our present knowledge of animal virus and vertebrate DNAs.

Furthermore, I consider it ineffectual to regulate on a local level research involving possible infectious entities. Unless there is national, and preferably international, regulation, local regulation would not serve to protect the inhabitants of that locale.

In addition, I have found the members of the Department of Biochemistry and Molecular Biology, Harvard University, conservative in respect to possible safety hazards from research with animal viruses and vertebrate cells.

As taxpayers and governmental officials, you have a responsibility to insure public health and safety, but you also have a responsibility to promote the public welfare. It is conceivable that the technique of recombinant DNA may lead to major benefits in terms of public health and welfare. Therefore, a balance must be made between the "zero" likelihood of harm and the possibility of beneficial results.

Howard M. Temin
University of Wisconsin Medical Center

I implore you to encourage the progress of the planned facilities for genetic research at Harvard and to do your utmost to foster a spirit which advances this exceedingly important direction in medical science.

The new NIH Guidelines to which these Harvard facilities and investigators will adhere go far beyond reasonable needs for personal and public safety. I assure you that the current hazards in many chemical, bacteriological, biological and physical laboratories in Cambridge, public and private, are far greater than those anticipated in recombinant DNA research.

I realize you have heard a different point of view from some Harvard and MIT scientists who have testified before you. I believe their views are not based on sound scientific evidence and are highly exaggerated. In my estimation, they represent a tiny fraction of the scientific community.

I implore you again not to suppress the serious and responsible search for new knowledge. If scientific inquiry is stifled in Cambridge, it will be done in Waltham, Palo Alto or Moscow. In 1976, please do not squander your most precious human resources.

Arthur Kornberg
Stanford University Medical Center

> **"And God blessed Noah and his sons, and said to them, 'Be fruitful and multiply, and fill the Earth. The fear of you and the dread of you shall be upon every beast of the Earth, and upon every bird of the air, upon everything that creeps on the ground and all the fish of the sea; into your hand they are delivered. Every moving thing that lives shall be food for you; and as I gave you the green plants, I give you everything.' "**
>
> *—Genesis*

One can hardly read the Guidelines, or the careful and sensitive statement by Donald Frederickson, the Director of NIH, on releasing them, and not be impressed with the goodwill and concern that animate them. Yet there is much in this enterprise and in the Guidelines themselves that troubles me greatly.

First and foremost: the very existence of the Guidelines begs the central question, whether this kind of research should proceed at all. The experiments are quite simple and straightforward. Can they be stopped? Perhaps they can. If one could neither publish the results nor exploit them commercially there would be little incentive to do them.

As for the Guidelines themselves, the first thing to understand is the context of utter ignorance of what to expect in which they had to be formulated. The Guidelines begin by saying: "At present the hazards may be guessed at, speculated about, or voted upon, but they cannot be known absolutely in the absence of firm experimental data — and, unfortunately, the needed data were, more often than not, unavailable."

Physical containment. The purpose here is to keep the recombinants from escaping the laboratory. The Guidelines list four levels of containment labeled P1 to P4; but in effect there are only two levels, a lesser—P3—and a greater—P4. This classification is itself deceptive, for it makes the prevalent P3 facility sound better than it is, three quarters of the way to the top, whereas in fact it is the lowest level of containment. P1 is just a laboratory, P2 the same laboratory with a warning sign on the door. A young woman demonstrating a P2 experiment at an open hearing before the Cambridge City Council made a point of putting on the prescribed laboratory coat; but she had long, loose, abundant hair that could have carried more bacteria or viruses than a dozen lab coats.

A P3 facility such as has just been authorized at Harvard employs various devices intended to minimize the escape of recombinants. Yet the reason proponents of the facility at Harvard gave for building it within our Biological Laboratories, close to the laboratories of prospective users — though the building is half a century old and infested with ants and cockroaches — was that workers in the facility would be the principal means of spreading contamination, and hence should have to move as short distances as possible. I think it is probably correct that the laboratory personnel will be the principal means of spreading any potential infection. But in that case, wherein lies the containment? Why the elaborate and costly precautions within the facility? — the small unit at Harvard is estimated to cost more than $800,000. And what matter whether distances between the labs are short or long? All these workers move freely throughout the building and the city; they meet with us, eat with us, and — most importantly — they teach classes of young students. I see no reason to believe that P3 containment, even if conscientiously enforced, can effectively contain.

Biological containment. One of the most unsettling aspects of present recombinant DNA research is that the host organism that receives the plasmids that carry foreign genetic material is almost always the colon bacillus, *Escherichia coli*, a constant inhabitant of the human bowel. To do potentially hazardous experiments, why pick an organism that lives in us? The reason is that we know more about *E. coli* than about any other living organism. Yet what is to keep some hybridized *E. coli* turned pathogenic from infecting its conventional human hosts? Or transferring those plasmids to human cells?

Hence the stress on the assurance that all recombinant experiments with *E. coli* will use the K12 strain, which, we are told, can exist only under special laboratory conditions and neither survives nor reproduces in the human gut. The use of this strain is the "biological containment."

In this connection Stanley Falkow of Seattle, Washington, submitted to the NIH Recombinant DNA Advisory Committee a highly informative report on the ecology of *E. coli*. According to Falkow, almost innumerable serologically distinct strains of *E. coli* inhabit the human colon from time to time, the population constantly changing. The more persistent (resident) strains last several months, other (transient) strains only a few days. The statement that the K12 strain does not survive in the human bowel rests primarily on observations by E.S. Anderson and H. Williams Smith that this strain "is a poor colonizer of the human alimentary tract." Smith found a mean survival time of about three days, Ander about six days. Anderson also found that it "multiplied to some extent in two of eight subjects." Hardly an impressive statistic! Furthermore he could detect plasmid transmission from K12 to other enteric flora when it was fed "in substantially high numbers."

Falkow confirms these observations, and adds another that is singularly important: Working with calves, he found that introducing certain plasmids into K12 increased its survival and multiplication in the gut many times over. He concludes that "it may not be too farfetched to suggest that some DNA recombinant molecules could profoundly affect the ability of this *E. coli* strain to survive and multiply in the gastrointestinal tract."

These are oddly inadequate data to carry such weight. We would like to know much more. How does K12 get along in persons whose colons are relatively empty of bacteria and hence offer it little competition? — such as newborn infants, or persons who have just been treated with sulfa drugs or antibiotics? So-called biological containment seems to me as problematical as P3 physical containment.

Enforcement. The Guidelines are just that, hence wholly voluntary. The only penalty now available for simply disregarding them is the possible withholding of federal research support. Obviously this applies only to research dependent on federal funds. It leaves out completely the rapidly growing industrial exploitation of recombinant DNA technology.

Benefits and risks. I have up to now said almost nothing of the potential benefits of this technology. I think that the most certain benefits to come out of it would be scientific: increased understanding of important biological phenomena, such as the mechanisms that turn specific gene activities on and off, that trigger cell multiplication and differentiation, that regulate cell metabolism. We are also offered the prospect of large practical benefits: teaching cereal plants to fix their own nitrogen from the air, new bacterial syntheses of drugs and hormones, the hope that increased understanding of cancer may lead to its cure. I cannot think of a single instance of such developments, scientific or practical, that does not also involve large potential risks.

Consider cancer. If indeed it turns out that recombinant DNA research will improve our understanding of cancer, that would still be far from showing us how to cure it. In spite of many statements, as vague as they are optimistic, that the cure of cancer lies in this direction, it is hard to see how that is to happen. Any such hope must be balanced against the real possibility that recombinant DNA experiments may induce new cancers. If right now I had to weigh the probabilities of either event I would guess that recombinant DNA research carries more and earlier risks of causing cancers than hope of curing them.

Add that about 80 percent of cancer in this country is now believed to be of environmental origin. The largest single cause of lung cancer is smoking, but one is free to smoke or not. About 40 percent of those environmental cancers happen in the work places, through involuntary exposure to a rapidly increasing variety of toxic materials in industrial use. If one were really concerned about cancer, there is the obvious place to attack it, with sure and immediate results.

Or consider a frankly industrial development. General Electric is reportedly trying to patent a newly assembled strain of *Pseudomonas* bacteria that can wholly digest crude oil. It was developed there by Ananda Chakrabarty by transferring plasmids from several strains, each of which could digest oil partially, into a single strain that can do the whole job. It is pointed out that this organism could be very useful for cleaning up oil spills. Very true; but how about oil that has not spilled? — oil still in the ground, or on the way, or stored? Can this organism be contained, kept from destroying oil we want to use? Or will we need to begin to pasteurize oil?

The corporate connection. As early as February 1974 *Fortune* magazine hailed the coming importance of genetic recombination in industrial developments. "The best microbes are freaks," it said and "many scientists see an important industrial role ahead for the powerful new methods of transferring genetic material from one cell to another." It named a number of them, including a few who are already directing corporate activities.

The industrial exploitation of recombination technology raises special problems, for in that, as any other business enterprise, the major goal is to maximize profits and, frequently in the past, public and worker safety and health have been subordinated to that end. Last May representatives of about twenty drug and chemical companies met with NIH Director Frederickson to discuss the proposed Guidelines. They expressed "general support," but made three points: (1) the fear that voluntary Guidelines might lead to enforceable regulations, (2) for reasons of competition, the companies could not afford to reveal what recombinant DNA experiments they were performing and (3) they found other features of the Guidelines onerous, for example the restrictions on large-volume experiments, which of course are less easily contained, but which they require in testing procedures for commercial feasibility.

The dilemma of the NIH. The recombinant DNA development faces NIH with an interesting predicament. Anything I say of this is said sympathetically, for under Donald Frederickson's perceptive leadership it is doing as well as could be hoped. Yet is it possible for the same agency both to promote and regulate? The old Atomic Energy Commission, set up originally to regulate, turned instead to promoting nuclear power, and that eventually destroyed it. It has been replaced by two separate agencies, one for research and development, the other for nuclear regulation.

NIH, on the contrary, set up to promote scientific and medical research, is now being forced into regulation. Its entire impulse, as that of all other institutions concerned with research, is to avoid regulation, to maintain full freedom of inquiry. Probably that is why it can bring itself only to promulgate voluntary guidelines. Surely it recognizes the previous history of ineffectuality of voluntary self-regulation in other areas. For the NIH Guidelines to be enforced, academically and particularly industrially, they would have to become regulations, backed by legislation, with adequate provisions for licensing, inspection and supervision. The NIH would like to avoid such measures and so, as a scientist, would I. Yet this situation seems to demand them, and I fear that scientists and science will eventually have to suffer because of them.

What to do. First, I think it essential to open a wide ranging and broadly representative discussion of the central issue: whether artificial exchanges of genetic material among widely different living organisms should be permitted.

Second, in consideration of the potential hazards and our present state of ignorance, I would confine all recombinant DNA experimentation that transcended species boundaries to one or a few national or regional laboratories where they can be adequately confined and supervised. There, every attempt should be made to define the hazards that are now only guessed at. If trouble should arise, I would expect it to involve first the workers in such laboratories and their families whose health should be carefully monitored. Until such trials have told us better what to expect, this kind of investigation should have no place in crowded cities or educational institutions.

Third, industrial research and development in this area need most of all to be brought under control. The usual secrecy that surrounds industrial research is intolerable in a province that can involve such serious consequences and hazards. The need for licensing, inspection and supervision will probably require national legislation. Hearings in the Congress should begin at once to consider these issues.

As I write these words, they trouble me greatly. I fear for the future of science as we have known it, for humankind, for life on the Earth. My feelings are ambivalent, for the new technology excites me for its sheer virtuosity and its intellectual and practical potentialities; yet the price is high, perhaps too high. We are at the threshold of a great decision with large and permanent consequences. It needs increasing public attention here and worldwide, for it concerns all humankind. That will take time, during which we can try to learn, as safely as that can be managed, more of what to expect, of good and ill. Fortunately there is no real hurry. Let us try, with goodwill and responsibility, to work it out.

"Crack out the liquid nitrogen, dumplings . . . we're on our way." 2/9/77

Reprinted courtesy of The Boston Globe.

Document 5.10

November 17, 1976
Carstens Letter to Watson

A. S. Carstens
5971 Avenida Chamnez
La Jolla, California 92037

Telephone
714-454-1847

November 17, 1976

Dr. James Watson
Cold Spring Harbor Laboratory
P. O. Box 100
Cold Spring Harbor, N. Y. 11724

Dear Dr. Watson:

We enclose one copy of our "package" which we have developed with Dr. DeNike and distributed quite widely. We do not believe that strengthening the Guidelines is a practical possibility, therefore, we will press forward to enact the Recombinent DNA Pre-exemption Act before the next Congress.

We expect to put this into proper legislative form with proper sponsorship.

Scientists in support should make their views known at the coming Hearings. We are in the process of attempting to form a real Environmental Coalition consisting of the Sierra Club, Friends of the Earth, Environmental Defense Fund, National Resources Defense Council, Wilderness Society,Audubon Society, National Wildlife Federation and other national groups.

We all admire your courage in standing up for your views and regret that the vast majority of scientists who might also help, do not have the courage to do so nor the vision the realize the ultimate consequences.

Sincerely yours,

A. S. Carstens

A. S. CARSTENS

ASC :rm

Hypothetical Scenarios Illustrative of Recombinant DNA Hazards

By: L. Douglas DeNike, Ph.D.

SCENARIO I.
A scientist, in a news conference, develops the following argument: "Right now, it is perfectly legal for anyone to develop, multiply and release to the environment the most deadly recombinant DNA-bearing organisms imaginable. Only if malice could be proved would there be any legal case against such a person. Moreover, we have no idea how easy or difficult it would be to do this, and the government has no program to ascertain said ease or difficulty. If recombinant DNA technology is allowed to proliferate in the absence of this information, and we then discover that making 'doomsday bugs' is relatively simple, then humanity is doomed. We must, therefore, know the worst *soon.* Therefore, I encourage molecular biologists to *activity engage in the attempt to fabricate exceedingly dangerous recombinants,* to give this research priority, and to widely publicize their findings."

SCENARIO II.
A noted molecular biologist of unquestioned competence, privately funded, holds a news conference and says, "I intend to devote my primary attention to the development of (1) a botulism-producing strain of *E. coli* viable in the human digestive tract, (2) a variety of *Pasteurella pestis* resistant to all widely-used antibiotics, (3) a plant rust capable of attacking all major grain crops, (4) a type of foot-and-mouth disease capable of eliminating cattle as a useful domesticated species. Every attempt will be made, within budgetary constraints, to keep these novel organisms and their prototypes in tight confinement. However, if they escape despite all precautions, then it is better for humankind to know the worst sooner rather than later."

SCENARIO III.
A molecular biologist stands before media microphones a few months later, and announces, "I have succeeded in implanting the gene for tetanus toxin into the currently common influenza strain, A-Victoria 75 virus. Extensive laboratory tests indicate that this modified microbe, if released, would be greater than 90% lethal during the next flu season, and might subsequently mutate into types essentially 100% lethal. Therefore, I ask the public for advice as to what I should do with the virus and the laboratory notes describing its development. Needless to say, the military and the CIA have importuned me for exclusive rights. Thus far, by means I prefer not to describe, I have kept them from pressing their demands. I have also been approached by a few fanatical population-controllers. They seem to believe in 'Super-Flu for the Super-fluous'."

SCENARIO IV.
Yet another molecular biologist, in his acceptance speech for the Nobel Prize, says, "Oh by the way. All of the nations now possessing nuclear weapons will totally dismantle and destroy them within six months. If not, then my lysogenic versions of the common rhinoviruses will make it fatal to contract a common cold. Sorry to be so blunt about this, old chaps, but the physicists' preparations for Armageddon have been going on for too long. It's time for some rationality, you know. In case you're wondering, the cultures and release mechanisms have already been distributed worldwide, and nothing which supervenes—other than general and complete nuclear disarmament—will deter their subsequent dispersal . . ."

SCENARIO V.
The head of a central African state, over satellite television, makes the following apology: "Past threats to our sovereignty made necessary the secret development of a deterrent 'ethnic weapon' germ which, if released, would cause eventual cancer in 96% of Caucasians. Now, we must apologize. It appears that a laboratory worker's error has resulted in the irreversible spread of this microbe among the local monkeys. There is no cause for immediate alarm, however, since the highly metastasizing tumors should not appear for about eight years . . ."

S. Harris
"CONGRATULATIONS— YOU'RE THE FIRST VICTIM OF RECOMBINANT DNA."

Sidney Harris © 1978. (This cartoon originally appeared in *Johns Hopkins Magazine*.)

Document 5.11

January 7, 1977
New Times (**8**:48), Excerpt

Genetic engineers are splicing together cells to create new forms of life. Are they on the track to a cancer cure—or on the path to biological disaster?

PLAYING GOD WITH DNA

By Arthur Lubow

It would be so much simpler if they would just say there's no risk. No danger of infection, no way a man-made microbe can crawl out of a lab and start an epidemic, no chance that manufacturing new organisms will derail the locomotive of evolution. So you ask hopefully: "Is there any risk?" And instead of an answer, you get an analogy.

"There is a possibility that a band of monkeys could type out the Preamble to the Constitution," says a Harvard Medical School professor. "But it is such a small probability that we don't even have numbers to express it." The monkey analogy is a good one. It appeals not to your knowledge of the simian mentality, but to your common sense. Who knows? Maybe if you put a dozen monkeys in Independence Hall, they would feel inspired. But you give the professor the benefit of the doubt; and he hopes you will be equally generous about his work with recombinant DNA. When he splices together two cells to create a new living organism, he is confident that nothing sinister will happen. Since you're watching over his shoulder, he wants you to be confident too.

An amateur celebrity, the researcher is nervous: he blinks like a bat in the light. He is accustomed to obscurity. Thousands of whitecoated scientists work quietly in their laboratories, writing articles with polysyllabic titles for arcane journals—and most of us could not care less. "I am not much interested," said a former secretary of defense, "in why potatoes turn brown when they are fried." But every once in a while a paper marked with chicken scratchings explodes with the power of a thousand suns. We look at the mushroom cloud over Hiroshima, and suddenly we want to know everything. A scientific question has become a political problem, and the man with a pencil or test tube is suddenly a public personage. The recondite paths of nuclear physics led to the perilous crossings of atomic power. And now molecular biology, the most exciting frontier of modern science, has brought us to a similarly troubling vista.

For the first time, the scientific establishment has announced that there are some nonhuman experiments that should be prohibited. For the first time, a community—Cambridge, Massachusetts, the home of Harvard and MIT—has declared that it will not permit certain scientific work to proceed within its borders. "People want to know what the hell they're doing in these laboratories," says the mayor of Cambridge. "And since they're spending United States tax dollars, we have a right to know." A scientist warns in the *New York Times Magazine* that recombinant DNA work may "endanger the future of mankind." Scientists have formed two camps, each savaging the other's intelligence and integrity.

The controversy is flaring on the front lines of science: molecular biology. Since the 1950s, molecular biologists have scooped up the lion's share of Nobel prizes in medicine and physiology. Buckets of federal money have irrigated the field. Compared to its sisters, chemistry and physics, biology is still an infant, but in the last two decades it has grown faster than the hyperpituitary boy in a Wonder Bread commercial. Scientists now have a fair understanding of the genetic process in primitive creatures such as bacteria, but they have never had a way to examine the setup in animals similar to man. They now have a way. It is called recombinant DNA.

"It is a revolution," says James Watson, who won a Nobel prize for his work on the structure of DNA. "It is *the* way to go. I would say that everyone who is seriously interested in the detailed structure of the gene is now using this technology." Just as splitting the atom accelerated the progress of physics, so recombinant DNA can speed the growth of modern biology. Researchers say their work may even disclose a cure for cancer. But theoretical risks accompany potential benefits, and the threat of recombinant DNA has been compared to the danger of nuclear radiation. Not since the congressional investigations of atomic energy in the fifties has science sparked such a heated political discussion. Scientists all over the world are watching the Cambridge saga. They know that the threat extends beyond Harvard and MIT. There is a domino effect. Already the United States Senate, the New York State Attorney General and the cities of San Diego, Princeton, New Haven and Bloomington, Indiana, are pondering the public health dangers.

Not only molecular biologists are worried. The mayor of Cambridge is turning his sights on other laboratories. The NIH (National Institutes of Health), which drafted guidelines for work with recombinant DNA, is considering similar restrictions on all experimenters with animal cells. A chink in the form of a question mark has cracked the scientific megastructure. There hasn't been such a fateful biological question since the Sphinx posed the riddle on the road to Thebes. But among all these scientists there is no Oedipus, and the question lingers unanswered in the air.

Is there any risk?

For the crucial questions about recombinant DNA, we have very few answers. The scientific technique, though, is disarmingly simple to understand. The genetic code for every form of life is contained in a giant molecule called DNA. The DNA molecule or a close relative is found in all cells, from amoebas to zebras. Of course every cell is full of molecules, but DNA is the king of the roost. It decides which proteins a cell will synthesize. By manufacturing the proper proteins, a cell can perform its individual function—producing insulin, transmitting nerve impulses, growing a blond hair. It is the DNA that decides whether a cell will be part of a man or a mushroom.

In molecular terms, sexual reproduction is merely a mingling of DNA. Each parent contributes half a DNA set to form a new cell. Obviously, the new cell is a recombination of the parents' DNA; but the word "recombinant" is reserved for something far less common—the recombination of DNA from different species.

The mouse who falls in love with an elephant is a likely subject for a droll fable. Everyone knows that animals from different species can't mate. In some cases, closely related species are able to have a go at it, but often the offspring are sterile: for every nectarine, there is a mule. But in biological laboratories, modern Dr. Frankensteins have found a way to create brand-new forms of life. Perhaps it's a sign of the times that the new creatures are not ten-foot monsters with spokes through their necks, but colorless cells invisible to the naked eye. The DNA of mice and elephants can now be joined—inside bacteria.

The DNA molecule in bacteria usually takes the form of a chromosome ring. However, many types of bacteria have snippets of DNA left over, which appear as smaller rings in the cell. These rings are called plasmids. When bacteria reproduce asexually, both the chromosome and the plasmids duplicate themselves, the cell divides, and *voilà!* Two cells. With some bacteria, this happens every 20 minutes.

A modest bit of merchandise, the plasmid, but it boasts two very useful features. It can often move from one cell into another. And it can be split in specific places by substances called restriction enzymes. For the recombinant DNA maestro, these properties make it worth considerably more than its weight in gold.

Using restriction enzymes as chemical cutters, a scientist can split the plasmid DNA and the DNA of some other organism—say, a mouse. He can then take a piece of the mouse DNA and insert it into the plasmid, seal the plasmid up with other enzymes, and put it back into the bacterium. When the bacterium divides, so does the plasmid, including the mouse

Photo by Rick Stafford, Stock, Boston.

Mark Ptashne: Speculation that researchers will create Andromeda-type germs is "not scientific talk—it's science-fiction talk."

DNA portion. In a day, the scientist can have millions of pieces of the mouse DNA he wants to examine.

But every silver lining has its cloud. Bacteria are wonderful reproduction machines, making limitless quantities of identical mouse, cat, fruit fly or other kinds of DNA to order. They are so useful because, even after you perform recombinant surgery on them, they are still free-living organisms. As it happens, the particular bacteria that scientists use for this transaction, *E. coli*, live quite happily in a wide range of locales, including the place they're named for, the human colon or bowel. If swallowed accidentally, they might take up residence in the human body. And so they pose—a risk.

In one dire vision, conjured up by critics of recombinant DNA research, some *E. coli* bacteria carrying foreign genes escape from the laboratory and perform untold mischief in the human body. To honor Michael Crichton, in whose novel, *The Andromeda Strain,* bacteria from the upper atmosphere wreak bloodcurdling havoc on earth, we shall call this the Andromeda-strain scenario. In another more mystical vision, the introduction of new forms of life upsets the master plan of evolution. Of special concern is the mongrelization of higher life forms with primitive organisms like bacteria. We shall call this the evolutionary-disaster scenario.

Our heads filled with nightmares, we run to the scientists for some answers. To our chagrin, we discover a Catch-22: the riddles posed by recombinant DNA technology can be solved only by experimenting with the technology. There are too many unknowns to give a flat answer. On the Andromeda-strain scenario, Charles Thomas of Harvard Medical School is skeptical: "The idea that one could paste together two segments of DNA so that that act would simultaneously confer a selective advantage on that organism and have it be dangerous is something I can't buy." Across the river at Harvard College, Mark Ptashne offers an extended metaphor: "If you tell somebody from Mars you're going to put a Ford motor on a lawn mower, he asks, 'Can it be dangerous?' You say it may be. He says, 'It may destroy the world.' You say that is very unlikely, but you can't prove it won't. It's not scientific talk. It's science-fiction talk. Every time you make a mutant of *E. coli*, you can't be sure what will happen, but all attempts to make the strain virulent have failed."

Talk about evolutionary disaster is even less concrete. "I don't think we are doing something which has never happened before," says James Watson. "But it's impossible to say. It's not a meaningful question." Unlike Watson, David Hogness of Stanford, a leading researcher in the field, does not dismiss the question. He argues that recombinations *are* part of evolution. He challenges the notion that, in nature, genetic material does not leapfrog across species barriers. He thinks there are natural mechanisms by which animal DNA can enter a bacterial chromosome: "People started in this field with the assumption that we were doing something new that had never been known in the world, but I no longer think that is true." Another pioneer in recombinant DNA work, Paul Berg of Stanford, is more forthright: "There isn't a single person that I know of who can document the existence of a barrier."

And so, in the absence of facts, the debate turns to personalities, to my word against yours, to imaginative scenarios. The debating platform moves from the musty pages of medical journals to the strobe-lit hearing rooms of politicians. And in the commotion, in the clouds of smoke above hypothetical fires, an uncrackable nut of a question remains unanswered.

Is there any risk?

Around Harvard Square, Cambridge Mayor Alfred E. Vellucci is known as the man who wanted to convert Harvard Yard into a parking lot. Just the thought of an ax touching one of those majestic elms is enough to choke the throat of any crimson-blooded alumnus. Few men besides the fun-loving mayor recall the birth of this legend. "In 1957, student cars were everywhere, and the people of Cambridge had no place to park their cars," Vellucci recounts in his thick Boston accent. "I wrote to Harvard and MIT and got no answer. I wrote again. No answer. So I proposed to the city council that we cut down the trees in Harvard Yard, pave it and turn it into a parking lot. You should have seen how fast they moved! They got their cars off the street immediately."

After the parking affair, Harvard appointed a special representative for community relations, but the relations are still not so good. "Oh, you know," Vellucci says, "like a couple of lovers—off again, on again." The mayor and Harvard are fighting these days over the construction of a medium-containment laboratory to be used, among other purposes, for work with recombinant DNA. The building would be nearly completed had Vellucci not read a June article in the Boston *Phoenix*, a local weekly. In the article, entitled "Biohazards at Harvard," professor of biology Ruth Hubbard was quoted as saying that the risks of recombinant DNA are "worse than radiation danger."

Vellucci read the article on the Monday morning it appeared. While he was discussing it with his aides, a secretary announced that George Wald and Ruth Hubbard were outside. "They came in and they talked to me about the thing and it was dangerous," Vellucci recalls. "So I said to them, 'If I call a special meeting of the city council to discuss this matter, will you come?' And they said, 'Yes.'"

Dr. Wald remembers it a little differently. A white-haired man with a theatrical personal style (as someone remarked, "When you talk to George, you always imagine there are 5,000 people standing behind you"), Wald won a Nobel prize in 1967 for his work on the visual pigments of the eye. More recently he has been celebrated for his speeches on behalf of the antiwar movement and other left-liberal causes. He and his wife, Ruth Hubbard, are both biology professors at Harvard. Wald says that even before their talk Vellucci had decided to call a meeting. But the Walds were instrumental. They gave the mayor what he needed—"people with credentials." Vellucci explains: "If I'm gonna take a stand against this goddamn thing, I need some people on *my* side. And since they said they would come, I was fortified, I was ready for a meeting, and that is the reason why we then flung the challenge at Harvard and MIT to send *their* scientists over here because I knew *I* had scientists on *my* side!" He pounds the table for emphasis.

As it turned out, there were two meetings, because so many people wished to speak. The meetings were staged before overflow crowds on hot summer nights, made even hotter

Mayor Alfred Vellucci: "I have an obligation to a whole lot of lay people. . . . And so I don't want to take a chance."

by the glare of television lights. In addition to the Walds, Jon King, an associate professor of biology at MIT, appeared as a major spokesman against the work. On what Vellucci likes to call "the Harvard team" there were several Nobel laureates and other distinguished scientists. To open the first meeting, a contingent from the Cambridge public high school sang "This Land Is Your Land." Who's in charge here: citizens or scientists? The point was clear.

"It's about time the scientists began to throw all their goddamned shit right out on the table so that we can discuss it," Vellucci growls. In his office hang framed portraits of past mayors, palpable presences admonishing him to safeguard the health of Cambridge citizens. "Who the hell do the scientists think that *they* are that they can take federal tax dollars that are coming out of our tax returns and do research work that we then cannot come in and question?"

In his cranberry doubleknit jacket and black pants, with his yellow-striped blue shirt struggling to contain a beer belly, right down to his crooked teeth and overstuffed pockets, Al Vellucci is the incarnation of middle-American frustration at these *scientists*, these technocrats, these smartass Harvard eggheads who think they've got the world by a string and wind up dropping it in a puddle of mud. And who winds up in the puddle? Not the eggheads. No, it's always Al Vellucci and the ordinary working people who are left alone to wipe themselves off. "They say, y'know, it's all right to dip apples into red dye, to dip candy into red dye, and don't worry about it, it doesn't affect you at all. And then you have the United States Department of Drugs come out and tell you, no, don't eat any more apples dipped into red dye, y'see, don't eat any more candy that's been dipped into red dye. And everybody's saying, don't take the pill because we're gonna tell you why the pill is no good for ya—y'know, these are all *scientists* who did all this work." Al Vellucci isn't buying any of it.

After the second meeting last July, the Cambridge city council voted a three-month ban on all medium-containment recombinant DNA work. The council later extended the moratorium through mid-December, to provide time for a council-appointed study group to prepare a review and recommendations. The council is closely split on the issue, and no one is predicting its final verdict, but Vellucci has already decided how he will vote. "I have learned enough about recombinant DNA molecules in the past few weeks to take on all the Nobel prize winners in the city of Cambridge," he says. "Regardless of what the review board says, I'm opposed to this work. If they say they're against it, fine, they're against it. If they say 'We're in favor of this thing with safeguards,' it is my intention to call up Dr. Wald, Dr. Hubbard and Dr. King and invite them into this office and have them sit down and go over the whole review with me. I have an obligation to a whole lot of lay people in this city who do not understand this whole goddamn thing. And so I don't want to take a chance."

The controversy over recombinant DNA has a particular poignancy for Paul Berg, because he was the first person to raise the question of risk. Six years ago, in his Stanford laboratory, Berg discovered a way to construct synthetic tails for DNA molecules. By adding these "sticky ends," he planned to insert a given DNA fragment into *E. coli* bacteria. The DNA he wanted to examine, SV 40, has been carefully studied by geneticists because it is able to incorporate itself into the chromosomes of the host cell. It also causes cancer in mice. With his newfound technique, Berg hoped to clone segments of SV 40—to multiply it by using *E. coli* as the host. He never thought of a possible danger: by transforming a cancer virus, he might be manufacturing a mutant with malignant powers.

In the summer of 1971, one of Berg's graduate students was attending a summer program at Cold Spring Harbor laboratory in New York, and she described Berg's experiment to an audience of biologists. The reaction was one of shock. "Several people called me up and pleaded with me not to do it," Berg recalls. "I went to talk to a number of scientists, and they all said not to do it—the unknowns were too great and the risks too significant. I could never convince myself or anyone else that I could reduce the risk to zero. So we turned that experiment off."

Two years later, at a summer conference of biochemists held in New Hampshire, Herbert Boyer of the University of San Francisco described his experiments with plasmids and restriction enzymes. Because natural substances—restriction enzymes—created the "sticky ends" that Berg had laboriously manufactured with synthetic tails, Boyer and others could now easily concoct hybrid plasmids, recombining the DNA of bacteria and other organisms. "It was obvious to everyone in the room that now you could do these things, that you didn't need anything fancy—all you needed were the restriction enzymes," says Berg. The conference authorized a letter, published in the journal *Science,* warning that since new organisms "with biological activity of unpredictable nature" could be created, "prudence suggests the potential hazard be seriously considered."

"Who the hell do the scientists think that *they* are," the mayor growls, "that they can take federal tax dollars that are coming out of our tax returns and do research work that we then cannot come in and question?"

Because Berg had grappled with these issues two years earlier, the National Academy of Sciences, to which the public letter had been addressed, asked him to form a committee. And so in the spring of 1974, a committee of a dozen leading molecular biologists met at MIT and drafted a second public letter. If there is anything in this world that reproduces faster than bacteria it is committees, so, naturally enough, one of the MIT committee's recommendations was the formation of a committee by the director of the NIH to develop guidelines for researchers of "potentially hazardous recombinant DNA molecules." The scientists also called for an international conference to discuss the problem; the conference was held in February 1975 at the Asilomar center in California. In anticipation of the conference, the MIT group asked researchers to refrain voluntarily from two types of experiments: the introduction of toxin-formation or antibiotic-resistance genes into bacteria that do not naturally contain such genes, and the addition of tumor virus or other animal DNA into bacteria or viruses—since such hybrid organisms might "possibly increase the incidence of cancer or other diseases."

By the time of the Asilomar conference, some scientists were beginning to have second thoughts. They feared that dramatizing the risks had been a mistake. Joshua Lederburg, a Nobel laureate and one of the giants of molecular genetics, pointed out that the MIT letter implied that scientists could not be trusted to police themselves. James Watson had another worry. "The scientists were a bunch of jackasses for raising the question of danger when they could not quantitate it," he says now. "I became very worried about the whole thing because in setting regulations I couldn't decide whether it was more dangerous to put yeast DNA or drosophila DNA or human DNA into *E. coli,* and I couldn't distinguish the difference. I just found the whole Asilomar thing irrational."

What was created at Asilomar was a graded-risk system of regulation. Some work would be banned outright for now—roughly, the sort of experiments described in the Berg moratorium letter—and other experiments could be conducted only under specified conditions. The threat was that organisms posing unknown dangers could escape from the lab; the procedures thrashed out at Asilomar would confine experiments believed to be potentially hazardous to more carefully controlled laboratories. Two sorts of containment procedures would be used. The first, physical containment, assigned experiments to workplaces with increasing safety provisions, ascending from P1 to P4. A second system of biological containment required for certain experiments the use of deliberately weakened *E. coli* which, even if they escaped, would be unable to survive outside a pampered laboratory environment. With some modifications, these regulations were adopted by the NIH (and later by the National Science Foundation). Although it has no police powers, the NIH provides funds for most of this research and it therefore wields a formidable carrot.

That is how the matter rests today. The NIH guidelines are in use. Most scientists are reconciled to them; a smaller group think they are needlessly strict but grudgingly follow them. However, there is a third group of scientists unhappy with the regulations, and although relatively small in number, they speak in a loud voice. For they think the guidelines are irresponsibly lax, and if experiments proceed, the public health and even the future of the planet are threatened.

In drafting guidelines, the scientists at Asilomar and NIH worried mainly about the threat of an epidemic—the Andromeda-strain scenario. But even with the recommended precautions, some critics fear the experiments are hazardous. Because no one understands the intricacies of cell genetics, no one can say with assurance that an apparently innocuous combination of genes will not produce a virulent strain of *E. coli.* The pathogens needn't be as flamboyant as the fictional Andromeda strain, which congealed the blood into paste. "What worries me," says Jon King of MIT, "are the organisms where you won't identify the symptoms. If you're making a really virulent agent, you'd catch it. We're talking about things that would affect you in very mild ways."

Organisms produced through recombinant DNA work, a scientific group warned, might "possibly increase the incidence of cancer or other diseases"

To pose a hazard once it escaped, an organism must do three things: survive outside the laboratory; beat out enough competitors in the already existing jungle of microorganisms to establish an ecological niche; and produce some dangerous effect, since the great majority of bacteria are harmless. Most scientists believe the odds against the accidental creation of such an organism are inexpressibly great, but Jon King is unconvinced. "Especially in a technological society that is filled with new chemicals, there is a very good chance that there are new niches that can be exploited," he says. "We do not understand why some microorganisms make you sick. These people are saying they can't possibly make a pathogenic organism, but they can't even tell the difference between existing pathogenic and nonpathogenic organisms. It's an argument you make because you want to refute the possibility of danger without exploring it."

Andromeda scenarists object vehemently to the use of *E. coli*, bacteria that live happily in our gut. "A few years of research would have led to an organism other than *E. coli*," says Erwin Chargaff, a retired Columbia professor who worked on nucleic acid composition in the 1940s. "This is really incredible. It's beyond imagination. Because *coli* is a symbiont of all of us, it is part of us, more than any other organism. Now the argument is we know more about *E. coli* than any other organism. But I think such clever people could learn in five or ten years enough about other organisms."

Experimenters with recombinant DNA stress that they are using a strain of *E. coli* that is not known to survive outside the laboratory. For research that seems especially hazardous, an even weaker variety of *E. coli* is required. "I recognize that *E. coli* represents a risk—we pointed it out in the first letter we wrote—but the insistence that we abandon this research until we find another organism is unrealistic," says Paul Berg. "We may never find another organism. We have an incredible background of information on *coli*. It would be necessary to find an organism that you can predict will be safe when you find it." James Darnell of Rockefeller University argues that because so much is known about it, the laboratory strain of *E. coli* "is not the most dangerous but perhaps the safest organism to use." "I would definitely feed recombinant DNA to my 16-year-old daughter if it were not considered idiosyncratic," says Charles Thomas of Harvard Medical School. "It goes without saying I'd drink it myself."

But disease is not the only concern. Even if no Andromeda-type germ is produced, Robert Sinsheimer fears, recombinant DNA work jeopardizes life on this planet. Sinsheimer, chairman of the Caltech biology department, is the leading spokesman for the prophets of evolutionary disaster, the second scenario. This Southern Californian, tan beneath his white tennis shirt, speaks with the fervor of a repented sinner, animated by the same passions that drove former Communists like Whittaker Chambers to lead the red-baiting crusades of the fifties. Only six years ago Sinsheimer was predicting that man could "rise above his nature to chart his destiny," and that in genetic engineering lay "the potential for worlds yet undreamt." He has had second thoughts.

Sinsheimer is now the most eminent opponent of the guidelines. He fears that even if no virulent recombinant is produced, merely the exchange of DNA among organisms that normally honor barriers as rigid as South African apartheid can have dire evolutionary consequences. In the evolution that Darwin described, new combinations of genes are constantly formed by reproduction within a species. Occasionally, a novel combination or a spontaneous mutation provides a natural advantage to an organism, and that organism is better equipped to survive. It lives longer, it reproduces more often and its genes have a greater likelihood of persisting. Over a time scale of aeons, more organisms with these genes survive, and the species gradually evolves in a particular direction.

Sinsheimer argues that an especially rigid barrier divides the genetic material of lower forms lacking a defined nucleus, such as bacteria, and higher forms such as animals and plants. This taboo is now being broken by man. "Genetic recombination among species has not been allowed in nature," he says. "I'm willing to concede it may happen occasionally, but there's no evidence for it and I have no idea of how to estimate the rate. You are changing in a short period of time the biological equilibrium that has been evolving over a very long period of time. The consequences are unpredictable. If these organisms do reproduce and find a suitable niche, there's no way we can get them back. We can't even find them. And how do we know what will evolve after they're out?"

To a scientist, the intuitiveness of Sinsheimer's argument makes it seem almost mystical. "I think Sinsheimer has allowed his imagination to crawl away with him," says Nobel laureate David Baltimore of MIT. Sinsheimer's argument evokes a passionate response for two reasons. First, he is an eminent scientist, and he can't be brushed off with talk of professional jealousy or pitiable ignorance. Second, he talks of unknowables and unquantifiables, so his opponents find themselves shadow-boxing with phantoms.

"One of the tragedies is that Bob Sinsheimer refuses to be specific and articulate his concerns," Paul Berg complains. "What is this barrier? What is the evidence that there is this barrier, and what would happen if it were breached? What he talks about is that it is unnatural. When they wanted to start immunizing people, that was also said to be unnatural. For one guy, no matter how eminent he is, to go up and say 'I fear,' and then to expect the whole world to stop—I don't care how eminent he is, it's not going to happen."

Still, Sinsheimer is undaunted. "The burden of proof belongs on the other side," he says. "I've thought about the problem of genetic engineering longer than most people and I'm not as sanguine as I would have been ten years ago." He favors phasing out the use of *E. coli* in recombinant DNA work "as fast as possible" and the confinement of all experiments "transferring DNA between species that do not normally exchange genetic material" to P4 (top security) facilities. He believes the NIH guidelines are too soft. "I don't think it's smart," he says. "I don't think it's sensible. We've got only one biosphere. The graded-risk concept doesn't make sense when you have only one thing to lose, as in the biosphere. I think the potential exists for some kind of biological catastrophe through the introduction of new kinds of organisms. You've got to be more careful. You can't just go blasting ahead when so much is unknown in an area of potential catastrophe."

"I would definitely feed recombinant DNA to my 16-year-old daughter if it were not considered idiosyncratic," one scientist says. "It goes without saying I'd drink it myself"

Slow down. That's what the critics say, and they think they're being quite reasonable. "What kind of disaster could happen from going slow?" asks Jon King. An aide to Mayor Vellucci says he is puzzled: "I don't quite understand what the enormous urgency to plough ahead is based on." If the research poses a serious risk, why not proceed under the tightest P4 conditions, until the light of knowledge wipes away these shadows of doubt? Why the rush?

But asking scientists to slow down is like asking hawks to walk. It goes against their nature. ...

In science, the slow of foot drop out of the race. Already European researchers working without guidelines have performed experiments that Americans are itching to try. "Essentially our work is crawling," moans Paul Berg. "This thing is slowed down now. I can't do the experiments I could do. We're not doing one-tenth of what could be done." David Hogness says if King and Sinsheimer had their way and assigned his research to P4 conditions, "I wouldn't do it. I wouldn't do experiments under Mars-shot conditions. It would be a serious question whether I would continue the research."

The most irritating thing about this controversy is it takes up so much time. Matthew Meselson of Harvard informs a reporter at 10 a.m. that this is his fifth call from the press that morning; as they speak, his secretary takes call number six. David Baltimore of MIT says, "At this point, 30 percent of my time is being taken up by talking on this issue." It is not just the hours subtracted, it is the diffusion of energy, the distraction from single-minded effort. To be a first-rank scientist, you must think science all the time. When you come home at night, you can't bring your test tubes with you, yet your mind lags behind in the laboratory. As James Watson puts it, "It's a world of total domination—you have to be manic."

Not since the Vietnam war has there been a distraction the size of the recombinant DNA controversy. From their perspective as scientists, the defenders of the research are horrified by the time lost. But, as political animals, they concede the dispute has had some good side effects. "I think about recombinant DNA and the ethical issues and the arguments on both sides," says Tom Maniatis, the man for whom Harvard is building its P3 (medium-containment) lab, "and in that sense it has represented a major distraction. But in many ways, I have found the situation for me has been very positive. It has forced me to think about the relationships between science and society."

One of the many ironies of the recombinant DNA controversy is that the scientists under attack are finding themselves on the unfamiliar side of a political issue. As a group, molecular biologists are extremely left-wing, even in a liberal university context. Because their political philosophy forces them to support public participation and control, they twist in uncomfortable contortions when responding to their leftist critics. "Increasing community control of science, which I used to support, now worries me," Mark Ptashne muses. "If you ask the average guy on the street, 'Do you want to hook up a mouse gene to a human gene?' he says, 'What good will it do?' You say, 'It may help our understanding of the gene.' He says, 'How much does it cost?' You say, '$50,000.' He says, 'Hell, no.'"

Times like these make a man appreciate the dictatorship of the proletariat. In the winter of 1970–71, Ptashne visited North Vietnam and to his pleasure found that the government of this socialist country supported advanced scientific work. One senses he would be quite happy in a worker state, guided by enlightened leaders who recognize Comrade Mark's achievements. "To me science is a cultural activity," he says. "One of the things that amazed me most about Hanoi was that in the middle of the bombing, they wanted to know about the lambda virus I was working on. They knew everything in the field up to the last year or so. I said, 'You guys are fighting for your lives—why do you care about this?' Even I was alienated from this work by that time. They said, 'We regard science as a basic form of human activity.' Science as a cultural endeavor makes more sense to them than it does to us."

Left-wing scientists endorse the ideal of community control, but, in practice, they are waiting for the ideal community. The communities they live in fall far short of the ideal. When Al Vellucci says 99 percent of the people of Cambridge would vote against recombinant DNA work, you can take his word for it. Cambridge citizens oppose the research because they can't see what good it will do them; they can only see the risks. "I don't think these scientists are thinking about mankind at all," Vellucci says. "I think that they're getting the thrills and the excitement and the passion to dig in and keep digging to see what the hell they can do." Vellucci's conception of the scientist jibes neatly with the definition given by the great sociologist Max Weber: "An inner devotion to the task, and that alone, should lift the scientist to the height and dignity of the subject he pretends to serve. And in this it is no different with the artist." That's what Mark Ptashne means when he calls science "a cultural endeavor."

The scientific world is no democracy. It is a hierarchical, aristocratic world, in which a man speaks with the authority of his accomplishments. Nobel laureates (in science, unlike literature, Nobel prizes reliably honor genuine achievement) constitute a nobility of the robe. You are what you do. Somehow the public senses that in the scientist's heart the outside world is irrelevant. In the search for knowledge, in the quest for a Nobel-size discovery, social concerns are a distraction.

Working in his dungeon laboratory, Dr. Frankenstein can't be bothered by intruders. He is a genius, he has uncovered the secret of life, and no one can stop his research. Only when his monster begins to destroy does he realize what he has done; and by then it is too late. In an interesting essay, Robert Brustein examined the horror-movie genre and found that Frankenstein is the public image of the research scientist "so obsessed with the value of his work that he no longer cares whether his discovery proves a boon or a curse to mankind." Whenever a "good scientist" appears in a horror movie to destroy the giant grasshoppers or the beast from 20,000 fathoms, he is always a "practical" scientist—a technician, really—and he is often opposed by the research establishment. The crewcut scientists who eventually conquer the Thing, for example, are hampered by a longhaired theoretician who wants to keep it alive "to find out what it knows." Eventually both he and the Thing are exterminated; Hollywood gives the public the ending it demands.

American scientists have always coexisted with the public in an uneasy symbiosis. Washington and Jefferson both hoped to establish a national endowment for the arts and sciences. But not until the administration of John Quincy Adams, a president aware that he was making a hopeless *dernier cri* for civilization, was a serious effort made. Our most erudite chief executive, Adams was the last of an aristocratic presidential line. He knew that he stood on the threshold of the age of the common man. After Jacksonian democracy drowned his proposals for a national university and an executive department to administer federal aid to the sciences, it took a hundred years and a total war to soften public antagonism toward paying the high costs of scientific research. Following World War II and the mystery revealed at Los Alamos and Hiroshima, scientists became the new high priests. They spoke a language as strange as Latin, they saved bodies instead of souls and, with society's money, they built monuments as expensive as cathedrals. If challenged by recalcitrant congressmen, they responded self-righteously that their work was necessary for a greater future. They had, after all, won the war.

Left-wing scientists endorse the ideal of community control, but in practice they are waiting for the ideal community

When money began to pour into biological research in the 1950s, the justification was the improvement of public health. The discoveries of antibiotics and a polio vaccine were almost as dramatic as a mushroom cloud. The budget of the National Institutes of Health quadrupled between 1950 and 1957, in a process well described in Daniel Greenberg's *The Politics of Pure Science*. Scientists may not have read de Tocqueville, but they had gotten the message: "In aristocratic ages science is more particularly called upon to furnish gratification to the mind; in democracies, to the body." Scientists vowed to gratify the body with a vengeance.

Most biologists, including experimenters with recombinant DNA, can be quite eloquent when asked to describe the potential medical benefits of their work. "Scientists profoundly believe that real benefits to mankind come out of basic research," says Walter Gilbert of Harvard; he talks of a time when bacteria implanted with insulin genes can cheaply and efficiently spin out insulin in test tubes. (Corporations, including General Electric, Lilly and Hoffmann-LaRoche, are already exploring such possibilities.) The shiniest golden apple is the hope of a cure for cancer. Since cancer is a fundamental disruption of controlled cell growth, any research on the mechanism of cell growth can be said to offer hope for a cancer cure. David Baltimore of MIT points out that even the current NIH regulations may be postponing the magical day of discovery. "Recombinant DNA would allow us to understand the cancer virus," he says, "but containment techniques are such that not much work can be done."

When public criticism becomes severe, researchers like to recall the patron saint of controversial biologists: Louis Pasteur. In 1884 Pasteur's neighbors wrote an angry letter, complaining that a proposed new laboratory "right in the middle of Villeneuve-l'Etang" could easily propagate "a horrible plague impossible to imagine." If their fears had been heeded, modern scientists point out a bit smugly, inoculation against infectious disease might still be a science-fiction fantasy.

Those who fear a slowdown of their research say the world is moving very fast, new problems appear every day and, if we stop the race of knowledge, we will be overwhelmed. "I think the danger of postponing is as speculatively high as continuing the work," says David Hogness. "We create many new problems through our intelligence. When the problem reaches a certain level, we attempt to solve it. We solve it with more intelligence. If you cut off these ways of solving problems, if you truncate this process, you should know exactly what you're doing."

You see the problem: although scientific work is motivated by curiosity, it must be justified to society. Over the last three decades, as government funding for science increased, the nature of research gradually changed. The struggling experimenter who hooks up wires in his basement is very different from the university professor filling out grant applications to finance a cyclotron. Not everyone thinks that bigger is better. One acerbic soul who yearns for yesterday and rubs his hands gleefully at the plight of recombinant DNA researchers is Erwin Chargaff, a white-haired biochemist whose empirical data provided evidence for the Watson-Crick model of DNA structure. A wit once said that Chargaff's idea of a scientist is Louis Pasteur played by Paul Muni. Chargaff hates molecular biology, and he hates what has happened to science. "If you had a National Music Institute that supported the writing of hundreds of symphonies a year, it would kill music," he says. "Scientific ideas come from heaven, just like melodies came to Mozart." Chargaff would like to turn the clock back, but he is wise enough to know that that won't happen, so he has resigned himself to being a gadfly, biting his adversaries with a sour venom. "It was quite clear that as soon as science was almost completely funded by the government, attempts would be made to push it into so-called useful channels," he says. "Into therapeutic uses, for genetic diseases, a form of eugenics. I am one of the few people who remember that the extermination camps in Germany began as experiments in eugenics. So I am almost congenitally opposed to improving the human lot. It begins with the do-gooders and it ends with the exterminators."

Chargaff fondly remembers the days when "scientists liked to observe nature without trying to change it." He scoffs at the notion that research must run to keep up with new problems. "The idea that science can make a better world is hubris," he says. "We have the arguments that this work is indispensable: we cannot exist without recombinant DNA because otherwise my grandson will die of cancer. I am old enough to remember before there were any cars or airplanes. I can very well imagine how we could live without many things that are now considered indispensable to daily life in America. I could live without genetic research. But since there are geneticists and they must have cigars, we must give it to them."

The bind for geneticists is that to earn their cigars they must prove their work is important; and once people agree, society will want to regulate the work. When J. Robert Oppenheimer remarked that scientists now knew sin, he could not have envisioned all the ramifications of the fall from innocence. Scientists relish their power but shun responsibility. Everyone wants to be a big shot. Michael Crichton captured the sentiment exactly when his scientist hero, summoned from a dinner party to work on a dangerous mission, tells his wife: "You see, I'm suddenly an important person." What whitecoated Walter Mitty doesn't dream of all the paraphernalia of an *Andromeda Strain* fantasy—the pink jumpsuits, red and silver keys, 40-2-5 nutrients and broad-spectrum suppositories? In his reverie, the scientist forgets that the government that gives him jumpsuits can also give him guidelines.

After World War II, scientists became the new high priests. They spoke a language as strange as Latin, saved bodies instead of souls and, with society's money, built monuments as expensive as cathedrals

Psychological longings and sociological imperatives have led scientists to underline their own importance. However, once the public acknowledges that research is useful and not frivolous, scientists are in trouble. We don't allow politicians or industrialists to police themselves. If scientists are that powerful, the public will think, they should also be regulated. Certainly they shouldn't be allowed to maintain their devil-may-care nonchalance. Oppenheimer defended his Los Alamos work by saying that he was only following political orders. Brecht's Galileo amplified this attitude: "I submit that as scientists we have no business asking what the truth may lead to. . . . I've written a book on the mechanism of the universe, that's all. What people make or don't make of it is no concern of mine." Brecht's first American audience in 1947 had the atomic bomb on its mind, but today's public may also think of genetic engineering. Paul Berg agrees with Galileo. "That's not my job, to worry about that," Berg says. "I can't worry that whatever I do will be used or misused for some malevolent purpose. I think it would be dangerous to try to anticipate where a given line of research is going and then head it off at the pass."

The specter of genetic engineering, with its baggage of test-tube babies and cloned supermen, has frightened humanists even before Aldous Huxley evoked a complete vision in *Brave New World.* That nightmare remains distant. Scientists have lots to do before those days are even possible, and Paul Berg, for one, hopes those days never arrive. But recombinant DNA poses a more immediate threat. "Only very recently have people begun to consider that the pursuit of truth could in itself be dangerous," says Robert Sinsheimer. "Nuclear physics was the first example. Doing an experiment can be a dangerous thing, not just to the experimenter—explorers have always been in danger—but to everyone around him."

Sinsheimer suggests that there should be "an agency for the protection of the future": someone must think about the risks of radioactive wastes, recombinant DNA bacteria and other technological marvels, he argues. Mayor Vellucci began to worry about recombinant DNA after the Walds compared the threat to a nuclear holocaust. That analogy surfaces repeatedly, in private conversations with scientists and at the two public Cambridge meetings. The Atomic Energy Commission, which both initiated and regulated nuclear research, provides a good example of the wrong way to administer science. The people doing the work, who have a vested interest in the progress of the research, should not have the last word on safety.

Although all the facts aren't in on the hazards of recombinant DNA research, the risks appear far more remote than the dangers of nuclear energy. Clearly, modern technology is creating other health hazards that pose a more certain and immediate threat to human life. "I don't like PCBs, I don't like hair dyes—they're a real story that would scare the shit out of you," says James Watson. "But all that the civilized English majors want to talk about is recombinant DNA." The risk of genetic research has captured the popular imagination because it extrapolates so easily into a science-fiction doomsday scenario.

"It is all too big and is happening too fast," says George Wald. It's an appealing lament. Man's scientific knowledge has outdistanced his social wisdom. The recombinant DNA controversy will be useful if it forces people to think about where we are going and what we should make out of our lives and our planet. Our technical capabilities are speeding ahead; our political institutions must struggle to keep up. Otherwise we will increasingly resemble idiots savants, who can multiply ten-digit numbers in seconds but can't find their own way home. ●

Document 5.12
May 1977
Bulletin of the Atomic Scientists (**33**:22)

The Cambridge Experimentation Review Board

How a citizens group helped a city council set safety standards for genetic research

Editor's note: The following report was filed on January 5, 1977, with the Cambridge, Mass., city manager by the Cambridge Experimentation Review Board, a six-member advisory group established to assist the city council in formulating regulations for the conduct of recombinant DNA research at Harvard University and the Massachusetts Institute of Technology. The board's recommendations were approved by the city council, with some further restrictions, on February 7.

The Cambridge Experimentation Review Board (CERB) has spent nearly four months studying the controversy over the use of the recombinant DNA technology in the City of Cambridge, Mass. The following charge was issued to the Board by the City Manager at the request of the City Council on August 6, 1976.

The broad responsibility of the Cambridge Experimentation Review Board shall be to consider whether research on recombinant DNA which is proposed to be conducted at the P3 level of physical containment in Cambridge may have any adverse effect on public health within the City, and for this purpose to undertake, among other studies, to:

- review the "Decision of the Director, National Institutes of Health to Release Guidelines for Research on Recombinant DNA Molecules" dated and released on June 23, 1976;
- review but not be limited to the methods of physical and biological containment recommended by the National Institutes of Health;
- review methods for monitoring compliance with applicable procedural safeguards;
- review methods for monitoring compliance with safeguards applicable to physical containment;
- review procedures for handling accidents (for example, fire in recombinant DNA research facilities;
- advise the Commissioner of Health and Hospitals on the reviews, findings and recommendations.

Throughout our inquiry we recognized that the controversy over recombinant DNA research involves profound philosophical issues that extend beyond the scope of our charge. The social and ethical implications of genetic research must receive the broadest possible dialogue in our society.

That dialogue should address the issue of whether all knowledge is worth pursuing. It should examine whether any particular route to knowledge threatens to transgress upon our precious human liberties. It should raise the issue of technology assessment in relation to long range hazards to our natural and social ecology. Finally, a national dialogue is needed to determine how such policy decisions are resolved in the framework of participatory democracy.

In the several months of testimony, we have come to appreciate the brilliant scientific achievements made in molecular biology and genetics. Recombinant DNA technology promises to contribute to our fundamental knowledge of life processes by providing basic understanding of the function of the gene. The benefits to be derived from this research are uncertain at this time, but the possibility for advancement in clinical medicine as well as in other fields surely exists.

While we should not fear to increase our knowledge of the world, to learn more of the miracle of life, we citizens must insist that in the pursuit of knowledge appropriate safeguards be observed by institutions undertaking the research. Knowledge, whether for its own sake or for its potential benefits to humankind, cannot serve as a justification for introducing risks to the public unless an informed citizenry is willing to accept those risks. Decisions regarding the appropriate course between the risks and benefits of potentially dangerous scientific inquiry must not be adjudicated within the inner circles of the scientific establishment.

Moreover, the public's awareness of scientific results that have an important impact on society should not depend on crisis situations. Many of the fears over scientific research held by the citizenry result from a lack of understanding about the nature of and the manner in which the research is conducted.

Members of the Review Board have made a determined effort to assess the risks to the Cambridge community of recombinant DNA research at the P3 level of physical containment. The National Institutes of Health, in issuing its guidelines, sought a balance between "stifling research through excessive regulation and allowing it to continue with sufficient controls." The function of the Review Board was not to repeat the long and careful deliberations of the National Institutes of Health, perhaps one of the most intensive biohazards studies in the history of biology. Our role was to examine the controversy within science. We called upon people from diverse fields to testify. We encouraged skepticism, and in doing so were able to determine the locus of the controversy.

Many of us felt that it was the role of the proponents of the research to justify that *no reasonable likelihood* exists in which the public's health would be compromised if the research is undertaken under the guidelines issued by the National Institutes of Health. We recognized that absolute assurance was an impossible expectation. It was clearly a question of how much assurance was satisfactory to the deliberating body, and in the case of the Cambridge Review Board that body was comprised of citizens with no special interests in promoting the research.

Decisions regarding the appropriate course between the risks and the benefits of potentially dangerous scientific inquiry...

The uncertainty we faced was not something fabricated in our community. It was expressed most eloquently by Donald Frederickson, NIH Director, when he issued their guidelines:

> In many instances, the views presented to us were contradictory. At present, the hazards may be guessed at, speculated about, or voted upon, but they cannot be known absolutely in the absence of firm experimental data—and, unfortunately, the needed data were, more often than not, unavailable.

Our recommendations call for more assurance than was called for by the NIH guidelines. We feel that under our recommendations, a sufficient number of safeguards have been built into the research to protect the public against *any reasonable likelihood* of a biohazard. For *extremely unlikely possibilities,* we have called for additional health monitoring, whereby appropriate personnel are responsible for the detection of hazardous agents, inadvertently produced, before they are able to threaten the health of the citizens in our community.

We recognize that the controversy over the use of the recombinant DNA technology was brought to the public's attention by a small group of scientists with a deep concern for their fellow citizens and responsibility to their profession. Many of these early critics are now satisfied that the potential hazards of the research are negligible when carried out under the NIH guidelines. There are also those scientists who continue to call for more stringent control over this technology, in many instances, against the majority view of their colleagues and amidst very strained personal relations. To them we owe our gratitude for broadening the context in which the issues are being discussed.

The willingness of scientists on both sides of the controversy to share their knowledge with us in our determination to arrive at a reasoned decision has been an inspiration.

The Cambridge Experimentation Review Board has spent over 100 hours in hearing testimony and carrying out its deliberations. Our decision is as unemotional and as objective as we are capable of offering. It provides a statement of conditions and safeguards that we deem necessary for P3 recombinant DNA research to be carried out in Cambridge.

The members of this citizen committee have no association with the biological research in question and no member of the Cambridge Review Board has ever had formal ties to the institutions proposing the research, with the exception of one member who has taught in unallied areas at both the institutions in question. Moreover, the City Manager in selecting a group of citizens representing a cross-section of the Cambridge community insured that the "empathy factor"—that is, the concern that the institutions proposing the research might lose valuable funds or that qualified researchers would leave in the event of a ban on the research—was never an issue in the deliberations.

In presenting the results of our findings we wish also to express our sincere belief that a predominantly lay citizen group can face a technical scientific matter of general and deep public concern, educate itself appropriately to the task, and reach a fair decision.

Board's Recommendations

Section 1

After reviewing the guidelines issued by the Director of the National Institutes of Health for Research Involving Recombinant DNA Molecules (issued June 23, 1976) it is the unanimous judgment of the Cambridge Experimentation Review Board that recombinant DNA research can be permitted in Cambridge provided that:

The research is undertaken with strict adherence to the NIH guidelines and in addition to those guidelines the following conditions are met:

I. Institutions proposing recombinant DNA research or proposing to use the recombinant DNA technology shall prepare a manual which contains all procedures relevant to the conduct of said research at all levels of containment and that training in appropriate safeguards and procedures for minimizing potential accidents should be mandatory for all laboratory personnel.

II. The institutional Biohazards Committee mandated by the NIH guidelines should be broad-based in its composition. It should include members from a variety of disciplines, representation from the biotechnicians staff and at least one community representative unaffiliated with the institution. The community representative should be approved by the Health Policy Board of the City of Cambridge.

III. All experiments undertaken at the P3 level of physical containment shall require an NIH certified host-vector system of at least an EK2 level of biological containment.

IV. Institutions undertaking recombinant DNA experiments shall perform adequate screening to insure the purity of the strain of host organisms used in the experiments and shall test organisms resulting from such experiments for their resistance to commonly used therapeutic antibiotics.

V. As part of the institution's health monitoring responsibilities it shall in good faith make every attempt, subject to the limitation of the available technology, to monitor the survival and escape of the host organism or any component thereof in the laboratory worker. This should include whatever means is available to monitor the intestinal flora of the laboratory worker.

VI. A Cambridge Biohazards Committee (CBC) be established for the purpose of overseeing all recombinant DNA research that is conducted in the City of Cambridge.

A. The CBC shall be composed of the Commissioner of Public Health, the Chairman of the Health Policy Board and a minimum of three members to be appointed by the City Manager.

B. Specific responsibilities of the CBC shall include:

- Maintaining a relationship with the institutional biohazards committees.
- Reviewing all proposals for recombinant DNA research to be conducted in the City of Cambridge for compliance with the current NIH guidelines.
- Developing a procedure for members of institutions where the research is carried on to report to the CBC violations either in technique or established policy.
- Reviewing reports and recommendations from local institutional biohazards committees.
- Carrying out site visits to institutional facilities.
- Modifying these recommendations to reflect future developments in federal guidelines.
- Seeing that conditions designated as I to V in this section are adhered to.

Section 2

We recommend that a city ordinance be passed to the effect that any recombinant DNA molecule experiments undertaken in the city which are not in strict adherence to the NIH guidelines as supplemented in Section 1 of this report constitute a health hazard to the City of Cambridge.

Section 3

We urge that the City Council of Cambridge, on behalf of this Board and the citizenry of the country, make the following recommendations to the Congress:

I. That all uses of recombinant DNA molecule technology fall under uniform federal guidelines and that legislation be enacted in Congress to insure conformity to such guidelines in all sectors, both profit and non-profit, whether such legislation takes a form of licensing or regulation, and that Congress appropriate sufficient funding to adequately enforce compliance with the legislation.

II. That the NIH or other agencies funding recombinant DNA research require institutions to include a health monitoring program as part of their funding proposal and that monies be provided to carry out the monitoring.

III. That a federal registry be established of all workers participating in recombinant DNA research for the purpose of long-term epidemiological studies.

IV. That federal initiative be taken to sponsor and fund research to determine the survival and escape of the host organism in the human intestine under laboratory conditions.

Section 4

In the event that the citizens of Cambridge, the members of the City Council or other interested parties wish to know how the Cambridge Experimentation Review Board carried out its charge to review P3 recombinant DNA research in the City, the final section of this report discusses the review process. In this discussion we include a brief chronology of events, some of the strategies undertaken by the Board for self-education and a description of its deliberation process.

On July 7, 1976, after having held two days of public hearings, the City Council of Cambridge voted a three-month "good faith" moratorium on all P3 level recombinant DNA research in the City and called for the establishment of a citizen review board to study the issue.

James L. Sullivan, City Manager of Cambridge, released the charge to the newly designated Cambridge Experimentation Review Board on Aug. 6, 1976, and issued the guidelines under which that body was to carry out its responsibilities. In addition, eight citizens and the newly appointed acting Commissioner of Health and Hospitals for the City were selected to constitute the Board. Members of the Board were chosen to reflect a cross-section of the Cambridge community. Of the eight citizen Board members, only three had ever met before. Seven of the eight had never had formal ties with either institution proposing the new research. The one individual who did have some formal ties with the universities has taught courses in structural engineering both at Harvard and M.I.T.

The Cambridge Review Board commenced its first meeting Aug. 26, 1976, and continued its hearings until the recommendations of the Board were issued to the Commissioner of Health and Hospitals on Dec. 21, 1976. Meetings were held twice weekly with each session lasting in excess of two hours.

At the Sept. 14 meeting, the Board arrived at a consensus on key policy issues related to the process of its inquiry. Dr. Francis Comunale, initially serving as chairperson, released the chair to the vice chairperson, Daniel Hayes. This decision was made to preclude any ambiguity or conflict of interest in having Dr. Comunale, the then acting Commissioner of Health and Hospitals in the role as chairman of the Board and the person to whom the Board advised on the matter in question. Dr. Comunale thereafter became an ex officio member of the Board. He attended meetings, without a vote, and excluded himself from the final deliberations leading to a decision.

At the same meeting the Board voted to request an extension of the moratorium for an additional three months, on the grounds that we needed the additional time to carry out the full scope of our charge, including a review of the Environmental Impact Statement, which at that time was not complete. The request for an extension of the moratorium was subsequently granted by the City Council and accepted by the institutions affected by the moratorium.

It was agreed that on all decisions undertaken by the Cambridge Review Board a consensus would be sought; if consensus could not be reached on an issue, the majority decision would prevail. Moreover, any Board member had the right to poll the entire membership on any issue requiring a vote. If consensus could not be reached on the final recommendation, then minority statements would be permitted in the Board's final report. The members agreed that Thursday meetings would be kept open for the public and the media, while Tuesday sessions would be held in private.

Among the more formidable problems facing this lay citizen board was its self-education. At the outset of the inquiry, the members of the Board were, for the most part, unfamiliar with the concepts, the basic scientific principles and the explanatory models underlying the recombinant DNA technology. The education of the Board members was carried out simultaneously with the inquiry process. We had to decide on the kind of information we would need to reach a decision as well as the kind of people who could provide us with that information.

There were several facets to the Board's information gathering and self-education strategies as exemplified in the following.

- Each Board member was provided with special technical documents on the controversy, including the NIH guidelines, the Environmental Impact Statement, and essays in journals such as *Science.* Along with technical materials, articles that were published in the more popular press and written for a wider readership were distributed to the Board members. As examples, the Board had articles from *Scientific American,* the New York Times Magazine, and *National Geographic.*
- A technical assistant to the Board, who had training in the biological sciences, offered help with translating technical concepts. The technical assistant also made available to the Board current articles, news analyses, and essays in leading journals relating to the controversy.
- Spokespeople who appeared before the Board were asked to reduce technical concepts to layman's terms, to present simplified models of bio-chemical events, and to draw upon analogies that helped foster

understanding whenever they were available.

• Members of the Board were witness to a forum on the recombinant DNA controversy in which proponents and opponents of the research presented their arguments and responded to questions from the audience.

• Two open-line telephone conversations were used to draw testimony from people outside the state. In one of these conversations, the NIH Director and a panel of experts responded to questions of the Board members.

• In a five-hour marathon session, the Board carried out a type of mock courtroom affair. Board members served as a kind of jury, while advocates on both sides of the issue presented their case, were given an opportunity to cross-examine one another, and responded to questions raised by the "citizen jury." This format enabled the Board members to evaluate how well scientists on each side of the controversy responded to the critical issues. Medical researchers and clinicians were also on hand to respond to testimony.

• Board members were taken through laboratories at Harvard and M.I.T. In one case a mock experiment was carried out which exemplified the various stages of the recombinant DNA process. Visiting the laboratories also helped the Board members concretize many of the specifications found in the NIH guidelines relating to physical containment.

Speakers appeared before the Board both on a voluntary basis and at the Board's request. The schedule of speakers called for fair representation of the views of opponents and proponents, as well as other persons who were called upon to broaden our understanding of the issues. Individuals on each side of the issue were heard from on intermittent weeks.

Some members of the Cambridge Review Board visualized the Board as a kind of "citizen jury" whose function it was to review and assess the significance of the recombinant DNA controversy within science. The use of the legal metaphor helped members of the Board clarify for themselves the role of lay citizens in this complex issue. The analogy was of only limited value since Board members functioned in a greater variety of ways than citizens called upon to jury duty. The Board determined the rules of its inquiry, called upon people to testify, listened to the arguments, cross-examined scientists and finally came out with its recommendations.

...must not be adjudicated within the inner circles of the scientific establishment.

The use of a "citizen court" in areas of controversy within science that have significant bearing on public welfare is quite new and untested. It encouraged discussions among Board members about where justification rests. At issue was whether the proponents of the research must prove that it is safe beyond all reasonable doubt or whether the opponents must prove that if recombinant DNA research were undertaken there would be significant potential hazards.

There was no clear consensus on the issue of who must justify what, and to what degree of satisfaction. However, the Board carried out its inquiry by seeking the strongest positions on both sides of the controversy, while simultaneously looking for weaknesses in the arguments.

Several intensive planning sessions were used to explore the Board's unresolved questions and to draw as wide a range of input from its citizen members as possible. The planning sessions were designed to overcome the factors that inhibit people from expressing their uncertainties. The aim was to eliminate any social hierarchies that could prevent full cooperation and participation from Board members. The success of full cooperation hinged upon the building of confidence for each individual member.

The planning strategy involved first covering the walls of a room with large sheets of paper. Then, a scribe wrote down suggestions from Board members, insuring that each individual completed his/her recommendations or queries before the issues were debated by the entire Board. Finally, the material on the sheets was reduced and synthesized by a technical assistant and sent out to the Board members for discussion at subsequent meetings. This method insured that each citizen member, whatever his/her stand on the controversy, and whatever his/her state of knowledge on the issues, had an unfettered opportunity for self-expression and participation.

Individuals appearing before the Board spent up to three hours discussing the issues and responding to questions. Members of the Cambridge ERB heard over 75 hours of testimony from more than 35 individuals representing both sides of the controversy. In addition, the Board spent over 25 hours in formal planning and deliberation as well as countless hours of reviewing related written material before arriving at our decision.

Finally, it is worthwhile noting that despite a considerable heterogeneity in the Board's makeup and differences in how its members initially perceived the controversy, we were able to reach a unanimous decision. □

Document 5.13

February 11, 1977
Science (**195**:558)

Gene-Splicing: At Grass-Roots Level a Hundred Flowers Bloom

For a research technique too new to have produced a single practical application, the recombinant DNA method of gene-splicing has evoked a perhaps unprecedented degree of public interest. Debate about the technique has raged through campuses, spilled over into city councils, and has now reached the attention of state legislatures.

Many of these bodies have made or are making their own reviews of the terms under which the research may proceed. So far all have accepted the guidelines issued by the National Institutes of Health last June, but usually with certain extra restrictions of their own.

With the exception of action being contemplated in New York State, these restrictions are of minor significance, so that in effect the NIH guidelines are being generally endorsed at the local level.

Yet public anxiety about the technique is so definite that even industry, in a change of position, is now, for reasons of self-protection, leaning toward having the government register and keep track of its gene-splicing activities.

Local involvement in the gene-splicing debate has included the following actions.

New York State. Having held public hearings on the gene-splice technique (*Science*, 12 November), the state attorney general's environmental health bureau has prepared a bill to control the research. The bill, which has not yet been introduced, would require everyone engaged in gene-splicing research or production to obtain a certificate from the state health commissioner, who would also specify training and health-monitoring programs. Deborah Feinberg, who drafted the bill, suggests in an accompanying report that all gene-splicing work should be done in P3 (moderate level) containment facilities.

An official of the New York State health department says his feeling is that all new laboratories should be equipped with P3 facilities, but that "we would not require everything to be done in P3 right away." But the department would probably upgrade some lower level experiments to P3 and P4 while endorsing the NIH guidelines in general, the official says.

California. Two committees of the state legislature are at present holding hearings, after which they will decide whether or not to introduce legislation. Marc Lappe, a special assistant in the health department who helped organize the hearings, says that the minimum likely requirement of such legislation would be to make the NIH guidelines applicable to everyone, particularly industry.

New Jersey. State attorney general William F. Hyland, whose interest in biomedical issues was manifested during his handling of the Karen Quinlan case, has been following the gene-splicing issue closely. His assistant on the subject, Dennis Helms, says his own feeling—Hyland has not yet come to a decision—is that state regulation is not a good idea for an issue that can be properly settled only on a national basis. There is no point in driving the research underground by excessive regulation, Helms believes, because "in the end we are going to depend on the responsibility of the individual scientist. But I can assure you the response will be electrifying if there is a bad accident. That will mean banning everything in the ridiculous fashion that always happens when you do things too fast."

Cambridge. The city council is in the throes of creating an ordinance on gene-splicing research. Though Mayor Vellucci would still like to ban all P3 and P4 research, the proposal of the citizens' review board—to endorse the NIH guidelines with added restrictions—will probably prevail in some form.

San Diego. Seeking to avoid a Cambridge-style confrontation, the University of California at San Diego informed city mayor Pete Wilson last year of its intention to build two P3 facilities. The mayor asked his quality of life board to set up a DNA study committee chaired by Albert Johnson, dean of sciences at UCSD. After hearing witnesses from both sides, the committee completed a report last week for submission to the mayor and council. The report endorses the NIH guidelines but in addition recommends that the council consider the desirability of confining all gene-splicing research to P3 facilities; that the university refrain from experiments requiring P4 facilities; that it notify the city of any P3 experiment requiring the highest degree of biological containment (EK3); and that an ordinance be passed to bring industry and others within the ambit of the guidelines. "Ideally, if this work is to continue, it should go forward with the full knowledge, understanding and approval of the people of San Diego as to its potential benefits and hazards," says the DNA study committee.

Madison. A resolution asking for a citizens committee and public debate on the issue will come before the city council in a few weeks. The University of Wisconsin set up a committee at the same time as the resolution was introduced. "There is now a little bit of jostling going on between the city and the university about whether there should be a public debate," says Philip Ball, an aide to the mayor.

Bloomington. The mayor's office, after following the events in Cambridge, heard rumors of gene-splicing research at the University of Indiana, and was later informed by the university of plans to build a P3 facility, where only P1 or P2 experiments would be conducted for the time being. The mayor's environmental commission has held one set of hearings and will hold another later this year, but has found no serious fault with the university's procedures. "People so far are pretty calm. For the most part the community is satisfied that the university is being responsible. We don't anticipate taking any action at the present time," says Deborah Mantell, an aide to the mayor.

Ann Arbor. City mayor Albert Wheeler, who happens to be a microbiologist, is taking no action. One reason may be that the University of Michigan has gone through a more intense debate about the technique than any other institution, Harvard and MIT included. Vigorous opposition to the research was mounted by Susan Wright, a historian of science. One of her supporters has said, and the proponents of the research agree, that the university will never be the same again.

Deciding that gene-splicing research was the wave of the future, the university planned to construct three P3 facilities. It set up a committee A to get the job done, a committee B to look at the ethical and social aspects of the research, and a committee C to assess biohazards. The nonbiologists on committee B concluded last March that the research should go ahead. The one dissenter, historian Shaw Livermore, opined that things were not so bad that we had to change the order of life. While the technique would help alleviate human distress, said Livermore, "I believe that the limitations of our social capacities for directing such a capability to fulfilling human purposes will bring with it a train of awesome and possibly disastrous consequences."

The university's board of regents held five meetings on the issue, deciding last May by a 6-to-1 vote that the research should be allowed to proceed. Robert Helling, head of one of the two groups using the technique, says the debate took up an enormous amount of time—he has done almost no research for a year—but has been well mannered. "There have been intense feelings at times but there has never been anything personal, we have tried to keep things civilized."

Industry. The industry attitude toward the control of gene-splicing research is about to undergo an important change. Apart from General Electric, all major firms known to be actively interested in the technique are drug companies, whose positions are coordinated by the Pharmaceutical Manufacturers Association in Washington. Until recently, the PMA has advocated voluntary compliance with the NIH guidelines, asking only that they be modified by lifting the restriction on large volumes of liquid and by protecting intellectual property rights.

Position Not Pragmatic

"We now realize that position was not pragmatic," says PMA scientific direction John G. Adams. "We are in a fishbowl on this. There are charges that we are doing something clandestine." Hence, for its own protection, the PMA is leaning toward having some agency of government keep a registry of industrial research; provided they were not subject to disclosure under the Freedom of Information Act, the industry could submit research plans and even results to the registry, Adams suggests.

All the major drug companies are interested in the technique but only six are actively engaged in it—Hoffmann-La Roche, Upjohn, Eli Lilly, Smith Kline and French, Merck, and Miles Laboratories. A smaller company whose activities have attracted a lot of attention is Cetus Corporation of Berkeley, California. Cetus has as consultants Stanley Cohen, who pioneered an important aspect of the technique, and Joshua Lederburg, also of Stanford. The company's present specialty is improving the genetics of industrial microorganisms, but gene-splicing "will be a very major aspect of our future output," says corporation president Ronald E. Cape. Cape believes that those suspicious of industry's involvement with the technique underestimate how conservative industry tends to be. Cetus is fully complying with the NIH guidelines but, Cape notes, "there is only so much curtailment that the United States, by regulation or moral suasion, can enforce on the rest of the world." At a recent meeting it was clear that the European scientists "were obviously able to move ahead in certain directions which were shut off, at least for the time being, to American scientists," Cape notes.

Another small company active in the field is Genentech, which is funding Herbert Boyer of the University of California, San Francisco, in a project to synthesize human insulin with the gene-splice technique. The human gene for the insulin precursor molecule has not been isolated, so the plan is to synthesize the gene chemically on the basis of the protein's known structure.

On the basis of work by Boyer and Cohen, Stanford and the University of California have applied for a patent on commercial uses of the gene-splice technique. The patent, which does not apply to academic or industrial research, would require commercial users to abide by the NIH guidelines. The Patent and Trademark Office announced last month that it would give accelerated processing to patent applications involving gene-splicing, "in view of the exceptional importance of recombinant DNA and the desirability of prompt disclosure of developments in the field."

The Patent Office action, stimulated by Betsy Ancker-Johnson, Commerce Department assistant secretary for science and technology, has been criticized by Senator Dale Bumpers (D–Ark.) for having weakened the NIH guidelines. The announcement allows firms to avoid disclosure of proprietary information if it would prejudice their foreign patent rights. Bumpers aide George L. Jacobson says that this in effect exempts industry from the disclosure provisions of the NIH guidelines, an action which preempts the discussions now going on elsewhere in government as to whether and in what form the guidelines should be made to apply to industry. Ancker-Johnson says that the announcement extends the guidelines to industry by requiring that they be followed if quick patents are to be granted.

Environmentalists. Several environmental groups have taken stances on the issue. The Environmental Defense Fund and the Natural Resources Defense Council have filed a petition with the Department of Health, Education, and Welfare asking for a public hearing to determine whether any gene-splicing research should be allowed and if so, under what conditions. Such a hearing, the petitioners state, would serve as "a broad-based public review of the existing NIH guidelines and would permit open debate on issues given little attention by the NIH Drafting Committee or the office of the director," such as whether the human gut bacterium *Escherichia coli* is a suitable host for gene-splicing work.

Friends of the Earth believes that a moratorium should precede the review: there should be a public debate "so that an official moratorium on recombinant DNA research can be imposed pending further public investigation." As for the Sierra Club, its board of directors decided on 9 January that, pending further information and discussion, "the Sierra Club opposes the creation of recombinant DNA for any purpose, save in a small number of maximum containment labs operated or controlled directly by the federal government."

What all these activities represent, at state and city council level, by industry and environmentalists, is an extended exercise in public education about the gene-splice technique and its implications. The calls to prohibit or slow down the research seem threatening and irrational to those scientists who first pointed to the risks and who believe, with some measure of justice, that there should be a presumption in favor of their estimate of how to deal with them. Yet the bottom line reached by most public bodies so far is to endorse the NIH guidelines with minor changes. Whatever further restrictions emerge from the present round of debate, the research will at least be proceeding on the basis of informed public consent, a desirable and probably inescapable condition for a technique of such consequence.—NICHOLAS WADE

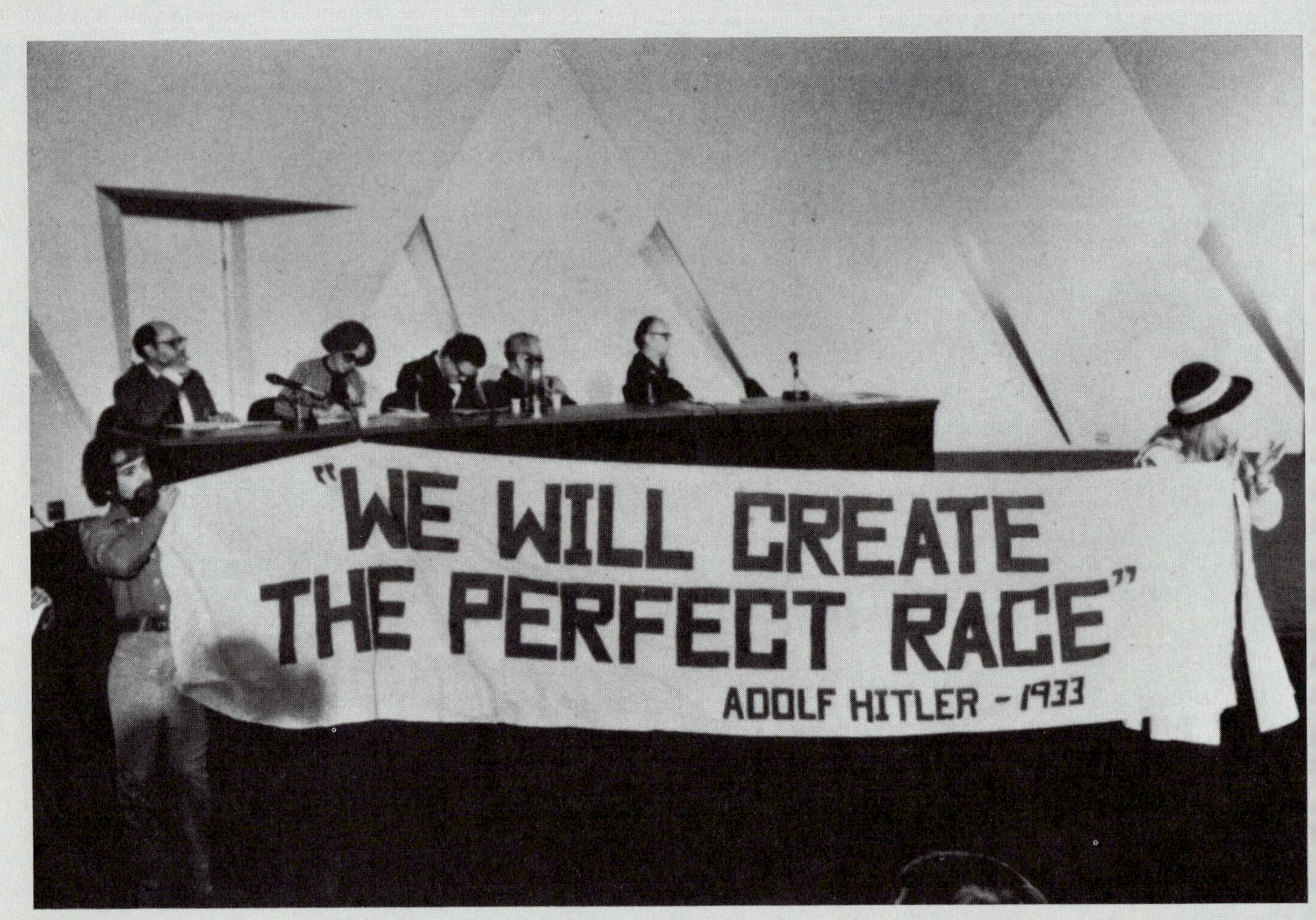

Demonstrators at the National Academy of Sciences meeting, March 1977.

Document 5.14

April 1977–1978

Dane County (Wisc.) Board of Supervisors Resolution, Excerpt

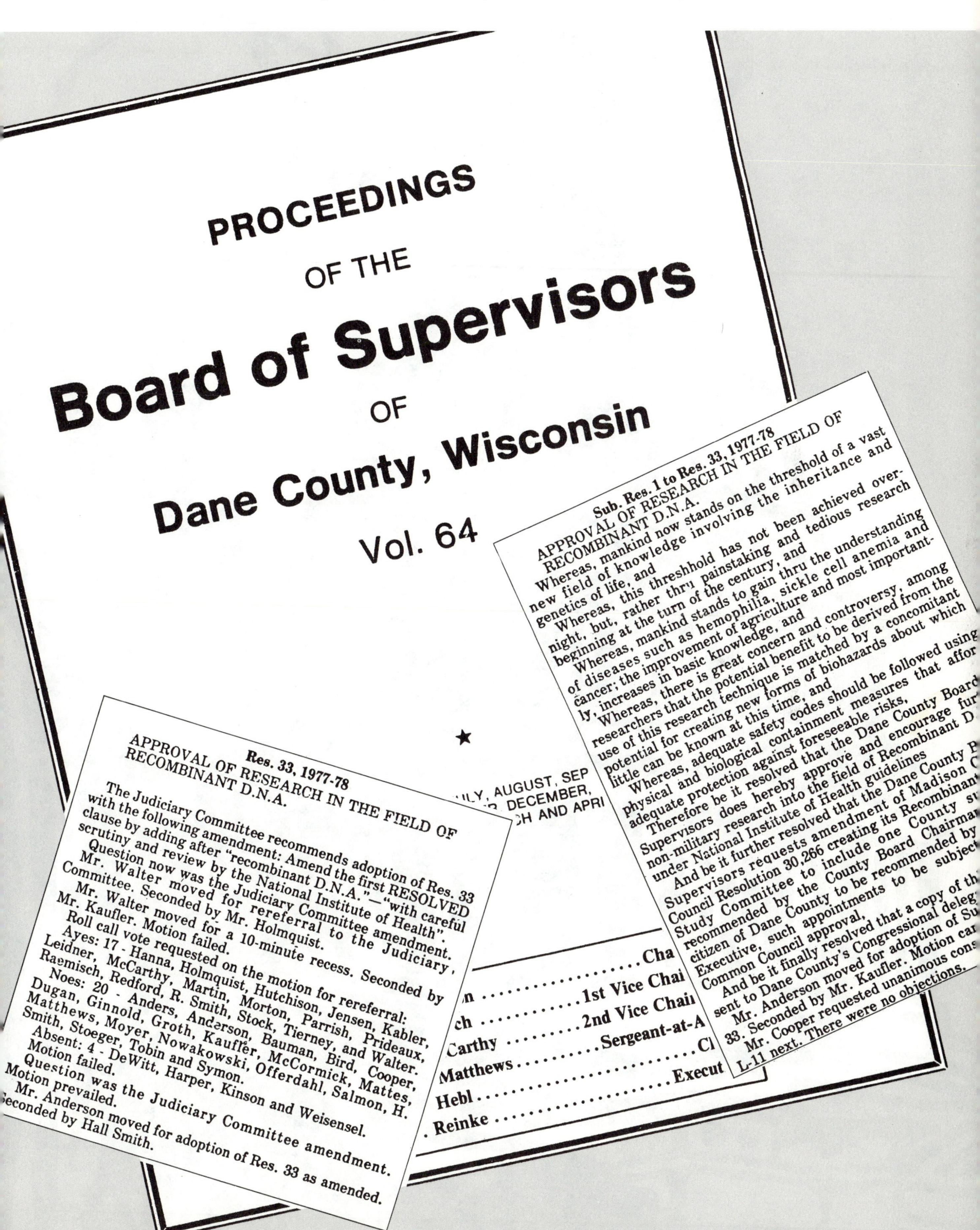

PROCEEDINGS

OF THE

Board of Supervisors

OF

Dane County, Wisconsin

Vol. 64

...LY, AUGUST, SEP
...R, DECEMBER,
...CH AND APRI

...n ....Cha
...ch ....1st Vice Chai
...Carthy ....2nd Vice Chai
Matthews ....Sergeant-at-A
Hebl Cl
. Reinke ....Execut

Res. 33, 1977-78

APPROVAL OF RESEARCH IN THE FIELD OF RECOMBINANT D.N.A.

The Judiciary Committee recommends adoption of Res. 33 with the following amendment: Amend the first RESOLVED clause by adding after "recombinant D.N.A."—"with careful scrutiny and review by the National Institute of Health".

Question now was the Judiciary Committee amendment.

Mr. Walter moved for rereferral to the Judiciary Committee. Seconded by Mr. Holmquist.

Mr. Walter moved for a 10-minute recess. Seconded by Mr. Kaufler. Motion failed.

Roll call vote requested on the motion for rereferral:

Ayes: 17 - Hanna, Holmquist, Hutchison, Jensen, Kabler, Leidner, McCarthy, Martin, Morton, Parrish, Prideaux, Raemisch, Redford, R. Smith, Stock, Tierney, and Walter.

Noes: 20 - Anders, Anderson, Bauman, Bird, Cooper, Dugan, Ginnold, Groth, Kaufler, McCormick, Mattes, Matthews, Moyer, Nowakowski, Offerdahl, Salmon, H. Smith, Stoeger, Tobin and Symon.

Absent: 4 - DeWitt, Harper, Kinson and Weisensel.

Motion failed.

Question was the Judiciary Committee amendment. Motion prevailed.

Mr. Anderson moved for adoption of Res. 33 as amended. Seconded by Hall Smith.

Sub. Res. 1 to Res. 33, 1977-78

APPROVAL OF RESEARCH IN THE FIELD OF RECOMBINANT D.N.A.

Whereas, mankind now stands on the threshold of a vast new field of knowledge involving the inheritance and genetics of life, and

Whereas, this threshhold has not been achieved overnight, but, rather thru painstaking and tedious research beginning at the turn of the century, and

Whereas, mankind stands to gain thru the understanding of diseases such as hemophilia, sickle cell anemia and cancer; the improvement of agriculture and most importantly, increases in basic knowledge, and

Whereas, there is great concern and controversy, among researchers that the potential benefit to be derived from the use of this research technique is matched by a concomitant potential for creating new forms of biohazards about which little can be known at this time, and

Whereas, adequate safety codes should be followed using physical and biological containment measures that affor[d] adequate protection against foreseeable risks,

Therefore be it resolved that the Dane County Board [of] Supervisors does hereby approve and encourage fur[ther] non-military research into the field of Recombinant D[.N.A.] under National Institute of Health guidelines

And be it further resolved that the Dane County B[oard of] Supervisors requests amendment of Madison C[ommon] Council Resolution 30,266 creating its Recombinan[t ...] Study Committee to include one County s[...] recommended by the County Board Chairma[n ...] citizen of Dane County to be recommended by [the County] Executive, such appointments to be subject [to] Common Council approval,

And be it finally resolved that a copy of th[is ...] sent to Dane County's Congressional deleg[ation].

Mr. Anderson moved for adoption of Su[b. Res. 1 to Res.] 33. Seconded by Mr. Kaufler. Motion car[ried.]

Mr. Cooper requested unanimous cons[ent ...] L-11 next. There were no objections.

THE ART OF LOBBYING

The NIH is an agency designed to fund and conduct medical and basic research. It is not a regulatory agency. To ensure compliance with the guidelines by individuals and institutions receiving the NIH's financial support, the agency can, of course, simply threaten to suspend that support. Such a threat is, however, of no consequence to private or industrial laboratories, many of which would undoubtedly do recombinant DNA research. The compliance of these laboratories with the NIH guidelines was voluntary, and they might well be reluctant to disclose full details of their research projects to the RAC, in order to avoid the risk of competitors' learning what they were up to. Industrial secrecy and propriety rights would be jeopardized. Even if industrial laboratories did abide by the guidelines, how could their performance be monitored? The need to urge private laboratories to comply with the NIH guidelines was one of the principal arguments used by advocates of new legislation to regulate recombinant DNA.

On July 19, 1976, Senators Edward Kennedy (Massachusetts) and Jacob Javits (New York) wrote to President Ford urging federal control of recombinant DNA research in industry (Doc. 6.1). Two months later, on September 22, the Senate Health Subcommittee, chaired by Senator Kennedy, devoted a day's hearings to the issue, which it had last discussed in a somewhat different context in April 1975 (see Chapter 4). In his opening statement the Chairman focused attention on the question of the compliance of industry with the guidelines (Doc. 6.2). Several microbiologists gave evidence, and Robert Sinsheimer expressed his general misgivings about the long-term consequences of recombinant DNA research (Doc. 6.3).

On October 21, 1976, other hearings were held in New York City, before State Attorney General Louis Lefkowitz. These hearings gave the East Coast opponents of recombinant DNA—Jonathan King, George Wald, Erwin Chargaff, and Liebe Cavalieri—an opportunity to repeat their calls for a total moratorium. However, the majority of scientists testifying felt that the NIH guidelines provided more or less adequate safeguards (Doc. 6.4), and only one (J. D. Watson) spoke out against any form of legislation (Doc. 6.5). In his report on the hearings, Attorney General Lefkowitz advocated a strict state bill that was duly drafted by his staff.

Against this background of government initiatives at city and state levels (Cambridge, Massachusetts and New York State), discussion in Congress continued behind the scenes and among the staffers, following Senator Kennedy's Health Subcommittee hearings. Meanwhile, on behalf of the administration, the Secretary of HEW with the approval of the President formed a Federal Interagency Committee on Recombinant DNA Research in November 1976 to address extension of the guidelines beyond the NIH to the public and private sectors. During the winter this Committee succeeded in persuading all government agencies which might possibly fund recombinant DNA research to adopt the NIH guidelines. It then closely examined existing legislation—such as TOSCA (Toxic Substances Control Act) and OSHA (Occupational Safety and Health Act)—to see if any could be used to regulate recombinant DNA research and ensure universal compliance with the NIH guidelines. With a dissenting opinion from P. B. Hutt, Chief Council of the FDA (Doc. 6.6), who cited Section 361 of the Public Health Act (42 USC 264) as an adequate statutory authority (if one assumes that recombinant DNA creates real dangers of "introduction, transmission or spread of communicable disease"), the conclusion of the Interagency Committee was negative. No existing federal agency had the necessary statutory powers, and a new bill would be needed, which the HEW began to draft.

Early in February 1977 while the Interagency Committee was reaching this conclusion, Senator Dale Bumpers (Arkansas) and Representative Richard Ottinger (New York) introduced identical bills that were referred to the Senate Committee on Labor and Public Welfare and the House Committee on Interstate and Foreign Commerce. The draft bills were "to provide for guidelines and a strict liability in the development of research related to recombinant DNA" (95th Congress: H.R.3191; S.621) (Doc. 6.7). These draft bills would give power to the Secretary of HEW to issue guidelines and licenses and have powers of inspection of premises. Anyone guilty of a misdemeanor would risk up to one year in prison or a fine of $10,000 per day for each day's violation, or both. But more than that, the bills had the following clause: "Civil Liability Section 7. Persons carrying out research involving recombinant DNA shall be strictly liable, without regard to fault, for all injury to persons or property caused by such research."

As the academic world suddenly realized, if such bills became law, virtually all recombinant DNA research would be halted. No institution could obtain or afford insurance against such liability. The alarm had rung. A bill designed specifically to ensure compliance of industry with the guidelines would not have stirred most academics, many of whom had an almost reflexive distrust of corporate activities and profits. But these bills were an enormous and direct threat and, some felt, an insult. The academic institutions saw the need to contact their Congressional representatives and organize themselves into a spontaneous lobby, part of which soon became more professional and known as Friends of DNA.

At the beginning of March 1977, Senator Metzenbaum (Ohio) introduced another bill into the Senate Committee on Human Resources (95th Congress: S.945), which was in some respects similar to the first bills. Senator Metzenbaum's bill, however, also had the novel proposal of a special thirteen-member "National Commission for the study of recombinant DNA research and technology" to be appointed by the Secretary of HEW. Among other things, this National Commission was to study the appropriateness of continuing recombinant DNA research (Doc. 6.8).

Across the country toward the end of March 1977, California's Governor Jerry Brown invited Halstead Holman (Stanford) and J. D. Watson to put the case for and against legislation (Doc. 6.9). The meeting provoked an interesting comment from Lance Montauk (Doc. 6.10) and diverse responses in the press (Doc. 6.11), but California did not proceed with legislation.

In Washington the Interagency Committee was meanwhile urgently drafting a bill of its own, which was completed at the end of March. It was introduced on April 6, 1977, by Secretary Califano at a meeting of Senator Kennedy's Subcommittee on Health and Scientific Research. The administration's bill envisaged licensing of facilities, registration and inspection, a penalty for misdemeanor of a fine of $5,000 a day, one year in prison, or both. Initially it provided for federal preemption and had a five-year sunset clause. Senator Kennedy's opening statement at the hearing revealed that he was decidedly against federal preemption of the right of local communities to insist on requirements stricter than the federal regulations (Doc. 6.12). After the hearing, Kennedy's staff drafted an amendment that replaced all the clauses of the administration's bill with others significantly different (95th Congress: S.1217; Calendar #334; Report #95359).

The resulting Kennedy bill (S.1217) (Doc. 6.13), like Senator Metzenbaum's draft, envisaged a National Recombinant DNA Safety Regulation Commission. It restored the fine to $10,000 per day, and, most importantly for the future of the bill, it specifically envisaged additional local regulations and eliminated federal preemption. A hearing held on July 22, 1977, revealed the strength of opposition to two fundamental parts of the Kennedy amendment: the establishment of a National Commission and lack of federal preemption (Doc. 6.14). An amendment from Senator Javits reintroduced a clause providing for federal preemption, and on August 2, 1977, Senators Nelson (Wisconsin) and Moynihan (New York) introduced an amendment to the Kennedy bill (Doc. 6.15) that would have both eliminated the proposed National Commission and restored federal preemption. After the summer recess Senator Kennedy withdrew support for his bill (Doc. 6.16 and Doc. 6.17).

Having traced this history of attempted legislation in the Senate, we return now to events in the House of Representatives. On February 28, 1977, Representative Harley Staggers (West Virginia) and Representative Paul Rogers (Florida) introduced a bill that was referred to the House Committee on Interstate and Foreign Commerce and on March 24 sequentially referred to the Committee on Science and Technology (95th Congress: H.R.11192). Initially this bill had the support of many academics because it envisaged federal preemption, relatively small penalties, and no National Commission. But as the bill went through subcommittees, federal preemption was weakened, penalties increased, and a National Commission was called for. By May 1977 Rogers was sponsoring a draft with a penalty of $50,000 per day of violation. As the bill became increasingly stringent, support from academics changed to opposition. By October 1977, the bill now designated H.R.7897 was discussed in the full Committee on Interstate and Foreign Commerce, and the scientific community's lobby was brought to bear on Chairman Staggers. Subsequently the bill—completely redrafted by Rogers' staff and now much more moderate and again having a strong federal preemption clause—was discussed in the Interstate and Foreign Commerce Committee. On March 14, 1978, another attempt to remove the federal preemption clause was watered down, and the House Committee on Science and Technology requested that the bill be sequentially referred to it. Both Senate and House bills (Doc. 6.18) languished after March 1978, neither body caring to continue the debate, and they died on December 31, 1978. The proposed revised guidelines published in the *Federal Register* on July 28, 1978, contain a useful tabular summary of all this Congressional activity (Doc. 6.19).

From early 1977 through the spring of 1978 when legislation seemed a real possibility, the scientific and academic communities, beginning as relative amateurs, mounted an increasingly professional and ultimately successful lobby—or set of lobbies, since there was inevitably more than one view. For example, the American Society for Microbiology, the Inter-Society Council for Biology and Medicine, and the Federation of American Societies for Experimental Biology, together representing a large proportion of the microbiologists in the United States, lobbied throughout 1977 for federal preemption and for a cautious and careful approach to legislation, which, in principle at least, they favored (Doc. 6.20 through Doc. 6.22). A group of thirteen members of the National Academy of Sciences drafted a resolution passed at the annual meeting on April 26, 1977, voicing their concern over the legislative developments (Doc. 6.23). A special delegation of the American Association for the Advancement of Science visited all pertinent legislators (the AAAS is the proprietor of the journal *Science*). Individual molecular biologists, including several of those who had signed the Moratorium letter in 1974, repeatedly gave evidence at hearings and played key roles in lobbying Senators and Representatives (Doc. 6.24). Groups of molecular biologists at scientific meetings signed petitions (Doc. 6.25), and as we discuss in Chapter 8, the signatories of the 1974 Moratorium letter also attempted during 1977 to draft a second joint letter opposing legislation.

When in the spring of 1977 the academic institutions were rudely awakened to the dangers both to themselves and to academic freedom of some of the proposed legislation, the lobby group called Friends of DNA quickly convened. Its membership included presidents, vice-presidents, senior administrators, and legal advisors of many of the most prestigious U.S. universities, as well as some government servants and individual scientists. Friends of DNA had the advantage of the services of two professional lobbyists working in Washington for Harvard University, Daniel Moulton and Nancy Nixon. Unfortunately we were denied permission to reproduce examples of the memoranda issued to Friends of DNA by Moulton, Nixon, and their colleagues, but they contained succinct analyses of the various pieces of legislation, voting records of members of Senate and House Committees, and advice as to which of those members to contact and the deadlines for such action. Friends of DNA received through their lobbyists a copy of the *Higher Education Daily* of October 31, 1977 (Doc. 6.26), which gives something of the flavor of the other memoranda.

Moves to pass federal legislation withered in the spring of 1978, leaving the situation unchanged regarding compliance with the guidelines by private laboratories, which was initially the chief reason for attempts at legislation. Such compliance remained purely voluntary. Local government, too, remained free to pursue parochial regulations, and two state legislatures, New York and Maryland, subsequently passed bills essentially requiring compliance with the NIH guidelines but imposing nothing stricter.

In New York this relatively happy outcome had required a considerable lobbying effort. In 1977 both houses of the state legislature passed the strict bill that State Attorney General Lefkowitz had drafted before it was vetoed by Governor Hugh Carey (Doc. 6.27). Elements of the environmentalist movement did not, however, drop the matter, and as a compromise a new and much less restrictive bill was passed in 1978 and signed by the Governor that summer. We give the last word in this chapter to Norton Zinder (Doc. 6.28), who played a leading role at state (New York) and national levels in lobbying against unreasonable and detrimental legislation.

Document 6.1

August 6, 1976
Science (**193**:468)

Recombinant DNA at White House

The issue of recombinant DNA research has been formally brought to the attention of the White House. In a letter of 19 July to President Ford, Senators Edward Kennedy and Jacob Javits urge him to make all such research, including that being conducted by industry, subject to federal control.

The implication of the letter is that if the White House fails to act, Congress will.

The research guidelines issued by the National Institutes of Health this June lack the full force of law and in any case apply only to NIH grantees. Kennedy and Javits, the chairman and prominent Republican member, respectively, of the Senate health subcommittee, are concerned that much recombinant DNA research would not be subject to any control.

"We urge you to implement these guidelines immediately wherever possible by executive directive and/or rulemaking, and to explore every possible mechanism to assure compliance with the guidelines in all sectors of the research community," runs the key passage of the senators' missive.

The proper scope of the guidelines is an issue that was raised at the public hearing convened in February by NIH director Donald S. Fredrickson. Fredrickson was urged by Peter Hutt, former general counsel for the Food and Drug Administration, to make the guidelines apply to everyone. If a significant loophole were left, Hutt implied, Congress would act to fill it.

One legal solution Hutt suggested was that the Secretary of Health, Education and Welfare should declare the guidelines universal in scope by invoking an obscure section of the Public Health Service Act, one which authorizes action to prevent the spread or introduction of communicable diseases.

In the event, Fredrickson decided to make the published guidelines applicable only to NIH, but recommendations have been sent up to the Secretary of HEW for extending their scope. The Secretary has not yet taken action, and there is some doubt that he will rush to do so. "In an election year, regulation of the private sector by government is not one of the things Republicans like to do," observes a Congressional staff aide.

The Kennedy-Javits letter praises the guidelines as such, saying that they are "a responsible and major step forward and reflect a sense of social responsibility on the part of the research community and the NIH." The "glaring problem" with them that caught Kennedy's interest, according to a staff member of the Senate health subcommittee, was their limited range of applicability. Kennedy's attention was first drawn to the problem by LeRoy Walters, a bioethicist at the Kennedy Institute, Georgetown University, who participated in the February hearing at NIH. Kennedy has been following the events in his home state of Massachusetts, where the city council of Cambridge recently resolved that there should be a moratorium on recombinant DNA research requiring physical containment conditions appropriate to experiments of high or moderate risk. Kennedy "doesn't disagree with the process going on at Cambridge," says a staff aide. The Senator feels, however, that what happens countrywide is different from legitimate local prerogatives, and that his concern with the issue should be on the national level.—NICHOLAS WADE

Document 6.2

September 22, 1976
Kennedy Opening Statement

from the office of

Senator Edward M. Kennedy

of Massachusetts

OPENING STATEMENT OF SENATOR EDWARD M. KENNEDY
AT A HEARING OF THE SENATE HEALTH SUBCOMMITTEE
ON RECOMBINANT DNA RESEARCH AND THE NIH GUIDELINES

Wednesday, September 22, 1976
9:15 a.m.
Room 4232 Dirksen Building

For Release: September 22, 1976
9:15 a.m.

This morning the Senate Health Subcommittee returns to the subject of genetic engineering. No area of scientific inquiry has generated the degree of public controversy which currently surrounds those scientists who have learned to create new forms of life in the laboratory.

I believe that the vigorous public debate over the potential benefits and risks of this work is much to the good. An important precedent is being set. It is that the implications of technological advances must be carefully considered early on, and must be considered in public processes with wide participation from as many diverse elements of society as is possible. Recombinant DNA research presents a prototype of the problems our society will face over and over again as technology develops. The issues go far beyond safety questions--in many ways, those are the easiest to answer. The real problem is to understand the social consequences of what science can now enable us to do. As long as the development of technology outstrips man's capacity to understand the implications of that technology, the chances for serious errors in judgment are increased and the possibility for serious societal disruption is very real.

Thus, I believe the debate over genetic engineering must go on. Scientists must tell us what they are capable of doing, but we as members of society must decide how it should be or whether it should be applied. Congress can not legislate an appropriate answer in this matter. But it can and should take the lead in assuring that these issues are discussed publicly and by as broad a segment of the population as possible. The plain fact is that genetic engineering has the capacity to change our society. How do we want it changed? What uses can we make of this knowledge? What degree of change is desirable and at what rate? What kind of society do we want to become? These are the questions that must be answered and the development of those answers must involve as many Americans as possible.

While these discussions take place, however, and while the research continues, we face the immediate, practical problem of assuring that the research is as safe as possible with the fewest risks to the fewest people.

Today we will focus on the guidelines that the National Institutes of Health have developed for recombinant DNA research. Although not perfect, the guidelines are generally considered to be an important step in the reduction of risk in this area. The problem is that many groups doing research <u>do</u> <u>not</u> fall under these guidelines.

It is my belief that at the very least, all federal agencies should comply with the NIH guidelines. These guidelines need not be viewed as fixed or final. But they should be observed while appropriate modifications are made. I also believe that industry research in this area should conform to NIH guidelines. The potential for industrial abuse is present in part because of the tradition of secrecy which surrounds potentially profitable research. There are ways to protect legitimate trade secret information and still have federal monitoring of DNA research in industry by NIH. I would hope that industry would voluntarily comply with the guidelines. If they don't, I will have to consider legislative action.

My concerns over industry compliance have been raised by the refusal of the general electric company to testify at this hearing. Their excuse was that their top scientist in this area is unavailable. They refused to send a senior corporate official to discuss General Electric's overall policy in this area. We wanted to know what research they are doing, what they contemplate, and whether they can comply with the NIH guidelines.

This Subcommittee is grateful for the cooperation of the Pharmaceutical Manufacturers Association in this hearing. But the actions of G.E. make it clear that we will have to have a complete accounting of <u>all</u> industrial research and development in this area. I have directed staff to vigorously pursue the matter as a high priority activity.

Document 6.3

October 15, 1976
Science (**194**:303)

Recombinant DNA: A Critic Questions the Right to Free Inquiry

When the issue of recombinant DNA came up last month before the Senate health subcommittee, the following exchange occurred between the senators and Robert Sinsheimer, chairman of the biology division at Caltech:

Kennedy: Do you agree that in terms of magnitude this is of as great significance as the splitting of the atom?

Sinsheimer: What this technology does is to make available to us the complete gene pool of evolution. We can take the genes of one organism and recombine them with those of others in any manner we wish. To my mind that is an accomplishment as significant as the splitting of the atom.

Schweiker: Are you saying that all that has gone before, we now have the power to change in some way—the evolutionary process?

Sinsheimer: Yes.

The senators did not follow up on the implications of the comparison they were drawing, but the analogy between nuclear energy and the recombinant DNA technique is one that Sinsheimer himself has raised. In a voice too gentle and well-mannered to receive much attention, he has been asking whether the scientist's claim of an absolute right to free inquiry should not sometimes be limited in the interests of society. Nuclear energy may yet turn out to be one such field that would better have remained forbidden territory. The recombinant DNA technique, he suggests, could prove to be another. "To impose any limit upon freedom of inquiry is especially bitter for the scientist whose life is one of inquiry; but science has become too potent. It is no longer enough to wave the flag of Galileo," Sinsheimer said in a lecture last year to the Genetic Society of America.

With the notable exception of Erwin Chargaff of Columbia, Sinsheimer has stood virtually alone in his doubts about the wisdom of going ahead with the recombinant DNA technique, a method of genetic engineering which in essence allows each gene in an organism to be manipulated, whether for study or practical purposes. Most biologists believe that the work should proceed under appropriate safeguards. That approach has prevailed, and is embodied in the guidelines for research issued by the National Institutes of Health this June. Most of the public debate about the technique has revolved around what particular level of safeguards is appropriate, and public attention now rests on the next logical stage in the approach, that of ensuring that the NIH guidelines are followed by other government agencies and by industry (see box).

The approach of the NIH guidelines is a reasonable and responsible first step which has the full endorsement of those who first drew attention to the possible hazards of the technique, including biologists such as Paul Berg, Maxine Singer, David Baltimore, and Norton Zinder. How can Sinsheimer both differ from such eminent authorities and have a case worth making? The answer, perhaps, lies with the difference of perspective in which the two sides view the recombinant DNA technique. Those behind the guidelines approach see the problem as being one of how to take the next step in a way that will bring the most benefits to science and society while keeping the risks to a minimum. In Sinsheimer's view, the question is whether, in the light of what we know of history and of the process of evolution, it is prudent to take that step at all.

Sinsheimer is a member of the National Academy of Sciences and editor of its *Proceedings*. Cracking the unusual structure of the virus ϕX174 is one of his feats. He has not always been skeptical of the fruits of scientific progress. As those who disagree with him on recombinant DNA find frequent occasion to recall, he was once an ardent advocate of genetic engineering. In an article of 1970 he looked forward to the advent of human genetic engineering as a way to escape the tyranny of heredity and improve man's intellect and other capacities.

Now Sinsheimer believes otherwise. In a talk given this June at the University of California, he warned of the dangers that may accompany new knowledge. " 'Know the truth and the truth will make you free' is a credo carved on the walls and lintels of laboratories and libraries across the land," Sinsheimer observed. But, he added,

> We begin to see that the truth is not enough, that the truth is necessary but not sufficient, that scientific inquiry, the revealer of truth, needs be coupled with wisdom if our object is to advance the human condition. . . .
>
> The twentieth century has seen a cascade of magnificent scientific discoveries. Two, in particular, have extended our powers far beyond prior human scale and experience. In the nucleus of the atom we have penetrated to the core of matter and energy. In the nucleic acids of the cell we have penetrated to the core of life.
>
> When we are armed with such powers I think there are limits to the extent to which we can continue to rely upon the resilience of nature or of social institutions to protect us from our follies and our finite wisdom. Our thrusts of inquiry should not too far exceed our perception of their consequence. There are time constants and momenta in human affairs. We need to recognize that the great forces we now wield might—just might—drive us too swiftly toward some unseen chasm.

Genetic engineering by the recombinant DNA technique is a thrust of inquiry that may be perilous in its consequences, Sinsheimer has come to believe. "I do fear," he said in the same talk, "that there are potentially grievous risks—of the spread of slow viruses or of cancer or of new pathogens, yet unborn, evolved from our inventions."

What has turned Sinsheimer from advocate to skeptic, from enthusiasm about genetic engineering to misgivings so grave as to set him on a different path from the mass of his colleagues? While in Washington for the Senate hearings on recombinant DNA, Sinsheimer explained in an interview some of the reasons for his change of mind.

In his earlier view of genetic engineering, he says, "I thought of very careful experiments to replace gene A with gene B—it never occurred to me that anyone would do a shotgun experiment [in which all the genes of an organism are manipulated more or less at random]." He was also more optimistic then that genetic engineering could be controlled.

Asked why more colleagues do not share his view, Sinsheimer replies that "I have been thinking about these things for longer than most of the people who are now more sanguine than I am. Scientists can be very insular, and to some degree they have to be. To be a good scientist takes an awful lot of dedication, and you have to really believe in it and believe that what you are doing is good and beneficial. It is such people who are less likely to entertain other points of view."

A certain narrowness of view is Sinsheimer's chief complaint with the NIH guidelines governing research on recombinant DNA. "This is a technology that was developed by scientists to solve their own problems, and they are still locked into that mode of thinking," he observes.

When he reviewed the guidelines, at the NIH's request, Sinsheimer found that they had dealt reasonably well with the immediate health hazards but "had given no thought to the evolutionary question." As to why the committee set up by the NIH overlooked this question, Sinsheimer remarks that it was "implicit for the guidelines committee to concern itself with health hazards—it simply was not constituted to cope with the larger issues."

By the "evolutionary question" Sinsheimer refers to the fact that some of the genetic manipulations made possible by the new technique may be of a type which evolution has been at pains to prohibit. Many recombinant DNA experiments require the insertion of genes from the cells of higher organisms, or eukaryotes, into prokaryotic cells such as bacteria. Sinsheimer conjectures that the apparent barrier to genetic interchange between eukaryotes and prokaryotes is one that is there for good reason, and that to transgress it by creating prokaryote/eukaryote hybrids in hundreds of laboratories throughout the world is to risk causing unpredictable—and irreversible—damage to the evolutionary process. For example, the barrier might be there to protect the genetic machinery of higher cells from prokaryotic take-over.

Sinsheimer has not been greatly impressed with the arguments brought forward against his barrier theorem. A common objection, raised for instance by Baltimore at last month's hearings, is that the prokaryote/eukaryote barrier is being broken all the time in nature, as for example when bacteria in the gut take up digested fragments of DNA. "That is an ad hoc argument. It is even worse than ad hoc—it is contrived," says Sinsheimer. Since there is no evidence that prokaryotes and eukaryotes do in fact exchange genetic information, those who say that microorganisms are always taking up eukaryotic DNA have to couple this assumption with the hypothesis that whenever it occurs, the organism dies out.

Photo by Floyd Clark/Caltech.

Robert L. Sinsheimer

Another argument is that all the genetic combinations that can occur have already occurred in the course of evolution. Sinsheimer feels intuitively that this is not the case. He observes too that many who make the argument also speak of the benefits of genetic engineering, a proposition which is predicated on the opposite assumption.

To those who contend that any new and harmful organisms accidentally created by the technique would not survive in nature, Sinsheimer replies that one could certainly design some quite fearsome microorganisms by the recombinant DNA technique. Can one be certain that such organisms, if they arose inadvertently, would always be at a disadvantage?

The intentional misuse of the recombinant DNA technique is another of the broader issues which Sinsheimer feels has been neglected in the guidelines approach. The guidelines deal with the immediate health hazards that scientists can foresee, but they don't take account of the hazards to other sectors of society, of the fact, for example, that the technique can be used by other sectors besides scientists, such as the military or terrorist groups.

Sinsheimer considers that deliberate misuse of the technique is a serious possibility. The problem is analogous to that of nuclear terrorism, he says. "It may well be that there are some technologies that you should not use, not because they can't work but because of the social dangers involved and the repression that would be necessary to prevent social danger."

The nuclear genie is now out of its bottle for good or ill, and the crucial time of grace for instituting control over the recombinant DNA technique is probably already over. Has a unique opportunity been missed? Sinsheimer returns to his theme of the consequences of new knowledge: "We have gone along for several hundred years with the belief that knowledge and the means for acquiring knowledge are always beneficial.

"The situation that first led anybody to question that assumption was the atomic bomb. I think that a lot of people wish there were a way to forget all about nuclear physics but there is not. For a while, many people hoped that that was an anomaly.

"But now here comes another one. How do you cope with this new observation that some kinds of knowledge and some kinds of technology can be very dangerous? We have no assurances that science will not lead us into a very dangerous world.

"How do you control that without interfering with a lot of the freedoms that people [scientists] have cherished? That is something we are only groping toward. . . .

"How do you make policies for an issue which may take 50 years to resolve? Our government, at least in the past, has not been ready to make long-term decisions.

"Some of my colleagues feel that it is the scientist's job to do science, and society's job to cope with what he does. I disagree with this in principle. The scientist must keep the public informed and involved because nobody else will.

"It is entirely possible, as Chargaff said, that the future may curse us [for the consequences of the recombinant DNA technique]. Really only the interests and concerns of the scientific community were involved in formulating the guidelines."

Those who formulated the guidelines have shown a curious reluctance to come out and debate Sinsheimer at his own broad level of argument. The pursuit of knowledge is held even by nonscientists to be a distinguishing value of society. Is that the answer to Sinsheimer's belief that the right of free inquiry should not be absolute in the case of recombinant DNA? If it is, nobody has rushed forward with it in any of the public documents intended to justify the NIH guidelines.

Sinsheimer believes that one step leads inevitably to another, that the recombinant DNA technique is the beginning of the genetic engineering of bacteria, of plants and domestic animals, and ultimately of man. "Do we want to assume the responsibility for life on this planet . . .? Shall we take into our own hands our own future evolution?" Sinsheimer has asked. If any of his opponents had heard the question, they might perhaps have answered to the effect that since man has now insulated himself from Darwinian pressures, some other means of genetic improvement must be found to assure his continued progress as a species. But Sinsheimer, who seems to have a virtual monopoly of long-range thought about the issue, has also provided an answer to the question. He says, in essence, that we aren't clever enough to know, so shouldn't yet try.

The recombinant DNA technique will clearly bring to birth a technology so potent that even its slightest deviations from the intended path may cause grievous perturbations in society at large. Historians half a century from now will no more blame the architects of the guidelines for failing to cope with every possible contingency than do their contemporaries blame Henry Ford for every highway casualty. Yet they may take a certain interest in the quality of the arguments being relied on for riding roughshod over the reservations articulated by Sinsheimer. Would they be very favorably impressed with what is on the record so far?—Nicholas Wade

"It is entirely possible, as Chargaff said, that the future may curse us [for the consequences of the recombinant DNA technique]. Really only the interests and concerns of the scientific community were involved in formulating the guidelines."

Document 6.4

May 1977

Bulletin of the Atomic Scientists (**33**:16)

Biologist Richard P. Novick: The moral-philosophical questions must be confronted.

Present controls are just a start

The public will not be protected until the NIH guidelines are strengthened, tightened and made universal

Richard P. Novick

In the statement that follows, I will address the following points.

- Controls for ribonucleic acid and deoxyribonucleic acid (R-DNA) research are necessary and should be universal; decisions affecting the conduct of this research should ultimately be enforced on a worldwide basis.
- Absolute containment of microbes is, like absolute zero or a perfect vacuum, an unattainable ideal.
- The NIH guidelines are an honest attempt, but one that falls short in a number of ways.

I would like to say, quite frankly, that my own personal interest in these experiments colors my view inevitably. In fact, given that I would like to do certain experiments involving R-DNA—experiments that I believe to be non-dangerous—I am quite unable to distinguish between the following two alternatives as the basis for this belief:

1. I am convinced they are not dangerous, and therefore it is okay to do them; or
2. I have convinced myself they are safe precisely because I want to do them!

And I would defy *anyone* whose self-interest is involved in a particular line of activity to make that distinction.

One can view the biohazards of R-DNA research on several levels, bearing in mind, first that it involves at present primarily microorganisms and, second, the basic underlying fact that microorganisms are self-reproducing and invisible. Therefore, the escape from the laboratory of an organism containing R-DNA, which I will refer to as a "novel biotype," must be regarded as irreversible. Survival and establishment of the escapee would depend solely upon its viability in the natural environment and would not be subject to intervention by man. Added to this is the possibility that even if the organism itself is unable to become established, it may donate its recombinant DNA molecules to others that are already established in the environment and are therefore viable.

At the simplest level, one should consider the consequences of purposeful or accidental dissemination of novel biotypes containing R-DNA. In such cases, either (a) the organism may have unpredictable

Richard P. Novick, M.D., is Chief of the Department of Plasmid Biology at the Public Health Research Institute in New York. He was chairman of the group of scientists known as the "plasmid group" that prepared the major preliminary proposal on DNA guidelines for the 1975 Asilomar conference.

and harmful properties; or (b) the known properties of the recombinant DNA may be harmful.

In this connection, I should like to mention that in the overwhelming majority of cases, the introduction of a viable organism into a new habitat has had harmful and usually unpredictable consequences. Therefore, I would argue that any modification of the biosphere through the release of a novel biotype constitutes a biohazard by definition.

At another level, one should consider the moral-philosophical question of how deeply man should go into interfering with natural heredity, especially his own—make no mistake, there are *no visible limits* to the exploitation of R-DNA technology.

For a number of reasons, I will confine my remarks to the first of these levels and I would like at this point to give an example of how a well-intentioned R-DNA experiment could go astray. I am giving this example because it involves a very specific experiment with very specific foreseeable consequences, rather than the highly speculative ones about which we have heard so much.

A Specific Experiment

Cellulose digestion. It has been proposed to develop a cellulose-producing *E. coli* for the purpose of enabling man to digest cellulose fibers. Presumably the gene for cellulose would come from one of the various cellulose-degrading bacteria. This experiment, under the NIH guidelines, could be performed under the physical containment level designated as P3, or P2 if it can be shown that the donor bacteria can exchange plasmids with *E. coli,* or if the DNA can be purified and shown not to contain "harmful" genes, that is, harmful *unknown* genes. But, is the gene for cellulose, which is, in this case, the desired gene, a harmful gene or not?

On developing the organism, the investigator would next wish to test it for its ability to permit humans to digest cellulose fibers and so benefit nutritionally. It is very possible that he might decide to go ahead on the assumption that the gene was probably beneficial and surely harmless. It is conceivable that the cellulose gene might give the organism an enhanced ability to survive in the human colon, or at least not hinder it. The human tests would then inevitably be tantamount to releasing the recombinant organism which might then become a widespread inhabitant of the human colon.

Assuming for the moment that the only new biological activity of this organism was that of cellulose digestion, it would seem worthwhile to ponder some of the possible consequences.

While cellulose digestion might enable us to obtain nourishment from non-nutritive sources, it might have possibly disastrous side effects. The human digestive system has evolved over many eons to deal with a certain type of natural diet. It has been shown that the presence of a substantial undigestible residue, mainly of cellulose fibers, is of major importance for the maintenance of normal bowel function.

Indeed, it has been shown that our highly refined, low residue diet is probably responsible for a number of non-malignant but rather unpleasant intestinal disorders such as constipation, irritable colon and ulcerative colitis, diverticulosis, adenomatous polyps, hemorrhoids, etc. And it has been suggested that such a diet is also responsible for the increased frequency of cancer of the colon and rectum that has been seen in our society in comparison with other groups that eat a more natural, high residue diet.

While the causal chain here is far from established, the possibility of trouble along these lines is a very real one. There is nothing in the NIH guidelines to prevent these experiments from being done, and it is perfectly possible that a scientist, either ignorant of this possibility or aware of it but deciding that the risk is acceptable, could go ahead with this line of experimentation and accidentally release the resulting organism.

I would grant that the NIH's guidelines could be tightened so as to include this experiment. I would argue, however, that they could not, *in principle,* be tightened or otherwise modified to include all such experiments. It seems to me inevitable that somewhere along the line we either will underestimate a possible danger of this sort or fail entirely to perceive it. Granted that this particular experiment may perhaps not now be undertaken, because of exposure and public commentary such as I have given here, the number of other possibilities is limited only by the imagination.

It should also be noted that I am able to predict these possible adverse consequences only because there happens to be information on the relation of diet to diseases of the colon. This information is relatively new, however, and it is more than likely that other such "beneficial" experiments will be envisioned for which similar information does not happen to be available.

The NIH guidelines were formulated on the basis of containment; the only consideration given to the question of performing one or another experiment was whether or not it could in theory be safely contained. I would like, therefore, to discuss briefly the question of safety with respect to containment. There are two questions here:

- Are the NIH guidelines good enough to ensure safety, given absolute compliance?
- Is absolute compliance likely, or even possible?

I should like to answer both questions with a qualified negative on the basis of the following considerations pertaining to the physical and biological containment systems described in the NIH guidelines.

Physical Containment

The NIH guidelines were formulated on the basis of matching what was perceived to be a graded series of risks to a correspondingly graded set of precautions, or containment levels, designated P1, P2, P3 and P4 (see box).

Among the latter, the only one

that is designed *in principle* to be absolutely effective is that designated P4. However, P4 containment (which was developed some time ago for the handling of ultra-dangerous pathogens at installations such as Fort Detrick, Md.) at its best does not provide perfect protection—laboratory accidents and infections, some fatal, have occurred and are probably unavoidable. This problem is underscored by the existence of laws prohibiting the importation of certain microorganisms into the United States *for any purpose whatsoever,* and by the existence in the guidelines themselves of a blanket prohibition of certain types of experiments.

The levels below P4 are increasingly less effective so that the containment levels constitute *in principle* a graded series of probabilities of escape of organisms. Aside from the unanswerable question of whether or not the assessments of relative risk for various experimental situations are valid, the fact remains that the system is *designed to permit a higher probability of escape* for organisms regarded as lower on the risk scale.

In other words, at its very best, the system is not designed to prevent *absolutely* the escape of any recombinant organisms except those perceived a priori to be the most dangerous. Moreover, one may with a great degree of certainty assume that in practice the containment systems will not operate at their optimum level of effectiveness.

On the one hand, microbiologists who are accustomed to working with dangerous pathogens will generally have gone through an elaborate training procedure and will be greatly concerned for their own safety. Such workers must, in general, be presumed to be very careful, indeed. Nevertheless, laboratory-acquired infections with serious pathogens occur with a disturbingly high frequency despite these precautions.

On other hand, most of those for whom these guidelines are intended are not now trained in microbiological safety techniques. In fact, most of them have chosen to work with *E. coli* and other relatively innocuous microorganisms *precisely* because with such organisms one didn't have to worry about contamination, dissemination and infection. One could, therefore, work with great rapidity, unhindered by any consideration of safety except that of protecting one's cultures from gross contamination by extraneous organisms.

I am not at all comforted by the thought of the bureaucracy that this set of proposals will engender; however, I am dissatisfied with the available alternatives.

Therefore, it is my belief that even with the best of intentions, it is extremely unlikely that all laboratory heads (who are the responsible authorities under the NIH guidelines) will be able immediately to master and to adopt the full range of safety precautions recommended by the NIH guidelines, and described in an extensive treatise on physical containment in Appendix D of the guidelines.

What is more, even if all of the procedures recommended in the guidelines were carried out to the letter, there would still be gross breaches of containment inherent in the system. This is, first, because accidents are inevitable and, second, because the guidelines themselves are seriously deficient and, in addition, leave much to the imagination and to the discretion of the individual investigator.

The problem here is that the average biochemist-molecular biologist is not likely to have the training and tradition of safety awareness either to perceive the various deficiencies or to exercise optimum discretion.

Let me give four simple examples in the area with which I am most familiar, namely handling of bacterial cultures with respect to the containment of aerosols: air suspensions of very fine (infectious) particles that are dispersed by air currents, are the major route of inhalation infection, and do not settle appreciably with time.

- Bacterial cultures are aerated either by shaking or by bubbling, both procedures that generate extensive aerosols. The only reference in the guidelines, including Appendix D, to these universal practices is on page 27913, where it is mentioned that "shaking machines" should be operated in a biological safety cabinet for containment levels of P2 or above. However, it is not entirely clear if "shaking machines" refers to aeration of bacterial cultures, nor is there any mention of the infectious aerosol caused by bubbling and of what to do about it.
- On page 27929 there is a serious misstatement regarding aerosols in connection with the use of a blender: "Before opening the safety blender bowl, permit the blender to rest for at least one minute to allow *settling of the aerosol cloud*" (emphasis added).
- A recommendation for opening ampoules on page 27928 is probably of unproven effectiveness and seems positively dangerous:

> The researcher uses an intense, but tiny, gas-oxygen flame and heats the tip of the hard glass ampoule until the expanding internal air pressure blows a bubble. After allowing this to cool, he breaks the bubble while holding it in a large low temperature flame; this immediately incinerates any infectious dust which may come from the ampoule when the glass is broken. Preliminary practice with a simulant ampoule of the same type actually in use is necessary to

develop a technique that will not cause explosion of the ampoule.

Here, there is, on the one hand, no proof that the low-intensity flame is effective. On the other hand, this technique is likely to lead to accidental explosion of ampoules, even in the most careful of hands.

• Wire innoculation loops are in near universal use for transferring cultures, and their use, especially their flame sterilization, has been shown to disseminate infected aerosols. There is no mention of this in the guidelines, and I would argue that the level of awareness of the problem is very low.

There are many instances of similar problems that could be quoted. I hope that these will suffice to make the point.

I think, therefore, that there will be a great deal of variability among laboratories and among individuals in the rigor with which the containment procedures are practiced. And there are two considerations here:

First, there are those who, despite the best of intentions, will have a difficult time observing the required discipline.

Second, and perhaps more important, there are those who believe that all of the present concern over the biohazards of novel biotypes is absurd: that is to say, there are *no* biohazards inherent in R-DNA research. There are a number of rationales for this belief, which I won't elaborate upon here. But I have very serious doubts that such people would be able, even if willing, to adhere rigorously to a prescribed set of elaborate and rather inconvenient procedures.

In addition, there is the question of natural human error, which is surely inevitable. I mean such things as spills, mis-labelings, various sorts of contamination, mechanical failures of physical systems, etc. In fact, to supervise the operation of a rigorous containment system even as outlined in the guidelines, for a laboratory of any size, would be an extremely time-consuming and, I might add, boring and unpleasant job.

It would be well worthwhile to avoid establishing a regulatory system that stood as an adversary to the scientific community.

Biological Containment

• *On the letter and the spirit of the guidelines.* One laboratory has been cloning fragments of eukaryotic DNA in an *E. coli* K12 recipient organism that contains a conjugative plasmid. Another laboratory has described the development as a cloning vehicle of a rather dangerous plasmid. This is a conjugative plasmid with the highest known degree of promiscuity (that is, the ability to infect very distantly related organisms). These experiments are certainly prohibited by the guidelines, although the prohibition is not spelled out precisely as such.

The relevant text, under biological containment (page 27915) reads: "The host is always *E. coli* K12, and the vectors include nonconjugative plasmids and variants of bacteriophage λ." I think it is very significant that the heads of both of these laboratories evidently consider the guidelines and all associated activities to be absurd and unnecessary.

• *Safe organisms.* A great deal of effort has been spent recently to develop host organisms and cloning vectors that have a greatly reduced chance of survival outside the laboratory. This effort has been highly successful on paper, but there are a number of potentially serious problems with it as a method of safety.

• *The human factor.* The bacterial host, χ1776, developed by Dr. Roy Curtiss and coworkers, is a very poor grower, is inconvenient to use, and requires constant monitoring to verify its many biological properties. Unless extreme care is exercised in its handling, it is very likely to be replaced in culture by a rapidly growing mutant, or by an exogenous contaminant.

Although the sentiment has been widely voiced that what is wanted is a fail-safe organism so that one can happily forget about physical containment, it is a certainty that χ1776 is not such a strain. In fact, I seriously doubt that it would be possible to produce one. The idea of a fail-safe biological containment system is an illusory and dangerous misconception.

• *Biological problems.* Along with the reduced viability of χ1776 there is its sensitivity to lysis (dissolution) under various conditions, notably the action of bile salts and detergents as well as mere growth outside the laboratory. Lysis results in the immediate release of all contained DNA. While there are some situations in which it is very likely that free DNA will be rapidly degraded, many others have not been tested. And I could quote at least two different examples of bacterial transformation experiments that were spoiled by the presence of plasmid DNA as a contaminant. In both cases, this contaminating DNA was taken up and expressed by the experimental organisms.

Finally, there is the possibility that crippled organisms may find some unsuspected escape route from their biological straightjackets. For example, the χ1776 strain is severely compromised in its ability to synthesize the envelope or wall that protects the bacterial cell. Some bacteria under such conditions undergo a transition to a stable, wall-less state referred to as an L-form. These L-forms are morphologically difficult to identify and are thought to be capable of inhabiting the tissues of a higher organism more or less indefinitely in a form that is very difficult to detect, identify and culture.

* * *

My major conclusion from this argument is that rather well-defined controls are entirely appropriate for recombinant DNA research. Parenthetically, such controls may be appropriate also for certain other types of genetic research, such as that involving chemical mutagens, and perhaps this will be considered in the near future.

The present control system for R-DNA research, based upon the NIH guidelines, constitutes a preliminary attempt which is good in many ways but which does not, in fact, provide adequate protection for two major reasons: First, the NIH guidelines themselves need to be strengthened and tightened, and second, their applicability needs to be extended.

At present, in my view, if the escape of a particular recombinant biotype is unacceptable, then that biotype should in general not be constructed, except in cases of overriding scientific necessity, in which case P4 containment should be used. Conversely, experiments involving R-DNA should ordinarily be undertaken outside of a P4 facility only if the possible escape of the novel organism is an acceptable outcome.

Precisely how and by whom the criteria of acceptability are decided upon and applied is to me the crucial question and the one to which a great deal of attention should now be given. At the very least, these criteria should be formulated by people with diverse backgrounds.

In the meantime, a glance at the guidelines suggests that those who framed the guidelines did, indeed, envision two classes of experiments: those where escape is deemed acceptable (that is, physical containment levels P1 and P2); and those where it is not (that is, P3 and P4). I believe that this division is appropriate, in general, if not in detail. However, I would like to see a number of improvements in the over-all control system:

- Broadening of the applicability of the controls. This could begin at the state level but must surely go beyond, for fairly obvious reasons.
- The strengthening and more uniform codification of the procedural guidelines.
- The development of training centers and courses for the handling of hazardous microorganisms and the requirement of a specific level of proficiency in this area for workers involved in R-DNA research.
- The consideration of enforcement and sanctioning, such as publication and patenting policies, and licensing of laboratories.
- The development of an international policy on the location and construction of protective facilities, especially P4.
- Worldwide acceptance of the principle of controls for this type of research and, ultimately, the development of guidelines on an international level.

I am not at all comforted by the thought of the bureaucracy that this set of proposals will engender. However, I am dissatisfied by the available alternatives. Practically speaking, I think the NIH guidelines should be used as a good first draft and should probably be enacted into legislation with a very liberal provision for modification and with a strong provision for an appropriate training program and licensing procedure.

Practically speaking, as has been pointed out by many, it would be well worthwhile to avoid establishing a regulatory system that stood as an adversary to the scientific community. While this ideal may, in practice, be difficult to achieve, it is to be hoped that the public and the scientific and industrial communities will exercise sufficient good faith to permit the establishment of a cooperative relationship in this respect. □

Levels of containment

The National Institutes of Health lists these four levels of physical containment for laboratory experiments involving recombinant DNA molecules:

P1 level (minimal)—A laboratory possessing no special engineering design features. It is a laboratory commonly used for microorganisms of no or minimal biohazard under ordinary conditions. Work is generally conducted on open bench tops. Public access is permitted.

P2 level (low)—Similar in construction and design to the P1 laboratory. It must have access to an autoclave within the building. Although the laboratory is not separated from the general traffic pattern of the building, access is limited when experiments of low biohazard are being conducted.

P3 level (moderate)—A laboratory having special engineering design features and physical containment equipment. It is separated from areas which are open to the general public. Separation is generally achieved by controlled access corridors, air locks, locker rooms or other double-door facilities. Access to the laboratory is controlled.

P4 level (high)—A facility designed to contain microorganisms that are extremely hazardous to man or may cause serious epidemic disease. It is a separate building or a controlled area within a building, completely isolated from all other areas of the building. Access to the facility is under strict control.

Further details of the four levels of containment are specified in the NIH guidelines, published in the *Federal Register*, 41:131 (July 7, 1976).

Document 6.5

May 1977

Bulletin of the Atomic Scientists (**33**:12)

An imaginary monster

The only danger we face is the specter of untested regulations

James D. Watson

I greatly welcome this occasion to speak about recombinant DNA research and the way it is likely to impact upon the citizens of New York State.

Through this research we can insert genetic material from one organism into a second organism with which it does not normally exchange genetic material. My view is that it offers a marvelous opportunity to probe more deeply the organization and functioning of human genetic material. From such studies we may also obtain new insights into how our normal developmental processes may fail, or how we may fall victim to degenerative diseases like arteriosclerosis or muscular dystrophy, or how we can fight the harmful effects of the disease carrying bacteria and viruses.

Toward these goals, recombinant DNA technology can provide a major step forward allowing us to do experiments which would be effectively impossible without it. Not surprisingly, many of the better scientific laboratories in the world are beginning such experiments, and we anticipate a rash of marvelous new biological facts to emerge over the next several decades.

However, the very novelty of this technique has raised the question whether we should worry about the creation of new organisms never yet tested by the evolutionary process. Might not these new creations multiply out of control and lead to diseases to which we have never yet been exposed?

In response to this dilemma a number of scientists, of whom I was one, suggested that we refrain from certain classes of experiments until we thought more seriously about whether there might be experiments that we might want never to be done. We further proposed an informal moratorium on such experiments until a much larger group of concerned scientists could meet together to argue the situation.

To my knowledge none of us then was deeply concerned, but since others had expressed worry, we thought the responsible course was to inform the public of the power of this new technique and what might be the potential consequences. Here it is important to note that we did not then propose a moratorium on most forms of recombinant DNA research—only on certain experiments involving the genes of certain disease carrying viruses and bacteria.

When some months later we assembled in California at the Asilomar conference, there was a great diversity of opinion as to whether we were dealing with a real problem. I for one, together with Joshua Lederberg of Stanford Medical School, thought that the drafting of any set of formal guidelines was unnecessary.

For example, we did not believe that the insertion of the genes of higher organisms or viruses into the commonly used K12 strain of the colon bacillus *Escherichia coli* posed any realistic danger. Under most circumstances these foreign genes are unlikely to function, and lacking any positive evolutionary selection would eventually be eliminated from their new hosts. And even if they did function, the K12 strain is well known for its inability to multiply in humans. So most recombinant strains if not protected by the experimental condition of the laboratory would quickly expire.

Even if genetic recombination, occurring in the gut, transfers some recombinant DNA molecules from dying K12 cells to *E. coli* cells adapted to growth in human intestines, I would not be concerned. Such transfers will never be a frequent event and, even so, there is no reason to believe that *in vivo* recombinants would ever have such a selective advantage that they would out-

James D. Watson, a molecular biologist and co-recipient of the 1962 Nobel prize for medicine and physiology, is director of the Cold Spring Harbor Laboratory in the state of New York.

grow the normal gut bacteria already highly evolved to multiply in our intestinal environment.

My thinking at Asilomar was strongly influenced by the relative ease with which DNA molecules can pass into cells and be incorporated in a genetically active form into the chromosomes of the new hosts. When this phenomenon was first discovered in 1943, it was thought to be restricted to the pneumococci bacteria. But over the years we have learned of more and more experimental cases where genetically active DNA has been transferred to a variety of different bacteria as well as to the cells of many higher organisms. That being the case, it seems probable that such transfers will also occasionally occur normally *in vivo.*

The bacteria in our body, for example, must constantly be exposed to DNA released from either sick human cells or from free DNA present in our food. If a person is on antacid pills, there is a good probability that his intestinal bacteria would be exposed to and take up a tiny fraction of the DNA present in, say, raw oysters or beef steak tartar. And even without the antacid, the very rare DNA molecule must pass through the digestive tract to reach not only our intestinal bacteria but, also, the cells lining the gut.

How infrequent such events are is not yet known. But we should assume that on an absolute scale the transfer of genetically active DNA is a very common event, and that it generally occurs with no harmful consequence to host cells already well adapted to their own genetic messages.

As a result, I see no reason to be apprehensive about any experiments in which we transfer genes of higher cells to bacteria. I even include the situation where we would transfer genes from viruses known to cause cancer in certain animals, say the SV40 monkey virus or the human Adenovirus 12. To start with, they will most likely be unable to function. But even if they could, the most probable consequence will be an upset in the normal controls over the host DNA synthesis leading to quick extinction of their respective bacterial host. And even if the recombinant DNA containing bacteria released a so-called oncogenic (tumor forming) product, I fail to see how it could lead to a human cancer. Conceivably a few cells might divide at the wrong time but this would not be a hereditary process.

As to the danger that such genes might escape from their lab-created

Biologist James D. Watson: A marvelous opportunity to probe more deeply.

bacteria and enter a human cell, here again I would not be concerned. For example, already when we are very young, we are infected by a variety of viruses, all of which have the capacity to make cells cancerous. Why we don't all come down with multiple cancers as a result of such exposure to infectious viruses is still very unclear.

In any case, the occasional rare exposure to a bacterial-derived "cancer gene" should have negligible impact compared to the massive assaults we receive every time we are infected with one of the many, many viruses with oncogenic potential. So the claim that recombinant DNA research poses a cancer threat to mankind is total nonsense.

Unfortunately, the majority of attendees at Asilomar were not prepared to say that the matter was becoming hopelessly overblown. Instead they argued that even though no one was able to give any quantitative arguments as to dangers, it might be prudent to repress the most paranoid of their bad dreams by creating the so-called "safe *E. coli* K12 strains." These would be strains so filled with mutant genes that they would have great difficulty in multiplying in the laboratory, much less in the human gut. The argument prevailed that even if we all believed that the ordinary laboratory variety of *E. coli* K12 was perfectly safe, why not give society another factor of safety?

The trouble, however, with this way of thinking, as Joshua Lederberg was the first to point out, was that by creating the so-called "safe strains" the general public might come to believe that our ordinary laboratory strains of *E. coli* were, in fact, dangerous. His caution, however, was to no avail, and over the past 18 months massive efforts have gone toward creating "safe strains."

In a real sense, by emphasizing the concept of the safe strain at Asilomar, the scientific community announced it was worried that it might get the public in trouble. So the National Institutes of Health had no choice but to come up with formal guidelines. But given the absence of any firm facts implicating real dangers, the guidelines necessarily have to appear capricious, if not self-serving, in their details, and so are not easily defensible on any quantitative ground.

So what started out as an attempt of the scientific community to appear responsible takes on increasingly the aspect of a black comedy. Not only do we have an ever growing set of NIH regulations, but we are all here today. Instead of focusing ourselves on potential major genetic or carcinogenic dangers to the inhabitants of New York—like the carcinogenic flame retardants in all our children's pajamas, the carcinogenic dyes most women use to color their hair, PCBs (polychlorinated biphenyls), etc. (the list is frighteningly long)—we are thinking of creating a 1976 version of the fall-out shelter debacle of the early 1960s.

I'm afraid that by crying wolf about dangers which we have no reason at all to worry about, we are becoming indistinguishable from my two small boys. They love to talk about monsters because they know they will never meet one. □

Document 6.6

February 20, 1976
Hutt Letter to Fredrickson

COVINGTON & BURLING
888 SIXTEENTH STREET, N. W.
WASHINGTON, D. C. 20006

TELEPHONE: (202) 452-6000
WRITER'S DIRECT DIAL NUMBER
(202) 452-6300

TWX: 710 822-0005
TELEX: 89-593
CABLE: COVLING

EDWIN S. COHEN
OF COUNSEL

February 20, 1976

Donald S. Fredrickson, M.D.
Director
National Institutes of Health
Building 1, Room 124
9000 Rockville Pike
Bethesda, Maryland 20014

Dear Don:

.....The most extreme possibility of regulatory control would be a requirement that the Federal government approve every recombinant DNA molecule experiment before it is conducted. I believe this type of rigid and complete control is wholly unnecessary, and should not be adopted. Far less stringent controls would appear to be sufficient to assure public protection. One possibility would be to license all laboratories and/or researchers conducting this type of experimentation. A second possibility would be to license the materials used in these experiments, such as the enzyme and the source of the foreign DNA, when they are intended for this specific use. Either or both of these possibilities could be accomplished without any change in existing legislation, under the provisions of Section 361 of the Public Health Service Act, 42 U.S.C. 264, which authorizes the Secretary of HEW to promulgate regulations necessary to prevent the introduction or spread of communicable disease.

Sincerely yours,

Peter

Peter Barton Hutt

Document 6.7

February—March 1977

Congressional Bills, Excerpts

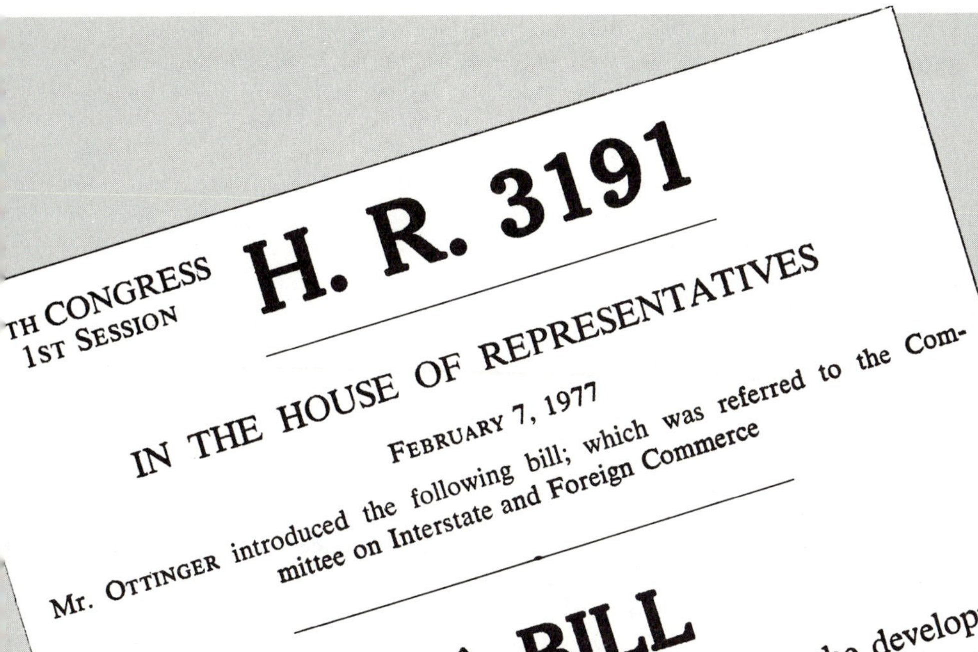

TH CONGRESS
1ST SESSION

H. R. 3191

IN THE HOUSE OF REPRESENTATIVES

FEBRUARY 7, 1977

Mr. OTTINGER introduced the following bill; which was referred to the Committee on Interstate and Foreign Commerce

A BILL

To provide for guidelines and strict liability in the developmen[t] of research related to recombinant DNA.

1 *Be it enacted by the Senate and House of Representa-*
2 *tives of the United States of America in Congress assembled,*
3 That this Act may be cited as the "DNA Research Act of
4 1977".

New York Post, March 8, 1977

Lawmaker urges halt to gene experiments

WASHINGTON (AP) — Rep. Richard Ottinger (D-N.Y.) has urged all U.S. scientific experiments in the transplanting of genetic material be suspended until controls are established to protect against potential threats to mankind.

Ottinger told the House subcommittee on health and the environment yesterday that uncontrolled research in genetics has implications similar to that of the splitting of the atom.

95TH CONGRESS
1ST SESSION

S. 621

IN THE SENATE OF THE UNITED STATES

FEBRUARY 4 (legislative day, FEBRUARY 1), 1977

Mr. BUMPERS introduced the follo... ch was read twice and referred to t... Public Welfare

CIVIL LIABILITY

22
23 SEC. 7. Persons carrying out research involving recom-
24 binant DNA shall be strictly liable, without regard to fault,
25 for all injury to persons or property caused by such research.

Document 6.8

March 8, 1977

95th Congress: Bill S.945, Excerpt

95TH CONGRESS
1ST SESSION

S. 945

IN THE SENATE OF THE UNITED STATES

MARCH 8 (legislative day, FEBRUARY 21), 1977

Mr. METZENBAUM introduced the following bill; which was read twice and referred to the Committee on Human Resources

A BILL

To regulate the conduct and development of research related to recombinant DNA, to establish a national commission to study recombinant DNA research and technology, and to provide citizens and governmental remedies.

1 *Be it enacted by the Senate and House of Representa-*
2 *tives of the United States of America in Congress assembled,*
3 That this Act may be cited as the "Recombinant DNA
4 Standards Act of 1977"...

17 ...TITLE II—STUDY OF RECOMBINANT DNA
18 RESEARCH AND TECHNOLOGY

19 PART A—NATIONAL COMMISSION FOR THE STUDY OF
20 RECOMBINANT DNA RESEARCH AND TECHNOLOGY

21 ESTABLISHMENT OF COMMISSION

22 SEC. 201. (a) There is established a Commission to be
23 known as the National Commission for the Study of Recom-
24 binant DNA Research and Technology (hereinafter in this
25 title referred to as the "Commission")...

1 ...(b) (1) The Commission shall be composed of thirteen
2 members appointed by the Secretary of Health, Education,
3 and Welfare. The Secretary shall select members of the
4 Commission from individuals distinguished in the fields of
5 medicine, law, ethics, theology, the biological, physical, and
6 environmental sciences, philosophy, humanities, health ad-
7 ministration, government, and public affairs; but six (and
8 not more than six) of the members of the Commission shall
9 be individuals who are or who have been engaged in recom-
10 binant DNA research and shall reflect diverse opinion on
11 safety and appropriateness. In appointing members of the
12 Commission, the Secretary shall give consideration to recom-
13 mendations from the National Academy of Sciences and
14 other appropriate entities. Members of the Commission shall
15 be appointed for the life of the Commission. The Secretary
16 shall appoint the members of the Commission within sixty
17 days of the date of the enactment of this Act.

18 (2) (A) Except as provided in subparagraph (B),
19 members of the Commission shall each be entitled to receive
20 the daily equivalent of the annual rate of the basic pay in
21 effect for grade GS–18 of the General Schedule for each day
22 (including traveltime) during which they are engaged in
23 the actual performance of the duties of the Commission.

24 (B) Members of the Commission who are full-time offi-...

1 ...COMMISSION DUTIES

2 SEC. 202. (a) The Commission shall carry out the
3 following:

4 (1) (A) The Commission shall (i) conduct a study
5 as to the appropriateness of continuing recombinant
6 DNA research; (ii) conduct a comprehensive investiga-
7 tion and study to identify the basic ethical and scientific
8 principles which should underlie the conduct of DNA
9 research; (iii) develop guidelines which should be fol-
10 lowed in such research to assure that it is conducted in
11 accordance with such principles; and (iv) make rec-
12 ommendations to the Secretary (I) for such administra-
13 tive action as may be appropriate to apply such guide-
14 lines to recombinant DNA research conducted or sup-
15 ported under programs administered by the Secretary,
16 and (II) concerning any other matter pertaining to
17 recombinant DNA research and technology.

18 (B) In carrying out subparagraph (A), the Com-
19 mission shall consider at least the following:

20 (i) The protection of researchers and the gen-
21 eral public from the dangers of recombinant DNA
22 research.

23 (ii) The role of assessment of risk-benefit
24 criteria in the determination of the appropriateness
25 of research involving recombinant DNA...

Document 6.9

Summer 1977

CoEvolution Quarterly (**14**:40)

Remarks on Recombinant DNA

BY JAMES WATSON

My position is that I don't regard recombinant DNA as a major or plausible public health hazard, and so I don't think that legislation is necessary.

My experience is partly as a scientist who has worked with DNA for . . . it's getting close to thirty years, and as an administrator, a director of a laboratory where I have moral and legal responsibility for our work with viruses, and I think about it.

When it was clear you could work with recombinant DNA, people initially had two worries. One was, a number of us were working with common viruses which can make cells cancerous – under some conditions – and they thought, well, what if you took the gene which is the cancer gene, put it into our bacteria – K-12, *e. coli* – maybe this cancer gene could then spread through the population.

The second worry was that you could take antibiotic-resistant markers in combinations which don't exist and by genetic manipulations produce a bacteria which is resistant to practically everything, and this then might spread through the population.

James Watson co-discovered, with Francis Crick, the helical structure of DNA, for which he received the Nobel Prize in 1962. He is director of the Cold Spring Harbor Laboratory in New York, which has been doing research on cancer, bacterial virus, molecular genetics, and protein synthesis. He has authored Molecular Biology of the Gene *(1968, 1970) and the famously candid account,* The Double Helix *(1968).*

In March of this year Dr. Watson was invited by Governor Brown's office to speak in Sacramento, California, on the subject of recombinant DNA research, which is being considered for restrictive regulation by the state legislature. Speaking at the same time was Dr. Halsted Holman of Stanford University, in favor of regulation.

Because most of the press on recombinant DNA has violently opposed further research, and those arguments are familiar, I am presenting here only Dr. Watson's views, which contain some co-evolutionary news. Casually polling my environmentalist acquaintances on recombinant DNA, I find the ones who are primarily political are very alarmed, the ones who do scientific work are not. Evelyn, while typesetting this, said that's a snobby remark. —*SB*

So a group of us, led by Paul Berg at Stanford, met one morning in April, 1974, at MIT and we said we would write a letter which said we shouldn't do those two types of experiments until a meeting was held in which we might discuss whether in fact these experiments posed any plausible hazard. We scheduled that meeting for February, 1975, at Asilomar in California.

As soon as the letter came out I began to get very queasy. A press conference was held, and the people who were organizing the meeting, of which I was not one, indicated that the scientists in renouncing their right to do particular research were doing something which hadn't been done before, and this was a very important step, and there was comparison made with the moral dilemmas which faced the physicists at the time that nuclear energy was being unleashed.

I always thought our action was rather minor.

Because of this publicity there was great expectations that something was going to happen which had never happened spontaneously – like a hydrogen bomb had never exploded spontaneously. And this was in a context which many geneticists had generated by saying that genetic engineering was around the corner. I've never seen the technology which would enable that, though I would love to be able to cure Tay-Sachs or sickle-cell anemia or any of these things by inserting genetic material.

A lot of people think that somehow you can enslave the masses by genetic engineering, whereas in fact the masses are already enslaved.

The general rule that seems to govern most of the people working with medical microbiology is that once you've domesticated a bacteria or a virus and got it to grow under conditions of a laboratory culture, it loses its virulence. It no longer has the ability to multiply in humans. If you could get the Legionnaire's Disease to grow well in a university laboratory, it probably wouldn't cause the disease. Biochemists generally live pretty long. (That's not true of organic chemists – they normally have a life span about five years shorter than other people, because those fumes that smell bad are bad.)

Anyway, Asilomar came up with what became the NIH (National Institute of Health) Guidelines. Working with the genes of *drosophila* (fruit flies) was safe, the toad got by, the mouse was in trouble, and humans were out. They got up at the end and said, "We've got a consensus." I got up and said that we got a consensus in the State Department about Vietnam, and certain members of the national press didn't like me. They made me out as someone who didn't care about children, the evil person who raced to get the Nobel Prize for DNA. If you were against regulation you were against motherhood.

Now, it was claimed that what we were now capable of was something uniquely new — in the sense that you would be able to assemble DNA material which doesn't normally exist. But in fact cells have an enormous capability to take up DNA through other means than the usual sexual recombination. You can change the heredity of plants by bacteria — the evidence is about 90% certain — plasmids pass into the plant cell and you have the bacterial gene functioning in a plant, and it causes galls. So DNA transfer I think is probably fairly common in nature.

There's a worry about exchange between bacteria. Could you create a new form of bacteria? Between two species there are often bacterial viruses or bacteriophages, which have a host range that allows them to grow in two sorts of bacteria. We know the virus can pick up genes from one and transfer them to the other. There's a very perceptive article published in **Bacterialogical Review** in October by a man from New Zealand, D.C. Reanney. He calls this "evolution by extra-sexual processes," or extra-chromosomal. He points out that maybe bacteria are just one species. They have relatively little DNA because if they need a gene for something, they're constantly being infected by viruses which will bring them the gene they want.

There are viruses which multiply in mosquitoes and will multiply in human cells. They will also multiply in plants. You can find cases in evolution — it's called convergent evolution — where the same things, such as pigments, appear in things which are apparently unrelated. This could very easily be a virus which has carried this DNA from one species to another.

We've been told there's a danger of *e. coli* getting a gene for cellulase via recombinant DNA. If there were a real selective advantage for *e. coli* to have cellulase, which would do away with our species, it would have it already — by viral transfer or other means. Evolution is constantly trying to seek out the best set of genes for a given organism to occupy a niche. I do not worry about "monsters."

Evolution hasn't stopped. Some of these people have made grandiose statements about "scientists interfering with God's world." The world is changing all the time. Every day there are new bacteria produced in unknown quantities. Viruses are produced. You feel punk one day, I feel punk, we don't know what's making us feel punk. The only thing that's new is some people are saying it's recombinant DNA.

I can fully understand why we're in this room. We're in this room not, I think, because the community in any sense wants to control science but because scientists said there might be a danger. And if the scientists say it, the public should listen. I think they made a great mistake in implying they could distinguish between the possible dangers and that there was any plausible set of guidelines. Some people have said the guidelines are capricious. I think they're totally capricious and totally unnecessary. We must have wasted $25 million on those precautions by now and it's on its way to $100 million. I think it's the biggest waste of Federal money since we built all those fallout shelters.

[Stanford medical professor Halsted Holman then gave his opposing view, that public debate and possible regulation on this matter is correct for the same reason it is right for a patient to know what is going to happen to him in surgery, what his chances are, and that nothing will be done without his "informed consent." The issue here is "informed consent by the public." In the questions following Dr. Holman's presentation Prof. Watson was prodded to reply. . . .]

Watson: Informed consent in this thing is meaningless because you can't tell people what kind of risk is going to occur. I think it's as difficult as trying to decide whether you should be a Jew, a Catholic, or a Protestant on some sort of logical grounds. It's like arguing religion, because you have the same number of solid facts. That's why we don't argue religion in polite company. People just get mad at each other, because they can't give each other any facts.

Question from audience: It's a new political principle if you say, "The less facts you have, the less the public has a right to make the decision." Up to now we've always heard, "The more facts you have, only scientists and specialists understand these facts, therefore the public shouldn't make the decision." If there's no facts, the public is no worse off than the scientists.

Watson: There's a tradition in western society of freedom. You generally let people do what they want unless it causes damage to other people. It's silly to control where there's no evidence of danger. I am totally agreed that the public should participate in any process where they can be given facts to think about. But the tradition is, you don't call fire until you see it.

The trouble with trying to assess recombinant DNA hazard is you can't count about it. For example, how would we know if recombinant DNA is bad for enfeebled patients? Well, I guess you would say a higher mortality at Stanford University Hospital, where there's recombinant DNA research, compared to a hospital in New York where there isn't. That may sound cynical, but you have to know that something bad exists before you can do anything. If suddenly the incidence of *e. coli* infections doubles in a matter of six months you would go into these people and see if they have one of these things produced in Stanley Cohen's lab. And if someone wants to sue Stanley Cohen for being irresponsible, they can do it.

You've got to smell a little smoke before you decide where the fire is. There's so many fires. Every newspaper we read has some new scare. Sanity demands a limited number of scares.

Maybe I'm just worried about too many other things to take this seriously. As I explained to Governor Brown, these flame retardants which are in all children's clothes — tris — are the most potentially dangerous carcinogens around. They threaten every child in California and in the whole country. Tris is <u>bad</u>. It's taken up through the skin. That's a real danger. ■

Document 6.10

March 18, 1977
Montauk Letter to Lackner

Lance Montauk
Room 4048
State Capitol
Sacramento
California 95814

18 March 1977

Jerome Lackner, Director
Department of Health
714 P St. Room 1253
Sacramento, California
95814

Dear Mr. Lackner,

This letter is in response to the presentation by Drs. Watson and Holman on the evening of March 15th.

During the discussion afterwards, it was not surprising that virtually everyone in the audience was in favor of some form of government regulation re recombinant research, since the audience consisted almost entirely of government regulators. In fact, the simple idea of government regulation in this area would not distress me very much - although the idea of no government regulation wouldn't distress me whatsoever. However, in today's context, these ideas are not simple.

One of my concerns was heightened at the end of the program, when I approached you with a query about the Western tradition of government staying out of the private lives and enterprises of citizens. Before you could respond, the Governor jokingly called you over to the news cameras with the question "Have you decided yet on whether to regulate DNA recombinant research?". Although the question was flippant, and your answer vague, the episode highlighted the dissonance between your statements and your actions when dealing with "elitism".

M.D./J.D. Department of Health Directors, and sons of Governors who are Governors themselves, are hardly in a position to decry the elitism of scientists. Statements about public access and involvement in decision making look shallow when I see the Governor query, what have you decided. Of course, your reply might be "but there have been public hearings, ample time for input from all citizens and interested parties, to let everyone make their views known". This does not change the fact that, in centralized decision making, relatively few people, and many of them appointed staff personnel at that, have the power to affect outcomes.

Now you might argue that those few people are not always elitists; that, despite their histories, they have populist perspectives. However, claiming that one truly represents the people, that one is expressing their consciousness, or that one most considers the public's best interest, does not make it so. Rather it assumes that those who disagree either a) don't have the people's best interests in mind, or b) are blind to the realities and thus confused in their conclusions. This is poisoning the well.

You had many examples of how the elitist military industrial research complex had impinged on our lives, such as Strontium 90, ozone depletion, and polychlorinated biphenyls. But now that the public is aware of the risks of halogenated hydrocarbons, have aerosol sales decreased? No, and instead we need legislation to curb their use. Do the underdeveloped countries of the world, containing the majority of passengers on spaceship Earth, want to abandon DDT? No. Would our safety and security have been enhanced if we had left all nuclear testing to the Soviet Union? Doubtful. Is per capita consumption of petroleum and other limited resources dropping, or do folks still buy big cars and drive over 55 miles per hour? Evidently, ordinary people are eager to enjoy the material wealth and security which science and industrial technology provide. As Pogo said, "We have met the enemy, and he is us."

You can suggest that this abundance was purchased through exploitation of the unfortunates around the world via U.S. imperialism; that it does not truly satisfy our important spiritual and human needs; that it is destroying our environment, our humanity, and eventually our species. However, the American public which you so enshrine gives little indication it wants to abandon materialism or creature comforts. Although I personally think our consumption should be greatly curtailed, that is an elitist thought - and we elitist affluents who urge abstinence could well ponder whether some of the wealth we forego is thereby lost forever to the poor who desperately need it.

In the midst of this all-American rapaciousness an unusually conscientious group of researchers in biochemistry and molecular biology asked themselves some penetrating questions a few years ago. As with all decent questions, there were no decent answers. They asked, "Is what we are doing unreasonably dangerous?" Even as they formulated answers, they found themselves politely summoned before inquisitorial bodies all around the country, which demanded satisfactory answers to these decent questions. But there were none, so these bodies threatened to make answers of their own. The result is that laws are relentlessly being promulgated by federal, state, county, and city governments.

If there was ever a time when you wouldn't want to even appear to be making life hard for concerned scientists, it would be the first time they timidly expressed their social conscience. But, politics being politics, this is just what happened. Thus, Watson wishes he had kept quiet. Next time he, or others, will remain silent. They are bright people who learn from experience, since that is what science is all about. What we have done is shut ourselves off from the open and honest appraisals, by researchers, of their own work, which we claim to want and need so much. And is it so absurd that these people trust themselves as much as they trust those institutions which brought them murders in Atascadero, genocide in Vietnam, and panic buying of saccharin because it is a rat carcinogenic?

Since these scientists know that drunks on the highway kill thousands yearly, that lead contamination is growing, and that every night tris(2,3 dibromopropyl)phosphate is stealthily migrating from our children's nighties into their tissues, is it any wonder that they question why government's top priority this month is to regulate DNA recombinant research?

The regulators' response is that some recombinant may kill us all. Well, people have been terrorized for centuries by new inventions and ideas - and, perhaps some day a new human venture will eradicate our species. But does anyone really believe that, through legislation, homo sapiens can be saved for its natural demise when the Earth falls into the sun? Surely, this is hubris.

I suggest that, usually, outside regulating is bad. In this case, it discourages scientific openness in the future, and it lowers both the risks and rewards in recombinant research by making the whole enterprise more tedious and costly. Perhaps generally the burden should lie on government to prove the necessity of regulatory intervention in the private sector's decision making.

The survivability of a species depends on the diversity of its gene pool. The more information, strategies, techniques, tools, and mechanisms a species can draw on in the face of myriad dangers, the more probable that some of its members will survive. Decision making is similar. Even the Energy Commission recently concluded it was unwise to put all your eggs into one fuel source.

The more varied the input contained in a decision, the more probability it will survive adversity. Now on the surface this may seem an argument for thorough, open, and complete centralized hearings, where "a decision" must be made. But if there are dozens of small decisions there can be more input than if there is one big decision. Instead of having one centralized decision covering a large population, let each subspecies, or institution, or family or individual, make their own decisions. Big mistakes come from big decisions. Big mistakes will not come from small decisions as readily, because real problems are not linear like chains which are only as strong as their weakest link - rather, they are multidimensional. They do not succumb to standard quantitative analysis. Instead of adding up the various factors in a huge balance in Sacramento or Washington, ending up with one plus or minus which inevitably ignores the factors and all their nuances, why not use local laboratory scales, each of which ignores less?

The more complex the problem, the less information can be adequately assimilated and unified by the deciding committee, commission, or board. (How many legislators know adenine from a lipoprotein?) But experimenters whose lives are immediately dependent on their own judgment clearly have most at stake. Like Madame Curie, they may inadvertently kill themselves. So might our PSA pilot, but basically we trust him more than the Civil Aeronautics Board, and for good reason.

A DNA researcher in her lab might some day destroy us all, or save us all. Some would prevent the destruction, others the salvation. The debate last Tuesday on this subject has raged since the kindly physician Dr. Guillotine provided his solution to the irksome problems typified by the mad research chemist Lavoisier. At least in California these issues are still alive.

Sincerely,

Lance Montauk

cc: Governor Brown
James Watson
Dr. Halsted Holman
Assemblyman Barry Keene
Stuart Brand (Governor's Special Consultant)
James Rote (Assistant Secretary for Resources)
Mark Lappé (Department of Health)
Sally Caleff (Senate Health Consultant)
Susan Brown (Office of Planning and Research)
Claudia Ayres (Office of Planning and Research)
Clyde Wilson (University of California, Berkeley)

Document 6.11

April 1977
Newspaper Articles

San Francisco Chronicle, April 4, 1977

Charles McCabe Himself

On Playing God

THOSE lovely people who gave us the atom bomb have another treat in store for us. Now they can create new forms of life, by jiggling about with genes.

I don't pretend to understand much about recombinant DNA, as the whole thing is called. When I encounter such sentences as the following, in a "simple" explanation, my head congeals:

"Researchers first remove plasmids from bacteria, using chemicals that cause the bacteria to split apart. The scientists then apply enzymes, which open the closed loops of the plasmids and allow insertion of small stretches of DNA that represent one or more alien genes."

But I do know what the act of creation is. When it is placed in the hands of men who by definition do not know what they are doing, it scares the daylights out of me. Jiggling with genes may cure cancer. Then again, it may cause outbreaks of new forms of cancer. Gene-splicing may clean up spills, and at the same time might defoliate half a state.

Who knows? The only thing that is certain about this business is that nobody knows. Even molecular biologists, whose racket it is, say that neither the putative good or harm of DNA recombinant research can be known for years.

Why, in the name of all that is sacred, can't we learn to let well enough alone? Why do we diddle ceaselessly with nature? Why will scientists persist in playing God?

San Francisco Chronicle, April 26, 1977

Editorials

Intruders in The DNA Labs

WE ARE WHOLLY out of sympathy with the headlong rush of certain California legislators to write laws and set up a bureaucratic commission to license the laboratories, public and private, of this state in which advanced biological research is carried out.

As best we can discover, these eager protectors of the public weal are proposing to round up all scientists engaged in genetic research, license them and handcuff them with rules of laboratory inspection and certification of laboratory personnel engaged in such genetic activities.

Experimentation with recombinant DNA molecules, the target of this legislative policing effort, is the technique for propagating hereditary material in bacteria and other cells. It was once thought to open up hazardous possibilities. Today, as more knowledge is gained, it seems to the scientists on the borderline of this unknown field to offer many fewer hazards.

Document 6.12

April 6, 1977
Kennedy Opening Statement

from the office of

Senator Edward M. Kennedy

of Massachusetts

OPENING STATEMENT OF SENATOR EDWARD M. KENNEDY, CHAIRMAN OF THE SENATE SUBCOMMITTEE ON HEALTH AND SCIENTIFIC RESEARCH, AT A LEGISLATIVE HEARING ON REGULATION OF RECOMBINANT DNA RESEARCH, WEDNESDAY, APRIL 6, 1977, AT 9:30 A.M. IN ROOM 6202 OF THE DIRKSEN SENATE OFFICE BUILDING.

For release: 9:30 a.m.
April 6, 1977

Today the Senate Health and Scientific Research Subcommittee holds a legislative hearing of truly historic importance. We are considering legislation to give the Federal Government the right to regulate a specific area of biomedical research—research with recombinant DNA.

The freedom of scientific inquiry is cherished in this country. Nonetheless, many scientists are themselves advocating that recombinant DNA research be regulated by the Federal Government. Scientists and lay persons alike have concluded that this research has the potential to significantly alter our society—for good or ill. The extent of that potential alteration is not known today—and cannot be known until more research is done. It may be that the risks are currently being grossly exaggerated. It may be that the benefits are overstated. But it may also be that the reverse is true. Because we don't know, and can't know at the present time, I believe we must err on the side of caution. I believe it is this feeling that has produced the unprecedented agreements that some form of legislation is necessary.

Whatever form the final legislation takes, it must build upon the unique process that has promoted the consensus which makes this regulation possible. For the first time a technical area of biomedical research has been the subject of widespread, indepth, public discussions. Scientists and lay persons have been talking to each other in all parts of the country. They have gained mutual respect. Scientists have learned that ordinary citizens can understand technical matters and make responsible decisions. Lay people have learned that the scientists can and will share information and that their proposals for self-regulation were not designed to slip something over on the public—but were responsible efforts at minimizing potential dangers.

To my mind this unprecedented dialogue between scientists and citizens must be built into the regulatory process. Because the risks, if there are any, will involve scientists and lay persons alike, decisions as to how to contain them must be made cooperatively. I believe a national commission, with a majority of nonscientists, would be the best vehicle to regulate this research. I would be interested in our witnesses' comments on this proposal. In addition, I would like their reaction to the following propositions:

1. All recombinant DNA research in this country should be subject to the same regulations.
2. The regulatory process ought to be flexible and reflect new knowledge of risks and benefits as it becomes available.
3. The ethical, legal and moral implications of this work should be given careful scrutiny.
4. Local communities should retain the right to be stricter than the Federal regulations—including the right to prohibit recombinant DNA research in their own communities.

The decision to have the Federal Government intervene and regulate a specific area of scientific inquiry is not to be taken lightly. It establishes a precedent whose consequences are not yet clear. It should make us all somewhat uneasy. What is reassuring, however, is that the impetus for this step has come from the ground up. It is not a case of Federal imposition of unwanted regulation on an uninformed and unsuspecting public. It is, rather an example of the regulatory process at its best. With the Government being responsive to the wishes of the people—scientists and lay persons alike. As such, there is great promise that this legislation will be accepted once enacted, and that this legislative precedent will be responsibly used in the future.

Document 6.13

April 1, 1977

95th Congress: Bill S.1217, Excerpt

Calendar No. 334

95TH CONGRESS
1ST SESSION

S. 1217

[Report No. 95–359]

IN THE SENATE OF THE UNITED STATES

APRIL 1 (legislative day, FEBRUARY 21), 1977

Mr. KENNEDY introduced the following bill; which was read twice and referred to the Committee on Human Resources

JULY 22 (legislative day, JULY 19), 1977

Reported by Mr. KENNEDY, with an amendment

[strike out all after the enacting clause and insert the part printed in italic]

A BILL

To regulate activities involving recombinant deoxyribonucleic acid.

Be it enacted by the Senate and House of Representatives of the United States of America in Congress assembled, That this Act may be cited as the "Recombinant DNA Regulation Act".

FINDINGS

SEC. 2. The Congress finds that

(1) work with recombinant DNA will improve the understanding of basic biological processes and offers many potential benefits,

(2) there exists, however, a possible risk that micro-

...

merce and regulations and licensure by the National Recombinant DNA Safety Regulation Commission are necessary and proper to effectively regulate such activities.

SEC. 3. The Public Health Service Act is amended by adding after title XVII the following new title:

"TITLE XVIII—NATIONAL RECOMBINANT DNA SAFETY REGULATION COMMISSION

"ESTABLISHMENT OF COMMISSION

"SEC. 1801. (a) (1) There is established within the Department of Health, Education, and Welfare a commission to be known as the National Recombinant DNA Safety Regulation Commission (hereinafter in this title referred to as the 'Commission').

"(b) The Commission shall be composed of eleven members. The President shall, within sixty days from the date of enactment of this title, appoint—

"(1) Six members of the Commission from individuals—

"(A) who are not and have never been professionally engaged in biological research,

"(B) who are qualified to serve on the Commission as members of the general public including persons who by virtue of their training, experience, or background in the fields of medicine, law, ethics,

...

and other necessary expenses incurred in the performance of such duties.

"DUTIES AND FUNCTIONS OF THE COMMISSION

"SEC. 1802. (a) (1) (A) The Commission is authorized—

"(1) to direct and supervise all personnel of the Commission;

"(2) to promulgate such rules and regulations as may be necessary or appropriate to carry out the duties and functions vested in it by this title;

"(3) to carry out the provisions of this title;

"(4) except as provided in section 1816(f), to utilize, with their consent, the services, personnel, and facilities of other Federal agencies and of state and private agencies and instrumentalities with or without reimbursement therfor;

"(5) except as provided in section 1816(f), to enter into and perform such contracts, leases, cooperative agreements, or other transactions as may be necessary or appropriate in the conduct of the work of the Commission and on such terms as the Commission may deem appropriate, with any agency or instrumentality of the United States, or with any state, territory or possession, or any political subdivision thereof, or with any public or private person, firm, association, corporation, independent testing laboratory, or institution;

"(6) to monitor compliance by the owners or operators of a licensed facility and persons authorized by a license to engage in recombinant DNA activities in connection with a licensed facility with the requirements of this title; and

"(7) to undertake such other activities as are incidental to enforcement of the provisions of this title.

"(b) The Commission shall encourage, through contracts, the development of effective epidemiological methods and safety monitoring technologies to identify and follow the

production and dissemination of recombinant DNA and the biological or chemical products thereof.

"(c) The Commission shall encourage on a continuing basis studies designed to assess the risks to human health and the environment which may be presented by recombinant DNA activities. The Commission shall insure that the findings of such studies shall be maintained and readily accessible to all interest persons.

"GENERAL REQUIREMENTS

"SEC. 1803. (a) Effective two hundred and sixty-five days after the date of the enactment of this title, no person may engage in recombinant DNA activities in the States or in

• • •

prior to the commencement of such project in accordance with section 1818.

"CIVIL PENALTIES

"SEC. 1809. (a) Any person who knowingly, willfully, or negligently violates a provision of section 1803 or section 1808 shall be liable to the United States for a civil penalty in any amount not to exceed $10,000 for each such violation. Each day such a violation continues shall, for purposes of this section, constitute a separate violation of section 1803 or section 1808.

"(b) A civil penalty for a violation of section 1803 or section 1808 shall be assessed by the Commission by an order made on the record after opportunity (provided in accordance with this subsection) for a hearing in accordance with section 554 of title 5, United States Code. Before issuing such an order, the Commission shall give written notice to the person to be assessed a civil penalty under such order of the Commission's or its delegates' proposal to issue such order and provide such person an opportunity to request, within fifteen days of the date the notice is received by such person, such a hearing on the order.

"(c) Any person who requested in accordance with subsection (b) a hearing respecting the assessment of a civil penalty and who is aggrieved by an order assessing a civil

• • •

civil action process may be served on a defendant in any judicial district in which a defendant resides or may be found. Subpenas requiring attendances of witnesses in any such action may be served in any judicial district.

"EFFECT ON STATE AND LOCAL REQUIREMENTS

"SEC. 1813. (a) It is declared to be the express intent of Congress to supersede any and all laws of the states and of the political subdivisions thereof insofar as they may establish or continue in effect with respect to recombinant DNA activities any requirement which is different from, or in addition to, any requirement applicable under this title, except as provided in subsection (b).

"(b) Upon receipt of an application by a state or by a political subdivision of a state and after notice and opportunity for a hearing on the record, the Commission shall, no later than three months from the date the application was received and in accordance with the provisions of paragraphs (1) and (2) of this subsection, exempt any existing or proposed requirement from subsection (a) of this section.

"(1) The Commission shall grant an exemption under this subsection only if it finds—

"(A) that the requirement of a state or political subdivision of a state applicable to recombinant DNA activities is, and will be administered so as to be, more stringent than a requirement under this title;

• • •

"(B) the reason for the requirement is relevant and material to the health and environmental concerns or comparable compelling local conditions of such state or political subdivision; and

"(C) compliance with the requirement will not cause such activities to be in violation of any applicable requirement under this title.

The Commission may not withdraw any such exemption for so long as it finds that such requirement remains more stringent than a requirement under this title and continues to be so administered.

"(2) The Commission shall not grant an exemption under this subsection if the Commission finds that such requirement is arbitrary and capricious.

"(c) An application for such an exemption may be accompanied by any materials gathered by the applicant in its legislative or administrative consideration of the existing or proposed requirement. Upon receipt of such an application and such materials, if submitted, the Commission shall publish such application in the Federal Register as a proposal, accompanied by a description of any supporting materials submitted therewith.

"(d) It is not the intention of the Congress that enactment of the Recombinant DNA Safety Regulation Act, promulgation of regulations thereunder, or compliance there-

• • •

Editors' note: The entire draft prepared by the administration was struck out by Senator Kennedy's amendment, parts of which are reproduced.

Document 6.14

July 22, 1977
Science (**197**:348)

Gene Splicing: Senate Bill Draws Charges of Lysenkoism

Considerable friction has been generated between Senator Edward Kennedy and part of the scientific community over the issue of recombinant DNA research. "It smacks of Lysenkoism," says a senior scientist of the legislation drafted by the staff of Kennedy's Senate health subcommittee. "We are being hassled out of existence for no reason at all," complains Walter Gilbert of Harvard. Kennedy's staff, on the other hand, says the bill establishes a minimum regulatory apparatus which is designed to wither away if scientifically unjustified.

Scientists' apprehensions about the bill have been amplified by Americans for Democratic Action, a liberal Democratic pressure group. On the initiative of a scientist member who cited the Kennedy bill, the ADA board recently adopted a resolution warning that Congress is "attempting to control specific activities through individual licensing and punitive action." Strict societal control of science, the resolution avers, has in the past preceded such excesses as Lysenkoism and "some of the inhuman practices in Nazi Germany."

The frustration behind these sentiments derives from fear that the impending legislation will set up a vast and cumbersome bureaucracy which will seriously impede research. Some scientists opposing the legislation consider it so restrictive as to constitute "prior restraint," a practice abhorred by civil libertarians in freedom-of-speech issues. Others fear that control of recombinant DNA research is only the tip of the iceberg, and that other techniques, such as cell fusion, will be next to be regulated. "It is clear that there are a whole bunch of regulators here who have discovered that we have been doing genetics for 30 years without permission. For a scientist that sounds hilarious, but they are dead serious," says an MIT biologist.

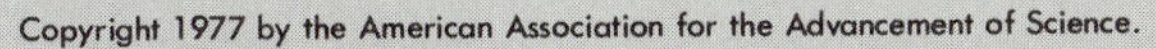

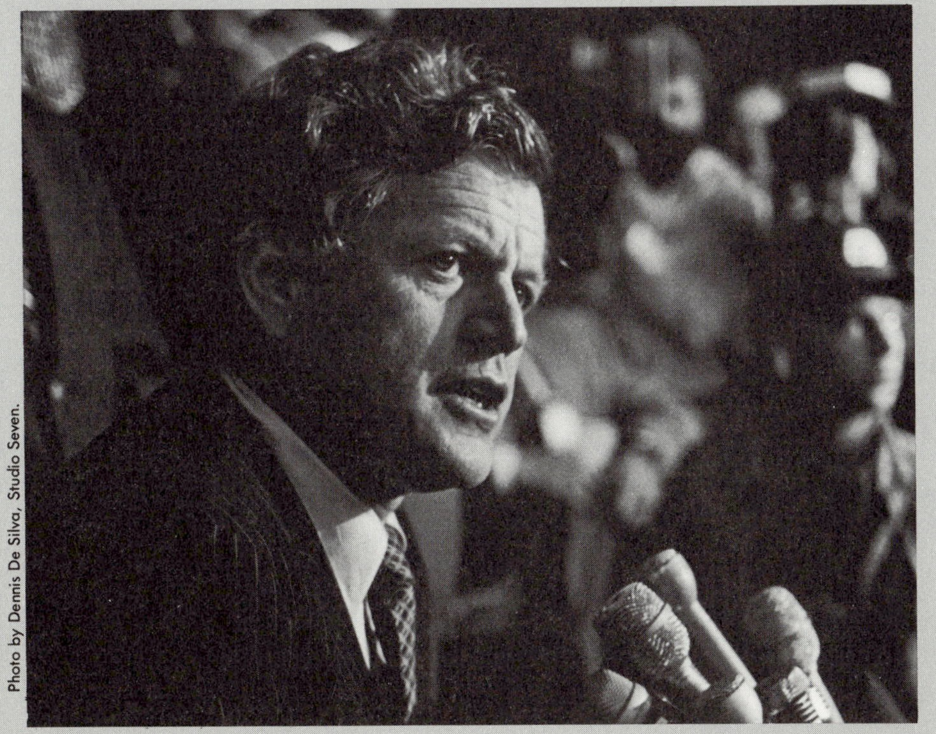

Photo by Dennis De Silva, Studio Seven.

Senator Edward Kennedy

Resentment of the Senate bill on gene splicing has been compounded by a separate development, the emergence of a belief that the originally perceived health hazards of the research, which the present NIH regulations are designed to address, have been overestimated. Though much of the knowledge underlying this evaluation has been available for several years, it seems first to have been brought together this April by an individual member of the NIH committee which drafted the regulations. The review is in the form of a widely circulated letter from Roy Curtiss of the University of Alabama to the director of NIH. It lays out the evidence which persuaded Curtiss to change his position on the possible health hazards of the research from one of greater to lesser concern.

"I have gradually come to the realization that the introduction of foreign DNA sequences into EK1 and EK2 host-vectors offers no danger whatsoever to any human being," except in very special circumstances, Curtiss writes: "The arrival at this conclusion has been somewhat painful and with reluctance since it is contrary to my past 'feelings' about the biohazards of recombinant DNA research."

"The Curtiss paper has had a big impact because he started from the other side and is a very credible guy," observes Alexander Rich of MIT. One important impact of Curtiss's palinode has been on the NIH Recombinant DNA Committee. At meetings held in May and June the committee recommended reducing the stringency of its guidelines in several respects (human shotguns to be permitted in P3 physical containment instead of P4; all P4 experiments to be permitted with only an EK1 host-vector). According to an account of the June meeting in the *PMA Newsletter*, the committee members "often mocked their own restrictions" with remarks such as " 'P4 is designed to prevent research,' 'P1 is a laboratory plus a bureaucrat,' and 'These high levels are political, not scientific.' "

The committee's recommended changes have yet to be approved by the NIH director; even if approved, it is not yet clear that they will be incorporated into legislation.

Another probable impact of Curtiss's letter was on the mood of delegates attending the Gordon Conference on Nucleic Acids in June this year. It was a letter from the members of the 1973 conference which first directed public attention to the possible hazards of gene splicing. But having heard an exposition from Alexander Rich of the pending legislation, 137 delegates signed a letter to Congress expressing worry that the regulatory machinery now being considered will be "so unwieldy and unpredictable as to inhibit severely the further development of this field of research." Much of the stimulus for the legislation seems to derive from exaggerations of the possible hazards, the letter adds (*Science*, 15 July, p. 208).

Another scientific group which has sought to persuade Congress is the Inter-Society Council for Biology and Medicine, a coalition of seven scientific societies. In a letter of 30 June to Congressman Paul G. Rogers, chairman of the House health subcommittee, the council emphasizes the "general acceptability" of the present House bill on recombinant DNA "as opposed to the current Senate version." The legislation produced by Rogers' committee, the letter says pointedly, "will permit free scientific inquiry while protecting the health of the public."

Scientific opponents of the Kennedy bill, and Kennedy's staff differ strongly on the interpretation of the bill's requirements. Essentially the bill establishes a presidentially appointed commission within the Department of Health, Education, and Welfare. The commission would license facilities to conduct gene splicing research, and would employ a team of inspectors to visit laboratories, examine records, and monitor compliance. In the event of infringements, facilities could have their licenses revoked, and researchers could be fined up to $10,000 a day per violation. The commission is to issue new regulations which are "no less stringent" than the present NIH guidelines. (The House bill essentially contains all the same features—licensing, inspectors, fines of $5000 a day, and new regulations—but with the major difference that enforcement is placed in the hands of local biohazard committees instead of a federal commission.)

Opponents predict that the damage caused to science by the legislation will be comparable to that done by Lysenkoism in the Soviet Union.

Opponents of the Senate bill complain that it makes the process of getting an experiment approved an intolerable struggle through layers of red tape. According to a staff member who helped draw up the legislation, but who declines to be identified, the bill simply requires that a researcher's facility be licensed, and his project registered with the commission; the only review and approval is by his local biohazards committee.

Opponents say the Senate bill creates an unwieldy bureaucracy which will spend some $25 million to regulate a mere $3 million of research. The Senate staff member says that the bureaucracy created by the bill comprises the president of the commission, who would be its only full-time member, and 50 inspectors. According to the congressional budget office, the cost of the regulatory apparatus will be less than $4 million a year.

Opponents claim that the Senate health subcommittee desires to regulate other aspects of biological research, gene splicing being only a first step. The staff member states that there is no basis for this claim, and that Kennedy has no such intention.

Opponents predict that the damage caused to science by the legislation will be comparable to that done by Lysenkoism in the Soviet Union. The staff member says he heard similar predictions about the creation of the Commission for the Protection of Human Subjects in Behavioral and Biomedical Research; now that the commission is about to expire, he says, the same people are urging that it be continued.

The Senate bill on recombinant DNA has been approved in committee (although Senator Gaylord Nelson is thinking of writing a minority report) and is likely to be taken up by the full Senate shortly. In the House, the bill prepared by Rogers' subcommittee is to be considered by the committee on science and technology before going to the House floor.—NICHOLAS WADE

Document 6.15

August 2, 1977

Congressional Record, Excerpts

Remarks by Senator Moynihan

Congressional Record*
S.13312 Aug. 2, 1977

I am pleased to join the Senator from Wisconsin in sponsoring a substitute amendment to S.1217.

Federal regulation of research into recombinant DNA is, indeed, a most complex and controversial issue. The research itself is being conducted at the very frontiers of science, where our knowledge is incomplete, but rapidly growing. While there are legitimate concerns regarding the safety of this research, we must take great care so as to not allow unjustified fears to cloud our judgement and cause us to set precedents which we may, at a later date, have reason to regret. This substitute amendment would, I believe, establish the necessary regulatory authority to control recombinant DNA research.

I find this proposal to be in all respects superior to S.1217. We must be very careful when creating such regulatory power that we examine its impact on private research institutions. In this case, the institutions that are affected are our universities and research laboratories, which have operated, and should continue to operate in the tradition of free inquiry into all research matters. Senator Kennedy's bill, S.1217, would establish unnecessarily stringent controls on the research activities of the scientific community.

The bill which we are introducing today provides the necessary regulation without forcing unnecessary intrusion into what is, legitimately, the academic domain.

*The full amendment was published in *Congressional Record*, August 2, 1977 (S.13312–S.13319). We reproduce here only excerpts of page S.13312, which substitutes the RAC for the National Commission, and pages S.13316 and S.13317, which restore federal pre-emption.

RECOMBINANT DNA ACT—S. 1217

AMENDMENT NO. 754

(Ordered to be printed and to lie on the table.)

Mr. NELSON.* Mr. President, I am today introducing an amendment in the nature of a substitute to S. 1217, a bill to regulate activities involving recombinant deoxyribonucleic acid (DNA) research.

S. 1217 has generated much new information and commentary since its introduction April 1, 1977.

The substitute proposal reflects new information and views, and differs from both the Senate bill (S. 1217) and from H.R. 7897, a bill pending before the House Interstate and Foreign Commerce Committee, with respect to a number of major issues: the nature and the extent of regulation necessary to combat potential risks from DNA research activities; the definition of recombinant DNA activities to be regulated; penalties for noncompliance; the authoritiy of the Federal Government to preempt State and local regulations. The proposal reflects concerns expressed in supplemental views filed with the report (No. 95-259) of the Human Resources Committee on S. 1217.

We are interested in the news of all interested parties on this important legislation.

I ask unanimous consent that the amendment be printed in the Record allowing these remarks, along with supplemental views on S. 1217, and an article in the New York Times, July 31, 1977, by Walter Sullivan, science editor.

There being no objection, the material was ordered to be printed in the RECORD as follows:

AMENDMENT NO. 754

Strike out all after the enacting clause and substitute the following:

SHORT TITLE

SECTION 1. This Act may be cited as the "Recombinant DNA Act".

FINDINGS

SEC. 2. The Congress finds that—

(1) research and other activities involving recombinant DNA will improve the understanding of fundamental biological processes;

(2) the knowledge gained from such research and activities may be of great benefit to medicine and agriculture and may provide many other benefits to society;

(3) there exist, however, uncertainties regarding the extent to which recombinant DNA or organisms or viruses containing recombinant DNA and activities involving recombinant DNA may present a risk of injury to health or the environment, and there is a risk that such organisms and viruses may spread quickly and without warning to persons, agricultural plants and products, and other items in or affecting commerce;

(4) the public interest requires that the health and welfare of the population of the United States be protected from such risk, and commerce in the United States is dependent upon such protection being provided; and

(5) to effectively accomplish such protection and consequently to effectively regulate commerce requires that the possession of recombinant DNA and any activity engaged in for its production (whether a research or commercial activity) be subject to control.

*For himself and Mr. Moynihan

"ADVISORY COMMITTEE

"Sec. 481. (a) There is established the Recombinant DNA Advisory Committee which shall advise the Secretary and make recommendations to the Secretary for the effective administration of this part.

"(b) The Advisory Committee shall consist of seventeen members appointed by the Secretary from individuals who by virtue of their training or experience are qualified to participate in the functions of the Advisory Committee. Of the individuals appointed as members of the Advisory Committee (1) one shall be representative of the interests of nonprofessional personnel engaged in recombinant DNA activities; (2) one shall be representative of the interests of entities engaged in such activities for commercial purposes; and (3) at least nine shall be individuals who are not actively engaged in any recombinant DNA activity and do not have a direct financial interest in such an activity. Of those nine individuals, at least four shall be scientists who are knowledgeable in the various scientific and medical disciplines necessary for an evaluation of the potential risks to health and the environment presented by recombinant DNA activities.

Receipt of a salary from an institution that sponsors recombinant DNA activity shall not qualify as a direct financial interest. One member of the Advisory Committee shall be a specialist in public health, one shall have background in the ethical concerns involved in biomedical research, and three shall represent the interests of the general public. The term of office of a member of the Advisory Committee shall be three years except that of the members first appointed to the Advisory Committee, six shall be appointed for a term of one year and six shall be appointed for a term of two years as designated by the Secretary at the time of appointment. Section 14 of the Federal Advisory Committee Act shall not apply with respect to the duration of the Advisory Committee.

"(c) members of the Advisory Committee (other than officers or employees of the United States), while attending meetings or conferences of the Advisory Committee or otherwise engaged in its business, shall be entitled to receive compensation at rates to be fixed by the Secretary, but not at rates exceeding the daily equivalent of the rate in effect for grade GS–18 of the General Schedule, for each day so engaged, including traveltime; and while so serving away from their homes or regular places of business each member may be allowed travel expenses (including per diem in lieu of subsistence) as authorized by section 5703 of title 5, United States Code, for persons in the Government service employed intermittently.

"EFFECT ON STATE AND LOCAL REQUIREMENTS

"Sec. 494. (a) Except as provided in sub-section (b), no State or political subdivision of a State may establish or continue in effect any requirement for the regulation of recombinant DNA activities.

"(b) Upon application of a State or political subdivision of a State, the Secretary may allow, by order promulgated after providing (in accordance with this subsection) notice and opportunity for an oral hearing on such application and after considering local conditions, exempt from subsection (a), under such conditions as may be prescribed in such order, a requirement of such State or political subdivision applicable to recombinant DNA activities if—

"(1) the requirement is more stringent than a requirement under this part which would be applicable to such activities if an exemption were not in effect under this subsection; and

"(2) the requirement is necessary to protect health or the environment and is required by compelling local conditions.

A State or political subdivision which submits an application under this subsection shall be given an opportunity for an oral hearing on such application to be commenced not later than sixty days from the date the application is submitted. The presiding officer at such a hearing shall upon conclusion of the hearing make a written recommendation to the Secretary respecting approval of the application upon which the hearing was held.

"(c) Within—

"(1) sixty days of the conclusion of a hearing held on an application submitted under subsection (b), or

"(2) one hundred and twenty days of the date the application was submitted, whichever occurs later, the Secretary shall either approve or disapprove such application. The decision of the Secretary shall be in writing, shall, if a hearing was held on the application, contain the recommendation made by the presiding officer at such hearing, and shall include a statement of the reasons for the decision of the Secretary.

Document 6.16

September 27, 1977
Kennedy Withdrawal Statement, Excerpt

REMARKS OF SENATOR EDWARD M. KENNEDY

AMERICAN MEDICAL WRITERS ASSOCIATION

NEW YORK CITY
SEPTEMBER 27, 1977

I am delighted to have the opportunity to address this convention of American medical writers. Your publications are the principal sources of information for the nation's health care community about the role and activities of Congress in the field of health. It is not an easy task, because Congress does not often speak with a single voice on vital issues like the organization, delivery and financing of health care.

The reforms of the Congressional committee system of the past few years have made some progress in the coordination of committee jurisdiction in other areas, but not in health care. We currently have two committees in the Senate and two in the House with jurisdiction over cost containment and health insurance. There is an additional committee in each body with overlapping oversight responsibilities in the science policy area.

The problem seems to be infectious, because the Executive Branch is beginning to suffer from the same confusing symptoms. The newly reorganized Department of H.E.W. is creating a separate health care financing administration, which will bring to the Executive Branch the same sort of division that has been so counterproductive in the Legislative branch.

You and your publications bear an especially heavy responsibility. Your readership—the nation's scientists, health care professionals and health care administrators—need to be more fully and accurately informed about policy developments in Washington. Only then can they participate in an effective way in the development of health and science policy. And only with their informed participation will any policy have a chance to succeed.

In three current areas, there is a great deal of confusion over present and future prospects for health policy—health manpower, recombinant DNA and national health insurance.

The Health Manpower Legislation enacted by the past Congress was an effort to walk a difficult line. Using the leverage of federal dollars, we tried to encourage the nation's medical schools to improve their social priorities, without compromising their integrity or imparing their tradition of academic freedom.

The premise of the legislation was that society, which pays the bulk of medical school bills, has an interest in the priorities and direction of health manpower policy. The legislation was complicated. In general, I believe that we succeeded in walking that difficult line. There were, however, some notable failures.

In the closing days of the Joint House-Senate Conference on the legislation, the House of Representatives insisted on a provision requiring American medical schools to admit all U.S. citizens studying at foreign medical schools. The only proviso was that the applicants must pass a qualifying examination.

The Senate initially opposed the provision. But we decided to agree to it, when we realized that the only alternative was the death of the entire manpower legislation.

Among the major interest groups, the Association of

American Medical Colleges supported the Senate decision, although they too approved the House provision. No one anticipated the full impact of the provision on the nation's medical schools, or the flood of criticism that it has produced.

I agree that the current law is an unwarranted intrusion by the federal government into academic medicine. It violates a basic academic freedom—the freedom of a medical school to select its own students in a nondiscriminatory way, according to its own publicized standards. It also discriminates against every American student who failed to be admitted to an American medical school, but who could not, for whatever reason, enroll in a foreign medical school.

It is a bad law which must be changed. It will be changed, and changed soon, and it should never have been enacted.

There are other deficiencies in the manpower law, but they will need careful study before they can be corrected. These include technical problems with the student loan program, and with the formula for determining places in primary care residency programs.

I am prepared to act immediately on the foreign medical school problem. But I would like to see these other areas addressed next spring, when the Senate Health Subcommittee will reevaluate the entire manpower legislation.

The second policy area where there is considerable controversy and confusion involves recombinant DNA. This unique biochemical technique has triggered an extraordinary national debate on the relationship between scientists and society. Town meetings have been held across the nation. Scientists and laymen have begun talking to each other, working together to understand the implications of the scientific experiments, trying to accommodate their different perspectives and values.

This DNA dialogue is the beginning of a new and more sophisticated interaction between science and society. For the first time, the principles of public accountability are being applied to the nation's science policy, just as those same principles are being appropriately applied in national defense and other critical areas of foreign and domestic policy.

The public deserves to participate much more fully in the evaluation, development and implementation of our national policy toward science and medical research. In its way, the change is a historic one. We are abandoning the blank check science policy of the 1950's and 1960's. As the recombinant DNA debate symbolizes, we are entering a new era of public participation in critical policy decisions in science.

I view this as a constructive development that should be welcomed by the scientific community. We have an opportunity to broaden public understanding of science. And we also have an opportunity to enhance public support for science, through the knowledge that our overriding public purposes and goals are shared by the scientific community.

This new concept of public participation in science policy has unsettled many members of the scientific community. They see it as a threat to the freedom of scientific inquiry, or as the forerunner of government controls over science.

But the truth is just the opposite. The greatest potential threat to science is the lack of public understanding and public participation. We cannot affort irrational, unthinking, unknowing intrusions into sensitive and highly emotional areas like scientific policy. The best safeguard against future error is a more informed and more participating public.

The experience with recombinant DNA shows that an informed public can make constructive contributions to the development of science policy. Laymen have been able to study and master complicated scientific details.

Nevertheless, in the historic context within which the recombinant DNA debate is taking place, the enactment of federal legislation is not a step to be taken lightly. Emotions are running high. Many thoughtful scientists are fearful of the proposed legislation, and many members of the public are equally fearful of the absence of legislation.

At times, the bitterness of the debate has obscured the constructive elements that are evolving. As we are learning, the scientific facts about recombinant DNA are in a state of flux. The information before us today differs significantly from the data available when our committee recommended the pending Senate legislation. The new work of Dr. Cohen at Stanford raises serious questions as to whether recombinant DNA can ever produce a "novel" organism. Dr. Cohen believes that by using this technique, scientists can only duplicate what nature can already do.

I am concerned about the fluctuating scientific data and the emotional atmosphere of the debate. Because this is an extremely sensitive area for legislation, I will suggest a compromise legislative approach next week.

I will propose legislation to extend the NIH guidelines to all parties conducting recombinant DNA research. It will be a one year bill. I will urge the Congress to enact it, and thereby defer consideration of permanent legislation for another year. At the same time, I shall help form a national recombinant DNA study commission, with equal representation of scientists and laypersons, to study the complex issues of recombinant DNA research. The commission will have a life of nine months, and it will be asked to recommend whether permanent legislation is needed. Members of the commission will be nominated by scientific and public interest organizations, and the final selections will be made by an independent group. David Hamburg, President of the Institute of Medicine, has agreed to chair that group. The full details of the process will be announced next week.

The successful establishment of this commission will be a constructive continuation of the current dialogue between scientists and laypersons. As such, it will reduce the risk of prematurely adopting a regulatory approach, no matter how flexibly that regulation might be structured. . . .

Document 6.17

September 29, 1977
Kennedy Letter to Thomas

United States Senate
COMMITTEE ON HUMAN RESOURCES
WASHINGTON, D.C. 20510

September 29, 1977

Professor Charles A. Thomas, Jr.
Harvard Medical School
Department of Biological Chemistry
Boston, Massachusetts 02115

Dear Professor Thomas:

Thanks for your note of September 20. I have indeed decided to simply extend the NIH Guidelines to cover all persons doing recombinant DNA research for one year instead of moving for a permanent regulatory structure. In order for that view to prevail, however, you and your colleagues will have to become active in support of it. The House of Representatives is still moving towards a permanent regulatory structure and Senator Nelson still persists in his desire to enact his permanent regulatory structure legislation.

Best wishes,

Sincerely,

Edward M. Kennedy, Chairman
Senate Subcommittee on Health
and Scientific Research

EMK:lhk

Document 6.18

October 1977

Newspaper Article and Headlines

New York Times, October 25, 1977

Congress Is Likely to Delay Until at Least Next Year DNA Research Regulations Once Thought Critical

By HAROLD M. SCHMECK Jr.
Special to The New York Times

WASHINGTON, Oct. 24—Legislation to regulate recombinant DNA research, which appeared to be an urgent issue last spring with leaders of health affairs in Congress, now appears likely to be delayed until next year in both houses.

There is even some sentiment for postponing beyond next year the legislation to govern this promising but controversial area of genetics research.

Even Representative Paul G. Rogers, Democrat of Florida, who is the chief architect of the legislation before the House Commerce Committee, conceded that it would be "very difficult" to get a bill through before Congress adjourned for the year.

Meanwhile, the office of President Carter's science adviser, Dr. Frank Press, has sent out a letter of query to all Federal agencies and departments interested in recombinant DNA research to survey their present views of the need for regulation.

Most Replies Received

Dr. Gilbert S. Omenn, associate director of the President's Office of Science and Technology policy, who is conducting the survey, said a majority of the replies had been received. He said he hoped to have at least a preliminary report on the survey ready for hearings next month on DNA legislation before a subcommittee headed by Senator Adlai E. Stevenson 3d, Democrat of Illinois.

Recombinant DNA research, sometimes called "gene splicing," has aroused much discussion and controversy among scientists and laymen alike in rencent years. There have been predictions of grave dangers as well as revolutionary benefits from the studies in which genetic material from widely different species can be spliced together to reproduce in living cells.

DNA (deoxyribonucleic acid) is the active material of the genes and chromosomes of all living things. Potentially, the gene-splicing research might produce some novel life forms different from anything nature has contrived.

In answer to a question, Mr. Rogers said there might still be time to complete the legislative process in the House if the current House-Senate conference on energy legislation should be completed promptly. Others seem less hopeful.

Senate Prospects Are Slim

On the situation in the Senate, one committee staff member said the prospects for action this year are getting "dimmer and dimmer."

The Congressional recess is expected to come sometime in November, although some members think the session could extend into December.

The conceivable benefits of the gene-splicing research include revolutionary new ways of augmenting the world food supply and curing disease. Conceivable risks include the loosing of bizarre, epidemic-causing plagues. Both benefits and risks are largely conjectural.

The first impetus for regulation of the research came from the scientists who were themselves pioneers in the work. By last spring a controversy over it seemed to be developing into a national debate that included fears for the genetic future of mankind and the view, heretical to most scientists, that some questions of biology were so potentially dangerous that the answers should not even be sought.

Some Guidelines Accepted

It was against this background of concern that committees of Congress took up the matter of legislating controls and safeguards.

The National Institutes of Health had already drawn up detailed and stringent guidelines for the studies. These were soon accepted as binding on all research done with Federal support. This, however, left no certain control over privately conducted industrial research and some concerns were known to be either interested or already actively engaged.

The major purpose of legislation, such as Mr. Roger's bill, was to draw all such research under regulations with adequate means of inspection and safeguards against violations.

A more sweeping bill, sponsored by Senator Edward M. Kennedy, Democrat of Massachusetts, would include establisment of a national commission that would have regulatory powers over the research. Many scientists objected to this on the grounds that it might create an unnecessary Federal bureaucracy and might seriously inhibit research.

In recent months many scientists, individually and in groups, have expressed the view that accumulating evidence shows the original fears concerning the research were excessive. A substantial number of other scientists still remain unconvinced of this. A few on both sides continue to hold extreme positions, either that all such research should be banned or that no regulations are needed.

Kennedy Backs Off

In the light of the view that past fears might have been excessive, Senator Kennedy last month withdrew support from his own bill. At present he favors establishment of a non-Government commission composed of laymen as well as scientists to study further the needs for regulation.

One person familiar with the situation in the Senate said it also seemed doubtful that Senator Kennedy would have the votes to get his original bill passed.

Another measure, drafted by Senator Gaylord Nelson, Democrat of Wisconsin, is similar in its main thrust to Mr. Rogers's bill. A spokesman for Senator Nelson said there would be no effort to push the Senator's bill this year unless the House acted on Mr. Rogers's proposed legislation.

Representative Harley O. Staggers, Democrat of West Virginia, chairman of the Commerce Committee, is known to be considering the introduction of an amendment to the Rogers bill that would postpone any regulatory legislation pending further study of the issues concerning recombinant DNA. It is understood that this study might take as long as a year.

Meanwhile, representatives of two of the nation's largest societies of biological scientists delivered a letter to the Commerce Committee last week expressing support for legislation that would bring the research under regulation and some degree of Federation surveillance. The letter from the American Society of ... ology and the Fed... Societies ...

Opposed by Scientists

October 3, 1977 THE CHRONICLE OF HIGHER EDUCATION

Kennedy Withdraws Bill to Regulate DNA Research

San Jose Mercury Wednesday, October 26, 1977

Danger less than feared?

theory on gene rese...

Document 6.19

July 28, 1978
Sequence of Legislation

	Bill	Chief sponsor	Date
House:			
H. Res. 131		Richard Ottinger, Democrat of New York.	Jan. 19, 1977.
H.R. 3191	Identical to S. 621	do	Feb. 7, 1977.
H.R. 3591	do	do	Feb. 16, 1977.
H.R. 3592	do	do	Do.
H.R. 4232		Stephen Solarz, Democrat of New York.	Mar. 1, 1977.
H.R. 4759		Paul Rogers, Democrat of FLorida	Mar. 9, 1977.
H.R. 4849	Identical to H.R. 4759	do	Mar. 10, 1977.
H.R. 5020	Identical to S. 621	Ottinger	Mar. 14, 1977.
H.R. 6158	Administration bill	Rogers	Apr. 6, 1977.
H.R. 7418		do	May 24, 1977
H.R. 7897		do	June 20, 1977.
H.R. 11192		Rogers and Harley Staggers, Democrats of West Virginia.	Feb. 28, 1978.
Senate:			
S. 621		Dale Bumpers, Democrat of Arkansas.	Feb. 4, 1977.
S. 945		Howard Metzenbaum, Democrat of Ohio.	Mar. 8, 1977.
S. 1217	Administration bill	Edward Kennedy, Democrat of Massachusetts.	Apr. 1, 1977
S. 1217	As reported; accompanied by Report 95-359.	do	July 22, 1977.
S. 1217	Amendment 754 in the nature of a substitute.	Gaylord Nelson, Democrat of Wisconsin.	Aug. 2, 1977.
S. 1217	Amendment 1713 in the nature of a substitute.	Kennedy	Mar. 1, 1978

Document 6.20

April 6, 1977

Halvorson Testimony, American Society for Microbiology, Excerpt

TESTIMONY

ON FEDERAL REGULATIONS OF RECOMBINANT DNA RESEARCH
BEFORE THE
SUBCOMMITTEE ON HEALTH AND SCIENTIFIC RESEARCH

HARLYN O. HALVORSON
PRESIDENT, *AMERICAN SOCIETY FOR MICROBIOLOGY*
DIRECTOR, *ROSENSTIEL BASIC MEDICAL SCIENCES RESEARCH CENTER*
PROFESSOR OF BIOLOGY, *BRANDEIS UNIVERSITY*

RECOMMENDATIONS

1. **Adoption of the NIH guidelines to govern both research and use of recombinant DNA, regardless of the source of funds.**

2. **Authority for licensing of facilities and determination of procedures to be regulated should be vested in the Secretary of Health, Education and Welfare or the Director of NIH.**

3. **Regulations should be flexible so that as further information becomes available the regulations can become less stringent, or if necessary, more stringent.**

4. **Uniform licensing procedures should be established nationally.**

5. **Licensing should be on an institutional basis. Local decision making should be in the hands of a local biohazard committee which will have public representation.**

6. **The licensing committee of the National Institutes of Health should be composed only of technically trained members who represent the subdisciplines of the sciences and are familiar with accreditation procedures.**

7. **Threat of license removal should provide the only necessary deterrent for compliance with the regulations.**

Thank you for the opportunity to appear before you today to present a summary of the views of the American Society for Microbiology regarding recombinant DNA.

My name is Harlyn Halvorson. I am a microbiologist from Brandeis University, Waltham, Massachusetts. Professionally I am Professor of Biology and Director of the Rosenstiel Basic Medical Sciences Research Center and, in addition, am Chairman of the Biohazards Committee at the University. I appear before you today as President of the American Society for Microbiology, an organization with both a great interest and expertise in this field.

To place these remarks in proper perspective, let me briefly describe the structure of ASM. The American Society for Microbiology has an active membership of over 25,000 individuals and is the largest, single, life science society in the world.

Just prior to the establishment of the NIH guidelines for recombinant research in June 1976, ASM was requested by Dr. Hans Stetten of NIH to review both the adequacy of the guidelines and their recommendations for containment facilities. As President of ASM I appointed a task force of experts knowledgeable in epidemiology, virology, infectious diseases and in isolation techniques to review the NIH guidelines. The committee unanimously recommended endorsement of the guidelines with minor modifications (see appendix). These recommendations were also unanimously endorsed by the Council Policy Committee of the American Society for Microbiology who expressed the conviction that these guidelines provide safe and proper procedures for the handling of recombinant DNA molecules.

We believe that the NIH guidelines provide an appropriate starting point for determining a national policy to govern both research on and the commercial use of recombinant molecules. The Department of Health, Education and Welfare is to be congratulated for the effective and rapid manner in which it has recognized the issues and dealt with them in the current guidelines, and in the procedures adopted for applying these guidelines in NIH sponsored research. This experience is clearly an example where, when the problem has been recognized, the scientific community, through the leadership of HEW, has provided a balanced analysis and an appropriate mechanism for policing its own sponsored activities.
If organisms carrying certain kinds of recombinant DNA should prove to be hazardous then this hazard exists whether the work is conducted in a research laboratory or in a commercial production firm. Since it is the organism that provides the potential hazard, not the mechanism of producing it, regulations should be directed to the control of the organism. At the present time the initial research that led to the creation of an organism carrying the recombinant DNA would be covered under the guidelines but the use of the organism would not. We therefore recommend that the guidelines be extended to include all research and use of recombinant DNA regardless of the source of funding. Moreover, protection against potentially hazardous research depends ultimately upon knowledge, judgment, and training in the use of hazardous microorganisms. We would support regulation containing licensing procedures, so long as the mechanism of licensing was not so cumbersome as to impede or stifle the research itself, and provided these regulations did not encompass existing research on infectious microorganisms not associated with *in vitro* recombinant DNA research or on clearly harmless recombinants.

Regulations should be enforced through licensing of laboratories and, if necessary, inspection. Such regulations should be *flexible* to permit rapid regulation of new potentially hazardous procedures and relaxation of regulations that prove to be unnecessary. Since this is a very complex field, *authority* for such licensing and the determination of which procedures should be regulated should be vested in the *Secretary of HEW or in the Director of the NIH.* For the hazards to be seen in their proper and realistic perspective, the agency responsible for applying these regulations must have extensive experience with bacterial epidemics. The most appropriate federal agency is the Center for Disease Control.

These same arguments make it absolutely necessary that any licensing procedures be nationally established. We, therefore, urge that the nation adopt a single system for DNA recombinant experiments. While this does not preclude local authority or responsibility in this issue it is apparent that recombinant molecules in microorganisms do not respect state or national boundaries. A uniform national procedure would preclude the expenses of establishing multiple levels of criteria, licensing, inspection and education. It would also be a step towards the goals of uniform international standards.

Local interests have an important and appropriate role to play on the institutional biohazard committee. We therefore recommend two stages of control: first, institutions should be licensed by an accreditation committee at CDC and, second, institutions should have a biohazard committee with local responsibility.

The essential question is how licensing can best be accomplished. Licensing of individual projects would be exceedingly cumbersome and unworkable because it would destroy the flexibility that is essential for creative research. We would favor licensing of an institution for work at a specified level of containment.

A local biohazard committee could play a leading role in assessing individual laboratories as to the adequacy of the safety procedures, the appropriateness of levels of physical containment and biologic containment, the provision of educational materials and review of the experimental design. In addition, quality control by local groups would be more knowledgeable of the strengths and weaknesses of any scientific colleague than governmental review agencies.

We have the following suggestions for procedures for local biohazard committees:

1. Eliminate non-hazardous experiments from licensure. Such an approach would permit concentration on more hazardous experiments and enhance cost effectiveness. For example, scientists have been recombining genes of *E. coli* in *E. coli* for years by established methods of genetic exchange such as transduction, transformation, and conjugation. The use of recombinant molecule technology to amplify and purify genes of an organism within the same species does not constitute an added risk. Such experiments, rated P1 EK1 in the present guidelines, are likely to be the most common ones nationally. P1 EK1 experiments could be approved by the biohazard committee but need not be certified.

2. Provide maximal local authority for decision making by biohazard committees within the boundaries of perceived risks of the experiments. This principle enables the quality control by local groups who will be more knowledgeable of the strengths and weaknesses of any scientific colleague than the governmental review agencies. In addition, this mechanism would overcome the problem of the lack of uniform educational procedures for Ph.D.'s and would address the issue of quality control. Thus the biohazard committee could review the facilities and biological containment for P2 EK1 experiments and possibly P2 EK2 experiments and recommend approval or disapproval based on a first hand examination of the project. The decision would be registered with the appropriate governmental agency.

3. All P3 and P4 experiments regardless of EK level should first be reviewed and examined by the local biohazard committee and then their decision forwarded to the appropriate governmental agency for final action. The governmental agency would be required to respond to the question of the adequacy of the institution to perform the proposed investigations within a specified period of time.

Examples of certification principles could be obtained from the National Board of Medical Examiners, The American Society for Microbiology, and the American Board of Pathology. The public has an important role to play. The appropriate public representation is on the local biohazard committee. We are concerned that federal legislation not impose unreasonable restraints on open inquiry and research. For example, we are deeply concerned over some of the sections of the recently introduced bill by Senator Bumpers, "DNA Research Act of 1977, S621." Section 7 of this bill dealing with civil liberties provides for liability without regard to fault for injury to persons or property covered by such research.

If there were great liability there would be extreme constraints on universities and individuals. An inability to protect from liability by an insurance mechanism could preclude undertaking this important research. Thus the effect of section 7 of the Bill S621, or related legislation, would be to discourage or forbid recombinant research in many of the academic institutions in this country.

One of our main concerns is the apparent intemperate rush to establish legislation in this field. We must not compromise these advances in medical science. Also we must not establish a dangerous precedent by moving so rapidly that legislation is drafted without the best possible information, provided through consultation with the appropriate qualified scientific and medical experts.

IT'S NOT NICE TO FOOL MOTHER NATURE!
DNA RESEARCH
CONRAD

Document 6.21

June 30, 1977
Acker Letter to Rogers

Inter-Society Council for Biology and Medicine
Washington, D.C.

Please reply to:
P.O. Box 34660
West Bethesda, MD 20034

30 June 1977

Honorable Paul G. Rogers
U.S. House of Representatives
Washington, D.C. 20515

Dear Mr. Rogers:

The Inter-Society Council for Biology and Medicine comprises chief executive officers of seven professional organizations*. The Council, along with a significant number of sister societies and prominent individuals (see listing enclosed), have joined in support of major provisions of Congressman Paul Rogers' version of recombinant DNA legislation.

The Council believes it is not wise at present, when so little is definitely known about the risks and benefits of recombinant DNA research, to enact legislation hastily. We favor the adoption of the National Institutes of Health guidelines and application of the guidelines to all who use recombinant DNA technology. All federal agencies have, in fact, adopted the guidelines as applicable to their own respective employees and laboratories.

The Council wishes to emphasize the general acceptability of H.R. 7897 as opposed to the current Senate version if the Advisory Committee can be structured according to principles set forth below. We are impressed that the legislation produced by Congressman Rogers will permit free scientific inquiry while protecting the safety of the general public.

If federal legislation is to be enacted, the members of the Council, the elected leadership of the societies which they represent (and their governing boards in many instances) support the following principles relative to federal recombinant DNA legislation, most of which are embodied in H.R. 7897.

1. That all responsibility for regulatory action relative to the discovery and use of recombinant DNA technology be vested in the

* American Institute of Biological Sciences
American Society for Medical Technology
American Society for Microbiology
American Society of Allied Health Professions
Association of American Medical Colleges
Federation of American Societies for Experimental Biology
National Society for Medical Research

Secretary of DHEW:

2. That an Advisory Committee should be established to advise and assist the Secretary of DHEW. The functions of the Committee should be advisory and not regulatory. The Committee's membership should include representatives of the public and a majority of scientists with appropriate technical expertise in this field;

3. That institutions and not individuals should be licensed;

4. That regulatory responsibility should be delegated to the maximum extent possible to local biohazard committees. Each committee should include representatives of the public and a majority of members with appropriate technical expertise in the activities conducted at the institutions in question;

5. That experiments requiring P1 facilities be exempt from federal regulation;

6. That license revocation is an effective and sufficient deterrent in obtaining compliance with regulations. Further that the Inter-Society Council is opposed to the bonding of scientists or to civil penalties;

7. That the Inter-Society Council goes on record favoring uniform national standards governing DNA recombinant activities;

8. That the Secretary of DHEW be granted authority and flexibility to modify the regulations as further scientific findings become available. Further that the legislation include a sunset clause;

9. That the Inter-Society Council continues to express its concern that the development of such important legislation governing research should proceed only after due and careful deliberation.

Sincerely yours,

Robert F. Acker

Robert F. Acker, Ph.D.
Executive Director
American Society for Microbiology

RFA:bmb

Enclosure

cc: House Committee on Interstate and Foreign Commerce
House Committee on Science and Technology
Senate Committee on Human Resources

Document 6.22

October 17, 1977
Hess and Acker Letter to Staggers

FASEB FEDERATION OF AMERICAN SOCIETIES FOR EXPERIMENTAL BIOLOG

9650 ROCKVILLE PIKE • BETHESDA, MARYLAND 20014

TELEPHONE: 301 — 530-7000 • CABLE ADDRESS: FASEB, WASHINGTON, D. C.

Member Societies

AMERICAN PHYSIOLOGICAL SOCIETY
AMERICAN SOCIETY OF BIOLOGICAL CHEMISTS
AMERICAN SOCIETY FOR PHARMACOLOGY AND EXPERIMENTAL THERAPEUTICS
AMERICAN SOCIETY FOR EXPERIMENTAL PATHOLOGY
AMERICAN INSTITUTE OF NUTRITION
AMERICAN ASSOCIATION OF IMMUNOLOGISTS

EUGENE L. HESS
Executive Director

JOHN R. RICE, C.P.A.
Comptroller

October 17, 1977

The Honorable Harley O. Staggers
Chairman, Committee on Interstate
and Foreign Commerce
2125 Rayburn House Office Building
House of Representatives
Washington, D.C. 20515

Dear Mr. Chairman,

With respect to legislation designed to regulate Recombinant DNA activities, there are some highly vocal scientists so concerned about the potential dangers of this research that no legislation could be sufficiently stringent to satisfy them. At the other end of the spectrum, there are highly vocal scientists who believe that no hazards exist and that regulations are unwarranted. The difficulty with both of these minority groups is that neither, on the basis of currently available evidence, can prove that it is right. Between the two groups is the mass of scientists who, along with the general public, are awaiting answers to questions concerning the potential risk of Recombinant DNA research.

In the meantime the federal government is trying to balance two obligations. One obligation is to protect the health and safety of the public; the other is to protect the right of free scientific inquiry.

The first step in balancing these two responsibilities of government was the development of the NIH guidelines which now govern all federally supported research in this area. These guidelines are in the process of being updated and will be revised periodically as new knowledge dictates or warrants. One feature of the Guidelines is that they apply only to federally funded research. Legislation will be required to extend their application to the private sector.

Legislation will also be required to establish the Guidelines as the single, uniform, national standard with right of preemption over State and local laws. In the absence of such federal preemption there will continue to be a proliferation of conflicting State and local laws creating chaos in the scientific community and a mounting number of court cases brought against the National Institutes of Health with potentially undesirable precedents. Also, without uniform national standards we lack a basis for seeking international agreement on guidelines.

Legislation will also be required to provide minimal necessary administrative arrangements, including the assignment of responsibility for administering the law, and to provide a legal basis for addressing such questions as disclosure of information.

In our view H.R. 7897, a bill before your committee, represents an adequate start in addressing the problems cited above and in striking a balance between the responsibility to protect the health and safety of the public and the responsibility to protect the right of free scientific inquiry. While not everyone is happy with every feature of the bill, we are prepared to rely on the legislative process to produce a final bill which will serve the interests of both science and the public.

Because time may be as important a factor as substance in this case, it is hoped that your Committee on Interstate and Foreign Commerce will have an opportunity to debate and act on H.R. 7897 before the end of this session.

Sincerely,

Eugene L. Hess, Executive Director
Federation of American Societies
for Experimental Biology

Robert F. Acker, Executive Director, ASM
on behalf of American Society for Microbiology

cc: Congressman Paul Rogers
Chairman, Subcommittee on
Health and The Environment

bcc: PAC
Society EOs
I-SCBM
Dr. Lewis, NSF

Document 6.23

May 1977

National Academy of Sciences *Letter to Members* (**7**:12), Excerpt

Academy Resolution on DNA Legislation Proposals

The Congress has before it numerous and diverse proposals for legislation that would subject research with recombinant DNA to some form of regulation. In some of these proposals, federal regulation would not exclude action by the local government. By the terms of at least two bills, under certain conditions, local governments could establish and administer regulations more severe than those established by the Federal Government, indeed could even ban such research. Institution of such locally derived regulations would require a finding by the Secretary of HEW that they are as stringent or more stringent than the federal regulations and will be so administered by the local government. Such local option provisions are suspect in that they would make possible a variegated regulation mix across the country; and they have implications that might be translated into other areas, of scientific research. Hence they are of concern to the Academy.

A group of 13 members submitted a resolution on the DNA regulatory situation for consideration at the recent Annual Meeting. The text of the resolution and the names of the proposers are set forth below. The resolution was passed, by voice vote, at the Business Session on April 26.

RESOLUTION

Three years ago the National Academy of Sciences played an important role in initiating responsible and informed discussion on the need for controls on recombinant DNA research. The subsequent debate, both inside and outside the scientific community, resulted in a set of guidelines formulated under the sponsorship of the National Institutes of Health. Now the Congress is considering several drafts of detailed and far-reaching legislation dealing with the regulation of recombinant DNA research.

The NIH guidelines are the result of careful deliberation and we favor their simple conversion into a uniform national set of regulations. However, much of the proposed legislation now before Congress would allow local communities or states to set their own stricter regulations, including even a complete ban on this type of research. It is reasonable for the local community to participate in supervising adherence to regulations. But the question concerns speculative rather than known dangers and these would not vary from one locality to another.

The research institutions of the country constitute an important national resource, and differing local options could subject that resource to arbitrary regulations. Overly restricting this type of research would severely degrade the capability of biomedical research and limit its contribution to the public welfare. In essence, it would allow a local community to affect a critical component of national policy. Above all, local option would set a dangerous pattern for the regulation of basic research in a manner that might deprive society of substantial future benefits.

Some of the legislation proposes to establish a national regulatory commission expressly to govern recombinant DNA research. Its broad powers for controlling basic research represent a wholly new and unfortunate departure. We are also concerned that it would set a precedent for the regulation of other areas of science. A much simpler and more flexible mechanism allowing for public participation could carry out those functions that may be needed.

THEREFORE, be it resolved that we voice concern about legislative developments in this area and ask the President of the National Academy of Sciences to relay these concerns to appropriate officials in the government.

D. Baltimore
B. D. Davis
H. Eagle
J. T. Edsall
C. Grobstein
D. M. Horstmann
R. Hotchkiss
R. J. Huebner
A. Rich
W. A. Rosenblith
F. N. Ruddle
R. L. Sinsheimer
E. L. Smith

(PROPOSED by the individuals listed above; PASSED by voice vote at the Business Session of the 114th Annual Meeting of the Academy, April 26, 1977.)

Document 6.24

October 7, 1977
Cohen Letter to Staggers

STANFORD UNIVERSITY MEDICAL CENTER
STANFORD, CALIFORNIA 94305

STANFORD UNIVERSITY SCHOOL OF MEDICINE
Department of Medicine

October 7, 1977

The Honorable Harley Staggers
Rayburn House Office Building
Congress of the United States
House of Representatives
Washington, D.C. 20515

Dear Congressman Staggers:

I appreciate the opportunity I had to discuss with you by telephone my concerns about the proposed legislation on recombinant DNA sponsored by Mr. Rogers (H.R. 7897). I am responding now to your request for more details about my views.

As you know, several years ago biologists learned to duplicate in a test tube the genetic recombinational process that normally is carried on by all living cells. This enabled them to propagate hereditary material from various biological sources in bacteria. My own laboratory at Stanford had an important role in the development of these techniques, and shortly after this discovery I joined with a group of scientific colleagues in publicly calling attention to the possibility that some gene combinations that could be made with these techniques might prove to be hazardous. Although there was no scientific basis for anticipating a hazard, because of the newness and the relative simplicity of these methods, it seemed reasonable and appropriate to proceed with caution. Further information was needed to enable us to assess more fully the implications of the research, and to determine whether there was in fact any risk. This was an opportunity in a new research area, we thought, to exercise care at the onset.

Our actions were voluntary, and were taken at our own initiative in the absence of legislation or government rulings. Unfortunately, our unprecedented attempt to guard against hazards that were not known to exist was so novel that it was widely misinterpreted by the press and others as implying we thought danger was likely. It has been inconceivable to most people that we would have taken such a step or that an extensive federal response would have occurred unless severe hazards were actually known, or unless there were at least sound scientific grounds for expecting danger.

During the past year or so, virtually all of the scientists who were among the first to express concerns about certain kinds of recombinant DNA experimentation have come to believe that our earlier concerns were greatly overstated. At the time this issue was first raised, the techniques were new. Although many other kinds of genetic manipulation (such as the creation of hybrid plants and hybrid animals) have been carried out for years, there was little experience in gene modification using these new procedures at the time this issue was first raised. However, while the anxieties of the public and of some members of Congress have been increasingly aroused, the work has proceeded without adverse consequences in many dozens of laboratories around the world. During the past four years, more than 200 scientific investigations involving recombinant DNA have been published, and literally hundreds of billions of bacteria containing many different kinds of recombinant DNA have been grown in the United States and abroad with no harm to humans or to the environment. Because the life cycle of bacteria is measured in minutes, it has been possible to study many thousands of generations of living cells containing recombinant DNA.

As scientists we deal with data, and the data obtained have persuaded us that our earlier concerns go far beyond any reasoned assessment. Many of us initially feared the spread of bacteria containing recombinant DNA and the possible conversion of harmless to harmful bacteria. However, experience has shown that the introduction of recombinant DNA into already weakened laboratory bacteria weakens them even further so that they are unable to compete successfully in nature and are at a disadvantage outside the special protective conditions of the laboratory. Moreover, even genetic material known to be associated with disease-producing traits fails to convert the laboratory organisms used for recombinant DNA experiments into disease-producing bugs. A recent NIH-sponsored conference of epidemiologists and infectious disease experts concluded unanimously that the danger of runaway epidemics from these bacteria has proved to be virtually non-existent. In addition, many of our earlier concerns stemmed from what we believed to be the "novel" nature of gene combinations that could be made using recombinant DNA techniques. However, recent experiments have shown that recombinant DNA molecules can arise in nature by reactions akin to those used in the laboratory.

Unfortunately, the Congress appears now to be reacting to our earlier very speculative concerns, which we raised before we had reassuring data of the type I've mentioned above. As Nobel laureate James Watson recently pointed out, "We suffer (from the results of) a massive miscalculation in which we cried wolf without having seen or even heard one." Legislation has been proposed to establish an extensive regulatory apparatus to control basic scientific research in an area where there is not one shred of evidence to support the view that special risks exist or special controls are needed. In other areas of research where biologists work with microbes that we know to be hazardous, the simple establishment of standard operating practices has been sufficient to protect laboratory workers and the public at large in the absence of legislation. Yet for recombinant DNA, a costly and cumbersome bureaucracy is proposed to govern the content and methods of scientific inquiry in the absence of any indication whatsoever of actual hazard or of any scientific evidence to even suggest that a substantial threat to the public health and welfare exists.

The passage of H.R. 7897 would lead to unprecedented and unwarranted government intervention into basic scientific research. Many knowledgeable scientists believe such intervention will seriously inhibit the solution of biological and medical problems, and will inevitably delay discoveries leading to the treatment and cure of diseases that really do exist and which do present a very real hazard to the health and welfare of the public.

I urge the tabling of H.R. 7897 and the establishment instead of a distinguished Study Commission consisting of both representatives of the public and scientists to examine in greater detail whether there is really a need for legislation in the light of newly available data. Maintenance of the present situation, which has resulted in no harm whatsoever, for another year or 18 months to enable a careful review would not seem to pose an unreasonable threat to the public. Certainly, the work will continue without analogous legislatively mandated controls in much of the rest of the world in any case, and microbes do not recognize national boundaries. It seems to me that the potential danger of establishing a mechanism for day-to-day control by the government of an area of basic research where increasing evidence indicates no unique regulation is required, is much greater than the entirely speculative risks from recombinant DNA.

I would be pleased to answer any specific questions you may have about these comments, or to provide documentation for the statements I've made.

With best wishes,

Sincerely yours,

Stanley N. Cohen

Stanley N. Cohen, M.D.
Professor of Medicine and
Professor of Genetics

SNC:seh

October 14, 1977
Zinder Letter to Staggers

THE ROCKEFELLER UNIVERSITY

1230 YORK AVENUE · NEW YORK, NEW YORK 10021

The Honorable H. Staggers
U.S. House of Representatives
Washington, D.C. 20515

October 14, 1977

Dear Congressman Staggers:

It has been brought to my attention that as Chairman of the Committee of Interstate and Foreign Commerce, you have been told that the scientific community is in favor of HR 7897. I want to assure you that this is not the case. While I can't speak for all my colleagues, I know that for example some of us spent the better part of the summer in what proved to be a successful effort to have our Governor veto the recombinant DNA bill passed by our State Legislature.

I should note that I was a member of the Committee of the National Academy of Sciences that in 1974 informed the scientists and the public of the possible consequences of the then newly discovered techniques of recombinant DNA. Still it is not paradoxical that I testified twice before the Senate Subcommittee on Health and once at a briefing breakfast with Congressman R. Thornton against legislative action.

What we did runs as follows. Given the state of knowledge then, we attempted to have the work proceed while blanketing it with a code of practice to protect the public. It was a most successful piece of self-regulation which would buy us time to have certain other experiments done to test the hazards if any. This process was to culminate, we hoped, in the N.I.H. guidelines of June, 1976.

Instead under the leadership of a very few but highly vocal dissenting scientists a significant opposition to this work appeared. As the argument became more shrill but far less substantive two things happened: the Congress embarked on legislation and the data started to come in that the experiments were safer than we'd thought. The scientific community had little to say about the structure of the bills written in the House and Senate. We were not listened to on the grounds that we were interested parties. So as the legislation appeared and feeling powerless to go back to our voluntary and reasonable codes, we supported in turn those bills we found least onerous. At first it was HR 7897 for compared to S1217 it was far less restrictive. With the appearance of the Nelson-Moynihan substitute amendment which we found still more favorable there was another switch in allegiance. When the community becomes fully aware (only three weeks old) of Senator Kennedy's total about face it'll follow him. This sounds dastardly but at all stages we have been pushing for less restrictive legislation and if possible a way to return to our voluntary regulation. Not being used to lobbying, we've probably made all kinds of mistakes.

Still your reticence to allow infringement of our freedom of inquiry and Senator Kennedy's turn about gives us hope that our message has gotten across. Believe me, sir, we neither want nor need legislation in this area. We'd like nothing better than to be able to return to our labs, stop going to hearings or awaiting the latest word from Washington.

If I can be of any service in any way please don't hesitate to call on me.

Yours sincerely,

Norton D. Zinder
John D. Rockefeller, Jr., Professor
of Molecular Genetics

Document 6.25

July 15, 1977
Science (**197**:208)

LETTERS

Recombinant DNA Research: Government Regulation

The following open letter to Congress represents a consensus of those who attended the 1977 Gordon Conference on Nucleic Acids. Discussions at the conference about the status of pending legislation proposed to regulate recombinant DNA research led to the formulation of this position, which was discussed and voted upon by the entire meeting. Subsequently, 137 individuals signed the letter, representing 86 percent of the members of the meeting. We are most concerned that the benefits to society, both practical and fundamental, that we foresee will not be forthcoming because legislation and regulation will stifle free inquiry. At the meeting this June, with a single exception, there was unanimous agreement that regulation beyond simple enforcement of the NIH Guidelines is unnecessary, and many expressed the view that less regulation would suffice to guard against any hypothesized dangers.

We are concerned that the benefits of recombinant DNA research will be denied to society by unnecessarily restrictive legislation.

Four years ago, the members of the 1973 Gordon Conference on Nucleic Acids were the first to draw public attention to possible hazards of recombinant DNA research. The discussions which started at that meeting resulted in the issuance in 1976 of the NIH Guidelines for the conduct of this research.

We, members of the 1977 Gordon Research Conference on Nucleic Acids, are now concerned that legislative measures now under consideration by Congressional, state and local authorities will set up additional regulatory machinery so unwieldy and unpredictable as to inhibit severely the further development of this field of research. We feel that much of the stimulus for this legislative activity derives from exaggerations of the hypothetical hazards of recombinant DNA research that go far beyond any reasoned assessment.

This meeting made apparent the dramatic emergence of new fundamental knowledge as a result of application of recombinant DNA methods. On the other hand, the experience of the last four years has not given any indication of actual hazard. Under these circumstances, an unprecedented introduction of prior restraints on scientific inquiry seems unwarranted.

We urge that Congress consider these views. Should legislation nevertheless be deemed necessary, it ought to prescribe uniform standards throughout the country and be carefully framed so as not to impede scientific progress.

WALTER GILBERT
Biological Laboratories,
Harvard University,
Cambridge, Massachusetts 02138

Document 6.26

October 31, 1977

Higher Education Daily

HIGHER ED SEES FEDERAL PRE-EMPTION AS CRUX OF DNA CONTROVERSY

With Congress immobilized by the debate over the hazards of recombinant DNA research, a number of communities and states across the country are taking steps to institute their own local controls.

The trend has alarmed colleges and universities and some scientists who say that a patchwork of local regulations would play havoc with the research, which involves altering DNA (deoxyribonucleic acid), the active material of genes and chromosomes. Putting aside the question of risk—the subject of a heated controversy within the scientific community—there is a growing realization that Federal legislation is needed, if only to keep states and local governments from imposing more stringent barriers.

According to the American Society for Microbiology, efforts to prohibit recombinant DNA research are underway in New Jersey and New York, and state controls are imminent in California and Massachusetts. Cambridge, Mass., has already adopted its own guidelines, as have the states of Maryland and Wisconsin.

"In the absence of any conclusive congressional action, communities that are concerned (with the dangers of DNA research) are going to move," predicts a Senate Human Resources Committee staffer.

The Senate. During the last several months, as the scientific debate on the research has intensified, the situation in Congress has become increasingly confused. Momentum for a DNA bill has dwindled, and it is now clear that legislation will have to wait until next year.

The Senate Human Resources Committee early this summer cleared S. 1217, which would have set up an independent commission to monitor and license recombinant DNA research. But Sen. Edward Kennedy, D-Mass., the sponsor of the bill, has withdrawn his support of the legislation. According to a Kennedy staffer, the senator changed his mind after reviewing new information which suggests the research risks are less than formerly thought.

The staffer said Kennedy plans to introduce next year legislation creating a study group to re-assess the problem and to extend to all researchers for one year research guidelines of the National Institutes of Health, which now apply only to researchers receiving Federal grants.

Sen. Jacob Javits, R-N.Y., is also preparing what a Javits staffer said would be "a whole other approach" to controlling the research. The staffer said Javits favors a strong pre-emption provision. Meanwhile, Sen. Adlai Stevenson III, D-Ill., the chairman of the Science, Technology and Space Subcommittee, has scheduled hearings beginning Wednesday on the safety of the research and the effects of regulation.

The House. On the House side, H.R. 7897 has been stuck in markup for weeks before the Interstate and Foreign Commerce Committee. The bill lodges authority for regulating the safety of DNA research with the Secretary of HEW.

Sources say that committee Chairman Harley Staggers, D-W.Va., may offer an amendment similar to Kennedy's new proposal.

After the Commerce Committee finishes with H.R. 7897, it will most likely be referred to the Science and Technology Committee.

Whatever happens, higher ed associations are supporting a tough Federal pre-emption provision. John Crowley, associate secretary of the Association of American Universities, says AAU is "strongly in support of the strongest pre-emption provision possible." Jerold Roschwalb, director of governmental relations at the National Association of State Universities and Land-Grant Colleges, calls pre-emption "the issue."

Roschwalb predicts researchers will flock to areas with the fewest restrictions if local jurisdictions are able to throw up their own barriers to the research.

Leslie Dach, a lobbyist for the Environmental Defense Fund, says that local governments should be able to impose control where there are valid concerns. "If scientists have to move, then they have to move," he says. "Their inconvenience seems to me to be secondary to the public's right to set up an ample margin of safety to protect their health."—CBL

Document 6.27

June 27, 1977
Pollack Letter to Watson

Department of Microbiology
School of Basic Health Sciences
HEALTH SCIENCES CENTER
State University of New York at Stony Brook
Stony Brook, New York 11794
516 444-2446

ALLIED HEALTH PROFESSIONS • BASIC HEALTH SCIENCES • DENTAL MEDICINE • MEDICINE • NURSING • SOCIAL WELFARE

June 27, 1977

Dr. J. D. Watson
Cold Spring Harbor Laboratory
P.O. Box 100
Cold Spring Harbor, N.Y. 11724

Dear Jim,

The New York State Senate has passed, and the State Assembly is considering, the attached act. In my opinion it would be a disaster to have this become law. As you know, I have been an early and earnest proponent of regulation of dangerous or potentially dangerous biological research, and I have argued forcefully for adoption of laws embodying the NIH-NSF Guidelines. Since my election as Chairman of the Institutional Biohazards Committee on the Stony Brook campus, I have tried to follow the spirit and the letter of the Guidelines as if they were law.

This act, however, completely bypasses the Guidelines. It discards the collective work of our colleagues, to vest total regulatory and punitive power in the hands of local Health Commissioners throughout the State. Dr. Stephen Harris, the Suffolk County Health Commissioner, would under this act cease to be a member of our local Institutional Biohazards Committee. Rather, he would become the sole arbitrator of our individual research efforts, with the power to levy a $5000/day fine if he and an investigator differed on any point of scientific procedure. A similar power would be in the hands of each county Health Commissioner. No appeal to the Recombinant DNA Directors of the NIH or of the NSF would be considered.

I am very discouraged and disturbed that the first local fruit of scientists' civic responsibility should be this medieval prescription for punishing witches and sorcerers. I would be glad to discuss this with you personally any time after July 11th.

Sincerely yours,

Bob

Robert E. Pollack, Ph.D.
Associate Professor of Microbiology

REP:cb
Encl.

LAWS OF NEW YORK, 1978

CHAPTER 488

AN ACT to amend the public health law, in relation to the certification of recombinant DNA experiments

Became a law July 20, 1978, with the approval of the Governor. Passed by a majority vote, three-fifths being present.

The People of the State of New York, represented in Senate and Assembly, do enact as follows:

Section 1. The public health law is hereby amended by adding a new article thirty-two-A to read as follows:

ARTICLE 32-A

RECOMBINANT DNA EXPERIMENTS

Section 3220. Findings.
3221. Definitions.
3222. Certification.
3223. Enforcement.

§ *3220. Findings. The legislature finds that new techniques of genetic manipulation which allow researchers to accomplish exchanges of genetic material between unlike organisms offer great potential for expanding human knowledge of genetics but also may present significant risks to public health, to the environment and to the health of research workers. These experiments are being conducted at research institutions in New York, including universities, hospitals and industrial facilities. Industry plans to mass-produce recombinant organisms should such organisms be found to have commercial uses. Such research and production should only be conducted under safe conditions as prescribed by the commissioner.*

§ *3221. Definitions. As used in this article:*

1. "DNA" means deoxyribonucleic acid.

2. "Recombinant DNA" means DNA molecules which;

(a) have been formed by joining together DNA segments in a cell-free system and which have the capacity to enter a cell and to replicate in such cell either autonomously or after they have become an integrated part of such cell's genome; or

(b) are the result of a replication of the DNA molecules described in paragraph (a) of this subdivision.

3. "Recombinant DNA activity" means the possession of recombinant DNA by any person and any activity undertaken by any person for the production of recombinant DNA.

4. "Person" means any individual, corporation, partnership, or legal entity of any kind, or a governmental agency, board or body.

5. "Cell-free system" means an environment outside of any cell or cellular organism.

§ *3222. Certification. 1. One hundred eighty days after the effective date of this article, no individual shall engage in a recombinant DNA activity unless such individual has been issued a certificate by the commissioner or such individual is affiliated with and is acting under the direction of and in a facility controlled by a person who has been issued such a certificate.*

2. Within one hundred eighty days of the effective date of this article, the commissioner shall prescribe regulations for the conduct of recombinant DNA activity which shall be the substantial equivalent of sections II (entitled "containment") and III (entitled "experimental guidelines") of the recombinant DNA research guidelines of the National Institutes of Health of the Department of Health, Education and Welfare published in part II of the Federal Register for July seventh,

EXPLANATION—Matter in *italics* is new; matter in brackets [ ] is old law to be omitted.

nineteen hundred seventy-six. If the National Institutes of Health guidelines are revised, the commissioner shall revise the regulations for the conduct of recombinant DNA activity accordingly.

3. Within one hundred eighty days of the effective date of this article the commissioner shall prescribe regulations for the:

(a) training and qualifications for individuals engaging in recombinant DNA activities;

(b) personnel health monitoring programs;

(c) establishment of institutional committees to oversee such activities; and

(d) periodic reports of the progress of such activities.

Regulations adopted pursuant to this section shall be reviewed periodically by the commissioner in light of current scientific knowledge to determine their continued adequacy and appropriateness.

4. The commissioner shall appoint an advisory committee which shall be composed of individuals representative of the public interest, laboratory personnel and of the scientific community to advise him in carrying out his duties under this article.

5. The commissioner shall by regulation, establish procedures for application for a certificate to conduct recombinant DNA activity. All proprietary information in applications or reports to the department by persons certified pursuant to this section, not available to the public or protected by a patent or copyright, shall be kept confidential.

6. The commissioner may approve or deny an application for a certificate to engage in recombinant DNA activity or may approve it upon such conditions as he shall prescribe.

7. The commissioner by regulation may prescribe reasonable fees for certification, not exceeding the cost of administrative services rendered by the department.

8. The commissioner shall, by regulation, provide for an abbreviated certification process for the conduct of recombinant DNA activity which is subject to, and which is in compliance with, policies and regulations promulgated by any agency of the federal government for the regulation of recombinant DNA activity.

9. No local authority shall enact or enforce any local law, ordinance, rule or regulation which would regulate or restrict recombinant DNA activity. Further, no local authority shall enact or duplicate any provision of this article as local law, ordinance, rule or regulation.

§ 3223. Enforcement. 1. The commissioner may conduct scheduled or unscheduled inspections of facilities where certified activities are being conducted.

2. The commissioner may suspend a certificate on the ground that the holder unreasonably refused to admit a department inspector to the premises where a recombinant DNA activity was taking place, or for any other violation of a regulation promulgated pursuant to this article. The commissioner shall hold a hearing within fifteen days following any such suspension to determine whether the suspension was reasonable. The commissioner may then annul the suspension, or if he finds the suspension to have been reasonable he may revoke the certificate or continue its suspension for a period not to exceed three months.

§ 2. This act shall take effect immediately.

The Legislature of the
STATE OF NEW YORK } *ss:*

Pursuant to the authority vested in us by section 70-b of the Public Officers Law, we hereby jointly certify that this slip copy of this session law was printed under our direction, and, in accordance with such section is entitled to be read into evidence.

WARREN M. ANDERSON — *Temporary President of the Senate*

STANLEY STEINGUT — *Speaker of the Assembly*

The Gene, The Scientists and The Law

Norton D. Zinder
Rockefeller University / New York, N.Y.

For a time now I've been feeling like an actor who opened in a play about five years ago to rave reviews. This play was re-reviewed about two years ago, and was taken apart by the critics, but the damn show just won't close. The show won't close because we keep drawing an audience. When the audiences stop coming it'll close and I hope it'll close soon.

My title has for my taste one too many nouns. You can guess which noun particularly disturbs me—the last one of "The Gene, The Scientists and The Law." For the last few years, many of us have been trying very hard to avoid getting Science involved with the Law. I'll try to tell you how and why.

Dr. Weissbach's complete description of a recombinant DNA experiment is terribly important because there is the myth abroad that *anybody* can understand all the scientific and the historical events that occurred which led to the polemic that exists relative to recombinant DNA research. Let me illustrate the kinds of errors that are made. I could give you "n" examples, but I'll describe just a few. I'm going to quote from the actual documents. I do so because although there are documents that exist, most people who write about this subject don't seek out the primary sources but misquote the misquote. I have been so badly misquoted on this subject that I'm rather sensitive and will try not to do the same.

My last outing in this polemic was a discussion on radio station WBAI, NY, with George Wald of Harvard, Marcia Cleveland of the N.R.D.C. and Bernard Talbot of the N.I.H. I was asked at the very last minute since someone else couldn't make it. I arrived at the studio and I picked up the WBAI program notes for that month's broadcasts.[1] This listing had been sent out as public information to the subscribers of WBAI. It's centerfold contained a story on the recombinant DNA program. As I read it, I almost wanted to cry. Why? Because this document contains one error after another even though it was written four years after the recombinant DNA discussion began. I'll just quote a few sentences to give you its flavor. "About three years ago, scientists at Miles Laboratories discovered a way to create restriction enzymes." That's what it says. There isn't a word in that sentence that's true except the connectives. It goes on about what these restriction enzymes can do—they don't quite have that right either. Then they close this paragraph in a statement that almost floored me, both as an individual and as a trained geneticist. It says, "Similarly, a gene resulting in a *racial* disease,"—I know of no racial diseases—"like sickle cell anemia, could surreptitiously be introduced into whole populations as a means of covert warfare, even by accident." The document goes on and on, and I don't think there are three words in it that are really meaningful, appropriate or true. This illustrates what is going out to the public and what the public is reading: these horrible, erroneous scenarios. Some like to believe that scientists and lay people are interacting in a new and productive way on this issue. Unfortunately, it just isn't true. The scientists don't know how to begin to correct all of the misinformation that has been disseminated.

Just a short time ago the final version of the

Environmental Impact Statement arrived from the National Institutes of Health. That was an interesting document as it came with an Appendix that has the letters contributed by those who commented on the Draft Environmental Impact Statement (DEIS).[2] I guess the people who wrote didn't know that the letters were going to be made public. The comments on the DEIS should have contained arguments with which I might disagree but would still respect for their cogency. As I flipped the pages I recognized a letter written on the same broken typewriter that I'd seen before (sort of the Pumpkin Papers). It came bearing the letterhead of The Sierra Club.[3] It contains absolute nonsense and goes on and on with many pages of the same. Letters from this individual led my colleague, Dr. James Watson—the Watson of Watson and Crick—to say that "after the letters from this gentleman, I am coming to the conclusion that Mineral King could best be saved by Walt Disney, Inc., rather than The Sierra Club."

Next there's a brief from the National Resources Defense Council,[4] a pro bono law firm involved in environmental issues. This should be better. Then I read, "In July of 1974, a committee of the National Academy of Sciences, chaired by Paul Berg of Stanford University, called for a voluntary moratorium on *all* recombinant DNA activities." That's not what we did. After all, I was there. There was no moratorium on all recombinant DNA experiments.

After being so brilliant about the history of the events, a lawyer now, mind you, with no training in science whatsoever, comments on the NIH guidelines and finds them to be too complicated: "The elaborate classification system of the current guidelines could be abandoned in favor of a much simpler system. NRDC suggests at most three classifications: Prohibited, Very Hazardous and Hazardous." Nice classifications if one doesn't mind using pejorative words. I do believe there are things in law—I'm not a lawyer and I would never attempt to tell lawyers how to act procedurally—but I do believe there are things called rules of evidence. Consider the last classification, "Hazardous." Why is this experiment "hazardous?" What evidence is there that this experiment is "hazardous?" There is no evidence. Somebody *said* it's hazardous. That's interesting. Somebody said it's *not* hazardous. Somebody said it *might* be hazardous. Now what is a lawyer supposed to do in such a situation? Since there is no material evidence one might conclude that it *might* be hazardous. Also the scientists who wrote the NIH guidelines should know a bit more about how to classify experiments than a lawyer. I've never found the law or legalese to be easy.

We move on to another letter from the N.Y. State Attorney General's Office.[5] I tend to become upset and angry when people who hold public office demean it by their actions. One of the important things was the date on this document: October 19, 1976. I read along, and it says, "As a result of the moratorium agreed to at Asilomar, no recombinant DNA research has been done, so it's not surprising that the dangers have not yet been encountered." Now the moratorium is called at Asilomar. I was there also and that's not what happened. Below I'll try to explain why I believe this kind of "error" recurs—it has a purpose.

I read on and come across another interesting point. Again, I'm glad Dr. Weissbach explained the three components of a recombinant DNA system, the host, the vector and the donor, as they are important. I quote, "In the description of the recombinant DNA experiments, the DEIS claims that the DNA vector is usually derived from the same species that will serve as the host. Again a false sense of security is created because the reader may assume that the host species will only receive the DNA of its own kind." At first I didn't know what they were talking about. It's only natural that a vector is derived from the host that it'll be put back into since one wants to use it to amplify some gene(s) that has been inserted into it. The vector must be able to grow in that host. Then I remember that I read another document that came from this same office sometime last Spring. It was a position paper justifying legislative regulatory action on recombinant DNA in NY State. There it says, "Any organism may act as a *vector donating* its DNA."[6] They don't know the difference between the vector and the donor.

You cannot read the guidelines, or the Draft Environmental Impact Statement believing that the vector is the donor and the donor is the vector, and make any sense out of them. It's like reading a manual on driving an automobile and believing that by depressing the accelerator the car will brake. The whole world is upside down. Moreover, the word 'vector' has a common well understood usage in the English language, meaning 'carrier.' Furthermore the DEIS has a glossary that includes the word 'vector.'

It's incredible that people can make such "mistakes." However the real tragedy is that this biased, erroneous statement written by two public servants was mailed off or at least dated two days before they held a public hearing on recombinant DNA in which they were supposed to become educated on the subject—at least that's my understanding of the purpose of a public hearing. The letter was dated October 19, 1976 and hearings were held on October 21, 1976. I don't believe that either the proponents or opponents of recombinant DNA research would have participated in such a sham if they had known that the minds of those in charge were already closed and there was no way to open them.

Somehow we've gotten to a position such that we're having a dialogue of the deaf. Nobody's listening, and everybody's talking past each other. Why? Some, such as Robert Pollock[7] of Stony Brook have suggested that the public just doesn't understand the peer review process by which the Academy rules itself. Peer review is used in awarding funds for research and in judging promotions. We have peer review without real conflict of interest. It's my own feeling, having participated in a number of peer review panels that while it's not a perfect system, I've never been a participant in any other discussion in which I felt so much equity was involved. There's an article in a recent *Scientific American*[8] by three sociologists who analyzed the National Science Foundation peer review process and described how well it works.

Still I don't believe this explains the depth of emotion on this issue. I'm not sure I can provide an adequate answer although I believe it is in large measure due to an anti-science and anti-intellectual trend in this country. These attitudes have been crystalized by a small but vocal band of dissident scientists who have found a receptive audience in those professional nay sayers, some so-called environmentalists who twitch at the words "new technology" without distinguishing the progressive from the regressive.

It all started back in 1972 when Paul Berg decided he wanted to put the DNA of a virus called SV40 (Simian virus 40) into the bacteria *E. coli* by a chemical procedure; a procedure that is not as simple as that provided by the current technology. This caused a bit of a stir in the scientific community and led to a meeting at Asilomar in January of 1973, to discuss biohazards. This was Asilomar I, where we found out how little we knew about laboratory biohazards. We collected data on the incidence and type of laboratory infections. The data we had was collated and led to a book published by the Cold Spring Harbor Press called *Biohazards*. This meeting had no particular impact with the press or the public but the volume produced is used as a guide by the scientific community. Things then began to move rather fast. In June of 1973 there was a Gordon Conference on Nucleic Acids; a special conference held every year in New Hampshire. At this Conference, Dr. H. Boyer announced that he had DNA restriction enzymes that led to sticky ended DNA. This meant that one could put any DNA together with any other DNA. The scientists at the meeting immediately saw the potential of this technique for making possible all kinds of new and important experiments. Concern was also voiced about the possible hazards in using this technique. Hence, being public minded they wrote a letter[9] to the National Academy of Sciences requesting that the NAS look into the matter. They didn't just mail the letter to the President of the Academy, they also published the letter in the journal, *Science*.

At that moment the issue of recombinant DNA went public and we entered the political process. None of us realized what would ensue and none of us realized that our lives would never be the same.

A committee was formed by the Academy, under the chairmanship of Paul Berg. It met at MIT on April 17, 1974. The people chosen to attend were those whom we felt would be most constructive in thinking through this situation. As a result of that meeting a letter to the scientific community [10] was written and published in *Science* and *Nature*. The letter asked the scientists to join us in a temporary moratorium on the performance of the two types of experiments we felt had the most potential for hazard. *Only two types of experiments were involved*!! We've all said this a thousand times but still people have misquoted that document and have found not what's there but what they'd like to be there. We intended only a temporary pause and only for certain experiments so that hopefully the whole subject wouldn't become pejorative. In fact, our so called leadership was really grass roots leadership. The experiments chosen were precisely those that the scientific community was most concerned about. Our tactics in handling the problem may have been right or wrong—history will judge—in asking for a pause in doing these experiments. We felt we were buying time for risk assessment experiments to be done. Even with the wisdom of hindsight and knowledge of how events were soon to overtake us, I still feel we did the right thing and chose the right experiments to defer. They're still listed as those with the highest potential hazard and require the highest level of containment.

We made two other recommendations which really took the issue public. The first was a request for an international conference—which turned out to be again at Asilomar—so that we could call on the expertise from various other scientific disciplines. There were only ten of us at the MIT meeting and we felt that the decision making required broader representation of the international scientific community. We also asked the NIH to set up a committee to establish codes of practice or guidelines for this work to be based on further more careful deliberation. Thus the scientific community recognized the potential hazards of this work and did not shirk it's duty but moved quickly to blanket this area with a code of practice. We said very clearly that even though the hazards were only potential—there's no evidence for any of them—we were going to be on the safe side. We said, in essence, 'those experiments that everyone's worried about—don't do,' as for the rest, consider them before proceeding. Because of simple peer pressure everybody cooperated.

In February of 1975 the Asilomar conference was held. It was attended by 90 American scientists, about 30-40 foreign scientists, 12 reporters and 4 lawyers. And what did we do at Asilomar? There we tried to broaden our thinking and see what kind of concerns there were. Some have said that the conference was too technical and didn't deal with all of the issues such as the moral and ethical ones. It was hard enough to burden this small area of science with the biohazard problem no less these other considerations. In addition it was necessary to bring all of the scientists up to date so that their thinking would reflect the latest information. The idea of biological containment was also discussed. It involves the use of host-vector systems that are so enfeebled that they can't live outside the laboratory. This was a technical problem which greatly attracted the scientists' efforts as it was a relief from trying to decide whether frog DNA was potentially more dangerous to use than mouse DNA. It was something they knew how to do. I've always felt that bringing biological containment to the fore was a strategic error on our part although it was the right thing to do. It had two consequences. First it implied that our standard bacterial strains were unsafe and second and more importantly it led to the belief that these experiments were more dangerous than any ever done before since a new form of containment was developed for them. Rather it was a very conservative approach to experiments with recombinant DNA but has not been understood as such.

The lawyers present at Asilomar would now and again shake us up and remind us that we could be sued. They advised us that only a sincere attempt at self-regulation would be of aid in preventing or mitigating such suits.

The reporters sat there with their tape recorders running, taking it all in and interviewing individuals at the intermissions. They had insisted on being allowed to come. Only one condition was required of them: they were not to write an article every day but agree to wait until the conference was over. Then they were, of course, free to do as they pleased. We didn't want any day's deliberations appearing prematurely in the press. Thus the reporters also stayed the full four days of meetings. They seemed to enjoy it. I recall a conversation with one science writer: he said, 'I always thought that most scientific conferences were boondoggles, but I've never seen a bunch of people work so hard.' That's also my memory of Asilomar. I don't think I had five hours sleep during that entire week.

On the basis of the deliberations we wrote provisional guidelines for the performance of recombinant DNA experiments.[11] It was voted on, paragraph by paragraph, by all of the 150 people there. Its provisions were more stringent than those in the Berg, et al. letter. The experiments were divided into essentially five categories: Forbidden, High-risk, Moderate-risk, Low-risk, Minimal-risk. This set the pattern for the ultimate NIH Guidelines. The categories were explicitly stated to be a first assessment to be modified up or down as further data would dictate. The document exists, despite this many people who have analyzed what we did and why we did it have not bothered to read it. Only last month I read an article in the *New England Journal of Medicine* by a bioethicist named W. Gaylin on recombinant DNA.[12] In referring to Asilomar he said: "Proud in their achievement at halting the major research, they engineered the most successful publicity and promotional package of any scientific meeting that I have ever encountered. Asilomar became the scientific version of 'Jaws' and the public was titillated but also frightened. In generating awe, however, they encouraged the fear that is the components of it. In their pride in their achievement, the researchers forgot that the public would also be aware that they resumed the research some short period later and more important it was "they," the scientists, who called both the halt and allowed the resumption. Precisely the opposite approach was needed." Why almost three years later can't Dr. Gaylin find out what really happened? Why did the conference become a jamboree? Very simple—there were 12 reporters there. After they left they wrote the story in great detail. We talked to them and later were called by others. The reporters wrote the story and we suddenly became public figures. Had there been no reporters present, as at Asilomar I, most people would probably never have heard of the conference. But it is his last statement which is indeed more important: ". . . more important it was 'they,' the scientists, who called both the halt and allowed the resumption." We now have the clue as to why the opponents of recombinant DNA research say that a moratorium on *all* recombinant DNA work was called for by either the Berg, et al. letter, or at the Asilomar Conference and then rescinded. For the opponents of this work to admit the truth, that the three steps from *Berg,* et.al. (1974), to the Asilomar statement (1975) to the NIH Guidelines (1976)[12] resulted in a continuing increase in the stringency of the self-imposed regulations, would destroy their thesis that we turned things off and on at our pleasure, without care or consideration for the public. Are these honest mistakes? I doubt it! There were *only* five primary documents [2,9,10,11,13] on this subject in 1976, all readily available and none of them except perhaps the NIH guidelines are difficult to read.

The NIH guidelines were promulgated about a year later.[13] They were based on the Asilomar guidelines but enlarged the categories and were more restrictive. With the NIH guidelines out all should have been well, but suddenly in June, 1976, everything started to fall apart. The argument changed in tone and became ad hominem. The first inkling, although few noticed it, was in a September, 1975 article by Erwin Chargaff, who wrote in *The Sciences* that "at the Council of Asilomar, the molecular bishops assembled to condemn the heresies of which they were the perpetrators."[14] By June of 1976 we were "feeble men, masquerading as experts."[15] In March of 1977 at the National Academy of Sciences Forum he complained bitterly about how the courageous small band of dissidents were subject to cruel and personal attacks but that he'd never stoop to such tactics.[16] It is too bad we all aren't such noble souls. Instead of being able to return to the laboratory and use this new tool to address the many basic and applied questions for which it is uniquely suited, we now found ourselves on the defensive.

The temper of the times changed. I feel that this change was a testimony to the power of George Wald. believe that a strange multibody system composed of George Wald, Mayor Vellucci of Cambridge, Mass., recombinant DNA and Harvard collided in a very distinctive way. Although Boston has many well-known members of Science for the People, who were opponents of this work all along, they could never have induced Mayor Vellucci to initiate the Cambridge hearings. During late '76 and early '77 everything seemed to go wrong. The media turned around and changed us from heroes to villains. All the while the most outrageous scenarios on the dangers of recombinant DNA appeared in the press. The public was justifiably frightened. There's one other curious coincidence in what happened last year. Senator Kennedy, who has the primary responsibility in the Senate for matters of health, also comes from the state of Massachusetts and found in the Cambridge, Mass., hearings a mythic paradigm for the interaction of lay people and scientists.

The polemic really heated up after the promulgation of the guidelines. Michael Rogers, Associate Editor of *Rolling Stone* magazine, in his book, *Biohazard*, described how some of us backed off, refused to answer our phones to reporters and were generally dismayed in the way our motives were questioned. Contrariwise the small group of alarmists turned to the media to spread their gospel in a way that made people think they represented majority view of scientists and so intensified their fears.[17]

In September of 1976 Professor James Darnell and I met with the people from the N.Y. State Attorney General's Office, and obviously failed in trying to explain to them the issues related to recombinant DNA research. In September, I testified before Senator Kennedy, Chairman of the Senate subcommittee on Health and Research. The hearings were ostensibly to celebrate the putting in place of the NIH Guidelines. At that hearing the Pharmaceutical Manufacturers' Association and several other associations, all of the Governmental Agencies, agreed to obey the NIH guidelines. However Sen. Kennedy kept referring to some company that wouldn't send a representative. Unfortunately, he didn't know and there's no way to tell somebody in a Senate hearing, that they weren't doing recombinant DNA experiments. I was depressed after I left the hearing because I felt that we were in real trouble. Some people felt, 'ok, it's over now, the Guidelines were in place and we're safe.' However the Winter of '76-77 was characterized by vitriol and invective on the part of both sides at the Universities, at town meetings and at governmental hearings.

In March of '77 a rash of bills was suddenly introduced into the Congress. Congressmen Ottinger of N.Y., Senator Bumpers of Arkansas, Senator Metzenbaum of Ohio, and Congressman Rogers of Florida all introduced regulatory legislation. Ultimately, there was a bill introduced by the Administration in the Senate through Senator Kennedy's subcommittee. The Senators rewrote it to make it far more stringent. Ultimately, of all the bills there was left only the Kennedy bill[18] in the Senate and the Rogers bill in the House.[19] It was said that legislation was needed because the NIH guidelines did not pertain to industry, etc. Since our twin goals had always been to protect the public and to continue the research under controls appropriate to our estimate of the potential hazards, I believe the scientific community would have accepted the simple codification of the NIH guidelines into the law of the land. These bills were far from that. They set up vast bureaucracies, cumbersome licensing, harsh penalties and tedious reporting procedures. Their rhetoric implied that scientists were guilty until proven innocent and hence the bills contained search and seizure provisions. They read like a narcotics bill. The House bill was somewhat better: it gave greater autonomy to local biohazards committees and had a less bureaucratic structure. The only concession we won was that both bills preempted local law. It is extremely necessary that science be protected from any arbitrary and capricious acts of local governments where more often than not extraneous issues enter the deliberations. We felt strongly that if nothing else there must be one national standard. We wanted to avoid a crazy-quilt of many layers of regulations that would provide not one iota more of safety but would both hamper the work and inevitably lead to inadvertent violations. If recombinant DNA experiments are dangerous, they're dangerous everywhere, not only in Cambridge, Mass. In addition, when the Federal government promulgates standards which by implication are inadequate to the task, it is derelict in doing its duty. So if we had to have a regulatory bill, we wanted preemption, and at least on this issue we prevailed last June.

In March the New York State Legislature received a recombinant DNA regulatory bill submitted by the State Attorney General's Office.[20] It was so poorly drafted and so harsh that it underwent considerable committee revision. The final product was still vague, particularly in the definition of what was to be regulated. The sentence containing this definition had six verbs.[21] After, 'recombinant DNA *means,*' (the first verb) it was impossible to determine which of the others was the operative verb. It either referred to everything in molecular biology or nothing; no one was sure. Fortunately, Governor Carey vetoed this bill.[22] He did it on principle: he said that there should be a national standard if there was any regulation of science and that laws that intrude on freedom of inquiry must be narrowly and precisely drawn. I believe Governor Carey has been very courageous. With his veto he has helped us all over the country for he showed other politicians that such action could be taken without dire political consequences. He also vetoed the laetrile bill, whose genesis I believe is part and parcel of the same kind of Know-Nothingism that led to the recombinant DNA bill. The scientific community that I represent owes the Governor a debt of gratitude.

In late June, '77 and early July, '77, a number of events occurred which began to turn the tide. The professional biological organizations began to lobby. As ammunition they had new data from experiments specifically designed to test whether the standard host for recombinant DNA, *E. coli* K-12, could become pathogenic by the introduction of genes known to affect pathogenicity in its bacterial relatives.[23,24] The experiments failed and hence it is almost impossible to believe that introduction of a few random genes into such strains could in any way make them pathogenic. Some experts said that it would take years of intensive genetic manipulation of these bacteria to make them pathogenic. In addition scientists at the very same Gordon Conference that had brought the question of recombinant DNA to the fore wrote an open letter deploring what they felt was overreaction to an issue that was well in hand in the laboratory. Thousands of experiments with recombinant DNA have been done during the past few years in hundreds of laboratories and there has been no ill effect on health or environment. The small coterie of alarmist scientists picked up no following, hence the media lost interest in them and began to present a more balanced picture. One headline for a *NY Times* article on recombinant DNA was, "No Sci-Fi Nightmare."

In the Senate, Senator Nelson of Wisconsin, a member of the Kennedy subcommittee, not only voted against the Kennedy bill but also introduced a substitute amendment.[26] He was joined bySenator Moynihan from NY, who as a former academic was concerned about the intrusive nature of the Kennedy bill. This amendment was the least restrictive legislation proposed but still had many faults. Parenthetically I might add that I was looking forward to a debate between Senator Kennedy and Senator Moynihan on the issue of academic freedom. It might have been quite a historic event. The Nelson-Moynihan bill did block the Kennedy bill in the closing days of the session this August. In September Senator Kennedy suddenly withdrew support of his bill citing the new data.[27] He said that he was going to introduce a one year interim bill and set up a Study Commission that would decide whether further regulation was needed. As of now (December), he's not done so. The House bill is totally stalled at this moment. It has yet to go to the full committee for discussion. The Chairman of the Committee on Interstate & Foreign Commerce, Congressman H. Staggers of West Virginia, feels the Rogers' bill to be too restrictive and favors something of a more interim nature. In any event at present there is no bill on the docket in the Senate and a stalled one in the House.

Thus it's not clear whether there will or won't be legislation, and if so what kind. When Congress returns I have the feeling some kind of bill will be passed. Perhaps one that only extends the NIH Giuidelines to industry. I hope it's a sensible bill because I've always believed that respect for regulations ultimately depends on one's perception of the wisdom of the rules or regulations. We all know that people tend to obey regulations when they respect them. They don't when they disdain them. In addition one of the problems is that law will replace peer review. Now the onus is on the scientists. That's a different psychological state from having the onus on some regulators. In addition law will justify the assumption of hazard, hazard justifies fear and fear can lead to irrational acts.

A few words in closing about the other concerns regarding recombinant DNA. There are the questions 'are we delving too far' (Sinsheimer), 'have we the right to counteract the evolutionary wisdom of millions of years,' (Chargaff). I don't consider these real problems. Man has been contravening evolution ever since he arrived on this planet. The "biggest break with nature," George Wald's favorite statement, was when man settled down, left the nomadic life, created agriculture and domesticated our plants and animals. As for the notion that there are things that we shouldn't know, I resent anyone telling me what I shouldn't know. Thomas Jefferson said it best: "there is no truth on this earth that I fear to be known."

Epilogue

The text above was derived from transcription of a tape of the talk which I gave from notes. Since the spoken word is so different from the written word, extensive editing was necessary. In the editing I did not try to achieve straight manuscript form but rather tried to retain as much of the substance as possible while not destroying the emotional tone that is clearly discernible listening to the tape. After three years of speeches, hearings at every level of government, interviews, briefing sessions and extensive correspondence and phone calls on this subject, I was angry, dismayed and feeling somewhat put upon in having to once again discuss this subject and realizing that the end was nowhere in sight.

Since I have not explicitly given my opinion of the safety of recombinant DNA research in the text, I now say that I believe it to be as safe as most other biological research and safer than that dealing with a number of very dangerous microorganisms such as *B. anthracis* or Marburg virus. Thus the NIH Guidelines are more than adequate and will, in time, I'm sure just whither away. They are, in fact, totally superfluous now.

References

1. W.B.A.I. New York, *NY Program Notes*, September, 1977.
2. Draft Environmental Impact Statement. Guidelines for Research Involving Recombinant DNA Molecules. National Institutes of Health, August 19, 1976. *Federal Register*, September 9, 1976.
3. Environmental Impact Statement on N.I.H. Guidelines for Research Involving Recombinant DNA Molecules, part 2, p. K172.
4. Ibid., p. K124.
5. Ibid., p. K80.
6. Report and Recommendations of the N.Y. State Attorney General on Recombinant DNA Research, February 8, 1977, p. 4.
7. Robert Pollock, Professor of Microbiology, S.U.N.Y. Stony Brook, personal communication.
8. Cole, S., Rubin, L. and J. Cole. Peer Review and the Support of Science (N.S.F.). *Scientific American* **237** (4), 34, 1977.
9. Singer, M.F. and D. Soll. Guidelines for DNA Hybrid Molecules. *Science* **181,** 1114, 1973.
10. Berg, P., Baltimore, D., Boyer, H., Cohen, S., Davis, R.W., Hogness, D.S., Nathans, D., Roblin, R.O., Watson, J.D., Weissman, S. and N.D. Zinder. Potential Hazards of Recombinant DNA Molecules. *Science* **185,** 303, 1974.
11. Berg, P., Baltimore, D., Brenner, S., Roblin, R.O., and M. Singer. Summary Statement of the Asilomar Conference on Recombinant DNA Molecules. *Science*, **188,** 991, 1975.
12. Gaylin, W., The Frankenstein Factor. *New England Journal of Medicine*, **297,** 665, 1977.
13. National Institutes of Health Recombinant DNA Research Guidelines. *Federal Register* 41 #131, 27902, July 7, 1976.
14. Chargaff, E. Profitable Wonders. *The Sciences*, **15,** 21, 1975.
15. Chargaff, E. On the Dangers of Genetic Meddling. *Science*, **192,** 938, 1976.
16. Chargaff, E. Potential Risks of Recombinant DNA Research. Research With Recombinant DNA. Pub. National Academy of Sciences U.S., p. 45, 1977.
17. Rogers, M., *Biohazard*. Alfred A. Knopf, New York, 1977, p. 187-196.
18. U.S. Senate. Recombinant DNA Safety Regulation Act S.1217. 95th Congress, 1st Session, 1977.
19. U.S. House of Representatives. Regulation of Recombinant DNA Activities H.R.7897. 95th Congress, 1st Session, 1977.
20. State of New York Senate (S.4009) & Assembly (A.6740D). 1977-78 Regular Sessions. An Act to Amend the Public Health Law, in Relation to the Certification of Recombinant DNA Experiments.
21. Ibid., N.Y. State Senate 4009D and A6740D.
22. Hugh L. Carey, Governor, New York. I Am Returning Herewith Without My Approval, Senate Bill Number S4009-D. August 5, 1977.
23. Curtiss, R. III. Letter to Dr. Donald Fredrickson, Director, National Institutes of Health. Available from N.I.H., April 12, 1977.
24. Gorbach, S.L. Letter to Dr. Donald Fredrickson, Director, N.I.H. re: Falmouth Workshop on Risk Assessment. July 14, 1977.
25. *NY Times*. "No Sci-Fi Nightmare," News of the Week in Review, July 24, 1977.
26. U.S. Senate. Recombinant DNA Act Amendment in the Nature of a Substitute to S.1217 Amendment #754. 95th Congress, 1st Session, 1977.
27. Edward M. Kennedy, Senator. Remarks to American Medical Writers Association, September 27, 1977.

SAFETY FILM 5063
PUBLIC DEBATE
NOW
PEOPLES BUSINESS COMMISSION
→36
26←
25A←
→1A
→2
→2A
KODAK SAFETY FILM 5063
5063
→7
→8
→8
KODAK SAFETY FILM 5063
KODAK SAF

CHALLENGING THE BASIC ASSUMPTIONS

Photos by Paul Conklin for the Academy Forum. Background photo courtesy Tom Broker, Cold Spring Harbor Laboratory.

Despite the stringency of the June 1976 NIH guidelines, which totally prohibited some experiments and prevented many more by requiring containment conditions that were simply not available, in the United States recombinant DNA research received increasingly adverse publicity during the second half of 1976 and the early months of 1977. Disaster scenarios became progressively more fantastic as uninformed imaginations were given a free rein and the editors of the mass media continued to find space for them. Cambridge Mayor Vellucci's letter to the President of the National Academy reflects well the atmosphere in some quarters (Doc. 7.1). As the demand for safety precautions became increasingly strident, the precautions themselves became increasingly stringent. In Congress as well as at various levels of local government, serious consideration was being given to drafts of legislative measures designed specifically to regulate recombinant DNA research. Some groups—in particular, Science for the People and the People's Business Commission—attempted further to confuse issues by linking recombinant DNA with a resurgence of a eugenics movement in the United States.

Serious discussion of the issue reached its nadir on the opening evening of a forum organized by the National Academy of Sciences in Washington, D.C. on March 7–9, 1977. Jeremy Rifkin, a spokesman for a small group known as People's Business Commission, disrupted the meeting and with sympathizers staged an extraordinary demonstration, the tone of which is clearly conveyed by a couple of photographs (see pp. 134, 202) and by Rifkin's final comment, "Let's open this conference up or close it down!" A few hours previous to the demonstration a press conference had announced the formation of a Coalition for Responsible Genetic Research (see Doc. 7.2) that included George Wald of Harvard, another Nobel Laureate, Sir MacFarlane Burnet from Australia, Lewis Mumford, the well-known writer on city planning and technology, American environmentalist groups, including Friends of the Earth, the Environmental Defense Fund, Natural Resources Defense Council, and Science for the People. The coalition's policy was the now-familiar call for a total moratorium on the research. After Rifkin's demonstration the forum proceeded as planned. The proceedings have been published by the National Academy of Sciences as "Research with Recombinant DNA"

(Washington, D.C., 1977), and we have not included excerpts here. In the words of biochemist Leon Heppel, dated March 21, 1977:

I was rather embarrassed that I couldn't think of anything to say at the DNA meeting, in view of the fact that several hundred dollars were spent by NIH to bring me to Washington. . . . I felt the way I would feel if I had been selected for an *ad hoc* committee convened by the Spanish Government to try to evaluate the risks assumed by Christopher Columbus and his sailors, a committee that was supposed to set up guidelines for what to do in case the earth was flat, how far the crew might safely venture to the earth's edge, etc.*

Amid such furor it was, of course, all too easy to lose sight of the crucial issue—namely, what evidence, necessarily circumstantial, might there be to suggest that producing *in vitro* new gene combinations poses any significant and novel biohazard. Indeed, the behavior of many of the opponents of recombinant DNA research forces the conclusion that for whatever reasons, their purpose was to widen, confuse, and cloud this central issue so as to avoid it and ensure the debate continued in a way that would allow them to challenge the social and political structure of science and technology in the United States (Doc. 7.3). Recombinant DNA had become, in effect, useful grist to their political and social mills.

A perceptive commentary on this situation was published in *New Solidarity* on June 26, 1977, by Carol Cleary. *New Solidarity* is a rather obscure publication linked, we believe, through the *Industrial Worker* to the International Workers of the World in Chicago. Unfortunately (but perhaps understandably) Ms. Cleary has denied us permission to reprint her article in full, and it is difficult in paraphrase to convey its pungency. The article roundly attacks Senator Kennedy for attempting to "railroad" legislation through Congress, expresses great skepticism about the environmentalist groups ("massive development of environmental terrorists"), states that a lobbyist of

*Victor McElheny drew our attention to S. E. Morison's *The European Discovery of America: Southern Voyages*, Vol. 2 (New York: Oxford University Press, 1974), which tells that Ferdinand and Isabella indeed set up a committee which correctly judged that Columbus' expectations of the distance westward to the Indies were far too optimistic. The earth's sphericity was never in question.

Friends of the Earth was paid by Senator Kennedy, and speaks of "manipulated local community opposition." It is a thoroughly scurrilous but entertaining piece of journalism. Obviously for some, exactly who speaks for "the people" is as controversial as recombinant DNA regulation was for others.

After the publication of the Ashby report (in Britain), events in Europe moved not only more slowly but much more dispassionately. It was easier for Europeans to avoid confusing the limited issue with more diffuse and emotional ones concerning the uncomfortable current relationship between science and society. For example, in its first carefully worded report dated February 1976, the Standing Advisory Committee on Recombinant DNA of the European Molecular Biology Organization (EMBO) (Doc. 7.4) drew attention to the long chain of events that would have to occur before any recombinant DNA experiment could lead to an accident, and it doubted whether in fact such sequences could ever actually happen. Almost exactly a year later Robin Holliday (NIMR London) published in *New Scientist* a more detailed analysis of the probabilities of these events and reached the same conclusion as the EMBO Committee (Doc. 7.5). Disaster scenarios were closer to science fiction than to science.

A few weeks earlier in the same magazine, however, Robert Sinsheimer had again expressed his partly scientific, partly ethical and social reasons for urging caution (Doc. 7.6). Spurred on by events during 1977 in Congress and in state and city legislatures in the United States, some molecular biologists who had been involved in the debate from its outset—Stanley Cohen, J. D. Watson, and Charles Thomas—as well as medical microbiologists such as Rene Dubos, and other respected biologists and commentators on science and science policy including Lewis Thomas and Sir Peter Medawar—attempted to reintroduce a calmer perspective (Doc. 7.7 through Doc. 7.12). They drew attention not only to the increasing lack of a sense of proportion regarding the conjectural hazards of the research, but also to the indisputable scientific, social, and political dangers inherent in attempts to censor, regulate, and legislate basic research. Their comments, "in defense of DNA," as Watson put it, had real impact. In some quarters—and in ones that mattered most—the necessity of such stringent regulations began to be questioned.

Document 7.1

May 16, 1977
Vellucci Letter to Handler

A PRIVATE COMMUNICATION

Letter to Members

NATIONAL ACADEMY OF SCIENCES

VOLUME 7 · NUMBER 4 MAY 1977

May 16, 1977

Mr. Philip Handler
President, National Academy of Science
2101 Constitution Avenue, N.W.
Washington, D.C. 20418

Dear Mr. Handler:

As Mayor of the City of Cambridge, I would like to respectfully make a request of you.

In today's edition of the Boston Herald American, a Hearst Publication, there are two reports which concern me greatly. In Dover, MA, a "strange, orange-eyed creature" was sighted and in Hollis, New Hampshire, a man and his two sons were confronted by a "hairy, nine foot creature."

I would respectfully ask that your prestigious institution investigate these findings. I would hope as well that you might check to see whether or not these "strange creatures," (should they in fact exist) are in any way connected to recombinant DNA experiments taking place in the New England area.

Thanking you in advance for your cooperation in this matter, I remain

Very truly yours,

ALFRED E. VELLUCCI
Mayor
City of Cambridge
Massachusetts

Document 7.2

March 7, 1977

Press Release and Position Statement

DR. GEORGE WALD,
NOBEL LAUREATE, WILL SPEAK:
PRESS CONFERENCE:
MONDAY, MARCH 7, 1:30 P.M.,
WATERGATE HOTEL,
Continental Room

COALITION FOR RESPONSIBLE GENETIC RESEARCH

INTERNATIONAL COALITION CALLS FOR PUBLIC INQUIRY AND MORATORIUM ON GENE TRANSPLANT RESEARCH; NOBEL LAUREATES, CLUB OF ROME FOUNDER, LEWIS MUMFORD AMONG SPONSORS

Citing the unprecedented hazards of gene transplant research, scientists, environmentalists, and public interest representatives announced the launching of the *Coalition for Responsible Genetic Research* at a press conference in Washington, D.C., coinciding with the National Academy of Science Forum on Recombinant DNA (gene transplant) Research.

The coalition includes sponsors from all over the United States, the Netherlands, Switzerland, England, Canada, Italy, France, New Zealand, and Australia who call for open public inquiry and a moratorium on gene transplant research and funding, pending policy formulation with broad public participation.

"The continuation of this research without public understanding and approval," states the coalition's position paper, "and, in fact, without a full comprehension of its potential by most of the involved scientists, poses a worldwide danger which is intensified by the fact that industrial investment in the developing genetic technology has already begun."

Among the hundreds of coalition sponsors are Sir MacFarlane Burnet, Nobel Laureate, Physiology and Medicine; Dr. George Wald, Nobel Laureate, Physiology; Dr. Aurelio Peccei, industrialist founder of the Club of Rome; and Lewis Mumford, author-philosopher.

Coalition for Responsible Genetic Research

Position Statement on Recombinant DNA Technology

Scientists, working toward the advancement of knowledge, have not concerned themselves in a substantial way with analyzing or evaluating the impact that their discoveries might have on society. Indeed, society as a whole has tended to acquiesce in the assumption that any advancement of scientific knowledge is a basis for progress, a potential source of future blessings of humanity. In spite of the pervasive effects of the atomic bomb and the increasing domination of so many aspects of life by technology, the assumption has been preserved by biologists who, to receive immediate public approval, have had to cite only the general relevance of their work to medicine.

For some years now, however, some scientific research has been creating products, information and techniques having an impact that may be more dangerous than beneficial (fluorocarbon-induced ozone depletion, radioactivity, dioxin, DDT, PCBs, vinyl chloride, photochemical smog, Mirex, etc.). In particular, genetic and molecular biological research is rapidly providing the basis for the practice of genetic engineering. The continuation of this research without public understanding and approval, and in fact without a full comprehension of its potential by most of the involved scientists, poses a worldwide danger which is intensified by the fact that industrial investment in the developing genetic technology has already begun.

Molecular biologists have now learned how fragments of DNA (the master hereditary chemical of all living cells) can be recombined or reassembled to produce new entities, "recombinant DNAs", which have the ability to enter cells and reproduce themselves therein. This type of agent has the theoretical capacity to alter fundamental biological processes in the biosphere. A primary danger is that the deliberate production of "desirable" changes in certain organisms might result in the disruption of the infinitely complex and delicate balance among living things, a balance which has evolved slowly in nature. A sudden disturbance of natural relationships could precipitate a disastrous and irreversible breakdown of our ecosystem. More immediately, current research itself involves serious hazards which inadvertently could result in the spread of cancer or the creation of new diseases. In addition, the possibility of applying genetic engineering to the evolution of the human species raises fundamental ethical questions which need to be brought out into the open.

Scientists involved in this research are not unaware of its dangers, and indeed were the first to call attention to them. But, caught up in the intellectual fascination of their work, they have engaged in only a limited debate. A few ill-defined public benefits of questionable feasibility have been proffered to justify the hazards, and containment guidelines have been formulated in an effort to minimize accidents; but the more profound social questions have been ignored by all but a handful of scientists.

It is time to end the isolation of scientists from society, before the technique of genetic engineering has become a fait accompli. Scientists need to be made aware of what they are doing, in the broadest sense, and -- perhaps even more important -- the public must be awakened to the broad powers which have been left in the hands of scientists and industrialists who are unqualified or reluctant to judge their import, because of suspected impact upon investment (e.g. the SST, aerosol propellants, taconite tailings, Kepone, etc.).

There is little or no precedent for organized consideration of the ends of scientific research, of any type, with respect to social, health and ethical values.

The Coalition for Responsible Genetic Research calls for:

1. ESTABLISHMENT AND IMPLEMENTATION OF DEMOCRATIC PROCEDURES that will ensure open discussion and public decision on the problems posed by gene implantation research and applications.
2. REVIEW OF THE IMPLICATIONS OF ALL DEVELOPMENTS in this field, particularly as they affect industrial and medical application.
3. ENCOURAGEMENT AND ASSESSMENT OF ALTERNATIVE TECHNOLOGIES that do not require the manufacture of novel organisms.
4. A VIGOROUS PROGRAM OF SYSTEMATIC RISK ASSESSMENT to keep pace with developments in this field.
5. PUBLIC PARTICIPATION IN INTERNATIONAL CONSIDERATION OF THE ENDS OF THIS RESEARCH with careful examination of ethical, health and social implications.

In order that decision-makers not be faced with a pre-formed policy of expansion, the Coalition calls for full assessment of the social and biological implications <u>before</u> decisions are taken that would set us on a course of proliferation.

Document 7.3
May—June 1977
Science for the People (9:14)

Recombinant DNA: Does the Fault Lie Within Our Genes?

Jon Beckwith

I have been doing research in bacterial genetics for the last 12 years at Harvard Medical School and I am a member of Science for the People. Over the last couple of years, we have been discussing in our laboratory how the recombinant DNA technique could make certain of our experiments much easier to do. However, as a result of these discussions we decided not to use this technique at all. This is not because the particular experiments we were talking about could be thought of as health hazards, in any way. Rather, my reasons were that I do not wish to contribute to the development of a technology which I believe will have profound and harmful effects on this society. I want to explain why some of us have arrived at this decision.

In 1969, a group of us in the laboratory developed a method for purifying a bacterial gene. We took that opportunity to issue a public warning that we saw developments in molecular genetics were leading to the possibility of human genetic engineering.(1) While we saw genetics progressing in this direction, we had no idea how quickly scientists would proceed to overcome some of the major obstacles to manipulating human genes. The reports on the use of recombinant DNA technology, beginning in 1973, represented a major leap forward. The result is that geneticists are now in a position to purify human genes. And proposals have already been put forward for the setting up of "mammalian DNA banks."(2) Further, techniques are being developed which will allow reintroduction of those genes into mammalian cells. These steps appear perfectly feasible.

In March 1977, the National Academy of Science held a forum on recombinant DNA in Washington, D.C. Several members of Science for the People, Jon Beckwith among them, spoke on the hazards of recombinant DNA. The Forum was marked by what the media have called "Vietnam era protest": several opponents of recombinant DNA from the People's Business Commission unfurled a banner which quoted Hitler: "We shall create a perfect race." On the same day that the Forum began, several organizations issued a joint statement in Washington opposing all recombinant DNA research.

There are still some barriers left to introducing genes into human cells, organs or embryos at the proper time or in the proper way. But these goals are not at all inconceivable and they may be achieved very rapidly (see reference 3). Whatever the current state of knowledge, to claim that the possibilities of genetic engineering of humans with this technique is far off is to totally ignore the history of this field.

In 1969, most scientists pointed to the impossibility of purifying human genes and claimed that such developments were *at least* decades off. In fact, they were four years off. Let's not be fooled again. Just as suddenly as recombinant DNA appeared on the scene, breakthroughs in "genetic surgery" may appear.

And when the day arrives in the near future when geneticists have constructed a "safe" vector for carrying mammalian genes into human cells, others will begin to use it for human genetic engineering purposes. There has already been at least one reported case in which there were direct attempts to cure a genetic disease in human beings with virus-carried genes(4) and in human cells.

But, why be concerned about human genetic engineering? There are, certainly many individuals and groups which have ethical or religious objections to any intervention of this kind in human beings. Possibly after widespread discussion within a society, those objections might predominate. I, personally, do not necessarily view all human genetic intervention as inherently to be opposed. But, I would rather point today to some concrete dangers of the development of recombinant DNA research by examining the scientific, social and political context in which it is proceeding. For that reason, much of what follows will speak to those issues rather than directly to recombinant DNA.

Scientific Developments

In the last 10 or 15 years, there have been advances in a number of areas of genetics which bring us to a situation today, in which *genetic engineering is already underway*. These include a variety of types of genetic screening programs in which it is possible to identify genetic differences between people by examining cells of individuals.(5)

The approaches are: 1) amniocentesis, where the cells of a fetus obtained from a pregnant woman can be examined for genetic variations. In a small number of cases, these variations are known to cause serious health problems and suffering may be eliminated by giving the parents the option of aborting such fetuses. 2) Post-natal screening — when infants are screened after birth for genetic differences. Again, in a small number of cases, those variations may cause disease and treatment may be provided. 3) Adult screening — where prospective parents can be advised of the likelihood of their bearing children who might carry particular genetic variations. While each of these programs has proved beneficial to some individuals, they have also encountered problems, been controversial, and, in some cases, caused suffering to those screened. In addition, all of these programs raise the basic question of who is deciding who is defective, or even, who shall live?(6)

There are other developments which have received much attention in the press — e.g. the possibility of cloning genetically identical individuals and the attempts to grow fertilized eggs in the test tube and then implant them in a woman's uterus.

At the same time that these developments in genetic technology were taking place, there was also a growth in studies in human behavioral genetics. In the last ten years, there has been a resurgence of supposedly "scientific" research which claims to explain many of our social problems as being due to genetic differences between people.(7) For instance, there are the attempts to say that the inequality which exists in this country or the lower achievement of various groups, particularly blacks, is due to inferior genes.(8) Or the proposals that criminality might be explained by genetic differences between the criminal and the noncriminal — the case of the XYY male.(9) (By the way, one of the reasons that I suggest that genetic engineering is already under way, is that XYY fetuses have been aborted after detection by amniocentesis.) (10) In both these cases, the scientific evidence has been shown to be nonexistent and, in some cases, fraudulent. In addition, there are the more recent attempts in the field of sociobiology to claim biological and genetic evidence to justify the lower status position of women in this society.(11) It is a disgrace that this government continues to support such shoddy, groundless and ultimately harmful research.

Socio-Political Context

These genetic theories and the problems with genetic screening programs did not arise in a social and political vacuum. They have followed a period of intense social agitation and social disruption in the United States. After blacks, other minority groups, the poor and women demanded a greater share of the wealth and power in this society, the response arose that such equality is genetically impossible. The ghetto uprisings and other violent confrontations which occurred during this period are explained as being due to people whose genes are "off." The demands of the women's movement are met with the answer that women are genetically programmed for the roles they now occupy.

Another more recent example of this genetic approach to social problems lies in the field of industrial susceptibility screening.(12) Arguments have been appearing in the scientific literature and elsewhere that occupational diseases, caused by pollutants in the workplace can be ascribed not to the pollutants themselves, but to the fact that some individuals are genetically more susceptible to the pollutants than other individuals. So the argument goes, the solution is not getting rid of the pollutants, but rather, for example,

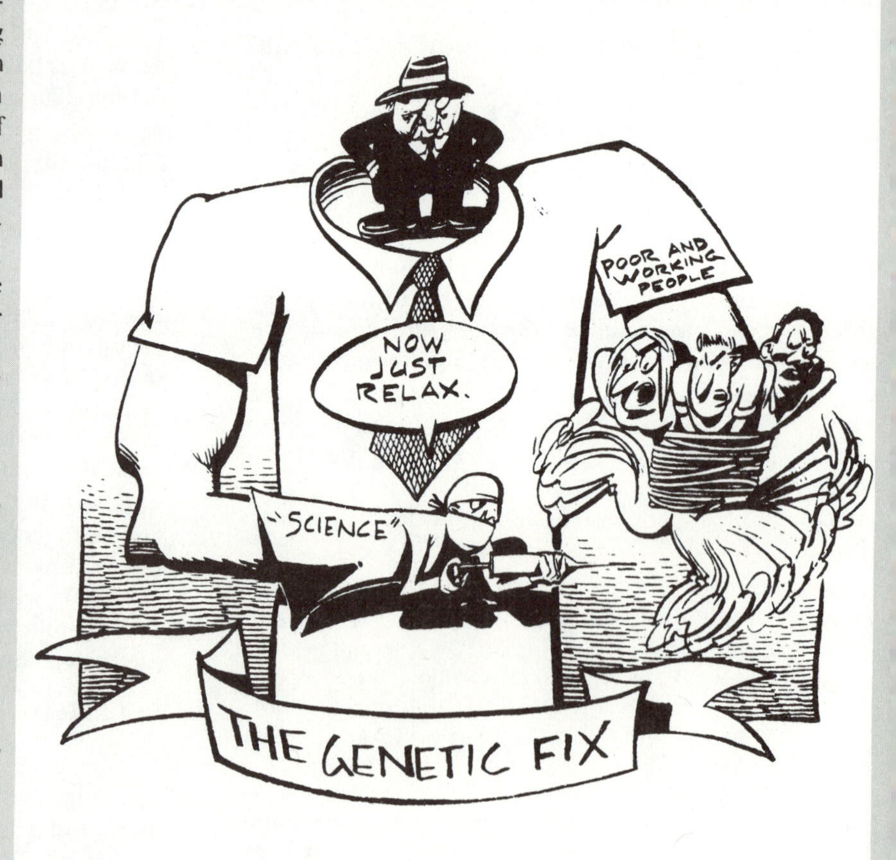

Graphic by Nick Thorkelson

Let me give you some examples of how we may move from the present technological fix to the genetic fix, once recombinant DNA techniques have provided the tools.

simply not hiring those individuals who are thought to carry the genetic susceptibility.

Now, clearly, whenever it is possible to warn someone of dangers he or she may face, that information is important. However, what is blatantly ignored by those promoting this area of research is that, in almost every case, nearly everyone in the workplace is at some degree of increased risk because of the exposure, for instance, to asbestos fibers. Yet, already, there are headlines in the newspapers such as the following: "Next Job Application May Include Your Genotype."(13) A Dow Chemical Plant in Texas has instituted a large scale genetic screening program of its workers.(14) Rather than cleaning up the lead oxide in General Motors plants, women of child-bearing age are required to be sterilized if they wish employment.(12,15) It is a genetic cop-out to allow industries to blame the disease on the genetically different individual rather than on their massive pollution of the workplace and the atmosphere. This is the epitome of "Blaming the Victim."(14a)

The end result of these genetic excuses for society's problems is to allow those in power in the society to argue that social, economic and environmental changes are not needed — that a simpler solution is to keep an eye on people's genes. And thus the priorities are determined. For example, major funding goes to genetics research and into viral causes of cancer and a pittance to occupational health and safety. This distorted perspective is reinforced by the emphasis and the publicity that recombinant DNA research has achieved with its claim for solving problems whose solutions are mainly not in the realm of genetics. Typical of the claims made by those promoting this area is a statement by biologist David Baltimore:*

> How much do we need recombinant DNA? Fine, we can do without it. We have lived with famine, virus and cancer, and we can continue to.(16)

This is not a neutral or apolitical statement. The sources of famine and disease lie much more in social and economic arrangements than in lack of technological progress. Aside from the incredible claims for the benefits of recombinant DNA, this statement essentially opts for the status quo. Social problems, such as famine and disease, are taken out of the arena of political action and sanitized behind the white coat of the scientist and the doctor. Of course, we might have both social and medical approaches to such problems going on at the same time. But given the current struggle over solutions to these problems, such statements can only provide weapons to those who would like to maintain present power relationships and profits. What is opted for are the technological fixes, in this case, the *genetic fix*.

Recombinant DNA — The Genetic Fix

Let me give you some examples of how we may move from the present technological fix to the genetic fix, once recombinant DNA techniques have provide the tools. In the United States over the last few years, approximately one million school children per year have been given drugs, usually amphetamines, by the school systems, in order to curb what is deemed disruptive behavior in the classroom.(17) It is claimed that these children are all suffering from a medical syndrome, minimal brain dysfunction, which has no basis in fact — no organic correlate. Now, clearly, there are some cases of children with organic problems where this treatment may well be important. But in the overwhelming majority of cases the problems are a reflection of the current state of our crowded schools, overburdened teachers and families and other social problems rather than something wrong with the kids. Imagine, as biochemical psychiatry is providing more and more information on the biochemical basis of mental states, the construction

*At the National Academy of Sciences Forum on Recombinant DNA, Baltimore responded to this section by acknowledging that proponents of the research had overdramatized the benefits. Their attitude had been that it was necessary to do this in order to justify the research to the public. Baltimore, by the way, has an interesting background, having been one of those most involved in opposing chemical and biological warfare research, having given one-half of the money from a prestigious award he received to Science for the People, and having supported Science for the People's struggle against XYY research in Boston. The divisions in the recombinant DNA struggle illustrate how political lines become sharpened when issues begin to hit close to one's own professional interests.

of a gene which will help produce a substance in human cells which will change the mental state of individuals. Then, instead of feeding the kids a drug every day, we just do some genetic surgery and it's over.

Don't forget that introducing genes into humans — genetic engineering — results in permanent changes. There is no way to cut the genes out. It's irreversible. At least, when protests were mounted in certain schools against the drugging of kids, the treatment could be stopped. That's not the case with the genetic solution. There's no going back.

Another example: A current idea, again without scientific foundation, is that aggression is determined by hormone imbalance. Males, it is said, are more aggressive than females because of the hormone testosterone or the absence of presumed female hormones. As a result, patients in mental institutions deemed aggressive are treated with the presumed female hormones.(18) But recently it has been discovered that there are genes in bacteria which will break down testosterone. Wouldn't it be a simpler, less costly approach to introduce such genes (in a functional state) into the "aggressive" patient? Maybe even social protest can be prevented that way. But what are the sources of aggression in this society? Isn't it possible that rather than hormone imbalance, it is social and economic imbalance — unemployment, racism, etc. — which spurs many people to "aggressive" behavior? And, while we're on the subject, would such genetic surgery be used on those in leadership positions in the society responsible for such atrocities as the Indochina War?

Similar approaches could be used to argue for gene therapy on fetuses, infants or on the workers themselves so that they can work in factories with high vinyl chloride levels. Given the sophistication of the new technologies, a new eugenics era may do even greater damage than the earlier eugenics movement (1900-1930).(19)

Conclusion

I would like to add a component to the benefit-risk discussion of recombinant DNA which has, for the most part, been ignored. This component is the risk of human genetic engineering to those without power in this society. Given the present social context, I believe these consequences are inevitable. It is not just the particular evils and damage to individuals I have mentioned in my scenarios that concerns me. The dramatic developments in this field, and the publicity it has received and will continue to receive, is already reinforcing the focus on the genetic fix. On the one hand, an atmosphere is being generated in which a variety of genetic approaches to social problems is accepted. And, as a corollary, social, political and economic changes are deemphasized. The priorities of the society cannot be allowed to be dictated by the technocrats and their technology. On the contrary, technologies must be developed only *after* social decisions that they are wanted and needed.

On this basis, I believe we should seriously consider whether recombinant DNA research should be pursued at all.☐

REFERENCES

Many resources on the subjects covered in this article can be obtained from Science for the People, 897 Main St., Cambridge, MA 02139.

1. J. Shapiro, L. Eron and J. Beckwith, Letter to *Nature*, 224, p. 1337 (1969).

2. F.H. Bergmann, "A Bank of Mammalian DNA Fragments," *Science* 194, p. 1226 (1976).

3. R. Pollack, "Tumors and embryogenesis," *Science* 194, p. 1272 (1976).

4. Shope Papilloma Virus described in T. Friedmann and R. Roblin, "Gene Therapy for Human Genetic Disease," *Science* 175, p. 949 (1972).

5. J. King, J. Beckwith and L. Miller, "Genetic Screening: Pitfalls," *The Science Teacher*, 43, No. 5, p. 15 (1976).

6. F. Ausubel, J. Beckwith and F. Janssen, "The Politics of Genetic Engineering: Who Decides Who's Defective," *Psychology Today*. 8 No. 1, p. 30 (1974).

7. Ann Arbor Science for the People, *Biology as a Social Weapon*, (to be published in paperback, Spring-Summer, 1977) Burgess. Minneapolis.

8. N. Block and G. Dworkin, *The IQ Controversy*, Pantheon, 1976. paperback.

9. J. Beckwith and J. King, "The XYY Syndrome: a Dangerous Myth," *New Scientist*, 14 Nov. 1974. 474: D. Borgaonkar and S. Shah. "The XYY Chromosome, Male — or Syndrome," *Prog. in Med. Genetics* 10, p. 135 (1974).

10. E.B. Hook, "Behavioral Implications of the Human XYY Genotype," *Science* 179, p. 139 (1973).

11. E.O. Wilson, "Human Decency is Animal," *N. Y. Times Sunday Magazine*, Oct. 12, 1975, p. 38.

12. T. Powledge, "Can Genetic Screening Prevent Occupational Disease?" *New Scientist*, 2 September, 1976, p. 486.

13. L. Timnick, "Next Job Application May Include Your Genotype Too," *Houston Chronicle*, Apr. 4, 1975.

14. D.J. Kilian, P.J. Picciano and C.B. Jacobson, "Industrial Monitoring: a Cytogenetic approach," *Ann. N.Y. Acad. Sci.* 269, p. 4 (1975). and abstract from the National Cancer Institute, *Carcinogenesis Program, 4th Annual Collaborative Conference*, Feb. 22-26, 1976.

14a. W. Ryan, "*Blaming the Victim*, Random House, N.Y. (1971).

15. E. Goodman, "A Genetic Cop-out," *The Boston Globe*, June 15, 1976. p. 37.

16. M. Yao, "Scientists Split on DNA Research," *The Michigan Daily*, March 4, 1976, p. 1.

17. P. Schrag and D. Divoky, *The Myth of the Hyperactive Child*. New York 1975).

18. J. Money, "Use of an Androgen Depleting Hormone in the Treatment of Male Sex Offenders," *J. Sex. Res.* 6, 165 (1970); D. Blumer and C. Migeon, "Hormone and Hormonal Agents in the Treatment of Aggression." *J. Nerv. Ment. Disease* 160, 127 (1975).

19. J. Beckwith, "Social and Political Uses of Genetics in the U.S.: past and present," *Annals of the N.Y. Acad. of Sci.*, 265, p. 47 (1976).

Document 7.4

February 14–15, 1976

Report of the EMBO Standing Advisory Committee, Excerpt

EMBO

EUROPEAN MOLECULAR BIOLOGY ORGANIZATION

Report on the First Meeting of the EMBO Standing Advisory Committee on Recombinant DNA held at London on 14/15 February 1976

Introduction

1. This committee was established by the Council of the EMBO with the endorsement of the EMBC, to advise, upon request, governments, organizations and individuals on the scientific and technical aspects of recombinant DNA research including the conditions under which this research should be done. At the request of the EMBO Council we have met to decide whether it is appropriate or necessary to elaborate guidelines, other than those currently available, for such research in Europe.

Considerations

2. We have considered, therefore:
the Ashby Report, the provisional report of the Asilomar Conference, the proposed guidelines for research involving recombinant DNA molecules of the National Institutes of Health Recombinant DNA Molecule Program Advisory Committee, drawn up at La Jolla on 4-5 December 1975 and currently being considered by the Director of NIH for adoption as official policy, various other documents of that committee, together with other relevant memoranda.

   We fully support a number of general principles enunciated in these several reports, namely that:
   a) recombinant DNA research promises great scientific and social benefits and, therefore, should proceed under appropriate safeguards.
   b) at present one cannot exclude the possibility that this work could entail undesirable biological side effects;
   c) therefore, physical and biological containment procedures should be instituted for certain types of experiments;
   d) the containment should be designed to match the best estimate of the possible risks involved.

3. This Committee feels that there are two categories of risk: one is the intentional creation of new genotypes containing combinations of genes known to specify harmful products or endow dangerous potentialities: for example, the introduction of genes for *Cl. botulinum* toxin into *E. coli*; the introduction of plasmids which increase the enteropathogenicity of *E. coli*; or the introduction of penicillin resistance into β-haemolytic streptococci or pneumococci. We believe that experiments in this category should in general be proscribed at the present time, as has been suggested in all previous draft guidelines drawn up by other committees.

4. The second category of risk arises from the possibility of adventitiously producing genotypes of unknown and undesirable properties. This risk depends upon chains of events, such as:
   a) unwitting introduction of a foreign genetic material with pathogenic potential into a micro-organism,
   b) the escape of this new micro-organism from the contained laboratory,
   c) the establishment of the new micro-organism in a natural environment,
   d) expression of this genetic information in the host micro-organism to the detriment of the ecosystem,
   e) or, the transfer of the harmful genetic material in the new micro-organism to the genome of some other organism to the detriment of the environment.

   Sequences of events, such as above, are in our opinion very unlikely to be realised.

5. During the past year, recombinant DNA research has proceeded under draft guidelines set out in the Ashby Report and the Asilomar Conference Report. For the most part the evaluation of risks cannot, at present, be based on adequate data, as the work of several other committees has already shown. The EMBO Committee does not feel, therefore, that it should engage in drawing up a new estimate of risks, since it is unlikely to be any more accurate than its predecessors.

6. The most detailed code of practice yet to be drafted is that which is currently being considered for adoption by NIH as official policy. This could, we believe, provide a basis for an international code of practice but details of implementation and the precise classification for particular experiments is something on which opinion may vary. The current NIH proposed guidelines are still under debate and until a final version is approved we believe more detailed comment would be premature. See letter to Dr. H. Stetten (Appendix I).

Recommendations

7. We feel that it is important to initiate an experimental analysis of the various parameters on which an objective assessment of the various putative risks might be based. To this end this Committee will strive to have experiments, such as those listed in Appendix II, performed as soon as possible. The results of such experiments should permit a more objective evaluation of possible problems.

8. Meanwhile experimental research should proceed with due caution and the national committees in Europe responsible for controlling this research should give serious consideration to the assessments of possible risks in various types of experiments and appropriate levels of containment that are proposed in the latest NIH proposed guidelines, which we believe represent the upper limit of stringency necessary.

9. Furthermore, we believe it is essential that those European national committees that are responsible for the control of recombinant DNA research should come into close contact, for instance through this EMBO Committee, through the EMBC and through the ESF to ensure that in Europe the conditions under which recombinant DNA research is allowed to proceed are as uniform as possible. We therefore recommend that procedures be established for the exchange of information and discussion of policy between European national committees.

10. This Committee is prepared to advise, on request, governmental and private institutions and individuals on scientific and technical aspects, including (a) the assessment of possible biohazards of particular experiments (b) the assessment of appropriate levels of physical and biological containment, so far as questions of principle are concerned. The EMBO Committee wishes to stress, however, that its role is only advisory, it has no juridical powers.

11. Drawing upon available expertise, this Committee intends as a matter of urgency to establish and implement procedures for evaluating biologically disarmed micro-organisms in relation to prevailing criteria of biological containment.

12. This Committee also intends to establish a voluntary registry of recombinant DNA research in Europe and to achieve this it will seek, within the limits set by confidentiality, the cooperation of individual scientists as well as research councils and other involved agencies.

13. We recommend that a small number of centres housing collections of strains of micro-organisms and animal cell lines used in recombinant DNA research be established in Europe.

14. Finally the Committee will continue to promote the organisation of courses on the safe handling and containment of pathogenic micro-organisms and on recombinant DNA technology.

Members of the Committee, present: Professor Ch. Weissmann, *Chairman*, Professor E.S. Anderson, Dr. K. Murray, Professor L. Philipson, Dr. J. Tooze, Professor H. Zachau

PRESENT on 14th February only: Professor W.F. Bodmer

ABSENT: Dr. S. Brenner, Professor F. Gros

PRESENT as ICSU observer: Dr. W.J. Whelan

EMBO EUROPEAN MOLECULAR BIOLOGY ORGANIZATION

EMBO European Molecular Biology Organization
Postfach 1022.40 6900 Heidelberg 1

DR. D. STETTEN
Deputy Director for Science
Office of the Director
National Institutes of Health, PHS
Bethesda
Maryland 20014
USA

Postfach 1022.40
6900 Heidelberg 1
West Germany
Tel. Heidelberg (0 62 21) 38 30 31

Heidelberg, 18th February 1976

Dear Dr. Stetten,

On behalf of the EMBO Standing Advisory Committee on Recombinant DNA, which was set up in January, I would like to thank you and your committee for the proposed guidelines for research involving recombinant DNA molecules which you have prepared and which, we understand, are now being considered by the Director of the National Institutes of Health for adoption as official policy.

We appreciate the extreme difficulty in attempting to formulate comprehensive and precise guidelines including assessments of putative risks, in view of the limited amount of available information. We support the broad principles of your draft, but prefer not to commit ourselves to all the detailed classifications, at least for the time being. We do feel, however, that any further tightening of your Committee's recommendations is unwarranted by the limited evidence that is at hand.

We enclose a copy of the report of the EMBO Committee, from which you will see that we have recommended that experiments specifically designed to provide information on which an objective assessment of possible hazards may be made, should be carried out as a matter of urgency. These experiments may partially duplicate some of the proposals discussed by your Committee.

We look forward to the publication of the official National Institutes of Health policy and we will keep you informed of the further developments in Europe with which the EMBO Committee is concerned.

Yours sincerely

Charles Weissmann
Chairman
EMBO Standing Advisory Committee

Document 7.5
February 17, 1977
New Scientist (**73**:339)

Should genetic engineers be contained?

The probabilities of the events in a hypothetical scenario of a man-made epidemic resulting from recombinant DNA research shows disaster to be closer to science fiction than to reality

Dr Robin Holliday FRS is head of the Division of Genetics, National Institute of Medical Research, London

Ever since Professor Paul Berg and his colleagues drew attention in the summer of 1974 to the potential dangers of certain experiments involving the formation of DNA molecules containing segments from widely different species, there have been numerous conferences and reports on the precautions which should be taken. Researchers agree that the hazards are difficult to assess; for this reason the supposed damages have almost always been referred to in a very general and nonspecific way. In this article I shall attempt to make a more specific assessment of the various accidents and biological events which would have to occur before human life was endangered. By doing this I hope to place in a more rational perspective the current debate on the wisdom of pursuing this research.

The experiments we are concerned with have been variously labelled as "genetic manipulation", "genetic engineering" or just "recombinant DNA". As genetic manipulation of chromosomes and the formation of recombinant DNA has been the basis of genetic experimentation for many decades, these terms are grossly misleading and have certainly lead to confusion among non-geneticists. Genetic engineering is more acceptable, but this is too general a term because it has been widely used in other contexts (for instance, the possibility of correcting human genetic defects by inserting normal DNA in place of mutant DNA). I prefer the term *heterogenetics:* the synthesis and study of replicating DNA molecules containing nucleotide sequences from unrelated organisms.

It is not of course possible to consider all experiments of this type so I simply discuss one standard "shotgun" experiment, which has been referred to as often as any, and which is listed in the most potentially dangerous category in the Williams working party on genetic manipulation. In such a shotgun experiment, DNA from a mammalian, perhaps human, source is fragmented with a particular restriction enzyme (specific endonuclease). The number of different pieces of DNA produced is extremely large, of the order of half a million. These pieces are inserted at random into bacterial plasmid DNA, and the plasmid is introduced into an *Escherichia coli* host. The plasmid replicates autonomously and can be transmitted by conjugation to other *E.coli* cells, or related bacteria, which do not already harbour a similar plasmid. In such an experiment, clones of bacteria can be recognised which contain plasmids with inserted DNA, and since the number of fragments of DNA is so large, almost every one of these clones is distinct. The investigator then has to find the particular clone in which he is interested (for instance, one containing a particular DNA sequence such as ribosomal DNA). This aspect of the experimentalist's problem need not concern us here.

We start our scenario with a standard microbiological laboratory without special containment facilities, which is using standard laboratory strains of *E.coli.* In the scenario I will not initially assign particular probabilities to each event, but simply refer to them as P_1, P_2, P_3 . . . etc. Similarly, I call the possible deleterious consequences C_1, C_2, C_3 . . . etc. At the end of the article I shall attempt to assess the actual probabilities involved, and so produce an overall value of risk. But one point of my article is to allow each reader to make his or her own judgements.

One careless scientist or technician, whilst pipetting by mouth, accidentally swallows a few thousand or million viable bacteria (Probability=P_1. We need not assign a different probability to other types of careless experimentation, for instance, the spilling of bacteria on hands or clothes or the contamination of food or drink). The cells survive and proliferate in his gut (P_2. Note that although *E.coli* is a component of the human intestinal flora, the laboratory strains are far removed from their original "wild type" ancestor). The plasmid the cells contain is transmitted to more natural bacteria of the flora (P_3). This plasmid still contains mammalian DNA and it is one of the axioms of those who worry about these experiments that such DNA in a foreign environment may behave abnormally. It may contain a latent or lysogenic mammalian virus which is induced to proliferate, perhaps because a stretch of DNA which keeps it in the lysogenic state in the mammalian cell has been replaced by bacterial DNA. Let us examine the probability of this (P_4): note that about half a million different clones of bacteria exist containing mammalian DNA, so if there is one latent virus per mammalian genome, $P_4=0{\cdot}000002$.

The route to an epidemic

There could, however, be 10 or 100 such latent virus genomes; let us be pessimistic and assume there are 50, in which case $P_4=0{\cdot}0001$. (In the context of this article "pessimistic" means assigning a high probability to a biological event). The plasmid containing this virus is present in many bacteria and in at least some of these it is induced to form virus particles. The virus is of mammalian origin, but it must be able to replicate its genome, produce coat proteins and so on, in a prokaryotic environment (the bacterial vehicle). I assign an overall probability P_5 for the induction, growth, maturation and release of virions. (Remember that so far no mammalian virus has so far been grown on any bacterium, although many people have attempted to do this). We then suppose that the virus is pathogenic (P_6) and that it infects the individual who is now its host. One consequence is that the individual dies (C_1); another is that he is a resistant carrier and transmits the virus to others by contact, or any of the usual channels of transmission (P_7). We suppose that some of these individuals are more susceptible than the original carrier (P_8) and become ill or die (C_2). Presumably infected individuals will be diagnosed as suffering from an unknown but potentially lethal disease, and would be isolated in hospitals. However, if only a proportion of infected individuals showed symptoms but were infective during a long incubation period (P_9), an epidemic is possible (C_3).

Another version of the scenario assumes that the virus is oncogenic (causes cancer). It might be reinserted in altered form into the chromosomes of cells of the human host and subsequently induce a malignant transformation (P_{10}). The individual originally infected would then die of cancer, but probably some years later (C_1). More insidiously, the virus could be both transmissible and oncogenic (P_{11}). In this case it would be capable of proliferating and spreading to other individuals, but would produce no outward symptoms. Nevertheless, it inserts itself into chromosomal DNA and eventually causes cancer. In this situation an epidemic of widespread cancer (C_4) might not occur until years after the original laboratory accident. It is almost certainly the possibility of consequence C_4, or perhaps C_3, which forms the basis of all the concern about "shotgun" or other related experiments.

This article first appeared in *New Scientist*, London, the weekly review of science and technology.

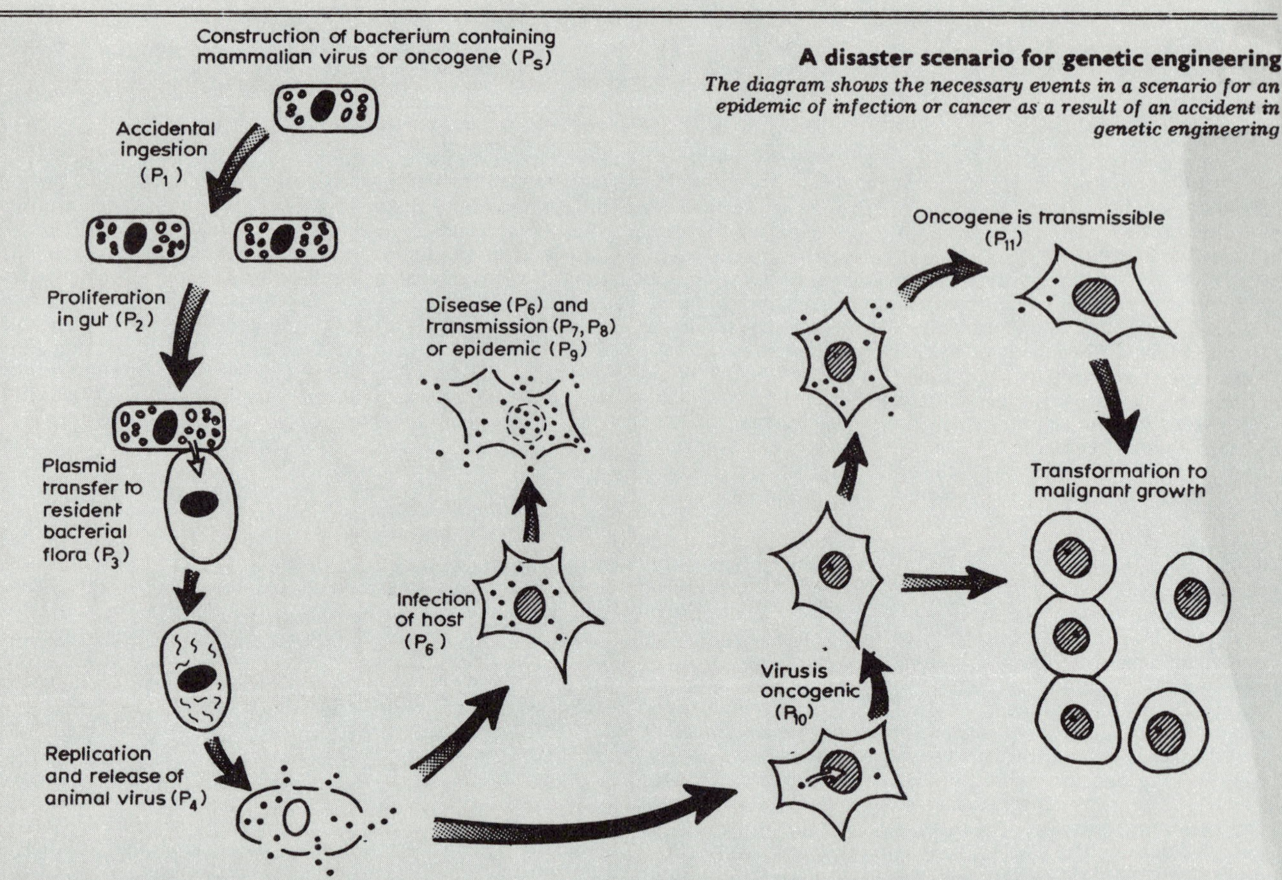

A disaster scenario for genetic engineering

The diagram shows the necessary events in a scenario for an epidemic of infection or cancer as a result of an accident in genetic engineering

The scenarios produce the following formulae:

First pathway of infection:

$P_1 \times P_2 \times P_3 \times P_4 \times P_5 \times P_6 = C_1$ (death by infection)

$P_1 \times P_2 \times P_3 \times P_4 \times P_5 \times P_6 \times P_7 \times P_8 = C_2$ (resistant carrier, causing others' death)

$P_1 \times P_2 \times P_3 \times P_4 \times P_5 \times P_6 \times P_7 \times P_8 \times P_9 = C_3$ (epidemic)

Second pathway of infection:

$P_1 \times P_2 \times P_3 \times P_4 \times P_5 \times P_{10} = C_1$ (death by cancer)

$P_1 \times P_2 \times P_3 \times P_4 \times P_5 \times P_{10} \times P_{11} = C_4$ (cancer epidemic)

To what extent can actual values be given to $P_1 - P_{11}$? I realise I run the risk of ridicule in attempting to assign such values in the following section. Nevertheless I think it is important to do so, and for other scientists to think about the probabilities they themselves would assign.

Most microbiologists working with non-pathogenic bacteria or other micro-organisms have at some time or other swallowed some viable cells, but I would be surprised if they did so more often than about once in 100 experiments, so the value of P_1 will be low, perhaps $\sim 0 \cdot 01$. For estimates of P_2 and P_3, I will rely on experiments carried out with volunteers by Professor E. S. Anderson, one of our experts in the field of plasmid transfer. He showed that *E.coli* K12 can survive in the gut for a few days and that a plasmid could be transferred to resident bacteria at very low frequency. The transferred plasmid did not spread; instead it soon disappeared. In these experiments with plasmids the volunteers ingested 10^{10} bacteria; Anderson concludes that the possibility of plasmid transfer would be remote if, under more realistic circumstances, someone swallowed just a few thousand bacteria (*Nature*, vol 255, p 502). Let us again be pessimistic and assume P_2 and P_3 together equal $0 \cdot 1$.

On the other hand, we have already seen that the probability of a plasmid containing a latent virus or oncogene must be low ($P_4 = 0 \cdot 0001$) simply because the mammalian genome is so large and fragments in the plasmid DNA are so small. I have already mentioned that no mammalian virus has ever been grown in a prokaryotic cell. Therefore, P_5 is likely to be extremely low; nevertheless I will assign a value of say $0 \cdot 001$. P_6, P_7, P_8 and P_9 are very difficult to assess. If the virus is pathogenic, it is likely to be transmissible, but extremely dangerous organisms such as the Green Monkey virus causing Marburg disease are fairly easily recognised and the people infected readily isolated in hospitals. I would describe P_6, P_8 and P_9 as unlikely, but P_7 as quite possible. I will however assign the fairly high probability of $0 \cdot 1$ to all four of these events.

Something can be said about P_{10}, since in spite of intensive efforts by many laboratories so far it has not been shown that any human cancer is caused by a virus. It is much more likely that natural and man-made chemicals in the environment are responsible for most human cancer (see, for instance, The Cancer Problem by John Cairns, *Scientific American*, vol 233, p 64). Both events P_{10} and P_{11} are therefore very unlikely but I again take the pessimistic view and give them values of $0 \cdot 01$. We can now tot up the probabilities of the various deleterious consequences:

$$C_1 = 10^{-2} \times 10^{-1} \times 10^{-4} \times 10^{-3} \times 10^{-1} = 10^{-11}$$
$$C_2 = 10^{-11} \times 10^{-1} \times 10^{-1} = 10^{-13}$$
$$C_3 = 10^{-13} \times 10^{-1} = 10^{-14}$$
$$C_4 = 10^{-10} \times 10^{-2} \times 10^{-2} = 10^{-14}$$

With the exception of P_4 and P_5, I believe I have assigned generally high values to all the other events which must take place before any biologically hazardous situation arises. Nevertheless, the cumulative values above represent extraordinarily small probabilities. Thus, if 10 scientists in each of a 100 laboratories carried out 100 experiments per year, the least serious accident (C_1) would occur an average once in a million years.

It could be argued that many of the probabilities have been misjudged. For instance, the lowest, P_5, could be said to be misleading because in screening for the required clone, many hundreds of thousands of bacterial colonies have to be picked and sub-cultured, and therefore the chance of one of these having a latent virus is very much higher. Although large numbers of clones may be screened, almost all the important prolonged experiments (the 100 experiments I referred to above) will be carried out with one or a few types of cloned plasmid. The essential point is that even if the product of the probabilities is underestimated by several orders of magnitude, the chance of an accident from studies in heterogenetics remains vanishingly small. Nevertheless, it can never be excluded.

Greater hazards

What I find extremely disquieting, is the degree of concern which has been generated in a world where far greater hazards should be everyone's concern. How can scientists accept a world bristling with nuclear weaponry, with the capacity of killing several times over every member of the human race, while at the same time demanding super-safety measures for shotgun or related experiments in heterogenetics? Some biologists would argue that nuclear technology and strategy is not their responsibility; others may even approve of it. To these I would point out that there are other types of biological experiments which are exceedingly dangerous, but which have over the years been carried out in many laboratories throughout the world without generating much concern. I refer to genetic studies with viruses or bacteria already known to be pathogenic to man. Induction of new mutations in these organisms, and the development of laboratory techniques for obtaining genetic recombinants from different mutant strains, seems to me to be potentially far more hazardous than anything discussed in the Williams Report. (The Williams working party does not consider experiments of this type, because research on known pathogens is covered by another White Paper, the Godber Report. However, this Report does not make the essential distinction between work on normal or "wild type" pathogens and the creation of abnormal new strains by genetic methods.) Pathogens evolve only if they have a balanced relationship with their hosts, but tampering with their genetic systems can easily upset this and could lead to an epidemic akin to that in rabbits when the myxomatosis virus first appeared.

Finally, I should make it clear that neither I, nor any of my immediate colleagues, have carried out or intend to carry out experiments of the type I have been discussing; therefore none of the present regulations governing containment facilities will have any effect on our research. I am, however, extremely concerned that some of the attitudes which have been taken up over this issue border on the irrational. For instance, a recent consultative document of the Government Health and Safety Commission proposed that "No person shall carry on any activity intended to alter, or likely to alter, the genetic constitution of any microorganism unless he has given to the Health and Safety Executive notice, in a form approved by the Executive for the purposes of these Regulations, of his intention to carry on that activity." (see *New Scientist*, vol 72, p 220). When attitudes like this prevail, it may only be a short step to the introduction of elaborate containment restrictions on conventional genetic research, which has been carried out since the turn of the century with a wide variety of organisms, and which scientists and laymen alike have long believed to be safe and beneficial to mankind. □

How can scientists accept a world bristling with nuclear weaponry, with the capacity of killing several times over every member of the human race, while at the same time demanding super-safety measures for shotgun or related experiments in heterogenetics?

Document 7.6
January 20, 1977
New Scientist (**73**:150)

An evolutionary perspective for genetic engineering

The US guidelines for controlling research in genetic engineering were drawn up by the National Institutes of Health. Here a prominent biologist claims that the guidelines are therefore formed around a public health framework and that they ignore the important evolutionary divisions between viruses, bacteria, and humans

Dr Robert Sinsheimer is chairman of the Division of Biology at the California Institute of Technology

We have come in our time to a transition point in the evolution of life on Earth. It is for that reason that I believe we need to think long and hard about the nature of the evolutionary process into which we would now intervene through genetic engineering, before we may, inadvertently, wreak great havoc. We can be sure that the recombinant DNA issue, as it is known, will not go away. There is already recombinant DNA in viruses and bacteria, and it is but a modest extrapolation to a future in which we will have recombinant DNA in plants, in invertebrates, in vertebrates, and no doubt in man.

I am not opposed to recombinant DNA research as such. I have said, and I still believe, there are some wonderful results to be derived from genetic engineering—and some that may literally be essential for the survival of our civilisation. But I see also a darker potential for biological and social chaos and I would hope to maximise the former and minimise the latter. It is for these reasons that I believe the policies to govern the recombinant DNA technology and genetic engineering need extensive thought and discussion, and that I believe the National Institutes of Health (NIH) Guidelines have been too narrowly conceived and are inadequate to the issue.

What, then, are my concerns? What would I do at this time? And, perhaps even more important, why has this issue become such a scientific contretemps? Why has it been so difficult to address and resolve?

We need to understand the significance of what has been accomplished up to this point. Recombinant DNA technology—the fruit of 25 years of extensive research in molecular genetics—makes available to us the gene pool of the planet—all of the genes developed in the varied evolutionary lines throughout the history of life—to reorder and reassemble as we see fit. We can look at the evolutionary tree tracing the development of each of the extant living species, each, as are we, the product of three billion years of evolution. That tree is a representation of the fact that evolution proceeds in a linear manner, by small increments, to produce gradually diverging species. Nature has, by often complex means, carefully prevented genetic interactions between species. Genes, old and new, can only interact within a particular species.

With recombinant DNA research we can now transform that evolutionary tree into a network. We can merge genes of most diverse origin—from plant or insect, from fungus or man as we wish. Most such combinations will, of course, be sheer nonsense, non-viable, and innocuous. A few will, by careful design, be valuable—if not, there would be little point to the whole enterprise. A few, however, may by design or inadvertence be deadly, in any of many ways.

The slow, almost measured pace of evolution permits the establishment, at any time, of quasi-equilibria among the various competing species. The balance is never static, but dynamic. Some species continue to find a suitable ecological niche; others die out. Most species that have lived have perished and have been replaced. We should perhaps recall

California Institute of Technology.

Dr Robert Sinsheimer

that the giant reptiles dominated the earth for 150 000 000 years—and then perished. Now we come, with our splendid science and our ingenuity, and we have now the power to introduce quantum jumps into the evolutionary process. But do we have the commensurate understanding to foresee the consequences to the currently established equilibria on which, quite literally, our life support systems depend?

Organisms evolve into an ecological niche which favours and permits their survival. They are where they are and what they are because of that evolution. Man likes to think that he is the exception, that we have made our own ecological niche. In part that *is* true, but in large part it is, at least as yet, a conceit. We rely literally on our fellow creatures; the plant world gives us our oxygen and our food; and the microbial world restores the planetary nitrogen and degrades our wastes.

Our resistance to disease, our susceptibility to disease, the severity of the symptoms caused by a disease are all reflections of our evolutionary adaptation into an available niche. For instance, there are in the United States some 25 deaths a year from botulism poisoning. One may think it very fortunate that botulism is not a contagious disease. Of course, this is not just due to good fortune. If botulism were a contagious disease, the human species could simply not be what it now is. Our ancestors would have had to find another niche.

My principal objection to the NIH Guidelines is that they do not reflect the evolutionary implications of what

This article first appeared in *New Scientist*, London, the weekly review of science and technology.

we are now about. They ignore the potential evolutionary consequences for ourselves *and* for our fellow creatures. The Guidelines were conceived to cope with the perceived, immediate medical hazards of recombinant DNA research. There were, I believe, reasons for this tunnel vision, to which I will return. As such, I believe their authors did a commendable job; they rank-ordered the hazards they envisioned and then, in a pattern of graded risk, imposed a graded set of containment provisions commensurate with the estimated risk. But it is clear that the authors of the Guidelines did not consider the transfer of genes across species, the introduction of quantum jumps into the evolutionary process, to be of *any* hazard, *unless* one could specifically pinpoint a gene of known toxicity. Thus, any DNA fragment from any invertebrate can be inserted into the bacterium Escherichia coli under moderate containment (P2, EK1) conditions. Any DNA fragment from any embryonic form of a cold-blooded vertebrate can be inserted into the E. coli organism under these conditions. Any DNA (from any source) that has been cloned (and is not known to code for a toxic agent) can subsequently be grown in the E. coli organism under P2, EK1 conditions.

Consider what is implied here. The DNA from an insect or an echinoderm can be cut with a restriction enzyme into some 20 or 30 or 50,000 fragments. Each contains some generally unknown cluster of genes. With another restriction enzyme one can produce a different set of 20 or 30 or 50,000 fragments. Any or all of these fragments can be inserted into E. coli and grown up into a clone. Somehow it is presumed that we know, *a priori*, that none of these clones will be harmful to man or to our animals or to our crops or to other microbes—on which we unthinkingly rely. I don't know that and, worse, I don't know how anyone else does.

Even more, this echinoderm DNA, for instance, may be prepared from organisms collected from Nature that live perhaps on a coastal shelf. Such organisms are surely not sterile preparations; they have their own, usually unknown, coterie of associated microbes and parasites which can include those deposited on the coastal shelf by our waste disposal systems, as well as more indigenous forms. When the DNA of the echinoderm is prepared and cloned one will inevitably prepare and clone, in some small proportion, the DNA of these small companions. That these small companions might include the spores of deadly bacilli, or the viruses of human waste seems to have received scant thought.

The Guidelines reflect a view of Nature as a static and passive domain, wholly subject to our dominion. They regard our ecological niche as wholly secure, deeply insulated from potential onslaught, with no chinks or unguarded stretches of perimeter. I cannot be so sanguine. In simple truth just one—just one—penetration of our niche could be sufficient to produce a catastrophe. I think there has been an inadequate realisation of the fact that we are here concerned with potentially irreversible processes; that living organisms, if they find a suitable niche, are self-perpetuating, and even more, are subject to their own future evolution, wholly beyond our control. This is a novel circumstance in the history of man-derived hazards. IF DDT or fluorocarbons prove to be unfortunate, their manufacture can be ceased and in time they and the hazard will vanish. Once released, self-propagating organisms will be with us, potentially for ever. A new pathogen need be created literally only once to cause untold harm. In time, no doubt, we could learn to cope with new pathogens as we have with the old; but as we know, the mortality and trauma associated with the adaptation to any new disease, or new strain of an old disease, can be most costly.

In the larger sense we have need to protect not only ourselves but the entire biosphere on which we depend and which is, in a sense, increasingly in our trust. We inherited and evolved in a marvellously balanced, self-sustaining world of life. Can we in truth predict what disruptions we may introduce into that world with our extraordinary inventions, our biological innovations not derived from the historic evolutionary processes?

‘The Guidelines reflect a view of Nature as a static and passive domain, wholly subject to our dominion. They regard our ecological niche as wholly secure, deeply insulated from potential onslaught, with no chinks or unguarded stretches of perimeter. I cannot be so sanguine.’

The concept of graded risk implicit in the Guidelines makes sense when one is concerned with large numbers at risk. If a few individuals, or even a considerable number, should perish because of our experiments, that would be tragic but not terminal. The concept of graded risk makes little sense if there is only one subject, as in one biosphere, the death of which would indeed be terminal. But the Guidelines are also, I suggest, deficient in another respect. The recombinant DNA technology was developed by scientists to solve their own scientific problems. The Guidelines were developed by these scientists to cope with the kinds of problems and hazards they could envision as emergent from the experiments they have in mind to do, and as I have said, primarily to cope with perceived, immediate medical hazards of such experiments. But as a technology the recombinant DNA techniques are in fact available to all sectors of society that have, or can purchase, the skills to use them. They are available to entrepreneurs, to flower fanciers, to the military, to subversives. Their application does not require vast resources. And the Guidelines have only in the most marginal way addressed the problems, the social hazards, which such applications may pose for our society. The potential for misuse is inherent in any scientific advance; it would seem to be particularly virulent here.

If I am not simply a Don Quixote, tilting at only imaginary dangers, how have we come to this parlous and perilous situation? I suggest there is a cluster of reasons which reinforce each other in such a way as to form, for many, an almost unpenetrable cloak or barrier to a new and wider perspective.

There is, of course, a wholly human tendency to worry about tomorrow's woe, tomorrow—an often, but not *always*, wise tendency based on the observation that many prophecies of woe are never realised. Here I would suggest such an attitude (and I *have* heard it expressed) overlooks the potential magnitude of *this* woe and especially overlooks the uniquely irreversible character of this enterprise. We are unaccustomed to thinking about irreversible steps. There is the very current tendency to believe that the human niche is now really quite impregnable. I suggest this is a very dangerous, if very human, conceit. But beyond these are a set of attitudes characteristic of science and scientists, and this issue has thus far been largely within the province of scientists. These are attitudes long established within science and deeply ingrained and if they are, as I suggest, no longer wholly appropriate we can expect that it will take a considerable time for many to mentally and emotionally readapt.

Among these we might mention the concept of freedom of inquiry, a hard won right, deeply treasured, earned by the trauma of our patron saints such as Galileo. If now this must, as I suggest, be at least partially restricted, acceptance will—and should—come grudgingly. Coupled to that issue is the shock inherent in the concept that science itself—the simple pursuit of truth, the exploration of Nature—could in itself be dangerous, not merely to the experimenter, but conceivably to the entire planet. We have perhaps become accustomed to the concept that some of the products of technology, exploited on a vast scale, can be dangerous on a vast scale. But the circumstance that science itself has become so powerful that the performance of a single experiment could have, directly, vast consequence is very new. We have not needed, and are therefore unaccustomed, to incorporate elaborate measures of safety into the design of our experiments.

I referred earlier to the seemingly curious fact that the Guidelines are addressed almost exclusively to the immediate medical hazards perceived in this research. I suggest there is a reason for this limited perspective. For the Guidelines were, after all, developed under the aegis of the National Institutes of Health whose purview is, primarily, medicine. It is conceivable that had the Guidelines been developed under a different aegis they might have reflected a different perspective. But this circumstance reflects a deeper problem; in fact, the Guidelines could not have been developed under any other aegis. Essentially all of the recombinant DNA research, indeed all of modern biological research in the United States, has been and is supported by the National Institutes of Health. I suggest that while the administration of the National Institutes of Health has been, truly, most enlightened, this predominant dependence upon an agency whose ultimate mission is health has distorted the science of biology in this country. It has, inadvertently I am sure, biased our values and limited our perspectives. And we are now beginning to see the cost.

There is yet another aspect which derives again from the changed circumstances of science. These changes arise in part from factors intrinsic to science: the very success of the scientific enterprise and the great powers now thereby in the scientists' hands. They also arise, however, from the changed social perception of science. We are now aware of the seemingly inevitable progression from scientific research to technological advance to social change. Aware of it and uncomfortable with it. And I suggest that major elements of our society are, determinedly, groping for means to reduce the apparent inevitability of this process, to bring it more under a conscious control.

> **'To do good science requires an intense dedication; one must live and breath science to do it well. And to do that one must really believe that the particular science one is doing is important and ultimately beneficial. To question that belief and still preserve that essential dedication will require a very considerable maturity.'**

In this process, the older concept that the scientists need have no responsibility for the social consequences of their discoveries is, I suggest, becoming increasingly untenable. The convenient fiction that there is a meaningful gap between the knowledge developed by scientific discovery and the uses society may make of that knowledge is seen to be just that—a fiction. It ignores the great social cost necessary in a free society, to maintain such a gap, even if it is possible.

And so, I suggest that scientists have—particularly in this issue—to begin to accept the responsibility to consider the social consequences likely to emerge from their research. And this must include the novel and painful conception that science *can* make the world a more dangerous place, that scientific advance could destabilise human society and could even imperil the human future. The atomic bomb was the first breach in our innocent conviction of the beneficence of science. My concern is that genetic engineering should not become the second.

This realisation is especially bitter. It is potentially demoralising, and therefore resisted. To do good science requires an intense dedication; one must live and breathe science to do it well. And to do that one must really believe that the particular science one is doing is important and ultimately beneficial. To question that belief and still preserve that essential dedication will require a very considerable maturity. All of this is a lot to swallow. These perceptions, while I believe they are true and probably essential to our very survival, bring no joy. Their recognition would, I suggest will, require a major wrench to the thinking of many if not most scientists. One could hardly expect such a transition to come quickly or easily.

Now, after all of this, what would I *do*? I would certainly like to exploit both the scientific and practical potentials of genetic engineering but with the minimum possible impact on, and interaction with, our existent biosphere. Ideally, I suppose, I would like to see such research and development done on the Moon! But more practically, I would like to see it confined to a small number of maximum containment facilities under careful and sustained supervision. And I would phase out, as soon as possible, the use of an organism such as E. coli, a ubiquitous species with a plethora of bacterial habitats, in favour of an organism with a very restricted range of habitats, far less likely to survive should it escape.

Some people with a misplaced paternalism attribute opposition to the Guidelines to a "childish fear of the unknown". It is not fear; it is *concern*. I am concerned because I have a realistic respect for the powers with which we toy. And I am concerned because I believe there is a very real possibility of a major cataclysm, an event that would mock all the striving, all the toil and effort, all the sweat and genius that has gone before.

I can state my objective very simply: the atomic age began with Hiroshima. After that no one needed to be convinced that we had a problem. We are now entering the Genetic Age; I hope we do not need a similar demonstration. □

'Son, I hear you failed genetics!'

This article is based on Dr Sinsheimer's contribution to the University of California Los Angeles, Distinguished Lecture Series, 1976-77

The Fanciful Future of Gene Transfer Experiments

Charles A. Thomas, Jr.
Department of Biological Chemistry,
Harvard Medical School, Boston, Massachusetts 02115

Science attempts to confront the possible with the actual.[1]

In preparing for my talk entitled "Some Future Possibilities for Gene Transfer Experiments," I planned to review some of the new vehicles and to illustrate how these might be used to study the sequential requirements for gene expression in higher plants and animals. However, much of this subject has already been covered or implied by the contributors to this Symposium. Moreover, this subject has already suffered from the unjustified extrapolations of earlier writers, and so I resolved not to add my own.

Much of this subject is related to the term "genetic engineering," which wisely was not used in the title of this Symposium, nor does it appear in the title of any contribution, save that of D.C. Reanney, where it is clearly used to mean "translocatable elements."

Nonetheless, in the public mind, much of what we have been hearing at this Symposium will be ineluctably confused with "genetic engineering." The public is already confused enough on this subject, and I believe some scientists are as well. This confusion is most evident in the recombinant DNA controversy that enmeshes so many scientists and so much science in this important area. It is a simple fact that the hazards thought to be associated with recombinant DNA are totally conjectural. No organisms of unusual pathogenicity have ever been formed from DNA recombined *in vitro*. And yet, the exploitation of this issue has met the special needs of many divergent groups – journalists, television reporters, government bureaucrats, university committees, politically motivated students and faculty, local and national politicians, and a few scientists. Why are all these people so concerned about something that has no scientific basis?

Perhaps there is no single answer to this question, and to prove that a given explanation is correct may never be possible. However, this question led me to reconsider the published remarks of various individuals dealing with the future possibilities of gene transfer experiments. I believe you will see that there is some reason to think the public has been preconditioned by incorrect and unwarrantable remarks that have been made by well-known scientists.

It was 33 years ago, in 1944, that Avery, MacLeod, and McCarthy demonstrated that DNA was responsible for genetic transformation in *Pneumococcus*. After a substantial delay, reinforcement was supplied by the Hershey-Chase experiment in 1952. The first symposium of which I am aware, on this subject, was held in 1963. It was organized by Tracy Sonneborn, who selected the title: "The Control of Human Heredity and Evolution."[2] (The title of this Symposium, "Genetic Interaction and Gene Transfer," is much more modest and appropriate.) Humans have always exercised considerable control over their own heredity and have been dominant factors in their own evolution, although these facts were given scarcely any attention by Sonneborn's panelists. The novel aspect was the idea of slipping in pieces of DNA – an idea that continues to bemuse us 14 years later. In spite of the alarming title, the individual contributors were for the most part reasonable, although ambivalent. For example, the first contributor, S.E. Luria, wrote (after a brief description of DNA, the code, gene expression, transformation, and transduction):

> From the standpoint of molecular biology we see today no clear and open paths to directed control of human heredity or, for that matter, of any heredity except possibly bacterial. Even at this level, the powerful tools of genetic change still suffer from . . . inefficiency.

and later,

> . . . nothing at present would justify either the prediction of a coming millennium of human betterment or the proclamation of an impending danger of genetic enslavement.

On the other hand, he recalled the rapid development of the atomic bomb from 1935 to 1945 and said:

> I would not think it premature . . . for the United Nations as well as the National Academy of Sciences . . . to establish committees on *the genetic direction of human heredity*. [Emphasis in original.]

Later he said,

> What has been foremost in my mind . . . has not been a feeling of optimism but one of tremendous fear of the potential dangers that genetic surgery, once it becomes feasible, can create if misapplied.

He went on to suggest that, if someone could infect people with a virus rendering them sensitive to CO_2 poisoning (analogous to a well-known virus in *Drosophila*), then

> Someone could gain a tremendous control over humanity by spreading such a terrible object, thereby holding the power of life and death over a large number of human beings.

There are two stories here. As a scientist Dr. Lauria expresses the opinion that there is no danger, yet as a panelist he generates an imaginative (if incomplete) story that is not unlike that found in *War of the Worlds*[3] by H.G. Wells! If you were a journalist reporting on that symposium, which one would you write about?*

---

*This same ambivalence can be found in Luria's book,[4] from which Eisinger[5] extracts the following text to justify his symposium on the ethics of human gene manipulation:

> In general it is impossible to foresee the consequences of potential new biotechnologies because there is no way to evaluate the interactions with the social settings in which they may become available and be applied. In a society based on cooperativeness and mutual helpfulness, genetic identity might add a biological substrate to the predominant values. Unfortunately, anthropologists have found that societies approaching such ideals exist only in a few isolated populations of South Sea Islanders. In a competitive, caste-ridden, power-dominated society, the ability to refashion human beings either by selection or by manipulations of eggs, sperm, and genes might become a tool to promote inequality and oppression. It might serve to create masses of obediently toiling slaves or to manufacture elites of identical rulers – in ancient Egypt, pharaohs married their sisters in order to generate successors as similar as possible to their own divine selves.
>
> Apart from such nightmarish speculations, the philosophical implication of experiments on human heredity done for eugenic rather than medical purposes are sobering. What would a world be like in which men – even a few men – were manufactured for some alien purpose, used as means rather than ends?

Perhaps the panelist seeming the most pessimistic about practical "genetic engineering" was Herman Muller. After pointing out the enormous difficulties, he stated that mate selection, eventually from frozen sperm banks, represented the most imminent possibility. In response to this, Tracy Sonneborn[2] said he was completely right – but then went on to say he was wrong:

> In spite of well-recognized technical difficulties, not insurmountable, the same semiblind selective approach [presumably cell transformation by unfractionated DNA] might be developed for human cells, even for cells that give rise to germ cells. Replacing such selected germ cell progenitors back in the body from which they were taken could lead to much control of human heredity.

This statement presents two problems: first, it conflicts with what is (and was) known about the differentiation of the embryonic gonad and the enormous inefficiency of such experiments stemming from the astronomical gametic wastage (particularly on the male side); second, the word "human" implies the human race rather than the individual developing from the hypothetical treated fetus. Therefore, this statement has a questionable scientific basis and is misleading.

Two years later Rollin Hotchkiss gave a lecture at the annual meeting of the American Institute of Biological Sciences in Urbana entitled "Portents for a Genetic Engineering."[6] A portent is an event or situation that presages evil. Not a cheery title.

> *Genetic Intervention.* By contrast, the new potential program, 'genetic engineering,' which is raised by the exploits of molecular biology, is a genetic intervention that could be practiced in private and in secret on individual genes of individual persons. I prefer to call it intervention since it may never be really engineered. It will be much more difficult to regulate, and legislation *against* it will seem like the same invasion of personal rights that legislating *for* eugenic measures appears to be. We have always been soft-hearted about man's right to make a fool of himself, as long as he does not become a malicious fool. Yet this kind of intervention has more than a little potential for altering the gene pool from which all future humans will draw their imprint. Let us try to assess how likely this genetic intervention is at present, and then consider how it is likely to be approached.

One cannot but be fascinated with kinds of genetic intervention which could take place "in private and in secret." Note also the reference to the human gene pool. But let us move on to even more exciting prospects:

> One can imagine farms of cultured human cells producing desired markers and delivering them up to viruses that can then profusely replicate them and infectiously transfer them to pregnant mothers, sick patients and other unlucky beings adjudged by the family doctor to be deficient in something he knows about. The wealthy and other royal families as always can even hope to purchase special advantages, such as determinants of musical ability, linkage groups providing skill in political oratory – or will they prefer skill in such gentlemanly pursuits as polo, or (somewhat less expensive) single factors enabling one to ride graceful and sure on an appropriately well-bred horse? It has been argued that these genes for genius and other categories of DNA cannot be pinpointed without elaborate selection tests, but I believe that the art of DNA identification and fractionation will soon bypass the laborious genetic cross. In fact, several workers, including L. Mindich and M. Roger in our laboratory, have already made small steps in this direction.

The author's style tends to charm the reader and confuse the subject. Is he actually proposing that a complex pattern of behavior can be conferred by the addition of a few genes to the embyro or adult? If so, these remarks conflict with everything that is known on the subject.

Consider again the next to the last sentence: does it have any scientific basis? Did it ever fit the known facts? Consider the last sentence: is it believable that steps have already been taken "in this direction?" Finally, Hotchkiss offers this definition of scientific responsibility:

> Let me stress the responsibility of the scientist to speak sometimes of what we don't know as well as what we do. I hope that this will soon again be earnestly inculcated in our young colleagues . . . [because] . . . in the case of genetic intervention, it can be a most serious matter.

While most of you have never read this article, I imagine that some have and have taken its message to heart. Let's have another look at that sentence: "Let me stress the responsibility of the scientist to speak . . . of what we don't know as well as what we do." This interpretation of the role of a scientist would not be accepted by many. Contrast it, for example, with a memorable line from Jacob's latest article,[1] "Evolution and Tinkering,"

> Science attempts to confront the possible with the actual.

Jacob goes on to describe the limited way in which science proceeds. But let me not become sidetracked with this altogether superb article. Let us continue with the "genetic engineers."

In 1966, Robert Sinsheimer spoke at CalTech's 75th Anniversary Conference. Perhaps the talk was a symposium lecture – not unlike this one. Sinsheimer, now freed from the customary constraints of experimental science, and undoubtedly reviewing previous articles mentioned above, supplied the press with some unmatched prose in an article[7] entitled "The End of the Beginning." This title* is in the grand tradition of *science fiction*. After creating a mood he writes:

> How will you choose to intervene in the ancient designs of nature for man? Would you like to control the sex of your offspring? It will be as you wish. Would you like your son to be six feet tall? seven feet? What troubles you – allergy, obesity, arthritic pain? These will be easily handled. For cancer, diabetes, and phenylketonuria there will be genetic therapy. The appropriate DNA will be provided in the appropriate dose. Viral and microbial disease will be easily met. Even the timeless patterns of growth and maturity and aging will be subject to our design. We know of no intrinsic limits to the life span. How long would you like to live?
>
> And in the end, after all those smaller steps to improve man's lot are taken, we may come to change man himself, his physique, his emotions, his intelligence, all of which are, in large part, the outcome of an inheritance pattern, which too can come under rational control. Not tomorrow. Perhaps not this century or the next. But it is only three centuries since Francis Bacon, and there are many centuries ahead.

Apparently, this kind of science fiction written by a real scientist met with some success because he reappeared again in 1968 with an even more disturbing title[9] – "Darkly Wise and Rudely Great." In the accompanying article he writes:

> The living organisms of today have had the benefit of two billion years of selective molecular evolution. Soon we shall have the cumulative ingenuity at our fingertips as well as *in* our fingertips, and with it not only the power to alter the natural world but also the power to alter our very selves.

There may be some among you who were surprised by Sinsheimer's reaction to recombinant DNA. All I can say is that you had not read your early Sinsheimer!

In 1969, for a symposium held by the American Association for the Advancement of Science at Boston University, Bernard Davis produced a manuscript entitled "Threat and Promise of Genetic Engineering."[10] This article is distinguished by its extreme seriousness. Davis apparently believed he was dealing with a serious, important, and very real issue. He dealt with it systematically and made a number of points relevant to the hypothetical situation. However, take for example his statement:

> And it is true that the possibility of altering or replacing specific genes would significantly broaden our control over the human gene pool, compared with our present dependence on the lottery by which each of us forms an enormous variety of germ cells.

---

*The possible precedent for this SciFi title could well be Asimov's *The End of Eternity*.[8]

We have all grown so accustomed to reading such statements in the press from journalists who don't know any better that we now accept them from scientists who do know better. Scientists (generally at a symposium with an outlandish title) supply the vocabulary to other scientists and journalists. I wonder if anyone has asked Dr. Davis how altering a specific gene in one or even 1000 sperm cells could have a perceptible effect on our control of the human gene pool. The question would be considered in bad taste; scientists speaking at certain kinds of symposia (or before TV or the press) are not supposed to be questioned about the scientific basis of anything they say.

However, the pace was quickening. In 1969 Shapiro and Beckwith isolated a segment of DNA from a transducing phage corresponding to a portion of the *lac* operon. They alerted the press and TV, and pressed their views that this kind of thing was another tool that could be used for "social controls," as Beckwith termed it in a later article.[11] Note the interesting parallel to Luria's suggestion[2] made 10 years earlier about the CO_2 sensitivity virus: "Someone could gain a tremendous control over humanity . . . ." Remember, with this kind of literature just as with certain art forms, one should never ask questions like "how, where, how many, etc." Such questions spoil the mood.

Amid these celebrations of the ethical consciousness of present scientists regarding future possibilities, a few restraining voices were heard.[12] Among the very few was that of Phil Abelson, who wrote[13] in July 1971:

> Talk of the dire social implications of laboratory-related genetic engineering is premature and unrealistic. It disturbs the public unnecessarily and could lead to harmful restrictions on *all* scientific research.

How right you were, Phil Abelson, and how well we understand this today!

But the bioethicists were now in the saddle and riding high. Avoiding the specific details, or the biological and economic realities of any genuine situation, they made grand pronouncements which took as a starting point a situation that never existed. For example, Leon Kass,[14] then the *executive secretary* of the Committee on Life Sciences and Public Policy of the National Academy of Sciences, drawing on his working paper for that committee, wrote:

> Recent advances in biology and medicine suggest that we may be rapidly acquiring the power to modify and control the capacities and activities of men by direct intervention and manipulation of their bodies and minds. Certain means are already in use or at hand, others await the solution of relatively minor technical problems, while yet others, those offering perhaps the most precise kind of control, depend upon further basic research.

He goes on:

> In contrast, biomedical engineering circumvents the human context of speech and meaning, bypasses choice, and goes directly to work to modify the human material itself. Moreover, the changes wrought may be irreversible.

and on:

> Indeed, both those who welcome and those who fear the advent of 'human engineering' ground their hopes and fears in the same prospect: *That man can for the first time recreate himself.* [Emphasis in original.]

Read these words, published in a major article in *Science*, as one who is unfamiliar with science – as a legislator, a lawyer, or a mother.

A reassuring article appeared in *The New York Times Magazine* entitled "The God Committee."[15] Reporting on the experiments of Merril (on λp*gal* curing galactosemic fibroblasts) *The New York Times* produced an article entitled "Altering the Cell – the Vistas Are Breathtaking."[16] Friedmann and Roblin wrote their serious long article in *Science* on gene therapy.[17] The National Institutes of Health sponsored a symposium at the Fogarty Center entitled "The Prospects of Gene Therapy."[18] I could go on: Jacques Monod and even Arno Motolsky became somewhat carried away by the "breathtaking vistas." The one sensible article known to me from this period was by Ben Lewin,[19] then located at the NIH. After summarizing the recent developments in a matter-of-fact manner, he concludes:

> The debate about genetic engineering is so far in advance of reality that attempts to establish ethical guidelines for genetic engineers are almost certain to be overtaken by whatever means for gene therapy may eventually emerge. Much of what has been said simply comprises safeguards, which should in any case be applied to any new medical treatment. The emotional response shown at the prospect of gene therapy is clouding a sober appraisal of the techniques which might be developed if the promise of recent research is fulfilled.

Nonetheless the torrent of alarming publications continued. Paul Ramsey, a Harvard Professor, published a book in 1970 entitled *Fabricated Man*,[20] again straight from *science fiction*. Woody Allen, in a film called *Sleeper*, developed a plot that turned on "cloning the nose." Kirk Douglas is producing a film based on *The Boys From Brazil*,[21] in which Hitler's fibroblasts were cloned by some nuclear transplantation scheme. All of this would be ridiculous humor except for the fact that it seems to be taken seriously by the public and by scientists who should know a great deal better.

More books: In 1973 *Genetic Fix*[22] by Amitai Etzioni (which incidently contains a more complete chronicle of this outrageous literature); Michael Hamilton's *The New Genetics and the Future of Man*,[23] one quarter of which is devoted to a long article by Leon Kass, whom I discussed earlier; and finally *Our Future Inheritance: Choice or Chance*,[24] a committee effort edited by Alun Jones and Walter Bodmer. This last is a fairly reasonable volume which received a good deal of journalistic attention and thereby distortion.

About 1972 Sinsheimer[25] wrote a chapter entitled "Prospects for Future Scientific Developments: Ambush or Opportunity" in *Ethical Issues in Human Genetics.* He opens with:

> Ours is a time of intense self-doubt, corroding confidence, and crippling resolve; a time of troubled present and ominous future; a time of strange clouds and sudden shadows seen in a fading light with cracking nerve. And hence it is not surprising that so great a triumph as man's discovery of the molecular basis of inheritance should provoke fear instead of joy, breed suspicion instead of zest, and spawn the troubled anguish of indecision instead of the proud relief of understanding.

By June of 1974, J. Eisinger[5] had organized a symposium for the Biophysical Society entitled "The Ethics of Human Gene Manipulation." At the risk of being tiresome, I emphasize that a nonbiologist reading this title and scanning the contribution by R. Roblin and another by Marc Lappé could easily conclude that human gene manipulation was a reality! Otherwise, why should these distinguished individuals be concerned about the ethical questions that arise from such manipulation? The facts are quite different: human gene manipulation (whatever one means by this) does not exist, and therefore no ethical questions have presented themselves.

In 1973-74 a group of accomplished molecular biologists, motivated by a deep sense of responsibilty, reversed Jacob's maxim and attempted to confront the "actual" with the "possible," then sounded the false alarm about potential* hazards of recombinant DNA.[27] Not a single experiment was done to substantiate the imaginary hazard, and, to my knowledge, none has been done to this day. Indeed, the epidemiologists and experts on water-borne infectious diseases have yet to make a significant contribution. Meanwhile, other scientists remained silent, even when presented with the most absurd nonsense, because they admitted that there were things they did not know . . . and proof of nonexistence of anything was not possible. In this process they exposed a political target of opportunity that invited attack by a wide range of individuals who had been fully equipped with a vocabulary to do so from

*One should note their incorrect use of the word "potential" – which may have caused some confusion in the public mind. The word "unsubstantiated" or "conjectural" would be more accurate. See also Singer and Soll.[26]

TV, SciFi, and biological scientists themselves. Is it any wonder that considerable public support was shown for those who wanted to ring the bell on these scientists about to plunge us all "irreversibly" into a Brave New World or violate the "wisdom" of two billion years of evolution?

Perhaps the most thoughtful and complete assessment of the role of molecular biology in medicine and genetics was supplied by MacFarlane Burnet[28] in his *Genes, Dreams, and Realities.* His conclusion is that molecular biology has the capability neither of helping nor of hurting the human condition for reasons that stem largely from the complexity of the systems involved. His views are not very popular because they would necessarily require biological scientists to assume a position of somewhat lowered importance. This is not what either individual scientists or our major government funding agencies want to hear – and consequently they do not hear it.

What are the prospects for therapy by gene transfer? From what I can see they are very limited, very costly, and largely unnecessary.

1. Only those genes that function exclusively in terminal stages of differentiation will be susceptible to complementation. They represent a very small fraction of the total genome.
2. The genes to be completed must correspond to non-autonomous functions; that is, they must produce diffusible gene products that act beyond the confines of the cell producing them. Otherwise each cell of the organ displaying the deficiency must be supplied with the gene – clearly a practical impossibility.
3. Samples of the individual's cells must be grown *in vitro*, where they must be infected and stably transformed by a transducing virus or chromosomal fragment.
4. The added gene must be fully dominant under these circumstances.
5. The transformants must be selected and grown to increase their number.
6. The transformed cells must be arranged to repopulate the appropriate organ – the spleen, liver, or bone marrow.
7. This entire process must be free of any untoward effects, such as tumor formation, even after a 20-year delay.
8. The entire process must be repeated for each patient in order to avoid histocompatibility problems, since inducing immune paralysis is known to be significantly dangerous.

There may be some to whom all this may seem easy. Indeed, I suspect that some scientists even justify the utility of their basic research in terms of visions of gene therapy. However, one should remember that applied research is more difficult than basic research. In basic research one only seeks understanding. In applied research one must get a new process to *work*, and that is much more difficult.

One of the realities of our life in the United States is the Food and Drug Administration and the body of public opinion and public interest that it represents. It is very likely that this agency would require so much prior testing over such a long time as to make any development of this kind so expensive that it would never be supported by the taxpayer for the benefit of a tiny minority of the population that would stand to benefit from the contemplated therapy.

There are other fatal objections to these speculations. Most genes function in a timed sequence during embryogenesis – only a few control terminal functions exclusively. Treatment of the gametes or early embryos (even if it could be done by overcoming the inefficiencies mentioned earlier) is warranted only if we believe that all couples have a "right" to have their own (repaired) children. There are many childless couples today – today more than ever. Fertility cannot be guaranteed by any government. Indeed, most of our scientific effort in this area has been devoted to promoting controlled infertility. There is already available a very cheap and often pleasant way of getting a healthy child: for example, by getting a healthy woman pregnant.

Therefore, in the absence of evidence that would refute *every one* of these objections, I would not bet one penny that gene therapy will ever be a practical reality, and all the talk to the contrary is wrong and misleads the public. Talk about genetic engineering and the manipulation of human genes is thus irresponsible.

Irresponsible and ill-considered remarks by scientists and bioethicists that have little or no scientific basis could be very dangerous to the scientific enterprise as a whole, especially today in 1977. Life is always more simple when we have clear villains or heroes, or when we have some driving ideology. At this point in our national life we are remarkably free of all of these: we have a popular president, no war to be for or against, and no dominant political or religious ideology. Under these circumstances Science could well be an attractive villain. Two generations of science fiction have laid the groundwork. Television and movies bring these same ideas into the consciousness of almost every American every day. And finally, our journalists (who are now more numerous than ever) prematurely publicize genuine scientific advances in terms that are consistent with those of science fiction. In this process, they often have the full complicity of the scientist himself. Who doesn't like attention? We should not be surprised to see scientists saying irresponsible things in order to enjoy public attention. It is human nature, and scientists are all too human.

In addition to TV and the press, I wonder if we should not look at the "Symposium" format with a more critical eye. When these assemblages are purely scientific, they promote scientific communication and are valuable for that reason. However, when they invite unchecked speculation on the part of the participants, or when a vote is to be taken on a pending resolution for communication to the press or to certain members of the legislature, then this process may encourage irresponsibility. Paradoxically, all this is done in an effort to be "responsible." A few years ago Arthur Koestler wrote a satire on symposia entitled *The Call Girls*.[29] This novel conveys extremely well the difficulties of symposia and should be studied by anyone who desires to improve this form.

REASONABLE EXPECTATIONS

After a review of some of the extravagent projections, it might be useful to mention some reasonable expectations for gene transfer experiments. There are some exciting possibilities. First of all, there is the practical possibility of producing valuable proteins in specifically designed *E. coli*. These proteins might be insulin, albumin, specific immunoglobulins, certain hormones, or antigens such as the capsid proteins of measles or flu. Estimates vary regarding the economic significance and practical value of this kind of biosynthesis, but most agree that it could be significant.

These new advances will be enormously important in elucidating the molecular basis of growth and development. Although much work must be done, it now seems possible to prepare specialized transducing DNA viruses containing a segment of DNA of completely known sequence and to work out just what features of this sequence are essential for gene function. Equally important is that it now appears possible to integrate these known sequences within the linear continuity of the chromosome. In this structural context, the inserted sequence will undoubtedly be subject to the same kinds of events that control other chromosomal genes. These controls are very likely to be those that are responsible for the stability of the differentiated state. To understand the molecular basis of the stability of the differentiated state, and the controlled manner in which it unfolds, would be a great human achievement; a goal we should not give up easily.

There is now virtual unanimity of opinion that cancer is, in the broadest terms, a developmental abnormality. Of course, it has a genetic basis; perhaps an epigenetic basis; but ultimately it is a developmental disease. About 1000 people die of cancer every day in our country. I can see that it is in the interests of the American people to avoid this particularly unpleasant way of dying. Therefore, I believe that it is right for them to support research in developmental biology. In general we are doing a fairly good job. Now, we are on the edge of some very important possibilities, and it would be a genuine tragedy to see this effort halted. However, we must remember that our activities take place in a larger context than just this room, or just the laboratory. The public is going to be heard and, given the confusing mix of misinformation they have received from biological scientists, I fear for what they may have to say.

SOME QUESTIONS

At the end of this talk, there was a brief exchange of comments with Dr. S.R. Kushner of the University of Georgia. While stating his agreement with much of what I said, he did want to come to the defense of some of the people who have spoken out against some of the things that may possibly be going on. He went on to say that scientists have not been that good at predicting what would happen in the future, and science fiction writers have been better; he added some illustrations. Therefore, while he agrees that gene therapy, based on what we know now, may not be possible, he feels it would be a mistake to rule out its ever happening. The same could be said about other scenarios produced by science fiction writers.

I replied that much of what science fiction writers have written has not come true (which was more of a retort than a reply). Kushner again said he agreed, but argued that we should not let our supposed rational approach to things inhibit our imaginations. In view of the incredible breakthroughs in the past 50 years, the next 50 years could render obsolete the kinds of approaches we are using right now and would open up entirely new vistas for dealing with genetic defects.

My reply was that our scientists are among the most imaginative people today; however, in addition to being imaginative, they must also be responsible.

I reproduce the essence of this exchange here because I believe Dr. Kushner has raised some important points that have been a source of confusion to both scientists and the public. Let me attempt to set down my own view below.

CODA

Science is not an imaginative subject. The reason is simple: for an idea to have scientific value, it must fit all the observable facts and also be consistent with a previously constructed theory; this theory being a useful algorithm that allows one to check rapidly whether a new idea is consistent with the total body of scientific knowledge. I agree with Jacob's view that science attempts to confront the possible with the actual. This is a more modest and limited conception, and I believe a more correct conception of what science is about.

However, this view itself is not in accord with the expressed opinions of many scientists. We often praise this or that individual as being a very "imaginative" or "creative" scientist. Others have said that science, underneath it all, is much like the arts – in that creative activity of the same kind is involved. This remark is based on a misunderstanding of the role of imagination in science. Imagination is essential to the elaboration of hypotheses – to dream up possibilities. As every experimental scientist knows, most possible explanations fail to fit the evidence. Imagination plays some role in the selection of experimental routes to test a given hypothesis. Usually there are many ways to go, and the trick is to pick a fast and sure route. But in the end the idea must fit the observations, and this must be demonstrated and repeated. In principle a scientific demonstration could be repeated by anyone, and there are well-developed statistical tests that guard against misinterpretation.

In spite of its achievements, science is a modest activity. It often spoils the fun. The most important and deeply held ideas or beliefs are often based on our imagination. This is well understood by every poet, lover, priest, and politician – and everyone else. This mental activity is the most distinguishing aspect of both man and mankind. Therefore, in view of the successes of science, it is tempting to elevate it into this higher, human realm. But as I said, science is not an imaginative subject, and attempts by scientists to make is so will lead to the destruction of science itself. One should remember that there is no assurance from heaven that science will go on forever. Since it can operate effectively only within an appropriate mode of civilization, it might well disappear again in a few hundred years, and reality once more would be viewed only in terms of our imaginations. Therefore, I would urge you to consider that the most responsible and socially useful role for a scientist today is to do science well; science fiction should be left to creative writers, who are doing their job very well indeed.

REFERENCES

1. Jacob, F., *Science* **196**, 1161-6 (1977).
2. Sonneborn, T.M., Editor, *The Control of Human Heredity and Evolution*, Macmillan, New York, 1965.
3. Wells, H.G., *War of the Worlds*, Penguin, New York, 1961.
4. Luria, S.E., *Life the Unfinished Experiment*, Scribner's, New York, 1973.
5. Eisinger, J., *Fed. Proc.* **34**, 1418-27 (1975).
6. Hotchkiss, R.D., *J. Hered.* **56**, 197-202 (1965).
7. Sinsheimer, R.L., *Eng. Sci.* (Dec. 1966).
8. Asimov, Isaac, *The End of Eternity*, Doubleday, New York, 1955.
9. Sinsheimer, R.L., *Eng. Sci.* (May 1968).
10. Davis, B.D., in *Identity and Dignity of Man*, Schenkman, Cambridge, MA, 1970.
11. Ausubel, F., Beckwith, J., and Janssen, K., Genetic engineering and the new eugenics, *Psychology Today* (June 1974).
12. Lederberg, J., Genetic engineering, or the amelioration of genetic defect, *The Pharos*, pp. 9-12 (Jan. 1971).
13. Abelson, P., Anxiety about genetic engineering, *Science* **173**, 285 (1971).
14. Kass, L.R., The new biology: what price relieving man's estate? *Science* **174**, 779-88 (1971).
15. Freeman, E., *The New York Times Magazine*, p. 84 (21 May 1972).
16. *The New York Times*, p. 1 (14 Oct. 1971).
17. Friedman, T. and Roblin, R., *Science* **175**, 949-55 (1972).
18. Freeze, E., Editor, *The Prospects of Gene Therapy, Fogarty International Center Conference Report,* DHEW Publ. No. (NIH) 72-61 (1971).
19. Lewin, B., New genes for old, *New Sci.* **54**, No. 792, 122-4 (20 Apr. 1972).
20. Ramsey, P., *Fabricated Man: The Ethics of Genetic Control*, Yale University Press, New Haven, 1970.
21. Levin, Ira, *The Boys From Brazil*, Random House, New York, 1976.
22. Etzioni, A., *Genetic Fix*, Maxmillan, New York, 1973.
23. Hamilton, M., *The New Genetics and the Future of Man*, W.B. Eardmans Pub. Co., Grand Rapids, 1972.
24. Jones, A. and Bodmer, W., Editors *Our Future Inheritance: Choice or Chance*, Oxford University Press, 1974.
25. Sinsheimer, R.L., in *Ethical Issues in Human Genetics*, pp. 341-51, B. Hilton and D. Callahan, Editors, Plenum, New York, 1972.
26. Singer, M. and Soll, D., *Science* **181**, 1114 (1973).
27. Berg, P. et al., Potential hazards of recombiant DNA molecules, *Science* **185**, 303 (1974).
28. Burnet, M., *Genes, Dreams, and Realities*, Basic Books, New York, 1971.
29. Koestler, A., *The Call Girls*, Random House, New York, 1973.

Document 7.8

February 18, 1977
Science (**195**:654)

Recombinant DNA: Fact and Fiction

Stanley N. Cohen

Almost 3 years ago, I joined with a group of scientific colleagues in publicly calling attention to possible biohazards of certain kinds of experiments that could be carried out with newly developed techniques for the propagation of genes from diverse sources in bacteria (*1*). Because of the newness and relative simplicity of these techniques (*2*), we were concerned that experiments involving certain genetic combinations that seemed to us to be hazardous might be performed before adequate consideration had been given to the potential dangers. Contrary to what was believed by many observers, our concerns pertained to a few very specific types of experiments that could be carried out with the new techniques, not to the techniques themselves.

Guidelines have long been available to protect laboratory workers and the general public against known hazards associated with the handling of certain chemicals, radioisotopes, and pathogenic microorganisms; but because of the newness of recombinant DNA techniques, no guidelines were yet available for this research. My colleagues and I wanted to be sure that these new techniques would not be used, for example, for the construction of streptococci or pneumococci resistant to penicillin, or for the creation of *Escherichia coli* capable of synthesizing botulinum toxin or diphtheria toxin. We asked that these experiments not be done, and also called for deferral of construction of bacterial recombinants containing tumor virus genes until the implications of such experiments could be given further consideration.

During the past 2 years, much fiction has been written about "recombinant DNA research." What began as an act of responsibility by scientists, including a number of those involved in the development of the new techniques, has become the breeding ground for a horde of publicists—most poorly informed, some well-meaning, some self-serving. In this article I attempt to inject some relevant facts into the extensive public discussion of recombinant DNA research.

The author is a molecular geneticist and Professor of Medicine at the Stanford University School of Medicine, Stanford, California 94305. This article is adapted from a statement prepared for a meeting of the Committee on Environmental Health of the California Medical Association, 18 November 1976.

Some Basic Information

Recombinant DNA research is not a single entity, but rather it is a group of techniques that can be used for a wide variety of experiments. Much confusion has resulted from a lack of understanding of this point by many who have written about the subject. Recombinant DNA techniques, like chemicals on a shelf, are neither good nor bad per se. Certain experiments that can be done with these techniques are likely to be hazardous (just as certain experiments done with combinations of chemicals taken from the shelf will be hazardous), and there is universal agreement that such recombinant DNA experiments should not be done. Other experiments in which the very same techniques are used—such as taking apart a DNA molecule and putting segments of it back together again—are without conceivable hazard, and anyone who has looked into the matter has concluded that these experiments can be done without concern.

Then, there is the area "in between." For many experiments, there is no evidence of biohazard, but there is also no certainty that there is not a hazard. For these experiments, guidelines have been developed in an attempt to match a level of containment with a degree of hypothetical risk. Perhaps the single point that has been most misunderstood in the controversy about recombinant DNA research, is that discussion of "risk" in the middle category of experiments relates entirely to hypothetical and speculative possibilities, not expected consequences or even phenomena that seem likely to occur on the basis of what is known. Unfortunately, much of the speculation has been interpreted as fact.

There is nothing novel about the principle of matching a level of containment with the level of anticipated hazard; the containment procedures used for pathogenic bacteria, toxic substances, and radioisotopes attempt to do this. However, the containment measures used in these areas address themselves only to known hazards and do not attempt to protect against the unknown. If the same principle of protecting only against known or expected hazards were followed in recombinant DNA research, there would be no containment whatsoever except for a very few experiments. In this instance, we are asking not only that there be no evidence of hazard, but that there be positive evidence that there is no hazard. In developing guidelines for recombinant DNA research, we have attempted to take precautionary steps to protect ourselves against hazards that are not known to exist—and this unprecedented act of caution is so novel that it has been widely misinterpreted as implying the imminence or at least the likelihood of danger.

Much has been made of the fact that, even if a particular recombinant DNA molecule shows no evidence of being hazardous at the present time, we are unable to say for certain that it will not devastate our planet some years hence. Of course this view is correct; similarly, we are unable to say for certain that the vaccines we are administering to millions of children do not contain agents that will produce contagious cancer some years hence, we are unable to say for certain that a virulent virus will not be brought to the United States next winter by a traveler from abroad, causing a nationwide fatal epidemic of a hitherto unknown disease—and we are unable to say for certain that novel hybrid plants being bred around the world will not suddenly become weeds that will overcome our major food crops and cause worldwide famine.

The statement that potential hazards could result from certain experiments involving recombinant DNA techniques is akin to the statement that a vaccine injected today into millions of people *could* lead to infectious cancer in 20 years, a pandemic caused by a traveler-borne virus *could* devastate the United States, or a new plant species *could* uncontrollably destroy the world's food supply. We have no reason to expect that any of these things will happen, but we are unable to say for certain that they will not happen. Similarly, we are unable to guarantee that any of man's efforts to influence the earth's weather, explore space, modify crops, or cure disease will not carry with them the seeds for the ultimate destruction of civilization. Can we in fact point to one major area of human activity where one can say *for certain* that there is zero risk? Potentially, we could respond to such risks by taking measures such as prohibiting foreign travel to reduce the hazard of deadly virus importation and stopping experimentation with hybrid plants. It is possible to develop plausible "scare scenarios" involving virtually any activity or process, and these would have as much (or as little) basis in fact as most of the scenarios involving recombinant DNA. But we must distinguish fear of the unknown from fear that has some basis in fact; this appears to be the crux of the controversy surrounding recombinant DNA.

Unfortunately, the public has been led to believe that the biohazards described in various scenarios are likely or probable outcomes of recombinant DNA research. "If the scientists themselves are concerned enough to raise the issue," goes the fiction, "the problem is probably much worse than anyone will admit." However, the simple fact is that there is no evidence that a bacterium carrying any recombinant DNA molecule poses a hazard beyond the hazard that can be anticipated from the known properties of the components of the recombinant. And experiments involving genes that produce toxic substances or pose other known hazards are prohibited.

Freedom of Scientific Inquiry

This issue has been raised repeatedly during discussions of recombinant DNA research. "The time has come," the critics charge, "for scientists to abandon their long-held belief that they should be free to pursue the acquisition of new knowledge regardless of the consequences." The fact is that no one has proposed that freedom of inquiry should extend to scientific experiments that endanger public safety. Yet, "freedom of scientific inquiry" is repeatedly raised as a straw-man issue by critics who imply that somewhere there are those who argue that there should be no restraint whatsoever on research.

Instead, the history of this issue is one of self-imposed restraint by scientists from the very start. The scientific group that first raised the question of possible hazard in some kinds of recombinant DNA experiments included most of the scientists involved in the development of the techniques—and their concern was made public so that other investigators who might not have adequately considered the possibility of hazard could exercise appropriate restraint. While most scientists would defend their right to freedom of scientific thought and discourse, I do not know of anyone who has proposed that scientists should be free to do whatever experiments they choose regardless of the consequences.

Interference with "Evolutionary Wisdom"

Some critics of recombinant DNA research ask us to believe that the process of evolution of plants, animals, and microbes has remained delicately controlled for millions of years, and that the construction of recombinant DNA molecules now threatens the master plan of evolution. Such thinking, which requires a belief that nature is endowed with wisdom, intent, and foresight, is alien to most post-Darwinian biologists (*3*). Moreover, there is no evidence that the evolutionary process is delicately controlled by nature. To the contrary, man has long ago modified the process of evolution, and biological evolution continues to be influenced by man. Primitive man's domestication of animals and cultivation of crops provided an "unnatural" advantage to certain biological species and a consequent perturbation of evolution. The later creation by man of hybrid plants and animals has resulted in the propagation of new genetic combinations that are not the products of natural evolution. In the microbiological world, the use of antimicrobial agents to treat bacterial infections and the advent of mass immunization programs against viral disease has made untenable the thesis of delicate evolutionary control.

A recent letter (*4*) that has been widely quoted by critics of recombinant DNA research asks, "Have we the right to counteract irreversibly the evolutionary wisdom of millions of years . . .?" It is this so-called evolutionary wisdom that gave us the gene combinations for bubonic plague, smallpox, yellow fever, typhoid, polio, diabetes, and cancer. It is this wisdom that continues to give us uncontrollable diseases such as Lassa fever, Marburg virus, and very recently the Marburg-related hemorrhagic fever virus, which has resulted in nearly 100 percent mortality in infected individuals in Zaire and the Sudan. The acquisition and use of all biological and medical knowledge constitutes an intentional and continuing assault on evolutionary wisdom. Is this the "warfare against nature" that some critics fear from recombinant DNA?

How About the Benefits?

For all but a very few experiments, the risks of recombinant DNA research are speculative. Are the benefits equally speculative or is there some factual basis for expecting that benefits will occur from this technique? I believe that the anticipation of benefits has a substantial basis in fact, and that the benefits fall into two principal categories: (i) advancement of fundamental scientific and medical knowledge, and (ii) possible practical applications.

In the short space of 3½ years, the use of the recombinant DNA technology has already been of major importance in the advancement of fundamental knowledge. We need to understand the structure and function of genes, and this methodology provides a way to isolate large quantities of specific segments of DNA in pure form. For example, recombinant DNA methodology has provided us with much information about the structure of plasmids that cause antibiotic resistance in bacteria, and has given us insights into how these elements propagate themselves, how they evolve, and how their genes are regulated. In the past, our inability to isolate specific genetic regions of the chromosomes of higher organisms has limited our understanding of the genes of complex cells. Now use of recombinant DNA techniques has provided knowledge about how genes are organized into chromosomes and how gene expression is controlled. With such knowledge we can begin to learn how defects in the structure of such genes alter their function.

On a more practical level, recombinant DNA techniques potentially permit the construction of bacterial strains that can produce biologically important substances such as antibodies and hormones. Although the full expression of higher organism DNA that is necessary to accomplish such production has not yet been achieved in bacteria, the steps that need to be taken to reach this goal are defined, and we can reasonably expect that the introduction of appropriate "start" and "stop" control signals into recombinant DNA molecules will enable the expression of animal cell genes. On an even shorter time scale, we can expect recombinant DNA techniques to revolutionize the production of antibiotics, vitamins, and medically and industrially useful chemicals by eliminating the need to grow and process the often exotic bacterial and fungal strains currently used as sources for such agents. We can anticipate the construction of modified antimicrobial agents that are not destroyed by the antibiotic inactivating enzymes responsible for drug resistance in bacteria.

In the area of vaccine production, we can anticipate the construction of specific bacterial strains able to produce desired antigenic products, eliminating the present need for immunization with killed or attenuated specimens of disease-causing viruses.

One practical application of recombinant DNA technology in the area of vaccine production is already close to being realized. An *E. coli* plasmid coding for an enteric toxin fatal to livestock has been taken apart, and the toxin gene has been separated from the remainder of the plasmid. The next step is to cut away a small segment of the toxin-producing gene so that the substance produced by the resulting gene in *E. coli* will not have toxic properties but will be immunologically active in stimulating antibody production.

Other benefits from recombinant DNA research in the areas of food and energy production are more speculative. However, even in these areas there is a scientific basis for expecting that the benefits will someday be realized. The limited availability of fertilizers and the potential hazards associated with excessive use of nitrogen fertilizers now limits the yields of grain and other crops, but agricultural experts suggest that transplantation of the nitrogenase system from the chromosomes of certain bacteria into plants or into other bacteria that live symbiotically with food crop plants may eliminate the need for fertilizers. For many years, scientists have modified the heredity of plants by comparatively primitive techniques. Now there is a means of doing this with greater precision than has been possible previously.

Certain algae are known to produce hydrogen from water, using sunlight as energy. This process potentially can yield a virtually limitless source of pollution-free energy if technical and biochemical problems indigenous to the known hydrogen-producing organisms can be solved. Recombinant DNA techniques offer a possible means of solution to these problems.

It is ironic that some of the most vocal opposition to recombinant DNA research has come from those most concerned about the environment. The ability to manipulate microbial genes offers the promise of more effective utilization of renewable resources for mankind's food and energy needs; the status quo offers the prospect of progressive and continuing devastation of the environment. Yet, some environmentalists have been misled into taking what I believe to be an antienvironmental position on the issue of recombinant DNA.

The NIH Guidelines

Even if hazards are speculative and the potential benefits are significant and convincing, wouldn't it still be better to carry out recombinant DNA experiments under conditions that provide an added measure of safety—just in case some of the conjectural hazards prove to be real?

This is exactly what is required under the NIH (National Institutes of Health) guidelines (*5*) for recombinant DNA research:

1) These guidelines prohibit experiments in which there is some scientific basis for anticipating that a hazard will occur. In addition, they prohibit experiments in which a hazard, although it might be entirely speculative, was judged by NIH to be potentially serious enough to warrant prohibition of the experiment. The types of experiment that were the basis of the initial "moratorium" are included in this category; contrary to the statements of some who have written about recombinant DNA research, there has in fact been no lifting of the original restrictions on such experiments.

2) The NIH guidelines require that a large class of other experiments be carried out in P4 (high level) containment facilities of the type designed for work with the most hazardous naturally occurring microorganisms known to man (such as Lassa fever virus, Marburg virus, and Zaire hemorrhagic fever virus). It is difficult to imagine more hazardous self-propagating biological agents than such viruses, some of which lead to nearly 100 percent mortality in infected individuals. The P4 containment requires a specially built laboratory with airlocks and filters, biological safety cabinets, clothing changes for personnel, autoclaves within the facility, and the like. This level of containment is required for recombinant DNA experiments for which there is at present no evidence of hazard, but for which it is perceived that the hazard might be potentially serious if conjectural fears prove to be real. There are at present only four or five installations in the United States where P4 experiments could be carried out.

3) Experiments associated with a still lesser degree of hypothetical risk can be conducted in P3 containment facilities. These are also specially constructed laboratories requiring double door entrances, negative air pressure, and special air filtration devices. Facilities where P3 experiments can be performed are limited in number, but they exist at some universities.

4) Experiments in which the hazard is considered unlikely to be serious even if it occurs still require laboratory procedures (P2 containment) that have for years been considered sufficient for research with such pathogenic bacteria as *Salmonella typhosa*, *Clostridium botulinum*, and *Cholera vibrio*. The NIH guidelines require that P2 facilities be used for work with bacteria carrying interspecies recombinant DNA molecules

that have shown no evidence of being hazardous—and even for some recombinant DNA experiments in which there is substantial evidence of lack of hazard.

5) The P1 (lowest) level of containment can be used only for recombinant DNA molecules that potentially can be made by ordinary biological gene exchange in bacteria. Conformity to even this lowest level of containment in the laboratory requires decontamination of work surfaces daily and after spills of biological materials, the use of mechanical pipetting devices or cotton plugged pipettes by workers, a pest control program, and decontamination of liquid and solid waste leaving the laboratory.

In other areas of actual or potential biological hazard, physical containment is all that microbiologists have had to rely upon; if the Lassa fever virus were to be released inadvertently from a P4 facility, there would be no further barrier to prevent the propagation of this virus which is known to be deadly and for which no specific therapy exists. However, the NIH guidelines for recombinant DNA research have provided for an additional level of safety for workers and the public: This is a system of biological containment that is designed to reduce by many orders of magnitude the chance of propagation outside the laboratory of microorganisms used as hosts for recombinant DNA molecules.

An inevitable consequence of these containment procedures is that they have made it difficult for the public to appreciate that most of the hazards under discussion are conjectural. Because in the past, governmental agencies have often been slow to respond to clear and definite dangers in other areas of technology, it has been inconceivable to scientists working in other fields and to the public at large that an extensive and costly federal machinery would have been established to provide protection in this area of research unless severe hazards were known to exist. The fact that recombinant DNA research has prompted international meetings, extensive coverage in the news media, and governmental intervention at the federal level has been perceived by the public as prima facie evidence that this research must be more dangerous than all the rest. The scientific community's response has been to establish increasingly elaborate procedures to police itself—but these very acts of scientific caution and responsibility have only served to perpetuate and strengthen the general belief that the hazards under discussion must be clear-cut and imminent in order for such steps to be necessary.

It is worth pointing out that despite predictions of imminent disaster from recombinant DNA experiments, the fact remains that during the past 3½ years, many billions of bacteria containing a wide variety of recombinant DNA molecules have been grown and propagated in the United States and abroad, incorporating DNA from viruses, protozoa, insects, sea urchins, frogs, yeast, mammals, and unrelated bacterial species into *E. coli*, without hazardous consequences so far as I am aware. And the majority of these experiments were carried out prior to the strict containment procedures specified in the current federal guidelines.

Despite the experience thus far, it will always be valid to argue that recombinant DNA molecules that seem safe today may prove hazardous tomorrow. One can no more prove the safety of a particular genetic combination under all imaginable circumstances than one can prove that currently administered vaccines do not contain an undetected self-propagating agent capable of producing cancer in the future, or that a hybrid plant created today will not lead to disastrous consequences some years hence. No matter what evidence is collected to document the safety of a new therapeutic agent, a vaccine, a process, or a particular kind of recombinant DNA molecule, one can always conjure up the possibility of future hazards that cannot be disproved. When one deals with conjecture, the number of possible hazards is unlimited; the experiments that can be done to establish the absence of hazard are finite in number.

Those who argue that we should not use recombinant DNA techniques until or unless we are absolutely certain that there is zero risk fail to recognize that no one will ever be able to guarantee total freedom from risk in any significant human activity. All that we can reasonably expect is a mechanism for dealing responsibly with hazards that are known to exist or which appear likely on the basis of information that is known. Beyond this, we can and should exercise caution in any activity that carries us into previously uncharted territory, whether it is recombinant DNA research, creation of a new drug or vaccine, or bringing a spaceship back to Earth from the moon.

Today, as in the past, there are those who would like to think that there is freedom from risk in the status quo. However, humanity continues to be buffeted by ancient and new diseases, and by malnutrition and pollution; recombinant DNA techniques offer a reasonable expectation for a partial solution to some of these problems. Thus, we must ask whether we can afford to allow preoccupation with and conjecture about hazards that are not known to exist, to limit our ability to deal with hazards that do exist. Is there in fact greater risk in proceeding judiciously, or in not proceeding at all? We must ask whether there is any rational basis for predicting the dire consequences of recombinant DNA research portrayed in the scenarios proposed by some. We must then examine the "benefit" side of the picture and weigh the already realized benefits and the reasonable expectation of additional benefits, against the vague fear of the unknown that has in my opinion been the focal point of this controversy.

References and Notes

1. P. Berg, D. Baltimore, H. W. Boyer, S. N. Cohen, R. W. Davis, D. S. Hogness, D. Nathans, R. Roblin, J. D. Watson, S. Weissman, N. D. Zinder, *Proc. Natl. Acad. Sci. U.S.A.* **71**, 2593 (1974).
2. S. N. Cohen, A. C. Y. Chang, H. W. Boyer, R. B. Helling, *ibid.* **70**, 3240 (1973); S. N. Cohen, *Sci. Am.* **233** (No. 7), 24 (1975).
3. If we accept the view that any natural barriers to the propagation of genetic material derived from unrelated species do not owe their existence to the intent of nature, we can reason that evolution has created and maintained such barriers because opportunities for genetic mixing occur in nature. Furthermore, we must conclude that limitations to gene exchange have evolved because the mixing of genes from diverse organisms is biologically undesirable—not in a moral or theological sense as some observers would have us believe—but to those organisms involved.
4. E. Chargaff, *Science* **192**, 938 (1976).
5. *Fed. Reg.* **41**(176) (9 September 1976), pp. 38426–38483.

Document 7.9
June 25, 1977
The New Republic (**176**:11)

We're going wild over bugs we've already eaten.

In Defense of DNA

by J. D. Watson

The fascination that the DNA molecule holds for so many has never, until recently, held the connotation of flirting with disaster. I, for one, have never given a moment's thought to whether my passion about the nature of the gene might be misplaced, much less a major danger to others or even to the future of mankind itself. Yet hardly a day now passes that I do not see the term "recombinant DNA" in some news article, and I am constantly being asked questions about the possible dangers of DNA research when I am in the company of students or even with our non-scientific friends. Answers to such questions are never simple, and now even less so. The vision of the hysterics has so peopled biological laboratories with monsters and super bugs that I often feel the discussion has descended to the realm of a surrealistic nightmare from which we will most surely soon awaken.

I am afraid, however, that it is not a nightmare from which we suffer, but a massive miscalculation in which we cried wolf without having seen or even heard one. Although from childhood we have been told that this is a silly way to proceed, we thought we were acting wisely because it was not an ordinary wolf we feared, one we could fight away with a club, but a creature so formidable that it could eat us up before we could even cry for help. The "super wolf" in this case is, of course, the new types of DNA molecules ("recombinant DNA") that can now be made in test tubes. The tools which have made such creatures possible is a newly discovered class of ordinary enzymes (the restriction enzymes) which make specific cuts in DNA molecules in a way that allows their genetic information to be specially rearranged. For example, the end product might be a new DNA molecule consisting of sections derived from both bacterial and human cells. Even more importantly, the means now exist to put these hybrid molecules back into bacteria, thereby creating new forms of bacteria capable of synthesizing human proteins together with their normal bacterial products.

If all goes as we now suspect it will, we should be able to produce, for example, human insulin in virtually inexhaustible quantities, as well as scores of other medically useful human proteins that are now in short supply. Naturally, many first-rate scientists have begun to play this game, for not only can we foresee a variety of important applied benefits, but recombinant DNA technology also provides a simple means to work out the detailed structure of the chromosomes of higher organisms. This opens up for the first time the possibility, for example, of detailed molecular characterization of the genes involved in the immunological response. This is a practical breakthrough of vast importance, and we can anticipate that in the next several years there will be a rash of discoveries which, without recombinant DNA, might not have been achieved within this century.

Simultaneously, however, with the arrival of this new technology, some of us began to wonder whether it might also have unexpectedly bad consequences, such as the creation of new types of organisms never yet subjected to the pressures of evolution, and which might have disease-causing potentials that we do not now have to face. In particular, we worried about (1) creation of bacteria selectively tailored to be resistant to all known antibiotics, or (2) insertion of the genes of tumor viruses into bacteria known to multiply in humans. Discussions first initiated at the 1973 Gordon Conference on Nucleic Acids were followed in April 1974 by the proposal by a number of molecular biologists, of which I was one, that there be a formal moratorium on these two types of experimentation until an international meeting could be held to discuss whether such experimentation did pose any plausible public health danger. It is important to stress that in making this proposal, we did not urge any restrictions

J.D. Watson is director of the Cold Spring Harbor Laboratory, a Nobel laureate and author of *The Double Helix*.

on most other recombinant DNA research because we believed it would be just as safe as experiments we now conduct routinely with many forms of viruses and bacteria.

Because those of us who signed the moratorium proposal were respected scientists, not known for environmental or political kookiness, we were taken seriously. Eleven months later, though, some 150 persons came together at the Asilomar Conference Center in California to make further proposals. Instead of discussing whether the limited moratorium should be continued, the debate focused on how widely it should be extended. The operative rule among participants was that while one's own experimental system was safe, there was no reason to assume one's neighbor's was. From the start I felt most out of tune, and almost no one took seriously my opening argument that since there was no effective way to regulate experimentation with the known pathogenic bacteria and viruses, how could we come up with any logical rules for assuring protection from dangers that we could measure even less well?

As we all know too well today, the result of the meeting was a near unanimous consensus that we must act as if virtually all forms of recombinant DNA technology were potentially dangerous and use devices for the physical containment of the new bacteria. There was also agreement that the specific bacteria (*E. coli* strain K12) to be used for such experiments should be especially tailored genetically so that even if they escaped into the outside world, they would have no chance to colonize the human digestive tract. All in all, most everyone, including the press, thought the affair was a smashing success, and the world of molecular genetics would gain infinite kudos for the deliberate self-restraint with which it had moved ahead.

I, however, left Asilomar feeling very uneasy. I believed the whole affair was a hasty rush into unjustified bureaucratic roadblocks that would set back the course of legitimate science, but my view was considered by some to be eccentric irresponsibility. I did not then, nor do I now, believe that all recombinant DNA research is necessarily totally safe. The future automatically entails risks and uncertainty, and no sane person rushes in directions where he anticipates harm to himself or others. Instead, we try to adjust our actions to the magnitude of risk. When no measurement is possible because we have never faced a particular situation before, we must not assume the worst. If we did, we would do nothing at all.

Much too late, I have come to believe that the two types of experiments under the original moratorium pose no real threat to the general public. Much too quickly, we concluded that it was dangerous to make bacteria which are resistant to many antibiotics or which *can* synthesize a deadly poison even though they normally don't. Instead, we should have focused upon the fact that most, if not all, bacterial species already exchange DNA with each other in nature—for example, through the infection process. Thus, if through recombinant DNA technology, we were to make an *E. coli* strain that, say, makes the cholera toxin, we are very likely repeating what nature has done many times in the past. There is every reason to believe that even if it did escape from the laboratory it would not pose any major public health threat. Even less convincing, especially in retrospect, were the arguments against putting the genes of tumor viruses into laboratory strains of bacteria. The argument here was that such cancer-gene-bearing bacteria might accidently colonize parts of the human body, releasing cancer-causing DNA that might pass into our cells and initiate a cancer. By the time of Asilomar, however, we already realized that many, if not all, of the so-called DNA tumor viruses were in fact ordinary animal viruses that routinely infect most of us early in life. By still unknown means, they remain latent in our bodies for the remainder of our lives, usually only expressing themselves as disease-causing agents under various physical conditions (such as Herpes virus-induced cold sores). So, the danger we face from our intestinal bacteria acquiring a little cancer virus DNA must be negligible compared to that we face every time we are infected with any of the innumerable DNA-containing viruses.

Why, then, did many first-rate scientists become upset at the thought that Paul Berg and his Stanford students wanted to place the genetic material from SV40 (a potentially cancer-causing monkey virus) within a strain of *E. coli,* called K12, known not to effectively colonize the human gut? I and many others, including finally Berg himself, took the matter seriously enough to initiate the chain of events that inexorably led to the regulatory mood at Asilomar. The best explanation for our intellectual sloppiness was suggested to me by Governor Jerry Brown of California who, after listening to me talk about recombinant DNA, asked me whether we had gotten into this mess because of "liberal guilt." It was this, and a paranoia about cancer, that affects most all of us.

The heart of our trouble comes not from silly left-wing agitation but from the fact that the molecular genetics establishment at Asilomar put the weight of its authority behind guidelines implying we could honestly predict that one form of recombinant DNA experimentation might carry more potential danger than another. In contrast, I, for one, saw no way to decide whether work on fruit fly DNA, or yeast DNA, or mouse DNA should be more or less restricted, if at all, and so found the Asilomar experience an exercise in the theater of the absurd. Particularly misguided was the placing of work with human DNA in the highest potential risk category, thereby restricting it to biological warfare-like facilities and insuring that almost no one in pure research could work with it. Yet this is a DNA to which our ordinary intestinal bacteria must be constantly exposed, for it is very hard to imagine that none of the human DNA released from the normal sloughing-off of dead gut cells does not occasionally enter neighboring bacterial cells and become integrated into their DNA.

The almost unanimous final consensus at Asilomar reflected oft-repeated speculation that human DNA might carry the genes of the so-called RNA tumor viruses. There is, in fact, no real understanding of why such cells harbor potential sources of trouble. No enlightened person, liberal or otherwise, would want responsibility for the slightest chance of increasing the incidence of human cancer. But, we should have considered that all vertebrate cells most likely carry many such genes as normal genetic components. Their real function must not be to cause cancer, and we should not have panicked about potential exposure to DNA sequences which we already possess.

Fortunately, there are now second thoughts about the matter, and there are efforts within the National Institutes of Health to downgrade the hysteria over test-tube-transferred human DNA. However, it will not be easy to convince people that anything has really changed, and the whole effort to rehabilitate human DNA must be seen by many as an attempt to place personal ambition ahead of the public good.

To say the least, we are in a rotten mess. We face the prospect that our miserable guidelines will be enacted into formal laws. And, their very existence creates fears that DNA research carries a profound threat to human existence. It does not help that some of our "best and brightest" say to each other that they now know that Asilomar was a tactical mistake, but then pop off to Washington to flatter legislators into thinking they are acting wisely in devoting massive attention to DNA.

Instead of going back to the question which should have dominated the Asilomar gathering—Is there a reason for formal guidelines?—our Congressmen are being asked to decide between two silly alternatives—(1) to either formalize the Asilomar nonsense at the level of the now-current NIH guidelines, and in so doing create another regulatory bureaucracy that will not only cost piles of money but legitimize a control over the direction of science that we have not seen since the Middle Ages, or (2) taking seriously the most extensive doomsday scenarios, effectively stop all recombinant DNA research. In fact, I do not know of any authority on infectious disease who takes the recombinant DNA doomsday scenario seriously.

So you might expect that our Congressmen would be deluged with mail from angry molecular biologists opposing either option. Instead, the dominant establishment attitude is that the matter is already so out of control that our only concern should be to contain our regulatory Frankenstein before it threatens all of biological research—before, for example, we have to obtain a "memorandum of understanding" every time we want to work with a new human cell line. This defeatist mood assumes that some federal law restricting recombinant DNA research is bound to come, and we should focus our efforts on seeing to it that it takes precedence over local laws. We should remember, however, that even a "reasonable" law that talks only about recombinant DNA is a perfect setup for a new round of local initiatives—say, to protect New Haven from unregulated work at Yale with potential human tumor viruses. And if that happens, as with the recombinant DNA story, we are likely to find that our opposition is for the most part led, not by individuals with any deep knowledge of or even fear of our work, but by persons who, for a myriad of reasons, do not like the fruits of science, if not of the intellect, and see us as the most vulnerable foe.

I thus believe that the ostrich-like approach of most molecular biologists to the events of the past two years can only lead to much more restrictive action against our freedom, not only to do good science, but to find the facts that may be necessary for the survival of our by now inherently advanced societies. While by now I find almost universal agreement among leading molecular biologists that the guidelines are a total farce, I sense that not one yet feels emotionally equal to the task of telling Senator Edward Kennedy and Representative Paul Rogers, chairmen of Congress's health subcommittees. Equally bad, by now so many top figures in our national scientific and health bureaucracy have defended the guidelines that, as with the Vietnam fiasco, none is willing to admit that he has been a naive fool. They are already being propelled by the momentum of their past nonsense to adopt their own form of the Cambodian escalation by trying to have the American guidelines enforced throughout the world.

Barring a miracle of enlightenment in Washington, in the White House if not in Congress, I fear that DNA research is condemned to be imprisoned in chains forged by research experts themselves. If the worst does occur in the United States, our only hope can be that Europe and the Eastern Bloc will not follow our trail of madness. Perhaps when we see that no Andromeda strain or doomsday bug has been produced abroad, but that progress continues to be made, the United States will resume free scientific inquiry.

Document 7.10

April 21, 1977
New York Times

Genetic Engineering

By René Dubos

Genetic engineering is not new. It began in 1930—almost half a century ago—when a technique was developed at The Rockefeller Institute Hospital to modify in the test tube the hereditary characteristics of the microbes that cause lobar pneumonia. This achievement led to the demonstration that the substance now known as DNA (deoxyribonucleic acid) is the carrier of heredity.

(Although I was present at the birth of genetic engineering, I never worked in the field and can only marvel at the new techniques of recombinant DNA that now make it possible to transfer genes from any kind of living thing to any other kind.)

In reality, it is misleading to state that "scientists are now able to create new forms of life." The genetic engineering of today means only the introduction into fairly simple microbes of a few genes derived either from other microbes or from higher forms of life, including man. However, the very fact that such genetic combination is possible has created hope and anguish—hope that the technique will be used to beneficial purposes, anguish at the thought that it might generate new diseases and lead to the manipulation of human nature.

For a long time, I had a quasi-religious hostility to experiments combining genes from different organisms, because I felt that this was countrary to the ways of nature. Like other biologists, I used to believe that gene exchange does not take place in nature, except during sexual conjugation between creatures of the same species. I now realize, however, that genetic exchange occurs frequently under natural conditions. For example, harmless bacteria readily incorporate gene fragments from other bacteria and thus become able to produce toxins, to resist antibiotics, to cause cancer in plants.

Contrariwise, virulent bacteria can lose the genes that make them dangerous. It is likely that bacteria also can incorporate DNA fragments from the animals and plants in which they reside.

Since gene exchange occurs widely in nature, I now feel that it is proper to do it experimentally under controlled conditions.

Even though human genes can be incorporated into bacteria and other organisms, changing human nature by genetic engineering seems to me impossible except perhaps for the correction of a few genetic maladies. Furthermore, my long experience in the field of infectious diseases has convinced me that laboratory techniques are most unlikely to produce strains of microbes that will start worldwide epidemics of new forms of disease. Billions upon billions of microbes exist everywhere in nature, constantly undergoing genetic changes, yet only very few can cause disease, and even then only when conditions are just right.

The great epidemics of the past and the increase in certain forms of cancer among us can be traced more to environmental factors than to new strains of microbes or viruses. For example, the severity of the 1918 flu pandemic was largely due to the conditions created by World War I.

I doubt that gene recombination in the laboratory will create microbes more virulent than those endlessly being created by natural processes. In any case, we know a great deal about handling dangerous microbes. I have been in daily contact with investigators working on smallpox, rabies, typhus, plague, cholera, tuberculosis, etc., yet have witnessed very few laboratory infections and never one responsible for starting an epidemic.

Like all human enterprises, genetic engineering may entail some unpredictable risk. I can only state that this is an acceptable risk because the potential benefits are large and the dangers purely hypothetical. Needless to say, countries will differ concerning what they regard as an acceptable risk. Some ban DDT and saccharin; others consider that the benefits derived from these substances justify their use.

The American people may decide that the safest course is a moratorium on DNA recombinant research, or regulations limiting it to a few institutions. Such policies would paralyze research in the United States but would not affect it in other countries.

DNA recombinant research can be carried out almost anywhere because it requires only simple equipment, unless put into a cumbersome straightjacket by unreasonable safety measures.

DNA recombinant research will go on in many places for several reasons. It enables even poor countries to engage in the most sophisticated field of biological science. It promises practical applications in medicine, agriculture and industry. It is one of the most exciting areas of knowledge, with large philosophical and scientific implications for the understanding of life.

René Dubos, a microbiologist and experimental pathologist, is professor emeritus of Rockefeller University.

"Somehow I was hoping genetic engineering would take a different turn."

Document 7.11

February 10, 1977
New England Journal of Medicine (**296**:324)

NOTES OF A BIOLOGY-WATCHER

The Hazards of Science

Lewis Thomas, M.D.

THE codeword for criticism of science and scientists these days is *hubris*. Once you've said that word, you've said it all; it sums up, in a word, all of today's apprehensions and misgivings in the public mind — not just about what is perceived as the insufferable attitude of the scientists themselves but, enclosed in the same word, what science and technology are perceived to be doing to make this century, this near to its ending, turn out so wrong.

Hubris is a powerful word, containing layers of powerful meaning, which is a peculiar thing when you consider its seemingly trivial history in etymology. It turned up first in popular English usage as a light piece of university slang at Oxford in the late 19th century, with the meaning of intellectual arrogance and insolence, applicable in a highly specialized sense to certain literary figures within a narrow academic community. But it was derived from a very old word, and as sometimes happens with ancient words it took on a new life of its own, growing way beyond the limits of its original meaning. Today, it is strong enough to carry the full weight of disapproval for the cast of mind that thought up atomic fusion and fission as ways of first blowing up and later heating cities, as well as the attitudes that led to strip-mining, off-shore oil wells, Kepone, food additives, SST's, and the tiny spherical particles of plastic recently discovered clogging the waters of the Sargasso Sea.

The biomedical sciences are now caught up with physical science and technology in the same kind of critical judgment, with the same pejorative word. Hubris is responsible, it is said, for the whole biologic revolution. It is hubris that has given us the prospects of behavior control, psychosurgery, fetal research, heart transplants, the cloning of prominent politicians from bits of their own eminent tissue, iatrogenic disease, overpopulation and recombinant DNA. This last, the new technology that permits the stitching of one creature's genes into the DNA of another, to make hybrids, is currently cited as the ultimate example of hubris. It is hubris for man to manufacture a hybrid, on his own.

This is interesting, for the word hybrid is a direct descendant of the ancient Greek word hubris. Hubris originally meant outrage; it was in fact a hybrid word from two Indoeuropean roots: *ud*, meaning out, and *gwer*, meaning rage. The word became *hydrida* in Latin, and was first used to describe the outrageous offspring from the mating of a wild boar with a domestic sow; these presumably unpleasant animals were, in fact, the first hybrids.

Since then the word hybrid has assumed more respectable meanings in biology, and also in literary and political usage. There have been hybrid plants and hybrid vigor, hybrid words and hybrid bills in parliament for several centuries. But always there has been a hidden meaning of danger, of presumption and arrogance, of risk. Hybrids are things fundamentally to be disapproved of.

And now we are back to the first word again, from hybrid to hubris, and the hidden meaning of two beings joined unnaturally together by man is somehow retained. Today's joining is straight out of Greek mythology: it is the combining of man's capacity with the special prerogative of the gods, and it is really in this sense of outrage that the word hubris is being used today. This is what the word has grown into, a warning, a code-word, a short hand signal from the language itself: if man starts doing things reserved for the gods, deifying himself, the outcome will be something worse for him, symbolically, than the litters of wild boars and domestic sows were for the ancient Romans.

To be charged with hubris is therefore an extremely serious matter, and not to be dealt with by murmuring things about anti-science and anti-intellectualism, which is what many of us engaged in science tend to do these days. The doubts about our enterprise have their origin in the most profound kind of human anxiety. If we are right, and the critics are wrong, then it has to be that the word hubris is being mistakenly employed, that this is not what we are up to, that there is, for the time being anyway, a fundamental misunderstanding of science.

I suppose there is one central question to be dealt with, and I am not at all sure how to deal with it although I am certain about my own answer to it. It is this: are there some kinds of information leading to some sorts of knowledge, that human beings are really better off not having? Is there a limit to scientific inquiry not set by what is knowable but by what we ought to be knowing? Should we stop short of learning about some things, for fear of what we, or someone, will do with the knowledge? My own answer is a flat no, but I must confess that this is an intuitive response and I am neither inclined nor trained to reason my way through it.

There has been some effort, in and out of scientific quarters, to make recombinant DNA into the issue on which to settle this argument. Proponents of this line of research are accused of pure hubris, of assuming the rights of gods, of arrogance and outrage; what is more, they confess themselves to be in the business of making live hybrids, with their own hands. The mayor of Cambridge, Massachusetts, and the Attorney General of New York have both been advised to put a stop to it, forthwith.

It is not quite the same sort of argument, however, as the one about limiting knowledge, although this is surely part of it. The knowledge is already here, and the rage of the argument is about its application in technology. Should DNA for making certain useful or interesting proteins be incorporated into *Escherichia coli* plasmids, or not? Is there a risk of inserting the wrong sort of toxins, or hazardous viruses, and then having the new hybrid organisms spread beyond the laboratory? Is this a technology for creating new varieties of pathogens, and should it be stopped because of this?

If the argument is held to this level, I can see no reason why it cannot be settled, by reasonable people. We have learned a great deal about the handling of dangerous microbes in the last century, although I must say that the opponents of recombinant-DNA research tend to downgrade this huge body of information. At one time or another, agents as hazardous as those of rabies, psittacosis, plague and typhus have been dealt with by investigators in secure laboratories, with only rare cases of self-infection of the investigators themselves, and none at all of epidemics. It takes some high imagining to postulate the creation of brand-new pathogens so wild and voracious as to spread from equally secure laboratories to endanger human life at large, as some of the arguers are now maintaining.

But this is precisely the trouble with the recombinant-DNA problem: it has become an emotional issue, with too many irretrievably lost tempers on both sides. It has lost the sound of a discussion of technologic safety, and begins now to sound like something else, almost like a religious controversy, and here it is moving toward the central issue: are there some things in science we should not be learning about?

There is an inevitably long list of hard questions to follow this one, beginning with the one that asks whether the mayor of Cambridge should be the one to decide, first off.

Maybe we'd be wiser, all of us, to back off before the recombinant-DNA issue becomes too large to cope with. If we're going to have a fight about it, let it be confined to the immediate issue of safety and security of the recombinants now under consideration, and let us by all means have regulations and guidelines to assure the public safety wherever these are indicated, or even suggested. But if it is possible let us stay off that question about limiting human knowledge. It is too loaded, and we'll simply not be able to cope with it.

By this time it will have become clear that I have already taken sides in this matter, and my point of view is entirely prejudiced. This is true, but with a qualification. I am not so much in favor of recombinant-DNA research as I am opposed to the opposition to this line of inquiry. As a long-time student of infectious-disease agents I do not take kindly the declarations that we do not know how to keep from catching things in laboratories, much less how to keep them from spreading beyond the laboratory walls. I believe we learned a lot about this sort of thing, long ago. Moreover, I regard it as a form of hubris-in-reverse to claim that man can make deadly pathogenic micro-organisms so easily. In my view, it takes a long time and a great deal of interliving before a microbe can become a successful pathogen. Pathogenicity is, in a sense, a highly skilled trade, and only a tiny minority of all the numberless tons of microbes on the earth has ever involved itself in it; most bacteria are busy with their own business, browsing and recycling the rest of life. Indeed, pathogenicity often seems to me a sort of biologic accident in which signals are misdirected by the microbe or misinterpreted by the host, as in the case of endotoxin, or in which the intimacy between host and microbe is of such long standing that a form of molecular mimicry becomes possible, as in the case of diphtheria toxin. I do not believe that by simply putting together new combinations of genes one can create creatures as highly skilled and adapted for dependence as a pathogen must be, any more than I have ever believed that microbial life from the moon or Mars could possibly make a living on this planet.

But, as I said, I'm not at all sure this is what the argument is really about. Behind it is that other discussion, which I wish we would not have to become enmeshed in. And I will tell you why.

I cannot speak for the physical sciences, which have moved an immense distance in this century by any standard, but it does seem to me that in the biologic and medical sciences we are still far too ignorant to begin making judgments about what sorts of things we should be learning or not learning. To the contrary, we ought to be grateful for whatever snatches we can get hold of, and we ought to be out there on a much larger scale than today's, looking for more.

We should be very careful with that word hubris, and make sure it is not used when not warranted. There is a great danger in applying it to the search for knowledge. The application of knowledge is another matter, and there is hubris in plenty in our technology, but I do not believe that looking for new information about nature, at whatever level, can possibly be called unnatural. Indeed, if there is any single attribute of human beings, apart from language, that distinguishes them from all other creatures on earth, it is their insatiable, uncontrollable drive to learn things and then to exchange the information with others of the species. Learning is what we do, when you think about it. I cannot think of a human impulse more difficult to govern.

But I can imagine lots of reasons for trying to govern it. New information about nature is very likely, at the outset, to be upsetting to someone or other. The recombinant-DNA line of research is already upsetting, not because of the dangers now being argued about but because it is disturbing, in a fundamental way, to face the fact that the genetic machinery in control of the planet's life can be fooled around with so easily. We do not like the idea that anything so fixed and stable as a species line can be changed. The notion that genes can be taken out of one genome and inserted in another is unnerving. Classical mythology is peopled with mixed beings — part man, part animal or plant — and most of them are associated with tragic stories. Recombinant DNA is a reminder of bad dreams.

The easiest decision for society to make in matters of this kind is to appoint an agency, or a commission, or a subcommittee within an agency, to look into the problem and provide advice. And the easiest course for a committee to take, when confronted by any process that appears to be disturbing people or making them uncomfortable, is to recommend that it be stopped, at least for the time being.

I can easily imagine such a committee, composed of unimpeachable public figures, arriving at the decision that the time is not quite ripe for further exploration of the transplantation of genes, that we should put this off for a while, maybe until next century, and get on with other affairs that make us less uncomfortable. Why not do science on something more popular?

The trouble is, it would be very hard to stop once this line was begun. There are, after all, all sorts of scientific inquiry that are not much liked by one constituency or another, and we might soon find ourselves with crowded rosters, panels, standing committees, set up in Washington for the appraisal, and then the regulation, of research. Not on grounds of the possible value and usefulness of the new knowledge, mind you, but for guarding society against scientific hubris, against the kinds of knowledge we're better off without.

If would be absolutely irresistible as a way of spending time, and people would form long queues for membership. Almost anything would be fair game, certainly anything to do with genetics, anything relating to population control, or, on the other side, research on aging. Very few fields would get by.

The research areas in the greatest trouble would be those already containing a sense of bewilderment and surprise, with discernible prospects of upheaving present dogmas. I can think of several of these, two current ones in which I've been especially interested, and one from the remote past of 40 years ago.

First, the older one. Suppose this were the mid-1930's, and there were a Commission on Scientific Hubris sitting in Washington, going over a staff report on the progress of work in the laboratory of O.T. Avery in New York. Suppose, as well, that there were people on the Commission who understood what Avery was up to and believed his work. This takes an excess of imagining, since there were vanishingly few such people around in the 1930's, and also Avery didn't publish a single word until he had the entire thing settled and wrapped up 10 years later. But anyway, suppose it. Surely, someone would have pointed out that Avery's discovery of a bacterial extract that could change pneumococci from one genetic type to another, with the transformed organisms now doomed to breed true as the changed type, was nothing less than the discovery of a gene; moreover, Avery's early conviction that the stuff was DNA might turn out to be correct, and what then? To this day, the members of such a committee might well have been felicitating each other on having nipped something so dangerous in the very bud.

But it wouldn't have worked in any case, unless they had been equally prescient about bacteriophage research and had managed to flag down phage genetics before it got going a few years later. Science can be blocked, I have no doubt of that, or at least slowed down, but it takes very fast footwork.

Here is an example from today's research on the brain, which would do very well on the agenda of a Hubris Commission. It is the work now going on in several laboratories here and abroad dealing with the endorphins, a class of small polypeptides also referred to as the endogenous opiates. It is rather a surprise that someone hasn't already objected to this research, since the implications of what has already been found are considerably more explosive, and far more unsettling, than anything in the recombinant-DNA line of work. There are cells in the brain, chiefly in the limbic system, which possess at their surfaces specific receptors for morphine and heroin, but this is just a biologic accident; the real drugs, with the same properties as morphine, are the pentapeptide hormones produced by the brain itself. Perhaps they are switched on as analgesics at times of trauma or illness; perhaps they even serve for the organization and modulation of the physiologic process of dying when the time for dying comes. These things are not yet known, but such questions can now be asked. It is not even known whether an injection of such pentapeptides into a human being will produce a heroin-like reaction, but that kind of question will also be up for asking, and probably quite soon since the same peptides can be synthesized with relative ease. What should be done about this line of research — or rather, what should have been done about it two or three years ago when it was just being launched? Is this the sort of thing we are better off not knowing? I know some people who might think so. But if something prudent and sagacious had been done, turning off such investigations at an early stage, we would not have glimpsed the possible clue to the mechanism of catatonic schizophrenia, which was published just this month from two of the laboratories working on endorphins.

It is hard to predict how science is going to turn out, and if it is really good science it is impossible to predict. This is in the nature of the enterprise. If the things to be found are actually new, they are by definition unknown in advance, and there is no way of foretelling in advance where a really new line of inquiry will lead. You cannot make choices in this matter, selecting things you think you're going to like and shutting off the lines that make for discomfort. You either have science, or you don't, and if you have it you are obliged to accept the surprising and disturbing pieces of information, even the overwhelming and upheaving ones, along with the neat and promptly useful bits. It is like that.

And even if it were possible to call most of the shots in advance, so that we could make broad selections of the general categories of new knowledge that we like, leaving out the ones we don't have a taste for, there would always be slips, leaks, small items of shattering information somehow making their way through. I have an example of this sort of thing in mind, a small item largely overlooked in its significance, a piece of news to match in importance, for what it tells us about ourselves and our relation to the rest of nature, anything else learned in biology during the past century. This is the astonishing tale — astonishing to my ears anyway — of the true nature of mitochondria and chloroplasts.

Between them, these organelles can fairly be said to run the place. They are, from every fair point of view, in charge. The chloroplasts tap the energy of the sun, and the mitochondria make use of it. Without them we might still have a world of microbes, but we could not have eukaryotic forms of life, nor metazoans, nor any of ourselves. Now, as it turns out, both of these can be viewed as living entities, organisms rather than organelles. The mitochondria live in our cells, and the chloroplasts in the cells of plants, as symbiotic lodgers. They replicate on their own, independently of nuclear division, with their own DNA and RNA, their own ribosomes, their own membranes, and these parts are essentially similar to the corresponding parts of bacteria and blue-green algae from which they are now believed to have descended. They are, in fairness, the oldest living inhabitants of the earth, and the least changed by evolution.

Well, this is the sort of knowledge I would call overwhelming, even overturning, in its implications. It has not yet sunk in, really, but when it does it is bound to affect our view of ourselves as special entities, as selves, in charge of our own being, in command of the earth. Another way to put it is that what we might be, in real life, is a huge collection of massive colonies of the most primitive kind of bacteria, which have adapted themselves for motile life in air by constructing around themselves, like a sort of carapace, all the embellishments and adornments of the modern human form. When you settle down to think a thought, you may think it is all your own idea, but perhaps it is not so. You are sharing the notion around, with more creatures than you could count in a lifetime, and they are the ones that turned the thought on in the first place. Moreover, there is more than a family resemblance, maybe even something like identity, between the mitochondria running your cells and those in control of the working parts of any cloud of midges overhanging a summer garden, or of seagulls, or the mouse in the basement, or all the fishes in the sea. It is a startling relationship, of such strange intimacy that none of us could have counted on before the facts began coming in. Would you prefer not to know about this? It is too late for that. Or would you prefer to stop it here and learn no more, leaving matters where they stand, stuck forever with one of the great ambiguities in nature, never to know for sure how it came out?

The only solid piece of scientific truth about which I feel totally confident is that we are profoundly ignorant about nature. Indeed, I regard this as the major discovery of the past 100 years of biology. It is, in its way, an illuminating piece of news. It would have amazed the brightest minds of the 18th-century enlightenment to be told by any of us how little we know, and how bewildering seems the way ahead. It is this sudden confrontation with the depth and scope of ignorance that represents the most noteworthy contribution of 20th-century science to the human intellect.

We are, at last, facing up to it. In earlier times, we either pretended to understand how things worked or ignored the problem, or simply made up stories to fill the gaps. Now that we have begun exploring in earnest, doing serious science, we are getting glimpses of how huge the questions are, and how far from being answered. Because of this, these are hard times for the human mind, and it is no wonder that we are depressed. It is not so bad being ignorant if you are totally ignorant; the hard thing is knowing in some detail the reality of ignorance, the worst spots and here and there the not-so-bad spots, but no true light at the end of any tunnel nor even any tunnels that can yet be trusted. Hard times, indeed.

But we are making a beginning, and there ought to be some satisfaction, even exhilaration, in that. The method works. There are probably no questions we can think up that can't be answered, sooner or later, including even the matter of consciousness. To be sure, there may well be questions we can't think up, ever, and therefore limits to the reach of human intellect that we will never know about, but that is another matter. Within our limits, we should be able to work our way through to all our answers, if we keep at it long enough, and pay attention.

I am putting it this way, with all the presumption and confidence that I can summon, to raise another, last question. Is this hubris? Is there something fundamentally unnatural, or intrinsically wrong, or hazardous for the species, in the ambition that drives us all to reach a comprehensive understanding of nature, including ourselves? I cannot believe it. It would seem to me a more unnatural thing, and more of an offense against nature, for us to come on the same scene endowed as we are with curiosity, filled to overbrimming as we are with questions, and naturally talented as we are for the asking of clear questions, and then for us to do nothing about it, or worse, to try to suppress the questions. This is the greater danger for our species, to try to pretend that we are another kind of animal, that we do not need to satisfy our curiosity, that we can get along somehow without inquiry and exploration, and experimentation, and that the human mind can rise above its ignorance by simply asserting that there are things it has no need to know. This, to my way of thinking, is the real hubris, and it carries danger for us all.

October 27, 1977
New York Review of Books (**24**:15)

The DNA Scare

Fear and DNA

P.B. Medawar

Playing God: Genetic Engineering and the Manipulation of Life
by June Goodfield.
Random House, 218 pp., $8.95

Biohazard
by Michael Rogers.
Knopf, 210 pp., $8.95

The Ultimate Experiment: Man-made Evolution
by Nicholas Wade.
Walker and Company, 162 pp., $8.95

I

It is the great glory as it is also the great threat of science that everything which is in principle possible can be done if the intention to do it is sufficiently resolute. Scientists may exult in the glory, but in the middle of the twentieth century the reaction of ordinary people is more often to cower at the threat.

Everybody will doubtless be dismayed to learn that it is possible in principle—and technically not even very difficult—to transform human beings into two sub-peoples: the one moiety brainy and comparatively beautiful—like the Eloi of H.G. Wells's famous journey into far future time—and the other moiety comparatively stupid but fitted by their docility and physical strength to do the dirty work and serve the others: Wells's Morlocks or Wagner's *Nibelungen*.

Why does not the mere possibility of this ultimate political prostration of mankind fill us with dismay? The reason is that the program I have just envisaged could have been embarked upon at any time in the past thousand years, merely by applying the most powerful of all forms of biological engineering—Darwinian selection—to a population—mankind—known by its open breeding system, lack of specialization, and rich resources of inborn diversity to be perfectly well able to respond to the empirical arts of the stockbreeder. The answer, in the form of a counter question, does something to explain why most biologists and laymen look rather coolly upon such attempts to curdle our blood: if these enormities have not been perpetrated or even seriously attempted hitherto by the comparatively straightforward and empirically well understood methods available for their execution, why should we now begin to fear that enormities as great or even greater will be executed by the much more costly and technically more difficult procedures of genetic engineering—by procedures which are conceptually well understood, to be sure, but are not yet anywhere near the level of proficiency in actual execution which the stockbreeder can command?

Nothing since the early days of atomic weaponry has caused so much dismay as the real or imagined threats associated with the development of genetical engineering and recombinant DNA research, the subjects of the books and papers under review.

At the root of all genetical engineering lies that which I described without qualification as the greatest scientific discovery of the twentieth century: that the chemical makeup of the compound deoxyribonucleic acid (DNA)—and in particular the order in which the four different nucleotides out of which it is assembled lie along the backbone of the molecule—encodes genetic information and is the material vehicle of the instructions by which one generation of organisms governs the development of the next. If the DNA message is altered, the effects of doing so are, in their context and of their kind, as far-reaching as the effects would be of altering the wording of congressional or parliamentary legislation or the wording of telegrams conveying diplomatic exchanges between nations. It is just such a process as this which in recent years has become possible by direct intervention—and to some degree at the experimenter's will—a situation quite different from the action of natural or artificial selection upon naturally occurring differences in the DNA messages characteristic of different organisms. The first process changes the genetic makeup of an organism, the second, acting upon naturally occurring genetic differences between organisms, changes the makeup of a population of organisms.

Introducing what has become the most talked about version of the first process—"recombinant DNA"—June Goodfield comments, "Very simply, it is the new technology that enables a scientist to take DNA from one organism and splice it onto DNA from another to create something absolutely new: new living molecules, new genes, and therefore new life."

The term "biological engineering"

need not of course be confined to that part of it which takes the form of an attempted manipulation of DNA. "Engineering" embraces all that accompanies and makes possible the translation of thought into action, and even if "thought" is too far-fetched a description of the acts of mind that underlie some of its manifestations, "biological engineering" can certainly be extended to include suspension of life in the deep-freeze, the attempt to rear babies to term outside the body, and other enterprises upon which the Medawars[1] have not thought "idiotic" too harsh a judgment.

Francis Bacon described the goal of the New Science of the seventeenth century as "the effecting of all things possible." The agents of this tremendous ambition were to be wise men and philosophers; he did not think there would ever come a time when people would do things merely because they *were* possible, yet that is exactly the mischief which the biochemist Erwin Chargaff, whom June Goodfield quotes, describes as the devil's doctrine: ***what can be done, must be done.*** It must have been some recognition of this source of temptation in themselves or in their weaker brethren that led to the remarkable resolutions of the Asilomar Conference of February 1975 in California at which scientists themselves proposed that certain types of experimentation with DNA should be abstained from. No literary folk have ever done as much. On the contrary: any suggestion that an author should not write exactly as he pleases no matter what offense he causes or what damage he does is greeted by cries of dismay and warnings that any such action would inflict irreparable damage on the human spirit and stifle forevermore the creative afflatus. Let us count it a mercy that we don't have to put up with this kind of talk from scientists; I mean, put up with the argument that the discovery of the truth is a complete justification for anything they may choose to do.

Although it was historically the most important, the Asilomar Conference of 1975 is not the only evidence of an awareness of possible evils acute enough to prompt scientists to accept guidance or impose upon themselves a censorship restricting their freedom to do exactly what they please. The National Institutes of Health have issued guidelines on the prosecution of recombinant DNA research and the British Medical Research Council has issued a cautionary document on genetic manipulation guided largely by the report of Lord Ashby's Working Party on this subject.[2] The Federation of American Scientists has issued a thoughtful and gravely worded public interest report[3] on the subject and the New York Academy of Sciences has devoted a symposium volume to a conference on the ethical and scientific problems raised by the human uses of molecular genetics.[4] At this conference Daniel Callahan asked, "How, then, are we to possess power without being possessed by it?" adding that this was the fundamental question underlying the problem of ethical responsibility in science. Lord Acton and others have pointed out that the same is true of political action. Callahan is not one to blame the weapon for the crime and he says that "if the quest for scientific knowledge is to be condemned because some of that knowledge may be misused, then so must the quest for all knowledge." Again, "there is no special responsibility applying to scientists that does not apply to others."

There was this difference though: scientists were now more fully cognizant than ever before of the way in which innocent-seeming and intrinsically inoffensive experimentation may lead to disastrous consequences. It was therefore, Callahan said, a special obligation upon a scientist to envisage what consequences of his work were *conceivable* and to share these misgivings with his colleagues. I believe that it is just this attitude which underlies the present unease of biologists about what the consequences of molecular genetic engineering may be.

In his book *Biohazard*, Michael Rogers does not plunge us right into the middle of things but explains carefully and intelligibly the classical researches that provided the conceptual foundations of modern genetic engineering, making special mention of Archibald Garrod, who first identified the so-called "inborn errors of metabolism" that occur because the body has a missing or defective gene, and of the classical experimental researches of Beadle and Tatum on the bread mold *Neurospora crassa* showing the connection between the action of genes and that of enzymes. Garrod's work and the *Neurospora* work represent some of the finest science of the twentieth century. From there he proceeds, justly and inevitably, to the dramatic and often recounted work on pneumonia bacteria by O.T. Avery and his colleagues in the Rockefeller Institute. These brilliant experiments first revealed that the genelike agent responsible for transforming certain bacteria from being non-lethal to lethal was none other than deoxyribonucleic acid—DNA for short—an abbreviation Rogers is sanguine enough to believe has now entered the vernacular. It is especially pleasing to see the prominence given to the name of a man, O.T. Avery, who deserves type as big and lights as bright as those of anyone who helped to tell the great story of DNA. Rogers, Wade, and Goodfield tell the same story of course: it is a good story and all three tell it well and in much the same way, though Goodfield's *aperçus* are the most personal.

It will now be helpful to take evidence from a variety of different well-informed sources.

Nature, the world's foremost scientific newspaper, has not stood aloof from the controversy. On the contrary, looking back over the "Recombinant DNA Debate Three Years On,"[5] an editorial declares that:

> ...information generated during

---

[1] *The Life Science*, P.B. and J.S. Medawar (Harper & Row, 1977).

[2] *Report of the Working Party on the Experimental Manipulation of the Genetic Composition of Micro-organisms*, Cmnd. 5880 (London: Her Majesty's Stationery Office, 1975), pp. 11-12.

[3] *FAS Public Interest Report*, Vol. 29, No. 4, Washington, DC, April 1976.

[4] *Ethical and Scientific Issues Posed by Human Uses of Molecular Genetics*, Annals of the New York Academy of Sciences, Vol. 265, January 1976.

[5] "Recombinant DNA Debate Three Years On," *Nature* (London), Vol. 268, July 21, 1977, p. 185.

the past three years indicates that the potential hazards associated with gene-splicing experiments may be more remote than once believed. For example, a special meeting of scientists and health experts, convened by NIH earlier this month, concluded that there is virtually no chance that recombinant DNA experiments could touch off an uncontrollable epidemic.

Nature goes on to cite Dr. Roy Curtiss, a respected microbiologist from the University of Alabama, as having written after much experimentation with the laboratory strains of the bacillus *E. Coli* that are being used in genetic research:

> I have gradually come to the realization that the introduction of foreign DNA into EK1 and EK2 host-vectors offers no danger whatsoever to any human being.

A more serious danger, maybe, is that the allegedly hazardous nature of the work may induce grant-giving agencies to impede the development of molecular biology or, more likely, to give molecular biologists seemingly valid reasons why their patrons should pull the purse strings together just when authentic supplicants are peering eagerly inside. A statesmanlike frown is accordingly directed by *Nature* at Senator Edward Kennedy's health subcommittee which is engaged in devising restrictive legislation that could possibly impede worthwhile research.

The Federation of American Scientists has a long record of service to the community, and the article "Splitting Atoms and Transplanting Genes," in its recent *Public Interest Report*, very properly reminds us of its stalwart services to the nation in making sure that the hazards of atomic energy became widely known. It now sees it as part of its function to do as much for recombinant DNA research, but far from holding up the profession to public obloquy, the FAS writes of it rather handsomely:

> The researchers have behaved with unprecedented restraint and caution. Raising the issue themselves; bringing it to public attention; urging the voluntary deferral of various experiments; and debating the hazards in full public view, represents four quite different and thoroughly commendable steps. In addition, most have, quite surprisingly, been able to come to agreement on a set of guidelines that have grown steadily more stringent—even while many of the researchers have grown more sanguine about the dangers. This is a tribute to the statesmanship of their leaders. It is no surprise that now they want to go ahead with research which all observers agree is filled with promise, and which promises tremendous assistance in understanding biology. They only ask a "yellow" light—the right to proceed with caution.

Among the hazards the FAS calls attention to is the accidental escape of potentially dangerous organisms as yet unknown in nature. The FAS seems to fear that the body's immunological system would be confounded by such unknown organisms. This fear of the unknown because it *is* unknown is not really justified. Human beings for example are perfectly capable of mounting immunological reactions against organisms new to them or even against chemical compounds which they have never met before—which, indeed, have not yet been invented. It is a misunderstanding of physiology to suppose that immunological-like neurological reactions depend at least in part on prior experience: there is after all always a first time we are confronted with any disease-causing organism but we do not necessarily succumb to it. The FAS goes on to say:

> The basic current hazard is the introduction into bacteria of genes which make the bacteria more dangerous. In the simplest case, such genetic changes might give one strain of bacteria the resistance to antibiotics that exists in other strains; thus some such antibiotic as penicillin might suddenly find that strains of bacteria that cause pneumonia had become resistant to its application.

It is notorious, though, that this process has been going on since penicillin was first introduced into medical practice and used more frequently and in larger doses than immediate needs called for. The appearance of antibiotic resistant strains of formerly susceptible bacteria is a typical evolutionary process. Although the existence of penicillin-resistant strains of bacteria is a major nuisance, it does not portend widespread disaster: rather it puts biologists on their mettle to find ways around the problem.

The FAS draws special attention to and endorses the main conclusions of the Working Party under Lord Ashby of the benefits and possible risks of genetic engineering. The Working Party's conclusions are worth setting out anew:

> We now have to declare our assessment of the potential benefits and practical hazards of using the techniques we have described. We reiterate our unanimous view that the potential benefits are likely to be great. The most substantial (though unpredictable) benefit to be expected from the techniques is that they may lead to a rapid advance in our detailed understanding of gene action. This in turn might add substantially to our understanding of immunology, resistance to antibiotics, cancer, and other medically important subjects.
>
> Furthermore, application of the techniques might enable agricultural scientists to extend the climatic range of crops and to equip plants to secure their nitrogen supply from the air. Another possible application is that segments of DNA, selected because they are templates for valuable products such as hormones, antigens or antibodies, might be produced in bulk by multiplying them in culture of *E. coli*: this would be of great benefit to medicine. And it is not inconceivable that the technique might ultimately lead to ways to cure some human diseases known to be due to genetic deficiency.
>
> In discussing the hazards of these techniques we have to distinguish between the risk to workers in the laboratory and the risk to the public. Many scientists are engaged on potentially hazardous research (using radioactive materials, or unstable chemicals, or pathogens). They and those who work with them are trained to take precautions; accidents are rare and they do not spread. But if the danger is one which might not be contained within the laboratory, the need for precaution is much greater and the public have a right to seek assurances that they are not at risk.
>
> Fortunately there are precedents for making such assurances. In the production of some vaccines, in public health and hospital laboratories, in research institutes for the study of infectious disease, it is essential to handle pathogenic organisms, some of them extremely dangerous. Accordingly, there is a well developed strategy of containment for these hazardous operations.... The dramatic response to any failure in containment illustrates how rare such failures are. A recent example of this is the enquiry which followed an outbreak of smallpox in London in 1973.

In short, the potential benefits of recombinant DNA research are great and the precautions the experiments call for must be commensurate with the magnitude of the risks involved.

I once had the pleasure of a lengthy formal discussion with the late Jacques Monod, at that time the Director of the Institut Pasteur, about a number of biological problems having to do with the threat and promise of genetical engineering—a subject upon which he was as well qualified as anybody in the world to express an authoritative opinion. We agreed that both the threats and the promises were greatly exaggerated and that the realization of both good and bad dreams was a very much more difficult exercise than it was commonly assumed to be, but at the same time Monod made an exception of cloning—the production of an indefinitely large number of replicas of some chosen human type.

Cloning was a definite possibility, he believed, as many others do too. The procedure has nothing to do with the recombinant DNA, however; it is biological engineering in the wider sense discussed above. To get it into perspective I should like to run over the procedure that would have to be adopted if cloning were to succeed. The first step would be to wash out from the fallopian tube a fertilized and therefore activated human egg—an egg developmentally ready to go. The egg would be stored in a cool, sterile nutrient medium outside the body and then manipulation could begin. If toads and newts are anything to go by, the egg's own nucleus could be replaced by a nucleus from an ordinary body cell (a lymphocyte nucleus, mayhap) from the tissues of the individual chosen for indefinite replication. The egg would then be maintained under conditions which allowed it to undergo a number of successive cell divisions—a process almost exactly analogous to twinning as it may sometimes occur *in vivo*.

That would only be the beginning of it, however, because for each such daughter egg to develop into a human being it would be necessary to find a woman whose uterus had been prepared by hormones in such a way that the daughter egg transplanted into it would continue with cell division and eventually attach itself to the uterine wall—"implantation" is the technical word. The embryo might or might not go to

term; if it did, it would necessarily have the same genetic makeup as the individual whose cell nucleus substituted for the nucleus of the original egg.

Anybody with any experience of experimental pathology—and the rival attraction of less exacting pursuits means that their number is getting less and less—knows that to carry through this program and to overcome all the misadventures that could so easily befail it would require a degree of organization that would make the mobilization and deployment of an army seem like running a Sunday school picnic. Even supposing a grant-giving agency composed mainly of wealthy simpletons could be found to support such a foolish enterprise, the very many misadventures known by all experimentalists to beset such a scheme would almost certainly prevent its being realized. We need not worry then about the difficulty of finding any one human being whose indefinite replication could be thought of with equanimity, for considered as a whole the enterprise is simply not on.

No appraisal of genetic engineering would be fair unless Erwin Chargaff were called upon. Chargaff was one of those who played a leading part in the discoveries that led to our modern understanding of DNA, and his part too, like Avery's, is not as well known as it ought to be. In writing "On the Dangers of Genetic Meddling"[6] Chargaff is very skeptical about the overflowing cornucopia of advances in medicine and human welfare which, it has been alleged, can grow out of the use of gene-splicing techniques—benefactions thought to include the repair of human genetic defects (a procedure very far beyond our present competence). Of this project I have said:

> It is mentioned in the same spirit as that in which a young biologist seeking funds to study the growth of sea cucumbers in a pleasant seaside resort urges his patrons to believe that such an investigation will throw a flood of light on the transformation from the normal to the malignant cell: it is a harmless form of window dressing that all grant-giving bodies understand and allow for.[7]

Chargaff declares that the genetic engineers are not nearly so proficient as they are given the credit for being about the splicing of eukaryotic DNA into DNA of microorganisms. (Eukaryotic DNA is the DNA of organisms, such as animals and higher plants, in which genetic material is marshaled and structurally organized in the form of chromosomes, the nucleic acid being combined with a basic protein to form a salt-like compound):

> Most of the experimental results published so far in this field are actually quite unconvincing.... It appears that the recombination experiments in which a piece of animal DNA is incorporated into the DNA of a microbial plasmid are being performed without a full appreciation of what is going on.

Chargaff is skeptical of the long-term efficacy of orthodox containment procedures for the possible escape of pathogens, and he asks why molecular geneticists have chosen as the subject of their experiments an organism *Escherichia coli*, the colon bacillus, which has for so many millennia been living in a state of symbiosis with man. "The answer is that we know so much more about *E. coli* than about anything else, including ourselves." He is right: so much knowledge and know-how is vested in *E. coli* that there is little likelihood of its being supplanted as a subject of experiment. In any event, so the patrons of *E. coli* argue, the laboratory organism has now been so modified in the course of prolonged culture outside the body that it no longer qualifies to be considered a regular member of the flora in our gut.

Clifford Grobstein,[8] well known for his sensible and temperate views, deplores the polarization of the recombinant DNA debate into an antithesis be-

[6] "On the Dangers of Genetic Meddling," Erwin Chargaff, *Science*, Vol. 192, June 4, 1976, pp. 938-940.

[7] "The Scientific Conscience," P.B. Medawar, *Hospital Practice*, July 1976, p. 17.

tween "best-case" and "worst-case scenarios." The worst-case scenario he envisages comprises: worldwide epidemics caused by newly created pathogens; the triggering of catastrophic ecological imbalances; the power to dominate and control the human spirit.

The last of these imagined dangers rather surprised me, for it seemed to me to be on all fours with H.G. Wells's Eloi/Morlock bad dream referred to earlier. Certainly nothing much more horrible can be envisaged than a procedure which not only fills the mind of man with untruth and misconception but leads to an active resistance to new learning and to anything that might conduce to improvement. Yet here again the technology that puts these grim possibilities within our power has also been known for five thousand years or more: it is known as "education," and it too has its brighter side, for whatever procedures may persuade us to approve evil can in principle also be used to make us reprobate evil and rejoice in and embrace the good.

Writing of the disquiet of the laity Grobstein makes it clear, though, that "the fear is not so much of any clear and present danger as it is of imagined future hazards." Grobstein fears that physical containment and the associated safety precautions reveal something of a Maginot Line mentality, for what is needed is research that will evaluate these hazards precisely, so that we know where we stand and shall not find ourselves standing still.

James D. Watson is well known to have a somewhat messianic conception of his role in the great revolution of molecular genetics, and it was hardly to be expected that he would remain silent amid the clamor of discussion on recombinant DNA. He says that the Asilomar Conference made him uneasy, and he now declares:

> I did not then, nor do I now, believe that all recombinant DNA research is necessarily totally safe. The future automatically entails risks and uncertainty, and no sane person rushes in directions where he anticipates harm to himself or others. Instead, we try to adjust our actions to the magnitude of risk. When no measurement is possible because we have never faced a particular situation before, we must not assume the worst. If we did, we would do nothing at all.[9]

I do not think Watson is being unduly sanguine and I specially applaud his choice of the word "sane."

II

Having now taken evidence from various quarters we may turn to the three works specifically under review. Nowadays laymen need not be told that "Cry havoc!" attracts more attention than the nightwatchman's reassuring "All's well, all's well." Happily none of these three books is disfigured by sensationalism; however there is something a little breathy about them all. None of them is definitive or pretends to be: these are interim reports: a definitive treatise could only be written from a height which none of the three authors can command.

Goodfield, though, turns her lack of inside knowledge to advantage by describing how she apprenticed herself to a laboratory in which recombination experiments were taking place. I liked specially her delighted description of the winding out of the exquisitely beautiful DNA fibers on a glass rod after they had been precipitated from solution by the addition of alcohol. It is not an essential part of her narrative, of course, but I sympathize entirely with her wanting to bring it in because when I myself prepared DNA for immunological purposes I can remember cruelly boring my colleagues by calling upon them to witness the very process June Goodfield describes.

Nicholas Wade might say that this episode illustrates his contention that "gene splicing is so simple a technique that for most present purposes it requires only a few dollars worth of special materials, all commercially available, and access to a standard biological laboratory." I think this is a misjudgment that reminds me of a prominent sociologist's published contention (I shall not say where) that the manufacture of atomic bombs now lies within the abilities of a high-school student. It could equally well be said that appendectomy is a remarkably simple operation requiring no more facilities than are available in a quite ordinary hospital. But its execution requires a knowledge and know-how—the biological or surgical equivalent of worldly wisdom—which puts it for all practical purposes far beyond the reach of any ordinary villain or casual mischief-maker—a villain who collected appendixes as others collect stamps.

June Goodfield's account has the merit of making it clear by implication why the conferment of antibiotic resistance is such a favorite exercise with genetical engineers. The reason is that it is not much good doing an experiment or modifying its procedure unless one knows whether the experiments work, or work better than before. When the conferment of antibiotic resistance is the transformation attempted, the organisms in which the transformation has been successful can be isolated very easily from a population that may be as diverse as the population of Times Square on a Saturday night (Goodfield's image).

Each of these three books is good and since there is general agreement in the nature of the promises and the threats it would be idle to single out any one of them; for each has special merits. They agree, too, on the history of discoveries bearing on DNA, though Rogers goes back as far as Miescher in the 1870s—the man who first extracted the stuff long called nuclein from pus (one good mark, if we were having a competition). This historical excursion will certainly earn him the contempt of those semi-literates who regard any work done earlier than in the past year or two as of merely antiquarian interest.

Writing of safety precautions in laboratories handling potentially dangerous materials such as tumor viruses Wade quotes W. Emmett Barkley, the biologi-

[8] "The Recombinant-DNA Debate," Clifford Grobstein. *Scientific American*, Vol. 237, No. 1, July 1977, pp. 22-33; "Recombinant DNA Research: Beyond the NIH Guidelines," *Science*, Vol. 194, December 10, 1976, pp. 1133-1135.

[9] "In Defense of DNA," J.D. Watson, *The New Republic*, June 25, 1977, pp. 11-14.

cal safety expert at the National Cancer Institute, in these terms:

> "In the majority of labs we visit we see things that ought to be corrected. The greatest offenders are university labs, not industrial labs. Most people working with tumor viruses have been exposed to some extent."

Barkley's is the cry of safety officers throughout the world—in factories no less than in laboratories. I offer it *gratis* to some graduate actuary or sociologist on the lookout for a PhD degree that he should study the life expectancy of safety officers in factories and laboratories. I suspect they die prematurely of diseases of stress.

Incorrigible though their clients seem to be, however, we must keep it firmly in mind that for every steel worker who falls into the blast furnace, and every cider maker who dissolves, boots and all, in raw apple juice (rich in frighteningly powerful enzymes), hundreds and hundreds do not. The parallel is not facetious, because no one is more gravely and immediately at risk of the hazards to which they are believed to be about to expose the public than the people who actually carry out supposedly hazardous experiments. I don't think the general public need take grave alarm until the inmates of institutions of genetic engineering themselves begin to fall by the way.

A further consideration that will influence the worldly wise is this: genetic engineers would very much *like* to confer upon microorganisms the ability to manufacture, in copious amounts, human insulin and the anti-viral agent called interferon, now being used in the treatment of some cancers.

When the engineers have demonstrated to everybody's satisfaction that they can do on purpose what they very much want to do, then will be the time to reappraise very critically the dangers consequent upon their inadvertently doing what they do not want to do anyway.

The large-scale manufacture of either human insulin or interferon would be a very great benefaction to mankind, for the trouble with interferon at the moment—so often judged therapeutically disappointing—is that there isn't enough to give it in dosages large enough for a clinical trial of adequate scale. Even penicillin did not finally triumph until it became possible to administer it in doses of the order of megaunits.

In Wade I came across for the first time the idea that nitrifying enzymes might conceivably be incorporated into plants that normally lack them, thus making it possible for them to capture from the atmosphere the nitrogen necessary for their growth and making them independent of added fertilizers (which are essentially compounds of nitrogen). The notion is not impossibly far fetched because some plants can be raised into whole organisms from single isolated cells. But here too I do very deeply sympathize with laymen and legislators who are trying to make sense of this whole strange farrago of pipe dreams and nightmares.

For their excess of fearfulness, laymen have only themselves to blame and their nightmares are a judgment upon them for a deep-seated scientific illiteracy which manifests itself in two ways.

In the first place the public deserve nothing but contempt for allowing themselves to be dupes of that form of science fiction which is our modern equivalent of the Gothic romances of Mary Shelley and Mrs. Ann Radcliffe; for being taken in, that is to say, by that trusty serio-comic character, the mad scientist, who to the accompaniment of peals of maniacal laughter cries out with a strong Central European accent, "Soon ze whole vorld vill be in my power."

The second reason for their excess of fearfulness is this: that because imaginative writing is the only form of creative activity most people know, even educated laymen have no idea of the width of the gap between conception and execution in science. A writer who hits on a good idea—or even a composer who thinks of or, like Sullivan, overhears a good tune—can take up pencil and paper and write it down; he does not have to sue for bench space in a laboratory or send in five copies of an application explaining what his poem is going to be about, how many sheets of paper it will occupy, what imagery it is going to be clothed in, or how mankind will benefit by its completion. But when a scientist has an idea he has merely reached the beginning of a long haul which will certainly involve an appeal for funds which he may easily not get. He cannot simply walk into his laboratory with a purposeful and dedicated look on his face and execute the idea he has in mind.

The existence of this large gap means in effect that the execution of recombinant DNA research depends very largely upon political decisions. I do not use the word "political" in the sense that it would depend upon congressional or parliamentary legislation but simply in the sense that the project and the means of executing it depend on decisions that are not the scientist's alone: they will depend at least in part upon peer judgment and on the policy decisions of an independent grant-giving body. But, it will be objected, many of those responsible for the decisions are themselves scientists; all right, but if one mad scientist is rare, a committee of scientists, all mad, is very much more improbable still. The existence of this very wide gap between conception and execution is that which allows the interposition of wiser counsels and restraining hands between the scientist's idea whether bright or foolish and the possibility of its being put into effect.

So much then for the etiology and cultural history of the forebodings that cause so much disquiet among laymen. To the professional scientist these suspicions of his competence and probity are most disquieting. In one of a number of wise discourses on civilization Sir Kenneth Clark remarked that all great advances in civilization are based upon *confidence*. Although I have tried to explain it, I find it difficult to excuse the lack of confidence that otherwise quite sensible people have in the scientific profession, among whom sanity is much more widely diffused than seems to be generally realized. Scientists want to do good—and very often do. Short of abolishing the profession altogether no legislation can ever effectively be enforced that will seriously impede the scientists' determination to come to a deeper understanding of the material world. □

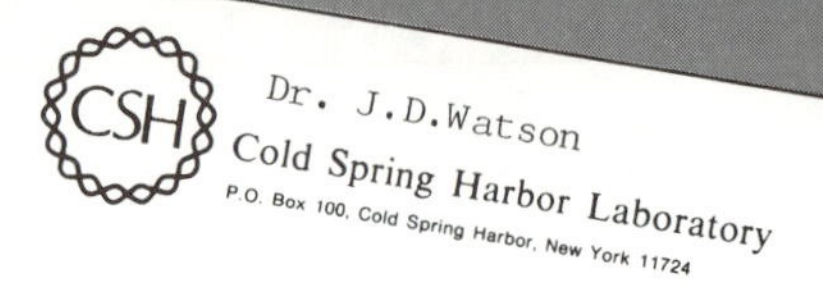

Dr. J.D.Watson
Cold Spring Harbor Laboratory
P.O. Box 100, Cold Spring Harbor, New York 11724

Dr. Norton Zinder
The Rockefeller University
1230 York Avenue
New York, N.Y

N. D. ZINDER
THE ROCKEFELLER UNIVERSITY
1230 York Avenue • New York, N.Y. 10021

FIRST CLASS

DR. STANLEY COHEN
UNIVERSITY SCHOOL OF MEDICINE
MEDICINE
94305

Dr. J.D.Watson
Cold Spring Harbor Laboratory
P.O. Box 100, Cold Spring Harbor, New York 11724

NEW YORK
SEP-19'77
N.Y.
U.S. POSTAGE
27

Dr. Stanley Cohen
Stanford University School of Medicine
Medicine

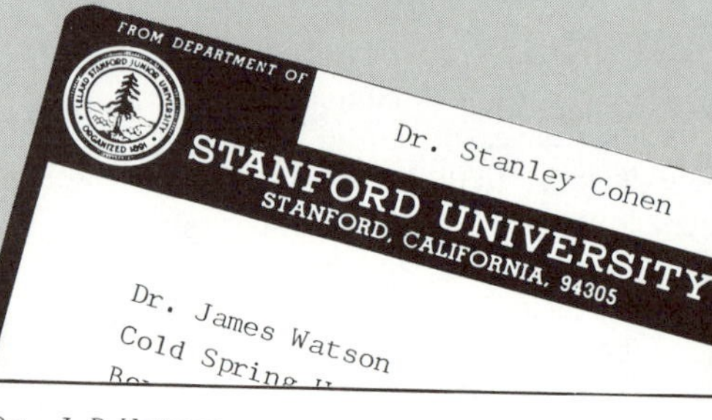

FROM DEPARTMENT OF
Dr. Stanley Cohen
STANFORD UNIVERSITY
STANFORD, CALIFORNIA, 94305
REMOVE TAB
Dr. James Watson
Cold Spring

STANFORD
AUG 11 '77
CALIF.
U.S. POSTAGE
.69

Dr. J.D.Watson
Cold Spring Harbor Laboratory
P.O. Box 100, Cold Spring Harbor, New York 11724

Dr. Paul Berg
Stanford University Medical Center
Department of Biochemistry
ford, Ca. 94305

Dr. N. Zinder
THE ROCKEFELLER UNIVERSITY
1230 YORK AVENUE · NEW YORK, N.Y. 10021

Dr. Paul Berg
Stanford University Medical Center
Department of Biochemistry
Stanford, CA 94305

A SECOND BERG LETTER

During the summer of 1977 when Congress seemed seriously to be considering the passage of specific legislation covering recombinant DNA research, some of those who had signed the Moratorium—or Berg—letter (Doc. 1.5) in the spring of 1974 decided to try to draft a second letter for publication. The new letter would give the reasons why, in view of new scientific information and the direction the public debate had taken, the original signatories now felt much less concerned about the conjectural hazards of recombinant DNA experiments than about the real hazards that the legislation pending in Congress might pose as a grave and unnecessary threat to intellectual freedom and scientific progress. As successive alternative drafts of a letter were circulated, it became apparent that agreement on a text was harder to achieve than it had been in 1974. The urgency was lost as the imminent threat of legislation diminished, and the attempt finally fizzled out in the spring of 1978.

Although a second letter was never published, we have included several of its drafts and some of the correspondence (Doc. 8.1 through Doc. 8.4), to show the differing opinions of various members of the original group of signatories. Norton Zinder's letter to the group (Doc. 8.2) is a particularly informative document.

Document 8.1

August 11, 1977
Cohen Letter to Watson
Watson's Response to Cohen

STANFORD UNIVERSITY MEDICAL CENTER
STANFORD, CALIFORNIA 94305 • (415) 497-5451

Stanford University School of Medicine
Department of Medicine

August 11, 1977

Dr. James Watson
CSH Labs.
P. O. Box 100
Cold Spring Harbor
New York 11724

Dear Jim,

Enclosed is my first try at revising Paul's draft. You may want to consider some of these additional points in preparing your own revision.

Best regards.

Sincerely yours,

Stan Cohen

SNC:seh

Enclosure

August 17, 1977

Dr. Stan Cohen
Department of Medicine
Stanford University Medical
Center
Stanford, California 94305

Dear Stan:

I like very much your proposed revise[d] draft and would gladly sign.

Here in New York we are most pleased b[y] Governor Carey's veto, especially pleasing was his reason – the bill would unnecessaril[y] restrain scientific endeavors! – not that the proposed national legislation would pro-tect against the possible new monsters. Norton was the key factor in mobilizing the letters, phone calls, visits to Albany, and we shall long remain in his debt.

Yours sincerely,

J. D. Watson
Director

cc: P. Berg
N. Zinder

REVISION

DRAFT OF BERG ET AL LETTER

In the spring of 1974 we advised the U.S. National Academy of Sciences and the Director of National Institutes of Health about the possible consequences of the indiscriminate use of recombinant DNA techniques. Although there was no firm basis for anticipating a hazard, we recognized that these new techniques enabled the construction of gene combinations that we believed were novel, and the properties of such combinations were not known at the time. The depth of our concern was reflected in the recommendation that certain scientifically valuable recombinant DNA experiments should be deferred until their potential hazards had been better evaluated or until adequate methods for preventing the spread of organisms carrying recombinant DNAs were developed.

Although guidelines for the safe conduct of research in other areas of science have long been available, no standard code of practice existed for the new area of recombinant DNA research at the time that we raised our initial concerns. However, our report and the subsequent Asilomar conference have since triggered widespread and detailed evaluations of the nature, probability, and magnitude of the postulated risks of recombinant DNA experiments. These discussions have now led to the formulation of codes of practice to guide research on recombinant DNA. Such guidelines are in many ways analogous to the standard operating practices employed in most other areas of scientific activity.

During the past three years, a number of developments have occurred that have altered our perspective and our estimate of the risks of recombinant DNA research; it is worth noting some of these developments:

(1) During the past four years, more than 200 scientific investigations involving recombinant DNA have been published, and literally hundreds of billions of bacteria containing a wide variety of recombinant DNA molecules have been grown and studied in the United States and abroad with no indication of harm to humans or to the environment. Because the life cycle of a bacterial cell is measured in minutes, it has been possible to study many thousands of generations of bacteria containing recombinant DNA molecules. Despite extensive efforts to detect some evidence of actual or potential hazard, none has been found to support our earlier concerns.

(2) Recent studies have shown that many recombinant DNA experiments simply recreate in the laboratory what occurs widely in nature. Combinations of genes indistinguishable from those constructed *in vitro* by recombinant DNA techniques are now known to be made intracellularly by the very same enzymes.

(3) Where they have been tested, organisms modified by recombinant DNA experimentation have been found to be at a disadvantage in their ability to compete in nature with the unmodified parental organisms.

(4) Substantial improvements in recombinant DNA methodology have greatly simplified experimental operations and have markedly reduced the likelihood of accidentally disseminating living organisms outside of the laboratory.

(5) Analysis by infectious disease experts and epidemiologists has led to virtually unanimous agreement that strain K12, the enfeebled laboratory variant of *E. coli* that is commonly employed for recombinant DNA experiments, is unable to colonize the normal human intestine or animal intestinal tracts. Whereas we were initially concerned about the use of an organism that is related to intestinal flora for this research, experts in infectious disease and epidemiology have concluded that there is little or no likelihood that *E. coli* K-12 strain can be transformed into an infectious or pathogenic organism, or even into a human intenstinal inhabitant, by the acquisition of a bit of foreign DNA. Genetically-modified derivatives of strain K-12 further reduce the likelihood of propagation or survival outside of specially devised laboratory environments. The use of vectors that are not self-transmissible and are not mobilizable to other host bacteria provides still another element of safety.

Considering the impressive progress made to date in the recombinant DNA field, the development of methods and practices to minimize the risk that is inherent in any area of research activity, and the increasing evidence that the special risks postulated earlier for recombinant DNA are non-existent, we regard the legislation currently before the United States Congress as unwarranted and unnecessary. In our view, the proposed introduction of prior restraints on scientific inquiry in the absence of any indication of actual hazard, and the spectre of harassment and punishment of scientists and institutions seeking answers to some of the most important biological and biomedical problems we face, pose a grave threat to future scientific progress in the United States. A review of the pending legislation in the Senate, S. 1217, and in the House of Representatives, H.R. 7897, has led us to the view that the greatest danger from recombinant DNA research is the legislation itself. These bills offer the prospect of unprecedented day-to-day control over scientific activities by the government in a well-intentioned attempt to protect the public, but in the absence of any evidence that the uniquely elaborate protective measures are necessary.

Although not each of us believes that the various strictures included in the NIH guidelines are required for public safety, we nevertheless agree that they represent a reasonable code of practice for recombinant DNA research. We believe that this generally accepted code of practice applies to the various segments of the research community in the same way that standard practices in other areas of scientific activity are applicable to both publicly and privately supported research. If a more specific mechanism is considered necessary to ensure public safety, we urge that reconsideration be given to the potential applicability of the Occupational Health and Safety Act to the bulk of this research, or alternatively, to revision of Section 361 of the Public Health Service Act. This latter act already provides the Department of Health, Education and Welfare with broad authority to regulate laboratory work with pathogenic micro-organisms. Certainly, the application of modification of mechanisms that currently exist to guard the public against definite hazards in other areas seems to be a more reasonable way of dealing with any residual anxieties about recombinant DNA research than is new legislation of the type proposed.

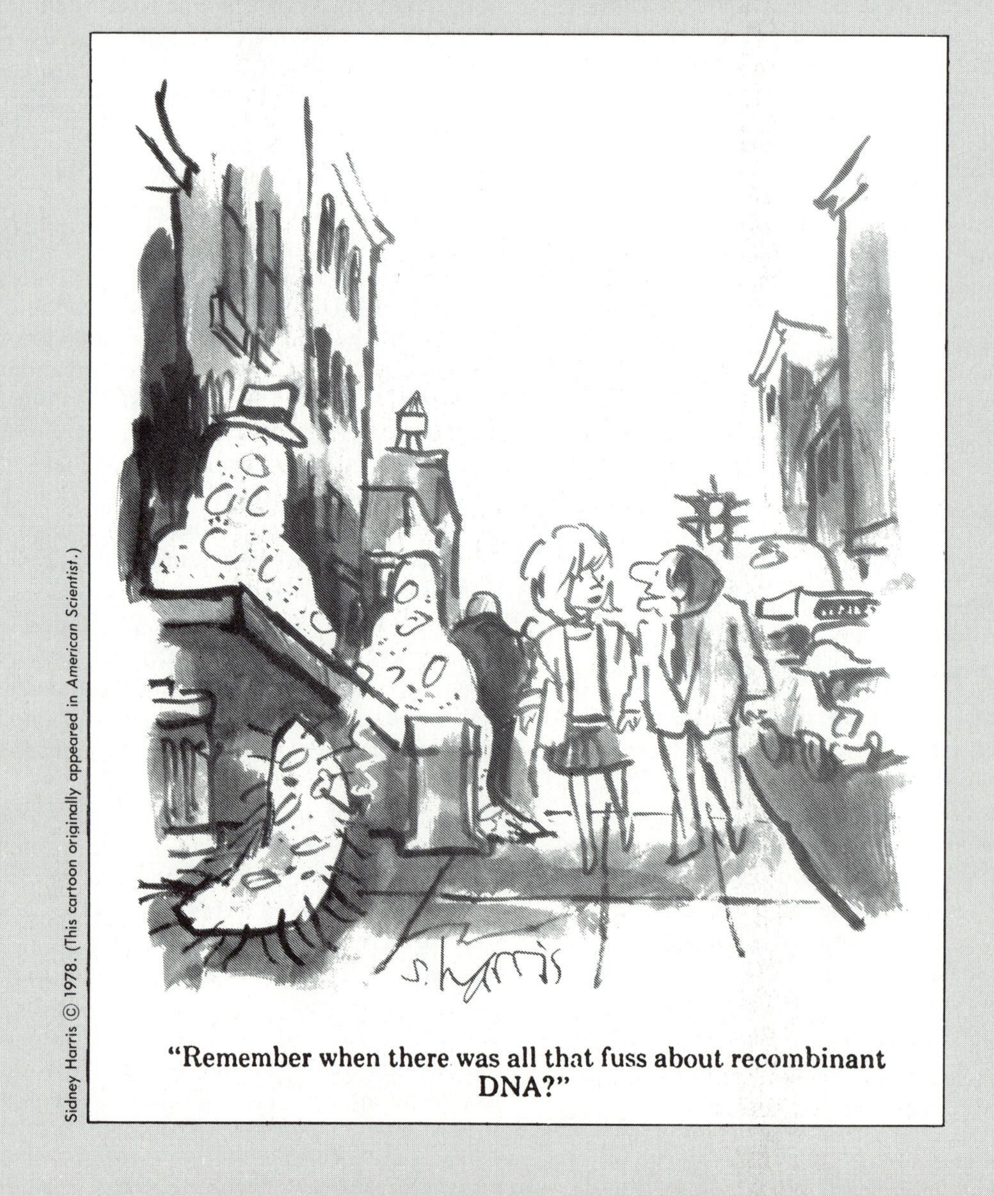

"Remember when there was all that fuss about recombinant DNA?"

Document 8.2

September 6, 1977
Zinder Letter to Berg et al.

THE ROCKEFELLER UNIVERSITY

1230 YORK AVENUE · NEW YORK, NEW YORK 10021

September 6, 1977

Dear Bergetal,

You have received or will receive from Paul copies of different versions of a letter to be issued by Berg et al. three years later.

For many reasons I believe that your consideration of these drafts and what to do with them can not be considered in a totally dispassionate manner (as if any of us are still dispassionate on the subject of recombinant DNA) but in what is to me at this moment a highly fluid political situation. I will recount as best I can the current political situation as I see it, although I'm sure it's incomplete.

The House: HR 7897, commonly known as the Rogers Bill, has been reported out of the Subcommittee on Health and Environment, but has not been given to the parent Committee on Interstate and Foreign Commerce. Rogers has asked Phil Handler to have an Academy Committee look over the need for legislation. The Academy Committee has met. It consists of competent individuals, but no one remotely working with Rec-DNA. It should deliver its report by the end of the month. A statement by us coinciding with theirs could be disastrous. In a preliminary version of the report that I've seen, they will come out reluctantly for legislation in order to apply to NIH guidelines uniformly and protect the science from capricious actions of local governments. It will state that Rec-DNA experiments as performed so far are quite safe and that the absence of legislation will lead to no catastrophe. They prefer the Nelson-Moynihan amendment to the Kennedy Bill S1217 or to HR 7897 as being closest to simply converting the NIH guidelines into law. (All this is not final and also <u>confidential</u>.)

The Senate: The bill S1217 has been reported out of the Kennedy Health Subcommittee and also the H. Williams parent Committee on Human Resources. There was only one dissenting vote, that of Gaylord Nelson, although in the committee report Eagleton and Chafee objected to the free-floating new AEC-like Commission to regulate Rec-DNA. This Commission is a direct result of pressure on Kennedy by H. Holman and L. Horowitz. We must note that at <u>every</u> hearing of the Kennedy subcommittee Holman was present, although he has no credentials whatsoever. (It's not only the Stanford people who know him as he was once a post-doc here at Rockefeller.) Kennedy seems adamant about this edifice, and his being talked to by Derek Bok, Jerry Weisner, the Inter-Society Council of Biology, etc., etc., etc. has not visibly moved him. Now, however, there are rumors of some dissension in the Kennedy camp. Horowitz's personality gets to everyone sooner or later. Unfortunately I've lost my contact in the Kennedy camp, but believe I can establish some contact in a few weeks. This friction may not be all to the good. I think Kennedy knows he's way out on a limb and may decide to take it out on us, rather than staff since he might feel we (the scientists) got him into this. In any event Nelson introduced his amendment, co-sponsored by Moynihan. It was a rushed job, piecing together with a few words earlier versions of HR 7897. I think they're still open to suggestion for modification. However, I also believe that we may be a bit naive in just what it takes to make the guidelines law. Boilerplate such as the OSHA-type whistle blowing laws are insisted on by the unions and while I believe "tenure" for technicians to be wrong, I'm not sure any bill relating to <u>hazards</u> to workers can exist without them. It's like being against motherhood. Licensing of P1 – P3 by local biohazards committees is about as convenient as filing an MUA. In any event, law means at the minimum licensing of facilities and registration of projects and probably penalties. The details of the application (hypotheses, etc.), I think, could be negotiated. The Nelson amendment already served one purpose. During the last week of the session L.H. managed to get 30 minutes for S1217 on the floor. It was ready for vote. With the amendment there, 30 minutes was too little time and hence it was put off.

My contacts on the Senate side say that they now doubt the bills will see the floor before January at the earliest. Congress comes back to the Energy bill, the Panama Canal Treaty, etc., and wants to adjourn mid-October for elections. On the House side it depends on what Rogers does after the NAS report. Thornton of the House Science and Technology Committee has asked for sequential referral as has the Judiciary Committee (depends on the Speaker of the House). It is my strong feeling that if Thornton ever gets it, he'll bury the bill unless we really ask him not to.

Local situation:

<u>California</u> – Kean Bill held off 'til January 1978

<u>Maryland</u> – Bill passed making NIH guidelines law. Dies if

Federal legislation passed. I know no details re: procedures

New York – Our noble governor vetoed the bill. There's lots of little things about which I'm sort of aware. The New York City Council has a resolution from the Friends of the Earth (well named, 'cause they ain't friends of people) prohibiting all Rec-DNA in New York City. I understand that since we are broke and run by state, only effect would be if Commissioner of Health got uppity, but the Hospital and Med School people have influence on him.

Princeton, U. of Washington, and and Cambridge, Mass. all have some modifield NIH plan.

Law suits: There are two suits that I know of.

1) FOE has one in Federal District of New York. It enjoins NIH from supporting, etc. Rec-DNA research. Suit based on NIH's failure to file Environmental Impact Statement and obtain clearance with Administrative Procedures Act and Council of Environmental Quality, etc. prior to issuing NIH guidelines. A technicality, but a well-drawn suit to this legal novice's eye.
2) Probably FOE instigated. Suit by man on behalf of 2-1/2 year old son in Frederick, Md. against construction of P4 facility and doing of Rowe-Martin experiment. Judge ruled facility can be built and will decide about experiment when time comes. Doubt it'll ever be done as at one time HEW lawyers wanted an Environmental Impact Statement (E.I.S.) for this experiment. Fredrickson realized what a precedent this would be if each experiment needed an E.I.S.

Us:

There is a corporate entity known as Bergetal (includes Maxine and probably Sydney) which is distrusted and disliked by part of the scientific community and some politicos. Its members provided the "proponents" for the recombinant DNA battles of this past year. So I don't think anyone doubts its position. Strangely this attitude to Bergetal doesn't extend to us individually. I guess that's because we're all so nice. A sure sign of this status is that many current statements that recount the history of rec-DNA skip from the Gordon Conference of 1973 to the Asilomar guidelines.

If you're still awake, by now you're convinced you've received a letter from Roy Curtiss. I do apologize for my long-windedness (unlike me) but the critical questions we must face are what do we hope to accomplish and how to do it considering the current milieu. I've discussed the matter with a number of political types and also some friends at Rockefeller, etc. They generally believe that Bergetal should keep its collective mouth shut as no matter what it says it'll incite public interest in what seems to be a cooling scene. They feel it is a time for other groups such as the Gordon Conference, Gorbach et al., the Biological Societies, the NAS, to speak out to the Congress on the new data that make what the Congress is doing overkill. Temperate articles such as we've had recently in the Washington Post, Washington Star, New York Times, even Science, and Time magazine are all to the good. The view is that while the Gordon Conference, which turns over, or any of us as individuals can change our minds, a statement by Bergetal is an incitement to riot. In any event, since we'll differ in detail if not in principle with the NAS Committee, above all we mustn't overlap their time frame. Perhaps we should see if we have some kind of consensus and be prepared, as the situation develops this fall, to move rapidly. I'd suggest that if we do, our message be directed to the Congress, not the public or the scientific community, even though it would diffuse out.

My guess as to our consensus runs as follows: what we did three years ago was appropriate to the state of scientific knowledge and also was necessary to calm down a somewhat hysterical molecular biology community (a fact conveniently forgotten these days). We were blanketing this area with a few principles of work extended at Asilomar and further extended by the NIH committee. These were all guidelines designed to be flexible so that they'd change as the data base changed. The public interest lagged behind about two years and now lags behind the new data base. The political dynamite of this issue escaped us!! I believe that those of us who attended the Academy Forum last March would agree that it was probably our nadir point. Structurally it was set up to make it appear that the scientists were split 50-50, while at the same time over a half dozen bills were being introduced into Congress. Returning home I was sure all was lost. Our current confidence just illustrates the volatility of the political process. I believe that we'd like the bills to go away and if necessary have the NIH guidelines extended as a code of practice by some kind of executive order. Is this possible? Paul said we got the veto in New York because we had to convince only one man. Ultimately that was true, but it took a lot of ohers to convince him. In Washington, I think it depends on Kennedy. I know that the aides on the House side are afraid of him. They feel even if they improve HR 7897 we'll lose in any conference with the S1217 as any compromise is our loss. Therefore, unless there's a sign of change of heart by

Kennedy I think we can only stop him with the counter bill by having the Societies and Universities lobby like crazy with the Congressmen from their States for the Nelson-Moynihan amendment.

I guess what I'm saying is the following. We can be purists and if we have a consensus put it out regardless of the political situation. As I told Paul over the phone, I've been busy so long calculating the results of moves – did I push too soon? too late? were the right people contacted? will he be angry at the truth? how far can I stretch "truth" without lying? – that I may have lost all perspective. Still I believe even if we make the purist statement or any other, I think we should await the return of Congress and try to find out through contacts and such people as the Harvard, MIT and Stanford lobbyists just what is the current status. We might go so far as to discreetly inquire what Rogers, Thornton, Nelson and even Kennedy, would think about a statement from us.

To be unequivocal, I favor our not doing anything now, but rather we try to ascertain whether we would agree to any legislation for whatever reason. The questions such as the changing data base have been all covered by others and we agree on this. I personally would accept a modified HR 7897 and Nelson-Moynihan bill with a sunset clause. I feel that if that's all we get, we've gotten out of this damn lucky. In fact, as some of you know, I am sort of committed to this. In order to get Moynihan to co-sponsor the Nelson amendment, I told him I would support the nine points of the Inter-Society Council. The decision had to be made fast and I felt his sponsorship was worth it. He's taking it seriously as he's sent all the Presidents of all the Universities in New York State his seconding remarks.

I've one last thought for us to consider. If we really believe that we have any collective power, perhaps we might consider a collective visit to Kennedy. Paul could write him and/or Stan could get in touch with L.H. I think he'd see us. If he'd see us and listen to us, it just might turn it all around.

Best,

Norton

P.S. Wally Rowe just called re: my hobby "cancer". Asked him about impact of statement by Bergetal. As the others, he saw nothing good, though was sympathetic to attempt to see Kennedy.

Document 8.3

September 19, 1977
Watson Letter to Berg

Dr. Paul Berg
Department of Biochemistry
Stanford University Medical School
Stanford, California 94305

Dear Paul:

I am happy with your latest version and would be most willing to sign it as I would have any of the earlier views - I liked the fact that it is getting shorter and less technical, for the audience we want to reach wants a simple message, not facts which they won't be able to evaluate.

Unlike Norton, I think it should go out soon. If it doesn't, the belief will persist that many prominent scientists are still in favor of legislation.

With best regards,

J. D. Watson
Director

REVISION

Three years ago we expressed our concern about the indiscriminate use of recombinant DNA techniques (Science, 185, p. 303, July 24, 1974). Although there was no evidence that recombinant DNA experiments were hazardous, we wanted reassurance that these novel experiments would be safe. Therefore, we recommended that certain recombinant DNA experiments be deferred until the question of potential hazards and how to deal with them could be better evaluated.

That evaluation, made independently in each nation where recombinant DNA research is performed, resulted in the formulation of guidelines or codes of practice to guide the use of recombinant DNA techniques. Most scientists believe that certain prescriptions in the guidelines cannot be justified by the scientific evidence we now possess. Nevertheless, investigators and their institutions, aware of the public's apprehensions have adopted the recommended procedures. Consequently, the possibility that experimental organisms will be hazardous or released is, in our view, exceedingly small.

More than three years of experience with recombinant DNA research has also led to a striking change in our assessment of the risks. Where it has been examined organisms modified by recombinant DNA methods are at a disadvantage in competing with their parental organisms; moreover, certain constructed DNA molecules, hitherto believed to be novel, can arise in nature by reactions akin to those used in the laboratory. There is also the virtually unamious agreement of experts in infectious disease and epidemiology that strain K12, the enfeebled laboratory variant of *E. coli* widely used for recombinant DNA experiments, is unable to colonize normal human or animal intestinal tracts. Based on recent experiments and existing data, these experts have concluded that there is little or no likelihood that strain K12 can be transformed into an infectious or pathogenic organism, or even into a human intestinal inhabitant by a bit of foreign DNA. Genetically modified derivatives of strain K12 and vectors that are not self-transmissable or mobilizable to other bacteria, provide a further measure of safety. Hence, our initial concern that the introduction and propagation of novel DNA elements on plasmids in *E. coli* would result in the dissemination of these new genetic combinations to intestinal bacteria was premature and, probably unwarranted.

On the other hand recombinant DNA methods have led to impressive scientific advances. Substantial improvements and innovations in the experimental operations have extended the utility of the technique and minimized the risks. New insights about the structure and organization of genes in higher organisms have emerged from such work. They promise important revelations about their function in health and disease. Isolation of the gene coding for insulin and the prospects for similar advances with genes coding for other therapeutic proteins brings closer the reality of practical benefits from DNA research.

In view of the rapid advance of scientific knowledge by recombinant DNA techniques and the entirely speculative nature of the hazards, we regard the enactment of legislation as unnecessary. Creating a costly, cumbersome bureaucracy to govern the content and methods of scientific inquiry would be unprecedented, unworkable and surely inhibit rather than foster basic research on important biological and medical problems. This scope of governmental intervention would be justified only if the research presented a clear and substantial threat to the public health and welfare. But in the absence of any indication of actual hazard, the waste of scarce public funds is unreasonable, unwise and is therefore, poor public policy. We believe that the provisions of Congressional bills S-1217 and HR-7897 will thwart basic biologic research and the achievement of its rewards for the public welfare. In our view the application or modification of already existing mechanisms that guard the public against known hazards is a more purdent way of dealing any remaining anxieties about recombinant DNA research.

Document 8.4

March 2, 1978
Zinder Letter to Berg et al.

TO: Berg, et al.

FROM: N.D. Zinder

I've revised Paul's statement for Berg, et al. and have made it stronger. The difficulty in the writing came because the statement made us sound either pompous, and/or silly and/or venal. Some of this remains.

It's difficult to know whether a statement by us will do good or harm, or be without effect. The situation has lost all the elements of having a scientific base and the legislation, etc., is purely motivated by different groups' perceptions of the politics of the situation. I've begun to feèl that work goes on in one universe and the polemic in an anti-universe. However with Paul I think these statements should be kept confidential until we decide if and what we do.

REVISION

Almost four years have passed since we called attention to the possible consequences of the indiscriminate use of recombinant DNA techniques. In the ensuing years these techniques have shown their power; power to solve important biomedical problems, power to provide insight into their own safety and unfortunately the power to engender a divisive polemic and pressures for restrictive legislation at every level of government.

Since our original action, in some measure, precipitated the unfortunate events, we feel obliged to speak out again. There now exists a body of evidence that leads us to conclude that research involving recombinant DNA is as safe as is any biomedical research. Hence it does not require and special regulation.

Why this conclusion and what are the real lessons of the last few years?

There have now been more than 250 scientific investigations involving the construction and propagation of many different kinds of recombinant DNA molecules. There has been no hint of harm to humans nor the least sign of invasion of any environment. When their ability to survive has been examined, organisms carrying recombinant DNA do not compete successfully with their parental or wild type organisms in the absence of specific selection. Furthermore, there are increasing indications that what were once thought to be novel recombinant DNA molecules can arise in nature by reactions akin to those used in the laboratory.

There is also virtually unanimous agreement by experts in infectious disease that even usual forms of strain K-12, the variant *E. coli* that is widely used as a host for recombinant DNA, can neither colonize human or animal intestinal tracts, nor be converted into an infectious or pathogenic organism by the addition of a bit of foreign DNA.

Moreover the predicted benefits from this technology have arrived more quickly than one could have hoped for. New insights have been obtained into the structure and organization of genes in higher organisms, radically altering our thinking about gene functions. Genes specifying insulin, growth hormone and somatostatin are now growing in *E. coli*; the last of these produces the natural gene product. We believe that similar accomplishments with genes coding for other valuable proteins such as those which could be used for production of anti-bacterial and anti-viral vaccines, or industrially important enzymes are not too far away. These benefits are no longer conjectural, they are here or on the horizon.

Moreover sophisticated techniques have been developed for the performance of experiments with recombinant DNA that in themselves reduce the possibility of any hazard. Further developments in this direction are readily predicted.

Therefore, we now conclude that research with recombinant DNA is safe. If we erred, as some have said, in being premature with our original statement, so be it. This is no excuse to continue the expenditure of time, effort and monies in pursuit of protection from a hazard that doesn't seem to exist. Continuation of such expenditures can only be detrimental to the public welfare. It is time for common sense and trust again to prevail.

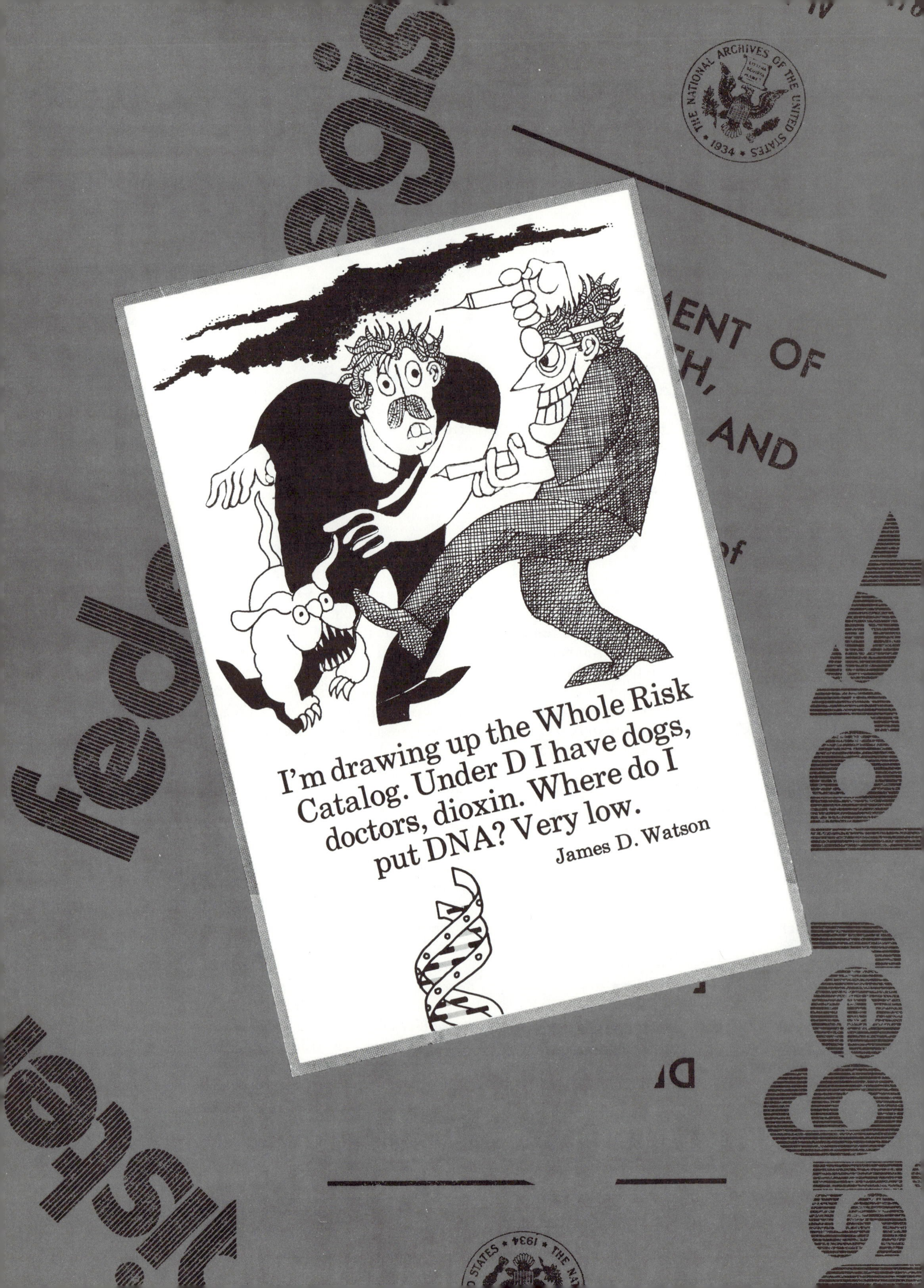
I'm drawing up the Whole Risk Catalog. Under D I have dogs, doctors, dioxin. Where do I put DNA? Very low.
James D. Watson

REVISING THE GUIDELINES: THE FIRST ATTEMPT

With Congressional committees debating legislation, states and cities taking initiatives, and claims and counterclaims for stricter or more lenient regulations reverberating, the RAC began in the spring of 1977 to consider the first review of the NIH Guidelines, which included a provision for at minimum an annual review. As part of this process the NIH, the National Institute of Allergy and Infectious Diseases (NIAID), and the Fogarty International Center convened, on June 20–21 at Falmouth, Massachusetts, a workshop to discuss the biology and ecology of *E. coli.* The participants at Falmouth included both molecular biologists and many medical microbiologists expert in the biology of wild strains of *E. coli* in their natural environments. This group came to the conclusion that the laboratory strain *E. coli* K12 is so genetically defective that there is no possibility of its being converted either by design or by accident into a pathogen capable of surviving and competing outside the laboratory with wild strains of *E. coli*, many of which are indeed pathogenic. This verdict—for *E. coli* K12 had in a sense been on trial at Falmouth—greatly influenced the RAC, which recommended relaxation of the containment conditions for many classes of experiments in the proposed revised guidelines that were published for public comment by Donald Fredrickson, Director of the NIH, on September 27, 1977.

We have not included excerpts from the *Federal Register* of September 27, 1977, since those revised guidelines were to be transformed by further amendments before they were finally issued by the Director of the NIH in December 1978. Document 12.4, however, includes a table comparing the required containment measures for particular classes of experiments in the original guidelines of 1976, the revisions proposed by the RAC in September 1977, and the revised guidelines finally adopted at the end of 1978. The table shows that as time passed the required safety precautions became increasingly less severe—a reflection of the progressive realization that initial concern and the responses to it had been exaggerated.

Returning to September 1977, the publication of the proposed revised guidelines elicited an immediate protest and a threat of legal action from Friends of the Earth (Doc. 9.1). That organization demanded that the NIH produce an Environmental Impact Statement under the National Environmental Policy Act of 1970 and provide public access to relevant information that had led to the proposal to reduce the stringency of containment measures. The proceedings of the Falmouth Workshop, which had decidedly influenced the RAC, were not available and unfortunately were not in fact published until May 1978, in the *Journal of Infectious Diseases* (Vol. 137).

Having criticized the NIH for its procedural failings, Friends of the Earth went on to request an extension of the prescribed deadline in order to digest all the material and formulate comments. The opposition to recombinant DNA research of some of the members of Friends of the Earth, and in particular their attempt through the courts to block federal financial support for the research, was not, however, universally welcomed by the supporters of that organization (Doc. 9.2). In a survey of the controversy in the September–October 1977 issue of its bimonthly journal *Science for the People*, that group again argued for a widening of the issue to encompass what were, in its opinion, interrelated areas (Doc. 9.3).

NIH Director Donald Fredrickson responded by arranging a public hearing on December 15–16, 1977, at which those in favor or against the proposed revisions testified (Doc. 9.4 and Doc. 9.5). A few days before the hearing the Environmental Defense Fund uncovered an alleged contravention of the guidelines in the laboratory of Charles A. Thomas at Harvard Medical School. Investigation led the NIH to order a halt to this research (Doc. 9.6). Earlier in 1977 a group in California had also been accused of breaking the NIH guidelines (Doc. 9.7). The atmosphere of academic research, already in this field fiercely competitive, was threatened with the pollution of informers, spies, and criminality.

Meanwhile, in California the synthesis in *E. coli* of a mammalian hormone, somatostatin, was announced by another group (Doc. 9.8). This was not only the first synthesis of a mammalian hormone in a bacterium but also the first synthesis of a functional protein from a gene that had been synthesized chemically. The practical and commercial benefits of recombinant DNA research were likely to be achieved much more quickly than anyone had initially dared to predict, as an article in *Business Week* pointed out (Doc. 9.9). This article also touched upon the sensitive issue of the compliance with the

NIH guidelines by industrial laboratories that were necessarily concerned with their patent rights.

From this bizarre constellation of events—the NIH proposing a modest relaxation of the guidelines; environmental groups arguing that the relaxation was premature and attempting to use the courts to block federal funding of this research; many molecular biologists urging much more extensive relaxations; lawyers at the NIH public hearing criticizing the NIH for faults in the procedures it had used in proposing the revisions; a laboratory at Harvard Medical School being ordered to halt its research because of an alleged infringement of the guidelines and a company in California announcing successful experiments with obvious commercial potential—one thing was abundantly clear. Revision of the NIH guidelines would be a slow and tedious process with much opposition. As we shall see, it took still another year to achieve, but the revision was then more radical.

Document 9.1

September 30, 1977
Hartzman Letter to Fredrickson

RICHARD M. HARTZMAN
ATTORNEY AT LAW
251 WEST 87TH STREET
SUITE 52
NEW YORK, NEW YORK 10024
212 - 866-8146

September 30, 1977

Dr. Donald S. Frederickson
Director
National Institutes of Health
9000 Rockville Pike
Bethesda, Maryland 20014

Dear Dr. Frederickson:

I am writing to you on behalf of Friends of the Earth in regard to the proposed revised NIH guidelines for recombinant DNA research.

We are deeply disturbed that these revisions have been proposed and published in the Federal Register for comment without the NIH having prepared the legally required environmental impact statement. The National Environmental Policy Act requires that an environmental impact statement "shall accompany the proposal through the existing agency review processes". Adequate public disclosure of the information base and analysis of the environmental impacts of the proposed revisions is not possible without a detailed impact statement, and hence informed and effective public comment is precluded. Nor can there be any assurance that the issues have been thoroughly considered and that the decisions being made are rationally based. The NIH action proposing revised guidelines is a continuation of the scientifically and legally defective processes used with respect to the entire recombinant DNA program and the promulgation of the original guidelines. If the NIH does not take immediate steps to comply fully with the law, Friends of the Earth will be constrained to initiate further legal action.

In addition, we find the 30-day comment period for the revisions wholly inadequate. Not only are the guidelines themselves highly complex, but the proposed revisions are profuse and often subtle. Nor would most interested commentators be able to obtain needed information concerning the basis for the revisions before the 30-day period had expired. A 60-day period for comment is much more reasonable for a complex matter such as this. In any case, the proposed revisions should be withheld from further consideration until an environmental impact statement has been prepared.

Finally, we are of the opinion that public hearings should be held before any final decisions are made on revisions to the guidelines. Merely holding open meetings of NIH advisory committees is not sufficient. This has been an issue of great public concern that has involved city councils and town meetings, state agencies and legislatures, federal agencies (such as the Environmental Protection Agency), and the United States Congress. It would be a serious omission if the key mechanism safeguarding recombinant DNA research, the NIH guidelines, were revised without public hearings that include an opportunity for interested groups and individuals to participate.

We are confident that, in the spirit of good government and concern for public health and the environment, you will take the necessary steps to comply with the law.

We await your prompt response.

Thank you.

Sincerely,

Richard M. Hartzman

jc

cc: Edward Kennedy
Paul Rogers
Joseph Califano
William Gartland
George Wald
Jonathan King
Philip Boffey
Nicholas Wade
Colin Norman

Document 9.2

Spring 1978
CoEvolution Quarterly (**17**:24)

Recombinant DNA, Paul Ehrlich, and Friends of the Earth

Lest it appear that we are trying to attack the wholly admirable Friends of the Earth organization, you should know that Ehrlich's letter criticizing FOE was sent to us by a loyal FOE staffer, Jim Harding.

The point and poignancy of the correspondence is how big simple issues, once joined, become unsimple; and how BAD/GOOD becomes goodbadgoodbadgoodbad. Many who would do good in the world become bruised and discouraged by that process. I wish they wouldn't. Detail work is what it's all about.

Paul Ehrlich, author of **Population Bomb** *and other population biology classics, is on FOE's Advisory Council. So was Lewis Thomas, author of* **Lives of a Cell** *and head of the Sloan Kettering Cancer Center, before he resigned in protest over FOE's lawsuit to have the NIH (National Institute of Health) recombinant DNA guidelines strictly enforced.*

Previously (Summer '77) we have printed DNA researcher James Watson's unalarmed views on this subject. Here appended is a recent paper by Bruce Ames on his lab's work with recombinant DNA in environmental testing for carcinogenic chemicals. How important is that? Well, according to a report in the 13 January 1978 issue of **Science** *the number of chemicals in the American Chemical Society's Chemical Abstracts Service is growing at an average rate of 6,000 new chemicals a week. Total in the registry is 4,039,907.*

Thomas H. Maugh II, in the **Science** *report, says: "Current estimates from EPA (Environment Protection Agency), indicate that there may be as many as 50,000 chemicals in everyday use, not including pesticides, pharmaceuticals, and food additives. EPA estimates that there may be as many as 1500 different active ingredients in pesticides. The Food and Drug Administration estimates that there are about 4000 active ingredients in drugs and about 2000 other compounds used as excipients to promote stability, cut down on growth of bacteria, and so forth. FDA also estimates that there are about 2500 additives used for nutritional value and flavoring and 3000 chemicals used to promote product life. The best estimate thus is that there are about 63,000 chemicals in common use. Small wonder then that determination of the safety of all commonly used chemicals is a massive project that may never be completely finished."*

To do a complete white-mice test for carcinogenicity of a single chemical costs about $150,000 and takes many months. A comparable Ames test costs $250 and takes two days. —SB

November 17, 1977

The Executive Committee
Board of Directors
Friends of the Earth
124 Spear Street
San Francisco, CA 94105

Dear Friends:

As a professional biologist, I have become increasingly concerned about the opposition to recombinant DNA research expressed by FOE and some other environmental groups. While there are valid reasons to be concerned about recombinant DNA technology, I think its hazards have been greatly exaggerated by some people. And some of the proposed action being taken against it could be counterproductive to the goals of the environmental movement in general and FOE in particular. Let me elaborate.

As you know, I have for a long time been interested in the social responsibility of scientists and how the results of scientific research affect public welfare. In the case of recombinant DNA research, I think the scientists involved have behaved admirably. Recognizing a possible serious hazard, they announced it to the public themselves and adopted voluntary restraints on their own research until the risks could be further examined.

As an evolutionist, I have been rather depressed by various features of the ensuing controversy. It has been stated that the research should be discontinued because it involves "meddling with evolution." *Homo sapiens* has been meddling with evolution in many ways and for a long time. We started in a big way when we domesticated plants and animals. We continue every time we alter the environment. In general, recombinant DNA research does not seem to represent a significant increase in the risks associated with such meddling – although it may significantly increase the rate at which we meddle. For example, if one wanted to create a bacterium that was highly resistant to all antibiotics, one would not do it by attempting to insert DNA from a tiger into its chromosome. The chances that recombinant DNA research might lead to the creation and escape of some kind of "super bug" seem extremely small. One must always remember that any laboratory creations would have to compete in nature with the highly specialized products of billions of years of evolution – and one would expect the products of evolution to have a considerable advantage. In addition, there is evidence that bacterial species have been swapping DNA among themselves for a very long time and perhaps even exchanging with eucaryotic (higher) organisms. Human investigators using recombinant DNA techniques thus are not necessarily performing the evolutionarily novel experiments that have been claimed.

I must add that I am less sanguine than some commentators on how rapidly certain benefits will accrue to humanity from recombinant DNA research. It is possible, for example, that cheaper insulin may be made available to diabetics. It will help improve systems for screening chemicals for carcinogenicity. Producing strains of nonleguminous crops that can fix their own nitrogen is a possibility, but what the other characteristics of such strains might be (i.e., how susceptible they would be to pests), and what sorts of yields they will produce remains to be seen. There is considerable reason for doubting that much can be gained by this route. And, large-scale intervention in human genetics will be possible, I suspect only in the distant future, if ever. It may never be socially desirable. In my view, the brightest promise of recombinant DNA is as a general research tool for molecular geneticists – a tool to use in the pursuit of the highest goal of science, understanding of the universe and of ourselves.

All of this is not to say that malign or careless use of recombinant DNA technology could not be responsible for a variety of disasters. But many other technologies – including those now in use in microbiology laboratories – are equally susceptible to misuse. As Sir Peter Medawar has pointed out, humanity has had a powerful technology for intervening in human genetics for a thousand years – selective breeding. People have decided not to put the knowledge of the stock breeder to such use, and hopefully society would be equally intelligent about the use in human genetics of any future recombinant DNA technology (the application of which is likely to be infinitely more complex).

It can be said with reasonable certainty that the results of virtually any pure scientific research have the potential for being turned against humanity. That applies, for example, even to the work our group does on the coevolution of butterflies and plants. If recombinant DNA research is ended because it could be used for evil instead of good, then all of science will stand similarly indicted, and basic research may have to cease. If it makes that decision, humanity will have to be prepared to forego the benefits of science, a cost that would be high indeed in an overpopulated world utterly dependent on sophisticated technology for any real hope of transitioning to a "sustainable society."

But even if an attempt were made to control some kinds of investigations directly, recombinant DNA research would be a peculiarly bad target for regulatory legislation and legal intervention. Workers in the field are now highly sensitized to possible dangers, are under careful scrutiny, and are funded by an agency that has already promulgated relatively restrictive guidelines on the research that can be done. Many other fields of research present greater hazards, promise fewer benefits, are populated by less responsible scientists, and would make better "test cases" for regulation. The genetic researchers who behaved responsibly are now paying a high price for their responsibility. They are being restrained from research that has considerable potential for benefitting the public. Of special interest to environmentalists and the FOE board is the work of Bruce Ames at Berkeley in developing tester strains of *Salmonella*, which are used in screening the products of the prolific synthetic organic chemical industry for carcinogens. Those of you who are familiar with it undoubtedly recognize that the "Ames screen" is the single most powerful technical tool that has been handed to environmentalists in decades. And yet Ames' research is being hindered by a complete – and pointless – ban on all recombinant DNA research with *Salmonella*. FOE is thus in the position of helping to block work on the most promising techniques for detecting environmental carcinogens.

Perhaps the most serious threat to recombinant DNA research is the FOE suit against the NIH. If it prevents NIH funding, it would effectively shut down all university research in molecular genetics in this country, closing the laboratories of our most competent and responsible geneticists. At the moment FOE is even blocking attempts to get further information on any potential risks of technology.

Certainly, FOE should maintain its interest in the general question of how science and society interface, but I think its position on recombinant DNA research is the result of overreaction rather than careful analysis of the difficult issues involved. I hope the whole question will be reexamined by the board and that you will get some input from scientists involved in the research. I think that I could persuade Paul Berg to meet with you. As you may recall, he was one of the original "whistle-blowers." You would find him a sympathetic and humane scientist who can give a well balanced presentation of the potential risks and benefits of recombinant DNA research.

Thank you for your attention,

Sincerely,

Paul R. Ehrlich

FOE's reply to Ehrlich was quite long and technical. It said in part:

December 2, 1977

Dr. Paul Ehrlich
Biology Department
Stanford University
Palo Alto, California 94305

Dear Dr. Ehrlich:

. . . FOE's position has not encompassed condemnation or discontinuance of this type of research per se. Rather, we have called for a "stop-to-think pause" to permit a systematic assessment of the possible benefits and of the risks relating to public health and the environment, and to gain effective public participation before widespread proliferation of the research. In addition, there are broader issues which need to be addressed: (1) commercial uses and patenting of life forms; (2) professional and governmental conflicts-of-interest; and (3) the social and ethical implications of genetic manipulation.

Development of recombinant DNA research has many striking parallels to the nuclear energy controversy. They both have been permeated by the assumptions that the work will go ahead; that the benefits outweigh the risks; that we can act now and learn later; and that any given problem has a solution. But the nuclear energy industry is foundering today because it was built on technological hopes and expectations rather than on technical solutions. FOE is concerned that premature development of recombinant research and its commercial application could well propel us, irreversibly, into a similar technological quagmire. We have the unique opportunity, now, before the intellectual and economic investment grows much greater, to assess the foregoing problems.

As you indicate in your letter, the genetic researchers who first called for a voluntary moratorium on certain types of recombinant DNA research did indeed perform an invaluable service. However, this new technology entails responsibilities beyond the initial call for caution, which are seemingly not yet recognized by these scientists and their colleagues. Among these responsibilities are the fulfillment of legal obligations, which are designed to protect public health and the environment, such as compliance with the National Environmental Policy Act (NEPA).

NEPA, which requires a detailed environmental impact statement, ensures that major federal projects are not undertaken *before* assessment of risks and possible benefits, exploration of alternatives, study of environmental impacts, full disclosure of the basis for action, and an opportunity for effective public participation in the decision-making process. NEPA litigation has been used effectively in other areas, such as nuclear energy, resulting in consideration of previously ignored problems. Cases in point are *SIPI v. AEC* and *Calvert Cliffs*. Preparation of impact statements have often led to beneficial modifications of projects.

Unfortunately, the NIH's violations of NEPA with respect to the funding of recombinant DNA activities and the development of guidelines have been flagrant and ongoing. The NIH has conceded that NEPA applies to the research guidelines which purportedly safeguard against hazards. Nevertheless, a draft impact statement on the guidelines was not prepared until two months *after* the guidelines were promulgated. It contained many glaring inadequacies, and was obviously not considered during the decision-making process. Only now, more than a year later, has a final impact statement belatedly been completed. Further compounding blatant disregard for the law, *revised* weakened guidelines were developed early this year and approved by the Recombinant DNA Advisory Committee in June. The revisions are now moving through further stages of the agency review process, even though an impact statement has not been prepared and the final statement on the *original* guidelines has only just come out.

The fundamental action that is subject to NEPA is the funding of the research, not the development of safeguards or guidelines. It is the research and its applications from which will come the environmental impacts, and each project poses its unique problems. Of course, some projects may be relatively harmless, but that must be determined in the NEPA process.

The New York State Attorney General has stated that

"Issuance of an impact statement after grants have been made to conduct recombinant DNA research and after publication of guidelines was a *per se* violation of the National Environmental Policy Act."

The NIH has created an adversary climate by failing to comply with NEPA, thus compelling Friends of the Earth to turn to the courts. Although we are forced to seek injunctive relief against recombinant DNA projects and the guidelines, the ultimate goal of this litigation is to place decision-making on recombinant DNA activities on a solid base and gain effective public participation. . . .

In the past several months there has been an effort to convince the public that the risks of recombinant DNA research are lower than previously thought. This coincided with the effort to kill legislation and promulgate relaxed guidelines. The claim that the risks are less than thought earlier is based primarily on the Falmouth Workshop on Risk Assessment, a letter from Roy Curtiss III, and a paper by Stanley Cohen. The new data which has come to light is mostly unpublished, meager, and limited to a consideration of the possibility of rendering E. coli K-12 pathogenic and the likelihood of plasmic transfer to other species. Although there is not much fear of E. coli K-12 becoming a pathogen, there is considerable concern over the transfer of genes to wild E. coli strains and other species. The Curtiss letter cites unpublished data showing a low probability of the transfer of non-conjugative plasmids. These numbers are reminiscent of those used in the nuclear debate regarding improbable events which have in fact occurred. It was the consensus at the Falmouth workshop, at which the Curtiss data was discussed, that more experiments are necessary to assess the likelihood of inadvertent transfer of recombinant DNA from E. coli K-12. . . .

The Stanley Cohen article has been circulated prior to publication, contrary to usual scientific canon, in order to accelerate its consideration in the formation of public policy. His conclusion that artificial recombinational events performed in the laboratory also occur in nature has been sharply criticized by scientific peers. . . .

The controversy over recombinant DNA research has not diminished. We are receiving a stream of comments from scientists who not only believe that there is insufficient evidence to relax the guidelines, but that the original guidelines were based on arbitrary assumptions. Compliance by the NIH with NEPA would bring into public focus the many unresolved concerns of this controversy, and help shape responsible public policy.

Regarding your suggestion that Paul Berg meet with the FOE executive committee, we do not believe that he would offer a balanced presentation. He has been an outspoken proponent of the research. However, we would both be happy to speak to him. An exchange of views may clarify our respective concerns and initiate a fruitful dialogue.

We hope this letter has explained the position of Friends of the Earth regarding the recombinant DNA controversy.

Very sincerely yours,

Francine R. Simring
Richard M. Hartzman, Esq.

Document 9.3

September–October 1977
Science for the People (9:28)

Dealing With Experts: The Recombinant DNA Debate

Bob Park & Scott Thacher

Molecular Biology Against the Wall

The proliferation of possibilities in recombinant DNA research has brought new excitement to molecular biology. Besides new vistas in "pure" research, remarkable applications and grim hazards have appeared on the horizon, and previously farfetched scenarios for genetic engineering seem much less distant.* The commercial aspects have aroused the curiosity not only of drug companies but of industry in general. Molecular biologists are invited to give briefings on Wall Street. A skirmish recently broke out in the Commerce Department when an official proposed accelerated patent procedures for recombinant DNA techniques. (So far G.E. holds three patents and both Stanford and University of California have applications pending.)

Simultaneously an unprecedented open debate has mushroomed on the control of this research. Numerous cities and towns, likely future hosts to recombinant DNA research, have joined the debate. For the first time, molecular biology has received local front-page coverage. No longer is the research a matter for "self-regulation" by scientists, through the good offices of the National Institute of Health (NIH) which funds most biomedical research. The issue has been catapulted to top-level policy-making involving the Secretary of Health, Education and Welfare (HEW), the Commissioner of the Food and Drug Administration (FDA), and an interagency task force which has recommended comprehensive legislation. Bills are now being formulated in Congress and in State legislatures.

*Applications include: industrial microorganisms which may transform chemical and pharmaceutical industry, production of biological materials not now available, plant varieties with unique abilities, e.g., nitrogen fixation. Potential hazards include: disease-causing bacteria never before encountered, ecological disruption, and new diseases of genetic regulation, e.g., cancer. For a more detailed discussion of the hazards, precautions, and alleged benefits, see paper entitled "Social and Political Issues in Genetic Engineering," by the Recombinant DNA Group of SftP, available from the SftP office: 897 Main St., Cambridge, MA 02139.

Harvesting the Culture of Elite Science

In recent years most working people have acquired a critical sense of the role of science and technology despite a tradition of science mystification and deference to elite authority. Many now recognize that unemployment, pollution, and disease are another side of the grand hype that science means automatic progress; they see that most of those white-coated experts are owned by business or government. Technology's record has fostered this disillusionment: e.g. PCBs, kepone, SST, Tris, nuclear power, occupational hazards, etc.

And so, in 1974, when molecular biologists themselves called for a moratorium on certain potentially dangerous experiments and asked that scientists discuss among themselves safeguards for this research, the news spread readily far beyond science to a quite interested public. Popular skepticism has been further stimulated by the disagreement increasingly visible among the experts themselves. But perhaps it was the prospect of actually engineering genetics — whether ours, someone else's, or that of plants or microbes — that finally cancelled the blank check of elite science, i.e. knowledge in the service of powerful institutions.

Open Debate on Usually Closed Issues

Debate on recombinant DNA research, both in and out of science, reveals that a Pandora's box has been pried open; social control of science is a live issue. Specific questions arise in three areas — the ostensible benefits, probable uses, and unintentional hazards. But we can go further and ask what underlies the disagreement among experts themselves and then ask how government policy in science could become the province of the people?

• One benefit promised from recombinant DNA technology is a breakthrough in world food production using new, specially engineered species of plants, which it is claimed would significantly reduce world hunger. This invites examination of the past effects of the Green Revolution—increased yields from selected hybrid varieties of rice, corn, and wheat. The results have not been

Bob Park and Scott Thacher are members of the Recombinant DNA Group of the Boston chapter of SftP. Bob has worked in clinical trials research in the drug industry and is planning to attend public health school. Scott is a graduate student in biophysics at Harvard, studying membrane biology.

to feed the hungry.(1) Predictions of new drug sources and supertherapies for intractible disease demand looking at the economic and social origins of most disease and health problems, questioning medical research priorities in general, and exposing what the high technology, "technical fix" approach to health care means.

While conceivably new therapies will be able to correct some of the non-controversial genetic defects known, there are many other conditions — virtually any characteristic with a claimed genetic predisposition — where the "correction" would amount to a form of genetic repression of individuals by society. Who decides when human variability becomes a genetic "defect"?(2) We need to spell out the implications — present and future — of emphasizing genetic fixes over giving society the treatment: they include declining social services, increasing channelling of individuals (IQ in education, occupational hazard vulnerability in employment), and ultimately suppression of deviance, dissent, unrest, and other "maladaptive" behavior.

While the ultimate uses of recombinant DNA technology are probably the gravest threat, it is on the immediate hazards of doing the research that the technical disagreements among the experts are most apparent.* The debate centers around the adequacy of containment for experimental organisms as well as the pretense that molecular biologists (or anyone else) know enough to guess at the broader ecological or evolutionary threats. How can supposedly objective experts** be in such disagreement? We think perceptions of "objective" reality are dependent on philosophical and ideological premises as well as on other immediate and material factors in people's lives. A large part of the benefit to risk estimate is speculative and thus is especially open to subjective valuation. For example, how one assesses benefits from recombinant DNA work is contingent on one's view of the social role of technology; predicting hazards depends on one's technological optimism.

Another source of subjectivity derives from one's own contribution to, or interest in, technology. For many in science, the value of their work depends to a considerable extent on how it contributes directly or indirectly to human betterment. In a society where institutions do not operate *a priori* to serve desirable social ends, there is an incentive to believe that better technology tends to shift the outcome in favor of serving those ends, that new knowledge has intrinsic positive value. Consequently, many medical researchers pursue answers to problems for which other solutions, such as changing social conditions, are lacking or are at least beyond their control. Some people, for this reason, may have an unduly optimistic outlook on recombinant DNA research. Others in science have careers whose success requires the rapid exploitation of scientific discovery. The advantages include publications, appointments, the realization of creative potential, esteem with family and colleagues, recognition by institutions and officials, and ultimately, entry into business and government circles. It is clear that in situations where advances are imminent, the personal benefits and risks of some scientists—as with investors—can very understandably differ from those of most working people.

Popular Critical Awareness on Technical Issues

Because technical issues cannot be resolved by reference to an "objective," neutral stance, it is especially vital that public policy* in science be determined by a process based on popular awareness, organization, and control. One form this could take would be labor unions with strong member participation and control, with extensive education programs, and with active involvement in defining and enforcing government policy and corporate behavior. Another avenue for popular control of science policy is community-based organizations watching over, for example, the health care system, medical research, and human experimentation.

Even without organization, however, public discussion, debate and criticism can have a major effect on the existing decision-making apparatus, as we are seeing. This process has not been encouraged by most prominent scientists. As Sidney Udenfriend, director of the Roche Institute for Molecular Biology** and member of the NIH advisory committee on recombinant DNA research, explained: "I'm afraid there's going to be some brush fires if we get communities involved in deciding biohazards. If we permit non-scientists to question our work in one area (DNA), we'll open ourselves up to all kinds of things...."(3)

How can good judgment on scientific issues be exercised by the "masses"? This, we propose, is analogous to the question: How do top government leaders and policy experts decide questions of science and tech-

*The concern arises from the use of the bacterium *E. coli* as a host because it is a normal inhabitant of the human G.I. tract (but occasionally causes serious disease). *E. coli* is used because it is the best known bacterium. But hybrid versions, created unknowingly when random samples of foreign DNA are spliced into its chromosome, could create a whole new class of disease-causing organisms.

**Definition of "expert": a person with extensive personal experience, both in theory and practice, in some area of technical knowledge, not necessarily certified by an academic degree. Being an expert, however, does not mean knowing the "truth" on a technical matter within one's expertise or better understanding the social implications.

*"public policy" — fundamental policies laid down by Congress or Executive branch on which government regulation is based.

**The Roche Institute is the "pure research" arm of Hoffmann-LaRoche, the most lucrative drug company in history, maker of Valium, Librium and others.

nology policy? *They rely on experts whom they believe to be credible.* The people, too, should be able to evaluate the credibility of experts. What are these experts' views on the general role of technology and on specific issues bearing on the people's interests? How have they contributed to dealing with the real problems of working people, and what are their stakes in these matters? Evaluating experts is an important task for any popular organization. Just as the rulers of the country can pick and choose between experts and the opinions that they espouse, so can the people.

Of course, the ability of the people to evaluate technical opinion would be considerably enhanced by their having more widespread technical knowledge and scientific understanding. This is a goal which progressive science workers and technical experts should facilitate, in contrast to what happens normally.

The Developing Controversy

In 1971 a scientist objected to a colleague's proposal to insert the virus SV40, which causes tumors in some animals, into the bacterium *E. coli* K12. It was feared the hybrid might escape from the laboratory, survive, and result in a new form of disease. The experiment was abandoned. The subsequent, self-imposed moratorium on certain gene-splicing research was partly intended to show that scientists could look after the danger of their own research. The first large scale discussion by molecular biologists of hazards took place in February, 1975, at Asilomar, Cal., where a rough consensus was obtained on how to deal with the safety question. However, the panel subsequently selected by NIH to write guidelines was made up mostly of scientists already using recombinant DNA techniques or planning to, and some advisors to the panel had direct commercial interests in it.(4) It was a foregone conclusion that the techniques would be developed and used extensively.

With minimal public participation, the NIH guidelines committee plunged forward (with occasional backsliding), buffeted on all sides by threatened feudal science chiefs. One early draft, available at the traditional Cold Spring Harbor phage* meeting in August, 1975, was sharply attacked by members of Science for the People and others as a retreat from earlier, more strict positions. Meanwhile, the debate went public.

The first large scale public confrontation on recombinant DNA took place at the Univ. of Michigan, Ann Arbor, in early spring of 1976, when the casual intentions of the university trustees to invest in a campus-

*Phage: a virus that lives in bacterial hosts, studied because of its relative simplicity.

These comments were written by Sheldon Krimsky, who was a member of Cambridge Experimentation Review Board (CERB).

As a result of the Cambridge experience we have a singularly important counter-example for those skeptics who would not believe that a group of citizens could grasp the issues of a technically complex debate, carry through an intense investigation of the issues and arrive at a decision that was sensible and thoughtful.

Basic science has just witnessed the end of its age of innocence. The events in Cambridge tell us that citizens are no longer willing to place their blind faith in research scientists who, in their eagerness to extend the boundaries of human knowledge, employ invasive technologies that have the capacity to alter significantly the world they wish to investigate. It is evident from CERB's recommendations that citizens recognized that academic science has become an industry. Researchers and their institutions compete for ever more scarce federal dollars.

CERB was sensitive to the fact that many of the claims scientists made about the risk-free nature of the research did not rest on hard empirical data. Proponents appealed to *a priori* assumptions, argued from analogy, deduced particular statements from evolutionary theory and made extravagant extrapolations from a narrow data base. It was the feeling of some board members that tests carried out under ideal conditions need not bear out under actual experimental conditions.

The main emphasis of the NIH guidelines was on the short-term risks of spreading biolgoical agents. CERB recognized the potential of releasing hazardous agents with long latency periods. The board recommended a national registry of those who are engaged in recombinant DNA research so as to make long-term epidemiological studies possible.

More than anything else, the report of the Cambridge Experimentation Review Board is a statement against elitism and self-regulation in one of the most carefully protected areas of scientific research. The following admonition was issued by CERB in its final report:

"Throughout our inquiry we recognized that the controversy over recombinant DNA research involves profound philosophical issues that extend beyond the scope of our charge. The social and ethical implications of genetic research must receive the broadest possible dialogue in our society. That dialogue should address the issue of whether all knowledge is worth pursuing . . . Knowledge, whether for its own sake or for its potential benefits to humankind, cannot serve as a justification for introducing risks to the public unless an informed citizenry is willing to accept those risks. Decisions regarding the appropriate course between risks and benefits of potentially hazardous scientific inquiry must not be adjudicated within the inner circles of the scientific establishment."

based recombinant DNA facility were unexpectedly dragged into the spotlight. The issue was raised by faculty member Susan Wright, with several other faculty and Ann Arbor SftP members joining in. It generated escalating interest on campus and within the surrounding community to such an extent that the university's Research Policies Committee felt compelled to arrange a full-dress forum, inviting a wide spectrum of experts from all over the country. It lasted two days and attracted a continuous attendance of over 600 people.

The outcome was that the two appropriate faculty committees gave near unanimous approval to proceed with the research, subject to the awaited NIH guidelines. However, far more significant was the effect of the debate locally in revealing the full depth of the criticism of the research, and nationally, in providing a stunning precedent for the growth of the controversy into a movement for popular control of science.

The Cambridge Experimentation Review Board

Just as final NIH Guidelines were about to be issued in June, 1976, Harvard University's plans to build a P3* facility came to light. Aware of Harvard's intentions, an interested City Councillor, Barbara Ackerman, attended a low-key "public" meeting called by Harvard's Committee on Research Policy to discuss the P3 plans. Simultaneously, the facility was announced in the lead article of a local alternative newspaper and immediately hazardous research in Cambridge became a burning issue, fanned by some local politicians running hard to catch up. They included Mayor Al Velluci who gained national attention for his efforts.** Thus recombinant DNA research became the focus of lengthy City Council meetings at which numerous opposing presentations were given and to which hundreds of people came, not all of them academically affiliated. An unprecedented 6-month moratorium on P3 and P4 recombinant research resulted, an act heard 'round the world, and equally startling, a citizens' review committee made up of non-experts was created to advise on the research hazard.

The experience of the Cambridge Experimentation Review Board (CERB) warrants close inspection as an example of public participation in making science policy. CERB, at the City Council's direction, was selected by the City Manager and consisted of people with neither personal interest in recombinant DNA research nor related professional interests, as with research scientists. Board members — all Cambridge residents, with an equal number of men and women — included a nurse, a social worker, two physicians, a businessman, a saleswoman, a university faculty member, a homemaker and an engineer. Taking its narrow assignment of dealing only with the immediate public health-safety issues, CERB met in both open and closed sessions biweekly for over 4 months and heard 75 hours of testimony ranging from NIH dignitaries and renowned advocates of the research to lab technicians and members of Science for the People. The board's final position allowed the research to proceed but with significantly stricter requirements than NIH. These included strengthening institutional biohazards committees, monitoring escape of vectors,* conducting local epidemiological studies,

*P3: the second highest level of laboratory "containment," ranging P1-P4, for keeping experimental organisms isolated and preventing their escape into the real world, from which they could never be recalled.

**The response of the politicians reflects more than just awareness within their constituencies of recombinant DNA issues. Cambridge has long been dominated by the imperial giants of Harvard and MIT, usually with cooperation from most city politicians, with effects which have included the removal of most of Cambridge's industrial employment and the constant encroachment on traditional working class neighborhoods by university expansion and housing for students, faculty, and the technological elite. In the 60's and early 70's, extensive industrial properties were bought up by the MIT-government-aerospace team to be transformed into an electronics, computers, and weapons research center. (Technology Square, for example, is a former site of numerous manufacturing plants.) The details of this process are contained in *Harvard, Urban Imperialist*, 1969, published by the Anti-expansion, Anti-ROTC committee at Harvard. The rent control law, finally passed in the late 60's with little help from most politicians, was a significant victory reflecting the widespread anger of the people against institutions like Harvard and MIT. The recombinant DNA issue was for the people of Cambridge but another example of imperial decision-making, and many politicians could not afford to let it pass.

*Vectors: organisms containing, in this case, hybrid DNA.

and setting up a city-wide biohazards committee. In addition, CERB recommended that the federal government extend the NIH Guidelines to cover industry, maintain a registry of workers in recombinant DNA labs, and fund health monitoring. CERB rejected assurances from Harvard and NIH scientists that the voluntary NIH Guidelines were a more-than-adequate protection against exceedingly improbable or inconceivable events. The CERB deliberations led to a city ordinance incorporating their recommendations and were in part responsible for the near-passing of another law banning P3 and P4 research indefinitely (defeated 6:5).

CERB's most important contribution was to show that non-experts could judge experts and make creditable public policy judgements.

The CERB report (5) revealed that public policy issues were not allowed to be obscured by the technical debates. This critical evaluation of the claims being made by experts is in sharp contrast to how the Science Court would function, as it has been proposed.*(6)

There were deficiencies in the CERB conclusions, but first let's examine how CERB was able to do what it did. CERB avoided becoming beholden to Harvard, MIT, or the science establishment in part because of the selection process that formed the board, but also because the development of an authority structure or hierarchy was minimized. For example, the original chairperson, who was also Acting Commissioner of Health and Hospitals in Cambridge, removed himself as a voting member on grounds of possible conflict of interest. In addition, all members were encouraged to take part in defining unresolved issues.(5) Finally, at least some members of the committee had a clear perception of political power and the people's interests, as well as an active commitment to working for those interests.

It is evident that the selection procedure which formed CERB cannot be counted on routinely in selecting citizens' boards since the success of this procedure depends on the orientation of the executive officers of, in this case, a municipal government. But even randomly selected committees of interested working people will not escape the problems of elitism, professionalism, and science mystification that affect all of us in contemporary society, unless some members have had experience in combatting this ideology.

The shortcomings of the CERB report reflect conditions which no citizens' committee could have easily overcome. It is unlikely that any representative committee (feeling the immense weight of world attention on its actions) could have strayed very far from the middle of the road in the absence of a visible migration of popular opinion on the issues. While there is considerable consciousness of the hazards possible in recombinant DNA research, very little organization or examination of the

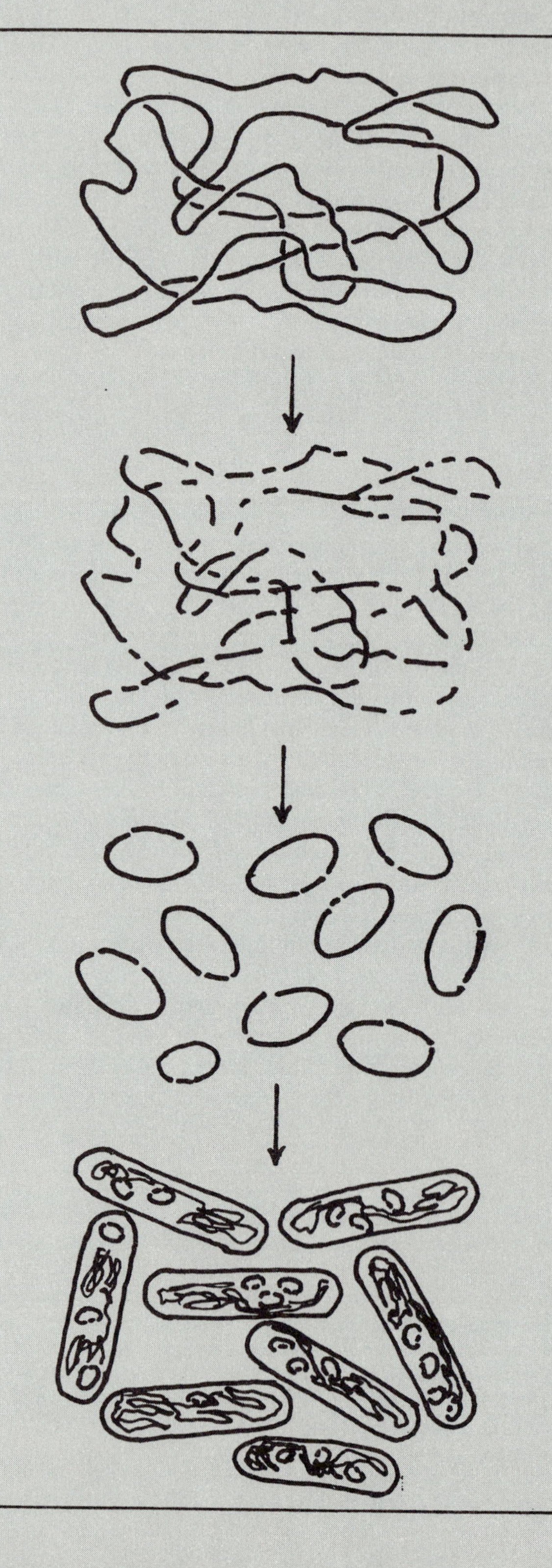

*In the science court concept for resolving disagreements among experts, as originally proposed by A. Kantrowitz, chairman of AVCO Everett Research Laboratory, a panel of scientific experts chosen in the usual manner of elite boards, would cross-examine technical claimants on the "facts," never venturing to examine broader questions of why who might believe what, and of course never similarly exposing themselves.

issues in political terms has developed on a mass scale. Thus it would be bizarre indeed if the committee had, at its own initiative, broadened the scope of its enquiry and pursued in depth questions we believe to be central: the likely specific uses of genetic engineering in *class* terms; the ecological or evolutionary dangers (in terms of infectious disease, soil ecology, and other specific areas); and benefits and risks in broad social terms — who really stands to gain, what are the indirect costs, who is at risk, and what alternatives are being ignored?

Actually, many Cambridge residents were suspicious and concerned over the proposed research at Harvard, according to two City Councillors. An outright ban on the research was favored by some. Had this awareness been better articulated and publicized, perhaps CERB would have taken a stronger stand. The progressive forces in the Cambridge debate could have been very effective in assisting communication between CERB and Cambridge residents.

A major factor in CERB taking a critical approach, aside from the nature of the committee itself, was pressure from a significant opposition minority within the local science "community" and the radical microcosm within Cambridge, both challenging the NIH/Harvard/MIT front. The availability of opposing experts — including technicians — allowed the committee to perceive the political nature of the debate on recombinant DNA research.

There are therefore two main lessons from CERB: 1) With some essential but rarely achievable prerequisites, a citizens' committee can acquire substantial critical expertise free of direct control by nearby institutions and can to some extent reject dominant and respected views. 2) Without a developed progressive movement concretely involved in similar or related issues locally, there are severe limitations to what even a well-selected citizen committee can do in forging an advanced position. This of course confirms the basic strategy of relying on "mass work" — going to, and being part of, the general populace rather than concentrating on influencing law makers, policy-level scientists, or other persons in high places.

A National Forum

Since the Ann Arbor and Cambridge excitement, there have been many smaller replications of the same debate(7). In March, the National Academy of Sciences sponsored a forum to end all forums on recombinant DNA, in Washington, DC. The NAS, the most select organization of elite science,(8) was probably concerned at the course the debate was taking and wished to present a moderate appraisal, especially for congressional staffers and the press. The panel of speakers was relatively balanced; the workshops were dominated by pro-recombinant forces, but the agenda was improved by the heavy turnout of counter-forces: members of the Peoples' Business Commission (formerly Peoples' Bicentennial), the Environmental Defense Fund, and the Coalition for Responsible Genetics Research. The only person at the NAS forum speaking for organized workers was an official of the Oil, Chemical, and Atomic Workers union, who pointed out that the NIH guidelines were ludicrous as far as protecting workers in industry is concerned.

Several developments were apparent. One was recognition of the extent of commercial inroads into recombinant DNA technology: a number of people argued that this technology, based on publicly funded research, should not be exploitable for profit. Another was the isolation of the most self-righteous and adamant proponents of the research from even mainstream, establishment scientists (who were a little embarrassed by this group). By then, in fact, the tide had already started to turn, and forces were being redeployed to the legislative field.

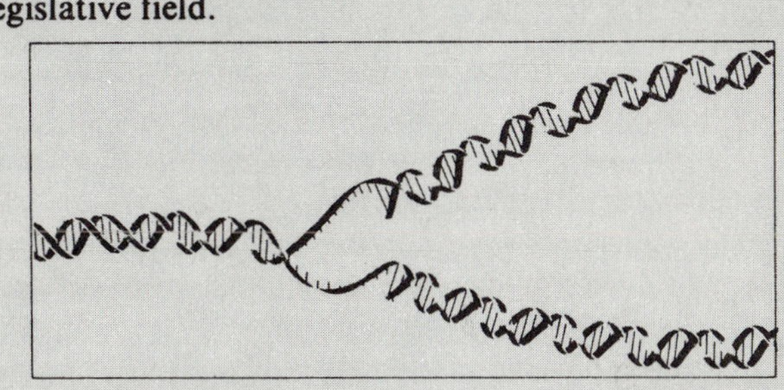

Legislative Shelter in a Storm

Some academic scientists and drug companies who previously had vigorously opposed legal controls on recombinant DNA research emerged in favor of national legislation at the NAS forum. Their position changed because they sought future protection from actions such as occurred in Ann Arbor and Cambridge. Many other people saw the legislation as necessary to cover industrial applications of recombinant DNA technology since the NIH gudelines applied only to government-funded research. As a result, California and New York are both considering legislation to cover the work. Two bills pending in Congress would essentially write the NIH guidelines into law with stiff penalties to enforce them.

The right of local communities to enact their own ordinances is an important issue. But the recent interagency report from the federal government emphasizes that national regulations must pre-empt local or state ones, and many scientists and pharmaceutical firms see this as the main value of the legislation.(9) The bill before the U.S. Senate, sponsored by Edward Kennedy (D., Mass.), gives local communities a real option to enact more strict legislation. Even Joseph Califano, Secretary of HEW, has felt the need to state publicly that he supports a local option.

While federal legislation will clearly give scientists the protection and sanction they need for recombinant DNA work, many are very resentful of the government's interference in their affairs. Philip Handler, president of the National Academy of Sciences, raises the spectre of "constraints that will swathe the research with bureaucratic complexities... and generally frustrate a career in research. If (regulation is) pursued yet further, science could be shattered."(10) A majority of the molecular biologists attending a Gordon Conference in June of this year were greatly aroused by the possibilities of arbitrary government interference in their affairs and stated publicly that earlier warnings by them and others concerning hazards had been exaggerated.(11) Nevertheless representatives of the Pharmaceutical Manufacturers Association concede that dealing with federal inspectors will be nothing new to them. Politically aware scientists at the NAS forum felt similarly. Donald Kennedy, newly appointed commissioner of the FDA and a former Stanford biology professor, went further and said, "Why should there be more regulation? The simple answer, I think, is because it is politically inevitable.... How much regulation are we going to have? Answer: As much as people insist on, in light of their own social value calculus." Biologist Clifford Grobstein, prominent in the debate in California, noted many at the NAS meeting who felt that "science has become too consequential to be left to the self-regulation of scientists or to be allowed to wear a veil of political chastity."(12)

Still Congress may give power to regulate the research to the same agencies — HEW and NIH — that provide most of the funding for the research. The "Recombinant DNA Research Act of 1977," introduced by Carl Rogers (D., Fla.) of the House Committee on Science and Technology, gives the Secretary of HEW full power to make regulations for the research and to license those who undertake it. Just as the Atomic Energy Commission was unable to both promote and regulate nuclear technology, so too, HEW, which runs the NIH, will have a conflict of interest.

The proposed federal regulations may frighten scientists, but it is doubtful they will eventually stymie research. Federal inspectors, according to Kennedy's bill, could examine any laboratory materials and could destroy or confiscate suspected dangerous recombinant organisms as well as recommend heavy daily fines, but enforcement would remain difficult. Inspectors would be hard pressed to see through the mass of laboratory paraphernalia in order to use their power meaningfully. As an alternative, Rogers' bill calls for local biohazards committees to be given the prime responsibility for enforcing the regulations, rather than federal inspectors. Such committees would have one third of their members from outside the regulated research institution and might possibly be more responsive to community concerns than a powerful federal bureaucracy.

Will federal legislation make the NIH guidelines more effective? The guidelines ask biologists to understand and follow relatively strict microbiological techniques which few have been trained in. Molecular biologists, especially, are used to treating the bacteria they study as harmless. Thus the guidelines are certain to suffer from much day-to-day negligence, especially from workers who are convinced there is no clear and present danger. (13) In one typical laboratory, the guidelines reportedly are often ignored.(14) Both congressional bills ask that employees who raise questions about safety be protected from loss of their jobs, but such a provision would be hard to maintain without strong local unions and safety committees.

The federal government is also trying to limit the liability of institutions doing the research. One bill, Rep. Ottinger's H.R. 3191, no longer under consideration by Congress, made it clear that institutions would be liable for an accident whether or not they had violated regulations. The federal task force on recombinant DNA research, however, concluded that if liability were unlimited, then the work might not proceed due to the costs of insurance. Already one contractor, Litton Industries, has bowed out of a government contract involving a high-containment P4 facility in Fredericksburg, Maryland, claiming it cannot get liability insurance.(15) Limiting liability would require legislation similar to the Price-Anderson Act which placed a ceiling on the liability of a power company for a nuclear power accident. Although the act was ruled unconstitutional recently in a federal court, similar provisions might still be written in the case of recombinant DNA research. At the moment, Kennedy's bill states that federal legislation shall not limit a citizen's right to sue over an accident.

Conclusion

Whether or not strong, meaningful laws are passed, requiring the slow, careful development of recombinant DNA technology — and whether they are enforced — depends on the critical consciousness of the people. The task of progressive science workers is to facilitate this process. Furthermore this objective makes sense only if it is broadened to include all interrelated areas, e.g., medical research priorities, occupational and environmental health, and genetic engineering uses. So too, the value of citizens' committees depends on informed popular opinion and agitation. Conceivably, legitimate citizens' committees could be arranged by coalitions of organizations in communities, independent of government, to help clarify technical disputes.

Evaluating experts is a political process. However, there is obviously no guarantee that politically progressive and responsible experts will necessarily have more reliable technical opinions and interpretations of fact. Ideally then, experts should be experienced in collectively defining positions and principles — participating with other, non-expert, working people. In this way the technical discipline and political sensitivities of experts will grow in good directions, along with everyone else's. Organizations are therefore needed in which both experts and non-experts can collaborate in non-elitist and anti-sexist practice toward progressive goals.

When working people begin to routinely and systematically evaluate the credibility of experts, the face of technology will change: governments and business will be less free to design our future against our interest.□

REFERENCES

1. H.M. Cleaver, "The Contradictions of the Green Revolution," *Monthly Review*, June, 1972; Nicholas Wade, "Green Revolution (1): A Just Technology, Often Unjust in Use," *Science*, 186, 1093, 1974.
2. Jon Beckwith, "Recombinant DNA: Does the Fault Lie Within Our Genes?," *Science for the People*, May-June 1977, p. 14.
3. *Drug Research Reports*, Dec. 3, 1976.
4. Francine Robinson Simring, "The Double Helix of Self-Interest," *The Sciences*, May/June, 1977, p. 10.
5. "The Cambridge Experimentation Review Board," (report of), *Bulletin of the Atomic Scientists*, May 1977, p. 22.
6. "The Science Court Experiment: An Interim Report," *Science*, 193, 653, 1976; Arthur Kantrowitz, "Controlling Technology Democratically," *American Scientist*, Sept.-Oct. 1975, p. 505.
7. Nicholas Wade, "Gene-Splicing: At Grass-Roots Level a Hundred Flowers Bloom," *Science*, 195, 558, 1977.
8. Nicholas Wade, "*The Brain Bank of America:* Auditing the Academy," *Science*, 188, 1094, 1975.
9. "Interim Report of the Federal Interagency Committee on Recombinant DNA Research: Suggested Elements for Legislation;" submitted to the Secretary of Health, Education, and Welfare. March 15, 1977.
10. *Chemical and Engineering News*, May 9, 1977, p. 3; editorial excerpted from Handler's annual report to the National Academy of Scientists.
11. "Recombinant DNA Research: Government Regulation," letter to *Science*, 197, 208, 1977.
12. Clifford Grobstein, "The Recombinant DNA Debate," *Scientific American, July, 1977.*
13. Richard P. Novick, "Present Controls are Just a Start," *Bulletin of the Atomic Scientists*, p. 16, May 1977.
14. Janet L. Hopson, "Recombinant Lab for DNA and My 95 Days in It," *The Smithsonian*, June 1977, p. 55.
15. "Scientist-critics less fearful now of DNA research," *The Boston Globe*, July 18, 1977.

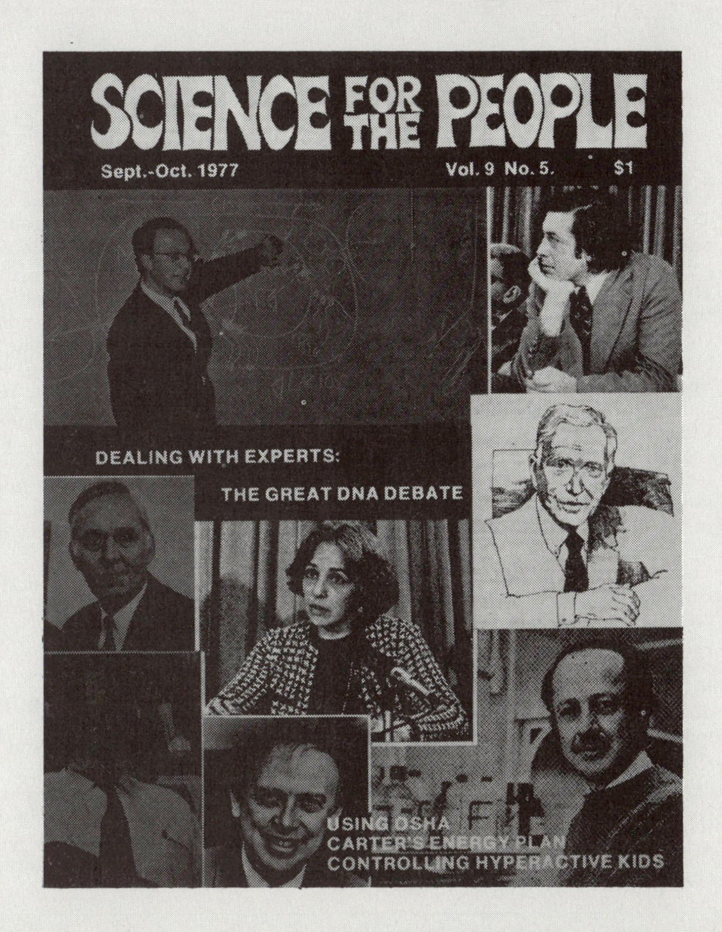

Document 9.4

December 17, 1977
Washington Post

But Some Colleagues Disagree

Discoverer Would Lift Curbs on DNA

By Victor Cohn
Washington Post Staff Writer

Dr. James Watson, the Nobel Prize-winning co-discoverer of the structure of DNA, the stuff of genes, surprised colleagues yesterday by calling for an end to government restrictions on DNA research as unneeded "nonsense."

He said "there is no evidence" that experiments in recombining DNA can do anyone any harm, the time and dollars wasted on maintaining restrictions is "enormous," and he and other scientists who recommended the precautions in the first place were "stupidly" wrong.

The emotional plea from the slender, intense Watson—who three years ago just as emotionally called for the restrictions—sent a shockwave through a National Institutes of Health advisory committee weighing changes in the restrictions.

But it did not make any of the advisers recommend abandoning them entirely without more evidence that some of the original fears of Watson and other scientists can be forgotten.

The advisors, in fact, ended a two-day public hearing with a discussion in which most of them endorsed most of NIH's new plans to modify the restrictions but keep them in effect for most researchers.

"Watson has put some questions in our mind that I think we need to consider very carefully," but "I don't think he's right—he's his usual extreme," said Dr. Walter Rosenblith, provost of Massachusetts Institute of Technology.

"Jim is speaking, I think, for a large group of scientists"—that is, a probable majority who now agree with him—and much of what he says may be correct, commented Dr. Alexander Rich of MIT, co-chairman of a National Academy of Sciences forum on the DNA problem this year. But the "prudent" and "correct" course, he said, is to modify the rules only as new evidence shows the research to be safe. It was the "extreme" Watson who, as an extremely emotional and highly original young man, joined Britain's Francis Crick to first describe the double-helix-shaped molecule which is the chemical of the genes in every organism.

Three years ago, Watson and a small group of others first warned that new experiments splitting and then "recombining" the DNA of various organisms could create new life forms having unknown powers.

The precautions have become a "disaster," Watson said yesterday, and the public has been misled into fearing "madman scenarios."

Various scientists' original fears were, mainly, that a cancer virus might be let loose, or other diseases or epidemics might be created, or that scientists might cause some subtle but disastrous changes in the genes of even lowly animals or organisms to somehow disturb the whole earth's ecology.

The first two fears were the main ones, Watson said, and laboratory experiences have now shown that, first of all, no one has gotten sick and, second, organisms in nature exchange DNA widely without disaster.

On a scale of things to worry about, he maintained, genetic research is "trival." And "if you worry about changing man's genetic nature," he said, "you should worry" about regulating sex. "Maybe we should have a committee there" or "require that all sex be performed at Fort Detrick," he said.

But some advisory committee members said genetic research is moving humankind toward genetic engineering, so a pause now for reflection is wise and not rash.

"God forbid" what the public would have thought if scientists themselves had not called attention to some of the dangers, said Patricia King, associate law professor at Georgetown University.

The scientific evidence on recombinant DNA's possible safety and the possible ubiquity of DNA combinations in nature is just starting to come in "and most of us haven't seen the scientific papers yet," said Dr. Robert Sinsheimer. A prominent biologist and critic of this research, he recently became chancellor of the University of California at Santa Cruz.

Restricting DNA Experiments

ONE OF THE TASKS Congress left undone this year was the passage of legislation to set regulations and limits on "gene-splitting" experiments. Its failure to act is understandable. The subject is new and usually difficult to understand. There is no consensus among scientists, or among those who observe science, on how the issues involved should be resolved. But a conference at the National Institutes of Health last week underlined, at least to us, the need for the congressional committees that began work on the subject early this year to get on with the job.

Gene-splitting is the term often used for the techniques involved in splicing together genetic material (DNA) from different species and inserting the result in bacterial cells. The experiments hold great promise for yielding knowledge about human life and ways to increase and improve it. But they also hold some dangers in the possible creation of new life forms and in the manipulation of the characteristics of future generations of human beings. Those dangers led NIH to establish safety guidelines for such experiments conducted by individuals and institutions receiving federal money. They have also encouraged several states and cities to conscder imposing safety rules of their own.

Many of the scientists who spoke at NIH believe that the existing rules are too stringent. Some want the rules abolished. They fear that this first step in the regulation of scientific experiments will lead to further government control over other areas of science. That, they believe, would be disastrous to the spirit of free inquiry on which basic science depends.

Ironically, those who make the loudest arguments against government regulation in this field provide their opponents with substantial ammunition. Underlying their rejection of any regulation is the suggestion that scientists know what is safe and that the rest of us must trust their judgment—not their collective judgment, that is, but their *individual* judgments. This is a dubious proposition, at best, and not one that is likely to be politically acceptable at a time when technological advance seems to be moving faster than the capacity of the public to understand or absorb it.

It may be that the rules NIH currently has in place are inadequate. For one thing, they do not cover the research being done by industrial scientists; for another, they do not seem to have quieted the fears of local governments. Congress should fill those voids with legislation that sets national safety standards for all research in the gene-splitting field. The scientific community will serve the country best by working with the appropriate congressional committees to ensure that those standards do not restrict research unduly but also do not expose individuals and communities to unnecessary risks.

Document 9.5

January 6, 1978
Science (199:30)

NEWS AND COMMENT

Gene-Splicing Rules: Another Round of Debate

Most biologists concerned with recombinant DNA consider that modest relaxations would now be justified in the National Institutes of Health rules governing the research.

But the agency has come under serious criticism for the way in which it has gone about revising its rules. At a public meeting held at the NIH campus on 15–16 December, Washington attorney Peter B. Hutt lectured the NIH for proposing the revisions with "undue, unnecessary and unseemly haste." Another attorney, Georgetown University law professor Patricia King, told the NIH that she was "terribly upset by the procedure for making the revisions."

Hutt's criticisms are significant because, as former general counsel for the Food and Drug Administration, he has acquired particular expertise in the regulation of biomedical issues. He is also a warm admirer of the scientific content of the NIH guidelines and of scientists' initiative in framing them.

At the NIH meeting, Hutt criticized not only the procedural basis of the present revisions but NIH's legislative and political strategy for handling the recombinant DNA issue over the last 2 years. He believes that instead of listening to those who didn't want to be regulated, the NIH should have followed the advice he gave in February 1976 to maintain the initiative and use already existing legislative authority to regulate gene-splicing research. "When NIH and HEW decline to use their legislative authority, they invite others to step into a regulatory void," Hutt opines. He regards the recent bills introduced in Congress as "the worst form of over-reaction" and Senator Kennedy's in particular as "an utter atrocity."

The authority which Hutt urged the NIH to use in 1976 is a statute which gives the Surgeon General sweeping power to control communicable diseases (*Science*, 27 February 1976). At last month's meeting Hutt said he had checked with other government attorneys and still felt strongly that the statute "provides ample legal authority." NIH director Donald Fredrickson, however, says the legal advice he has always received is that the statute is insufficient and that attempts to use it would be challenged in the courts. The Pharmaceutical Manufacturers Association has advised the NIH that it would accept the statute, even though it would have to be "stretched" a little, as a basis for regulating industry.

The purpose of last month's meeting was to consider the guideline revisions prepared by the NIH recombinant DNA advisory committee in May and June 1977. The revisions, which constitute generally minor relaxations of the present guidelines, are chiefly inspired by the realization that *Escherichia coli* K12, the standard bacterium used to propagate recombinant DNA molecules, is a much safer host than was originally believed. A principal fear at the time the guidelines were framed was that new genes spliced into *E. coli* K12 might accidently convert it into a pathogen, perhaps of epidemic potential. K12 now seems to be even more enfeebled than was thought; moreover, deliberate attempts to convert it into a pathogen have been unsuccessful, suggesting that accidental conversion is rather unlikely.

The data and arguments supporting this view were marshaled at a conference held in Falmouth, Massachusetts, this June. The conclusions of the Falmouth conference were not seriously challenged at the NIH meeting, but what upset the attorneys and others was the way they had been used. Once it was decided to use the conference for public regulatory purposes, Hutt explained, "there was an obligation to make that data available to the public." That and much other unpublished data on which the revisions depend should have been issued in time for public comment. Hutt listed five other examples of what he regarded as procedural flaws in the proposed revisions. He gave high marks to the science, but the regulatory quality of NIH's work struck him as "pedestrian at best." The guidelines could "probably be loosened even further, but you can't do that if you don't have public procedures built in. You have to recognize that we live in a participatory democracy," Hutt believes.

A notable dissenter from the scientific consensus at the NIH meeting was Robert Sinsheimer, now chancellor of the University of California, Santa Cruz. He restated his previous belief that the present guidelines are "extraordinarily anthropocentric," in being focused on the threat to human health rather than to the environment in general. Noting the "extensive reliance on unpublished data," and the difficulty of interpreting experiments he hadn't seen, Sinsheimer said he was "hesitant to endorse the numerous changes which correspond to a decrease in the level of containment."

The NIH guidelines were criticized from the opposite direction by James Watson of Cold Spring Harbor. Watson, a member of the National Academy group that called for a moratorium on certain experiments in July 1974, had already decided by the time of the Asilomar conference in 1975 that all restrictions on gene splicing should be lifted, and has been saying the same ever since. His extempore oration at the NIH meeting seemed to strike a chord even with those who didn't entirely agree with him.

The gist of his remarks is as follows:

> As one of the signers of the original moratorium, I apologize to society.
>
> One fear we had was cancer; the other, on the part of left wing liberals like myself, was of the CIA.
>
> The second fear I never thought much of because I was once invited to Fort Detrick and they had got nowhere. As for the Rube Goldberg scenario of a tumor gene getting into *E. coli* and thence to people, that just never could have happened.
>
> When the public heard of it, you got this fear of illegitimate sex, of mixing things that

didn't ordinarily mix up. Upon reflection, this was very silly. I think DNA moves around a lot. It doesn't always do good or bad, but it is going to happen.

The whole thing is that you just don't know. But we don't live in a risk-free society.

I'm drawing up the Whole Risk Catalog. Under D I have dogs, doctors, dioxin—where do I put DNA? Very low.

We have had enormous attention paid to people having evaded the guidelines. We should be careful at the penalties we impose on people who cheat at tiddliwinks.

This is nonsense, this whole hearing is nonsense. There are lots of things that scare the shit out of me, like Tris, but recombinant DNA, no!

This is supposed to be a great dialogue between scientists and the public. I can't think of a worse subject, because there is nothing to discuss.

The question now is, what is the best way to get out of this political mess.

Science is good for society. We are being attacked by everyone who doesn't have the guts to go ahead.

The dangers of this thing are so slight—you might as well worry about being licked by a dog.

The upshot of last month's meeting was to give NIH a reasonably coherent message, to the general effect that the science behind the proposed revisions is fine but the regulatory aspects need more work. Fredrickson agrees that Hutt "has some legitimate criticisms," and concedes that NIH is not expert in regulatory matters. But, he adds, recombinant DNA is not like a routine FDA problem: "We are not regulating Campbell's soup."

The NIH director believes the agency has the responsibility to propagate standards for gene-splicing research but he is unhappy about enforcing them as well. He would like to see some other agency, such as the Center for Disease Control, assume that task. "I think it is a conflict of interest for the NIH to be both the sponsor, conductor and regulator of this kind of research," he told the meeting. "My own belief is that it would be to the maximum advantage of the country for a very simple legislative package to be passed extending the existing guidelines to everyone."

Fredrickson's decision on whether to adopt the proposed revisions is formally independent of whatever action Congress may take during the next session. Last sessions' attempts to frame legislation failed to reach the floor of either the House or Senate, but Congress has not yet lost interest in the issue.

—NICHOLAS WADE.

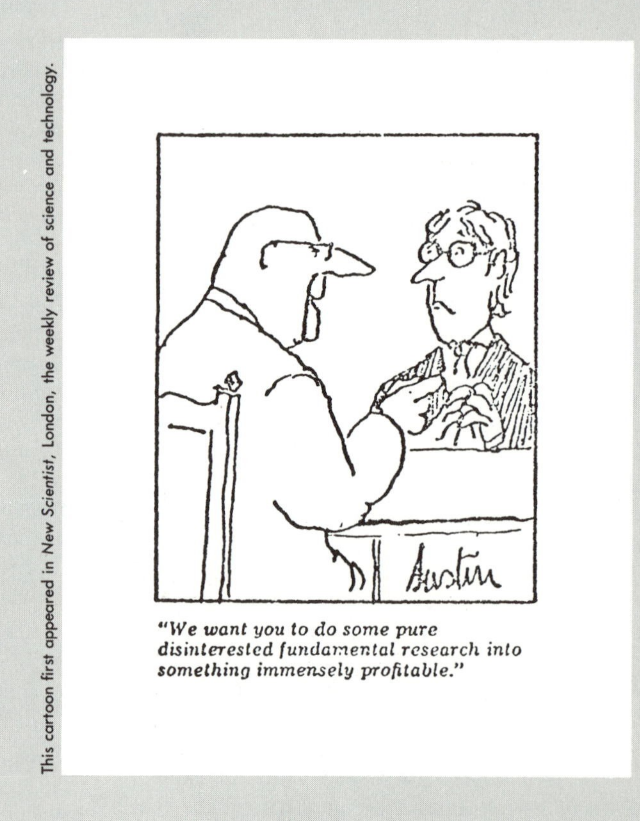

"We want you to do some pure disinterested fundamental research into something immensely profitable."

This cartoon first appeared in *New Scientist*, London, the weekly review of science and technology.

Document 9.6

January 6, 1978
Science (**199**:31)

Harvard Gene Splicer Told to Halt

Harvard Medical School biologist Charles A. Thomas has been instructed by the National Institutes of Health to put a temporary halt to his research with recombinant DNA. The halt is to continue until anonymous allegations that the NIH guidelines on gene-splicing research were broken in Thomas's laboratory have been checked out. Thomas was until recently a member of the NIH committee which framed the guidelines.

The basis for the NIH action is a purely procedural oversight which may well turn out to be the mistake of others besides Thomas—the failure to file with NIH a document required by the guidelines and known as a Memorandum of Understanding and Agreement (MUA). Thomas declines to discuss the matter but a friend of his, MIT biologist Alexander Rich, says the absence of the document "appears to be a three-way error" on the part of the NIH, the Harvard biohazards committee, and Thomas. An NIH official says NIH was not to blame, because Thomas's grant application—the continuation of an existing grant—did not mention recombinant DNA.

The allegations were indirectly outlined to the NIH around the beginning of December. A Freedom of Information request filed on 6 December by Environmental Defense Fund staffer Leslie Dach prompted the NIH immediately to dispatch to Harvard a three-man team headed by NIH investigator James Schriver.

The team discovered the matter of the missing MUA. Schriver, on a second visit to Harvard, has asked the university to take responsibility for assessing other allegations. One is that Thomas used his lab, rated as a "P2 level" containment facility, to do an experiment requiring the P3 level of containment. Rich says his information is that the experiment in question was completed before the NIH guidelines came into effect in June 1976, and that all Thomas's subsequent research has been at the P2 level. A press statement issued by Thomas declares that all research in his laboratory "has been done in strict accord with the NIH guidelines."

On present evidence the whole episode may amount only to the misplacement of a piece of paper, a slipup doubtless in part attributable to the teething troubles of setting up a new system. Yet in sending its team to Harvard and asking Thomas to postpone research, NIH has taken unusually strong steps, and there may be more to follow. "If we find out Thomas has been doing recombinant DNA research without an MUA, then some appropriate action is going to have to be taken, but what that is I don't know," says an NIH official.

The NIH's regulatory vigor probably has much to do with an uncomfortable morning spent by agency officials before Senator Adlai Stevenson's science and space subcommittee. At the November hearing, Stevenson repeatedly pressed NIH director Donald Fredrickson to explain the agency's tardiness in investigating the breach of its guidelines which occurred early this year at the University of California, San Francisco, and to say what sanctions NIH proposed to exercise against the institution. NIH's reaction to the Harvard situation seems designed to deny Stevenson a second occasion for the same criticisms.

The allegation that a pharmaceutical company, Miles Laboratories, is supporting research not permissible to NIH grantees under the present guidelines was made at last month's meeting at the NIH. The project apparently referred to is an attempt by Frank Young of the University of Rochester, New York, to develop a cloning system involving *Bacillus subtilis* instead of the usual *Escherichia coli*. The allegation is baseless, however, because no cloning has yet taken place on the Miles project. When a cloning system has been developed, Young says, he will submit it for approval to the NIH before further use even though, as an industry-supported project, NIH approval is not required.—N.W.

Document 9.7

September 30, 1977
Science (**197**:1342)

NEWS AND COMMENT

Recombinant DNA: NIH Rules Broken in Insulin Gene Project

A breach of National Institutes of Health rules on gene splicing occurred earlier this year in the Department of Biochemistry and Biophysics at the University of California, San Francisco, one of the leading centers for practice of the new technique. No hazard resulted, but the episode underlines some of the difficulties experienced by research laboratories in adapting to the new rules.

The breach was the use of a biological component, or "vector," before it had been certified by the NIH director. The researchers, a team engaged in isolating the rat gene which codes for insulin, say they destroyed the experiment as soon as they realized their mistake.

The experiment was repeated in a certified vector and published in *Science* on 17 June. It received considerable attention because the researchers had achieved, much earlier than expected, the first step toward the goal of isolating the human insulin gene and using it for the manufacture of insulin protein. The UCSF team was in competition with a group at Harvard which was known to be working with a better source material.

UCSF's preeminence in the gene-splicer's art has brought it some mixed blessings. Because of the practical implications of what its researchers are doing, a company called Genentech has established a relationship with Herbert Boyer, one of the pioneers of the technique. Members of the insulin team have set up a nonprofit corporation, the California Institute for Genetic Research. These commercial developments are a tribute to the department's success, but have also created internal stresses. "Capitalism sticking its nose into the lab has tainted interpersonal relations—there are a number of people who feel rather strongly that there should be no commercialization of human insulin," says UCSF microbiologist David Martin.

Another mixed blessing is fame, which has attracted press attention not only to the department's achievements but also to certain internal tensions. A lengthy and circumstantial article in the June issue of the *Smithsonian* called into question the respect accorded to the NIH safety rules by UCSF researchers, and in particular by the younger, postdoctoral workers who perform most of the experiments. Written after a 3-month internship in Boyer's lab by Janet L. Hopson, formerly a reporter for *Science News*, the article observed that "half of the researchers here follow the guidelines fastidiously; others seem to care little. . . . Among the young graduate students and postdoctorates it seemed almost chic not to know the NIH rules," Hopson noted. In a letter to the editor criticizing the article, Boyer stated that "In practice [the NIH rules] are followed seriously."

The stresses of both commercial success and media attention came together this May when the insulin team announced their production of the rat gene. Other researchers resented not only the intrusive presence of the press but also the fact that they were hearing of their colleagues' success for the first time. The team had worked in unusual secrecy, which many regarded as inappropriate in an academic setting as well as disruptive. "People would stop talking when you came into the room, or change the subject if you tried to make conversation about how the insulin project was going," says UCSF biochemist Brian McCarthy. The insulin team say that no secrecy was intended, and that it was the speed of obtaining results that occasioned surprise.

The secrecy and suspicion surrounding the insulin gene experiment, together with perhaps a touch of resentment, helped to fan rumors within the department alleging that the NIH rules had been broken and even that the experiment published in *Science* might not have been performed as described. The focus of the rumors was the obvious fact that the whole intricate experiment had been completed only 3 weeks after the NIH had certified the vector which the researchers used. Even experienced gene splicers were surprised by the rapidity of execution. "It is conceivably possible to do such an experiment in three weeks if everything works perfectly the first time, but you know as well as I do that science never works as well as you hope," a member of the department remarks.

"Well, what can I say? It did work well. We were all set to go," says a member of the insulin team queried on the speed of the experiment. Yet members of the team concede that an earlier experiment took place, but say it was aborted half way through and before any pertinent information had been gained.

The episode of the earlier experiment illustrates the problems experienced by research laboratories both in acclimatizing to the NIH's gene-splicing rules and in devising ways of enforcing them. "We can't run a policing service," says David Martin, until recently the chairman of the UCSF biosafety committee. "What we do is to try to raise the consciousness of the individuals involved to make them respect the guidelines." Martin spent 2 days investigating the episode and concluded that a breakdown in communications was responsible. Precise details remain obscure because the minutes of the biosafety committee record only Martin's conclusion, and a lengthier account prepared for the minutes by the present chairman, James Cleaver, is far from complete.

The episode revolves about the use of the vectors—virus-like entities known as plasmids—which are used to carry genes of experimental interest into bacterial hosts. Two particularly useful plasmid vectors, known as pBR322 and pMB9, had been developed by Boyer and others. But before either could be used, it had to go through a two-stage administrative process, the first of being approved by the NIH recombinant DNA committee and the second of being certified by the NIH director. Vectors, in other words, are off limits to researchers until certified.

Plasmid pBR322 was tentatively approved by the NIH committee on 15 January, finally approved on 23 June, and certified for use on 7 July. Plasmid pMB9 was certified on 18 April.

According to both Martin and two members of the insulin team, the episode of the earlier experiment was as follows. Early this year, sometime in February, an attempt was made to produce copies of the rat insulin gene with pBR322 as the vector. The gene was linked to pBR322 and the plasmids inserted into NIH-certified bacteria designated EK2 hosts. Some of the bacteria were successfully colonized by the plasmids. To verify that the clones of colonized bacteria contained the rat insulin gene it would have been necessary to carry the experiment to completion by extracting and analyzing the DNA sequence of the genes. This step was not performed because at that moment, on or around 1 March, the team say they learned that the plasmid had not been certified for use. They decided to destroy all their clones. Further attempts were made to clone the gene with an already certified vector, known as pCR1, but without success. When pMB9 was certified, team members say, they had all their materials ready to go, and repeated the experiment from scratch in the new plasmid. Martin says that he inspected the team's records and has "complete confidence" in their statement that the entire experiment was done after the approval of pMB9. "When the manuscript appeared so soon, people said 'How in hell can you do it?' But you can do it—the experiment was very straightforward," Martin said from Great Milton near Oxford, England, where he is at present on sabbatical. "I think they realized they would be questioned and that if there were any shenanigans they would be really jeopardizing a hell of a lot, not only their own careers but the whole advance of science through recombinant DNA technology."

In the categories of the NIH rules, the experiment with pBR322 was assigned to the highest available containment level short of going off campus to a specially secure laboratory. But it is clear that the experiment posed no issue of public health since it was performed in the required type of laboratory—a "P3" facility—and with a vector which has now been certified as safe.

The experiment also took place at a time when the NIH rules were still new and local procedures for implementing them were still in a formative stage. "This was one of the inevitable things that happened as we tried to evolve a new system," comments biosafety committee chairman Cleaver.

Not wholly plain from the Martin-Cleaver accounts of the episode is how the insulin team came to believe it was all right to go ahead with pBR322 prior to certification. Howard Goodman, chief of the laboratory which did the cloning and sequencing part of the experiment, is out of the country, as he has been for most of the past year. One of his postdoctoral colleagues says there was great confusion at the time about the status of the plasmid but the reason for the confusion is "sort of cloudy now."

According to Martin, it was clear that everyone knew that pBR322 had been approved but not certified, but the NIH, Martin says, was advising researchers that certification was imminent and that they should go ahead. According to the minutes of the 20 May meeting of the UCSF biosafety committee, Martin reported that the researchers "had been verbally informed that the certification of an approved EK2 vector was imminent and to proceed with its use."

This version is strenuously denied by William Gartland, director of the NIH Office of Recombinant DNA Activities, and by his only assistant, Daphne Kamely. Both say that they would never have advised use of any vector before certification. Gartland notes that the team "must have got the vectors from Boyer, who certainly knew they were not certified" because Boyer had complained repeatedly to the NIH of the delay in certification. According to Boyer, the insulin team "kept on asking" if the plasmid had been certified and he told them it had not. Boyer states that he never encouraged anyone to go ahead prior to certification. Thus the source of encouragement for the team to go ahead prior to certification remains obscure.

A different account from Martin's is given by William Rutter, a member of the insulin team and chairman of the UCSF Department of Biochemistry and Biophysics. In its memorandum filed with the UCSF biosafety committee, the insulin team had said it would use as vectors pCR1 and any other vectors that might in future be approved by the NIH recombinant DNA committee. When the NIH committee approved pBR322 on 15 January, Rutter says—(the committee gave tentative approval on 15 January and full approval on 23 June)—he therefore assumed that pBR322 was sanctioned for use, since he was not then aware of the NIH distinction between approval and certification.

The UCSF biosafety committee did not learn until May that the pBR322 experiment had taken place. Researchers doing recombinant DNA experiments are required to file a description of the experiment for committee approval. But, as Rutter has said, the memorandum filed by the insulin team did not mention pBR322 specifically. Researchers using the P3 laboratory at UCSF are also required to sign a logbook describing their experiment. Yet Martin says that when he inspected the logbook at the outset of his investigation, he found "nothing recorded." Those in charge of the facility "were not being compulsive enough in seeing that people were filling in the logbook"—a situation which has now been corrected, Martin adds.

Cleaver, Martin's successor as biosafety committee chairman, told *Science* that two entries from the insulin team are recorded in the logbook between 1 February, when the logbook was instituted (the P3 lab officially opened on 9 November 1976), and the end of April. An entry on 1 February notes in the column headed "vector" that pCR1 will be used, and the second entry on 23 April gives pMB9 as the vector. The pBR322 experiment, according to Rutter, took place after 1 February. The vector was not mentioned, he says, because the 1 February entry referred to the general experiment already described in the memorandum filed with the biosafety committee, in which the team had said it would use pCR1 and other approved vectors, and the logbook has to be signed only for each experiment, not for each use of the laboratory. "The signing in of the logbook meant for the insulin cloning experiments in general. The vectors were not designated specifically, and that was just human error," says Rutter. Rutter's laboratory provided the insulin gene for the experiment; it was members of Goodman's laboratory—Goodman was away until mid-April—who performed the cloning experiments.

How should a local biosafety committee respond to an incident of this sort? "I would have expected that the biohazards committee to have investigated the whole thing," says Gartland. Martin did conduct an investigation, and he took action both with the NIH and the insulin team. "I felt comfortable we had resolved the question and eliminated the possibility of it happening again," he says. But in fact, written documents of the committee record criticism only of the NIH.

The pBR322 experiment raises no question of hazard but it does raise the possibility that the insulin team might have gained an unfair advantage over other researchers who had abided by the NIH rules. Another team at Harvard is also working on the same problem. Members of the UCSF team say that they gained no information from the pBR322 experiment which was helpful to the later experiment with pMB9. As it happens, the Harvard team was not neck-and-neck with UCSF because it has not even now published any results.

As far as is known the pBR322 experiment is the only occasion on which the NIH rules governing recombinant DNA research have been broken. The researchers say that the breach was the result of innocent error, a statement not refuted by the available evidence. The experiment presented no hazard to public health nor, in the event, was any unfair advantage gained over competitors. As for the UCSF biosafety committee, its response included action to ensure against repetition of the incident, although not a full public account. The committee's discussion of the experiment, as reflected in the minutes of its 20 May meeting, is confined to an attempt—unsupported by available evidence—to ascribe the error to confusion generated by NIH. But both the experiment and the biosafety committee's response to it occurred in circumstances to which researchers were then still adapting, and for which there were few, if any, precedents.—NICHOLAS WADE

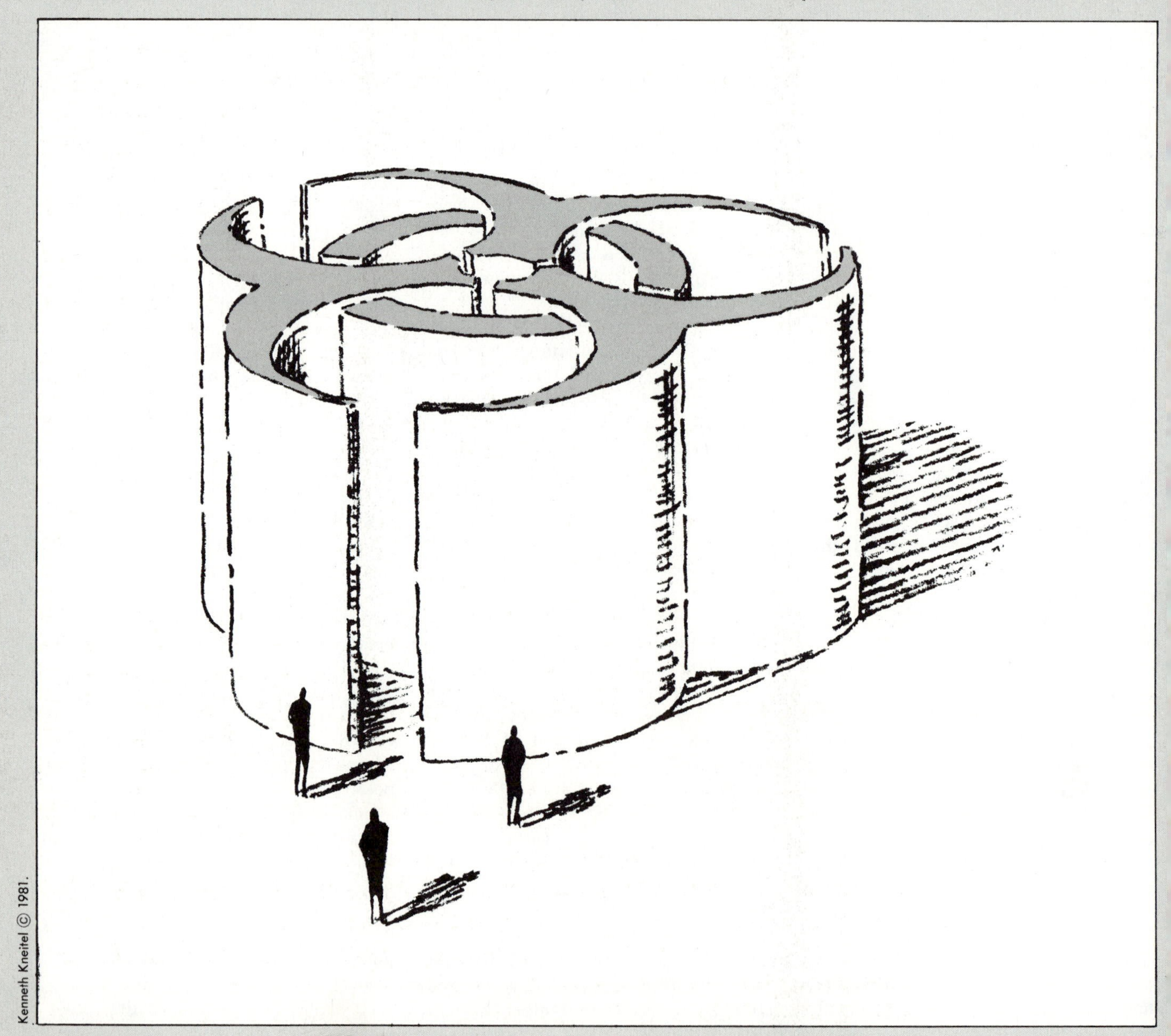

Document 9.8

December 9, 1977
Science (**198**:1056), Excerpt

Expression in *Escherichia coli* of a Chemically Synthesized Gene for the Hormone Somatostatin

Abstract. *A gene for somatostatin, a mammalian peptide (14 amino acid residues) hormone, was synthesized by chemical methods. This gene was fused to the* Escherichia coli *β-galactosidase gene on the plasmid pBR322. Transformation of* E. coli *with the chimeric plasmid DNA led to the synthesis of a polypeptide including the sequence of amino acids corresponding to somatostatin. In* vitro, *active somatostatin was specifically cleaved from the large chimeric protein by treatment with cyanogen bromide. This represents the first synthesis of a functional polypeptide product from a gene of chemically synthesized origin.*

The chemical synthesis of DNA and recombinant DNA methods provide the technology for the design and synthesis of genes that can be fused to plasmid elements for expression in *Escherichia coli* or other bacteria. As a model system we have designed and synthesized a gene for the small polypeptide hormone, somatostatin (Figs. 1 and 2). The major considerations in the choice of this hormone were its small size and known amino acid sequence (*1*), sensitive radioimmune and biological assays (*2*), and its intrinsic biological interest (*3*). Somatostatin is a tetradecapeptide; it was originally discovered in ovine hypothalamic extracts but subsequently was also found in significant quantities in other species and other tissues (*3*). Somatostatin inhibits the secretion of a number of hormones, including growth hormone, insulin, and glucagon. The effect of somatostatin on the secretion of these hormones has attracted attention to its potential therapeutic value in acromegaly, acute pancreatitis, and insulin-dependent diabetes.

The overall construction of the somatostatin gene and plasmid was designed to result in the in vivo synthesis of a precursor form of somatostatin (see Fig. 1). The precursor protein would not be expected to have biological activity, but could be converted to a functional form by cyanogen bromide cleavage (*4*) after cellular extraction. The synthetic somatostatin gene was fused to the lac operon because the controlling sites of this operon are well characterized. ...

The amount of somatostatin synthesized was variable and about a factor of 10 less than the maximum predicted yield. This variability could be interpreted in several ways. Protein degradation by endogenous proteases, the inability to fully solubilize the chimeric protein, and the selection of altered plasmids could all be contributing factors to the variability in yield. Although recombinant DNA experiments with chemically synthesized DNA are inherently less hazardous than those with DNA from natural sources, consideration should be given to the possible toxicity of the peptide product. A major factor in the choice of somatostatin was its proven low toxicity (*3*). In addition, the experiment was deliberately designed to have the cells produce not free somatostatin but rather a precursor, which would be expected to be relatively inactive. The cloning and growth of cell cultures were performed in a P-3 containment facility.

KEIICHI ITAKURA
TADAAKI HIROSE
ROBERTO CREA
ARTHUR D. RIGGS
Division of Biology, City of Hope National Medical Center, Duarte, California 91010

HERBERT L. HEYNEKER*
FRANCISCO BOLIVAR†
HERBERT W. BOYER
Department of Biochemistry and Biophysics, University of California, San Francisco 94143

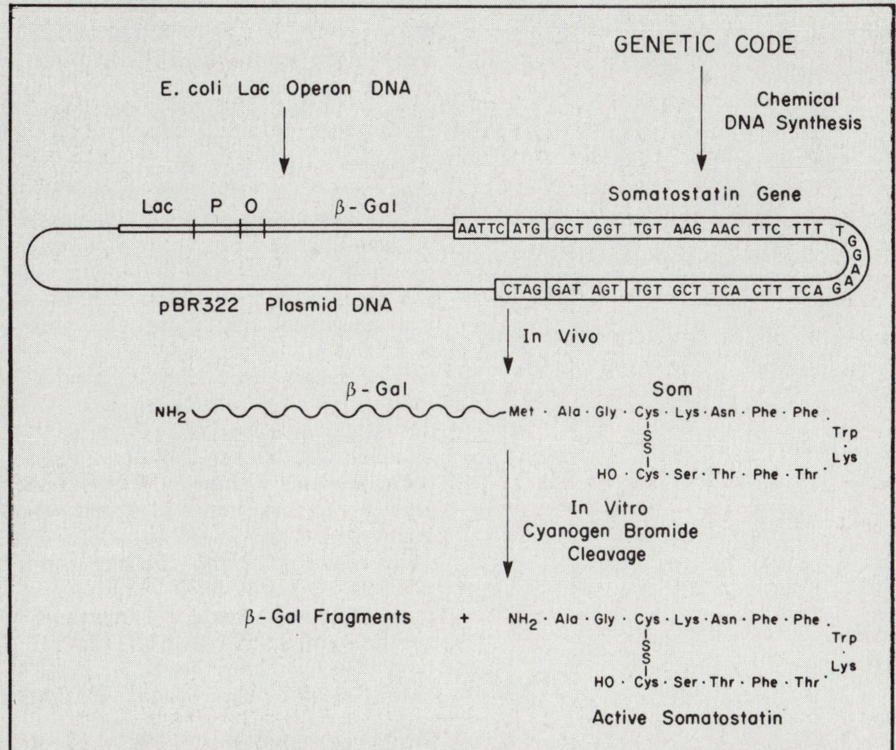

Fig. 1. Schematic outline of the experimental plan. The gene for somatostatin, made by chemical DNA synthesis, was fused to the *E. coli* β-galactosidase gene on the plasmid pBR322. After transformation into *E. coli*, the chimeric plasmid directs the synthesis of a chimeric protein that can be specifically cleaved in vitro at methionine residues by cyanogen bromide to yield active mammalian peptide hormone.

Document 9.9

December 12, 1977
Business Week

RESEARCH

A commercial debut for DNA technology

A tiny San Francisco company, just two years old, has scored a biomedical research coup that may have left its competitors in the dust. Genentech Inc. will get the patent rights to a new means of producing a brain hormone called somatostatin. But the exciting news is that scientists for the first time have employed controversial recombinant-DNA (gene-splicing) technology and the young science of artificial gene synthesis to produce the hormone. In addition, somatostatin has potential both as a research tool and as a medicine, and variations on its structure might well open the way for a whole new family of drugs capable of treating diseases that today defy medicine's best efforts.

The scientific breakthrough came at the University of California at San Francisco, where researchers—along with the City of Hope Medical Center in Duarte, Calif., and the Salk Institute—had been pursuing the new technique since mid-1976. "Molecular biology has reached the point where it can become involved in industrial applications," says Herbert W. Boyer, leader of the research team and a co-founder of Genentech, who now serves as a consultant to the company. "Our strategy," says Robert A. Swanson, Genentech's 30-year-old president, "is to concentrate solely on recombinant DNA and to manufacture and market products to major medical, pharmaceutical, and industrial companies."

Lots of competition. Genentech's connection with UC-San Francisco has led to unease among scientists in Boyer's lab—a feeling that is shared by some science policy advisers within the White House. And the advance comes at a time when many scientists and citizens still worry about recombinant-DNA research and its potential for harm.

Nevertheless, there are nearly 300 recombinant-DNA research programs now under way in the U. S., most of them funded by the National Institutes of Health (NIH), which oversee the safety of such experiments. Though Genentech followed NIH guidelines, because of the unique arrangement covering its research, the company will be first to exploit the somatostatin results commercially. Once production is under way—perhaps by the middle of next year—UC-San Francisco will share in the royalties, along with the City of Hope where the gene synthesis work was done.

Funding such research is expensive: The somatostatin experiments alone cost several hundred thousand dollars. But Swanson claims to have raised nearly $1 million in backing so far from sources such as International Nickel Co. and his former employer, the venture capital firm of Kleiner & Perkins. Despite the obvious risks of exploiting an unproven technology, Swanson insists that "our investors have deep pockets."

Though Genentech seems to have a clear headstart, it is by no means alone in its determination to cash in on the potential of recombinant-DNA technology. Across the bay in Berkeley, six-year-old Cetus Corp. is also opening a recombinant-DNA facility to complement its work on conventional chemical and radiological means of mutating bacteria. "This is the hottest area in biology today," says Peter J. Farley, Cetus' executive vice-president. Two months ago, Standard Oil Co. (Indiana) bought one-fifth of Cetus for about $10 million.

Elsewhere, Upjohn Co. will soon open its own recombinant-DNA lab. According to Joseph E. Grady, head of Upjohn's infectious disease research, the company expects to develop marketable applications within five years. Abbott Laboratories is just now beginning work on recombinant DNA, while Miles Laboratories Inc. is becoming the major supplier of the so-called restriction enzymes that scientists use to cut strands of DNA for recombination. Altogether, between 10 and 15 industrial labs are now pursuing

Genentech: Planning to market a synthesized brain hormone, somatostatin

Biomedical breakthrough: Artificial genes able to produce body chemicals

recombinant-DNA experiments.

Trying for insulin. The Genentech research began with the construction of an artificial gene by the team at the City of Hope under the leadership of molecular biologist Arthur D. Riggs. The scientists chose to construct the gene for somatostatin because the hormone's chemistry, worked out at the Salk Institute, is reasonably well-known, and because sensitive tests are available to measure whether it is actively working within a cell. More important, somatostatin seems to play an important role in regulating body growth and inhibiting the production of insulin in the pancreas. Thus, it and other hormones now under study seem to have wide possible application in treating diseases such as diabetes. Today, somatostatin costs around $30,000 per gram to synthesize chemically, but Genentech believes it can bring the cost down to $300 or less.

Once it had an artificial gene, Boyer's team at UC-San Francisco used restriction enzymes to cut open a ring of DNA known as a plasmid in the cells of a special strain of *Escherichia coli,* the human gut bacteria most commonly used in recombinant-DNA work. The strain the team used, called K-12 bacteria, had been specially mutated so that it could not survive outside laboratory conditions. The gene was then stitched into the plasmid, and the combination was introduced into another K-12 bacterium, which accepted the foreign genetic material as its own. Then, for the first time anywhere, the artificial gene not only replicated itself but also instructed the bacteria to produce somatostatin.

The experiment is the third "first" registered by UC-San Francisco scientists this year. Earlier, they successfully inserted into *E. coli* a rat gene responsible for the production of insulin. But while the work was hailed as an early proof of recombinant-DNA's potential value to man, one early phase of the experiment had also involved the first violation, albeit accidental, of NIH safety guidelines. The scientists had to destroy the earlier experiment.

$100 million market. Such miscues only heighten fears that recombinant-DNA research might lead to the production of lethal organisms. Scientists who discount the danger have launched an effective campaign to calm the worriers, and this summer they successfully headed off congressional control of their work. Now the Carter Administration is urging industrial labs to comply voluntarily with the NIH guidelines. But an even more effective means of review may emerge from a recent court decision in favor of Upjohn that allows man-made organisms to be patented. Thus, the companies would have legal protection for their discoveries while their lab procedures could be scanned through the patent application process. Yet another check would be possible through the Food & Drug Administration, which would pass on the introduction of any new medicines.

Genentech has already filed for patent protection for its somatostatin technology and will similarly cover the expected breakthrough for insulin—an eventuality made more likely by the new research. Says Irving Johnson, vice-president of research at Eli Lilly & Co., currently the largest producer of insulin: "Commercial results are more imminent than thought." Swanson and Boyer expect their company to compete effectively for the $100 million insulin market, and they see all sorts of future applications from hormones to antibiotics and even enzymes.

For now, though, they will have plenty of business producing products already in demand. Swanson points out that "missionary marketing" of new substances is not in Genentech's development plans. "The field is opening up rapidly," adds Boyer, "and we have the flexibility to move." ■

FRIDAY, MARCH 31, 1978
PART III

DEPARTMENT OF HEALTH, EDUCATION, AND WELFARE

National Institutes of Health

■

U.S.—EMBO Workshop to Assess Risks for Recombinant DNA Experiments Involving the Genomes of Animal, Plant, and Insect Viruses

REVISING THE GUIDELINES: CLONING VIRAL GENOMES

The environmentalist organizations were unhappy with the proposed revised guidelines, arguing that a revision was premature; many molecular biologists were equally unhappy because the proposed relaxations did not, in their view, go far enough. Animal virologists in particular felt that they were being especially penalized in both the original and the proposed revised guidelines by containment requirements for recombinant DNA molecules including parts of viral chromosomes that were far in excess of those needed when they handled the intact, infectious viruses. In July 1977, Lennart Philipson (Uppsala) and Pierre Tiollais (Paris) drew attention to this anomalous and scientifically nonsensical situation in an article in *Nature* (Doc. 10.1).

The difficulties U.S. virologists now faced if they wished to do recombinant DNA experiments with the genomes of animal viruses are apparent from Document 10.2. The RAC decided to permit staff of the NIH to perform, under the highest possible levels of containment (an EK2 host-vector system and a P4 physical containment laboratory), one of the experiments then explicitly prohibited by the guidelines, namely, the introduction into *E. coli* of the whole genome of polyoma virus, which can cause tumors in rodents under experimental conditions. This experiment was to be performed with the express purpose of obtaining data directly pertaining to the assessment of the conjectural risks of recombinant DNA research. A private citizen living near the P4 laboratory at Fort Detrick, Maryland, who was receiving moral support from environmentalist groups and other individuals, sought in vain a court injunction to block the experiment.

On November 26–27, 1977, the EMBO Standing Advisory Committee on Recombinant DNA met in London, principally to discuss the proposed revised NIH guidelines. The EMBO Committee, of which Lennart Philipson was a member, welcomed the proposed relaxations but felt they did not go far enough. There also was particular concern about the excessive stringency of containment for cloning viral DNAs, and the Committee decided to establish an *ad hoc* group of virologists to consider the issue further. The EMBO report (excerpted in Doc. 10.3) was sent immediately to the director of the NIH, as well as to 17 European governments and the 350-odd scientist members of the EMBO. A member of the EMBO committee (John Tooze) gave evidence at the December 1977 hearings at the NIH, dwelling particularly on the question of recombinant DNA research with viral DNAs.

The outcome of all this was the so-called U.S.–EMBO Ascott Workshop, the report of which was published in the *Federal Register* (Doc. 10.4). The Ascott Workshop, like the Falmouth Workshop, proved to be one of the turning points in the debate about guidelines for this research. It pointed out that studying the chromosomes of dangerous viruses was to be done more safely in *E. coli* by recombinant DNA methods than in animal cells by conventional virological techniques. It recommended massive and wholesale reductions in containment requirements for experiments with most viruses, P2 instead of P4 physical containment and EK1 instead of EK2 host-vector systems. The endorsement of these recommendations by *ad hoc* committees set up by the NIH to review the Ascott Workshop led to their incorporation, with a few changes, into the second version of the revised guidelines. This move precipitated a critical reexamination of other sections of those proposals. If, for example, animal virus DNA was to be used at the P2 level of containment with an EK1 host-vector system, why would it be necessary to use P3 physical containment and an EK2 host-vector system for primate DNA? The answer was that such stringent conditions were totally unjustifiable, and the guidelines were in the end more generally relaxed (see Chapter 12).

July 14, 1977
Nature (**268**:5616)

Rational containment on recombinant DNA

The US National Institutes of Health has recommended that all recombinant DNA experiments, involving any animal virus, should be subject to the highest containment conditions, even though other work with the intact virus may require less stringent conditions. Lennart Philipson, of the Department of Microbiology, Uppsala University, Sweden, and Pierre Tiollais, of the Institut Pasteur, Paris, put forward the case for categorising recombinant DNA experiments involving animal viruses more rationally.

THE development of molecular genetics has depended largely on the detailed analysis of bacterial viruses, the bacteriophages, and their use to transfer bacterial genes between bacteria which do not normally recombine. Animal viruses now promise to provide a similarly detailed insight into the molecular mechanisms of gene expression in animal cells, which will undoubtedly contribute to an understanding of man and his domestic animals both in health and disease and, it is hoped, contribute to the development of medical and veterinary science.

Restriction enzyme mapping of the viral genome enables regions of interest to be excised and analysed. In many cases the full development of this work requires the use of recombinant DNA techniques to purify and clone the resulting portions of the viral genome. Animal viruses can also be used as vectors for introducing other eukaryote DNA into animal cells in which its expression may be studied. At present, however, the guidelines on recombinant DNA research laid down by the US National Institutes of Health (NIH) require almost all recombinant DNA experiments involving animal viruses to be carried out under the highest containment conditions. In practice these requirements preclude most experiments.

Animal viruses as a class are unfortunately regarded by the general public and the uninformed scientific community as rather mysterious entities which cause pandemics, epidemics and isolated cases of dreadful infectious diseases against which there is no remedy. Scientists themselves have coloured this picture by suggesting that viruses may cause cancer in man, implying to the layman that cancer is an infectious disease. In reality, animal viruses are extremely diverse; some are known to be highly infectious and pathogenic. Others seem to be harmless. It should therefore be mandatory to discriminate between different viruses according to their pathogenicity to man and other mammals when formulating guidelines for the use of viral genomes in *in vitro* recombinant DNA research.

The conjectural hazards envisaged arise from the (hypothetical) risk that virus genomes, when introduced into bacteria or when used as vectors for introducing DNA into mammalian cells, may give rise to "new infectious entities" or may transfer their pathogenic or tumour-producing capacity from the new host back to man or other mammals. Both types of risk depend on chains of events with low probabilities.

Against these arguments it should be considered, first, that the transfer of all or part of the viral genome to bacterial cells may not be an entirely new event. For millions of years prokaryotic organisms have been exposed to eukaryotic DNA and several microorganisms have efficient systems of transferring DNA over recombinant barriers. Second, the development of a 'new infectious entity' would probably require expression of the viral genes in the prokaryotic cell. This must have a low probability since most genes from eukaryotes and viruses, which have been transferred up till now, are not faithfully expressed as RNA or protein in bacteria[1-4]. Even if the inserted viral genes are expressed the resulting risk would not be greater than that from production of viral proteins in cell cultures, since only fragments of the viral genomes will be inserted in the experiments proposed.

The conjectural hazards may therefore be confined to the transfer of viral nucleic acids from the modified bacterium to an animal or human host. This would require a chain of events, each of low probability: the bacterium containing hybrid DNA must first escape from the laboratory and then establish itself in a new host. The viral nucleic acid must then enter the cells of the new host, and there express its tumorigenic or other harmful properties. Even without considering containment we are probably discussing probabilities in the range 10^{-20}–10^{-28} for the transfer of harmful genes into the new host.

Against this background it is obviously necessary to prohibit experiments with viruses such as Lassa fever virus or Marburg virus, which are considered to be high risk viruses in diagnostic laboratories; but it is not clear that viruses classified as low risk, such as adeno, SV40 and polyoma viruses, need such high containment. SV40 virus has already been injected accidentally together with polio vaccine into millions of human subjects without any registered harmful effects[5,6], although antibodies against SV40 appear in the serum. Some adenoviruses infect humans readily and about 50–80% of the population develop antibodies to adenovirus types 1 and 2 from the age of 5–10 years[7]. Even the adeno–SV40 hybrid viruses which were originally isolated when adenovirus vaccines were developed in monkey kidney cells have been introduced into large groups of military recruits as an inactivated or live vaccine without any detrimental effects[8,9]. Careful studies have failed to find an association between adenovirus transcripts, T antigens or other adenovirus products with tumours in humans[10,11].

Normal biochemical work with adeno and SV40 viruses is considered to fall into the low risk category under guidelines published by the National Cancer Institute and those from the Center for Disease Control, Atlanta, Georgia. These regulations allow the use of large amounts of virus. Milligram quantities of viral DNA are handled regularly by investigators and technical personnel in several biochemical laboratories. Provided that expression of virus DNA is prohibited in the bacterial cell, which is likely (see above), it must be more risky for personnel to handle viral DNA than to insert it in a bacterial vector. The NIH guidelines request P4 and EK2 conditions to handle partial or intact viral DNA in a prokaryotic host although as we have pointed out the risks envisaged are comparable to or lower than those encountered when working with intact virions. It is possible to claim that virions present a higher risk since as long as the protein coat is present the virion may possibly escape from the laboratory and infect humans. Thus, there seems to be a disproportionately high containment requirement for work with animal virus genomes in bacteria.

It is also difficult to understand the containment requirements for work with eukaryotic vectors. In the case of SV40 it has been established that SV40 DNA will form hybrids with cell DNA when cells are infected at high multiplicity[12,13]. Such experiments are in essence shotgun experiments, but since no *in vitro* recombination is involved it is considered a low risk experiment requiring no containment facilities. If a similar experiment is carried out *in vitro* between DNA from the same cell and SV40 virus DNA (probably with lower efficiency), it is labelled as P4 according to the NIH guidelines. Insertion of SV40 sequences into adenovirus DNA also occurs frequently when the two are cultivated together, as exemplified by the adeno–SV40 hybrid viruses[14,15] and new hybrids recently developed (J. Sambrook and G. Fey, personal communication). To take advantage of the natural recombination between SV40 and adenoviruses requires only a moderate risk containment facility, but the NIH guidelines absolutely forbid these experiments using recombinant DNA techniques. Here there is also a distinct discrepancy between the guidelines and other regulations for the study of animal viruses.

Most of the confusion is probably due to the fact that natural and artificial recombinant DNA reserch has not been considered as a whole. Only the new *in vitro* recombinant DNA technique has formed the basis for developing the NIH guidelines, although guidelines for work with dangerous microorganisms have been in existence for a long time. Consistent rules for all recombinant DNA research are therefore needed.

A good definition of recombinant DNA research might be that suggested by the Standing Advisory Committee on Recombinant DNA of the European Molecular Biology Organisation (EMBO): recombination of DNA molecules of different biological origin by any methods that overcome generally recognised natural barriers to mating, infection and recombination, to yield molecules that can be propagated in some host cells and the subsequent studies of such recombinant DNA molecules. *In vitro* recombinant DNA research is only a subsection of this definition. Within the broader meaning of the term there are several experiments now carried out in low risk containment according to generally accepted guidelines, which must be analysed and compared with the *in vitro* techniques before any regulatory guidelines are issued. It is, for example, possible to transfer the genome of several animal viruses into the capsid of unrelated viruses[16–19], and thereby increase the host range of the viral genes. It is also possible to fuse animal cells, which may involve undesirable transfer of viral genomes to new hosts[20,21]. Many of these experiments involve conjectural hazards comparable to those implied by the *in vitro* recombinant DNA technique.

But in these cases the experiments also probably carry only a low risk if they involve low risk viruses, since similar events probably occur in nature. Therefore we would like to propose that the entire area of recombinant DNA research should be re-evaluated. Experiments which are considered to involve real and proven risks should be prohibited at present. The transfer, by any method, of high risk viral genomes, potent bacterial toxins and antibiotic resistance to gonococci and streptococci fall into this category. Conjectural hazards which may or may not become real should rapidly be evaluated under secure conditions. Meanwhile *in vitro* recombinant DNA research can probably proceed in several areas including the transfer of low risk viral genomes to prokaryotic cells and the development of eukaryotic vectors from low risk viral genomes.

In closing, it should be emphasised that both mutagenisation and deletion of a multitude of genes in the laboratory have, to the best of our knowledge, never provided a selective advantage for an organism in the natural environment. Some genetic manipulation with plant cells may be an exception. It is necessary for the opponents of recombinant DNA research to provide examples of such selective advantage otherwise this debate will focus more on faith than science. □

1 Morrow, J. F. *et al. Proc. natn. Acad. Sci. U.S.A.* **71**, 1743 (1974).
2 Kedes, L., Chang, A. C. Y., Housman, D. & Cohen, S. N. *Nature* **255**, 533 (1975).
3 Chang, A. C. Y., Lansman, R. A., Clayton, D. A. & Cohen, S. N. *Cell* **6**, 231 (1975).
4 Tiollais, P., Perricaudet, M., Pettersson, U. & Philipson, L. *Gene* **1**, (in the press).
5 Sweet, B. H. & Hilleman, M. R., *Proc. Soc. exp. Biol. Med.* **105**, 420 (1960).
6 Magrath, D. I., Russell, K. & Tobin, J. O. H. *Br. med. J.* **2**, 287 (1961).
7 Fox, J. P. *et al. Am. J. Epidemiol.* **89**, 25 (1969).
8 Rapp, F., Melnick, J. L., Butel, J. S. & Kitahara, T. *Proc. natn. Acad. Sci. U.S.A.* **52**, 1348 (1964).
9 Melnick, J. L. in *Viral and Rickettsial Infections of Man* (eds Horsfall and Tamm) IV ed. (Lippincott, 1965).
10 McAllister, R. M., Gilden, R. V. & Green, M. *Lancet* i, 831 (1972).
11 Gilden, R. V. *et al. Am. J. Epidemiol.* **91**, 500 (1970).
12 Aloni, Y., Winocour, E., Sachs, L. & Torten, J. *J. molec. Biol.* **44**, 333 (1969).
13 Lavi, S. & Winocour, E. *J. Virol.* **9**, 309 (1972).
14 Rapp, F., Melnick, J. L., Butel, J. S. & Kitahara, T. *Proc. natn. Acad. Sci. U.S.A.* **52**, 1348 (1964).
15 Lewis, A. M., Levin, M. J., Wiese, W. H., Crumpacker, C. S. & Henry, P. H. *Proc. natn. Acad. Sci. U.S.A.* **63**, 1128 (1969).
16 Rapp, F., Butel, J. S. & Melnick, J. L. *Proc. natn. Acad. Sci. U.S.A.* **54**, 717 (1965).
17 Choppin, P. W. & Compans, R. W. *J. Virol.* **5**, 609 (1970).
18 Závada, J. *J. gen. Virol.* **15**, 183 (1972).
19 Huang, A. S., Besmer, P., Chu, L. & Baltimore, D. *J. Virol.* **12**, 659 (1973).
20 Gerber, P. *Virology* **28**, 501 (1966).
21 Weiss, M. C. & Green, H. *Proc. natn. Acad. Sci. U.S.A.* **58**, 1104 (1967).

Ferdinand J. MACK, Jr., Plaintiff,

v.

Joseph A. CALIFANO, Jr., et al., Defendants.

Civ. A. No. 77-916.

United States District Court, District of Columbia.

Feb. 23, 1978

Preliminary injunction was sought to prevent an experiment testing the biological properties of polyoma DNA cloned in bacterial cells. The District Court, John Lewis Smith, Jr., J., held that requested injunction would not issue since record reflected that National Institute of Health had carefully considered potential risks of the experiment under its guidelines and had taken necessary precautions, experiment was designed to provide important and needed information on possibilities of recombinant DNA technology and none of plaintiff's affidavits established that the experiment was likely to cause harm to human health or the environment.

Motion denied.

Health and Environment 25.15(3)

Government scientists would not be preliminarily enjoined from testing biological properties of polyoma DNA cloned in bacterial cells where experiment would employ an E coli designed to "self destruct," weight of scientific opinion was that recombinant DNA research in accordance with National Institute of Health guidelines would have no adverse environmental or public health consequences, experiment was designed to provide important and needed information on possibility of recombinant DNA technology and it appeared that compliance with NIH guidelines would insure that no recombinant molecules would escape from the laboratory environment. National Environmental Policy Act of 1969, §§ 1 et seq., 102, 42 U.S.C.A. §§ 4321 et seq., 4332.

Ferdinand J. Mack, Rockville, Md., for plaintiff.

L. Mark Wine, Dept. of Justice, Washington, D.C., for defendant.

OPINION

JOHN LEWIS SMITH, Jr., District Judge.

Plaintiff seeks a preliminary injunction to prevent an experiment testing the biological properties of polyoma DNA (deoxyribonucleic acid) cloned in bacterial cells. The experiment is to be conducted in Building 550, Frederick Cancer Research Center at Fort Detrick, Maryland. Also before the Court is defendants' motion to vacate a voluntary stay.

Defendants are Joseph A. Califano, Jr., Secretary of Health, Education, and Welfare, Donald S. Frederickson, Director of National Institutes of Health, and John E. Nutter, Chief Officer of Specialized Research and Facilities, National Institute of Allergies and Infectious Diseases, National Institutes of Health.

On May 31, 1977 plaintiff, an infant resident of Frederick, Maryland, filed a motion for temporary restraining order and preliminary injunction to enjoin defendants from undertaking the experiments or constructing facilities at Fort Detrick to be used for the research. On July 18th a stipulation was entered into by the parties staying all proceedings pending finalization by the defendants of an Environmental Impact Statement and providing for 30 days notice to plaintiff of any experiments to be conducted at Fort Detrick after such finalization. In accordance with the stipulation, defendants advised plaintiff that an Environmental Impact Statement (EIS) became final on November 28, 1977 when the Council on Environmental Quality published notice of its receipt in the Federal Register. Plaintiff contends that the statement does not comply with the requirements of the National Environmental Policy Act (NEPA), 42 U.S.C. § 4332(2)(C) and other statutes.

A motion of the American Society for

Microbiology for leave to file a brief as Amicus Curiae with respect to the public health consequences of the proposed research was granted. Counsel for the Society participated in oral argument and submitted a brief. Dr. Naum S. Bers, Rockville, Maryland, appeared individually as a concerned citizen and was granted permission to file a statement.

Plaintiff asserts that defendants are planning to conduct experiments with polyoma, a virus known to cause cancer in mice. He states that the nature of the organisms to be created by the research is such that even a miniscule quantity, if released, in the environment would represent a threat to life and health. He further contends that the Fort Detrick experiments are to be conducted by defendants without determining the applicability of NEPA and according to the very guidelines of the Department of Health, Education, and Welfare (HEW) classified as "prohibited".

Defendants on the other hand take the position that the EIS and NIH (National Institutes of Health) guidelines reflect the cautious manner in which the scientific community and NIH have considered the new technology involving recombinant DNA molecules. They further state that the final EIS was completed after extensive public comment and discussion of alternatives. Much of plaintiff's concern, they state, is based on an apparent misunderstanding of the nature of the materials to be used in the experiment. Plaintiff's affidavits are based on the belief that the experiment here in question will be conducted utilizing a common strain of escherichia coli (E coli) as the host-vector for the planned studies. Significantly, the NIH guidelines "prohibit certain kinds of recombinant DNA experiments which include virtually all the known hazards—for example, those involving known infectious agents."

The research is now restricted by these guidelines to implanting any new genes into enfeebled strains of E coli, a human gut bacteria that has been modified even further to make it safe as the new DNA's laboratory host. In the planned experiment a derivative of E coli K-12, which has been specifically designed to "self destruct", will be employed. E coli K-12 is unable to colonize within the human intestinal tract and causes no known human or animal disease. *See* EIS at page 73. This EK2 host-vector system will not survive passage through the intestinal tract of animals and will "die" because of its dependency on chemicals not found in nature.

Defendants further point out that the complete experiment will be conducted in P4 physical containment laboratories which have been shown to safely contain microbes presenting a known and demonstrable hazard to man. For each certified EK2 system, "Appendix H, page 10 of the EIS", NIH reviews extensive scientific data to determine that the system meets the standards for safety. EIS at 81. *See* NIH guidelines, Appendix D, page 15. It is evident, therefore, that there is actually a two step distinction between the common strains of E coli which "do live in people" and the EK2 host-vector system which will be used in these experiments.

Counsel for the American Society for Microbiology states that the weight of scientific opinion now considers that recombinant DNA research in accordance with the NIH guidelines will not have adverse environmental or public health consequences. He contends that the present guidelines are more conservative than necessary and that certain restrictions in these guidelines could be safely modified. He further asserts that these guidelines are *not* in fact the very guidelines of HEW classified as "prohibited" as was asserted by plaintiffs. In the opinion of the Society, the proposed Fort Detrick experiment will specifically advance the public interest and present no risk of harm to the environment.

The research involves dividing and then rejoining the heredity-carrying material of various organisms—deoxyribonucleic acid, or DNA—to make recombinant hybrids that carry some of the traits of two unrelated forms. It is contended that the value of such work is that it may create new medicines, vaccines, industrial chemicals or crops. The risk, some scientists claim, is that it could create unexpectedly dangerous new ailments or epidemics. Many scientists are of the opinion that exaggerations of the hypothetical hazards have gone far beyond any reasoned assessment. They take the position that the experience of the last four years, including many laboratory experiments, has shown no actual hazards.

Recently the Supreme Court has summarized the limited role of the courts in determining whether the agencies have complied with NEPA.

> The only role for a court is to insure that the agency has taken a 'hard look' at environmental consequences; it cannot 'interject itself within the area of discretion of the executive as to the choice of the action to be taken'. *Kleppe v. Sierra Club*, 427 U.S. 390, 410 n. 21, 96 S.Ct. 2718, 2731, 49 L.Ed.2d 576 (1976) citing *Natural Resources Defense Council v. Morton*, 148 U.S.App.D.C. 5, 16, 458 F.2d 827, 838 (1972).

The EIS does represent a "hard look" by NIH at recombinant DNA research performed in accordance with its guidelines. It appears that compliance with the NIH guidelines will insure that no recombinant DNA molecules will escape from the carefully controlled laboratory to the environment.

Plaintiff requested an extension of time to furnish additional evidence. He submitted supplemental affidavits from four of his previous affiants which reaffirmed their previously expressed opinions. None of the affidavits established that the experiment is likely to cause harm to human health or to the environment. The recombinant DNA Research Guidelines represent an effort by many scientists to evaluate the hazards and provide safe methods for their control. The record reflects that NIH has carefully considered the potential risks of this experiment under the guidelines and has taken the necessary precautions.

The experiment is designed to provide important and needed information on the possibilities of recombinant DNA technology. Important scientific information relative to the possibilities of this technology would be delayed if a preliminary injunction were granted.

Accordingly, plaintiff's motion for a preliminary injunction is denied. Defendants' motion to vacate stay is granted.

FINDINGS OF FACT

1. Plaintiff's complaint, motion for temporary restraining order and preliminary injunction were filed on May 31, 1977.

2. On July 18, 1977 a stipulation was approved by the Court providing that all matters would be stayed pending the finalization of an Environmental Impact Statement (EIS), that defendants would give plaintiff thirty days notice before proceeding with the experiment, and that no such experiment would be conducted prior to the finalization of an EIS.

3. The National Institutes of Health (NIH) guidelines and final EIS were the result of a lengthy administrative process. The scientific community had operated for several years under voluntary restraints limiting the kinds of recombi-

nant DNA research that could be undertaken.

4. A public meeting was held by NIH on February 9th and 10th, 1976 which was announced to the public in the Federal Register. Thereafter NIH published the "decision of the Director, NIH and NIH guidelines on recombinant DNA research on July 7, 1976. At that time NIH announced that it was preparing a draft environmental impact statement. 41 Fed.Reg. 27902 (1976). On September 9, 1976 the draft EIS was published in its entirety in the Federal Register. 41 Fed.Reg. 38426 (1976). The final EIS was prepared and notice of its availability published in the Federal Register on November 28, 1977. 42 Fed.Reg. 6588 (1977).

5. The NIH guidelines govern all facets of NIH-funded research using recombinant DNA techniques. The guidelines provide detailed requirements for both physical and biological containment designed to insure that recombinant DNA molecules will pose no threat to man or the environment. The experiment is to be conducted in accordance with these guidelines.

6. The laboratory at Fort Detrick is a P4 laboratory with extensive safeguards built into its design. The experiment is to be conducted under P4 physical containment requirements—the highest level of physical containment. P4 facilities are governed by rules limiting access, providing for change of clothes before entering and leaving, and numerous other safety features. All recombinant DNA materials are handled in gas-type safety cabinets and removed only after sterilization.

7. The experiment is to be conducted using EK2 host-vector systems. In the planned experiment, a derivative of E coli K–12 which has been specifically designed to self-destruct if removed from the controlled laboratory environment will be used. E coli K–12 itself is safe and has been used for years without known harm to the laboratory workers or to the environment.

8. E coli K–12 is unable to colonize in the human intestinal tract and causes no known human or animal disease. The EK2 system uses a K–12 derivative that must have special chemicals found only in an artificial laboratory setting in order to survive and is safer than ordinary K–12. If these chemicals are not present, the EK2 is designed to self-destruct.

9. Recombinant DNA research has already become a valuable aid in progress against illness. Benefits include applied medical advances and an accelerated understanding of the genetic and biochemical basis of the disease process.

10. The experiment is designed to provide important knowledge concerning recombinant DNA technology.

11. The experiment poses no substantial risk to human health or to the environment because (1) there is little likelihood the materials will escape from the maximum containment of the P4 facility; (2) if such an escape did occur, the recombinant DNA molecules would not survive but would self destruct outside the laboratory environment; and (3) the particular virus being used has never been implicated in human disease.

12. Plaintiff has offered no evidence to show that he will suffer irreparable injury or that there is any significant possibility that the experiment will have an adverse impact on the environment.

CONCLUSIONS OF LAW

1. Plaintiff has not shown that he would be irreparably injured unless a preliminary injunction is granted.

2. Plaintiff has not sustained the burdens imposed upon him by *Virginia Petroleum Jobbers Association v. FPC* 104 U.S. App.D.C. 106, 259 F.2d 921 (1958) in that he has failed to demonstrate that he would be irreparably injured in the absence of the issuance of an injunction, that he is likely to prevail upon the merits of the controversy, and that the public interest lies in granting the requested relief.

Illustration by Barbara Thomas.

Document 10.3

November 26– 27, 1977, Excerpt

CEBM/77/11 E
Date: 30 November 1977
Original Language: English

EUROPEAN MOLECULAR BIOLOGY CONFERENCE

Eighth Ordinary Session (Second Part)

Report of the 4th meeting of the EMBO Standing Advisory Committee on Recombinant DNA held at Heathrow Airport (London) on 26–27th November 1977

The Conference is asked to take note of the report of the 4th meeting of the EMBO Standing Advisory Committee on Recombinant DNA research.

3. Proposed Revisions to the NIH Guidelines

3.1 General Comments

In June 1976 the NIH of the USA issued *Guidelines for Research Involving Recombinant DNA Molecules*, it being understood that periodic revisions would be made to the guidelines in the light of further knowledge and experience. During the summer of 1977 the NIH Advisory Committee began the process of revising the guidelines and proposed revised guidelines were published in the first issue of the NIH's *Recombinant DNA Technical Bulletin*. The EMBO Committee has discussed in some detail the revisions that are proposed and has reached the conclusions reported in the following paragraphs. The EMBO Committee will convey by letter its comments to the NIH Advisory Committee and to the director of the NIH.

3.2 The revisions proposed represent a significant relaxation of the containment conditions required for most classes of recombinant DNA experiments and they also provide greater flexibility in attaining particular levels of physical containment. The EMBO Committee considers that these changes are justified for the reasons discussed below and suggests that the levels of containment for certain classes of experiments should indeed be reduced somewhat further than is proposed by the NIH Advisory Committee.

3.3 The principal reasons for recommending significant relaxation of the containment measures for recombinant DNA experiments in general, and those which involve the use of *E. coli* K12 as the host organism in particular, are two-fold. First, we are becoming increasingly aware that in nature there is a considerable potential for exchange of genes that does not depend upon extensive homologies between DNA sequences. Exchange of genes between bacteria and higher plants, for example, has now been shown to occur in nature. Moreover, in the laboratory it has been shown that *E. coli* producing restriction enzyme not only take up DNA but also incorporate it into their chromosomes by an *in vivo* process that involves the endogenous restriction enzymes. In short the sorts of gene combinations that can be generated in the laboratory by *in vitro* recombinant DNA methods can apparently also be generated naturally *in vivo* albeit on a lesser scale. Second, new data about the biology of *E. coli* K12 indicate that (1) *E. coli* K12 does not colonise the human gut (2) *E. coli* K12 is not pathogenic (3) *E. coli* K12 cannot be made communicable or pathogenic even by introducing genes for toxins and other pathogenic properties from other strains of *E. coli* by standard genetic methods (4) the probability of transmission in the animal gut of non-conjugative plasmid vectors from an *E. coli* K12 host to some other bacterium is less than 10^{-16} per bacterium per day. In short *E. coli* K12 is so genetically defective that it is incapable of becoming a pathogen. The strains of *E. coli* K12 that have been further disabled by the introduction in the laboratory of many mutations which further restrict the conditions in which the organism can survive (biologically contained *E. coli* K12: EK2 hosts) present no threat to man or his environment.

3.4 When the original guidelines were drawn up this information was not available; at that time it was suggested by some that *E. coli* K12 might be readily converted into a pathogen by the introduction of foreign DNA. Moreover, natural barriers to the exchange of genetic material between species seemed greater than we now know them to be. The new evidence dispels many of these fears and so the containment measures initially and presently required are now recognized as being too stringent and should be relaxed.

3.5 Specific Comments

Physical Containment Specifications

The EMBO Committee welcomes the more detailed descriptions of physical containment measures given in the revised NIH guidelines, noting that they are substantially in accord with the proposals and recommendations of a joint NIH-EMBO workshop held in March (Annex 2). The separate description of the three components of each level of physical containment, namely the laboratory design, the laboratory practices and the special containment equipment, is particularly useful. Concerning these specifications the EMBO Committee has only a few comments: (a) it believes that mouth pipetting should be prohibited in the P1 laboratory, as it is prohibited in P2-P4 laboratories, and it recommends that simple air exhaust cabinets, such as a conventional fume cupboard, be used in the P1 laboratory for those manipulations likely to produce large amounts of aerosols; (b) in the specification of the P3 laboratory, which may be equipped with running water, or located beneath or next to laboratories with running water, precautions against the consequences of flooding or contamination of the mains water supply are not discussed.

3.6 Proposed "Trade offs"

The proposals to increase the flexibility of containment measures by allowing certain alternative combinations of safeguards are welcomed by the EMBO Committee. The EMBO Committee supports the principle that, for certain experiments, if biological containment is increased by one step a reduction of one step in special containment equipment can be allowed, and conversely, that biological containment can be reduced by one step when the physical containment equipment is increased by one step. These changes are envisaged by the NIH Committee for P3 and P4 physical containment and for EK1 and EK2 biological containment. The EMBO Committee proposes that they be extended to P2 and P3 physical containment such that a combination of a laboratory designed to P2 specifications with a class III cabinet exhausting to the outside air, be considered equivalent to a laboratory designed and equipped to P3 specifications.

3.7 Categorization of Safety Measures for Particular Classes of Experiments

The EMBO Committee taking into consideration the new data mentioned above in paragraph 3.3, believes that for "shotgun" experiments involving tissue from healthy organisms the basic combinations of biological and physical containment should be those shown in column 1 of Table 1. For comparison column 2 of Table 1 shows the containment combinations proposed in the revised NIH guidelines. The EMBO Committee's recommendations relax the containment combinations further than the proposals of the NIH Advisory Committee. The former does not believe primate DNA to be more hazardous than the DNA of other mammals and birds, and none of the existing European guidelines make a distinction between primate and other mammalian DNA so long as it is obtained from normal healthy tissue.

TABLE 1

Source of DNA*	COLUMN 1 EMBO Standing Advisory Committee's proposals	COLUMN 2 NIH Advisory Committee's proposals	
All Mammals and Birds	P2 + EK2	Primates	P3 + EK2
		Others	P2 + EK2
All other eukaryotes	P1 + EK2	invertebrates	P1 + EK2
	P2 + EK1	or	P2 + EK1
		plants	P2 + EK1
		or	P1 + EK2
Prokaryotes	P1 + EK1	P2 + EK 2 (not well characterized)	
		P1 + EK2 (well characterized)	

*Applies only to non-pathogenic species, to DNA from normal healthy tissue of such species, and to *E. coli* K12 host vector systems.

3.8 The EMBO Committee believes that DNA of amphibia, reptiles and fish poses less of a conjectural hazard than that of mammals and primates and, therefore, requires a lower level of containment.

3.9 The EMBO Committee has not discussed in detail the containment conditions for experiments with the DNA of organisms, of the various phylogenetic classes, that are pathogenic or are the hosts of pathogens. The containment conditions for these, and for experiments with DNA from pathological tissues from any organism should be decided case by case, taking into consideration factors such as the production by the pathogen of a polypeptide or non-polypeptide toxin, the host range of the pathogen, the precise nature of the pathological tissue and the disease, etc. Irrespective of the taxonomic position of the donor organism, all experiments involving the DNA of pathogens or pathological tissues should be referred to the national committee before they are started. In general the EMBO Committee recommends that in these cases either the physical or the biological containment measures be increased by at least one step.

3.10 Prohibited Experiments
The EMBO Committee accepts in principle the proposal made in the NIH guidelines that certain recombinant DNA experiments should not at present be initiated; however, the lists of particular pathogenic organisms that fall under the prohibition, in particular agents in Class 5 of the *Classification of Etiologic Agents on the Basis of Hazard*, may not necessarily be appropriate for all European countries. For example several viruses of domestic animals that are not present in North America are endemic in parts of Europe, and it would be unreasonable to absolutely prohibit working with them in those countries. The decision as to which pathogenic organisms should be classified as too dangerous to use must be the responsibility of national or regional authorities.

3.11 Cloning Viral DNA
The EMBO Committee believes that the containment categorization of experiments with animal viral DNAs which is proposed by the NIH Advisory Committee is too indiscriminate and excessively stringent considering the proposed classification of experiments with other classes of DNA and the longstanding, accepted safety precautions for handling intact virus particles and viral nucleic acids. The EMBO Committee believes that there is no justification for placing shotgun experiments with DNA of all viruses of warm blooded and cold blooded animals in one single category (P4 + EK1 or P3 + EK3) because their host range, virulence and pathogenicity vary markedly. It seems unreasonable to require equally stringent containment conditions for example for DNA of mouse minute virus on the one hand and feline leukaemia virus and Mason Pfizer monkey virus on the other. The EMBO Committee proposes that it would be more reasonable either to consider experiments with viral DNA on a case by case basis or to produce a detailed set of recommended categories for experiments with specific viral DNAs. Since the latter is beyond its competence the EMBO Committee hopes in the near future to establish an *ad hoc* international group of virologists to draw up such proposals.

3.12 Viruses as Vectors
In experiments involving the introduction of foreign DNA into cultured cells of animals using DNA viruses as vectors, biological containment is assured by the very restricted permissive conditions for the host cells; the only routes by which the recombinant molecule might escape are by chance infection of a contaminating microorganism or within a viral capsid and the size of the recombinant molecule may well preclude its encapsidation. The EMBO Committee believes that the categories of physical containment required should depend principally upon the taxonomic relationship between the donor of the DNA being cloned and the natural host species of the vector virus.

A detailed set of categories based upon the following principles is called for:

donor DNA from a species that is a host for the vector—low level physical containment

donor DNA from an eukaryote not susceptible to the vector—medium level physical containment

donor DNA from a prokaryote—higher level physical containment

For example cloning of mouse DNA using polyoma virus as a vector and mouse cells as host should not require precautions more stringent than those routinely used for many years in laboratories studying polyoma virus infection of mouse cells and mice.

The EMBO Committee finds the proposals for this class of experiments in the revised NIH guidelines not sufficiently discriminating because they would impose unnecessarily high levels of physical containment for experiments with many eukaryotic DNAs.

Document 10.4

March 31, 1978

Federal Register, Excerpt

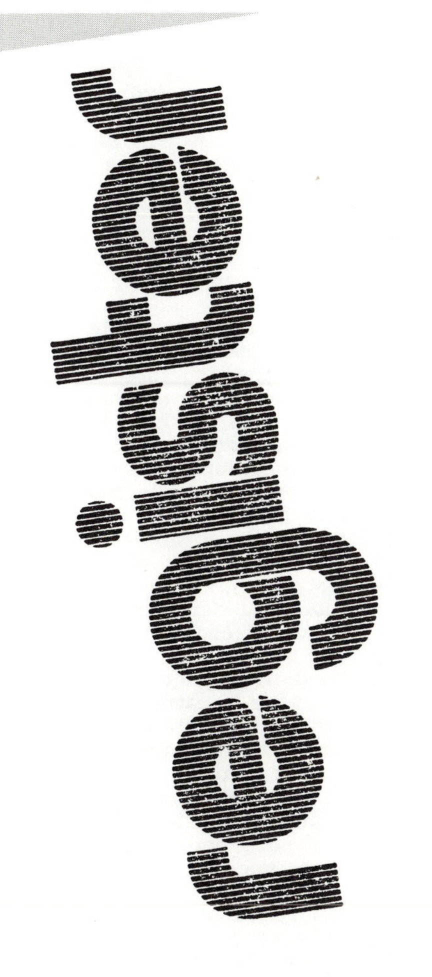
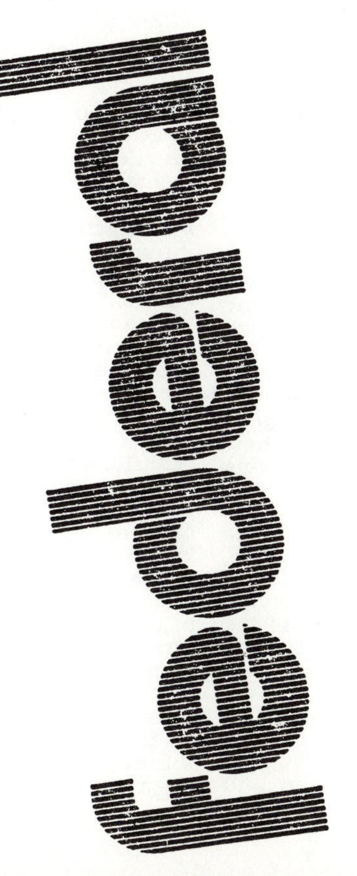

FRIDAY, MARCH 31, 1978
PART III

DEPARTMENT OF HEALTH, EDUCATION, AND WELFARE

National Institutes of Health

■

U.S.—EMBO Workshop to Assess Risks for Recombinant DNA Experiments Involving the Genomes of Animal, Plant, and Insect Viruses

[4110-08]

DEPARTMENT OF HEALTH, EDUCATION, AND WELFARE

National Institutes of Health

U.S.—EMBO WORKSHOP TO ASSESS RISKS FOR RECOMBINANT DNA EXPERIMENTS INVOLVING THE GENOMES OF ANIMAL, PLANT, AND INSECT VIRUSES

Report

On September 27, 1977, proposed revised Guidelines for Research Involving Recombinant DNA Molecules were published in the FEDERAL REGISTER (42 FR 49596 et seq.) for public comment and consideration. The proposed revised Guidelines and comments received were considered by the Advisory Committee to the Director, National Institutes of Health (NIH), at its meeting on December 15-16, 1977.

In response to discussion concerning viruses at the Director's Advisory Committee meeting, a joint U.S.—EMBO Workshop to Assess Risks for Recombinant DNA Experiments Involving the Genomes of Animal, Plant, and Insect Viruses was held in Ascot, England, on January 26-28, 1978. The workshop was attended by 27 scientists from the United States, the United Kingdom, West Germany, Finland, France, Sweden, and Switzerland. The participants were invited because of their scientific expertise and not as representatives of any government or of any policymaking group.

The primary purpose of the meeting was to conduct a scientific and technical analysis of possible risks associated with cloning eukaryotic viral DNA segments in *E. coli* K-12 host-vector systems and with the use of eukaryotic viruses as cloning vectors in animal, plant, and insect systems. In addition, there were general discussions of the possible importance of recombinant DNA technology for the solution of problems in basic and applied virology and of the classification of viruses with respect to the hazard that laboratory research with them might pose to the laboratory worker or to the community.

To provide further opportunity for public comment and consideration, the Workshop report is presented below. This report will be considered by the Recombinant DNA Molecule Program Advisory Committee at its meeting on April 27-28, 1978. On the basis of comments received and review by the Recombinant Advisory Committee, the Director, NIH, will subsequently issue revised Guidelines for Research Involving Recombinant DNA Molecules accompanied by a decision document explaining the modifications.

Please address any comments on this report to the Director, National Institutes of Health, 9000 Rockville Pike, Bethesda, Md. 20014. All comments should be received by April 15, 1978. Additional copies of this notice are available from:

Director, Office of Recombinant DNA Activities, Building 31, Room 4A52, National Institutes of Health, 9000 Rockville Pike, Bethesda, Md. 20014.

Dated: March 22, 1978.

DONALD S. FREDERICKSON,
Director,
National Institutes of Health.

REPORT OF U.S.-EMBO WORKSHOP TO ASSESS RISKS FOR RECOMBINANT DNA EXPERIMENTS INVOLVING THE GENOMES OF ANIMAL, PLANT, AND INSECT VIRUSES

This is the report of a joint U.S.-EMBO Workshop held in Ascot, England, January 27-29, 1978, which was convened to discuss the possible risks of recombinant DNA experiments involving the DNA's of animal, plant, and insect viruses. The 27 scientists in attendance (see attached roster) had expertise in clinical infectious disease; public health, medical and diagnostic virology; the biology of virus infection; biochemical virology; and plant, insect, and veterinary viruses. Five of the participants are actively engaged in recombinant DNA experimentation. A consensus statement of the discussions in the areas of pathogenesis and epidemiology of viral diseases, potential benefits of recombinant DNA experiments involving eukaryotic viral DNA, viral hazard classifications, and cloning in prokaryotic and eukaryotic systems is presented below. The group's conclusions, with respect to possible risks of recombinant DNA experiments involving viruses are based on the best available scientific data derived from publications, knowledge of current activities in the field of virology, and first-hand experience in the virology laboratory.

INTRODUCTION

Viral disease is a complex process that involves a series of critical steps; these include entry of the virus particle into the host, infection of specific cells at the portal of entry, replication of the virus in the infected cells, and usually, the spread of the progeny virus particles within the infected host to other susceptible cells. Depending upon the nature of the particular viral agent, the deleterious effects for the host, if any, may result from cytolytic activity, cellular transformation, chronic cellular dysfunction, or the provocation of an injurious immunological response. Viruses contain 5 to 150 or more genes and their coordinated functioning is required for viral growth and, consequently, for survival of the virus in nature. Even though we do not generally understand the precise role of each viral gene product, it seems clear that viral infection and disease production requires proper functioning of most, if not all, viral genes and, in general, is not a consequence of any single viral gene product. In the case of oncogenic papovaviruses, transforming retroviruses and possibly adenoviruses, individual viral genes are thought to be responsible for the transforming properties of the virus.

Recombinant DNA experiments have already yielded new information about the structure and control of expression of genes in higher organisms that could not have been obtained by conventional techniques. DNA cloning provides unparalleled opportunities to explore the basic biology of animal and plant viruses. Virologists will be able to probe more deeply into the control of viral gene expression and discover phenomena of general cell biological significance; techniques will be more readily available to elucidate the sequence of viral nucleic acids, to shed light on the role of viral gene products in pathogenicity, and eventually, to understand the molecular biology of animal and plant viruses to the extent that some bacteriophages are now understood. It seems apparent that this new information will lead to a deeper understanding of viral diseases and to new ways of combating them. In the immediate future the ability to obtain useful amounts of pure viral genomes and subgenomic fragments that cannot be obtained by other means will provide scientists and physicians with invaluable and inexpensive diagnostic protein and nucleic acid reagents. In the more distant future it should be possible to use gene cloning techneques to obtain large amount of viral proteins; one practical benefit from such developments might well be effective and safe vaccines for control of diseases caused by hepatitis viruses, herpesviruses and influenza viruses, and many other viruses, both known and, as yet, unknown.

In addition to being able to clone viral genes in bacteria we are now able to envisage using certain animal viruses as vectors for the propagation of foreign genes in animal cells; a similar system for exploiting a plant virus, cauliflower mosaic virus, to clone foreign genes in plant cells may shortly become available. The chief importance of animal and plant virus vectors is that they can be used to carry genes into cells in which they may be fully expressed as well as propagated. By using specifically designed viral vectors it may eventually prove possible to deliver a specific gene to specific target cells; such techniques have obvious medical, economic, and agricultural applications but their realization will depend upon a great deal of basic research.

Viral Classification Systems and Recombinant DNA Experimentation

The group extensively discussed the current safety procedures for holding and handling in the laboratory certain animal viruses, in particular those likely to be used in cloning experiments in the foreseeable future, either as vectors or as the sources of the nucleic acids to be cloned. Inevitably the recommended safety measures for using animal (and plant) viruses in research vary from country to country. In the context of recombinant DNA research, which involves a novel set of circumstances, none of the available classifications of viruses according to the risks they pose is entirely satisfactory. A list of animal viruses was therefore prepared and the viruses were ranked according to their known hazard on the basis of: (a) the severity of human disease that they can cause, particularly in persons exposed in the laboratory; (b) their potential for infecting laboratory workers; (c) the risk that a laboratory infection might result in spread to the community; and (d) the impact such spread might have on the community or environment (Table 1). In this list four bacteria and one rickettsial agent with different pathogenic potentials have also been included as a frame of reference for the 22 viruses identified. Because of time constraints, many animal viruses, particularly those of agricultural and veterinary importance, were not included in the table. Although not comprehensive, the list contains most of those animal viruses that have been previously mentioned in the context of recombinant DNA experiments.

This list can also serve the very useful function of a reference scale, familiar to both microbiologists and clinicians, for expressing the degree of concern that a given conjectural hazard may engender, by comparison to a known biohazard.

Cloning Viral DNA's in E. Coli K12

The cloning of viral DNA's and cDNA's in *E. coli*K12 using EK1 and EK2 plasmid and lambda phage vectors was discussed in light of the conclusions of the Falmouth meeting that *E. coli* K12 is not pathogenic and does not efficiently colonize the vertebrate digestive tract (Gorbach, 1978). Not for want of trying, the participants were unable to envisage a sequence of events which could occur with significant probability that would allow *E. coli* carrying either whole DNA genomes of certain viruses or sub-genomic fragments of virtually any virus to lead to disease. The question was also raised as to whether or not, in the extremely remote possibility that all of the biological and physical containment barriers broke down, intestinal bacteria carrying cloned whole viral genomes might bypass the natural barriers to infection by the virus particle. As summarized in the following section, the group concluded that the probability that K12 organisms carrying viral DNA inserts could represent a significant hazard to the community was so small as to be of no practical consequence.

RECOMMENDATION

Based on these considerations the participants of the U.S.-EMBO Workshop concluded that the use of P2 (NIH Guidelines) or CI (Williams Report) containment measures, in conjunction with an EK1 host-vector system should provide adequate containment for cloning any viral genome or fragment thereof and recommended this as the minimum containment levels for recombinant DNA experiments involving eukaryotic viral DNA inserts. However, if the virus itself must be handled at higher levels of physical containment it seems prudent at the present time to use the more stringent containment conditions. It was emphasized that containment practices must include adequate training and the use of high quality microbiological technique.

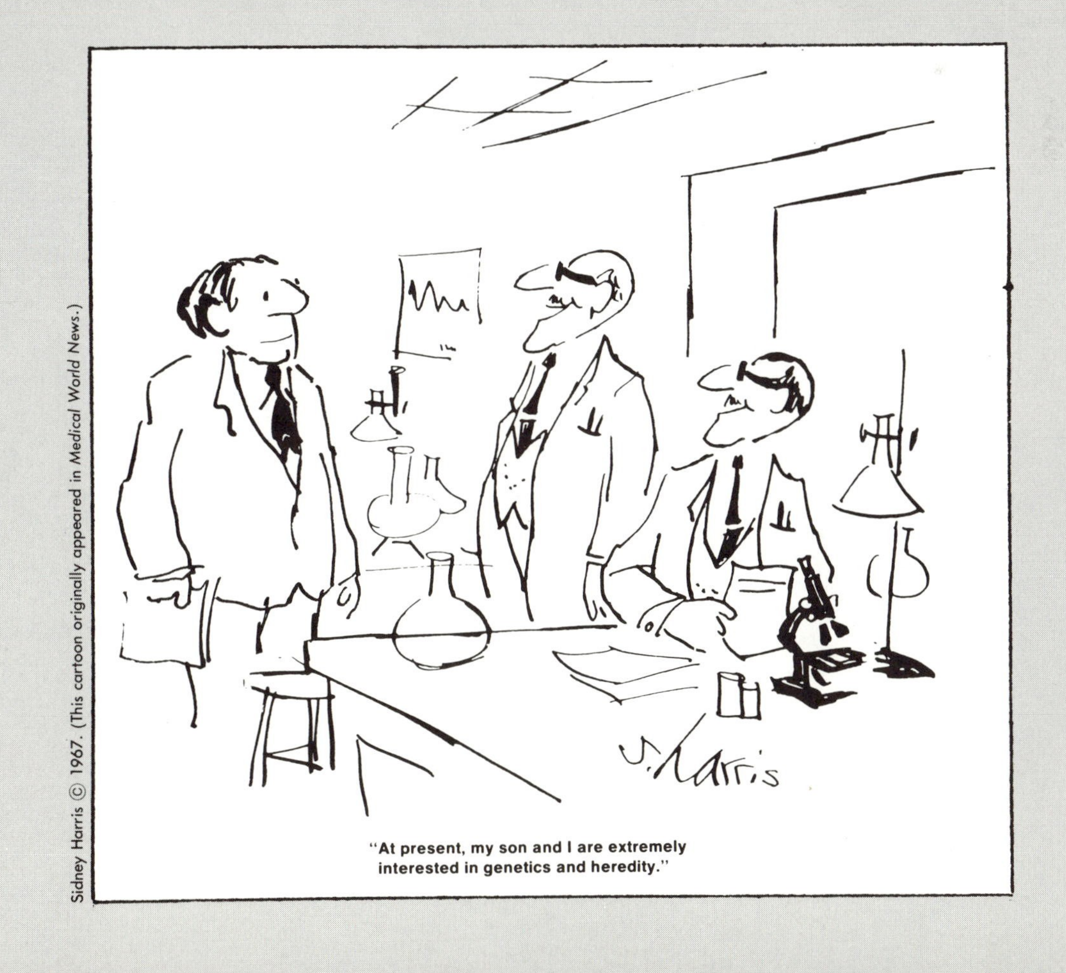

"At present, my son and I are extremely interested in genetics and heredity."

EUROPEAN SIDE SHOWS

If the U.S. circus filled the big top, Europe provided the side shows. In most European countries one or more national committees were set up in the months or years following Asilomar. The issue was also tackled by regional organizations such as the EMBO, already mentioned, the European Science Foundation (ESF), and the Commission of the European Economic Community (CEEC). By and large, however, the debate in Europe was more leisurely, and with a few exceptions the national discussions and decisions were (and remain to this day) highly derivative of those in the United States (Doc. 11.1). One exception was France, another was Britain.

After Asilomar two committees were set up in France. One was to consider general ethical questions; the other, under the sponsorship of the *Délégation Générale de la Recherche Scientifique et Technique* (DGRST), was responsible for deciding guidelines and containment conditions for experiments. The DGRST's *Commission de Contrôle* gradually set about drafting national guidelines to supercede the 1976 NIH guidelines, which were being used in the interim.

During 1977 the French Control Commission, however, used the NIH guidelines as guidelines rather than as immutable regulations and, for example, approved an experiment proposed and subsequently performed at the Pasteur Institute by Lennart Philipson and Pierre Tiollais involving recombining fragments of adenovirus DNA with a host-vector system, developed at the Pasteur Institute, in a P3 laboratory. At that time such experiments were impossible in the United States because of the stringency of the required containment conditions.

By December 1977—and the date is crucial—the French Control Commission had drafted a very short and simple set of guidelines whose general containment conditions were very similar to those recommended by the EMBO Standing Advisory Committee in November 1977 (compare Doc. 11.2 and Doc. 10.3). The French draft was made available to the NIH before its public hearing was held on December 16–17, 1977 (see Chapter 9). The fact that in some European countries it had already been possible and would in the future increasingly be possible to do experiments under less rigorous containment conditions than in the United States, where the required con-

Background photo courtesy Tom Broker, Cold Spring Harbor Laboratory.

tainment was prohibitive, was not without its effect on the discussion of the proposed revised guidelines both at the NIH public hearing in December 1977 and subsequently during 1978.

In Britain following the prompt report by the Ashby Working Party in January 1975 (see Doc. 1.12), seven months elapsed before a further Working Party was established by the British government. Under the chairmanship of Sir Robert Williams, the Working Party was to draft a code of practice; it reported a year later in August 1976, by which time there was local impatience (Doc. 11.3).

Many technical aspects of the Williams report (excerpted in Doc. 11.4) paralleled the 1976 NIH guidelines. Four categories of physical containment were defined, the first three (CI–CIII) being somewhat more exacting than the P1–P3 counterparts in the United States. Biological containment was advocated, but the specifications of biologically contained host-vector systems were left vague. No attempt was made to give specific criteria, nor were the NIH criteria recommended, and eventually a special subcommittee had to be set up in Britain to validate host-vector systems. A small set of examples of suggested combinations of containment for particular classes of experiments was included. The Williams report was, however, intentionally less comprehensive than the NIH guidelines; it did not try to cover every possible class of experiment. Instead it advocated experiment-by-experiment decisions, to build up a body of case law, leaving the decision-making task to a central advisory body, the Genetic Manipulation Advisory Group (GMAG), which it recommended be established. This approach, it was hoped, would allow more flexibility, and up to a point that proved to be the case. For example, Joseph Sambrook and Michael Botchan of the Cold Spring Harbor Laboratory travelled to the Imperial Cancer Research Fund Laboratories in London during the spring and autumn of 1978 to carry out with colleagues there recombinant DNA experiments with adenovirus and simian virus 40 under Category III containment conditions, stipulated by the GMAG. The 1976 NIH guidelines, still in force at that time, demanded such stringent containment conditions for those experiments that they could not be fulfilled at the Cold Spring Harbor Laboratory or anywhere else in the United States.

On the other hand, the specifications of Category III containment in the United Kingdom were significantly more stringent than the P3 specifications in the United States, and that certainly hindered research in many laboratories in Britain.

But returning to the chronological sequence of events, from the outset it was envisaged that all British laboratories, not just those receiving public funds, should comply with the regulatory system. Although initially a voluntary system was advocated, the possibility of the government's acquiring statutory power by issuing a regulation under existing law, the Health and Safety Work Act, was clearly foreseen. In the vacuum between the publication of the Williams report in August 1976 and the establishment in December 1976 of the GMAG, an event occurred in the United Kingdom that underlined the dangers and possible consequences of a rampant but uninformed bureaucracy. The U.K. Health and Safety Commission and its Executive, an independent body of civil servants set up under the Health and Safety at Work Act and responsible for its implementation, circulated in September 1976 a consultative document. The Commission not only proposed to make notification of genetic manipulation experiments compulsory (which was eventually decided but not until August 1978) but also devised a definition of genetic manipulation so broad as to include conventional genetics and much medical and veterinary practice. The crucial and offending paragraph read as follows:

> Notification of intention to carry on genetic manipulation: No person shall carry on any activity intended to alter, or likely to alter, the genetic constitution of any micro-organism unless he has given to the Health and Safety Executive notice, in a form approved by the Executive for the purposes of these regulations, of his intention to carry on that activity.

The British scientific community was united in its protest (Doc. 11.5 is one example) and the consultative document was rightly howled out.

Meanwhile, in October 1976 a Committee of the European Science Foundation (ESF), an organization of the European research councils and academies, made a comparison of the NIH and the U.K. guidelines and recommended that the latter should be adopted in Europe (Doc. 11.6). The recommendation was, however, not universally followed.

On December 8, 1976, the British government announced the names of the first members of the GMAG. Scientists involved in recombinant DNA research were in a very small minority, unlike the situation in the NIH's RAC; the lay public interest was to be represented, as were the chemical and pharmaceutical industries. Interestingly there were no trade-union representatives named at this time, but four were soon to join the Committee. One trade union, the Association of Scientific, Technical and Managerial Staffs (ASTMS), had two representatives on the GMAG; that trade union, with many laboratory technical staff and some academics among its members, subsequently played an active role in the British recombinant DNA debate (Doc. 11.7).

Inevitably the disclosure of information by industry to the GMAG periodically gave rise to some severe difficulties over privacy of proprietary information and possible jeopardy of patent rights (Doc. 11.8). During 1978 it was decided by the British government that the statutory powers foreseen in the Williams report were indeed needed to ensure universal compliance in the United Kingdom with the national procedures for regulating recombinant DNA activities. On August 1, 1978, Britain established a precedent by becoming the first and still only country in which it is a legal obligation to report recombinant DNA experiments to a central body (Doc. 11.9 and Doc. 11.10). The decision and GMAG's role in it were not universally approved (Doc. 11.11).

We should mention here that elsewhere in Europe the question of enacting specific legislation for recombinant DNA research was raised periodically. In the Netherlands, for example, from time to time some molecular biologists advocated a new law. Their reason was the same as that of some U.S. molecular biologists in 1977 (see Chapter 5)—a new law would preempt extremely restrictive local government initiatives that were threatening. In the Federal Republic of Germany the Ministry of Science and Technology drafted at least two versions of a bill over a three-year period, but the draft has not yet been discussed in Parliament. The Commission of the

European Economic Community in Brussels also over a period of several years considered drafts of a directive that would compel the nine EEC states to make registration of recombinant DNA work a legal obligation. Fortunately the pace of events outstripped the painfully slow bureaucracy of the Commission, and this directive was downgraded to a recommendation in the summer of 1980. A final decision on this question has still not, however, been reached by the EEC.

Document 11.1

March 3, 1977

Nature (**266**:2)

Heading for harmony?

The debate prompted by genetic manipulation developments is intensifying in Europe. **Chris Sherwell** reports

IN PARIS on Tuesday a committee gathered quietly for its routine monthly meeting to discuss research involving genetic manipulation using recombinant DNA technology. In Brussels last Thursday another committee lodged deep in the labyrinth of the European Commission of the EEC met to discuss the same subject. On Thursday of this week Britain's equivalent body gathers for its own regular monthly session. Within the next week a commission on genetic manipulation in the Netherlands is expected to report its long-awaited findings. And on 15 March a new and potentially crucial group within the European Science Foundation (ESF) convenes its first meeting in Strasbourg.

Plainly, the pace of developments relating to recombinant DNA research in Europe has quickened. The debate is now consuming the time of government officials, science administrators and molecular biologists in at least nine different countries and no less than five international organisations. Indeed, the aforementioned meetings are but the tip of an enormous iceberg, and if activity is symptomatic of decisiveness, the energy now being spent discussing the subject ought inevitably to produce results. For those researchers actually wanting to do the work, however, it has meant long periods of delay, uncertainty and confusion.

The debate itself stems from the development over three years ago of elegant new genetic engineering techniques, the implications of which prompted widespread concern. This spawned the growth of guidelines, specifying conditions under which various types of experiment using the techniques should be done, and the growth of committees to administer them. The result is that the issue has spread well beyond the bounds of the research community immediately involved.

In the United States for example, where the pace has been hottest, the debates freest and public involvement greatest, the National Institutes of Health (NIH) finally published a complex set of guidelines last June. Failure to broaden these voluntary controls now threatens legislation to make them stick (see box), a development which could complicate the European debate.

So far the European debate has proceeded unsensationally, mostly at the national level; progress is varied. International discussions have gone quietly ahead, both informally among individuals and formally through those bodies which have perceived a role for themselves. The chances of them continuing without rancour could diminish, however, particularly if national differences emerge strongly in certain international forums.

Country catalogue

Most European countries have in fact trodden broadly similar paths. The examples of the United States and Britain have been important, the latter adding a new word to the English vocabulary: 'geemag'. The Genetic Manipulation Advisory Group (GMAG) was established in Britain last year following recommendations from the Williams working party on genetic manipulation.

Precautions contained in the Williams guidelines, like those in the NIH guidelines, involve both physical and biological containment. The standards of physical containment are more strict under Williams, however, while biological containment receives greater emphasis in the NIH guidelines. Among other differences, the NIH guidelines provide an extensive list of experiments and appropriate containment measures; under Williams GMAG decides matters case by case. GMAG itself had early problems over trade union representation, but later began business and by this week was meeting again (see box).

At least eight other countries in Europe have sought to involve themselves in the same sort of process. directly involved. One, called an 'ethics committee' and having under 10 members, meets spasmodically to discuss the whole subject of genetic manipulation. The other, known as the *Commission de Controle*, was set up in 1975 under the auspices of France's main research body, the *Delegation Générale de la Recherche Scientifique et Technique* (DGRST). It has a membership of about a dozen, most of them with interests in recombinant DNA and related research. Industry and trade unions are not directly involved, though this could change; both are evidently prepared to abide by the *Commission* line.

The *Commission* meets regularly once a month to assess the conditions under which experiments in France are to be performed. It began by using the Asilomar guidelines, the product of the Asilomar conference in California in 1975, and subsequently used the NIH guidelines. More recently it has used both the Williams and NIH guidelines.

Over a period of some 18 months it has considered some 50 applications. Less than 10% have involved the more stringent containment conditions equivalent to the P3 and P4 categories of the NIH guidelines.The number of facilities in France which could now provide such containment is no more than a couple, and those would probably be at the P3 level.

Germany. The more fretful search for West German guidelines has produced four drafts, the latest of which was due to be distributed by the end of February, with the hope of producing a final draft by the spring. The government department involved, the Ministry of Research and Technology (BMFT), took over after a somewhat unsuccessful start under the German Research Council (DFG) using NIH guidelines.

The new national guidelines are expected to follow much the same lines as the NIH and Williams precedents. There will probably be an equivalent to Britain's GMAG with mixed representation. Guidelines will be voluntary, and it is hoped that industry, for the inclusion of which there are apparently no immediate plans, will follow them. There are no national P4 facilities in Germany, and few that would qualify as P3 facilities.

Switzerland. The Swiss Academy of Medical Sciences created a standing committee known as the Commission on Experimental Genetics. Headed by Werner Arber, Professor of Microbiology at the University of Basel, it has a membership of about a dozen, consisting of experts in the field and government officials from the health and science and technology ministries. Industry is ostensibly as keen as the universities to see controls implemented, and is represented on the Commission.

The Commission decided recently to follow the NIH guidelines, but this could still change in the future. It has been agreed with government agencies that no legislation is necessary. Researchers applying to the Swiss grant-giving body (*Suisse National Fonds,* SNF) and supplying relevant details would go before the Commission only when the SNF is unhappy; but they will probably be invited to register their work.

Sweden. An 11-man "committee concerning research with recombinant DNA" was set up last spring under the auspices of the Natural Science Research Council (NFR), the Medical Research Council (MFR) and the Swedish Cancer Society (RmC). Its chairman, Professor Peter Reichard of the Karolinska Institute, is the appointee on the committee of the NFR, which also appoints two lay representatives—in this case two MPs. The MFR, RmC, the Council for Forestry and Agricultural Research, the Board of Health and Welfare and the Association for the Pharmaceutical Industry each appoint one member; one member jointly represents the Academy of Sciences and the Academy of Engineering Sciences, another jointly represents the National Defence Research Institute and the Board for Technical Development; the Central Organisation for Salaried Employees appoints a representative for technical staff.

The committee will help granting agencies and government authorities to determine safety conditions for the experiments which they fund, and will probably work in the private sector as well. The committee will also advise researchers over safety precautions and help in risk classification, for which a working group has just been appointed. Researchers will be expected to submit experimental protocols to national advisory committees, which are responsible for specifying the containment measures. The containment procedures proposed are those of Williams, but experiments prohibited under the NIH guidelines, it is suggested, should not be carried out. Local safety committees would be responsible for supervising the measures.

So far the committee has received no applications and no recombinant DNA work is proceeding. The only group in Sweden which has worked in the field is at Uppsala under Professor Lennart Philipson, the MFR's appointee on the committee. His work is in abeyance, and he has yet to submit an application. The group has applied to have two new P3 laboratories built, but finance for these is not yet assured.

Denmark. Denmark has two committees. One, headed by Dr Kjeld Marcker from Aarhus, is an ad hoc committee established by the country's research council. It is examining work being done in Denmark and hopes to decide by the summer whether there ought to be a special research programme. The other is a Committee of Registration with 10 members representing research councils and industry but not trade unions. Its concern is with safety aspects, and it has been operating only a few months. The precise course it follows depends on the outcome of the first committee's work.

Holland. In January of last year, after consultation between the Royal Dutch Academy of Science and the National Health Council on the one hand, and the Ministry of Education and Science and the Ministry of Health on the other, a "Commission on Genetic Manipulation" was formed consisting of experts in the field and chaired by Professor Bootsma of Erasmus University, Rotterdam.

Its task was to compile an inventory of recombinant DNA research being done in the country and to advise laboratories on safeguards and the authorities on controls. The commission is expected to report its findings within the next two weeks, but there are differences of view about whether these should actually be published. The Dutch began by opting for the NIH guidelines but then switched to the Williams guidelines; now, however, it seems possible that national guidelines will be proposed which are more strict than either.

So far industry has not shown any public interest in the guidelines. Applications have come for six projects, from four universities. Three have come from the University of Amsterdam, and one from the Free University of Amsterdam, the University of Leiden and the University of Groningen.

Israel. A committee of the Israel Academy of Sciences and Humanities, chaired by Professor Leo Sachs of the Weizmann Institute, has recommended the establishment of a national safety committee for recombinant DNA research and of special safety committees at universities and research centres. The local committees would recommend appropriate safety precautions for every research proposal and then pass the matter to the national body, which has the final word.

The Academy decided that Israeli researchers would in general follow the NIH guidelines but would take into consideration recommendations from other bodies.

International catalogue

For Europe's researchers, worry as well as confusion is the product of the growing involvement of an increasing number of international bodies. But the important issue—whether or not there should be harmonisation of guidelines—already seems settled. There were benefits in allowing each country to follow its own path and arrive at its own conclusions, but different precautions in different countries threatened to encourage a concentration of researchers, so harmonisation looked preferable.

Moreover, with the subject a matter for public debate, there was the possibility that if one country applied more stringent controls than another, the difference itself could become an issue. Harmonisation could pose problems, though, if US researchers are faced with legally enforced regulations and there is public pressure for similar regulations in Europe. Creating a disinterested peer group competent to judge colleagues' work could be difficult enough without the added (if necessary) complication of drafting and enforcing legislation.

The case for harmonisation does not,

however, solve the problem of which body has the authority to encourage the move towards it. A look at the organisations involved reveals the difficulty.

The World Health Organisation (WHO). The WHO has two bodies actively involved, the WHO Special Programme on Safety Measures in Microbiology and the WHO Environmental Programme. Being global and not confined to Europe, and concerned with the public health and safety aspects of the research, the WHO interest is peripheral as yet, although it has begun consultations, with, for example, Professor S. W. Glover, chairman of the International Microbial Genetics Commission, which has a genetic engineering sub-group.

The International Council of Scientific Unions (ICSU). ICSU has recently created a sub-committee known as COGENE, headed by the US microbiologist Bill Whelan. Its first official meeting is in Paris at the end of May. Because it too is part of a global agency, and because it represents international commissions in various disciplines more than national councils, ICSU can acquire an important role outside North America and Western Europe.

It can, for example, involve the East European countries and also the USSR, which is apparently keen to participate in the research under agreed guidelines—the USSR has a committee under Professor Bayev which hopes to produce its own guidelines by the middle of the year. ICSU can also involve important Third World countries, such as India and Nigeria, which cannot be prevented from pursuing the research but which might participate in any programme organised by ICSU on their behalf to help ensure that they take the necessary precautions.

The European Molecular Biology Organisation (EMBO). EMBO is based at Heidelberg, where a P4 facility for international use is under construction. It was the first international European body to become involved in guidelines for recombinant DNA research, and has a Standing Advisory Committee on Recombinant DNA chaired by Professor Charles Weissmann. This was set up in January 1976 and held its first meeting in London the following month to discuss whether it was worth elaborating guidelines, other than those available, for Europe as a whole.

At its second meeting, in London last September, the committee compared the NIH and Williams guidelines. It suggested the establishment of national advisory groups to specify containment measures for each experiment on the basis of a detailed protocol submitted to it. The committee specifically recommended against the idea of using some combination of the procedures from the two sets of guidelines. And it advised that experiments forbidden under the NIH guidelines not be carried out.

The European Science Foundation (ESF). Established in November 1974, the ESF is made up of 45 national research councils and academies from 18 European countries and aims to create a close-knit community of science and research in Europe. Sweden quickly suggested to the founding committee that it should consider the whole question of genetic manipulation, including its social, legal and ethical aspects. With the EMBO committee able only to provide advice on request, and then only about scientific and technical aspects, the ESF decided in October 1975 to broaden a preparatory working party into an ad hoc committee on Recombinant DNA which could propose through ESF members whether and what action should be taken at European level.

The committee met three times in 1976 under the chairmanship of Professor Povl Riis of Denmark; members included molecular biologists, physicians and lawyers. Their brief was broad, and they concluded that the recommendations and code of practice of the Williams report should be adopted as the guidelines for recombinant DNA research in Europe. They also recommended that national registries of research should be established and that laboratories should be legally obliged to declare their work to it; laboratories would adhere to agreed guidelines voluntarily, however, and supervision and monitoring would be a national responsibility. National variations, it suggested, should be minimised.

The ESF has now created a new committee made up of representatives of the geemags of its members with the aim of proposing guidelines for Europe. This European Committee on Recombinant DNA will meet for the first time in Strasbourg on 15 March. It will note differences in the practices of various countries and consider prescribing measures for the future.

The European Commission. The European Commission, which has a dual role in the EEC of both initiating and implementing Community legislation, finally jumped into the fray in January. Spotting the opportunity provided by a potential need for Community-wide legislation and harmonisation, Directorate General XII (Research, Science and Education), headed by Dr Gunter Schuster, called geemag heads to Brussels on 21 January for "informal consultations". This offended some sensibilities, not least because Dr Schuster was seeing members of the new ESF committee on which he is himself the EEC's representative.

The Commission is apparently contemplating a directive, the device by which it can request member states to modify and harmonise their legislation. In the case of recombinant DNA research this might involve asking the Nine to ensure that they take the same precautions, but not interfering with the operations of individual geemags. At the January meeting there was no stern objection to the idea, provided a directive was not too specific or detailed.

The worry is not chiefly about Commission interference, although its record in science is less adequate that it might be. Most people recognise that it possesses the authority both to hasten the necessary harmonisation and to incorporate research done in the private sector a common framework. The worry for the moment is related more to the style and timing of the Commission's involvement, which could be self-defeating if it breeds resentment among researchers.

The outcome of the January meeting was presented last week at a meeting in Brussels of the Medical Research Committee, a sub-committee of CREST, the Commission's Scientific and Technical Research Committee. According to the office of the director of the biology programme, the meeting reached no firm conclusions and is due to meet again only in June. That may mean that EEC invlovement will remain peripheral for a while yet. If so, the immediate burden of recommending a path for recombinant DNA research in Europe now lies with the ESF. □

A TRANSLATION OF

French Guidelines on Recombinant DNA Experiments

13 DECEMBER 1977

Introduction

B) SAFETY FACTORS:

1) Physical containment

The laboratories are defined as P1, P2, P3 and P4 in the same way as in the N.I.H. Guidelines (June 1976 section II), except that, in a P3 facility, the exhaust air must be filtered through Hepa filters, and a double door autoclave must be available. This type of facility will be called P3*.

The level of protection can be increased by the use of safety cabinets located inside these laboratories. It is the laboratory+cabinet system which determines the level of containment. These levels are called L1, L2, L3 and L4. If an experiment is not entirely done inside the safety cabinet, the level of containment depends on the laboratory alone.

The cabinets considered here are of two types: laminar-flow (class II cabinets, N.I.H. 1976, section II) and isolators, which are defined as class III cabinets (N.I.H., 1976, section II) possibly improved by additional secondary safety systems.

Laboratory-cabinet systems are listed in Table I. They must be certified before utilization.

2) Biological containment

The containment brought by the biological system is specific for each host-vector system. The levels of safety depend on both the vector and the host and are called B0, B1, B1* and B2.

For the experiments using *E.coli* K12, its plasmids and its phages, levels B1 and B2 correspond to levels EK1 and EK2 (N.I.H. Guidelines, June 1976, section IID).

Level B1* corresponds to a B1 system carrying several mutations (for example, dap$^-$ or streptomycine-dependent bacteria). Level B0 corresponds to systems which are less safe than B1 (for example, transmissible plasmids).

Cloning systems involving *B. subtilis* 168 as the host cell and non-transmissible plasmids are classified as B1*. If the host cell is thy$^-$, spo$^-$, the system is

*Such systems will be certified by the National Committee only after an appropriate scrutiny.

classified B2 (see footnote).

If non-transmissible plasmids, which form unstable recombinants (loss of 1% per generation in the absence of any selective pressure) are used, B1 and B1* systems become B1* and B2, respectively (see footnote).

3) Training of personnel

The scientists and technicians involved in these experiments should be familiar with the biological and physical containment they are using. This comprises a competence with: a) the host-vector system so that the personnel are able regularly to monitor the genetic characteristics on which the security of the system depends; b) the system of physical containment and the emergency measures in case of accidents (including fire); c) in the case of L3 and L4 experiments the training of investigators should be analogous to that required for handling pathogenic microorganisms.

The control of the training of the laboratory personnel will be done by local safety and health committee (see Rapport d'Activité 1975-1976 de la Commission REG).

TABLE I: Levels of physical confinement

	Laboratory	Cabinet
L1	P1	None
L2	P1	Laminar flow hood
	P2	
L3	P2	Isolator[(a)]
	P3*	
L4	P3*	Isolator[(a)]
	P4	

[(a)] See text.

TABLE II: Classification of experiments[(a)]

DNA origin		Classification	
Mammalian		L3	B1
	or		
Avian		L2	B2
Amphibian		L2	B1
Reptiles	or		
Fish		L1	B2
Other		L2	B1
	or		
Eukaryotes		L1	B2
Prokaryotes		L1	B1
Viruses	see note 1		

[(a)]The cloning of DNA from pathogenic organisms and from cells producing viral genomes will be analyzed on a case by case basis. In general these experiments will require increased physical and/or biological security.

Note: The list of publications which have formed the basis for these Guidelines is available from D.G.R.S.T.

*Such systems will be certified by the National Committee only after an appropriate scrutiny.

C) CLASSIFICATION OF EXPERIMENTS:

Experiments involving the formation or the propagation of an unlimited number of recombinants in volumes lower than 10 liters are classified in Table II.

NOTES TO TABLE II:

1) The diversity of the known viruses and of their interactions with host cells makes impossible a classification which can be used in all cases. Therefore, each case will be classified according to the following considerations: a) cloning DNA from oncogenic or from highly pathogenic viruses is judged on a case basis (at least L3 B2); however, b) if the localization of oncogenic or highly pathogenic genes is known, and if they are eliminated before cloning by fractionation, the classification will be based upon the efficiency of the fractionation proposed in the experimental procedure.

2) If the number of clones is sufficiently low (see text), the Committee can authorize using a level of physical *or* biological containment decreased by one degree (this does not hold true for viruses). This number is set as lower than 100 for mammalian DNA fragments having an average molecular weight of one million and will be decreased for fragments from less complex genomes.

3) If the DNA fragment, whatever its origin, has been cloned, characterized and is known not to be pathogenic, it is possible to use levels of physical *and* biological containment decreased by one degree each. If, on the other hand, the recombinant to be isolated carries a pathogenic DNA fragment, levels of biological and physical containments will be increased by one degree each.

4) B0 and B1 host-vector systems will be judged on a case by case basis.

5) The addition of chemicals susceptible to inactivate the bacteria carrying recombinant plasmids, and/or recombinant phages, allows the experiment (and specially extraction of DNA) to be carried at a level of physical containment reduced by one degree.

6) The Committee should be notified of any genetic drift occurring during the propagation of recombinant DNA molecules.

Document 11.3
July 22, 1976
Nature (**262**:244)

Why are we waiting?

Although Mr Mulley, UK Secretary for Education and Science, has had the report of the Williams Working Party on Genetic Manipulation of Micro-organisms on his desk for well over a month, it is due to remain there for at least a "few more weeks". Until then scientists will be in the dark over the details of the code of practice that will apply to recombinant DNA research in Britain. There are several possible excuses for Mr Mulley's tardiness. None of them is good enough.

The Williams Working Party had its origins in a statement published by Professor Paul Berg and his colleagues in July 1974 in which they expressed concern at some of the possible consequences of the research they themselves had pioneered. The improbable pathway to disaster that they foresaw involved the escape from laboratory containment of bacteria whose genes had been experimentally recombined with genetic material from another organism, the survival of those bacteria in the outside world, their colonisation of the human intestine and the expression of their foreign genetic component to the detriment of the human host.

That scenario was taken seriously enough on both sides of the Atlantic that most scientists have since held to a voluntary moratorium on suspect research. American concern, latterly under the auspices of the National Institutes of Health, culminated four weeks ago in the issue of a complex series of guidelines. The British wheels were set in motion in July 1974 with the Ashby Working Party, convened by the Advisory Board for the Research Councils, which reported in December of that year. Seven months later Mr Mulley set up the Williams Working Party to produce a code of practice and to consider the establishment of a central advisory service for laboratories carrying out the procedures in question.

The Williams Working Party report is unlikely to contain many surprises. It will not recommend the proscription of any particular experiments but will suggest the precautionary measures that are appropriate for various categories of research. Few specific experiments will be quoted and no attempt to quantify their dangers will be made. The report is expected to endorse Ashby's suggestion of a central advisory service and to recommend how it might operate.

The scientific community anxiously awaits the report and Mr Mulley's reactions to it. The longer they have to wait for a go-ahead on experiments which in some cases have been on ice for two years, the worse (whatever the precautions suggested) will become the atmosphere of frustration and suspicion that has gradually built up. Inevitably moratorium-breaking has already occurred, and the consequences for those involved have on occasion been unpleasant.

Why then is Mr Mulley keeping us waiting? One possibility is that he is reserving the right to modify his reactions in the light of the response at home and abroad to the American guidelines. If so the delay could well be lengthy in view of the recent clash between Cambridge (Massachusetts) City Council and Harvard University. The trouble arose when the Mayor of Cambridge attempted to block plans to build a high containment laboratory within the Harvard Biological Laboratories; at the moment there is a three-month moratorium during which the council will review the position before deciding on the future of Harvard's genetic engineers. Similar problems are expected in other American cities. Although there is a reasonable chance that such clashes will be avoided in Britain, the longer Mr Mulley delays, the more likely they are to occur.

A second reason for delay may simply be the need for extensive briefing and consultation between departments. If so Mr Mulley must lavishly oil the cogs of bureaucracy. Most likely the delay is due to his consideration of the introduction of statutory controls of recombinant DNA research. Although the Working Party is thought to have been reluctant to recommend statutory controls, they may have made some suggestions in that direction because of forceful representation from the unions. The same pressures, which clearly must be respected since the unions represent those most likely to be directly carrying out the research, may now be holding Mr Mulley back.

The British code of practice, with or without statutory backing, could be very influential. In contrast to the American code, which applies only to NIH-supported research, the British code will apply to all academic institutes and probably also to industry via the Health and Safety Executive. If that breadth is matched by an authoritative depth and practical recommendations, the code could well be adopted by other European countries. That could happen through the European Molecular Biology Organisation which meets to consider the matter on August 12. It would be a great pity if the British code had not emerged by then. □

Document 11.4

August 1976

Williams Working Party Report (HMSO Cmnd. 6600), Excerpts

REPORT OF THE WORKING PARTY ON THE PRACTICE OF GENETIC MANIPULATION

Presented to Parliament
by the Secretary of State
for Education and Science
by Command of Her Majesty
August 1976

LONDON
HER MAJESTY'S
STATIONERY OFFICE
50p net

Cmnd. 6600

5. CENTRAL ADVICE AND CONTROL

5.1 The Ashby Report (paragraph 6.5(d)) suggested that "as an initial step a widely publicised advisory service, perhaps offered by public health laboratories, would help to safeguard the interests of the public and of those engaged in the experiments". The Government has already accepted that it has a responsibility to ensure that authoritative advice and guidance are available to laboratories using the techniques available for genetic manipulation and we were asked to make recommendations for the establishment of a central advisory service.

5.2 We were also asked to consider the practical aspects of applying in appropriate cases the controls advocated by the Working Party on the Laboratory Use of Dangerous Pathogens, which recommended that initially control should be on the basis of voluntary acceptance by laboratories of the advice of a centrally appointed Dangerous Pathogens Advisory Group (DPAG), but that various existing legal powers which could be invoked to give statutory force to such advice should "be consolidated so that the Departments of Health and Agriculture can act with full authority, without delay and with uniform principles."*

A voluntary system of advice and control

5.3 We recommend the establishment of a Genetic Manipulation Advisory Group (GMAG). Since a central advisory service will need to command the respect of the public as well as of the scientific community, including scientists in industry, the membership of the GMAG should include not only scientists with knowledge both of the techniques in question and of relevant safety precautions and containment measures but also individuals able to take account of the interests of employees and the general public. We hope that the Government will agree to the establishment of the GMAG on this basis at an early date so that work which is scientifically desirable may proceed quickly and safely.

5.4 The main functions of the GMAG should be to advise on the category into which a particular experiment would fall, taking into account the factors discussed in Section 2 above and on the application to particular cases of the code of practice recommended in Section 3 above. To do this the GMAG will need to maintain records of the facilities available in different laboratories and the qualifications of Biological Safety Officers: in time it should in effect establish a register of approved laboratories. It should also review experimental protocols regularly as part of a continuing assessment of precautions which may need to be changed as the subject develops. In particular, the GMAG should assess any new methods of physical or biological containment that may be developed by laboratories and consider whether they would justify major modifications of practice. Such assessment would need to be on the basis not only of submitted documents but also of independent technical evaluation and validation and all interested laboratories should be kept informed of such developments. A procedure for the acceptance by the GMAG of new methods of containment and of disabled strains might be introduced. The GMAG should publish an annual report and be ready to advise on general matters connected with the safety of genetic manipulation, including health monitoring, for which it should be able to call upon the experience of epidemiologists, and the training of staff.

5.6 We envisage the following stages in the consideration of a proposal for an experiment involving genetic manipulation:

i. discussions within the laboratory both of scientific merits and of potential hazards. The laboratory's Biological Safety Officer and a properly constituted and representative safety committee should have key roles to play and the discussions should lead to provisional conclusions about the desirability of conducting the experiment and

---

* Report of the Working Party on the Laboratory Use of Dangerous Pathogens, paragraph 61.

about the containment category into which the experiment should fall;

ii. if as a result of these discussions there is no doubt that the proposed experiment belongs to Category I or II the GMAG should be notified immediately, but work could proceed under the appropriate conditions as specified in the code of practice;

iii. if the discussions suggest that the proposed experiment falls into category III or IV, reference must be made to the GMAG for advice before a final decision to undertake the experiment is taken;

iv. a proposal referred to the GMAG will be examined for the detail of the proposed experimental protocol and of the physical facilities and safety measures at the laboratory concerned. After such consideration the GMAG would either:

(a) advise that there would be no objection to the work proceeding as proposed; or

(b) advise that there would be no objection if specified precautions were adopted; or

(c) advise that the experiment could not be undertaken safely.

5.7 It will be important for rapid assessments to be made of category I and II protocols which are reported to the GMAG so that any inconsistencies in a local decision can be quickly corrected. The GMAG will therefore need a scientific secretariat able to react quickly to protocols (consulting a member of the GMAG as necessary) and to deal directly with a laboratory if it seems necessary to question a local decision and to ask for delay pending consideration by the GMAG. Such cases may be rare but the possibility emphasises the need for speed and flexibility in the procedures for transmitting advice to laboratories.

5.8 The protocols of experiments likely to fall into category III or IV will need more thorough consideration by the GMAG, possibly at a regular meeting of the whole group, taking into account:

i. the nature of the experiment, with special reference to the biological factors referred to in Section 2;

ii. the facilities at the laboratory concerned. An inspection may be necessary for this purpose in the early stages until a register of approved laboratories (paragraph 5.4 above) is established;

iii. the experience, ability and training of the research workers and technicians and of the Biological Safety Officer; and

iv. the arrangements for monitoring the health of staff.

5.9 Formulation of advice on protocols for experiments in category III or IV may take some time and could in general proceed in parallel with a laboratory's preparations and planning. But if a voluntary system is to maintain the confidence and co-operation of the scientists concerned, it is important that the time taken should be kept to a minimum and should only rarely exceed three months.

5.10 The efficacy of the advisory machinery recommended above will depend on the willingness of laboratories to accept and act on central advice. Our consultations with both academic and industrial scientists convince us that scientists will in fact welcome and be ready to comply with authoritative guidance from the centre.

Statutory control

5.11 On the basis in paragraph 5.10 above, the advisory system we describe would in practice amount to a system of control broadly comparable to that now operating through the DPAG for laboratories working with dangerous pathogens. Some of our witnesses urged that there was a need for specific statutory powers similar to those advocated by the Working Party on the Laboratory Use of Dangerous Pathogens (paragraph 5.2 above). We carefully considered these views and set out our conclusions below.

5.12 We noted that the Health and Safety at Work Act lays a clear duty on the employer to protect his employees and also to avoid hazard to the public. We were advised that in the event of legal action a court, in considering whether an employer had taken all reasonably practical steps, would be likely to give great weight to whether he had sought or taken advice from the Genetic Manipulation Advisory Group once it was set up. The Health and Safety Executive also has powers of inspection to ensure compliance with requirements for the safety of workers and the public. These existing powers and duties arising from the Health and Safety at Work Act already provide a safeguard to the public. We were advised that as an additional measure, regulations could be made under the Act to require laboratories to submit experimental protocols and appropriate supporting information to the GMAG and we recommend that this should be done. Given such a requirement, it seems very unlikely that the advice of the GMAG would be disregarded.

5.13 The Health and Safety at Work Act does not however cover hazards to the plant and animal populations and it seems unlikely that existing powers available to the Agriculture Departments could be invoked to provide any necessary statutory controls in this field. In any case, if statutory control is to be envisaged it may be desirable, on the analogy of the recommendations of the Working Party on the Laboratory Use of Dangerous Pathogens referred to in paragraph 5.2 above, to envisage some consolidating legislation. Such specific legislation might be directed simply towards compulsory consultation with the GMAG or perhaps extend to a system of licensing for laboratories. A practical difficulty is that a definition of the work to be controlled (for example, on the lines in paragraph 1.3 above) would almost inevitably become outdated as the science developed and as new techniques emerged. Such a difficulty could be met by a provision that the work subject to control should be specified in regulations which could be amended as necessary more readily than major legislation.

5.14 We recommend that the system of voluntary control we have described should be established as quickly as possible since we believe this could provide immediate and effective control of the hazards while permitting valuable work to proceed safely. The operation of this system will enable the Government to consider the desirability and practicability of introducing specific consolidating legislation at a later date.

6. RECOMMENDATIONS

We recommend that:

i. experiments in genetic manipulation, conducted in appropriate conditions of physical and biological containment, should be encouraged (1.3);

ii. further work should be done on the development and characterisation of disabled organisms and that any which are developed should be made freely available to all workers in the field (1.6);

iii. no genetic manipulation experiment should be undertaken in containment conditions less stringent than those used for work with common pathogens (2.8);

iv. the code of practice in Appendix II should be adopted as a basis for the conduct of these experiments (3.1);

v. every laboratory conducting these experiments should have a safety committee and a Biological Safety Officer (3.4);

vi. appropriate training should be made available and be required for all research workers, technicians and biological safety officers in genetic manipulation laboratories (4.2);

vii. a Genetic Manipulation Advisory Group (GMAG) should be established to advise on appropriate precautions for the conduct of these experiments (5.3);

viii. the GMAG should be separate from the Dangerous Pathogens Advisory Group (DPAG) although there should be liaison between the two groups (5.5);

ix. a system of voluntary control should be established as quickly as possible (5.14);

x. regulations should be made under the Health and Safety at Work Act to require laboratories to submit experimental protocols to the GMAG (5.12);

(Numbers in brackets refer to paragraphs in the report)

7. CONCLUSION

We see the system of advice and control we have proposed as providing a framework within which progress can be made in an exciting and important new field of science that offers great potential benefit. Provided that the system operates flexibly so that advice to laboratories can be made available quickly and so that there can be a rapid response to new developments (with a view for example to modification of precautions if necessary), we believe that scientists will welcome and act on authoritative guidance from the centre. We think that it may be necessary for the Government to consider the introduction of specific statutory powers to control genetic manipulation but that a decision on this should be deferred until there is experience of the operation on a voluntary basis of the system which we recommend.

Suggested categorisations for some typical experiments in Categories I, II, III & IV

(These examples all assume standard biochemical manipulations)

Source of nucleic acid	*Specification of nucleic acid sequence*	*Vector/Host System*	*Category*
Mammals	Random	Phage or plasmid/bacteria, not disabled	IV
	Random	Phage or plasmid/bacteria, disabled	III
	Purified*	Phage or plasmid/bacteria, not disabled	III
	Purified*	Phage or plasmid/bacteria, disabled	II
Amphibians and reptiles	Random	Phage or plasmid/bacteria, not disabled	III
	Random	Phage or plasmid/bacteria, disabled	II
	Purified*	Phage or plasmid/bacteria, not disabled	II
	Purified*	Phage or plasmid/bacteria, disabled	I
Plants and invertebrates and lower eukaryotes	Random	Phage or plasmid/bacteria, not disabled	II
	Random	Phage or plasmid/bacteria, disabled	I
	Purified*	Phage or plasmid/bacteria, not disabled	I
Mammals Amphibians and reptiles Birds	Random	Virus capable of infecting man or growing in tissue culture cells	IV
	Purified*	Virus capable of infecting man or growing in tissue culture cells	III
Viruses pathogenic to vertebrates	Random	Phage or plasmid/bacteria, disabled	IV
	Purified*	Phage or plasmid/bacteria, disabled	III
Animal viruses, non-pathogenic to man	Random	Phage or plasmid/bacteria, disabled	II
Bacteria specifying toxins virulent to man	Random	Phage or plasmid/bacteria, disabled	IV
Plant pathogenic bacteria	Random	Phage or plasmid/bacteria, not disabled	II
Plant viruses	Random	Phage or plasmid/bacteria, not disabled	II
Bacteria or fungi non-pathogenic to man, animals or plants	Random	Phage or plasmid/bacteria, not disabled	I

* The term "purified" means fractions with little chance of including any unrecognised extraneous sequences (see paragraph 2.6.i). It is of course possible to have sequences selected because of their pathogenicity and these would raise the level of containment required.

3. CODE OF PRACTICE

3.1 We were asked to draft a central code of practice for laboratories undertaking experiments involving the techniques of genetic manipulation. We have done this on the basis of the four levels of physical containment (I – IV) summarised in paragraph 2.8 above; the code of practice is set out in full in Appendix II, together with a table illustrating the major differences between the four containment levels. The following headings are used in the code, as far as these are applicable to the particular level of containment being described:

- Laboratory (premises and facilities)
- Biological Safety Officer
- Staff—selection
  - training
  - supervision
  - protective clothing
  - health
  - discipline
- Packaging and transport of samples
- Security
- Special requirements of experiments involving laboratory animals or plants.

3.2 In drawing up the code of practice, we have taken account of the code set out in the Report of the Working Party on the Laboratory Use of Dangerous Pathogens and in view of the similar requirements of our category IV containment level and the category A level for dangerous pathogens, we have tried as far as possible to follow the provisions and wording of that Working Party's code of practice. We also noted that a code of practice covering three containment levels is being prepared by Sir James Howie's Working Party on the Prevention of Infection in Clinical Laboratories.

3.3 As indicated in the introduction to the code of practice, we have not attempted to deal with all the technical queries that may arise on various aspects of the recommended containment levels: we have sought rather to provide guidelines for the GMAG to build on in drawing up the detailed specifications that may be necessary for assessment of individual laboratories.

Document 11.5

October 13, 1976
Subak-Sharpe, Williamson, and Paul
Letter to British Secretary of State

DEPARTMENT OF BIOCHEMISTRY

UNIVERSITY OF GLASGOW
GLASGOW G12 8QQ.
TELEPHONE: 041-339 8355 EXT.

13 October 1976

ARW/SS
Mrs. Shirley Williams
Secretary of State
Department of Education and Science
Westminster
LONDON

Dear Secretary

Re: Genetic Manipulation: Report of the Williams Committee Command 6600 and the Health and Safety Commission Consultative Document entitled "Compulsory Notification of Proposed Experiments in the Genetic Manipulation of Micro-Organisms'

We draw your attention to the wider implications of two particular documents which have been occasioned by the public discussion of recombinant DNA technology. In the evaluation of this new technology by CMD 6600, both the considerable potential benefits and the conjecturally associated hazards have been stressed and Safety rules and Safety precautions have been proposed which should nullify such hazards. The real and foreseeable hazards of recombinant DNA technology are those normally associated with the handling of pathogenic micro-organisms, and for these there are tried and tested safety measures. The Report of the Working Party under the Chairmanship of Professor Sir Robert Williams was clearly concerned not only to see that sound, adequate and tried safety precautions were brought to bear on experiments with recombinant DAN techniques, but also to draft regulations which would lead to the progressive evolution of genetic safeguards against conceivable hazards that might arise in certain recombinant DNA experiments. It should be appreciated both that many of the conceivable but unproven hazards, imagined and projected onto the public debate may not really exist; and that some connected hazards could exist which have not yet been perceived. Many identified benefits are expected to result from recombinant DNA technology, but there are probably numerous important advantageous uses not yet foreseen.

In these circumstances there is clear need for carefully framed regulations which can cover all of the conceivable experiments and at the same time ensure that progress continues. It is a truism that Progressive Scientific Research inevitably means contending with the unknown. The Williams Committee perceived that the logical, practical and effective way to regulate recombinant DNA technology is by a flexible system which keeps abreast of progress in the field. Their proposal to set up the Genetic Manipulation Advisory Group is therefore highly commendable. It follows the general procedures for the responsible self-regulation of basic research by the acknowledged experts in the field. Such a review system, administered like the Grant Review system by the Research Councils, would have the greatest force and respect from our scientists, International Science and from industry, without need for the intervention of any other statutory body.

In contrast to the closely reasoned and highly responsible Report and recommendations of the Williams Committee, the consultative document of the Health and Safety Commission comprises and hasty and microbiologically ill-advised attempt to write sweeping rules and assume extensive powers. These, if practised literally, could restrict and impede an enormous range of advanced bio-medical research, ongoing at the present time and separate from the new recombinant DNA technology. For example, the proposed restriction on any activity intended to alter, or likely to alter, the genetic constitution of any micro-organism (regulation 2), directly proscribes the whole of microbial genetics, much pharmacology and even medical and veterinary practice (e.g. every use or administration of any antibiotic). Had such regulations been enforced earlier in this century, our present highly effective treatment and control of infectious diseases could not have evolved. It must be appreciated that genetic recombination is a natural phenomenon continuously ongoing at all levels, and found widespread in nature, and that conventional microbial genetic work is adequately covered within present legislation.

The Health and Safety document allows for exemptions from the HSC regulation, but this would presumably require that every standard procedure despite its previous safety record would need to be re-examined and revalidated. For this purpose, as well as for the licensing of workers in the use of all microbial genetic techniques including the new recombinant DNA technology, the Health and Safety Executive would need to set up a further expert Committee as well as substantially expand their Inspectorate. A very high level of microbiological, molecular biological and genetic expertise would be necessary for the competent operation of such inspection and control. The extra administrative load and expense involved in implementing the Health and Safety Commission document would of necessity be enormously greater than that involved in setting up and running the Genetic Manipulation Advisory Group. This expense could not be justified since in our view implementation of the HSE proposals would not serve any purpose beneficial to persons covered by the Health and Safety at Work Act or to the public at large.

Can valid criticism be levelled at the Genetic Manipulation Advisory Group by arguing that the scientists on that Group would have a vested interest in the continuation of experiments using recombinant DNA technology? We think not, for it was only the action of responsible scientific experts that initiated public discussion of the need for caution and for safety procedures to remove conceivable hazards arising in recombinant DNA technology. This public spirited action, which could only have been taken by scientists at the forefront of their field, should be recognized by continued public trust in the body of scientists who would serve on the Genetic Manipulation Advisory Group specifically to ensure that safe progress is made.

One predictable reaction can be envisaged against the sweeping and restrictive regulations proposed by the Health and Safety Commission - some scientists might be dissuaded in the future from discussing publicly the full implications of their research. It is/the public interest that scientists are not discouraged from freely speculating on the conceivable hazards, however remote, which might result from their research.

We therefore request that you make the Williams Working Party Report, which was accepted on behalf of the Government by your predecessor, the basis for a voluntary scheme based upon a Genetic Manipulation Advisory Group. We particularly request your support in ensuring that the Health and Safety Commission consultative document is not written into Law via the Interpretation Act, 1889.

We are also most uneasy about the HSC attempt by regulation 4 to set a precedent for changing the meaning of 'work' - We deplore their undemocratic proposal to gain new all embracing powers over the public without this even being properly debated by our elected representatives in parliament.

Yours faithfully

Professor J H Subak-Sharpe, BSc, PhD, FRSE
Virology Department Glasgow University
Hon Director, MRC Virology Unit

Professor A R Williamson, BSc, PhD, FRSE
Gardiner Professor of Biochemistry
Glasgow University

Dr J Paul, MBChB, PhD, FRCP (Edin. Glasgow)
FRC Path. FRSE
Director, Royal Beatson Memorial Hospital
Glasgow

October 1976

EUROPEAN SCIENCE FOUNDATION

1, quai Lezay-Marnésia
67000 STRASBOURG (FRANCE)

Recommendations concerning Recombinant DNA research

October 1976

RECOMMENDATIONS OF THE EUROPEAN SCIENCE FOUNDATION'S AD HOC COMMITTEE ON RECOMBINANT DNA RESEARCH (GENETIC MANIPULATION), ADOPTED BY THE ESF ASSEMBLY ON 26 OCTOBER 1976

I. Preamble

At its meeting on 10 September in Amsterdam the Committee discussed the final version of the guidelines for research on Recombinant DNA molecules drawn up by the United States National Institutes of Health (NIH) and the United Kingdom Report of the Working Party on the Practice of Genetic Manipulation. It noted the differences between the two reports, both with regard to the supervision of Recombinant DNA research in order to protect the public and the environment, and to the scientific assessments.

One such difference was that the NIH system of safeguards relied to a greater extent on biological containment than the United Kingdom system, which was based essentially upon physical containment.* Another was that considerably more detailed provisions were written down in the American guidelines than in the UK recommendations and code of practice. A further difference was that the NIH administrative system was designed primarily for research funded by the NIH, whereas under the UK system **all** laboratories would come under the aegis of a national central advisory committee.

In the committee's view, each set of proposals formed a coherent system, relying upon interlocking provisions for physical and biological containment and for personal and institutional responsibilities regarding their implementation. The Committee felt that to attempt to make a conflation of the two systems would be neither feasible nor desirable. Among the reasons for this opinion were the obvious dangers of confusion, and of the emergence of a lowest common denominator for the safety precautions proposed for different classes of experiments.

After discussion of the two series of proposals the Committee took the view that there were a number of advantages in the proposals for an advisory system and a code of practice for physical containment, which had now been adopted by the government of the United Kingdom, and decided to recommend its adoption in other European countries. The reasons for this opinion included the following :

— as mentioned above, the advisory system and code of practice proposed

* Biological containment means the use of disabled experimental organisms with a severely limited capacity to survive outside artificial conditions provided in the laboratory. Physical containment means physical facilities and safe handling techniques designed to prevent the organism from infecting laboratory staff or escaping from the laboratory.

in the United Kingdom report are designed to cover all laboratories carrying out Recombinant DNA research, public or private, whatever their sources of finance — and this offers greater safeguards to the public than any system in which some of this research does not come under supervision ;

— for broadly comparable categories of experiments, the UK system requires a slightly higher level of physical containment, which involves tried and tested procedures, rather than specified levels of biological containment, which, at the present time, is difficult to achieve, and is also of somewhat debatable reliability. However, the Committee stressed the importance of developing and testing host-vector systems designed to provide biological containment, as any additional safety measures for Recombinant DNA research were clearly to be welcomed ;

— the Committee noted that the advisory system was supported by the law in the form of the British Health and Safety at Work Act, under which a regulation was to be made requiring laboratories to submit protocols stating the conditions of all experiments with Recombinant DNA to the central advisory committee. The Committee emphasized strongly the importance of such legal support to ensure the effectiveness of an advisory service. It assumed that similar support existed in comparable legislation in other European countries, or could be provided.

It also stressed the advantages of flexibility in decision-making in such a rapidly developing field. A system combining the use of general categories of experiments and individual consideration of protocols for all experiments carrying an appreciable level of risk, potential or conjectural, was felt to provide this flexibility at the same time as satisfactory safeguards for the public. The Committee recognised, however, that in those countries where Recombinant DNA research was less developed the provision of expert advice on all experiments might not always be possible. This led the Committee to express the strong hope that those advisory bodies who did possess sufficient expertise and experience to advise on doubtful cases would, together with the EMBO * Standing Advisory Committee on Recombinant DNA research, be willing to give assistance to other advisory bodies who desired it. One forum through which such experience and expertise could be exchanged would be a European committee composed of representatives of national advisory bodies ; the establishment of such a forum forms the substance of recommendations 8 and 9 below.

The Committee agreed that further views and advice in particular upon the scientific and technical aspects of the two series of recommendations would be valuable. It therefore asked the EMBO Standing Advisory Committee on Recombinant DNA to compare the provisions of the American guidelines and the British code of practice and to advise the ESF Committee of its findings, in particular those relating to scientific matters. The report of the EMBO committee was completed in October and was circulated at the meeting of the ESF Assembly.

II. Recommendations

We recommend that :

1) subject to appropriate safeguards, research on Recombinant DNA

---

* European Molecular Biology Organisation.

molecules should be promoted and further developed in Europe.

2) adequate measures be taken to ensure the protection of the public and its members as well as of animals, plants and the environment. These measures are set out in recommendations 3 to 6 below.

3) guidelines for such research, designating in sufficient detail the appropriate safeguards for the different kinds of experiments with Recombinant DNA, be rigorously followed by all researchers and all laboratories carrying out such experiments.

4) in the first phase, the recommendations and code of practice in the United Kingdom Report of the Working Party on Genetic Manipulation be adopted as the guidelines for Recombinant DNA research in Europe, *provided that the European Committee which is referred to in Recommendation 9 below is set up immediately in order to ensure that the same levels of containment are used in the different European countries for the same categories of experiments.**

5) national advisory bodies with responsibilities for interpreting the recommendations and code of practice for Recombinant DNA research referred to above, for advising researchers in their use, and for supervising their implementation should be established forthwith in those European countries which have not already set up such a body.

6) in European countries national registers of laboratories carrying out Recombinant DNA research be established and that all laboratories, whatever their source of financial support, be legally required to declare relevant aspects of their programmes of Recombinant DNA research to a national register. We envisage such registers as being non-classified documents and we advise that they should be exchanged among the national bodies referred to in Recommendation 5.

7) in each country the necessary administrative and supervisory measures be devised, in connection with appropriate existing national legislation, to ensure that the guidelines adopted be strictly adhered to by all laboratories — governmental, industrial or private — or in universities or other research establishments.

8) the national bodies in Europe referred to in Recommendation 5 should keep in close contact with each other, with the EMBO Standing Advisory Committee on Recombinant DNA, with the appropriate committees of the NIH in America and with committees for Recombinant DNA research in other countries. We wish to emphasize the importance, both for the safety of the public and for the development of the science, of ensuring that the same experiments are not classified as differing in level of risk, and hence level of containment, in the different countries.

9) to give effect to Recommendation 8 above, a European Committee of representatives of those national bodies for Recombinant DNA research *referred to in Recommendation 5 above,* * of the EMBO Standing Advisory Committee on Recombinant DNA research and of the European Medical Research Councils *and with representation of Agricultural research* * should be established under the aegis of the ESF. It should come together at an early date and hold regular meetings, at frequent intervals in the first phase. ** These meetings will provide an opportunity fo~ mutual information, consultation and advice on general

* The words in italics are amendments which were proposed and agreed during the Assembly.

** It is proposed that this committee should meet first in January 1977 and at three-monthly intervals during the year.

policy for Recombinant DNA research and for the discussion of decisions on specific experiments, *in particular concerning biological containment.* * This process should lead to the building up of common body of experience in all questions concerning Recombinant DNA research, at European level. A further task of this Committee will be to keep the guidelines under continuous review and to consider whether and, if so, how they should be revised in the light of the development of the science and of safety measures.

Members of the European Science Foundation's ad hoc committee on Recombinant DNA research

Professor P. Riis	Denmark (Chairman)
Professor D. Bootsma	The Netherlands
Professor P. Chambon	France
Professor E. Deutsch	Fed. Rep. of Germany
Professor L. K. Dunican	Ireland
Professor W. Fiers	Belgium
Professor H. L. A. Hart	United Kingdom
Dr. B. Harvald	Denmark and EMRC
Professor I. Kreft	Yugoslavia
Professor B. H. Lindquist	Norway
Professor J. F. Miquel	France
Dr. D. de Nettancourt	Belgium
Professor F. Pocchiari	Italy
Professor A. Portoles	Spain
Professor J. R. Postgate	United Kingdom
Professor L. Raiser	Fed. Rep. of Germany
Professor P. Reichard	Sweden
Professor A. Tissières	Switzerland
Dr. J. Tooze	EMBO
Professor H. Tuppy	Austria
Professor H.-G. Zachau	Fed. Rep. of Germany and EMBO

Document 11.7

August— September 1978
Medical World, Excerpt

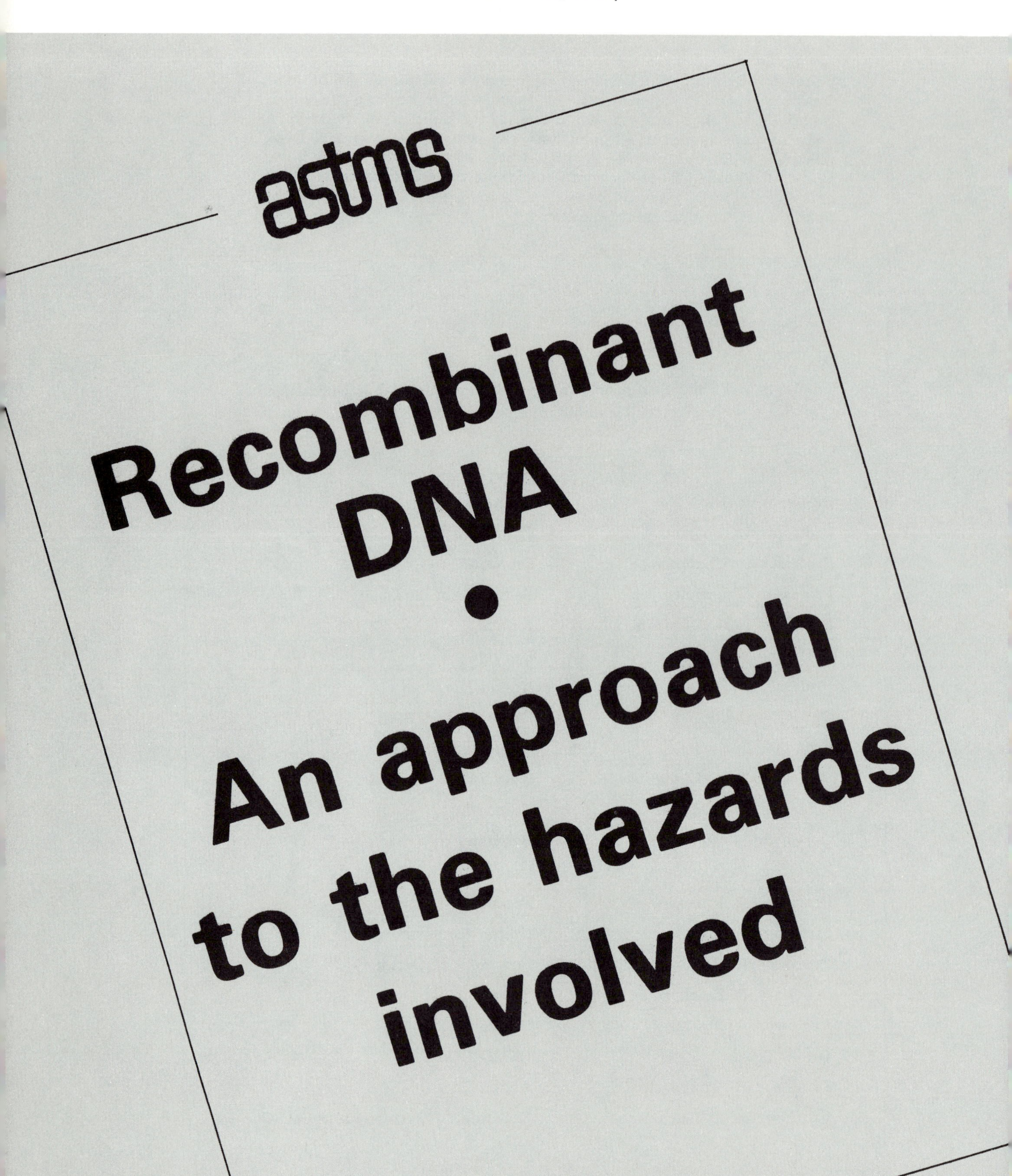

GMAG-seeking out ways to rein in the dangers

by Donna Haber, Divisional Officer and member of GMAG

In 1976 a working party, set up by the Government, recommended that research in genetic engineering (recombinant DNA) go ahead, with a number of safeguards to ensure the safety of the staffs working on the research, and of the community. This was an important event, given the controversy that was going on in some other countries — especially the USA — about this research. The report of this Williams Working Party was accepted by Parliament and attracted very little notice at the time.

One of the specific recommendations of the working party's report was that a Genetic Manipulation Advisory Group be set up to advise on appropriate precautions for the conduct of these experi ments. This in itself was interesting and important for a number of reasons.

For one thing, the Genetic Manipulation Advisory Group (or GMAG, as it has come to be known) could ensure that the system for the application of safety precautions would not be a rigid one. GMAG would (and does) consider the precautions necessary for each experiment individually.

The advice that we give is based on the guidelines in the working party report, but, because of the way we operate, it is possible to apply those guidelines with reasonable flexibility. It is also possible to have regard, in the advice we give, to any new information which becomes available on the risks involved. This is most important, as this is a new and changing field.

Discussed

The way in which this happens is that when an experiment is first proposed, it is discussed in the laboratory concerned. The key people involved in these discussions would be those who wish to do the experiment, the Biological Safety Officer of the establishment, and a 'properly constituted and representative safety committee'.

In their discussions they will come to the conclusion that the proposed experi ment falls within a certain category. There are four categories to choose from, category I being the least potentially hazardous, and category IV being the most potentially hazardous.

If their conclusion is that the experiment falls into category I or II, they can proceed with the work, but they must notify GMAG immediately. If it falls into category III or IV, then they must notify GMAG and wait for advice before beginning work.

Laboratory safety committees are therefore at least as important as GMAG for ensuring that this work is done safely. By categorising an experiment as category I or II, they can determine that the work starts immediately, even before GMAG advice is received.

Additionally, because GMAG acts as an advisory body and not a policing body, it is up to the local safety committees to monitor the work as it goes on and ensure that it continues within the appropriate guidelines. They can always approach GMAG for additional advice and, of course, ASTMS safety representatives have access to Bob Williamson and myself (the two ASTMS nominees on GMAG), the ASTMS Health and Safety Officer and Health and Safety inspectors for further advice and help.

But basically it will be the safety committees and, most importantly, the union safety representatives, who will be the ones who know what is going on in the laboratory, and thus the only ones in a position to give an alarm, should one be necessary. It goes without saying that ASTMS representatives should be involved in this process.

Once a proposed experiment has been approved by the local safety committee, it comes to GMAG **with the comments of that safety committee.** The more extensive and knowledgeable those comments are, the easier is Bob's and my task on GMAG, as we are then advised of our members' views in that situation. Members can write to us separately if they feel that they have concerns that are not being adequately expressed by their safety committee.

Demonstrate

Comments from the local safety committee also demonstrate to the whole of GMAG how much reliance they can place on the capability of that particular safety committee.

GMAG then has three choices: (1) we can advise that the work can proceed in the category proposed; (2) we can advise that there should be additional precautions taken and/or that the work falls

Photo by Jack Blake.

Genetic engineering requires staff who are trained in safety as well as a high standard of scientific competence.

Reprinted from *Medical World*. Published by the Association of Scientific, Technical, and Managerial Staffs, London.

into a different category; (3) we can advise that that particular experiment could not be undertaken safely.

This advice is given to the people involved locally, and to the Health and Safety Executive, which accepts our advice as safe practice under the Health and Safety at Work Act. Any enforcement is then up to the Health and Safety Executive through its inspectors and this could lead ultimately to prosecutions under the law in unsafe working situations.

Another important aspect of GMAG is that its membership includes not only scientists, but representatives of industry, the trade unions and the general public. In this sense it is unique, and something of an experiment. It was constituted in this way, according to the working party report, to command the respect of the public as well as the scientific community.

I would think that in saying this they were taking notice of debates taking place in other countries, where there was public concern being expressed, but no channels available for it to be fed into the official and governmental bodies which were discussing guidelines for this work, and perhaps anticipating the same kind of concern arising in the UK.

Public concern — and indeed in some places alarm — about genetic engineering has arisen, I believe, for a number of reasons. First, there is the problem that we do not know for sure whether or not this research is potentially dangerous. If there are dangers, then they would be very great.

Accumulated

Evidence now being accumulated seems to suggest that the more horrific outcomes which some people once predicted are most unlikely, and that the majority of this research can be done quite safely — but there is still no absolute proof one way or the other.

Secondly, there has, in some places, been a growing distrust of science and scientists. This is centred around such issues as the use of nuclear power, the development (or attempted development) of biological and chemical warfare, the use of medical and scientific knowledge for the torture of political prisoners, and technological pollution of the environment.

Although nobody would want to blame the whole of the scientific community for such abuses, there is no doubt that scientific knowledge is involved in them.

Thus, when we have a new field such as genetic engineering, the possibilities of which for good and/or ill are unknown, and which could have a huge impact on the whole community — then it is not enough to delegate the control of it only to scientists. There must also be an input from representatives of the community who will be affected by the research.

The establishment of GMAG is obviously an attempt to take account of these factors, and to provide a solution which will help this work to go ahead at the same time as allaying public anxiety about it. But does it work?

I feel on the whole that it does, even given some of the criticisms which have been made of it, and which I believe in some cases have been justified.

The main criticisms coming from some scientists — although by no means all of them — seem to be that they resent non-scientists being involved in the way their work is done, and that the whole process of notification of GMAG takes too long and consequently their work is unjustifiably delayed. The view has also been expressed by some scientists that this work is really not hazardous at all — or at least less hazardous than much other work going on in laboratories.

However, this is a separate issue from those which I am attempting to put forth in this article and will be discussed more fully at the public conference we are organising in October (see advert on page 8). I am, rather, proceeding here on the assumption that there is some hazard in this work, and trying to look at whether or not GMAG works in this context.

As to the first criticism, I disagree with it totally. I think that the public has a right to be involved in the control of work which could affect them so greatly — scientific work or non-scientific work.

On this issue, the involvement of the trade unions is very important, not only because the TUC (who nominated the four trade union members of GMAG) represents the interests of so large a part of the public — and specifically the interests of those employees who would be most at risk from this research — but also because the trade unions are powerful independent organisations with the capability of being extremely effective in areas in which they demonstrate their concern.

I would also ask that scientists with this view look at the other side of this coin, as it is certainly something that the non-scientists involved have discussed; that is, if all of us together take the responsibility for advising on safety, then, again, it would be all of us — and not just the scientists — who would have some responsibility if the advice were to be found to be wrong.

Also, I would contend that the queries and views of people not involved in a particular type of work can introduce new and fresh ideas and viewpoints into arguments, and can sometimes spotlight concerns overlooked by people too closely involved. The fact that we raise on GMAG concerns that are felt by non-scientists, and that these concerns are then discussed and dealt with, is most important in ensuring that this work is carried on in a way which will not alarm the general public.

Finally, there are issues dealt with on GMAG (such as the composition of safety committees) which need no scientific background to discuss — but which, in fact, benefit from a different type of expertise. At the risk of sounding immodest, I certainly feel that non-scientists have made an important contribution to the work of GMAG.

Justified

As to the criticism concerning delay, I feel that this has been justified at times, but I also think that the situation is improving. GMAG has now been in existence for approximately one and a half years and with experience, with getting to know one another, and with having settled among ourselves some fundamental issues, we are now able to function quicker and more efficiently.

I don't see how the initial delays could have been avoided in such a new kind of group and with a membership consisting of such diverse interests. It is regrettable, but unavoidable and, as I have said, the situation is now improving.

There are other issues which could be raised and discussed in this context. I will not do so because of a lack of space, but I hope that others reading this article will, and will take them up either in the Readers' Letters section of *Medical World*, or at our conference in October, or both.

Document 11.8
December 15, 1977
New Scientist (**76**:683)

GMAG falls foul of privacy constraints

Roger Lewin

When the Genetic Manipulation Advisory Group (GMAG) meets tomorrow its most important task will be to attempt to resolve what one committee member describes as "a rumbling area of discontent": that of confidentiality over industrially sensitive information. GMAG's chairman, Sir Gordon Wolstenholme, has drawn up what he told *New Scientist* is a form of declaration of confidentiality" which, with some modification, may provide the compromise necessary to satisfy the different requirements voiced within GMAG and outside.

Ever since GMAG first met exactly one year ago tomorrow to begin its task of monitoring genetic engineering in the UK, it has become steadily clearer that confidentiality is a key issue: in the words of one person on the committee, "confidentiality is the single most difficult issue that GMAG has faced". Scientists anxious to forge ahead in an exciting and rapidly advancing area are somewhat unhappy that their research proposals will be scrutinised by yet another committee manned in part by potential intellectual competitors.

The question of competition becomes even sharper where there is the possibility of commercial exploitation. Industrial laboratories are concerned about two things: first, that their plans should not become known to other companies; and secondly, that in having their research proposals processed by a public body they do not endanger possible patents through "prior disclosure".

All members of GMAG (which is made up of representatives from the scientific, industrial, and trade union communities) accept that the possibility of endangering potential commercial exploitation has to be taken very seriously. Nevertheless, representatives of trade unions have a duty to report to their constituency any matter they feel is relevant to safety. Cloning human insulin genes, for instance, would bring these two responsibilities into sharp conflict. In an attempt to reconcile these different interests Wolstenholme set up a confidentiality subcommittee.

One input to this subcommittee was representation from companies pursuing genetic manipulation research. A solution that would have satisfied this interested group was that a "mini-GMAG" should be set up within the main committee that would deal with commercially sensitive proposals. Although some GMAG members accepted the idea, the union representatives, according to Wolstenholme, were very unhappy about it. Over a period of several months, punctuated by several stormy meetings, a new compromise was worked out; and this is the one involving Wolstenholme's "declaration of confidentiality".

Currently GMAG members are bound by the Official Secrets Act, but strictly speaking, this covers only government property and it is a moot point as to whether proposals from research laboratories (academic or industrial) handled by the committee can be classified as government property. When, in July next year, GMAG achieves full legal status its members will be covered by section 28 of the Health and Safety at Work Act (1974), a provision that allows disclosure of the committee's business if health is at risk.

Neither of these provisions was sufficient to satisfy industry that its commercial interests were safeguarded, which is why it was necessary to have what Wolstenholme describes as "a very serious exercise to give industry a genuine assurance of our responsible behaviour". In addition to the "declaration of confidentiality" to be signed by every GMAG member, Wolstenholme has tackled the "leakage" of information between companies by asking GMAG members to declare their industrial connections, such as consultancies and substantial stockholdings. Whenever a commercially sensitive proposal comes before the committee Wolstenholme can consult his secret list and invite the appropriate member(s) to absent themselves during the scrutiny.

It now remains to be seen whether these provisions will satisfy industry. So far no commercially sensitive applications have been submitted to GMAG. There are many awaiting submission once conditions are right. □

UK extends the law to genetic engineering

From 1 August, 1978 any scientist intending to carry out experiments in genetic manipulation in the UK will be obliged by law to notify both the Genetic Manipulation Advisory Group (GMAG) and the Health and Safety Executive (HSE). Notification will be made very much as it is under the recent voluntary system, which has been operated by GMAG for the past year and a half and which has resulted in almost complete cooperation. Application of the new regulations should therefore be painless.

The new Genetic Manipulation Regulations have been made under the Health and Safety at Work Act. They arise directly from a recommendation of the Williams Working Party report of 1976 and were first drafted in August 1976. That draft, particularly in its provisional definition of genetic manipulation, was so criticised, if not ridiculed, that almost two years have been spent in wording the final regulations. Or, as John Dunster, deputy director general of the HSE, put it "Consultation is a very real process".

The final definition of genetic manipulation is said to have been widely approved and is in line with that accepted by the EEC and by several other countries. It reads: "the formation of new combinations of heritable material by the insertion of nucleic acid molecules produced by whatever means outside the cell, into any virus, bacterial plasmid, or other vector system so as to allow their incorporation into a host organism in which they do not naturally occur but in which they are capable of continued propagation".

The other stumbling block experienced by those drafting the regulations was how to define 'work'. The problem is that not all students consider it to be a four letter word; indeed some carry out research work and yet are not employed and consequently have not come under the Health and Safety at Work Act. The solution in the Genetic Manipulation Regulations has been to extend the meaning of work to encompass those who are 'non-employed' such as students.

The compulsion encoded in the regulations begins and ends with the notification of both GMAG and the HSE of intention to carry out genetic manipulation. It is, in theory at least, a voluntary matter whether or not any advice offered by GMAG (HSE will be relying on GMAG for that) is followed.

In practice, however, failure to comply with GMAG's suggestions is almost certain to lead to prosecution under the Health and Safety at Work Act. Information leading to such actions is expected to come either from the HSE's 'intelligence network', particularly safety officers in laboratories or via GMAG which includes two assessors from the HSE. The presence of these assessors has already been persuasive on rare occasions when there was reluctance to meet GMAG's advice.

The Genetic Manipulation Regulations will not be the last official word on the subject in the UK. Already a governmental Select Committee on Science and Technology is in the process of looking at genetic manipulation. Meanwhile, Leo Abse, a member of parliament, has criticised the new HSE regulations as being totally inadequate and providing no effective, enforceable control. Lastly, if the HSE considers that scientists are not complying with the suggestions of GMAG, it intends to incorporate the suggestions into an enforceable Approved Code under the Health and Safety at Work Act.

Peter Newmark

Document 11.10

May 1978
British Health and Safety Commission, Excerpt

Regulations and Guidance Notes | Health and Safety at Work | Health and Safety Commission

Genetic manipulation

Preface

This booklet contains

(a) The Regulations on Genetic Manipulation (SI 1978 No. 752) which are operative from 1 August 1978.

(b) Guidance Notes.

The Regulations provide that persons should not carry on genetic manipulation unless they have previously notified the Health and Safety Executive and the Genetic Manipulation Advisory Group.

The Regulations extend the meaning of work in Part I of the Health and Safety at Work etc Act 1974 to include any activity involving genetic manipulation. Non-employed persons engaged in genetic manipulation will have the same duties under the Act as self-employed persons.

The Guidance Notes offer practical guidance on the Regulations and contain a specimen of the form approved by the Health and Safety Executive for notification of intention to carry out activities involving genetic manipulation, as defined in the Regulations, to the Health and Safety Executive and to the Genetic Manipulation Advisory Group.

Regulations

Citation and commencement

1 These Regulations may be cited as the Health and Safety (Genetic Manipulation) Regulations 1978 and shall come into operation on 1 August 1978.

Interpretation

2 (1) In these Regulations, unless the context otherwise requires –
"genetic manipulation" means the formation of new combinations of heritable material by the insertion of nucleic acid molecules, produced by whatever means outside the cell, into any virus, bacterial plasmid, or other vector system so as to allow their incorporation into a host organism in which they do not naturally occur but in which they are capable of continued propagation;

"the Genetic Manipulation Advisory Group" means the central advisory group so known, comprising members appointed by the Secretary of State, the setting up of which was announced to the House of Commons on 5th August, 1976[a].

(2) The Interpretation Act 1889[b] shall apply to the interpretation of these Regulations as it applies for the interpretation of an Act of Parliament.

LONDON: HER MAJESTY'S STATIONERY OFFICE 1978

Enquiries regarding this publication should be addressed to:
Area offices of the Health and Safety Executive, or the general information and enquiry point, Health and Safety Executive, 1 Chepstow Place, London W2 4TF. Tel 01-229 3456.

Meaning of "work" and "at work"

3 For the purposes of Part I of the Health and Safety at Work etc Act 1974 the meaning of the word "work" shall be extended to include any activity involving genetic manipulation and the meaning of "at work" shall be extended accordingly.

Modification of section 3(2) of the Health and Safety at Work etc Act 1974

4 Section 3(2) of the Health and Safety at Work etc Act 1974 shall be modified, in relation to any activity involving genetic manipulation, so as to have effect as if the reference to a self-employed person included a reference to any person who is not an employer or an employed person in relation to that activity.

Notification of intention to carry on genetic manipulation

5 No person shall carry on any activity involving genetic manipulation unless he has given to the Health and Safety Executive and to the Genetic Manipulation Advisory Group notice, in a form approved by the Executive for the purposes of these Regulations, of his intention to carry out that activity.

(a) Official Report, 5th August 1976, vol. 916, col. 1014.
(b) 1889 c.63.

Exemptions

6 The Health and Safety Executive may exempt any person or class of persons from the requirement of Regulation 5 above in respect of any activity or class of activities but the Executive shall not grant any such exemption unless it has consulted the Genetic Manipulation Advisory Group.

Service of notice on the Genetic Manipulation Advisory Group

7 Any notice required by Regulation 5 above to be given to the Genetic Manipulation Advisory Group shall be deemed to have been duly given if it is given to the secretary of that Group in accordance with the provisions of Section 46(2) of the Health and Safety at Work etc Act 1974.

Guidance notes for Health and Safety (Genetic Manipulation) Regulations 1978

1 The Genetic Manipulation Regulations 1978 made under the Health and Safety at Work etc. Act 1974 (HSW Act), specify that no person shall carry on activities involving genetic manipulation unless notice of their intention to do so has been given to the Health and Safety Executive (HSE) and to the Genetic Manipulation Advisory Group (GMAG).

2 GMAG is an advisory body (set up under the aegis of the Department of Education and Science) whose main functions are:

- to review and assess the conjectured hazards of any activities involving genetic manipulation as defined;
- to categorize individual experiments according to the appropriate level of physical and biological containment;
- to advise on the application of the code of practice recommended in the Report of the Working Party on the Practice of Genetic Manipulation.

3 The Regulations define genetic manipulation as:

"the formation of new combinations of heritable material by the insertion of nucleic acid molecules, produced by whatever means outside the cell, into any virus, bacterial plasmid, or other vector system so as to allow their incorporation into a host organism in which they do not naturally occur but in which they are capable of continued propagation".

This definition is intended to cover all activities involving genetic manipulation which could present a hazard to the health and safety of workpeople and/or the general public, and should be taken to include, *inter alia:*

(a) the introduction of plant or animal virus nucleic acid, intact or reduced, into either a prokaryote or a primitive eukaryote;
(b) the introduction of prokaryote viral or plasmid nucleic acid, intact or reduced, into a eukaryote cell;
(c) co-infection with potential vector nucleic acid and prepared DNA, relying on the cell to carry out the ligation step.

4 Enquiries concerning the scope of the definition of genetic manipulation for the purposes of the regulations should be addressed in the first instance to HSE (at the address given in para 6) which will keep closely in touch with the Secretary of GMAG.

5 The regulations follow the recommendation of the Working Party on the Practice of Genetic Manipulation (Command 6600 HMSO), that Regulations should be made under the HSW Act, requiring laboratories carrying out activities in genetic manipulation to submit experimental protocols and appropriate supporting information to a central advisory group before beginning any relevant work. It was decided that, because of HSE's general responsibility for enforcing the HSW Act, notification should be to HSE as well as to the Central Advisory Body, GMAG.

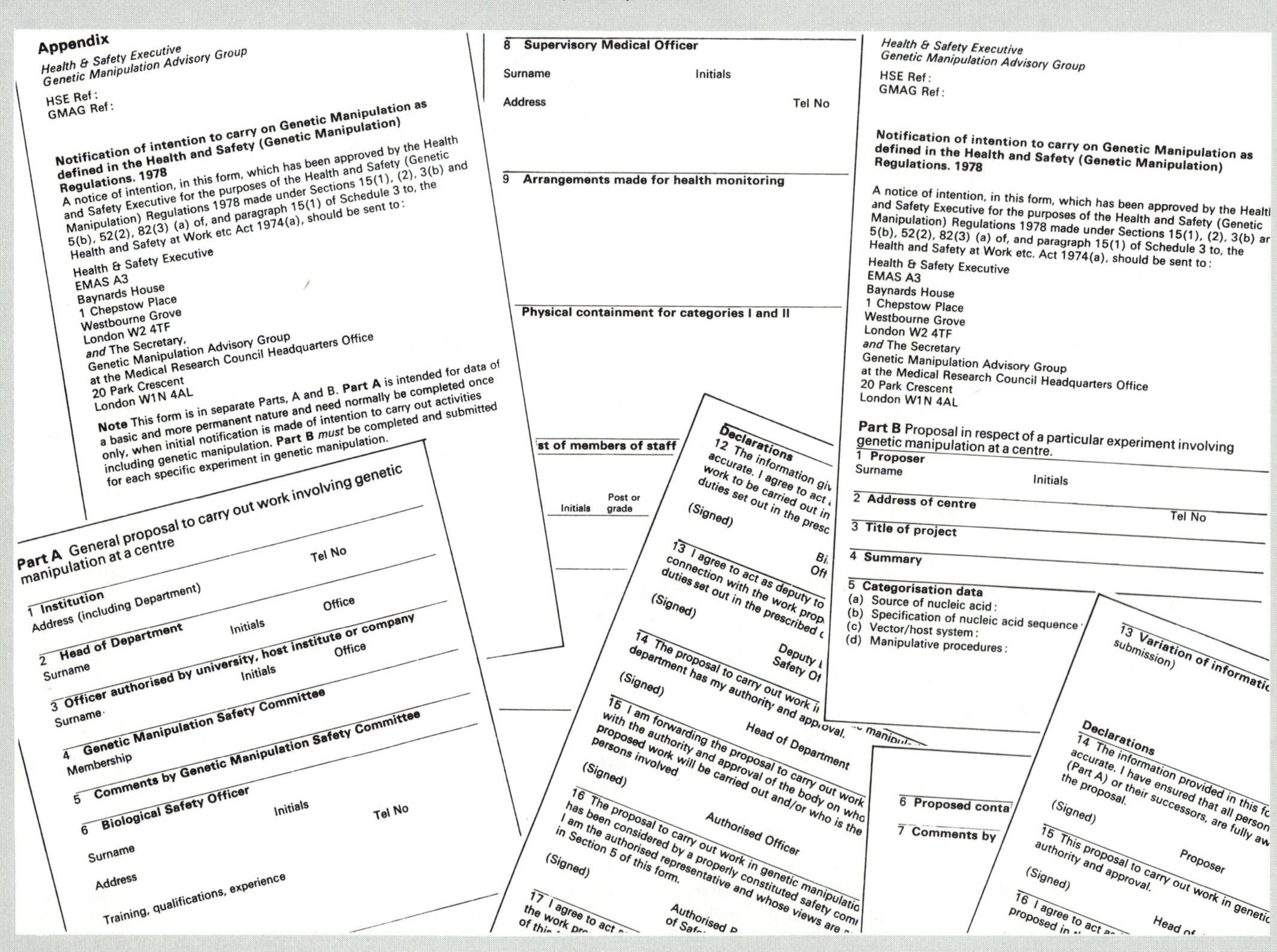

Appendix
Health & Safety Executive
Genetic Manipulation Advisory Group
HSE Ref:
GMAG Ref:

Notification of intention to carry on Genetic Manipulation as defined in the Health and Safety (Genetic Manipulation) Regulations. 1978
A notice of intention, in this form, which has been approved by the Health and Safety Executive for the purposes of the Health and Safety (Genetic Manipulation) Regulations 1978 made under Sections 15(1), (2), 3(b) and 5(b), 52(2), 82(3) (a) of, and paragraph 15(1) of Schedule 3 to, the Health and Safety at Work etc Act 1974(a), should be sent to:
Health & Safety Executive
EMAS A3
Baynards House
1 Chepstow Place
Westbourne Grove
London W2 4TF
and The Secretary,
Genetic Manipulation Advisory Group
at the Medical Research Council Headquarters Office
20 Park Crescent
London W1N 4AL
Note This form is in separate Parts, A and B. **Part A** is intended for data of a basic and more permanent nature and need normally be completed once only, when initial notification is made of intention to carry out activities including genetic manipulation. **Part B** *must* be completed and submitted for each specific experiment in genetic manipulation.

Part A General proposal to carry out work involving genetic manipulation at a centre
1 **Institution** Tel No
Address (including Department)
2 **Head of Department** Initials Office
Surname
3 **Officer authorised by university, host institute or company** Office
Surname Initials
4 **Genetic Manipulation Safety Committee**
Membership
5 **Comments by Genetic Manipulation Safety Committee**
6 **Biological Safety Officer** Initials Tel No
Surname
Address
Training, qualifications, experience

8 **Supervisory Medical Officer**
Surname Initials
Address Tel No
9 **Arrangements made for health monitoring**
Physical containment for categories I and II
...st of members of staff
Initials Post or grade

Declarations
12 The information giv... accurate. I agree to act... work to be carried out in... duties set out in the presc...
(Signed)
13 I agree to act as deputy to... connection with the work prop... duties set out in the prescribed...
(Signed) Deputy ... Safety Of...
14 The proposal to carry out work i... department has my authority and approval.
(Signed) Head of Department
16 I am forwarding the proposal to carry out work with the authority and approval of the body on whic... proposed work will be carried out and/or who is the persons involved
(Signed) Authorised Officer
16 The proposal to carry out work in genetic manipulatio... has been considered by a properly constituted safety comm... I am the authorised representative and whose views are ... in Section 5 of this form.
(Signed) Authorised ... of Saf...
17 I agree to act ... the work pro... of this...

Health & Safety Executive
Genetic Manipulation Advisory Group
HSE Ref:
GMAG Ref:

Notification of intention to carry on Genetic Manipulation as defined in the Health and Safety (Genetic Manipulation) Regulations. 1978
A notice of intention, in this form, which has been approved by the Health and Safety Executive for the purposes of the Health and Safety (Genetic Manipulation) Regulations 1978 made under Sections 15(1), (2), 3(b) an... 5(b), 52(2), 82(3) (a) of, and paragraph 15(1) of Schedule 3 to, the Health and Safety at Work etc. Act 1974(a), should be sent to:
Health & Safety Executive
EMAS A3
Baynards House
1 Chepstow Place
Westbourne Grove
London W2 4TF
and The Secretary
Genetic Manipulation Advisory Group
at the Medical Research Council Headquarters Office
20 Park Crescent
London W1N 4AL

Part B Proposal in respect of a particular experiment involving genetic manipulation at a centre.
1 **Proposer**
Surname Initials
2 **Address of centre** Tel No
3 **Title of project**
4 **Summary**
5 **Categorisation data**
(a) Source of nucleic acid:
(b) Specification of nucleic acid sequence
(c) Vector/host system:
(d) Manipulative procedures:
6 **Proposed conta...**
7 **Comments by...**
13 **Variation of informati...** submission)

Declarations
14 The information provided in this fo... accurate. I have ensured that all person... (Part A) or their successors, are fully aw... the proposal.
(Signed) Proposer
15 This proposal to carry out work in genetic... authority and approval.
(Signed) Head of...
16 I agree to act a... proposed in...

Document 11.11
June 29, 1978
Nature (**273**:696)

Recombinant DNA is safe*

Professor R. H. Pritchard argues that there are no real hazards in recombinant DNA research.

THE publication of the first report by the Genetic Manipulation Advisory Group has been the occasion for a number of articles in the press giving muted praise for the unprovocative, unemotive and constructive approach of we British to the problems posed by the attempt to control research with recombinant DNA. The Group has had an unenviable task and I would not wish to minimise the difficulties it has had in attempting to formulate rules that are acceptable to everyone concerned. Nevertheless, the occasion must not be allowed to pass without comment on the less commendable features of its Report and on the present level of debate on this issue in Great Britain.

The only conceivable justification for the expensive and growing bureaucracy which now controls recombinant DNA research would be that it posed an identifiable and distinct hazard to human health or society. Consequently it cannot be emphasised too often that no such hazard has ever been identified. Nor have I heard a sustainable theoretical argument to suggest that such hazards are real. Recent comment in the press and Parliament indicates a total misunderstanding of this point as witnessed by the fact that the so-called hazards have almost invariably been compared with those surrounding the nuclear energy industry. The difference between the two is unambiguous. The hazards of radiation are known from experience of them. They are quantifiable, and appropriate control can therefore be based on reasoned argument.

GMAG and other official bodies that have preceded it must take substantial responsibility for allowing such a widespread and fundamental confusion to exist and to continue to exist. Indeed, it is most disturbing that the GMAG report, if only by default, leaves the impression that there is broad agreement about the basic premises upon which its existence is based and that it need only concern itself with the definition and execution of regulations based upon these premises. Discussions I have had with many biologists indicate that some regard the recombinant DNA debate as the longest running and most expensive farce in town. Others accept the need for caution and control in the execution of certain kinds of experiments but believe that the GMAG guidelines are far too sweeping. Others believe that by confining their attention to *in vitro* recombinant DNA experiments the Group is operating in a logical vacuum with the result that if some of the conjectured hazards are real then the horse will probably have bolted before ever it enters their stable door. I have yet to meet a biologist who approves of the proposals as they stand. None of this divergence of views would be apparent to those readers of the GMAG Report who are not biologists, although it is they who will be most influenced by it.

The total absence of reasoned argument based on observations has led to the most disturbing aspect of the whole debate. This is the continual resort to authority rather than to evidence as a justification for control. GMAG summarising the background to its birth tells us "In July 1974 a group of distinguished biologists drew public attention to these problems" (the hypothetical risks of recombinant DNA research). And later (para. 4), "The question of hazards was raised by responsible scientists and we feel it essential that, however strong an individual worker's intuitive feeling may be that the hazards have been grossly overestimated, we have a right to expect of him, until we know more of the reality of the situation, that he should do the work either with scrupulous care or not at all". Mere workers, apparently, who disagree with "distinguished" and "responsible" scientists can only have "intuitive" reasons for doing so.

I do not wish to question the distinction of the scientists concerned, but clearly it is they who acted upon intuition. It is my personal opinion that their intuition was unsound, that they were ill-informed and consequently, in view of their distinction, it is they who acted irresponsibly. The soundness of the suggestions made by these scientists has never been rigorously discussed by those in this country who have been asked to do so. The Working Party chaired by Lord Ashby was unable to identify the hazards of recombinant DNA research. The following Working Party under Sir Robert Williams nevertheless felt able to divide experiments into four categories of increasing risk. Now we find GMAG debating in its report whether DNA from birds should be categorised in terms of its hazard level with mammalian DNA (because birds are warm blooded) or with amphibian DNA (because of the closer evolutionary relationship). One is tempted to wonder why the group didn't consult the I Ching for advice.

Practising biologists will not find it easy to adhere to restrictive and time-consuming rules, and ensure that others do too, when there is so little confidence that these rules have a rational basis. Consequently, the question that now concerns us most is on what evidence and by what means will the controls be relaxed. On this question as on others surrounding the recombinant DNA debate there seems little ground for optimism.

Consider, for example, the suggestion by the distinguished scientists that the creation of novel antibiotic resistant phenotypes among bacterial species by *in vitro* recombination is a potential hazard. Antibiotic therapy has been in widespread use for decades. A consequence has been the development of resistant strains of pathogens and non-pathogens. Such strains can become the prevalent ones in hospital environments and are a hazardous nuisance. I am not aware, however, of a single example in all this time of *any* bacterial species in which resistance to *any* antibiotic has become the prevalent phenotype except in an antibiotic-contaminated environment. Is there a reasonable alternative to the supposition from these observations that there is a price to be paid for antibiotic resistance by a naturally sensitive species and that this price is too high for resistant strains to predominate without positive selection in their favour? If resistance confers no selective advantage, what is the cause for concern except the one which has always been with us? This is how best to employ antibiotics so as to take maximum advantage of the natural selection pressures *against* resistant organisms to minimise their prevalence. Had the resources devoted to the recombinant DNA debate been devoted instead to investigation of this latter question, an important medical problem might have been solved. Instead a bureaucracy has been created.

But to return to relaxation of controls. Is it likely that GMAG, or any other body, will ever have in its possession more convincing evidence than it has now that *in vitro* transfers of antibiotic resistance determinants present no hazards? I suspect not and therefore wonder how relaxation of the blanket controls of this kind of activity are ever to be relaxed if they are not to be removed immediately. Moreover, if relaxation here proves a problem, then how much greater a problem will it prove to be in the case of the more esoteric areas of conjecture?

Great Britain will, from 1 August, have the doubtful privilege of being the first country to bring recombinant DNA research under legislative control. I find it hard to understand why GMAG has not publicly and firmly advised that this legislation be deferred until there is evidence, or even broad agreement, that the postulated hazards do indeed exist. □

*Editors' note: R.H. Pritchard wishes to point out that this title was not his but was added by the editor of *Nature*.

BIOHAZARD
CAUTION possible
BIOLOGICAL HAZARD

REVISING THE GUIDELINES: SUCCESS AT LAST

After the public hearings at the NIH in December 1977, the Director referred the proposed revised guidelines back to the RAC for reconsideration. During the first half of 1978 this task proceeded, influenced by many events including the Falmouth and Ascott Workshops (see Chapters 9 and 10), the decisions in other countries (Doc. 12.1), the loss of impetus in Congress for any legislation, and the progressively changing assessment by the scientific community of the probability of recombinant DNA research causing a significant hazard (Doc. 12.2 and Doc. 12.3). All of these considerations pointed to a more radical relaxation of containment measures than had been proposed in September 1977. On July 28, 1978, the Director of the NIH was able to publish in the *Federal Register* for a further round of comments three related documents—a decision to publish revised guidelines, the proposed revised guidelines, and an environmental impact statement. This time there would be no opportunity for criticism or for obstruction based on the lack of the impact statement. Together these documents occupy 137 pages of the triple-column text from which we have reprinted a very small sample here (Doc. 12.4). The introductory pages explain and justify in some detail the decision to relax the guidelines, while the tables compare the containment conditions for particular experiments required in the 1976 guidelines, the RAC's proposed revision of September 1977, and the revised guidelines, which were finally published in December 1978 (see below). A glance at the table shows the progressive relaxation.

As the NIH's proposals were being digested, more commentators began to acknowledge that for some the recombinant DNA issue had become a political pawn (Doc. 12.5). Retiring Chairman of the RAC DeWitt Stetten wrote in a valedictory his astute analysis of events from Asilomar onward (he likened Asilomar to a religious revivalist meeting) and produced his single-sentence guideline (Doc. 12.6). Simultaneously a former member of the RAC, Waclaw Szybalski, protested about the layers of bureaucracy that now surrounded recombinant DNA research (Doc. 12.7 and Doc. 12.8).

On September 15, 1978, a one-day public hearing was held by HEW in Washington with Peter Libassi, the General Council of HEW in the chair, to receive comment on the proposed revised guidelines. Representative Richard L. Ottinger and some public-interest groups focused their criticism on the administrative sections of the guidelines (Docs. 12.9 through 12.11), and, as we shall see, the NIH responded to the criticism. The environmentalist movement had by 1978 become the firmest opponent of recombinant DNA research, which led J. D. Watson to publish his rebuttal of their position (Doc. 12.12).

Throughout this period Senator A. E. Stevenson kept alive the question of federal legislation and compliance by industry (Doc. 12.13). His colleague, the former lunar astronaut Senator H. Schmitt (New Mexico), became on the other hand the champion of those opposed to any legislative move (Doc. 12.14).

After these hearings, the NIH and the HEW continued to prepare the revised guidelines. Some opponents of the revision again pointed out their fears of possible immunological diseases arising from bacteria producing animal and human proteins, but other experts discounted them (Doc. 12.15). In November the NIH held a three-day workshop for chairmen of Institutional Biosafety Committees, in part to brief them about life with the new guidelines. In response to the comments that had been made at the hearing and by letter, some changes in procedural matters were introduced, and the important decision to increase the size of the RAC from 11 to 25 members was taken. The new members were selected to include persons knowledgeable in law, public policy, ethics, the environment, and public health, and among those appointed, perhaps as a gesture of compromise, were recent critics of the guidelines (Doc. 12.16). The revised guidelines that were to come into force on January 2, 1979, were published in the *Federal Register* on December 22, 1978, and on December 29, the NIH announced in a press release the new appointees to the RAC.

The revision had taken over 18 months but had at last been achieved—and marked a significant change in attitude. Donald Fredrickson, Director of the NIH, whose patience must have been severely tried during that period while his agency was repeatedly criticized from all sides, summed up the situation in a paper presented for him by William Gartland, Secretary of the RAC, at a meeting organized by COGENE (the Committee on Genetic

Experimentation of the International Council of Scientific Unions) in England on April 1–4, 1979 (Doc. 12.17). Fredrickson was unable to attend as planned, because potential health risks from the nuclear power-plant accident at Three Mile Island, Pennsylvania, prevented his leaving the United States.

Document 12.1

1978

Genetic Engineering (p. 279)

HARMONIZING GUIDELINES: THEORY AND PRACTICE

John Tooze
European Molecular Biology Organization,
Postfach 1022.40, 69 Heidelberg,
(Federal Republic of Germany)

Since it was first raised in the USA, the issue of the conjectural biohazards of recombinant DNA research has been publically debated with more or less acerbity in virtually every country in which molecular biological research is pursued to a significant extent. Society does not usually preoccupy itself with protracted, repetitious discussions about activities that are only conjecturally hazardous since there are many really hazardous or potentially hazardous activities, some with long histories, that have a greater claim on public attention. Recombinant DNA research, however, has been an exception. In due course historians may be able to identify the manifold reasons why, but in the meantime we must live with the consequences of these debates. Perhaps in the long run the most significant, detrimental one will prove to be an increased distrust of laymen for scientists, but undoubtedly the chief, immediate consequence is, of course, that special safety precautions involving additional expenditures of money, energy, and time must be taken by anyone doing this research.

As we all know, following the lead given by the USA, in country after country, with alacrity or reluctance national committees have been established to draw up sets of safety guidelines, or adapt those of other countries, in which a graded series of containment measures is matched against a classification of gene cloning experiments based upon estimates of the degree of their conjectural biohazard. (While the principle of being innocent unless proved guilty was abandoned at the outset, the notion that the punishment should fit the crime has survived.)

Devising the specifications for scales of physical containment laboratories ranging from little more than a notice on the door and what should be universal elementary laboratory hygiene, to a laboratory in which the investigator and the environment are completely protected from the experiment, or vice versa, has not proved to be controversial since there is considerable practical experience of such facilities gained from many years of work with really dangerous pathogens. The notion of biological containment, the idea that the host and vector organisms used in recombinant DNA research should be so mutated as to have a totally insignificant probability of surviving outside the special permissive environment of the experiment, which was elaborated at the Asilomar Conference, has also been widely accepted by a grateful scientific community.

By contrast the attempts to estimate the degree of conjectural risk of individual or particular classes of recombinant DNA experiments have proved contentious and have resulted in sets of national guidelines that are not identical, are sometimes not entirely selfconsistent, and in some cases contradict common sense. That this should be the case is hardly surprising because the committees that were asked to devise guidelines had to answer such questions as, to what extent does a population of biologically disabled *E. coli*, each cell carrying a nontransmissible plasmid with a random 1×10^6 dalton piece of the genome of a fox, constitute a biohazard and to what extent would populations of the same host-vector system carrying similar sized fragments of the genomes of frogs, fishes, flies, figs, and flu virus constitute a greater or lesser biohazard? A medieval scholar used to debating the sex of angels would have felt at home with the dialectic if not the premises.

Although the history of the drafting of the U.S. National Institutes of Health (NIH) guidelines during 1975 revealed that consensus conjectures about the relative magnitude of the conjectural biohazards of different classes of shotgun experiments were not easy to obtain and changed from month to month as pressure from one group or another in one direction or the other was applied to the fragile fabric, the principles eventually enshrined in the NIH guidelines greatly influenced discussions elsewhere. The cardinal one was that the closer the phylogenetic relationship between the donor of the DNA to be cloned and the species considered to be at risk—inevitably the latter has usually been mankind—the greater the conjectural biohazard. The rationale for this was that the more closely related two species are the greater the homology between their DNA sequences, and, therefore, the greater the chance of a recombination event between the two DNAs. Furthermore it was argued that since the host range of

Reprinted by permission from *Genetic Engineering*, Boyer and Nicosia, eds., Elsevier/North-Holland Biomedical Press.

many viruses appears to have a phylogenetic basis, the possibility of inadvertently activating a latent virus of the donor species that could infect the species at risk was greater the closer the phylogenetic relationship of the two species. This possibility of activating latent viruses also led the NIH Committee to conclude that embryonic DNA poses a lesser conjectural hazard than the corresponding DNA from adults.

Such arguments are, however, more ethereal than they at first appear. Illegitimate recombination between DNAs possessing little homology is a well documented phenomenon, and as we learn more about the genetics of bacteria we realize that species barriers to gene exchange in nature are not so insurmountable after all. Moreover the life cycle of most viruses does not include a latent phase in which the viral nucleic acids become associated with the host cell DNA. These considerations notwithstanding, every set of national guidelines so far produced is based on phylogenetic considerations.

Another influential decision of a quite different nature was made by the NIH Committee. It decided to adopt an encyclopaedic approach which involves trying to envisage every possible category of recombinant DNA experiment and laying down for each of them appropriate combinations of physical and biological containment. The motives for this decision were entirely worthy: the aim was to avoid bureaucracy by decentralizing the procedures for implementing the guidelines. In a country the size of the USA, with a large number of laboratories doing this work, a decentralized system in which the investigator, a local safety committee and the customary peer review committees of national funding agencies make the decisions, by reference to encyclopaedic guidelines, seemed particularly attractive.

Encyclopaedic guidelines have, however, an inherent disadvantage; they involve making sweeping judgements based on, at best flimsy evidence, and the result is often too discriminate. When dealing with a subject that is riddled with conjecture—indeed is nothing but a pyramid of conjectures—a system which is flexible and allows a maximum of discrimination to be exercised has obvious advantages. Moreover, in the USA a variety of political factors, but notably the present trend towards an ever more populist society in which any and every issue—even a session of the local university's safety committee—is cause for a town meeting, has forced many in the U.S. scientific community not only to accept but even actually to welcome more central governmental control. Faced with the prospect of a series of differing parochial guidelines, each more proscriptive than the last, many of those wishing to do this research, as well as those responsible for administering it, have become advocates of pre-emptive federal legislation.

Table 1. Guidelines

Shotgun Experiment	Initial NIH	Revised NIH	French	German	Dutch	British
primate	P4 EK2 P3 EK3	P4 EK1 P3 EK2	P3 EK1 P2 EK2	P4 EK1 P3 EK2	P4 EK2	P4 EK1 P3 EK2
mammalian	P3 EK2	P3 EK1 P2 EK2	P3 EK1 P2 EK2	P4 EK1 P3 EK2	P3 EK2	P4 EK1 P3 EK2
invertebrate	P2 EK1	P2 EK1 P1 EK2	P2 EK1 P1 EK2	P2 EK1	P2 EK1	P2 EK1 P1 EK2

After the unanimity of Asilomar and the period during which most European countries followed the NIH guidelines, discrepancies between the different national guidelines are now beginning to increase as the above table shows. The situation in France and the Federal Republic of Germany illustrates this most vividly. Both countries formally adopted national guidelines during the last 6 months, those of France, which are the least restrictive of all national guidelines, reflect the growing belief that the fears of recombinant DNA research have been exaggerated and that the initial NIH guidelines were considerably too stringent. Opinions which, incidentally, are shared by the Standing Advisory Committee of the European Molecular Biology Organization. By contrast the Federal German guidelines are conservative and for some experiments, including the very important class of mammalian shotgun experiments, are more stringent than the revised NIH guidelines. The source of such discrepancies seems to me to be that, while an increasing majority of biologists are coming to the opinion that the containment requirements laid down 18 months or 2 years ago are quite unnecessarily severe, different national committees differ in their assessment of the rate and extent of presently acceptable relaxation. All too often one hears in the corridors that this or that committee believed on scientific grounds one thing, but decided to say another for political reasons. Political considerations are overruling scientific ones and what is acceptable in one society may be, or may be believed to be unacceptable at present in another.

Those who wish to see the potential scientific and practical benefits of this technology realized, must be thankful that no rigorous attempts to impose some international set of guidelines replete with detailed lists of containment categories for each conceivable type of cloning experiment have yet been made. Activities that pose only conjectural hazards are surely not fit subjects for national legislation, let alone international agreements or laws. That does not mean that meetings between those responsible for drawing up and implementing guidelines or regulations in different countries are a waste of time. On the contrary as experience within the European Science Foundation, for example, shows they can be very useful so long as they restrict themselves to exchanges of information and do not seek to impose any particular set of guidelines or procedures.

The history since Asilomar clearly indicates, to me at least, that any attempt during those 3 years to have drawn up international guidelines would have had deleterious consequences. We are dealing with a very rapidly developing area of science. As far as we know, it is certainly no more hazardous than many other, now routine, biological techniques whose development—unheralded by well meaning but nevertheless alarmist public statements by those who invented them—rightly excited no concern amongst the general public and entailed no dangers for it. Any move to international guidelines for recombinant DNA would not only further impede research which, as this Symposium has emphasized, promises enormous benefits to mankind, but would also seem to substantiate the unfounded fears of uncomprehending public opinion, which all too often is fed sheer drivel by newspapers, television, and radio, whose editors ought to be more jealous of their columns and broadcasting hours.

For example in Heidelberg, where the European Molecular Biology Laboratory is building a P4 containment facility for recombinant DNA research, one leftwing newssheet a few weeks ago had the headline "Das Monster im Stadtwald"; the conservative local paper published a long article in favour of the P4 laboratory but riddled with so many inaccuracies that its opponents must have been overjoyed, with friends like that who needs enemies; while a Stockholm newspaper published a photograph of the Heidelberg Laboratory (Sweden is one of the 10 states paying for the EMBL) with a caption translatable as "The most dangerous laboratory in the world". One of the ironies of such rubbish is that for the study of really dangerous pathogens recombinant DNA technology offers a very great safety factor.

Although there are indications that reason is beginning to reassert itself over hysteria, we have landed ourselves in a mess. Tangled in a web of regulations and bombarded by public opinion the prospects of restraints on this research eventually fossilizing into law is daunting. Why did it happen? There are in the revised version of the NIH guidelines two particularly revealing sentences tucked away in the introductory paragraphs which provide part of the answer. They read as follows: "During this period (i.e. since June 1976) the Committee has also become better informed about the general ecology and epidemiology of infectious microorganisms. Of particular relevance has been the information received from many medical microbiologists, including data from experiments with *Escherichia coli K-12*". Here we have the confession that while molecular biologists know a great deal about the molecular genetics of laboratory strains of *E.coli*, they are, or were in 1974, 1975, and 1976, ignorant of the epidemiology, ecology, pathology, and plain biology of naturally occurring strains of *E.coli*, and one suspects, the other microorganisms they study. In other words those who drew up the first NIH guidelines, which had such pervasive influence throughout the world, were not fully conversant with the then available and pertinent body of knowledge accumulated over many years by medical microbiologists. The latter, I think it is true to say, knowing something about the biology and epidemiology of naturally occurring strains of pathogenic microorganisms, have never taken the conjectural hazards of recombinant DNA research as seriously as those who invented the technique. I do not mean to imply that all those familiar with real pathogenic microorganisms would necessarily dismiss the conjectural biohazards of every conceivable recombinant DNA experiment as entirely insignificant but I feel certain that, had they been fully consulted from the outset, they would have put and kept these conjectural risks in perspective and prevented some of the hysteria, not to mention the staggering waste of time and energy, that has ensued.

It is, of course, easy today to criticise the group of most eminent molecular biologists for publishing in 1974 the so-called, and widely misinterpreted "moratorium" letter, without having involved from the outset those who were less ignorant of pathogenic microorganisms than themselves. No doubt they did not imagine the consequences. But with the easy wisdom of hindsight we can now see that there was a trap into which the unwary molecular biologists, without realizing it, fell headlong; the trap, of course, was that while they know so much about the laboratory organism *E.coli* K12, more about it than it is known about any other organism, they know so little about *E.coli*, the large family, one hesitates to use the word species, of bacteria that occur in soil, rivers, gastrointestinal tracts and elsewhere. Great expertise in the former caused them and their audience to believe that they were more than amateurs of the latter. The result was the set of very stringent requirements laid down in the 1976 NIH guidelines and an attempt to censor the directions of scientific inquiry which, like the dialetical discussions of consensus conjectures about the relative extents of conjectural risks, harked back to the medieval ages.

What began as a sincere, albeit tactically naive attempt by a group of scientists aware of their social responsibility and anxious to demonstrate an ability to self-regulate the direction of their research has led to the very antithesis of self-regulation, to imposed restrictions and regulations which may yet acquire the majesty of the law. Molecular biologists feared that they might unwittingly create a miasma and they have, but of verbiage, paper, red-tape and bureaucracy rather than noxious microorganisms. The formidable bureaucracy that now envelopes recombinant DNA research will not, I suspect, quickly and quietly go into voluntary liquidation; the money that has been and will continue to be spent on maintaining the committees, and on the redundant containment facilities cannot be recouped, while the temptation to insist that experiments continue to be done in those facilities, if for no better reason than to justify the decisions to build them, may well prove irresistible for a long time to come.

There are many lessons to be learnt from the episode and a vast accretion of archival material—a sort of instant history in the shape of documents by the ton and tapes by the mile—through which those with the patience may in the future ferret. Some of the most salutary lessons are already obvious. Clearly a substantial section of the lay public (apparently more strongly convinced that basic research is the fount of technological invention than some of the politicians and administrators responsible for funding it) is deeply suspicious of basic science and its practitioners and is anxious somehow to reduce the rate at which knowledge is acquired and exploited. Ignorance to them is less frightening than knowledge, since the latter increases the burden of choice and demands and forces society to change and evolve at a faster rate than they believe possible, comfortable or beneficial. These pessimists found in recombinant DNA research a new issue to focus upon, that apparently confirmed their gravest suspicions of science, for here was a set of discoveries the power of which apparently frightened even the scientists who had made them. There seemed to be an analogy with atomic physics and the progression from the conjectural hazards of recombinant DNA research by way of potential hazards, and hazards to outright horrors was as quick as it was irresistible. We can now only counteract this antiscience philosophy by fully exploiting recombinant DNA techniques to acquire new insights into the structure and function of genomes and by devising practical applications for the new knowledge. To achieve this we must work towards the dismantling of the unnecessary impediments.

Another important and painful lesson is that scientists should be as rigorous judges of the limits of the range of their expertise as they are of each other's experiments. When they fail to recognize these limits they are liable to act in ways that sooner or later, and rightly, are bound to deepen the layman's distrust of the scientist's own assessment of the implications and consequences of science and hence invite imposed censorship. The recombinant DNA issue has shown that scientists are conscious, in this particular instance perhaps overconscious, of their responsibility to society. We should now try to ensure that the community of molecular biologists is not castigated further for its errors of strategy and tactics and that the lay public, rather than becoming increasingly resentful and mistrusting, comes to appreciate not only the benefits that can flow from this research, but also the honourable motives of those who first raised the issue of conjectural hazards and why their good intentions led to such unfortunate consequences.

The opinions expressed above are those of the author and are not necessarily those of either the European Molecular Biology Organization or the European Science Foundation.

Document 12.2

April 19, 1978
Falkow Letter to Fredrickson

UNIVERSITY OF WASHINGTON
SEATTLE, WASHINGTON 98195

School of Medicine
Department of Microbiology and Immunology, SC–42

April 19, 1978

Dr. Donald Fredrickson
Director
National Institutes of Health
Bethesda, Maryland 20014

Dear Dr. Fredrickson:

I am writing you following my return from Europe after attending the International Symposium on Genetic Engineering held in Milan March 29-31, 1978 and the Plasmid Workshop held in Berlin from April 1-5, 1978. During the course of the meetings I consulted with a number of scientists about your request to prepare a documented list of microorganisms which are known to readily exchange genetic material under laboratory conditions or in Nature. I am glad to tell you that this material is being collated for transmission to the Recombinant DNA Advisory Committee.

On March 30, 1978 Professor E. S. Anderson, Professor M. H. Richmond and I held a meeting (our second) on Rick Assessment under the auspices of the Committee on Genetic Experimentation (COGENE) which was formed by the International Council of Scientific Unions. It was our unanimous concensus that we now have available sufficient data in man and livestock to show that:

a) the survival of E. coli K-12 sublines in man and animals is exceedingly low;

b) the frequency of transfer of self-transmissible plasmids from K-12 to resident gram negative flora of man occurs very rarely indeed;

c) that the frequency of transfer of plasmids from the resident flora to fed E. coli K-12 strains is extraordinarily low and;

d) that the frequency of mobilization of non-conjugative plasmids (the type actually used in DNA recombinant experiments) from K-12 to other enteric flora is not detectable under normal circumstances.

The combined monitoring data of several laboratories over a three year period has failed to demonstrate the colonization (indeed not even the detection) of K-12 strains in research workers and their families. Moreover, monitoring of the laboratory environment has not detected significant numbers of K-12 cells and in reconstructed accidental spills, the viable count of K-12 cells were found to decrease by 10 million fold in 4 hours if simply left to dry or was instantly sterilized if exposed to the mildest germicidal agents.

I wish to emphasize to you that all of the monitoring, survival studies, and transfer experiments were performed with ordinary E. coli K-12 rather than the certified EK-2 strain, X1776. Hence, since studies with ordinary E. coli K-12 are found to be negative, X1776 must represent a rather good case of "overkill". Analogous EK-2 strains have been developed in Great Britain by Sidney Brenner.

Upon my arrival in Bristol, England in June, 1978 to begin my sabbatical leave from the University of Washington, I shall correllate the data from participating laboratories into a single paper that will be submitted for publication by September, 1978. I am glad to be able to tell you that the risk assessment monitoring has proved to be so uniformly negative. We would be pleased, of course, to learn whether you or others feel that further monitoring data is required. Of course, further risk assessment experimentation is required but hopefully this will address other questions of safety.

Aside from these positive aspects of my travel, I confess that, in another vein I found the meetings most distressing. It is painfully obvious that because of the very restrictive nature of the NIH guidelines, as well as the beauracratic wall that the guidelines have spawned, American biologists can no longer expect to keep pace with either Western European or East European science. That is not to say that other countries have not adopted research guidelines. They most certainly have. Yet, the guidelines adopted by the European community and the USSR retain a flexibility and a scientific reality that is absent from our own. Nor do I believe that the European or Soviets are any less concerned about safety than we. It seems, however, that they learned valuable lessons from our mistakes. It is difficult for me to convey to you the intense discomfort many of us felt when we heard the results of reasonable, totally safe experiments being described by Western and East European scientists that are now literally forbidden to U.S. scientists.

Perhaps the adoption of the pending revised NIH guidelines will rectify, in part, these discrepencies. It is unfortunate that the Congress has received a good deal of inaccurate information about the history of the guidelines and were subjected mostly to the two extreme views of recombinant DNA technology.

Many of us will continue to work towards a rational solution of the recombinant DNA controversy. Yet, there is but so much time that one can afford to spend on such "political" matters without giving up science and our obligations to our students and families. I believe I told you that most of us who were involved in the Asilomar Meeting and in drafting the first set of NIH guidelines could be described as "walking wounded". I thought many of my wounds had healed but, after this trip, I fear my wounds have been re-opened. I believe that if Congress passes legislation regulating recombinant DNA research, American biological science will be adversely affected for several years to come.

Although we are falling behind, it is a tribute to our younger scientists that they persevere in the face of beauracratic adversity and continue to perform brilliant, innovative experiments. But these young scientists are beginning to lose their spark of excitement and they do not flourish. It is a pity.

I am glad on the one hand to send you news of our positive progress in one area; I regret that my news is not so encouraging in another (equally important) quarter. I am taking the liberty of sending copies of this letter to my congressional representatives and to several legislators considering recombinant DNA legislation. I doubt if it will have any impact but I shall sleep the better for it.

Sincerely,

Stan Falkow

Stanley Falkow, Ph.D.
Professor of Microbiology
and Medicine

csj

cc: Senator Henry Jackson
Senator Warren Magnuson
Senator Edward Kennedy
Congressman Joel Pritchard
Congressman Paul Rogers
Congressman Harley Staggers
Mr. John Stewart, Science Technology and Space

Document 12.3

April 1978

Clinical Research (**26**:113)

EDITORIAL

Trying to Bury Asilomar

When I was a boy in Chicago, the scientist was a poorly paid absent-minded dreamer, very bright, if not a genius, and culturally destined never to lead the masses much less have anything to say to any public official more important than a high school principal. Then came Oppenheimer and the atomic bomb, and physicists were men of affairs that we and the generals could not live without. Biologists were another matter. No one thought they posed a threat to anyone, and they would grow better hybrid corn or even someday cure cancer if they stayed by themselves and did not take on roles where they would only get in someone else's way.

I became one of them, and for 20 years it was the best of all possible worlds. Lots of money was dug up by Mary Lasker and her crowd and we were accountable only for the quality of our science. Even better, no one got mad at us when we didn't soon hit plenty of practical medical jackpots. That is, until Vietnam came, and our teach-ins told our students that the White House had the collective brains of a goose, a belief reinforced when Lyndon Johnson then went out to NIH to give the word that we troublemakers had better help the electorate or the gravy train would end.

Being a brand new Salk, however, is not that easy. It is not that we do not want to be do-gooders but if it was a cinch to come up with a new vitamin that would make everyone live to 100 years without Social Security, it would have happened long ago. Invariably, great challenges mean we don't know what to do next. Consequently most of our research efforts are holding operations which we hope will generate unexpected observations. With further luck, they may give us a fresh outlook on some pressing human problem like too many mosquitos or god-awful arthritis. So while we generally didn't go to work thinking what next can we do for the masses, we never lost sleep worrying whether we were in cahoots with the idle rich who preach the virtues of hard work by others and then live off of more pollution. Far better to be thought unworldly and of conscience clean.

Now, however, we are in a strange mess. The word is about that we are closet big businessmen who need to be kept in check by massive regulations, if not the threat of jail. All because some people have spread the word that DNA, the stuff which makes up our genes, might do us all in. I find this assertion total nonsense and cannot think of any potential environmental pollutant which worries me less. And everyone I know who works with DNA now feels the same and mere mention of "NIH Guidelines" or "Memorandums of Understanding" makes our mouths froth. Under ordinary circumstances our response would be obvious—head down to Washington and tell Congress that these scaremongers are an odd coalition of spaced-out environmental kooks and leftists who see genetics as a tool for further enslaving the masses.

The heart of our trouble, however, does not lie with the far-out doomsday scenarios these sad souls joyfully peddle. It was we totally up-to-date molecular biologists who first gave the DNA scare its legitimacy. To be sure we never said we were deeply worried for none of us were. I don't remember even five seconds of mild anxiety before I signed the call for the moratorium. This document, a semiofficial response to fears of genetic engineering at the service of CIA types, urged restraints from the making of certain recombinant genomes until we convened a gathering of international bigwigs on the Monterey Peninsula at Asilomar. There we were to discuss whether the 1973 breakthrough that gave us recombinant DNA was likely to generate dangerous new bugs. I didn't know how we could clearly answer this dilemma, but saw no harm for a little scientific restraint in the then post-Watergate mood of self-confession. Most of us were liberal-left types from birth, and our fellow McGovernites would see that we were not always out for quick fame, and placed the general good ahead of our immediate scientific goals.

Here we were terribly naive—the real world

never stops anything it wants unless it has already smelled smoke. Though everyone at Asilomar kept saying "potential risk," experienced newspaper addicts thought differently. No one would invent phrases like P3 or EK2 unless they could measure the monster's size. The very act of setting up committees to pinpoint the most risky experiments could only magnify the public fear that we biologists now had our own diabolical form of the bomb.

By this stage, I knew we were in way over our heads and felt that somehow this senseless hysteria should be nipped in the bud. So I argued, to the surprise of most at Asilomar, against the forming of any official guidelines that spelled out how we should work with recombinant DNA. I thought, by necessity, any regulations had to be capricious, and in the absence of even the slightest reason for being afraid, we were very silly to be making all this fuss. Only Joshua Lederberg and Stanley Cohen were publicly on my side, and those who privately agreed with me thought I was risking my hide by appearing so indifferent to the general good. Almost everyone reasoned that the public would not dump the matter and went along with Sidney Brenner's eloquent plea that we must make a show of self-restraint or the lawyers and ethicists would take over and give us stupid laws that could really shut us down.

Unfortunately, the high hopes of many that Asilomar might defuse the public concern were soon dashed. Within weeks Congress got in the act, and the guys who gave us the guidelines eventually were in front of Senator Kennedy. Given the importance ascribed to the topic, he saw no reason to accept the argument that a matter which might affect the life or death of our society should be left in the scientists' hands. Just the opposite—it seemed the perfect occasion for the public to be asked whether they were to be the next guinea pigs for our experiments. Here the molecular biologists quickly found they had little room to maneuver. Not being that afraid, they could only say you must not get that concerned and there is no reason for specific laws to control us. On the other hand, they had to say we needed the guidelines, for this was the Asilomar message. So they were left to argue how safe we would be if everyone followed the guidelines and did their "potentially most risky" experiments using specially disabled bacteria and elaborate containment facilities designed to prevent the escape of organisms bearing recombinant DNA. Not surprisingly this message failed to reassure Kennedy who could not believe that any sensible person would propose diverting so much scarce research funds or creating a brand new bureaucracy unless he were much more scared than he publicly let out.

Then every subsequent defense of the guidelines led automatically to how we would deal with sloppy lab practices, if not downright cheating. No one could propose that mistakes will never occur or that molecular biologists are so unworldly that they cannot be dishonest. This is a point on which Congress needs no education, and once the issue was raised on Capitol Hill it became very likely that some form of punitive legislation would be enacted to ensure strict compliance with the guidelines.

If, instead, the recombinant DNA community had been divided as to whether we needed any guidelines, the final outcome would have been unclear and might have been decided by whether the Senate and House had the gumption for still another regulatory commission. But with virtually every first-class molecular geneticist then arguing that we need their protection, Congress would have been accused of shirking their duty if they did not seize the initiative. The question then became of how to get a law where we, not the professional demonstrators, control the action.

Up through today we have barely squeaked through, and that three years have passed since Asilomar without legislation reflects more the inherent chaos of Congress than the skill with which we lobby. And though there have been moments of panic, the situation now is less out of our control. The most chilling early proposal from Senator Kennedy's baliwick was that final power should reside in a committee where by law the scientists who know what DNA is must be in the minority. The vision of a "Friend of the Earth" bearing a large wig, righteously telling us what not to do, could be a scene from *MASH.* And Representative Rogers' Health Subcommittee dallied with the thought that the way to keep us clean was the taint of a felony and $50,000 a day fines, penalties which most certainly would make docile puppets of every last one of our nation's bankers.

Through these ups-and-downs, we increasingly had the hope that recombinant DNA would finally so bore Congress that no one would do the work to see a bill through to its enactment. I don't think, however, that we shall ever win this way. As long as most responsible scientists are thought to favor perpetuation of guidelines, every new example of supposed cheating will rekindle some new hysterical outburst. Only if the scientific community strongly says the guidelines can go, do we have a fighting chance to weather such incidents without the possible creation of a truly imbecilic law.

I fear, however, that we academics do not have the skilled politician's knack of suddenly reversing his course without the slightest suggestion that he is soft in the brain. So the prevailing approach today is to urge the enactment of a much watered-down bill that will both keep all control over the recombinant DNA within NIH and preempt the power of more local bodies from making their own decisions as to which experiments should be done. And there are dreams that we might get a sunset clause which would limit the legislation to a several-year period after which everyone could bury the guidelines without acute embarrassment. If all this were to come about, we would breathe better and stop despairing that we may have closed the door in this country to many of the better experiments that we can do on the origins of cancer. For the moment, however, we must remain on guard. The final form of any legislation is often imprinted at the last moment in conference, and I see no reason for expecting Senator Kennedy to give way on preemption. So we had best face up to the prospect that the local governments will justify their own bans by citing federal law as proof that recombinant DNA poses a realistic threat.

We may thus hope that those scientists whose first foray into the public arena gave us Asilomar will ask themselves whether if Asilomar had never occurred, would they now see any compelling public health reason to propose formal restrictions on recombinant DNA research. If their answer is no, they should ask Congress for another chance to be heard. They should not delude themselves that Congress's main advice has come from the likes of George Wald. Their role has been much more of noise than of substance. Any legislation that we finally get will reflect Congress's belief that they have sensed the mood of our most down-to-earth experts, and the guidelines are what they want. I do not believe this, but I can effectively only speak for myself. Now is the time for my fellow DNA workers to come forward and speak their minds.

J.D. WATSON
Director
Cold Springs Harbor Laboratory
Cold Springs Harbor, New York

Illustration by Barbara Thomas.

Document 12.4

July 28, 1978
Federal Register, Excerpt

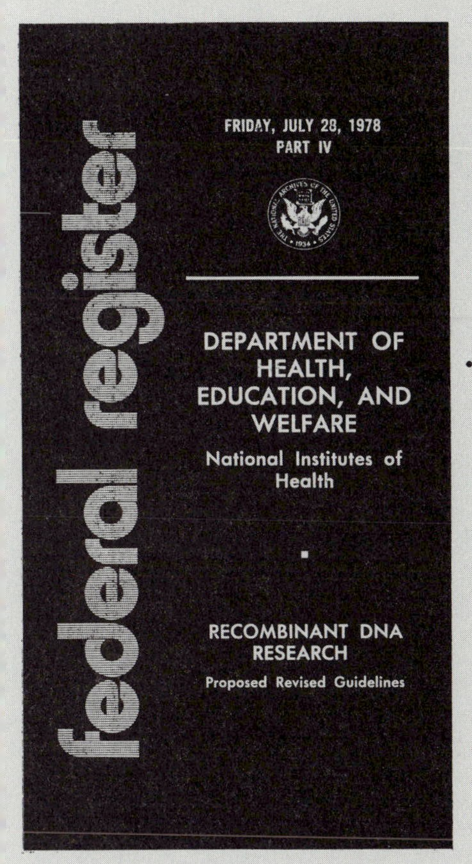

[4110-08]

DEPARTMENT OF HEALTH, EDUCATION, AND WELFARE

National Institutes of Health

RECOMBINANT DNA RESEARCH

Proposed Revised Guidelines

STATEMENT BY JOSEPH A. CALIFANO, JR., SECRETARY OF HEALTH, EDUCATION, AND WELFARE

The Director of the National Institutes of Health is publishing proposed revisions to the NIH guidelines on research involving recombinant DNA molecules.

The original guidelines are being updated in light of NIH's experience operating under them and in light of our increasing knowledge about the potential risks and benefits of this research technique. As experience accumulates, we should review and evaluate the evidence to assure that the restrictions imposed are appropriate to potential risks—strengthening restrictions where needed, relaxing regulation where justified...

In publishing these proposals for public comment, I recognize the extraordinarily difficult challenge that developing sensitive but effective regulations in this field poses for NIH, for the research community, and for the concerned public.

Necessarily, this task poses difficult questions that we will never be able to answer with complete certainty. I hope that those concerned will analyze the proposed revisions with care and give us their views on the strength of the evidence that supports the proposed revisions, the specific scientific and research containment procedures that the proposed revisions require, and the procedures and the standards they establish for future changes in the guidelines and for the exercise of discretion under them.

...We particularly seek comment on the sections in the proposed revisions that establish the mechanisms for administering and revising the guidelines. For example, do the proposed revisions strike the proper balance in establishing:

The procedures for permitting otherwise prohibited experiments and for exempting classes of research from the guidelines;

The standards for the exercise of administrative discretion under the guidelines;

The composition of the Department's Recombinant DNA Advisory Committee and of the institutional biohazard committees.

To review the comments on the proposed revisions, I am establishing a departmental review committee, consisting of Mr. Peter Libassi, the Department's General Counsel (Chairperson); Dr. Donald Fredrickson, the Director of NIH (Vice-Chairperson); Dr. Julius Richmond, Assistant Secretary for Health; and Dr. Henry Aaron, Assistant Secretary for Planning and Evaluation.

I have asked this committee to hold a public hearing to insure full and complete opportunity for comment. In order to hold this hearing and to issue the revised guidelines on a reasonably prompt schedule, no extension of the 60-day period for public comment will be possible.

In preparing these revisions, the National Institutes of Health have already held 19 hours of public hearings and have received continual advice from the scientific community and from the public. I want this open process to continue. I urge those concerned to help us find the proper balance by providing us with their comments on NIH's proposed revisions.

Dated: July 19, 1978.

JOSEPH A. CALIFANO, JR.,
Secretary.

DEPARTMENT OF HEALTH, EDUCATION, AND WELFARE

NATIONAL INSTITUTES OF HEALTH

RECOMBINANT DNA RESEARCH

PROPOSED REVISED GUIDELINES—NIH

This will introduce three related documents that the National Institutes of Health (NIH) is publishing for public comment: (1) a *Decision of the Director, NIH,* to publish revised NIH guidelines for research involving recombinant DNA molecules, (2) the *Proposed Revised Guidelines—NIH,* and (3) an *Environmental Impact Assessment* of the proposed action. The Secretary of Health, Education, and Welfare has approved the release of these documents for public comment.

As stated in the Secretary's preface, we are particularly concerned that public comment be invited on the scientific and procedural aspects of the proposed guidelines. A public hearing on these proposed revisions will be held in late September at the Hubert H. Humphrey Building, Washington, D.C. All comments received on or before September 25, 1978, will be considered, and no extension of the comment period will be granted. Within 45 days after the comment period, final guidelines will be promulgated with a notice in the FEDERAL REGISTER.

The events leading to the proposed revisions are described in the "Introduction and Overview" to the *Decision* and in the "Foreword" to the *Assessment.* This preface will orient the reader to other NIH documents that are directly related to the present publications. One is the current *Recombinant DNA Research Guidelines,* effective since June 23, 1976 (published in the FEDERAL REGISTER, July 7, 1976). Another is the *Environmental Impact Statement* (EIS) on those guidelines, published in 1977. The EIS, which contains a copy of the guidelines, is available from the Government Printing Office (stock No. 017-040-00413-3) and in GPO depository libraries throughout the country. A third related document is *Proposed Revised Guidelines on Recombinant DNA Research,* which the NIH Recombinant DNA Advisory Committee (RAC) recommended to the NIH Director on September 1, 1977. The RAC proposal was published in the FEDERAL REGISTER, September 27, 1977, for public comment.

The RAC-proposed revisions were discussed at a public meeting of the Advisory Committee to the Director (DAC), held at NIH on December 15-16, 1977. The DAC and special consultants heard witnesses from environmental groups, the scientific community, industry, etc. All correspondence from the public in response in the FEDERAL REGISTER publication was available to those present and will be published, with the transcript of the meeting and other documents, as part of a continuing public record of NIH activities concerning recombinant DNA. The Director, NIH, acting in light of the 2-day discussion, all commentaries, and the DAC's recommendations, has arrived at the present proposal—the *Proposed Revised Guidelines* (NIH)—and offers it for public review.

The current (1976) guidelines contain an appendix entitled "Supplementary Information on Physical Containment." This has been omitted from the present proposed guidelines but, in revised and expanded form, will be available on request as "Laboratory Safety Monograph—A Supplement to the NIH Guidelines for Recombinant DNA Research." For a copy, write to the Office of Recombinant DNA Activities, Building 31, Room 4A52, National Institutes of Health, Bethesda, Md. 20014.

The *Environmental Impact Assessment* is based on an intensive analysis of the current guidelines, the RAC-proposed alternative, and the present NIH alternative. The conclusion of this analysis is that there would be no adverse impact of the NIH-proposed changes upon the environment.

The guidelines as presently proposed are designed to discharge the continuing obligation of NIH to insure that recombinant DNA research goes forward under standards of safety reflecting the latest scientific knowledge, so that the public and the environment are protected from any hazards while deriving the full benefits of the recombinant DNA technique.

Written comments and inquiries concerning the *Proposed Revised Guidelines* should be addressed to the Director, National Institutes of Health, Bethesda, Md. 20014. All comments received will be available for public inspection at the Director's office on weekdays (Federal holidays excepted) between the hours of 8:30 a.m. and 5 p.m.

Dated: July 19, 1978.

DONALD S. FREDRICKSON,
Director,
National Institutes of Health.

DECISION OF THE DIRECTOR, NATIONAL INSTITUTES OF HEALTH, TO ISSUE REVISED GUIDELINES FOR RECOMBINANT DNA RESEARCH

JULY 19, 1978...

...INTRODUCTION AND OVERVIEW

Today, with the concurrence of the Secretary of Health, Education, and Welfare, and the Assistant Secretary for Health, I am proposing revisions to the NIH Guidelines for Recombinant DNA Research.(*1*) These Guidelines were first issued on June 23, 1976. The proposed revisions result from a continuing process of scientific and public exchange similar to that of the 1976 edition. This overview sketches the background for proposed revisions and summarizes the proposed changes. It references accompanying documents and other pertinent sources of information.

The probable risks and benefits of recombinant DNA research—the larger subject of which the NIH Guidelines are a part—have been discussed in numerous forums since first addressed in 1973.(*2*) Congress has held multiple hearings on related issues, including proposals to convert the Guidelines to Federal regulations(*3*) and redefine recombinant DNA research to narrow the range of experiments subject to regulations. Early in 1977 the NIH Recombinant DNA Advisory Committee (RAC), the scientific and technical committee responsible for proposing revisions to the Guidelines, began its task of identifying changes needed in the Guidelines and forwarded suggestions to me for consideration. In order that public comments could be heard on the RAC-proposed revisions, published in September 1977,(*4*) a meeting of the Advisory Committee to the Director, NIH (DAC), the committee responsible for public oversight, was held in December. The extensive record of this hearing bears witness to almost unanimous agreement that the original Guidelines badly need updating, and suggests numerous directions in which revisions might move.(*5*)

Much of the discussion at the December 1977 meeting of the DAC affirmed the need for continuous reevaluation of the scientific premises underlying the original Guidelines. Since Asilomar,(*2*) growing evidence has suggested that other experts ought to review the concerns of the molecular biologists who first raised questions about the safety of recombinant DNA research. Scrutiny from experts in infectious diseases, epidemiology, virology, botany, ecology, laboratory safety, and other disciplines has been needed. NIH sponsored a workshop for this purpose at Falmouth, Mass., 6 months before the DAC meeting. Here, old and new information about *E. coli* K-12—the host most used in recombinant DNA experiments—was interpreted carefully. From this came a consensus that the chances of this host being convertible to an epidemic pathogen are negligible.

Those attending the December DAC meeting also heard complaints that containment levels were set too stringently for recombinant DNA work on viruses and plants. This applied both to the original Guidelines and to the revisions proposed in September 1977. A decision was made to address the issues through workshops without delay.

One of these workshops, held at Ascot, England, dealt specifically with viruses.(*7*) Here, experts from several countries, most of whom had no stake in recombinant DNA experiments, reached an unequivocal opinion that the risks of cloning viral DNA in a bacterium like *E. coli* K-12 are not greater, and are usually much less, than the risks of handling the parent virus alone. They also stressed that defective viruses pose little risk of infection when used as vectors for cloning DNA in eukaryotic cells, since the cells cannot survive outside permissive laboratory conditions and the virus cannot escape in a viable form. A second working group, meeting in Bethesda,(*7*) then reviewed the conclusions. It agreed that the original Guidelines imposed stricter containment on use of viral DNA or of viruses as vectors than could be justified by any available fact, and recommended changes.

A group of agricultural scientists met in Washington, D.C., in March 1978(*8*) to consider the containment conditions for incorporation of DNA from plant pathogens into *E. coli* K-12 and for use of viral and other vectors in plants. An important concept discussed at this workshop is the lack of evidence that *E. coli* K-12, or any other strain of this bacterium, is capable of acquiring an ecological niche in plants and thus infecting them.

On April 27-28, 1978, the RAC considered the recommendations concerning viruses and plants and agreed with most of them. With a few changes, they are a part of the proposed revision.

At the December 1977 hearings before the DAC, other aspects of the Guidelines evoked requests for revisions.(*5*) There was overwhelming sentiment for exempting from the Guidelines experiments involving recombination of DNA within the same strains or from pairs of organisms that transfer genes in nature. Discretion to exempt such experiments is not provided in the original Guidelines; nor does one find there the flexibility to permit other experiments for purposes of risk assessments needed to determine the merits of particular standards, or how these should be revised. Other criticisms of the Guidelines stressed the delays and confusion created by excessive centralization of administrative control, and it is evident that implementing procedures must be changed.

Certain background elements merit comment. Shortly after the NIH Guidelines were published, the British guidelines appeared.(*9*) Subsequently other national guidelines have been issued, most recently those of the Soviet Union. Many international aspects of DNA research and its regulation have been reviewed elsewhere.(*9*) The December 1977 DAC meeting directed attention to instances where the NIH Guidelines either exclude experiments that have been conducted abroad or make them far more difficult to do.(*10*) Moreover, factual bases for the greater stringency of the U.S. (NIH) Guidelines cannot be shown.

Five years have passed since concerns were first raised about the hypothetical hazards of laboratory experiments with recombinant DNA. The thousands of individual applications of such techniques have produced much useful knowledge, but no evidence has come to light of a product created by these techniques that has been harmful to man or the environment. Foreign genes inserted into prokaryotic host-vector systems have been faithfully replicated and produced in quantities valuable to science. On the other hand, prokaryotes generally have not been able to translate eukaryotic genes into biologically active proteins. No new facts or unconsidered older ones have emerged to support the fears of harmful effects, and one prominent early proponent of guidelines has repudiated his support for them.(*11*) At the least, there is growing sentiment that the burden of proof is shifting toward those who would restrict recombinant DNA research.(*12*)

Although clearly the time has come to revise the original NIH Guidelines for Recombinant DNA Research, it is not the time to conclude that they are being altered in preparation for their early abandonment. Understanding of gene regulation and expression is increasing inexorably and at an awesome pace. We may predict that ways will be found to achieve and control the translation of foreign genes by a variety of hosts.(*13*) As the barriers to translation are dropped, some of the larger promise of recombinant technology will be realized. In some proportion to the harvest of positive results, a capability must be maintained for observing any capacity of these experiments to yield harmful products, and for communicating this to all who have an interest in similar experiments.

In preparation for this next phase of recombinant DNA research, several shifts in NIH guidance are necessary. Experiments posing no threat to safety must be exempted from the Guidelines; and provisions must be made to remove others as soon as their harmlessness becomes evident. Any universal rules imposed on this kind of activity derive validity from continual modification dictated by results of the experimentation they govern.

Primary responsibility for comliance with the rules must be located where the work is done. There it must be shared fully by principal investigators, those who work in their laboratories, institutional biosafety committees, and the institutional leaders. The NIH Office of Recombinant DNA Activities (ORDA) should be relieved of its burden of obligatory prior approval of certain experiments, so that it can better carry out, along with the RAC, two central functions. These are the continuing synthesis and interpretation of the Guidelines, and the maintenance of full communication among all who must use them.

To recapitulate, these new proposed Guidelines arose from a proposal made to me by the RAC in September 1977. Numerous amendments have been made on the basis of public comments received at the December 1977 hearing, in extensive correspondence before and after that, and recommendations of special expert workshops whose reports were then assessed by the RAC in April 1978. The proposal and the amendments have been the products of long and intense participation by numerous persons representing many points of view. I now summarize the more important proposed changes. The basis for decision on each element of revision is provided in detail in subsequent sections of this document.

SCOPE AND APPLICABILITY OF THE GUIDELINES

Recombinant DNA containing synthetic sequences is now explicitly part of the definition of what is included under the Guidelines. The standards of the Guidelines now apply to all recombinant DNA experiments conducted in an institution that receives any support from NIH for recombinant DNA research. This includes a registration requirement.

The original Guidelines contain a number of prohibited experiments. There was little sentiment for the removal of all the original prohibitions—although it has been noted that the U.S. (NIH) Guidelines are the only national guidelines to stipulate "prohibited" activities. The original prohibitions, with one modification and a necessary "flexibility" clause,(*14*) are therefore retained in the proposed revision. They immediately precede a new section called "Exemptions"—a juxtaposition chosen to emphasize that the prohibitions still override.

The first exemption from the guidelines covers the handling of DNA outside a host organism or virus. Such "naked DNA" has been handled in laboratories for years and is rapidly inactivated in nature.

The exempted experiments of the second class consist essentially in rearranging, or deleting from, molecules of nonchromosomal or viral DNA. No foreign DNA is involved. An example would be the introduction of a DNA molecule formed from pieces of SV40 virus into eukaryotic cells in tissue culture. Since there is little if any basis for presuming such "rearrangement" or "deletion" experiments to be hazardous, they are now excluded.

A third class of exemptions are experiments called "self-cloning," in which DNA found naturally in a host may be reinserted into that host. These are reproductions in the laboratory of events that occur in nature.

Similarly, provision is made in the proposed guidelines for exemption of a fourth class of experiments that involve donor-host pairs that normally exchange DNA. Such genetic exchange is known to occur widely between various species of bacteria and is generally mediated by certain plasmids or viruses. Experimental recombinations of this type are only an imitation of what nature is able to accomplish handily in the absence of Federal regulation. A list of donor-host pairs to be exempted is begun in this revision and will be expanded periodically as knowledge grows, The initial choice from several possible lists submitted to me by the RAC is a conservative one, restricted to pairs of organisms for which there is documented evidence of natural exchange.

Finally, a fifth exemption is provided for removal of other recombinations when they are shown to be safe. The last two exemptions create some of the discretionary power for modifying the guidelines that was so lacking in the original. Provision will be made for public input to such decisions, either by announcement of proposed exemptions prior to consideration by the RAC or before a decision by the Director becomes effective.

CONTAINMENT

I have made one decision that will not be regarded with equal pleasure by all engaged in recombinant DNA research. P1 containment previously permitted mouth pipetting. In accord with a previous recommendation by the European Molecular Biology Organization (EMBO), its virus Working Group strongly recommended prohibiting this practice; and so did NIH safety advisors. The RAC at its meeting on April 27-28, 1978, recommended that mouth pipetting be prohibited only for those P1 recombinant DNA experiments involving viral DNA. Rather than create two separate classes of P1, and in recognition of the present availability of excellent mechanical devices for pipetting, I am proposing that mouth pipetting no longer be permitted in P1 containment. Since it is already prohibited in P2-P4 containment, this bans the use of mouth pipetting for any experiment covered by the Guidelines.

CONTAINMENT GUIDELINES FOR COVERED EXPERIMENTS

The recommendations of the RAC, arising from the Ascot-Bethesda workshops, represent the first realistic appraisal of any hazards that might lie in the use of viral vectors or the cloning of viral DNA. Recombinant techniques offer access to ares of viral biology that are vitally important. Such studies should not be impeded unnecessarily. I have accepted the April 1978 recommendations of the RAC in this area with minor amendments. The revised guidelines emphasize the current dictum that any hazards of working with viruses in recombinant DNA experiments are maximal at the first stage, when the virus itself with its full genomic complement is handled.

The RAC unanimously approved modest changes in containment for plant experiments. I have also approved them provisionally, contingent upon concurrence by the Department of Agriculture.

A new sentence has been added to the guidelines giving much needed flexibility in the setting of containment levels.*(15)*

ROLES AND RESPONSIBILITIES

Two years' experience with the guidelines has offered valuable tutelage in the limits of external (Federal) control of laboratory experimentation. Scientists and their co-workers have long experimented with pathogenic organisms, poisonous plants and animals, and hazardous chemicals. The laboratory is not among the more notorious occupational settings for accidents or illness, and damage to community or environment by basic laboratory research is almost unknown. Control over the use of radioisotopes in the laboratory, long a Federal preserve, is not comparable to use of recombinant DNA techniques; for the risks of using radioisotopes are calculable and mistakes are easily measured. Thus realistic and durable standards can be set. Without a base for the setting of such standards, conventional regulation is difficult at best, and at worst can be preposterous.

In the case of recombinant DNA technology, we are in the midst of a search for any risks, and thus for applicable standards. The scientists who raised the possibility of risks also realized that the only effective safeguards lay in a maximum enhancement of the collective nature of the scientific process. The usual communications networks of science had to be augmented and the evaluation of results and the reaching of consensus accelerated. These actions, is was reasoned, would help establish a set of initial rules, and there was the added assumption that they could and should be kept up to date. All using the new techniques would sign a "memorandum of understanding" to the effect that, until things became clearer, the basic communality of scientific inquiry would be especially emphasized in any work with recombinant DNA techniques.

The power of the Government to require such discipline of its grantees was an attractive reason for the scientists to request Federal intervention. And the Federal capacity to achieve the essential communication and consensus-building has been one of the most positive results of this experiment in adminstration. But the price of Fedeal intervention includes a heavy tax of formalism. In the instance of these guidelines, diverse pressures have made difficult the appropriate balancing of substance and procedure. I have already alluded to one of the undesirable results—a chilling inflexibility of the original guidelines—and its proposed correction by revision.

Prior NIH clearance is mandatory for new NIH grants and contracts involving recombinant DNA techniques and for all projects in P4 facilities. In the proposed revised guidelines, prior NIH clearance is no longer required for changes at the P1–P3 levels. These changes must be approved by the institutional biosafety committee (IBC), and NIH will then review the IBC actions. This proposal reverses an October 1977 issuance stating that changes in ongoing projects require prior NIH clearance. The requirement resulted in numerous delays in projects which could not be justified on grounds of safety.

The proposed guidelines would strengthen institutional responsibilities and authorities in determining compliance. A full partnership with all investigators and their institutions is intended. The role of the IBC is particularly enhanced through delegation of some discretionary powers that were previously reserved for NIH and the RAC. To better meet these obligations, an institution using P3 or P4 containmen is required under the proposed guidelines to have a qualified biological safety officer.

Experience gained in the past 5 years in explaining recombinant DNA technology has shown how valuable can be a community's activities. At least one member of the IBC is to be a "public member"—i.e., one who has no financial connection with the institution. Further, to ensure opportunity for public particiation at the national level, procedural are set forth, as explained in Part IV of the decision, that provide public notice and solicit comment on the major actions of NIAH.

Another stipulation of the revised guidelines is that failure of compliance can lead to suspension of NIH support for recombinant DNA research.*(16)*

Provision is now made for the private sector to register voluntarily its recombinant DNA activities with NIH. Also, other consulting services, including certification of host-vector systems, will be provided. The service will be accompanied by protection of proprietary data as mandated by law.

NIH issued a draft environmental impact statement on the guidelines in September 1976. This was revised after public comment and issued in final form in October 1977. It concluded that the activities covered by the guidelines had no predictable impact on the environment, since all the risks discussed were hypothetical. The EIS was examined by a Federal district court in 1978.*(17)*

In parallel with the process of revising the guidelines, NIH has conducted an environmental impact assessment, including an analysis of how current experiments supported by NIH will be affected by this revision. Again, the activities covered by the revised guidelines deal only with hypothetical risks, and thus the assessment reveals no predictable impact on the environment. Its content is published herewith in a companion document.

ORGANIZATION OF THE REMAINDER OF THIS DOCUMENT AND ABBREVIATIONS USED

The Recombinant DNA Molecule Program Advisory Committee is somnetimes referred to below as the Recombinant DNA Advisory Committee or Recombinant Advisory Committee or RAC.

The meeting of the Advisory Committee to the Director, NIH, which took place in December 1977 is sometimes referred to below as the meeting of the Director's Advisory Committee or of the DAC or the December 1977 public hearing.

The "NIH Guidelines for Research Involving Recombinant DNA Molecules" as issued on June 23, 1976, and publishes in the FEDERAL REGISTER on July 7, 1976, are sometimes referred to below as the original guidelines or the 1976 guidelines or the current guidelines.

The proposed revised guidelines prepared by the RAC and published in the FEDERAL REGISTER on September 27, 1977, are referred to below as the PRG–RAC.

The proposed revised guidelines which are being proposed now by NIH are referred to below as the PRG-NIH.

The remainder of this document is divided into four parts corresponding to the four parts of the guidelines; i.e., I. Scope of the Guidelines; II. Containment; III. Containment Guidelines for Covered Experiments; and IV. Roles and Responsibilities.

Within each of these four parts there are two subsections; i.e., Review of RAC-Proposed Guidelines and Review of Comments and NIH-Proposed Guidelines. The first subsection describes how the PRG-RAC differs from the 1976 guidelines; the second describes (1) the public comments received both before and after the December 1977 DAC meeting, concerning the PRG-RAC, and (2) the changes which have been made in response to these comments leading to the PRG-NIH.

FOOTNOTES TO INTRODUCTION AND OVERVIEW

(*1*) In addition to the proposed revised guidelines and this "Decision Document," there is also being released an Environmental Impact Assesement, including numerous appendices.

(*2*) The capability to perform DNA recombinations, and the potential hazards, had become apparent to scientists at the Gordon Research Conference on Nucleic Acids in July 1973. At their behest the National Academy of Sciences created a committee that organized an international conference held in February 1975 at Asilomar Conference Center, Pacific Grover, Calif. Approximately 150 scientists, of whom a third were from foreign countries, were present. The committee also called on the National Institutes of Health to establish an advisory committee to draft guidelines for the conduct of this research. Temporary guidelines were issued at Asilomar pending issuance of NIH guidelines.

In response, the NIH Recombinant Advisory Committee (formally "NIH Recombinant DNA Molecule Program Advisory Committee") was established in October 1974 to advise the Secretary of HEW, the Assistant Secretary for Health, and the Director of NIH to accomplish these tasks. The several meetings at which the Recombinant Advisory Committee developed its proposed guidelines in 1975 were announced in the FEDERAL REGISTER and were open to the public. The committee, after preparing several draft versions of guidelines, reached agreement on a recommended revised version, which was referred to the NIH Director for review in December 1975.

A special meeting of the public advisory Committee to the Director, NIH, was convened in February 1976 to review these proposed guidelines. In addition to current members of the committee, a number of former committee members as well as other scientific and public representatives had been invited to participate. There was ample opportunity for comment and an airing of the issues, both by the committee members and the public witnesses. All major points of view were broadly represented.

The proposed guidelines were reviewed by the Director, NIH, in the light of comments and suggestions made at the public hearing as well as extensive written correspondence received after the meeting. When the final guidelines were released in June 1976, an accompanying decision paper described in great detail all relevant public comments and the reason for accepting or rejecting specific recommendations in preparing the final guidelines. The NIH guidelines and the Decision of the Director, NIH, were published in the FEDERAL REGISTER on July 7, 1976. In addition, copies of the guidelines were widely distributed to foreign embassies, medical and scientific journals, NIH grantees and contractors, and professional research societies.

(*3*) The following committees have held hearings and/or markup sessions on Recombinant DNA legislation:

House—The Subcommittee on Health and the Environment and its parent, the Committee on Interstate and Foreign Commerce; the Subcommittee on Science, Research, and Technology and its parent, the Committee on Science and Technology.

Senate—The Subcommittee on Health and Scientific Research and its parent, the Committee on Human Resources; the Subcommittee on Science, Technology, and Space. Its parent, the Committee on Commerce, Science, and Transportation, has not held any hearings or markup sessions on this topic.

The following bills on recombinant DNA technology have been formally introduced:

	Bill	Chief sponsor	Date
House:			
H. Res. 131		Richard Ottinger, Democrat of New York.	Jan. 19, 1977.
H.R. 3191	Identical to S. 621	do	Feb. 7, 1977.
H.R. 3591	do	do	Feb. 16, 1977.
H.R. 3592	do	do	Do.
H.R. 4232		Stephen Solarz, Democrat of New York.	Mar. 1, 1977.
H.R. 4759		Paul Rogers, Democrat of FLorida	Mar. 9, 1977.
H.R. 4849	Identical to H.R. 4759	do	Mar. 10, 1977.
H.R. 5020	Identical to S. 621	Ottinger	Mar. 14, 1977.
H.R. 6158	Administration bill	Rogers	Apr. 6, 1977.
H.R. 7418		do	May 24, 1977
H.R. 7897		do	June 20, 1977.
H.R. 11192		Rogers and Harley Staggers, Democrats of West Virginia.	Feb. 28, 1978.
Senate:			
S. 621		Dale Bumpers, Democrat of Arkansas.	Feb. 4, 1977.
S. 945		Howard Metzenbaum, Democrat of Ohio.	Mar. 8, 1977.
S. 1217	Administration bill	Edward Kennedy, Democrat of Massachusetts.	Apr. 1, 1977
S. 1217	As reported: accompanied by Report 95-359.	do	July 22, 1977.
S. 1217	Amendment 754 in the nature of a substitute.	Gaylord Nelson, Democrat of Wisconsin.	Aug. 2, 1977.
S. 1217	Amendment 1713 in the nature of a substitute.	Kennedy	Mar. 1, 1978

(*4*) The Recombinant Advisory Committee considered its proposed revisions at meetings throughout 1977. The version proposed to the Director, NIH, in September 1977, appeared in the FEDERAL REGISTER on September 27, 1977.

(*5*) This meeting of the Director's Advisory Committee took place in Bethesda on December 15–16, 1977. A summary of the meeting appeared in the recombinant DNA technical bulletin, and the complete record will shortly be published by NIH in vol. 3 of the series recombinant DNA research.

(*6*) The NIH-sponsored meeting at Falmouth, Mass., on June 20–22, 1977, was chaired by Dr. Sherwood Gorbach. A complete record of this meeting appears in the "Journal of Infectious Diseases" (May 1978).

(*7*) The "U.S.-EMBO Workshop to Assess Risks for Recombinant DNA Experiments Involving the Genomes of Animal, Plant, and Insect Viruses" was held on January 26–28, 1978, in Ascot, England. It was attended by experts on viruses from the United States, Britain, and other European countries, a majority of whom were not engaged in recombinant DNA research. The primary purpose of the meeting was to conduct a scientific and technical analysis of possible risks associated with cloning eukaryotic viral DNA segments in *E. coli* K-12 host-vector systems and with the use of eukaryotic viruses as cloning vectors in animal, plant, and insect systems. The report of the workshop was published in the FEDERAL REGISTER on March 31, 1978, and appears as appendix E to the accompanying environmental impact assessment. The results of the Ascot meeting were then reviewed by another group of U.S. virologists who converted them into recommendations for revision of the guidelines. This working group was chaired by Dr. Harold Ginsberg and met on April 6–7, 1978. Its report appears as appendix F to the accompanying environmental impact assessment. The report was considered by the Recombinant Advisory Committee at its April 27–28, 1978, meeting.

(*8*) The "Workshop on Risk Assessment of Agricultural Pathogens" was held on March 20–21, 1978, in Washington, D.C., under the auspices of the National Science Foundation, the Department of Agriculture, and the National Institutes of Health. A copy of the report of this workshop appears as appendix G to the accompanying environmental impact assessment.

(*9*) The United Kingdom guidelines, also known as the "Williams report," were issued in August 1976. A fairly comprehensive review of the international aspects of recombinant DNA research, including issuance of national guidelines, is contained in the "Report of the Federal Interagency Committee on Recombinant DNA Research: International Activities," November 1977. This is available from the Office of Recombinant DNA Activities, National Institutes of Health, Bethesda, Md. 20014.

(*10*) Under the NIH guidelines, experiments using prokaryotic hosts other than *E. coli* K-12 are severely limited whereas such experiments are proceeding in Europe, especially with *Bacillus subtilis*. Certain other categories of experiments require, according to the NIH guidelines, either P4+EK2 or P3+EK3 containment. Since no EK3 system has as yet been certified and since the first P4 facility has only recently been certified, these experiments were effectively forbidden. The same experiments require significantly lower containment under some European guidelines.

(*11*) Prof. James Watson, in testimony at the December 1977 DAC meeting and in print, has sought repentance for his earlier activities in support of special precautions for recombinant DNA research.

(*12*) The report, "Science Policy Implications of DNA Recombinant Molecule Research," March 1978, of the Subcommittee on Science, Research, and Technology of the Committee on Science and Technology, U.S. House of Representatives, says, "The burden of proof of safety factors should not be borne exclusively by proponents of recombinant DNA research; opponents must assume a corresponding burden."

(*13*) Significant differences exist between prokaryotes and eukaryotes in the ways proteins are synthesized under genic direction, and these account for limitations on the apparent success of many recombinant DNA experiments to date. A major thrust of current recombinant DNA research is in the direction of overcoming these differences. There is every reason to believe that this research will succeed. At my invitation, Dr. Malcolm Martin of NIH has drawn up this brief analysis of the state-of-the-art:

The potential use of recombinant DNA techniques to produce biologically useful reagents is predicated on: (a) the faithful replication of a segment of foreign DNA in a new host cell; (b) the synthesis of messenger RNA (mRNA) complementary to the inserted DNA; and (c) the efficient translation of the mRNA into a polypeptide. In nearly all cases that have been examined to date, DNA, from both eukaryotic and prokaryotic sources, has been amplified in prokaryotic host-vector systems. The fidelity of this entire process (a, b, and c) has been verified in several instances in which prokaryotic DNA segments have been cloned in *E. coli* and resulted in the synthesis of new polypeptides. Thus, in such cases, the informational content contained in the inserted prokaryotic DNA is expressed as evidenced by the synthesis of mRNA and novel proteins.

With few exceptions (some yeast inserts) the expression of eukaryotic DNA in the form of biologically active or biochemically detectable polypeptides in prokaryotes has not been demonstrated using chromosomal DNA inserts and unmodified vectors. In nearly all cases where the system has been rigorously examined, it has been shown that eukaryotic DNA has replicated in *E. coli;* in some instances, RNA complementary to the inserted eukaryotic DNA has been identified.

Messenger RNA synthesis and function in E. coli. The synthesis of messenger RNA (mRNA) in a prokaryote, such as *E. coli,* proceeds in a linear fashion along the DNA template of individual gene segments or groups of related genes. In nearly all cases examined, the mRNA molecules are the faithful colinear transcripts of prokaryotic genetic information and can be used in an unmodified form to direct the synthesis of prokaryotic polypeptides. The informational content of mRNA corresponds directly to the nucleotide sequence of DNA in such systems (i.e., *all* nucleotides present in a prokaryotic gene are transcribed into messenger RNA which, in turn, programs the synthesis of a corresponding protein). Control of this phase of gene expression appears to be solely at the level of RNA synthesis.

In prokaryotes (and eukaryotes), nucleotide sequences preceding the sequences corresponding to the actual genes play a major role in determining (a) whether a given DNA sequence will be transcribed into RNA and (b) whether the RNA so synthesized will efficiently bind to ribosomes, a prerequisite for protein synthesis. For example, certain DNA sequences interact with regions of RNA polymerase and thereby participate in the initiation of RNA synthesis; they are not represented in the final RNA product. DNA sequences specifying binding to ribosomes are physically located between those for initiation of RNA synthesis and sequences encoding the amino acids of a particular protein (the gene) and are also contained in the functional mRNA molecules.

Messenger RNA synthesis and metabolism in eukaryotes. Our understanding of gene regulation and expression in eukaryotic cells has increased markedly during the past 10 months. A common feature of all systems that have been carefully evaluated is that the initial, faithful RNA copy of the DNA is extensively modified to produce a functional form of mRNA. The final mRNA contains only a fraction of the sequences present in the original RNA product. That is to say, portions of large RNA molecules are removed by mechanisms that are, at present, poorly understood and the remaining segments of the primary RNA transcript are then rejoined to one another. In nearly all cases an RNA segment containing a ribosomal binding site is joined to a segment coding for a polypeptide; in addition, larger gene segments are often joined together. This process was first observed in animal virus systems (1, 2) where it was shown that viral mRNA, containing the information for a product which had been previously mapped to a specific locus on the viral genome, was complementary to regions of the viral DNA which were separated by more than a thousand nucleotides.

Support for the concept of complex modification leading to functional mRNA in eukaryotic cells has recently come from recombinant DNA experiments in which chromosomal DNA has been cloned in *E. coli.* When individual cloned eukaryotic genes are carefully analyzed, intervening DNA sequences which interrupt the actual sequence of the gene in chromosomal DNA have been identified. To date, such intragenic DNA has been detected in ovalbumin (3, 4), β globin (5, 6), immunoglobulin (7), and even tRNA genes (8). In one instance it has been clearly shown that the intervening DNA sequences, present in the primary RNA transcript of β globin DNA, are absent in β globin mRNA (5). These mechanisms presumably function in some regulatory fashion to modulate eukaryotic gene activity.

Implications for Recombinant DNA Research

A. The discovery of the existence of complex processes involved in the maturation of mRNA eukaryotic cells and the demonstration of intragenic DNA in several eukaryotic genes suggests that: (1) cloning of chromosomal DNA in *E. coli* DNA (shotgun or purified) will pose little, if any, risk since the maturation mechanisms have never been observed in prokaryotes; and (2) investigators who wish to develop prokaryotic cloning systems for the purpose of synthesizing useful biological products will utilize cDNA copies of functional mRNAs or synthetic DNA with a nucleotide sequence derived from a known amino acid sequence as DNA inserts.

B. Vectors are currently being "engineered" to ensure efficient transcription and translation of DNA inserts. Using slightly different approaches, groups in San Francisco and at Harvard (9–11) are preparing DNA segments which: (1) contain the sequences necessary for interaction with *E. coli* RNA polymerase linked closely to (2) sequences which encode a bacterial ribosome binding site. Such DNA segments can then be added to a prokaryotic cloning vector next to the site into which a foreign DNA will be inserted. This arrangement will facilitate the transcription of the inserted DNA and enable the mRNA so synthesized to bind to bacterial ribosomes. This embellishment has already been used to maximize the expression of a bacteriophage gene and human somatostatin DNA in a plasmid vector system (10, 11).

References for Footnote 13

1. Berget, S. M., Moore, C., and Sharp, P. A. (1977) Proc. Nat. Acad. Sci., USA *74*, 3171–3175.

2. Aloni, Y., Dhar R., Laub, O., Horowitz, M., and Khoury G. (1977). Proc. Nat. Acad. Sci., USA *74*, 3686–3690.

3. Breathnack, R., Mandel, J. L., and Chambon, P. (1977) Nature *270*, 314–319.

4. Weinstock, R., Sweet, R., Weiss, M., Cedar H., and Axel, R. (1978). Proc. Nat. Acad. Sci., USA *75*, 1299–1301.

5. Jeffreys, A. J. and Flavell, R. A. (1977). Cell *12*, 1097–1108.

6. Tilghman, S. M., Tiemeier, D. C., Seidman, J. G., Peterlin, B. M., Sullivan, M., Maizel, J. V., and Leder P. (1978). Proc. Nat. Acad. Sci., USA *75*, 725–759.

7. Tonegawa, S., Brack, C., Hozumi, N., and Schuller R. (1977). Proc. Nat. Acad. Sci., USA *74*, 3518–3522.

8. Valenzuela, P., Venegas, A., Weinberg, F., Bishop, R., and Rutter, W. (1978). Proc. Nat. Acad. Sci., USA *75*, 190–194.

9. Backman, K., Ptashne, M., and Gilbert, W. (1976). Proc. Nat. Acad. Sci., USA *73*, 4174–4178.

10. Itakura, K., Hirose, T., Crea, R., Riggs, A., Heyneker H., Bolivar, F. and Boyer, H. (1977). Science *198*, 1056–1063.

11. Backman, K., and Ptashne, M. (1978). Cell *13*, 65–71.

(*14*) Prohibition (i) in the original guidelines forbids experiments with "oncogenic viruses classified by NCI as moderate risk." The absence of evidence that use of these viruses will lead to formation of agents harmful to man and the potential for obtaining useful new knowledge, relevant to carcinogenesis in particular, and genetics in general, supports the removal of the prohibition. The removal of this prohibition was proposed in the report of the Working Group on Viruses which met on Apr. 6–8, 1978, and endorsed by the RAC at its Apr. 27–28, 1978, meeting. The reasoning behind this is that recombinant DNA experiments with pieces of these viruses cloned in *E. coli* K-12 pose no more risk, and actually appear to pose clearly less risk, than work with the whole infectious virus itself. Since NCI recommends that work with these whole viruses not be prohibited, but rather be performed under containment conditions similar to P3, there is no scientific reason to prohibit recombinant DNA work with these viruses. The "flexibility" clause refers to power of the Director, NIH, to waive prohibitions when the public interest may be served by such action.

(*15*) The Guidelines say at the beginning of Pt. III, "Given below are containment guidelines for permissible experiments. Changes in these levels for specific experiments (or the assignment of levels to experiments not explicitly considered in this section) may be expressly approved by the Director, NIH, on the recommendation of the Recombinant DNA Advisory Committee." Insertion of such language into the guidelines was recommended by the RAC at its Apr. 27-28, 1978, meeting. It recognizes that the classification of experiments given in Pt. III will necessarily be imperfect, as investigators in the future devise new ways to conduct recombinant DNA experiments not currently foreseen and therefore not explicitly considered in the guidelines. Also, new data may become available showing that certain particular experiments currently assigned a particular containment level are, indeed, clearly more (or less) safe than envisioned at this time.

(*16*) See App. C to the guidelines and Part IV of this "Decision Document."

(*17*) In May 1977, a resident of Frederick, Md., brought suit in the U.S. District Court for the District of Columbia to enjoin a proposed risk-assessment experiment which was about to be undertaken in a maximum containment facility (P4) located at the Frederick Cancer Research Center. (*Mack* v. *Califano*, Civil Action No. 77-0916). On February 23, 1978, the court issued a decision refusing to grant the injuction. In so doing, the court observed that the environmental impact statement on the original guidelines constituted a "hard look" at recombinant DNA research performed in accordance with the guidelines. The court further noted that compliance with the guidelines, it appeared, would insure that no recombinant DNA molecules would escape from the carefully controlled laboratory to the environment, and that the guidelines "represent an effort by many scientists to evaluate the hazards and provide safe methods for their control."

The plaintiff appealed (Appeal No. 78-1156), and on Mar. 8, 1978, the Court of Appeals for the District of Columbia upheld the district court decision.

Appendix A-1

Table: Comparison of Containment Levels

The following table compares the containment levels for all permissible types of recombinant DNA experiments. It designates the levels under the Guidelines now in effect (since June 1976), under those proposed by the RAC (September 1977), and under the present NIH-proposed revisions.

A dash (—) indicates that the category is not classified in the edition of the Guidelines under which the dash appears.

It should be stressed that the table is not definitive, since the containment levels of the Guidelines have been redefined and other requirements modified.

Appendix A

COMPARISON OF THE CONTAINMENT LEVELS OF THE 1976 GUIDELINES AND THE PROPOSED REVISED GUIDELINES

INSERTED DNA	CONTAINMENT LEVELS					
	JUNE 1976 GUIDELINES		RAC PROPOSED REVISED GUIDELINES		NIH DIRECTOR'S REVISED GUIDELINES	
	PHYSICAL	BIOLOGICAL	PHYSICAL	BIOLOGICAL	PHYSICAL	BIOLOGICAL
Shotgun Experiments						
1. Primate	P4 or P3	EK2 EK3	P4 or P3	EK1 EK3	P2 —	EK2[1] —
2. Primate DNA from uninfected cells	—	—	P3	EK2	—	—
3. Embryonic primate	P3	EK2	—	—	—	—
4. Mammals (other than primates)	P3	EK2	P2	EK2	P2	EK2
5. Birds	P3	EK2	P2	EK2	P2	EK2
6. Cold blooded vertebrates	P2	EK2	P3 or P2	EK1 EK2	P2 or P1	EK1 EK2
7. embryonic and germ line	P2	EK1	P2 or P1	EK1 EK2	—	—
8. producing potent toxins	P3	EK2	—	—	—	—
9. producing potent polypetide toxins	—	—	P3	EK2	P3	EK2
Lower eukaryotes						
10. producing toxins, or are pathogens	P3	EK2	—	—	—	—
Shotgun Experiments (Cont.)						
11. producing potent polypeptide toxins, or are pathogens	—	—	P3	EK2	P3	EK2
12. producing non-polypeptide toxins	—	—	P3 or P2	EK1 EK2	P3 or P2	EK1 EK2
13. producing toxins affecting invertebrates or plants.	—	—	P3 or P2	EK1 EK2	P3 or P2	EK1 EK2

INSERTED DNA	CONTAINMENT LEVELS					
	JUNE 1976 GUIDELINES		RAC PROPOSED REVISED GUIDELINES		NIH DIRECTOR'S REVISED GUIDELINES	
	PHYSICAL	BIOLOGICAL	PHYSICAL	BIOLOGICAL	PHYSICAL	BIOLOGICAL
14. remainder of species	P2	EK1	P2 or P1	EK1 EK2	P2 or P1	EK1[2] EK2
15. Plants	P2	EK1	P2 or P1	EK1 EK2	P2 or P1	EK1 EK2
16. carrying pathogens or making dangerous products	P3	EK2	—	—	—	—
17. carrying pathogens or making polypeptide toxins	—	—	P3	EK2	—	—
18. making potent polypeptide toxins	—	—	—	—	P3	EK2
19. making non-polypeptide toxins	—	—	P3 or P2	EK1 EK2	P3 or P2	EK1 EK2
Prokaryotes exchanging genetic information with E. coli						
20. Class 1 CDC agents	P1	EK1	—	—	P1	EK1[3]
21. Class 2 CDC agents	P2 or P2	EK1 EK2	—	—	P2	EK1[3]
22. Prokaryotes not exchanging genetic information with E. coli	P3 or P2	EK1 EK2	See Prokaryotic DNA recombinants (Categories 24-26)		P2 or P1	EK1 EK2
23. if pathogenic species	P3	EK2			P3	EK2
Prokaryotic DNA recombinants						
24. source of DNA is extensively characterized	—	—	P2 or P1	EK1 EK2	—	—
25. source of DNA is not extensively characterized	—	—	P2 or P3	EK2 EK1	—	—
26. source of DNA is pathogenic species	—	—	P3	EK2	—	—
Characterized clones						
27. Species that exchange information with E. coli	P1	EK1	—	—	See Characterized clones (Category 50)	
28. Species that do not exchange with E. coli	P2	EK1	See Characterized clones (Category 50)		See Characterized clones (Category 50)	
29. Purified DNA other than Plasmids, Bacteriophages, and Other Viruses:	One step reduction in physical or biological containment from that used in corresponding shotgun experiments		Same		Same, except for primate DNA (See Footnote #1)	
Plasmids, Bacteriophages and Other Viruses						
30. Animal viruses	P4 or P3	EK2 EK3	See Categories 32-41		*	
31. when clones shown to be free of harmful regions	P3	EK2	See Categories 32-41		*	
Viruses of Warm-Blooded Vertebrates						
32. DNA viruses + transcripts of retrovirus genomes	—	—	P4 or P3	EK1 EK3	*	

Document 12.5

September 7, 1978
New Scientist (**79**:672)

Recombinant DNA as a political pawn

Molecular biologists across the US will be watching Washington DC with anxious eyes during the next month as Congress decides finally whether to legislate on genetic engineering

Dr Roger Lewin
reports from Washington DC

The US Congress reconvenes this week after its short summer break, and it has just four short weeks in which to decide whether or not to pass a law governing recombinant DNA (genetic engineering) research. Both the House of Representatives and the Senate have bills before them which—if they so choose—they could bring to the floor for debate and thence to the statute book (given that certain little differences between the bills are ironed out on the way). The pathway to this apparently simple position over recombinant DNA legislation is strewn with argument and counter-argument, advice and counter-advice, and a tangled wreckage of numerous previous bills that have been proposed (often with unseemly haste), championed (frequently with unbridled enthusiam) and finally unceremoniously scrapped. The position on Capitol Hill is, however, anything but simple, with the unmistakable sound of astute lobbying and skilful politicking echoing endlessly around every facet of this much-troubled issue.

Whether or not Congress passes a bill before the end of the session depends not so much on the scientific and technical questions involved, or indeed on the social issues that arise immediately from them, but rather on how strong a political pawn it can be in the intricate power game that is played out on Capitol Hill. James Watson, who in 1974 was one of the small group of scientists that formally suggested a possible hazard in the new research, is so exasperated at the way recombinant DNA has become entangled in the political process that he has said that "The question now is, what is the best way out of this political mess?". Political expediency may, however, come to Watson's aid, for 1978 is election year for many members of Congress, and as genetic engineering has apparently lost the vote-pulling power it once seemed to have, there has been a distinct diminution in congressional commitment to passing recombinant DNA legislation. If the existing bills are not brought up for debate before 7 October, they will automatically expire with the end of the 95th Congress.

Supposing October comes and goes without the birth of a new law to control recombinant DNA research—what then? There are three possibilities. The first—and least likely—is that Congress will take up the issue once again in the new session. But, with the debris of close on a dozen bills littering both houses, and with the mountains of testimony that have resulted from nine separate hearings in four different committees, there is unlikely to be much enthusiasm for beginning it all once more. Unless, of course, there is some kind of hazardous outbreak resulting from the research! The second possibility is that appropriate federal agencies, such as the Federal Drugs Administration (FDA) and the Occupational Safety and Health Administration (OSHA), will be asked by the executive branch of government to extend their existing powers so as to encompass recombinant DNA activities. And the last is that genetic engineers will ply their trade unencumbered by the law, provided they adhere voluntarily to the guidelines established by the National Institutes of Health.

The recent saga over containing the potential hazards inherent in genetic engineering (such as the possibility of accidentally creating dangerous new organisms) has been dominated by two main points. The first is how to ensure that *all* research—both in universities and industry—is covered adequately by a regulatory umbrella. And the second concerns the degree to which local communities can have their own say in how research in their area is controlled. Right from the very beginning of discussions in Congress, it has been this last point that has divided the two houses. The Senate subcommittee on health, chaired and inspired on this issue by Edward Kennedy, has consistently favoured extensive public intervention: draft bills have had direct public participation built into the proposals, and there has always been provision for local government to enact its own legislation if it so chose. By contrast, the principal House bills, which came from Paul Rogers's health subcommittee, were much more appealing to the scientific community: power was more centralised, the public was much more at arm's length, and there would be federal preemption (local laws would not be allowed).

Many scientists were concerned that, given the (apparent) inevitability of legislation, the law should include federal preemption so that researchers would be protected from possible idiosyncratic actions of local legislators: the success of Alfred Vellucci, mayor of Cambridge, Massachusetts, in halting for a while recombinant DNA research at two of the country's most prestigious universities—Harvard and Massachusetts Institute of Technology—set pulses racing in many a laboratory around the country. Kennedy, who happens to be the senator for Massachusetts, is of course a self-professed champion of participatory democracy, so it is not surprising that he wishes the public to have its say on this issue. He is also ever anxious to put scientists in their place.

Simple legislation favoured

During the 18 months since the first recombinant DNA bills were introduced into Congress the repeated revisions have made the provisions less and less draconian, the result, no doubt, of intense lobbying by the American Society of Microbiologists and by individual scientists themselves. The current bill before the House, for instance, is merely an interim measure giving the force of law to the NIH guidelines and proposing a commission to review the supposed hazard of the research. This bill, which was sponsored by Paul Rogers and his main committee chairman William Staggers, was introduced on the last day of February this year, just 10 weeks after Donald Fredrickson, director of NIH, testified to an NIH hearing on the guidelines, saying, "My own belief is that it would be to the maximum advantage to the country for a very simple legislative package to be passed extending the existing guidelines to everyone." Shortly after the Rogers and Staggers bill appeared, Kennedy followed suit with his—very similar—version which, of course, excluded federal preemption. Of the two, Fredrickson and Joseph Califano, secretary of the Department of Health Education and Welfare (HEW), favour the House bill.

Given Kennedy's open enthusiasm for throwing a precisely tailored legislative cloak over this particular branch of the scientific community, it came as something of a surprise when in May he declared that, after all, he considered new legislation to be unnecessary. A Kennedy aide told me that "The senator has shifted his position

This article first appeared in *New Scientist*, London, the weekly review of science and technology.

because the bulk of evidence collected during the past two years suggests that the research is not as risky as was once perceived." He also said that Kennedy balks at the idea of setting a precedent for legislative intervention in basic scientific research. Referring to the problem of federal preemption, the aide said that "Kennedy particularly does not want to impose legislation 'because the science is risky', and then not allow people exposed to the risk to have any say in how it should be handled."

It seems likely that if Kennedy could be sure of getting his bill through the Senate he would have pushed on with it. But, although politically he is very powerful, philosophically he is pretty much on his own over blocking federal preemption. So, rather than go ahead and be defeated, he has dropped the bill altogether. There is, however, a possibility that he will switch positions yet again: he could take up the bill once more, and even be persuaded to accept some degree of preemption—if the political climate requires it.

Bills pass through Congress in parallel: one through the Senate and the other through the House of Representatives. Eventually, however, a single bill has to be produced to become law, and this means that differences between the two versions have to be settled in conference. Such disparities may be removed by rational debate, but horse trading—one point for another, one bill against another—is not unknown. Currently, the House and Senate health subcommittees have 12 going to conference between them, and on at least four there are substantial differences.

It is quite possible, therefore, that Rogers could encourage Kennedy to give some ground on the recombinant DNA bill (supposing it got to conference) in exchange for concessions on one of the other bills. Possible, but not very likely. Rogers retires from political life this session and, despite the energy he has invested in the recombinant DNA debate, it is certainly not his highest priority. The House bill, having passed through the health subcommittee, the Committee on Interstate and Foreign Commerce, and the Committee on Science and Technology, is currently with the Rules Committee, an influential body that decides when—and if—a bill will reach the floor of the House for debate and a vote. If the committee gives the bill a ruling within the next two weeks, there will still be time for the Senate to act before the end of the session.

In dropping his recombinant DNA bill in May, Kennedy did not abandon interest in the issue. He is proposing that, instead of enacting new legislation, existing law—under the FDA, OSHA and HEW—could be invoked. One regulation that has been much touted as a possible provision for controlling genetic engineering is Section 361 of the Public Health Service Act: this allows the secretary of HEW to "make and enforce such regulations as in his judgement are necessary to prevent the introduction, transmission, or spread of communicable disease. . . ." Kennedy was one of six senators who signed a letter on 1 June to Joseph Califano which suggested that existing regulations, including Section 361, should be invoked to cover recombinant DNA research. They gave three reasons for their suggestion: first, "the need for new legislation is less clear than it was one year ago when the initial bills were introduced"; second, "the existing regulatory deficiencies relating to Federally supported research can be remedied by executive action"; and third, "the heavy legislative schedule may preclude action in this session of Congress". This last point is a revealing comment on the political priority now accorded to recombinant DNA legislation by the senators.

Califano's reply, which has only just been dispatched to the senators, firmly rejects the manoeuvre: it states that "The department does not intend to invoke existing statutory authorities to regulate DNA activities at this time and continues to support legislation if it embodies the moderate approach of [the House bill]". Rogers and his aides can take comfort from the clear endorsement of their bill. If the House were to pass Rogers's bill there would clearly be a great deal of pressure from the Administration on the Senate to follow suit with a very similar piece of legislation.

It is also clear from Califano's letter, however, that enacting new legislation is not HEW's *first* choice alternative. Although in the last paragraph of the letter he restates his liking for Rogers's bill—"Should the Senate choose to act, I would strongly urge adoption of an approach similar to [the House bill]"—he also says "We are pleased with the progress made in the absence of legislation and believe that invocation of existing authorities, however appropriate, would not contribute materially to our objectives. . ." The implication is that, supposing everyone agrees with the notion, HEW would be happy to continue with a voluntary system based on the NIH guidelines. Fredrickson, Califano's chief advisor on this matter, has said publically that using existing regulations would be the worst of all possible alternatives.

Kennedy and his five co-signatories are not the only people who are currently raising the suggestion of employing existing laws. For instance, on 4 May Harrison Schmitt, a member of the Senate Science, Technology and Space Subcommittee, wrote an exploratory letter to Califano along these lines. The secretary's reply, again, was to reject the notion, saying that the Federal Interagency Committee, which reported in March last year, took the view that "in order for section 361 to be applicable to recombinant DNA research, there would have to be a reasonable basis for concluding that the products of such research cause or may cause human disease". There is no such basis, Califano says. Indeed, in its report, the interagency committee concludes that "no single legal authority or combination of authorities currently exist that would clearly reach all research and other uses of recombinant DNA techniques and meet all requirements." And yet Kennedy and others are attempting to resurrect the idea.

Perhaps Kennedy is simply using the suggestion as a way of holding the stage in the Senate. If he were to lose interest totally it would be possible for a House-type bill to be pushed through by someone else, and that would mean that he would have lost on the preemption issue. The invocation of existing regulations, or even the continuation of a voluntary system under the NIH guidelines, would allow local governments to enact laws if they chose to. And this is Kennedy's prime aim. So, all Kennedy has to do is discourage movement on the House bill and he's achieved his goal. There is a strong feeling that he no longer regards the possibility of hazard as either important or real.

A last flurry of activity on the legislation front is due to come from Senator Adlai Stevenson's Science, Technology and Space Subcommittee. In November last year the committee held three days of oversight hearings, taking testimony from more than 20 witnesses. A Stevenson aide told me that the hearings report, due now, recommends legislation along the lines of the Rogers's bill, but with a somewhat toned down preemption provision so as to achieve a compromise between the two houses of Congress. The report argues that there is a need for more than voluntary compliance with guidelines, particularly in view of the probable expansion in commercial exploitation. Despite the report's recommendation, Stevenson's office is not optimistic about the chances of legislation this session. And nor is Gilbert Omenn, a staff member of the President's science adviser's office: "The prospects for legislation are negligible, and the need for legislation is negligible", he says. He is content that a voluntary system should continue, adding that "there is a better understanding of the whole issue now"—he clearly feels that prospects of the scientific community being threatened by rampaging Mayor Velluccis are also negligible now!

During the past two years of intense debate over recombinant DNA Fredrickson and his colleagues at NIH must frequently have wished that the whole problem would just go away—one NIH official told me wearily that the issue has consumed as much as 70 per cent of the organisation's energies. The problem has of course not conveniently vanished, but it has become greatly muted during that period. A continued absence of any demonstrated hazard, together with *some* evidence—though not as much as is frequently claimed—that the bacterial host for experiments (*Escherichia coli* K12) is unlikely to survive outside the laboratory long enough to be dangerous, has convinced many people that the research is after all not the science fiction monster that was once feared. Some critics claim that the diminution of concern over possible hazards has been carefully orchestrated by researchers who want to get on with their work, and that the potential danger remains just what it was two years ago.

Be that as it may, the upshot of the present mood is that at the end of July the NIH published its latest set of revised guidelines which have significantly eased restrictions on a number of types of experiments. In the preamble to the revisions Frederickson goes so far as to say that "there is a growing sentiment that the burden of proof is shifting toward those who would restrict recombinant DNA research". And to further make the point about a significant change in attitude in the Administration, the name of the local groups designated to oversee the way research is carried out has been switched through 180 degrees from bio*hazard* committees to bio*safety* committees.

But the most significant alteration of all in the guidelines is extension of their constituency from just federally-funded researchers to *all* researchers, including those in private industry, a shift that must be seen as a key ploy to avoiding either new legislation or invocation of existing regulation. Califano referred to the (then) impending revision in his recent letter to Senator Schmitt, and said that it had come about partly as a result of "the absence of legislation". The new guidelines provide for a national registry of all recombinant DNA research, makes available NIH expertise and advice to private industry and—a vital issue—guarantees the confidentiality of commercially sensitive information. Sanctions against federally-funded researchers if they break the rules will be a loss of their support; and industry will be able to patent and market recombinant DNA products only if it can demonstrate that it has followed the guidelines throughout its activities. The whole system is voluntary, but with incentives for compliance both in academic and industrial laboratories.

Paul Berg, who chaired the National Academy of Sciences group that first spelled out the potential dangers of recombinant DNA research in 1974, says "I think things are moving in the right direction". He regards the hazard of the new science as much less than he did four years ago, "partly on the basis of direct evidence, and partly through a clearer perception of what is involved in terms of mechanisms of molecular biology". By contrast MIT biologist Jon King, a consistent proponent of caution in this area, is not so happy: "There simply isn't the evidence to enable us to change our minds yet about the potential hazard", he insists. He points out that there will be stiff commercial competition and this will encourage people to cut corners. "History consistently tells us that guidelines are not followed without enforcement", King says; "Think of what happened with asbestos, analine dyes, pesticides and coal mining—there are lots of examples showing that commercial pressure often leads to human illness—unless it's properly controlled." □

Illustration by Barbara Thomas.

Document 12.6

July 1978
Gene (**3**:265)

Invited Editorial

VALEDICTORY BY THE CHAIRMAN OF THE NIH RECOMBINANT DNA MOLECULE PROGRAM ADVISORY COMMITTEE*

I am taking a Chairman's prerogative to invade the printed agenda*. I should like at this time to share with you the reasons why I have felt impelled to resign my chairmanship of this Committee. Shortly after our last meeting of November 1977, I asked the Director, NIH, to accept my resignation and find a replacement for this chairmanship. He asked me to assist in the selection of a new Chairman and I have provided to him the names of candidates from which he is soon to make a choice. I am certain that you will be pleased with the name of my successor, and that the Committee will give the new Chairman the same devotion and industry which it has given to me.

There were, of course, personal reasons for my resignation. I am four years older than I was when I was first appointed, I fatigue more easily, and, as you are all aware, my visual acuity has continued to decrease until I am able to read only a very small fraction of the large amount of paper which passes over my desk in relation to this function. In addition, I have had a growing unhappiness with some of the directions which the recombinant DNA program has taken over the past four years. From my conversations with members of the Committee, I believe that this unhappiness is shared by some of you, and this may be a good opportunity to verbalize this discontent.

Prior to the Asilomar meeting of February 1975, I had had only modest contact with nucleic acids and with genetics. I had worked in the laboratory with lipids, polysaccharides, and proteins, but had never handled any nucleic acids. I had never worked on a genetic problem, and had certainly never engaged in microbiological research. Except for some briefing which I secured from members of the intramural NIH family, I came to Asilomar cold.

It has taken me several years to analyze and unscramble the experience of the Asilomar meeting. I now understand it more fully than I did at the time. It had many elements of a religious revival meeting. I heard several colleagues declaim against sin, I heard others admit to having sinned, and there was a general feeling that we should all go forth and sin no more. The imagery which was presented was surely vivid, but the data were scanty. I recall one scientist presenting information on the difficulty of colonizing the intestinal tract with *Escherichia coli* K-12, but his presentation was given little attention. We were all, in effect, led down to the river to be baptized and we all went willingly. I, for one, left the meeting enthralled. I had never been to a scientific meeting which had so excited me. On my return to Bethesda, I was asked to summarize the events at Asilomar before a meeting of the generally staid NIH Institute Directors and I believe I was able to transfer to them some of my excitement. Over the succeeding months, the Recombinant DNA Molecule Program Advisory Committee met and, by July 1975, it drafted a set of guidelines at Woods Hole, Massachusetts, which I at the time thought to be reasonably satisfactory. They did not conform to my prior notion of guidelines exactly, since they bordered on the encyclopedic. Nonetheless, I felt that we had successfully compromised most of the burning issues over which the Committee was initially strongly divided. When these guidelines were distributed, however, they elicited vigorous and often emotional responses, and among these responses

*Presented at the meeting of the Recombinant DNA Molecule Program Advisory Committee, April 28, 1978, NIH, Bethesda, Maryland.

there was one which I recall vividly. It charged our Committee with having violated the "spirit of Asilomar." At the time this expression did not catch my attention, but on consideration I was struck by the fact that despite the many, many meetings which I had attended at Atlantic City, I had never heard a reference to the "spirit of Atlantic City." This charge, in fact, pinpointed for me the notion that the experience at Asilomar was essentially a spiritual one rather than an intellectual one. It was, in the usual sense, not a scientific meeting at all. Whatever its purpose may have been in the minds of its initiators, a result was to fire the imagination, first, of the newspaper correspondents who were abundantly represented, and then of a substantial segment of the newspaper-reading public.

By December 1975, our Committee, meeting at La Jolla, again assembled a set of guidelines. Whereas up to that time I had insufficient confidence in my own judgment to hold a firm opinion on this issue, and found myself swayed by the views most recently presented, it was about the time of the La Jolla meeting that I began to wonder whether, indeed, any of the postulated hazards of recombinant DNA molecule technology were likely to materialize.

The La Jolla guidelines served as the basis for a discussion at a meeting of the NIH Director's Advisory Committee early in 1976, and this, in turn, was followed in July by the publication of the official NIH guidelines. In this last transformation, something happened which I found disturbing.

The mission of NIH is, I believe, very simply stated. It is to conduct and to support the very best biomedical research that it can find to conduct and support. Similarly, the mission of our Committee and of the guidelines which it drafted was to provide assurance that research in the area of recombinant DNA molecules would be conducted in such a fashion as not to jeopardize the laboratory, the community, or the environment. Both missions, it should be noted, are stated positively. It is the purpose both of NIH and of this Committee to encourage, to promote — not to forbid or to impede. The legal profession represented at the Director's Advisory Committee meeting was critical of the concept of guidelines, which in my judgement are designed to provide *guidance* to the investigator and to those who review his proposal. We were informed that what was needed was regulation, not guidance. This was exemplified by the recommendation that our instruction, written largely in the subjunctive mood (the investigator *should* . . .) be replaced by the more peremptory language of regulations (the investigator *shall* . . .). I recall arguing against such change in vain.

My reasons were very simple. It is my interpretation of the history of science and indeed of all culture that *regulation is antithetical to creativity*, and creativity is the most important component of scientific advance. From this, it follows that the best regulation for the flowering of science is the least regulation — that is, the least regulation compatible with the needs of society. Furthermore, I feared and my fears were, I think, justified that regulation might lead to legislation with a specification of sanctions, i.e., punishment, for those who were in violation of the regulations. Whereas the so-called regulatory agencies of Government must from time to time adopt a punitive posture, this is, I believe, a poor posture for a research agency such as the National Institutes of Health.

Against what hazards were we proposing to draft regulations? With the passage of time, the hazards that had been pictured at Asilomar seemed to recede. Whereas a great number of positive and useful scientific results are being published based upon the technology of recombinant DNA molecules, to the best of my knowledge no adverse results have been noted. Indeed, I believe that there is at this time not one iota of acceptable evidence, i.e., data publishable in a scientific journal, to indicate that the recombinant DNA molecule technology has ever enhanced the pathogenicity or the toxigenicity of any microorganism. This, of course, does not mean that it never will do so, but it does cause one to wonder whether all of the present fuss is truly justified. It places the hazards in this area in the same category as those in many other areas for which we have no positive evidence. To clarify this point, let me offer you an analogy. Ever since the Middle Ages, it has been suspected that the ghosts of those who died by suicide are more menacing than ghosts in general. This anxiety, once implanted in the minds of the people, led to some interesting containment practices. The bodies of victims of suicide were excluded from traditional burial places, lest their ghosts pollute or otherwise disturb the more peaceful ghosts of those who died of natural causes. They were doomed to be buried in the crossroads, and in order to ensure that the ghosts not escape from the tomb, a stake was driven through the body of the victim into the underlying soil, thus pinning the ghost into its grave. This containment practice continued for many centuries and was ultimately abandoned only in the 18th century. Experience since that time has justified the conclusion — that the hazard which had earlier been postulated was either of very small magnitude or possibly nonexistent. We may yet prove to be wrong about the safety of unpinning the ghosts of suicide victims, but I should be surprised if this were so.

How long do we wait, in the absence of any positive evidence, before we decide that the hazards in a particular area of research are at a socially acceptable level? To this question I have no specific answer. Soon we may come to the conclusion that the manipulations of recombinant DNA technology do not of them selves add significantly to the dangers inherent in the conduct of microbiological research. Then we can replace our complex and, I repeat, encyclopedic guidelines by a very simple statement. This might take the following form: "The conditions of containment appropriate for any recombinant DNA experiment are those which are dictated by the most virulent or dangerous organism entering into that experiment." Is anything more really required?

I hope that none of you will construe any of my critical remarks as being personally directed. They are not. I have thoroughly enjoyed and been stimulated by my contacts with the many members of the Committee. I hope that I have established enduring friendships with many of you, and I shall certainly follow your further deliberations with great interest and concern. I should like particularly to express my appreciation to the several members of the NIH staff who have worked so hard and so loyally to keep this project afloat: Dr. Leon Jacobs who, from the beginning, has served as Co-Chairman of this Committee, Dr. Bernard Talbot, who has worked enormously hard and valiantly, Dr. William Gartland, Director of the Office of Recombinant DNA Activities, his small but energetic staff — Dr. Kamely and Dr. Goldberg. Then, there is Ms. Betty Butler, who not only made certain that all the paper flowed in the right directions but also nursed us through our several tortured meetings. To work with all of these people has been a very rewarding experience.

I wish you well in your future meetings.

DeWitt Stetten, Jr.

Deputy Director for Science
National Institutes of Health
Bethesda, MD 20014 (U.S.A.)

Trends in Biochemical Sciences

Dangers of regulating the recombinant DNA technique

Waclaw Szybalski

Bureaucratic regulation of the recombinant DNA technique is not justified because no practical risks exist and the benefits are very great. The true dangers of imposed regulations include: (1) creating novel or aggravating existing hazards; (2) political dangers; (3) depleting and misdirecting valuable resources; (4) fostering bureaucracy; (5) denying or delaying benefits to society; and (6) misdirecting environmental concerns and other beneficial measures.

The recombinant DNA technique, which permits inserting fragments of practically any DNA into specially developed host-vector systems, offers abundant scientific and practical benefits, several of which have already been realized. However, the very success of this method has created apprehensions based on intuitive feelings that benefits must also be accompanied by risks.

As with some other important new scientific discoveries, the benefits and the hypothetical risks have been widely debated during the past few years, and various highly unlikely but dramatic scenarios of biological hazards that might affect man and the environment have been contrived. To counter these imaginary risks several kinds of regulations have been proposed, including various forms of guidelines, differing from country to country, administrative regulations, ordinances and even federal legislation.

This regulatory zeal is based on the erroneous assumption that regulation of risks, even imaginary and disproven ones, could only benefit laboratory workers and society in general. However, little thought has been given to the possibility that regulations might create novel and unnecessary risks of their own, including introducing cumbersome and harmful laboratory procedures, all in the name of 'safety'.

This contribution (based on earlier articles [1,2] containing extensive reviews of the pertinent literature) will consider the dangers and detriments created by imposed regulations, as exemplified mainly by the official guidelines and regulations of recombinant DNA activities developed in the United States.

Who knows the hidden dangers that await us all?

Creating novel hazards and aggravating existing ones

The physical P1 to P4 containments prescribed by the N.I.H. Guidelines [3] require progressively more elaborate laboratory equipment and procedures to ward off the speculative dangers of recombinant DNA. Both the equipment and the cumbersome procedures conceivably might create new dangers to laboratory personnel. For instance, even the simplest P1 containment requires daily decontamination of the table tops with germicidal agents [3], which in turn pollute the surroundings and might endanger the health of personnel through chronic exposure to toxic chemicals given off by large laboratory surfaces. Also, the unnecessary increase in the scale and frequency of autoclaving does magnify the risk of burns or even explosions besides wasting energy and effort. Many other examples of novel dangers, often created by adapting biochemical techniques to confined spaces within biocontained laboratories, could be cited.

Political dangers

The political dangers include among others: (1) loss of constitutional freedoms; (2) designing laws that invite disrespect; (3) misleading the public; and (4) politicizing science and creating an atmosphere of mistrust and witch hunting.

Freedom of inquiry is guaranteed under the First Amendment to the U.S. Constitution, as interpreted by the U.S. Supreme Court [4]. This concept, that no scientific ideas or notions should be restricted out of fear of the ideas themselves, has been developing for several centuries. As early as 1894, the Board of Regents of the University of Wisconsin espoused that 'whatever may be the limitations which trammel inquiry elsewhere, we believe that the Great State of Wisconsin should ever encourage that continual and fearless sifting and winnowing by which alone the truth can be found' [5]. It was also stressed that in science we need to accept no constraints other than those absolutely essential, well documented, 'serious and decisive', since 'the burden of proof in favor of legislation lies with the public and those who would do the regulating' and because 'if one introduces the principle of regulating research, more will follow' [6].

Laws that invite disrespect are self-defeating, and regulations designed to avert non-existent dangers would not be obeyed unless there was strong enforcement bordering on intimidation. History is replete with examples of unreasonable laws and it would be tragic to create new ones.

Misleading the public and creating unfounded concerns should certainly be avoided. Official adoption of safety regulations indicates to the layman that real danger exists, even if there is uniform agreement among knowledgeable specialists that there are no practical risks in random cloning of any DNA in *E. coli* K-12 host-vector systems [7–9]. It is very difficult for the layman 'to obtain a sense of where the bulk of informed opinion really lies, to distinguish statements of facts from assertions and indeed from science fiction' [10]. And some scientists, even prominent ones, did not help by making irresponsible and nonsensical statements as, e.g. implying that bacteria developed to bio-oxidize oil spills on water surfaces could also destroy oil deep in storage tanks or oil wells [11], where oxygen is absent and bacteria cannot grow because of the deficiency of water and nutrients [12].

Politicizing of science represents another danger, since it leads to creation of a new kind of adversary relationship between scientists and the general public. It fosters a system of informers and leads to accusations, mistrust and adverse publicity in the name of hypothetical risks or political beliefs, with examples already accumulating [1,2,13,14]. Historical precedence includes the Holy Inquisition, witch-hunting, the tribulations of Copernicus and Galileo Galilei and, more recently, the infamous Scopes trial and the Lysenko affair.

Depleting and misdirecting valuable human and fiscal resources

The various regulatory and administrative activities connected with recombinant DNA technique result in loss of the productive time of affected scientists, who now have to fill out a miriad of forms, defend the safety of their research, and participate in unnecessary committee meetings and various bureaucratic activities [1,15]. This could lead to another more long-term danger, namely that young scientists would tend to avoid research areas fraught with such stifling regulations, and their intellectual contribution to this important and beneficial area of human endeavour would be permanently lost.

Balkanization of research is another danger which would be caused by a variety of regulations, differing from country to country and even city to city [14]. This would cause migration of scientists, as is already happening, and the so-called 'brain drain' from those countries with unreasonably restrictive regulations, with a resulting decline in the scientific research base.

Since fiscal resources are limited, regulations would result in diminished financial support for research. Many millions of dollars would be diverted to administering and enforcing the regulations (see estimates in [16,17]) and to the construction of unnecessary but expensive physical containment facilities.

Fostering bureaucracy and policing basic scientific research

The regulation of research employing the recombinant DNA technique might add several new dimensions to the bureaucratic process, namely licensing, evaluation and approval of each experiment, inspections, seizures and destruction of recombinant DNA, court procedures, fines and general policing of daily laboratory procedures [17]. Obviously this is not only distasteful but also unworkable for the very sensitive creative process. There are true dangers of not so surrealistic scenarios, in which eager inspectors are trying to identify and control the imaginary risks while enforcing the regulations by 'seizing and destroying' [17] the 'dangerous recombinant DNA' and reporting the lax researchers to the judicial authorities. Bureaucracy has a tendency to expand rapidly. For instance, the simple requirement for a scientist to state in writing that his research will conform to the official guidelines has become an elaborate procedure of many months duration, even for the simplest, totally safe P1-EK1 type of experiment (see Table I), where *E. coli* or coliphage genes are cloned in an *E. coli* K-12 host-vector system. According to a current proposal [18], this kind of experiment should not even be included under the guidelines. The M.U.A. application procedure for the cloning of heterologous and especially eukaryotic DNA often takes many more months or years and the M.U.A. frequently is not approved for trivial or procedural reasons.

Denying or delaying benefits of research to society

Application of the recombinant DNA technique is already speeding up progress in basic biological research, has resulted in development of procedures for synthesis of a few human polypeptide hormones, and has the potential to supply many substantial medical, agricultural, technological, ecological and other benefits [19]. Denying or delaying any of these would exact a high price in lives needlessly shortened or lost and in the prolongation of human misery. Even the present guidelines [3] prohibit or make excessively cumbersome the current highly beneficial research directed toward developing life-saving proteins, including safer vaccines, for example, for Herpes II virus [2], specific antibodies, many human proteins and polypeptide hormones [19]. Again, there is infamous historical precedence in the attempts by several local communities to interfere with the research carried on by Louis Pasteur, which might have totally blocked or delayed the development of several life-saving vaccines [20].

Misdirecting legitimate environmental concerns and other beneficial measures

The uncontrolled increase in the human population and in industrial activities seriously threatens our environment and hence the quality of life. Effective measures to protect our environment are of the utmost importance. The recombinant DNA technique would be of considerable value in solving certain environmental problems; microorganisms capable of degrading specific pollutants could be constructed and 'dirty' technology could be converted into 'clean' microbiological and enzymatic processes. In the latter context, any improvement in the efficiency of biological nitrogen fixation would decrease our reliance on the energy-wasting chemical synthesis of nitrogen fertilizers, which contribute to the pollution and eutrophication of lakes and other waters.

It is, therefore, erroneous and regrettable that several respectable environmental movements are misdirecting their funds and efforts toward opposing the beneficial research employing the recombinant DNA technique and advocating restrictive regulations [2,13,14,21,22]. This is an example of how the lay public and environmental groups were misled by erroneous scenarios implying that *E. coli* strains which carry recombinant DNA could somehow pollute the environment and change the ecological balance. In fact, *E. coli* strains in general are not endowed with the capacity to propagate effectively in nature outside the gut and a few other body cavities, and *E. coli* K-12 used in recombinant DNA research has lost the capacity to propagate even in the gut and requires specially adapted laboratory media [7,8,19]. An important additional fact, which is often overlooked, is that natural selection plays a crucial role in epidemiology and the environmental impact of microorganisms, since even if one assumes that *E. coli* K-12 could inadvertently acquire foreign DNA able to produce a 'hazardous product', it would not become capable of spreading through the natural environment, which is already saturated with well-adapted microorganisms [23,24]. Only if *E. coli* K-12 concomitantly acquires a whole spectrum of additional properties that make it as well suited for a given environment as the pre-existing well-adapted organisms, would it have a chance to establish itself, but only for a limited period, until its mutants (which would probably dispose of some adaptively useless recombinant DNA, for instance, that coding for some 'hazardous product') or variants of pre-existing microorganisms would evolve into still better adapted organisms and overgrow it. This well known evolutionary phenomenon of *periodic selection* [25], when new better-adapted variants created within the bulk of the wild-type population overgrow and thus replace this population and all former spontaneous or added variants (including recombinant DNA mutants), is at odds with the scenarios of epidemiological and ecological dangers of recombinant DNA.

Conclusions

The early proposals to regulate the recombinant DNA activities were motivated by well-meant and honest epidemiological and environmental concerns, which were only intuitive and not based on knowledge of the epidemiology and ecology of *E. coli* K-12 and the host-vector systems. The possibilities of some risks constitute a typical working hypothesis which should be either proven or disproven by rigorous application of already known facts and well-designed experiments. However, the mass media, the public and the regulatory bodies have misconstrued this working hypothesis as proof that true dangers do exist and that some action is required to protect the public. This regulatory zeal was probably also well meant, although the proponents have unfortunately forgotten or not realized that the usual regulatory procedures are not compatible with the delicate creative process of basic scientific research. We are now beginning to realize that, whereas the dangers of regulations are real and might inflict serious long-term harm on society, the risks of recombinant DNA activities are non-existent from the practical point of view and even hard to define. Thus, this example of unjustifiable regulations shows how 'good intentions' of zealous proponents can result in much harm and little good.

There is hope, however, that reason will prevail and on the one hand the present N.I.H. Guidelines will become greatly relaxed or eliminated, while on the other hand no legislation will materialize. There are indications both predictions might be true. The most recent July 1978 draft of

TABLE I

Example of bureaucratic procedure to obtain approval of a Memorandum of Understanding and Agreement (M.U.A.)[1] for a P1-EK1 experiment[2]

Date	Actions
4 Jan.	Telephone inquiry about the procedure for securing M.U.A. directed to the N.I.H. Office of Recombinant DNA Activities (O.R.D.A.). Instructed to contact the N.I.H. Granting Agency from which grant was obtained.
5 Jan.	Telephone inquiry to the N.I.H. Granting Agency. First advised to contact O.R.D.A. but, after explaining this was already done (see above), asked to call again when a more knowledgeable person would be present.
6 Jan.	Another telephone inquiry, but no information obtained.
9 Jan.	Another telephone inquiry results in a promise that instructions and forms will be obtained from O.R.D.A. and then mailed.
2 Feb.	Received Recombinant DNA Research Guidelines of 7 July 1976 and N.I.H. Guide for Grants and Contracts of 17 October 1977.
13 Feb.	M.U.A. is prepared, typed, duplicated and signed.
16 Feb.	Received from N.I.H. new N.I.H. Guide of 15 Feb. 1978.
20 Feb.	M.U.A. is modified and retyped according to new 15 Feb. N.I.H. Guide, and sent to the Institutional Biosafety Committee.
25 Feb.	M.U.A. is returned as unsatisfactory because it did not contain '*specific* statements as to what containment practices would be followed' according to the Institutional Biological Safety Committee Memorandum of 3 November 1976, and because it stated 'I agree to abide by the provisions of the current N.I.H. Guidelines (including shipping) as long as the new Guidelines are not approved and as far as they do not violate the common sense principles of good, safe and efficient laboratory practices'.
28 Feb.	New M.U.A. is prepared, typed, duplicated and signed, as to conform with both the N.I.H. Guide of 15 Feb. 1978 and the Memorandum of 3 November 1976 (see above, 25 Feb. 1978). It had to contain the following statement 'I am familiar with and agree to abide by the provisions of the current N.I.H. Guidelines, the notice contained in the N.I.H. Guide for Grants and Contracts of 15 February 1978, and other specific N.I.H. instructions pertaining to the proposed subject. I agree to comply with the requirements specified by the Guide pertaining to shipment and transfer of recombinant DNA materials'.
1 Mar.	New M.U.A. is sent to the Institutional Biosafety Committee. A copy of the N.I.H. Guide of 15 Feb. 1978 is included since the Biosafety Committee never received this Guide.
7 Mar.	Institutional Biosafety Committee approved the M.U.A., and signatures of the Chairman of the Committee and the Subcommittee on Recombinant DNA are affixed.
8 Mar.	Signature of the Director of Laboratory is secured.
10 Mar.	Signature of the University Director of Research Administration is secured.
16 Mar.	The M.U.A. and covering letter are sent to the Division of Cancer Research Resources and Centers, N.C.R., N.I.H. In the covering letter it was indicated that 'up to now, about 50 man-hours were spent on this application . . . . . , at a cost of about $600, ...'.
23 Mar.	Received a phone call from the Assistant Program Director, Division of Cancer Research Resources and Centers, asking to prepare a new revised M.U.A., because the number of the N.I.H. grant was not included. After explaining that 'illustration page one' in the N.I.H. Guide of 15 Feb. 1978 did not specify to provide the grant number, and offering to provide the grant number by phone, the N.I.H. official stated that the M.U.A. anyway has to be retyped and resubmitted. A confirming letter specified that 'fresh signatures of the relevant institutional officials will be required' in addition to preparing a new revised M.U.A.
14 Apr.	After calling the Office of Recombinant DNA and obtaining additional instructions, a revised M.U.A. with the grant number is prepared.
17 Apr.	These revised M.U.A. copies are mailed with a covering letter to the Division of Cancer Research Resources, N.I.H.
15 May	Received a letter dated 10 May and an M.U.A. approval from N.I.H. dated 26 April 1978 and valid only for 15 days until 30 June 1978, when the application procedure will probably have to be repeated because of the N.I.H. grant renewal and change in the grant number.

[1] Procedure as specified in N.I.H. Guide for Grants and Contracts 7 (1978) No. 3, pp. 5–11, and in current N.I.H. Guidelines.

[2] Lowest containment experiment; in this specific case the cloning of phage λ DNA in *E. coli* K-12 host-vector systems.

the Guidelines [26] is a great improvement upon the 1976 version [3], although many of its aspects, including the bureaucratic regulations, still leave much to be desired. The federal legislation seems to have become a political football with responsibility for the regulations being passed back and forth between the Senate and the Secretary of H.E.W., with no vote at present, and irreconcilable differences on the pre-emption provision between the House version and Senator Kennedy's draft [27,28].

References

1 Szybalski, W. (1978) in *Biomedical Sciences and Public Responsibility* (Fudenberg, H. H. and Melnick, V. L., eds), pp. 97–141, Plenum Press, New York
2 Szybalski, W. (1978) in *Genetic Engineering* (Boyer, H. W. and Nicosia, S., eds.), pp. 253–275, Elsevier/North Holland, Amsterdam
3 Recombinant DNA Research Guidelines (1976) *Federal Reg.* 41, 27907–27943
4 Grisvold v. Connecticut (1965) *U.S. Supreme Court Rep.* 381, US 479–531 (see p. 482)
5 Report of the Investigating Committee (18 Sept. 1894) in *Papers of the Board of Regents*, University of Wisconsin, Madison, WI, U.S.A.
6 Callahan, D. (1978) in ref. 13, pp. 95–98
7 Gorbach, S. L. (1977) *Recomb. DNA Tech. Bull.* 1, 19–23
8 Risk Assessment of Recombinant DNA Experimentation with *Escherichia coli* K-12. Proceedings from a Workshop Held at Falmouth, MA, 20–21 June, 1977 (1978) *J. Infect. Dis.* 137, 609–714
9 U.S.-EMBO Workshop to Assess Risks for Recombinant DNA Experiments Involving the Genomes of Animal, Plant and Insect Viruses (1978) *Federal Reg.* 43, 13748–13755
10 Handler, P. (1978) in ref. 13, pp. 4–34
11 Wald, G. (1976) in ref. 22, pp. 239–245 (see p. 244)
12 Burris, R. H. and Szybalski, W. (1978) in ref. 13, p. 385
13 Regulation of Recombinant DNA Research (1978) *Hearing before the Subcommittee on Science, Technology and Space of the Committee on Commerce, Science and Transportation, U.S. Senate, 95th Congress*, 2, 8 and 10 November 1977, Serial No. 95–52, pp. 1–432, U.S. Government Printing Office, Washington
14 Science Policy Implications of DNA Recombinant Molecule Research (1978) *Report, Subcommittee on Science, Research and Technology, Committee on Science and Technology, U.S. House of Representatives, 95th Congress*, Serial X, 21–754, pp. 1–78, U.S. Government Printing Office, Washington
15 Nelson, G. (1977) in ref. 17, pp. 57–63
16 Recombinant DNA Act (1978) *Committee on Interstate and Foreign Commerce Rep.* No. 95–1005, Part 1, (pp. 1–44) U.S. Government Printing Office, Washington
17 Recombinant DNA Safety Regulation Act (1977) *95th Congress, Senate Rep.* No. 95–359, pp. 1–63
18 Recombinant DNA Research. Proposed Revised Guidelines (1977) *Federal Reg.* 42, 49596–49609
19 Boyer, H. W. and Nicosia, S., (eds.), (1978) *Genetic Engineering*, Elsevier/North Holland, Amsterdam
20 Eisenberg, L. (1977) *Science* 198, 1105–1110
21 Minutes of the Board of Directors Meeting, Sierra Club (8–9 Jan. 1977) pp. 24–29
22 Genetic Engineering, Human Genetics and Cell Biology – Evolution of Technological Issues – DNA Recombinant Molecule Research (1976) *Subcommittee on Science, Research and Technology of the Committee on Science and Technology, U.S. House of Representatives, 94th Congress*, Ser. KKK, pp. 1–259, (Supplemental Report II; 80–497) U.S. Government Printing Office, Washington
23 Davis, B. D. (1976) *Science* 193, 442
24 Davis, B. D. (1977) *American Scientist* 65, 547–555
25 Atwood, K. C., Schneider, L. K. and Ryan, F. J. (1951) *Cold Spring Harbor Symp. Quant. Biol.* 16, 345–354
26 Recombinant DNA Research. Proposed Revised Guidelines (1978) *Federal Reg.* 43, 33042–33178
27 Wade, N. (1978) *Science* 200, 744, 1368
28 Schmitt, H. (1978) *Science* 201, 106–108

Waclaw Szybalski is at the McArdle Laboratory for Cancer Research, University of Wisconsin Medical School, Madison, WI 53706, U.S.A.

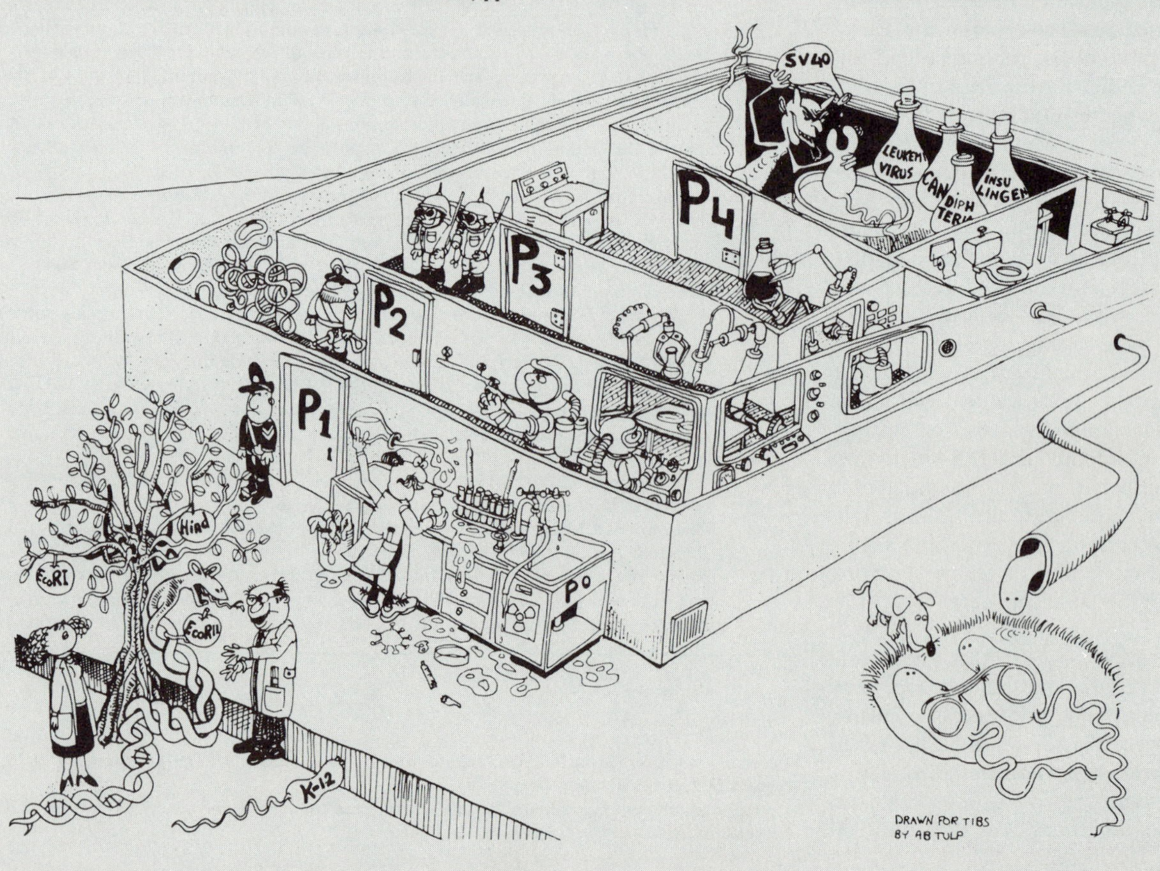

Document 12.8

November 22, 1978
Lyddane Letter to Udry

DEPARTMENT OF HEALTH, EDUCATION, AND WELFARE
PUBLIC HEALTH SERVICE
NATIONAL INSTITUTES OF HEALTH
BETHESDA, MARYLAND 20014

November 22, 1978
Watson, James D.
"Cold Spring Harbor Laboratory
Cancer Research Center"
5 P01 CA 13106-08

Mr. William R. Udry
Administrative Director
Cold Spring Harbor Laboratory
Box 100
Cold Spring Harbor, New York 11724

Dear Mr. Udry:

The above referenced application for a grant-in-aid has been identified as one involving recombinant DNA molecules. Such a project requires that your institution submit a Memorandum of Understanding and Agreement (MUA) containing the information described in the NIH Guide for Grants and Contracts, Vol. 7, No. 3, February 15, 1978, which should be available at your institution.

The original and two copies of the MUA signed by the appropriate designated official, along with the certification statement, should be forwarded to the executive secretary of the initial review group (study section or committee). This information is on one of the self-addressed cards contained in the kit. This card is returned to the applicant after assignment of the proposal for review. The application will be considered incomplete and may not be reviewed until the requested information is received.

The attached illustration is provided for your convenience.

Sincerely yours,

Irene G. Lyddane

Irene G. Lyddane
Grants Clerk
Referral Branch
Division of Research Grants

IGL/fr

Enclosure

cc: Dr. James D. Watson

Cold Spring Harbor Laboratory
P.O. Box 100, Cold Spring Harbor, New York 11724 (516) 692-6660

MEMORANDUM

TO: Terri Grodzicker

FROM: Steve Kron

DATE: December 8, 1978

With regards to the attached letter concerning our submission of updated MUA's, I have enclosed a copy of the latest NIH guidelines concerning Recombinant DNA Research. I have outlined in yellow those items which I feel are appropriate to our current submission.

As I informed you the other day, NCI will not process our continuation application for the Cancer Research Center (C.R.C.) grant until we submit these updated MUA's.

It is my understanding that we must submit updated MUA's (see page 3 of guidelines) for every project on the C.R.C. grant which involves recombinant DNA and which will be performed during 1979. This would include the MUA's on the list I gave you Wednesday since none of these MUA's had a date later than October 1, 1978.

I will be forwarding other information to NCI next Friday, (December 15, 1978) and would appreciate it if you could have these updated MUA's ready for mailing by that date. I would not like to delay their submission to any later date, as I fear a delay in the receipt of an award notice from NCI with a resultant delay in receipt of funds by the lab.

cc William R. Udry
Dr. Watson

DOCUMENTATION REQUIRED FOR PROPOSALS INVOLVING RECOMBINANT DNA

This information supersedes the article entitled "DNA Recombinant Research" in the October 17, 1977 issue of the *NIH Guide for Grants and Contracts*, and the "Important Notice" concerning recombinant DNA in the Public Health Service research grant and training application kits. Information contained herein is excerpted from the *NIH Guide for Grants and Contracts*, Vol. 7, No. 3, February 15, 1978. Any additions or modifications to these requirements will be reported in future issues of the *NIH Guide for Grants and Contracts*. Described herein are (1) requirements for notations on research and training grant applications; (2) a restatement of the requirements for the Institutional Biohazards Committee; (3) refinements to the MUA; (4) new notations that will appear on the NIH Notice of Grant Award forms for grants involving recombinant DNA; (5) procedures for requesting prior approval in funded recombinant DNA projects; (6) specifications for shipping and transfer requirements for recombinant DNA materials; and (7) a reminder about foreign grant applications.

I. NOTATION ON RESEARCH AND TRAINING GRANT APPLICATIONS

Application forms are under revision to include a check block indicating whether or not recombinant DNA research is involved. Until such time as these forms are available, applicants should specify in capital letters at the bottom of the first page of the application "THIS APPLICATION DOES/DOES NOT INVOLVE

RECOMBINANT DNA." Labelling the face page of the application will assist in expediting the processing of the application.

II. INSTITUTIONAL BIOHAZARDS COMMITTEE (IBC)

Each institution where research involving recombinant DNA technology is being or shall be conducted must establish a standing biohazards committee. Suggestions for the composition of such a committee are discussed under Section IV of the Guidelines, which also discusses the roles and responsibilities of principal investigators and institutions. A roster of the members of the Institutional Biohazards Committee must be submitted to the NIH.

The minimum information must include the names, addresses, occupations, and qualifications of the chairman and members of the committee. This information must be submitted to:

Office of Recombinant DNA Activities
National Institute of General Medical Sciences
National Institutes of Health
Bldg. 31, Room 4A52
Bethesda, MD 20014

The composition of Institutional Biohazards Committees is subject to review by the Office of Recombinant DNA Activities for compliance with recommendations stated in the Guidelines. It is the responsibility of each grantee institution to update this information at least annually. As stipulated in the Guidelines, the Office of Recombinant DNA Activities shall assist in the formation of an Area Biohazards Committee (ABC) when this is appropriate. Such an Area Committee shall be necessary when additional expertise from outside a given institution is necessary for the Biohazards Committee to fulfill its functions.

III. CONTENTS OF MEMORANDUM OF UNDERSTANDING AND AGREEMENT (MUA)

Applications for the National Institutes of Health involving recombinant DNA research, as defined by the Guidelines, must be accompanied by a *proposed* MUA with the statements shown in the attached illustration. Because the information provided is captured by a data management system for NIH use, applicants are urged to follow the sequence and format of the illustration as closely as possible.

Incomplete MUAs render the application incomplete. An application without a *proposed* MUA is incomplete and will not be reviewed until a properly executed MUA is provided. Once an MUA has been submitted for the proposed project, changes desired to that project must be made by submitting a revised MUA. The revised MUA should contain either a copy of the original MUA, showing the desired changes, *or* a statement explaining how the original MUA should be modified.

The *proposed* MUA must contain:

A. A description of each proposed series of experiments that involves recombinant DNA molecules and the individual investigator responsible for each experiment, if other than the principal investigator. Do not submit a separate MUA for each experiment.

   Descriptions should include a summary of the research project and should indicate the sources of DNA, nature of inserted nucleic acid sequences, hosts, and vectors. Descriptions must be of sufficient detail to provide information about the experiments without need for reference to the application. Descriptions should provide for each recombinant DNA experiment an indication of the approximate time of initiation after the start date of the project period (e.g., first year, second year, third year,). The time of availability of the required facilities should also be provided in the description. Ordinarily no more than two pages of description for each experiment are acceptable.

B. An assessment of the level(s) of physical and biological containment for each experiment as required by the current NIH Guidelines for these experiments.

C. A description of the facilities and specific procedures that shall be used to provide the required levels of containment. Each performance site must be identified with the name of the organization, city, and state.

D. A specific brief statement by the principal investigator agreeing to abide by the provisions of the current NIH Guidelines and the requirements contained in this Notice concerning shipment and transfer of recombinant DNA materials (see Section VI).

The principal investigator must also attest to the accuracy of the information in A through D of this document.

E. Information concerning Institutional Biohazards Committee review:

   1. When facilities are in existence, a *certification* is required indicating that the Institutional Biohazards Committee has reviewed the proposed project for recombinant DNA experiments and found adequate and in compliance with the NIH Guidelines, this Notice, and other specific NIH

instructions pertaining to the proposed project, the (a) procedures, (b) project and facilities personnel in place at the time of review, and (c) facilities. The date of the review must be specified.

2. When facilities are proposed or are under construction or renovation at the time of the application, an *assurance* in lieu of a certification must be provided. The assurance is signed by the appropriate institutional official(s) to indicate that the Institutional Biohazards Committee has reviewed the proposed project for recombinant DNA experiments and found adequate and in compliance with NIH Guidelines, this Notice, and other specific NIH instructions pertaining to the proposed project, the (a) procedures, (b) project and facilities personnel in place at the time of review, and (c) plans for facilities proposed or under construction or renovation. The assurance includes a statement that recombinant DNA experimentation shall not occur until the completed facility has been reviewed by the Institutional Biohazards Committee and a revised MUA, with *certification*, has been approved by NIH and research is authorized by issuance of a revised award with a Footnote 2 (see Section IV).

F. A statement by the appropriate institutional official that the Institutional Biohazards Committee shall monitor, throughout the duration of the project, the facilities, procedures, training, and expertise of the personnel who are working on the project and for the facilities.

G. The signature of both the institutional official(s) and the principal investigator.

H. The date of signature by the institutional official of the applicant institution. This shall become the date of the proposed MUA for future reference.

Multiple Sites: When recombinant DNA research is proposed at multiple sites, the proposed MUA must specify items A through C above by each site. When recombinant DNA research is proposed at sites governed by other than the applicant institution, signatures of the appropriate officials at the applicant institution *and* the institution(s) where the recombinant DNA research is to be conducted are required. The signatures shall indicate that the Institutional Biohazards Committees of the institutions where the research is to be performed have given the certification and/or assurance required in item E of the MUA, and that the other information is complete and accurate concerning the research to be performed at the site under the jurisdiction of the signer's institution.

MUAs with Noncompeting Applications: All noncompeting continuation applications involving recombinant DNA research projects must be accompanied by a *proposed,* updated MUA that incorporates a statement that the facilities, procedures, and project and facilities personnel in place at the time of review have been reviewed by the Institutional Biohazards Committee prior to the submission of the application, and that the project continues to be in compliance with NIH Guidelines.

Fellowship Applicants, Research Career Development Award Candidates (RCDA), Research Career Awardees (RCA), and Program Directors for Institutional Research Training Grants:

1. Fellowship applicants, RCDA candidates, RCAs, or Program Directors for Institutional Research Training Grants may attach to the application a copy of the MUA(s) *approved* by the Office of Recombinant DNA Activities when the proposed research or training is part of a funded NIH project involving recombinant DNA.

The fellowship applicant, RCDA candidate, RCAs, or Program Director (if other than the principal investigator) must sign the MUA copy under the signature of the principal investigator, indicating that he/she has become familiar with and agrees to abide by the provisions of the current NIH Guidelines, this Notice, and other specific NIH instructions pertaining to the proposed project. The principal investigator and the appropriate institutional official must also initial and date the approved MUA copy to indicate that the copy is current and accurate and that the research or training proposed for the fellowship applicant, RCDA candidate, RCAs, or Program Director is consistent with the approved recombinant DNA project.

2. If any recombinant DNA work is proposed other than that indicated in the *approved* MUA(s), a separate *proposed* MUA must be submitted to NIH.

IV. APPROVAL OF MUAs AND AWARD PROCEDURES

This section applies only to grants awarded by NIH. NO PROJECT INVOLVING RECOMBINANT DNA RESEARCH CAN BE FUNDED WITHOUT AN MUA *APPROVED BY THE NIH*. NOTE THE CONCEPTS OF A *PROPOSED* MUA VERSUS AN MUA *APPROVED* BY THE OFFICE OF RECOMBINANT DNA ACTIVITIES, AND AN *ASSURANCE* VERSUS A *CERTIFICATION* ON AN MUA. AUTHORIZATION FOR USE OF FUNDS TO CONDUCT RECOMBINANT DNA EXPERIMENTS

CAN ONLY BE EFFECTED BY THE ISSUANCE OR REVISION OF A NOTICE OF GRANT AWARD. All NIH awards made for approved applications which propose recombinant DNA experimentation shall carry one of the following footnotes:

A. *Footnote 1*: "Funds from this award may not be used to conduct recombinant DNA experiments."

B. *Footnote 2*: "Recombinant DNA experiments must be conducted in compliance with NIH Guidelines and approved MUA dated --/--/--."

C. *Footnote 3*: "No funds from this or future awards may be used for recombinant DNA experiments until a revised MUA is approved by NIH and authorized on an award document."

Footnote 1 shall be used in those instances where recombinant DNA research was originally proposed by the applicant but, for whatever reason, an MUA has not been approved by the NIH at the time of award.

Footnote 2 shall be used in those instances where recombinant DNA experiments shall begin during the awarded budget period and an approved MUA, with proper *certification*, is on file with the NIH at time of award.

Footnote 3 shall be used in those instances where, at time of award, an approved MUA is on file with the NIH containing: (1) an assurance that adequate facilities shall be available at some time in the future, or (2) certification that adequate facilities exist for experiments to take place during a future *budget period*. It should be noted that noncompeting applications must include a proposed MUA recertifying existing facilities, procedures, and personnel.

V. PRIOR APPROVAL REQUIREMENTS FOR CHANGES IN CURRENTLY FUNDED PROJECTS

Grantees engaged in active projects supported by NIH who wish to modify their existing projects with respect to recombinant DNA research must obtain prior approval from the NIH in the following instances: (1) initiation of recombinant DNA experimentation not previously approved by NIH; (2) acceleration of the schedule for recombinant DNA experimentation to the current budget period; (3) change of experiments to different physical or biological levels; (4) changes of host, vector, or source DNA; (5) cloning other than originally approved DNA segments; (6) change of physical location of the experiments; and (7) change in principal investigator.

The institution and the principal investigator must apply to the awarding Bureau, Institute, or Division for permission before proceeding with the proposed changes. The request to conduct such experiments must be accompanied by a new MUA; a revised MUA; or in the case of minor modifications to approved recombinant DNA experiments, a letter, signed by the principal investigator and the appropriate institutional official, requesting an amendment to the current MUA. A revised MUA should contain a copy of the current MUA, showing the desired changes, *or* a statement explaining how the current MUA should be modified. The signature of the institutional official shall signify that the change has been cleared with the Institutional Biohazards Committee.

For approved changes requested via a new or revised MUA, the Bureau, Institute, or Division shall issue a revised award citing the date of the approved MUA. For approved changes requested via a letter, the Bureau, Institute, or Division shall issue a revised award citing the approved MUA date and specifying the approved changes. The Office of Recombinant DNA Activities may authorize the Bureau, Institute, or Division to inform by letter, the appropriate institutional official of the approval of minor changes in lieu of issuing a revised award.

NOTE: No changes in recombinant DNA experiments in currently funded projects may be initiated prior to obtaining written approval from the NIH.

VI. SHIPPING REQUIREMENTS

All MUAs submitted with competing and noncompeting applications involving recombinant DNA research must indicate that the principal investigator (program director, fellow, or candidate) agrees to comply with the NIH Guidelines, this Notice, and other specific NIH instructions pertaining to the proposed project. Included in the provisions are the following pertaining to shipment or transfer of recombinant DNA materials:

A. Prior to shipment or transfer of recombinant DNA materials to other Federally funded investigators within the United States, the sending laboratory shall obtain a letter from the requesting laboratory stating that:

1. Research involving recombinant DNA molecules shall be conducted in compliance with the NIH Guidelines, this Notice, and other NIH instructions, and that the requesting laboratory shall not transfer

the recombinant DNA materials to other laboratories;

2. The requesting laboratory has been reviewed by its Institutional Biohazards Committee which has certified that facilities, procedures, and the training and expertise of the personnel involved are adequate;

3. An approved MUA with a certification is on file with the funding agency of the requesting laboratory;

4. A copy of this letter is on file with the requesting laboratory's Institutional Biohazards Committee.

B. Prior to shipment or transfer of recombinant DNA materials to non-Federally funded investigators or institutions within the United States, the sending laboratory shall obtain a letter from the requesting laboratory stating items 1, 2, and 4 under A above.

C. Prior to international shipment of recombinant DNA materials, the sending laboratory shall obtain a statement from the requesting laboratory stating that research involving recombinant DNA molecules shall be conducted in accordance with the containment levels specified by the NIH Guidelines, or applicable national guidelines if such have been adopted by the country in which research is to be conducted, and that the requesting laboratory shall not transfer the recombinant DNA material to other laboratories.

D. The sending laboratory shall maintain a record of all shipments of recombinant DNA materials and shall provide NIH with a complete list of such shipments in the annual progress report for NIH grants and contracts.

VII. FOREIGN GRANT APPLICATIONS

Applicants for NIH awards in foreign countries first should contact the Office of Recombinant DNA Activities for information on NIH policies and procedures for recombinant DNA research projects to be conducted outside of the United States.

MEMORANDUM OF UNDERSTANDING AND AGREEMENT

Description (To be supplied)

Levels of Physical and Biological Containment (To be supplied)

Facilities and Procedures for Containment (To be supplied)

The information above is accurate and complete. I am familiar with and agree to abide by the provisions of the current NIH Guidelines, the Notice contained in the *NIH Guide for Grants and Contracts* of February 15, 1978, and other specific NIH instructions pertaining to the proposed project. I agree to comply with the requirements specified by the Guide pertaining to shipment and transfer of recombinant DNA materials.

____

Principal Investigator Date

I certify that the Institutional Biohazards Committee (IBC) has reviewed on (date) the proposed project for recombinant DNA experiments, and found adequate and in compliance with NIH Guidelines, the Notice contained in the *NIH Guide for Grants and Contracts* of February 15, 1978, and other specific NIH instructions pertaining to the proposed project, the (a) procedures; (b) project and facilities personnel in place at the time of review; and (c) facilities. I agree to comply with the requirements specified in the Guide pertaining to the shipment and transfer of recombinant DNA materials.

AND/OR

I assure that the IBC has reviewed on (date) the proposed project for recombinant DNA experiments, and found adequate and in compliance with NIH Guidelines, the Notice contained in the *NIH Guide for Grants and Contracts* of February 15, 1978, and other specific NIH instructions pertaining to the proposed project, the (a) procedures; (b) project and facilities personnel in place at the time of review; and (c) plans for facilities proposed or under construction or renovation. Recombinant DNA experimentation shall not occur until the completed facilities have been reviewed by the IBC and an MUA with certification has been approved by NIH and research authorized by issuance of a revised award citing the date of the approved MUA.

I agree that the IBC shall monitor throughout the duration of the project the facilities, procedures, and training and expertise of the personnel who are working on the project and for the facilities.

____

Applicant Institutional Official Date

____

Institutional Official Date
(Additional Performance Sites, if applicable)

STATEMENT OF HON. RICHARD L. OTTINGER ON RECOMBINANT DNA RESEARCH REGULATION September 15, 1978

Mr. Chairman, members of the panel, I appreciate the opportunity to appear before you this morning.

I would just like to make a few brief remarks this morning about the urgency with which I am convinced HEW must act to regulate recombinant DNA research.

As the first Member of Congress to have introduced legislation on this matter nearly two years ago, I have watched with escalating concern the increase in commitments, particularly in the commercial sector, to recombinant DNA research.

Quite simply, the whole matter is rapidly getting out of control, and it is at this juncture appropriate for HEW to take the reins and protect the public from any further loss of control until Congress acts.

The failure of the Congress to act is one which troubles me, and I am personally very disappointed. But that failure is not one upon which HEW should now base any excuses for inaction. Rather, it is directly *because* Congress has not acted and cannot act during this session that I come before you today to urge your action.

The failure of the Congress resulted from some of the most vigorous and, in many instances, distasteful, lobbying I have ever seen on any issue. When the scientific establishment gangs up on Congress, touting its expertise as the only thing going and claiming to be capable itself of protecting the public, it seems clear that Congress is hard-pressed to resist. Those sorts of assertions must now be resisted by HEW.

Recombinant DNA research has the potential for bringing us some of the greatest horror stories since we dropped the atomic bomb in Japan. The recent transport of smallpox from England to the United States is illustrative of the kind of problem we might face with DNA research. We were told a year ago by the World Health Organization (WHO) that smallpox had been eradicated and was only a concern for historians. It is now clear such optimism was premature. While smallpox may have been confined to labs, mostly for archival purposes, the threat to public health was not entirely eliminated.

The virus in England escaped through a "faulty filter," and contaminated workers in and around a laboratory at Birmingham University. A medical photographer fell victim to this virus and died. The director of the lab committed suicide. The threat from this outbreak even spread to the U.S. when an individual exposed in England flew to North Dakota.

This episode with smallpox parallels and highlights several concerns about research in the area of genetic manipulation:

1. Labs cannot provide foolproof containment of dangerous organisms;
2. Biohazard outbreak, unlike any other type of threat, can rapidly spread to virtually any neighborhood on earth;
3. The more over-optimistic the picture painted by the scientific community, the more vulnerable the public is to health disasters, especially from unprecedented, genetically-hybrid forms, against which nature may have evolved no immune mechanisms.

I realize that stringent safeguards on this type of research may be expensive and/or have an inhibiting effect on the progress of research. But we treat the threats casually, at our own peril.

HEW must now act, and I would urge that you do so under Section 361 of the Public Health Service Act. It may not be the perfect instrument, but at least it provides some authority for action. I do wish I could assure you that the Congress would quickly provide you more precise authority, but I cannot. Even if we were to act early next year, and that's doubtful, it would take an additional year to prepare regulations.

We cannot afford to wait any longer. Imperfect as the world is and as Section 361 is, waiting is going to make matters much worse.

STATEMENT OF THE ENVIRONMENTAL DEFENSE FUND ON THE PROPOSED REVISED NATIONAL INSTITUTES OF HEALTH GUIDELINES FOR RECOMBINANT DNA RESEARCH SEPTEMBER 15, 1978

Good morning. My name is Leslie Dach. I am a Science Associate with the Environmental Defense Fund (EDF). EDF is a non-profit organization with over 46,000 members. EDF undertakes legislative, judicial and administrative actions to minimize human exposure to toxic chemicals. EDF has been deeply involved in the national debate on recombinant DNA. In 1976, EDF petitioned the Department of Health, Education and Welfare to use its authority under §361 of the Public Health Services Act to regulate recombinant DNA activities. EDF submitted comments on the current NIH recombinant DNA guidelines and has testified before Congress, and the Advisory Committee to the Director of the National Institutes of Health on the issue of recombinant DNA.

EDF thanks Secretary Califano for holding this public meeting. Because of the limited time allotted to each speaker, I will only be able to touch on a small number of EDF's concerns. EDF intends to submit extensive written comments on the proposed guidelines.

Today, I would briefly like to discuss the following five aspects of the proposed revised guidelines:

1. the need to rewrite sections of the guidelines to eliminate ambiguity.
2. mechanisms for assuring adequate public participation in NIH activities mandated by the guidelines.
3. membership of the Recombinant DNA Advisory Committee (RAC)
4. membership and responsibilities of the Institutional Biosafety Committees (IBCs).
5. confidentiality

EDF is concerned that the poor quality of the drafting of the guidelines will result in confusion and compliance failures. As presently written, instructions for persons conducting recombinant DNA activities are spread throughout the proposed guidelines themselves, the Director's decision document, the Environmental Impact Assessment (EIA) and appendices to these documents. Often, these documents contradict each other. But only the guidelines will have the indisputable force of law. Moreover, the guidelines are sometimes so vague that their intent is unclear, and problems are sure to arise.

The section of the *Federal Register* package dealing with certification of host-vector systems exemplifies these problems. The guidelines state that certain new host-vector systems "may not be used unless they have been certified by NIH." But there is no indication of what office within NIH supplies this certification. The person seeking certification does not know if he or she is free to use the host-vector system after approval by the Recombinant DNA Advisory Committee or if they must wait for approval of the NIH Director. Just this sort of ambiguity was cited by the University of California at San Francisco to explain their violation of the existing guidelines.

Another example of poor drafting is in the section of the guidelines dealing with exceptions from prohibited experiments. The Director's decision document indicates that the rationale for allowing exceptions is to provide for experiments for which there are compelling social or scientific reasons (43 *Fed. Reg.* 33048). Yet the guidelines themselves (43 *Fed. Reg.* 33070) merely indicate that weight will be given in the decision making process "both to scientific and social benefits and to potential risks." Clearly, the standard for excepting experiments is different in the two documents.

When differing interpretations of the same issue are given in different sections of the *Federal Register* package, it is impossible for EDF or a person conducting a recombinant DNA activity to ascertain which is the correct one. In addition, it is unrealistic to assume that the reader will scour all the documents in the *Federal Register*. Most important, the EIA and the Director's decision document do not have the indisputable force of law. EDF therefore maintains that all information necessary for compliance be included in the guidelines themselves. Finally, ambiguities such as the one described within the guidelines should be clarified.

The second area of concern I will discuss today is the failure of the guidelines to specify procedures for public notification and public comment. The guidelines often indicate that a decision will be made "after appropriate notice and opportunity for public comment." (43 *Fed. Reg.* 33070). No clarification of this is given. For example, the section of the guidelines dealing with certification of new host-vector systems makes no mention of procedures for public notification or public comment. (Interestingly, the Director's decision document states (43 *Fed. Reg.* 33057), "I agree that prior notification to the public should in the *Federal Register* be given when the RAC considers applications for certification.")

The procedures for public notification and comment must be described, in detail, in the guidelines. Failing this, the public will have no assurance that opportunity for public comment will be provided. If pressure for a quick decision is strong or NIH wishes to avoid public scrutiny, the guidelines enable it to make decisions in secret. In addition, valuable resources are likely to be wasted fighting out the mechanism of "appropriate opportunity for public comment," each time the opportunity is given.

As an example of how public participation mechanisms should be constructed, EDF proposes the following mechanism for decisions concerning exceptions from prohibited experiments.

Within 10 days of receipt of an application for an exception from a class of prohibited experiments, NIH should publish in the *Federal Register* notice of the receipt, details of where the material submitted in support of the exception can be obtained and the closing date of the public comment period. At a minimum, the comment period should be 45 calendar days. Final notice of agency action should also be published in the *Federal Register*. All material submitted to NIH should be available to the public.

Our third area of concern is the membership of RAC. The guidelines (43 *Fed. Reg.* 33086 and 33087) describe the functions of RAC. They include recommending revisions in the guidelines, exceptions from prohibited experiments and exemptions from the guidelines. These decisions involve weighing both social and scientific risks and benefits. This is especially true for allowing exceptions from classes of prohibited experiments. Such decisions are clearly not purely scientific ones. As such, RAC must include non-scientist members who adequately represent the interests of the general public. The guidelines do not provide for this. They do not contain any provisions for selecting RAC members. (Again, the Director's decision document mentions this issue. It indicates that two non-scientists are currently RAC members and that more public members may be added.)

Our written comments will provide more details on this issue. Requirements similar to the ones I have outlined are contained in HEW's proposed regulations for the governing bodies of Health Systems Agencies (43 *Fed. Reg.* 22858).

EDF maintains that at least one-third of the RAC be composed of individuals who are not engaged in biomedical research and who can reasonably be expected to represent the interests of the general public. Such individuals would include representatives of labor, public interest groups and elected or appointed public health officials. A subcommittee composed of a majority of RAC members who represent the interests of the general public should be given authority to make recommendations to the Director of NIH concerning exceptions to prohibited experiments and exemptions from the guidelines.

I will discuss our final two areas of concern, confidentiality and the membership and responsibilities of the Institutional Biosafety Committees simultaneously. IBCs have primary responsibility for insuring compliance with the guidelines.

I do not have time to describe the membership and responsibilities of IBCs as outlined in the proposed guidelines. I will therefore proceed immediately to a description of the changes EDF feels are necessary.

One-third of the membership of each IBC should be composed of individuals who have not been affiliated with the institution for at least a year prior to their service on the IBC. At least one of these individuals should be a non-doctoral person from a laboratory technical staff. This person should be elected by the institution's technical staff. At least one person should represent the health department of the local government. This person should be selected by the health department. The remainder of the non-affiliated members should be persons who may be reasonably expected to represent the interests of the community.

The membership ratios EDF suggests are necessary to insure that IBCs will satisfactorily perform their oversight and regulatory roles. It is well known from studies of medical licensing and disciplinary boards and from the events surrounding the violations of the recombinant DNA guidelines at Harvard Medical School and University of California at San Francisco that peer review does not provide adequate public protection. The proposed guidelines requirement for one non-facility member does not address this issue. There is no provision that this individual represent the public interest. The person could very well be a recombinant DNA scientist from a neighboring institution.

Turning to confidentiality, EDF maintains that all IBC meetings should be announced and open to the public. All MUAs and project registrations from institutions that must comply with the guidelines should also be available to the public. EDF's reading of applicable statutes (5 U.S.C. §552 and 18 U.S.C. §1905) indicates that such information, submitted to IBCs or NIH, is not exempt from the provisions of the Freedom of Information Act (5 U.S.C. §552). In light of their legal authority under the guidelines, IBCs are agencies within the meaning of 5 U.S.C. §552. 5 U.S.C. §552 gives NIH the *discretion* to withhold trade secrets from the public. 18 U.S.C. §1905 provides a criminal penalty for violation of an explicit prohibition against revealing trade secrets or confidential statistical information. No such prohibition exists here. Moreover, the courts have held that the research designs and protocols of non-commercial scientists are not trade secrets and therefore must be revealed to the public under 5 U.S.C. §552. Washington Research Project v. Department of Health, Education and Welfare 504 F.2d 238 (1974). This decision applies to all recombinant DNA activites regardless of whether the investigator plans to seek a patent on his research. In addition none of the information EDF maintains should be available to the public meets the test for confidential commercial information contained in National Parks and Conservation Association v. Morton (498 F.2d 765 (1974).

EDF maintains that no changes in ongoing recombinant DNA projects subject to the guidelines should be allowed without NIH approval. The error rate of IBC's has been variously estimated at 4-15% by NIH's Office of Recombinant DNA Activities (ORDA). The turn around time for ORDA review of an IBC assessment of containment requirements is 4-5 days. For these two reasons, EDF maintains that no changes should be allowed without ORDA approval. To speed up the turn around time, EDF suggests that ORDA be given additional staff. There are presently only two people reviewing IBC assessments of containment requirements.

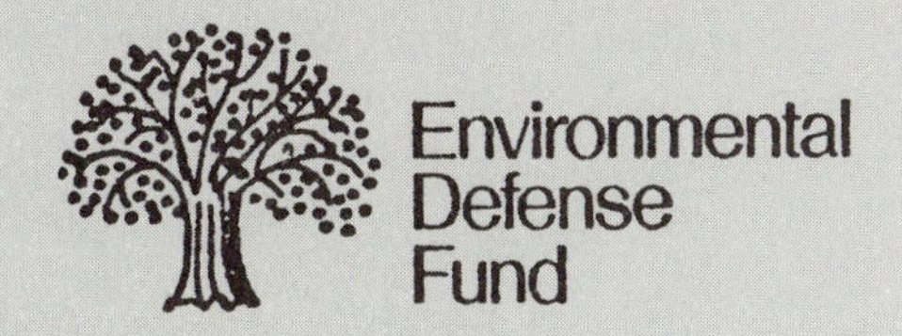

September 15, 1978

FRIENDS OF THE EARTH

620C Street, S.E., Washington, D.C. 20003

(202) 543-4313

David Brower, *President*

September 15, 1978

TESTIMONY OF PAMELA LIPPE, ASSISTANT LEGISLATIVE DIRECTOR, ON THE PROPOSED REVISED GUIDELINES FOR RECOMBINANT DNA RESEARCH

I. INTRODUCTION AND OVERVIEW

We in the public interest community look at recombinant DNA as a powerful new technique *and* technology with unprecedented potential for benefit and harm. Contrary to recent scientific pronouncements, we do not want to stop recombinant DNA research or scientific progress; we simply wish to achieve the benefits without incurring unnecessary risks.

The scientific community looks at recombinant DNA as a powerful tool which will allow them to answer questions beyond their previous understanding, and to do things that stretch the limits of their imagination.

While it is understandable that scientists more clearly perceive the beneficial aspects of their efforts, a quick survey of recent scientific and technological breakthroughs illustrates the time bomb and elusive nature of many of science's most cherished dreams.

As the Stevenson subcommittee pointed out in its recent oversight report, "At issue is the extent to which research scientists should be entrusted with responsibility for their own conduct, individually or through peer review, or be subjected to external control and scrutiny."

Our position is that if the situation is serious enough to develop guidelines for the conduct of this research, then it is serious enough for the guidelines to be enforced. Self-regulation has never worked in the past. There is no reason to expect recombinant DNA is to be any different.

In general my comments will focus on the procedural aspects of the guidelines as requested by Secretary Califano. I hope that you find these comments and recommendations useful.

II. SCOPE AND APPLICABILITY

In general, we support efforts to further extend the guidelines to the private sector such as those in the proposed revisions. It is unfortunate that no more comprehensive mechanism exists to transform the guidelines into full fledged regulations—legal authority extending to industry and academia alike, and to both effectively.

We would still urge that HEW use Sec. 361 of the Public Health Service Act. Congressional inaction on this issue does not in any way diminish the need for meaningful regulation in this rapidly developing new field. It was "hard ball" politics, not new data that killed both the Senate and House bills.

The new data, miraculously discovered and/or developed in only one year's time offers assurance about one small part of the potential danger—e. coli k-12. We are already considering approval of new host vector systems, deliberate release into the environment, and exempting research with wild type e. coli. What little we know now, will become even less as we continue to open broad now biological and ecological vistas.

We urge you to reconsider your decision regarding Sec. 361 in light of the inadequacy of the existing NIH proposals as pointed out in today's testimony, and the need for legally enforceable regulations due to increasing interest in the private sector. Even if Congress acted next session, which is highly uncertain, it would be a *minimum* of two years before regulation would be in place. I hope you will agree with us that the drawbacks of that course of inaction far outweigh the use of Sec. 361.

Recommendations:

1. Promulgate the guidelines as regulations under existing authority.

2. In the absence of full regulation under existing authority, could NIH-funded investigators require an approved MUA with certification on file with the NIH, prior to shipment or transfer of recombinant DNA materials.

Exemptions

The requirement for the fourth class of exemptions is that they "exchange DNA by known physiological processess." There is no longer a requirement that they must be "natural" physiological processes. "Known" is a much broader class of exemption. How does the NIH define it? More significantly, there is no requirement that that particular class of exemptions be "safe."

Recommendations:

1. All exemptions must be made on basis of proven safety based on experimental data (i.e., no significant risk.)

2. The list of exchangers should be by species, not genus, and added only on the basis of safety as well as exchange.

3. There should be full compliance with NEPA.

Exceptions

The guidelines point out that weight must be given to both "scientific and societal benefits and potential risks," acknowledging that these decisions are not only technical, but are political. This situation demands much broader representation in the decision-making. Appropriate notice and comment is simply not adequate; the decision is already made. The decision makers can and do ignore public comment.

Exceptions are easements of prohibitions. These experiments were not prohibited lightly; the ban should not be lifted without careful consideration of all possible consequences drawing on the full range of expertise, especially the public.

Recommendations:

1. All exceptions from the guidelines should be undertaken only after a full environmental impact statement to allow for assessment as well as better public input and awareness. In the case of a single risk assessment experiment, where containment will not be breached, an EIA may be satisfactory.

2. Waiver on large scale culture size should only be granted after full scale risk assessment studies on the organism, and an EIS on the large scale production process have been done.

3. The guidelines presently state that prohibition of deliberate release into the environment should be lifted only "if the requirements for a waiver are met, and if the requirements of NEPA are considered." What are these requirements for a waiver? And aren't the requirements of NEPA going to be met, not only considered?

Certification of Host-Vector Systems

New host-vector systems raise a whole new range of ecological and biological issues. The projections of safety being made from e. coli k-12 are no longer germane. These decisions must be made on the basis of as complete knowledge as possible. The NEPA process will help to inform the decisions.

Recommendation:

1. A full EIS must be required before approval of any new host-vector system.

III. RESPONSIBILITIES OF THE NIH

Due Process Considerations

1. The reference to "shifting the burden of proof" should be stricken from the guidelines. As discussed earlier, it is clearly premature. For an excellent discussion of where the burden of proof lies, I urge you to read Peter Hutt's March 3, 1978 letter to Dr. Fredrickson p. 2.

2. "Appropriate notice and opportunity for public comment" is an inadequate public participation mechanism. By that point, the important decisions have already been made. Our comments in the past have too often been ignored. The public perspective must be introduced at a much earlier point in the decision-making.

3. In that regard, the Director's Advisory Committee (DAC) does not adequately substitute for public involvement on the Recombinant DNA Advisory Committee (RAC). Once again the decisions are made by the time the "public" gets input. It may be possible to change the details, but the basics are set in concrete. It is this kind of public participation that has led to distrust of the policies of the NIH.

4. Public membership must be substantially increased on the RAC. As the Stevenson subcommittee concluded, "the technical and social policy aspects of even physical and biological containment standards or host-vector approval cannot be readily distinguished." Scientific decisions cannot and should not be separated from their social and political context. In attempting to do so, the NIH has insured that the decisions were simply made in a less representative social and political context.

Recombinant DNA Advisory Committee

The RAC has turned out to be much less conservative

than the Director, NIH. In many cases, they went further in the PRG-RAC than the Director finally allowed in the PRG-NIH. Advisory Committees have often been expected to counterbalance internal conflicts of interest, and in this case ensure that the public health and safety is being considered adequately. Unfortunately, the RAC does not seem to be fulfilling that function.

Recommendations:

1. EPA, OSHA, FDA, and CEQ should have full voting membership on the RAC. This is essential now that the NIH is accepting voluntary compliance by industry. In order to protect proprietary information, the RAC will have to hold meetings not open to the public. We need to have some assurance that our concerns will be represented. Perhaps more importantly, the RAC could use the advice of experienced regulators, particularly as they begin to deal with industry.

The Interagency Committee is not adequate participation and input for these agencies. It has not met once this year. The advice of EPA and OSHA, in particular, will always be "appropriate" (p. 33051), they should be fully involved in the decision-making.

2. The majority of open spaces on the RAC should be filled with individuals nominated directly by the public. Until we have members who share our perspective, and whose judgement we trust, the decisions of the RAC will always be suspect. This trust is in the long term interests of the NIH, as well as ourselves.

Penalties

HEW must find some credible mechanism for ensuring enforcement of the guidelines; withdrawal of NIH funding is totally inadequate. The sanction is so severe that the NIH will probably never use it, except for the most gross violations, witness the recent experience with UCSF and Harvard.

HEW must find some mechanism by which penalties can be assessed appropriate to the infraction, and then use them as required.

IV. RESPONSIBILITIES OF THE INSTITUTION

Health Monitoring

Institutions are only required to determine the necessity of doing health monitoring, not to actually do it. There is nothing in the guidelines that requires medical monitoring, data collection, maintenance of records, or for that matter good experiments designed to determine what if anything is happening with recombinant organisms inside human bodies.

Very little study has been done on the epidemiology of recombinants in humans. And although we've heard repeatedly that no one has been injured by this kind of research, they're still saying that about commercial nuclear reactors. It is important to note that the most dangerous kinds of experiments have not been done. Almost all work has been executed with the safer recombinants, primarily because of the lack of high containment labs. But with the new guidelines, what used to be done in P-4 and P-3, can now go on in P-3 and P-2. New host-vector systems, work in wild-type e. coli because of exemption, and large scale processes will make what little we know essentially inapplicable.

Recommendations:

1. Make mandatory requirements for data collection and record keeping in a unified manner, obviously allowing for easier accumulation and analysis.

2. Encourage and support epidemiological studies of recombinant DNA laboratories to develop the information needed to make decisions on the basis of experimental data.

3. Carry out an assessment in a year or two to see if a centralized collection of this data is required.

Institutional Biohazard Committees

The NIH has delegated considerable authority to the institutional biohazards committees:

1. to allow research to proceed prior to NIH approval on changes in P1 through P3 protocols.

   a. This policy will lead to confusion and discrepancies between institutions. Approximately 10-15% of all MUA's are disapproved by the NIH on non-trivial grounds. This creates ample opportunity for problems. Recent conversations with ORDA indicate that processing of MUAs is down to 24-48 hours.

Recommendation: Require prior NIH approval on non-trivial changes on P-2 through P-4. If significant delays develop, hire more staff, fund less research. (one staff person worth)

2. to approve single step reductions in containment levels for experiments with purified DNA and characterized clones.

   a. authority for single step reductions in containment levels, as well as the previous provision, will lead to a patchwork of local regulation which was decried by the scientific community in regards to federal pre-emption of stricter standards, but is apparently welcomed for lowering them. Furthermore, there is no requirement that these fully "characterized clones or purified DNA" be safer.

Recommendation: prior approval by the NIH should be required on P-2 through P-4 protocols and only granted on the basis of proven safety (i.e., no significant risk). The function of the insert as well as its quantitative purity should be taken into account.

3. to "review and oversee" all recombinant DNA projects because of the "impossibility of Federal surveillance to enforce these standards externally," and because it is an "effective and relatively inexpensive administrative mechanism."

a. It is the above section which underlies the basis for our concern about the whole regulatory approach that the NIH has developed. The last year has shown the inadequacy of *local* surveillance to enforce these guidelines as well as Federal. As chairman of the Harvard Biohazard Committee Bernard N. Fields has pointed out in the Boston Globe, "Monitoring of compliance is on an honor system." That is hardly an "effective administrative mechanism."

The proposed revision requiring one public member and recommending one non-doctoral staff person, chosen by the institution, is absolutely inadequate to ensure a balancing of perspectives. One public member on a committee of 9-15 will be intimidated, ignored, and outvoted continually.

Recommendations:

One third of the members of an IBC should represent the public and worker interests. The use of institutional scientists and environmental health and safety personnel has already been discussed in the Laboratory Monograph. This lessens the need for overwhelming scientific expertise on the committee. The make-up and selection process should be specified in great detail in the guidelines. For example:

one or more non-doctoral staff (nominated by that population or workers.)

one public health official and/or elected official.

one or more "volunteers" representative of the public, well-suited to represent the public interest.

The public member should volunteer to insure continuing interest. Notification of openings should be made in local media. Selection should be made by the local governing body or health department.

One third of the members should represent scientific disciplines related to risk assessment (e.g., epidemiologists, environmental health scientists, physicians with both research and clinical experience in infectious disease, microbial ecologists.)

One third should represent disciplines relevant to recombinant DNA technology, biological safety and engineering.

The IBC must have all meetings publicized and open to the public except for those portions specifically dealing with confidential information. This should be required in the guidelines.

Conclusion

Until there is adequate representation of diverse interests on the decision-making bodies, oversight and regulation of recombinant DNA research is a sham. The present proposed revisions do not adequately protect the public, the worker, and in the final analysis, the recombinant DNA researcher himself.

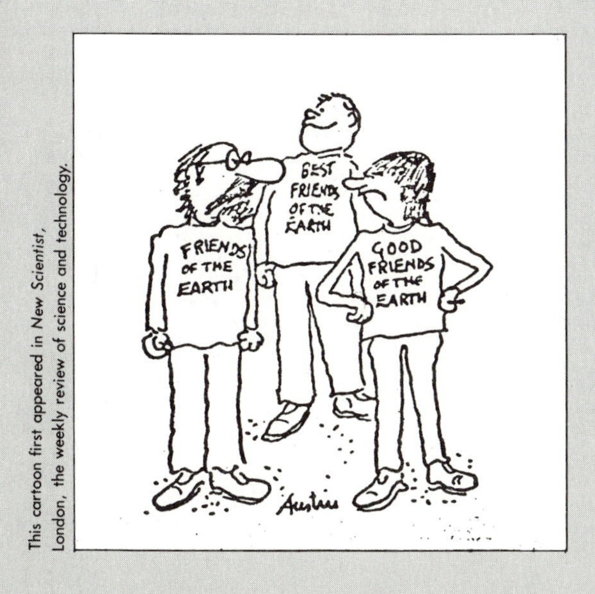

This cartoon first appeared in *New Scientist*, London, the weekly review of science and technology.

Document 12.12

May 14, 1978

The Washington Post OUTLOOK *Columnists / Editorials*

SUNDAY, MAY 14, 1978

The Nobelist vs. the Film Star

By J. D. Watson

UNTIL THE LAST YEAR, I never thought much about my allegiances. My parents were for Roosevelt and against the spoilage of our land by senseless land spectators or industrial giants who put steel mills where there had been sand dunes and the prairie warbler had nested. People who went on bird trips or camped in the national forests and wanted to save Mineral King were the right sort, while those who owned big yachts or stripped the rolling fields of Ohio for coal were the bad guys whom we must get laws to stop. So it was natural to make out a modest check whenever Robert Redford or some equally fine fellow asked you to help him defend the environment and fight the polluters who would give us more cancer.

Now, however, I must confess that I didn't respond to Robert Redford's latest appeal. It is not that I am against him as a folk hero, but, though he must be unaware, he and I are, for practical purposes, real enemies. For some of the money he raises for the Environmental Defense Fund is being used to try to stop the experiments we do with "recombinant DNA."

This test-tube-made genetic material now provides an incredibly powerful means to find out what human genes are like. And in so doing it will give us important new ways to think, say, about our immune systems, or how our blood cells are made or the nature of the genes that go out of control when cancer arises.

This being so, I most certainly am a Friend of DNA and want work with recombinant DNA to go as fast as possible. In the old days, this impulse would generally be viewed as good for the earth. Now, however, there exist highly vocal groups who think I'm a danger to the world. The Friends of the Earth, the Sierra Club and the Natural Resources Defense Council, as well as the Environmental Defense Fund, all say that our experiments pose a realistic threat to our way of life and must be constrained by their new breed of environmental lawyers.

All this initially surprised me, since I had always regarded environmentalists as among our most intelligent public groups and thought that the original rules for work with recombinant DNA which had come out of the 1975 Asilomar Conference should more than reassure them. Particularly since I found those guidelines a terrible overkill and probably not at all necessary.

My fellow DNA workers wanted, however, to act more than clean and certainly to give the impression of being responsible citizens. So they suggested that we largely work with specifically enfeebled organisms that would not live well outside our test tubes. And when, after Asilomar, the matter was taken up by the National Institutes of Health, they in turn wanted to look like the perfect guardian of our health, and so the guidelines which we now have to live with became more than tough. In fact, they effectively blocked most of the better experiments that directly relate to cancer.

See DNA, Page D2

Watson, director of the Cold Spring Harbor Laboratory in New York state, won a 1962 Nobel Prize in medicine for his work on the structure of DNA.

DNA Restrictions Attacked

DNA, From Page D1

As a result, the DNA community is now very unhappy working under harsh rules we do not believe necessary and which waste vast sums of sorely needed research funds. We now want to relax greatly the guidelines we imposed upon ourselves.

Unfortunately, we find this task to be much more complicated than their original drafting. Our main problem is that in our original statements about recombinant DNA, we kept referring to "potential dangers." Instead we should have said "conjectural dangers," since there was, and still is, not a trace of evidence that any of the experiments pose a threat to those who do them, much less to the general public.

♦

IN BEING SO linguistically sloppy, we gave a long awaited opening to two groups which were out to embarrass us. The first consists of disgruntled long-out-of-productive-science biochemists, who use any opportunity to say bad things about how the effects of modern science are carried out. The other is a tiny, though noisy, group of Boston-based academic leftists who fantasize that the rich will finally subjugate the masses by giving them bad genes manufactured by recombinant DNA methodologies. This is a mad idea which I suspect they are too intelligent really to believe. It must be a tactical move in their zany campaign to convince the Boston poor to rise up against the elitist imperialism of MIT and Harvard.

We never expected, however, that we would be branded as polluters by the environmental movement. For until recombinant DNA came along, we always thought we were on their side.

After all, who wants to see our planet not fit for our children to inherit? When they went to court to block DDT or keep the skies of Monument Valley blue, we could only applaud. So why now are we on opposite sides? Can we have on blinders, and can our self interest as scientists not allow us to see how indifferent we are to the harm we may do? Might, in fact, the professional environmentalists present arguments that we just can't face up to?

I feel strongly this is not the case. Compared to almost any other object which starts with the letter D, DNA is very safe indeed. Far better to worry about daggers, or dynamite, or dogs, or dieldrin, or dioxin or drunken drivers, than to draw up Rube Goldberg schemes on how our laboratory-made DNA will lead to the extinction of the human race.

The strains of viruses and cells we work with in the laboratory generally are not pathogenic for man, and all we know about infectious diseases makes it unlikely that the addition of a little foreign DNA will create any danger for those who work with recombinant DNA-bearing bacteria. Even if no special guidelines existed, and we only employed the standard microbiological practices of routine sterilization, we should have no reason to be concerned for our health. Equally important, we should not worry that our experiments will profoundly alter evolution by creating bizarre life forms unlike any seen before. DNA is frequently carried from one species to another by viruses, and the global evolutionary impact of our experiments must be negligible compared to naturally occurring DNA transfers.

If this is so, how can we explain the enthusiasm with which so many professional environmentalists wish to shut us down?

The answer, I fear, is that such groups thrive on bad news, and, the more the public worries about the environment, the more likely we are to keep providing them with the funds that they need to keep their organizations growing. So if they do not watch themselves, they will always opt for the worst possible scenario.

For the short term this may give them more recruits, but I worry about the long-term effect. No one will benefit if we perceive the credibility of our environmental movements to be no better than that of the most troglodytic of our industrial firms.

If what they say about DNA is nonsense, do we have any compelling reason to listen to them when they come out against pesticides that give us shiny apples or tell us that the waters of the Mississippi are likely to give us cancer? I would like someone to set me right on such matters, but whom to trust now is not that clear.

Document 12.13

October 14, 1978

Congressional Record, Excerpts

Statement of Senator Adlai Stevenson

THE STATUS OF RECOMBINANT DNA RESEARCH

• Mr. STEVENSON. Mr. President, a year ago I cautioned the Senate not to rush legislation regulating recombinant DNA research. In retrospect, perhaps my remarks were taken too seriously. On the surface, remarkably little has changed in the interim, and the job of passing needed legislation must now await the 96th Congress.

Additional hearings by the Senate Subcommittee on Science, Technology, and Space and the House Committee on Science and Technology have brought the total to nine separate sets of congressional hearings in 3 years. The Recombinant DNA Safety Regulation Act (S. 1217), the bill introduced by Senator KENNEDY and reported by the Human Resources Committee in July 1977, is still on the Senate Calendar. Although a substitute amendment has been introduced, no further action has been taken.

Earlier in the current session, the House Commerce and Science and Technology Committees reported with minor differences the Recombinant DNA Act (H.R. 11192), introduced by Congressmen ROGERS and STAGGERS; but it has not been brought to the House floor. Secretary Califano, in the meantime, has confirmed his Department's opposition to using existing Public Health Service Act authority to apply the National Institutes of Health research standards to recombinant DNA activities in the private sector. The proposed NIH guidelines revisions, originally published in September 1977, are still under consideration by HEW.

In many respects, the appearance of inactivity is misleading. In the past year a great deal of progress has been made in both the politics and the science of recombinant DNA. Although these developments have not obviated the need for congressional action, legislation must take into account the circumstances that now prevail.

The enormous power of the technology has been amply demonstrated, not only in increasing understanding of basic biological phenomena but also in producing what are likely to be useful products. The number of NIH research projects using recombinant DNA techniques has more than doubled to nearly 500, even under restrictions that many scientists consider excessive.

There have been additional experiments to assess the possible risks of recombinant DNA work to human health, as well as discussions of these and other environmental risks among a wider range of biological and health scientists. Taken together, the results of these experiments and conferences have diminshed somewhat the original concerns of scientists that led to the voluntary moratorium on certain experiments, the Asilomar Conference in 1975, and the issuance of the NIH guidelines in 1976. These new judgments are reflected in the changes in the standards proposed by NIH and now in the final stage of HEW review.

The debate over safety continues, but it is better informed and much less emotional. The serious issues are more procedural than substantive. Sponsors of the House and Senate bills seemingly remain divided over the question of Federal preemption, but there is less anxiety in the academic and industrial communities about excessive and inconsistent regulation by State and local governments.

The November 1977 hearings of the Subcommittee on Science, Technology and Space revealed a surprising degree of consensus about the elements of a regulatory system applicable to both public and private activities and commensurate with the remaining uncertainties about the safety of the research. The subcommittee has summarized this consensus in a report, "Recombinant DNA Research and Its Applications," published in August 1978. The report also includes the first systematic effort to examine the issues that are likely to arise in the regulation of large-scale commercial applications of recombinant DNA techniques. Certain commercial developments appear to be more imminent than even the subcommittee's most optimistic witnesses predicted less than a year ago.

FUTURE LEGISLATION

The question remains whether enactment of Federal legislation is still advisable, primarily to bring privately supported research within the scope of the NIH guidelines. Let me explain why I believe that Congress should act.

First. In spite of increasing confidence in the safety of current recombinant DNA research, the cautious approach that has been taken to the use of this technology is reasonable and will continue, whether or not there is legislation. The failure in the past several years to observe any case of illness or other harm does not prove the safety of all recombinant research, it merely tends to support the consensus that permissible experiments conducted on the whole under prescribed physical and biological containment standards do not pose a significant risk. By the same token, activities under the revised guidelines must be monitored in order to determine the correctness of the judgments on which they are based. For example, NIH observes in its Environmental Impact Assessment:

> Few recombinant DNA experiments have been conducted with viral DNA, since the overly stringent containment levels of the current guidelines greatly inhibited their use. Under the (proposed revised guidelines), such work would be carefully monitored to insure that any new information on safety or risk were quickly reviewed and any appropriate amendments to the guidelines were made.

And further:

> The (proposed revised guidelines) focus on areas of experimentation that need special attention for the possibility of potential hazard. Work in progress that is expected to yield valuable new information will need to be monitored—for example, experiments in which "engineered" systems should permit intentional expression of genetic functions. The current revisions intend to remove as a focus of attention the type of project that does no more than mimic nature and to permit serious attention to new developments that would futher the expression of new genetic functions.

Both the workshop on viruses and the workshop on agricultural pathogens noted areas of ignorance about the possible effects of certain experiments even with E. coli K-12 and for that reason urged further investigation. Results of the important risk-assessment experiments with polyoma virus being conducted at the NIH P-4 facility at Fort Detrick have yet to be published. Subject to unusual case-by-case exceptions, the proposed guidelines retain the prohibitions on certain presumably hazardous categories of experiments, the release into the environment of organisms containing recombinant DNA, and the production of certain large cultures.

Future research, moreover, will not be confined exclusively or perhaps even primarily to E. coli but may involve cloning in a variety of organisms such as soil microbes, lower eukaryotes, fungi and yeast. NIH concedes that knowledge of their characteristics is inferior to the information acquired over many years about E. coli.

It may be necessary to devise entirely new criteria in order to evaluate the safety of these new systems. The Director of NIH assures us nevertheless, that the "same considerations of safety and risks associated with the use of E. coli K-12 will also apply to any new host-vector system to be certified." In other words, Federal review, approval, and monitoring of the use of new host-vector systems will be necessary for the foreseeable future, even if physical containment procedures eventually become routine practice as a result of training and habit and other requirements of the guidelines are dispensed with on the basis of new evidence.

A more thorough program of risk assessment is needed, but it cannot establish once and for all that recombinant DNA research is without risk. Such knowledge will be acquired gradually over the course of years, and it will require Federal support of a substantial research effort simply to keep pace with the development of recombinant DNA techniques. As Dr. Frederickson observed in his introduction to the revised guidelines:

> It is not the time to conclude that (the guidelines) are being altered in preparation for their early abandonment. Understanding of gene regulation is increasing inexorably and at an awesome pace. We may predict that ways will be found to achieve and control the translation of foreign genes by a variety of hosts . . . In some proportion to the harvest of positive results, a capability must be maintained for observing any capacity of these experiments to yield harmful products.

Second. In these circumstances, it is no less anomalous now than it was a year or two ago to restrict one class of researchers but no another, merely on the basis of their location and source of funds. Moreover, the alternatives to new legislation are deficient or have been rejected.

For a brief time this year there was a revival of interest in using existing statutory authority, specifically section 361 (Control of Communicable Diseases) of the Public Health Service Act, to regulate research in the private as well as public sectors.

In June I joined several members of the Human Resources Committee in inquiring whether, in view of recent scientific developments and the heavy congressional agenda, HEW would reconsider its reluctance to invoke section 361. After an inordinate delay, Secretary Califano replied in September that he would do so only in the unlikely event of a demonstrated emergency. There remains some question whether the Food and Drug Administration will require a firm seeking the agency's approval of the end product of recombinant DNA research to certify that it has complied with the guidelines; but it is clear that such action, though useful, will not suffice to put all research activities on the same regulatory footing.

In the meantime, NIH proposes to extend the guidelines' coverage to privately supported research in institutions receiving NIH funds for recombinant DNA work and to establish a voluntary system under which industry can register projects and consult with NIH. Both are steps in the right direction, but voluntary registration does not invalidate the conclusions of the subcommittee's report "that a program of voluntary compliance by the monitoring of industrial research activities is insufficient." In recognition of industry's concern about the disclosure of trade secrets, the new guidelines advise firms to "consider applying for a patent before submitting information to DHEW." So long as the patentability of RDNA products is in doubt, companies may be reluctant to register even completed experiments.

Third. So far those private companies which have acknowledged that they are sponsoring or conducting recombinant DNA research have agreed informally to comply with the guidelines' prescriptions. But voluntarily adherence will become increasingly problematic as the research proliferates, more positive results are achieved, and the line between research and development is crossed.

As I noted earlier, NIH would continue to ban the production of certain cultures in excess of 10 liters and the release of recombinant DNA organisms into the environment, although specific exceptions could be made for experiments in institutions subject to the guidelines. Industry will have a greater interest than academic researchers in conducting developmental activities but may be deterred by proprietary considerations from full participation in voluntary registration. Even if a company agreed to inform NIH of all research and development projects, there is no provision in the revised guidelines for granting it an exception to the prohibitions.

The problem is not merely theoretical. In recent comments on the new guidelines, Eli Lilly Co. states:

> Recombinant DNA technology is progressing rapidly from a laboratory phenomenon to developmental activities. The necessity for conducting scale-up experiments (greater than 10-liter volumes) is imminent.

Further, Lilly urges that procedures for providing NIH with information about private recombinant DNA activities not "subject the participants to a possible loss of confidential information and consequent competitive disadvantages through Freedom of Information requests." Lilly's concerns about the status of its activities under the revised guidelines are reasonable. It seems to me that the only satisfactory way to resolve these problems, as well as to insure compliance with the guidelines, is to apply the rules universally and uniformly.

Fourth. In addition to placing similar obligations on all researchers, legislation is needed to achieve effective implementation of the standards. Authority to regulate recombinant DNA research should be vested in the Secretary of HEW and responsibility for monitoring and enforcing compliance should be delegated to an appropriate agency of the Department apart from NIH. The NIH Director himself has stated repeatedly that there is "an inherent conflict of interest" in the Agency's both sponsoring research and acting as its "policeman." The NIH Office of Recombinant DNA Activities lacks adequate staff to perform this necessary oversight function.

And NIH's lack of experience in regulatory activities is indicated by the ambiguity of the guidelines' procedural provisions, by their failure to establish clearly the responsibilities of institutions, institutional committees and investigators, and by the absence of any mention of procedures to investigate and correct violations. While the proposed revisions make some improvements in administrative procedures, they retain or create a number of serious ambiguities and weaknesses. In the letter to Secretary Califano which follows this statement, I have recommended that HEW correct these problems before the new guidelines become effective. In the long run, however, the solution is to relieve the National Institutes of Health of the administrative burden and permit them to devote their considerable resources to standard setting.

For these reasons, I am pleased that Secretary Califano, according to his letter of September 12, continues to support moderate legislation. In January I will introduce a bill incorporating the elements suggested by HEW as well as other recommendations of the Subcommittee on Science, Technology and Space. These include provisions clarifying the authority of the Secretary, facilitating the issuance of administrative regulations, permitting exemptions for activities that pose no significant risk, instituting uniform national standards while allowing States and localties to take other actions to assure their citizens that the Federal regulations are being observed, and insuring the accountability of institutions and investigators conducting research.

I trust that we will have the active and sustained support of the Executive in passing such legislation early in the next session. The issue of recombinant DNA research is too important to accept an abdication of responsibility by either branch. Next year we have the obligation to conclude this important science policy debate by taking moderate, informed, and constructive action.

Document 12.14

September 26, 1978
Schmitt Letter to Watson

HARRISON SCHMITT
NEW MEXICO

United States Senate
WASHINGTON, D.C. 20510

September 26, 1978

Dr. James D. Watson
Director, Cold Spring Harbor Laboratory
Cold Spring Harbor
Long Island, New York 11724

Dear Dr. Watson:

Due to your previous interest in legislation impacting on recombinant DNA research, I would like to share with you the enclosed Report of the Senate Commerce Committee on the hearings held November 2, 8 and 10, 1977, by its Subcommittee on Science, Technology and Space.

Since I did not agree with many of the conclusions and recommendations made in the Report, I have filed minority views which appear starting on page 50 of the Report and, with supporting documents, continue through page 80. I hope you will find this Report useful and that it will bring you up to date on the latest activity in Congress on the subject of recombinant DNA.

Thank you for your continuing interest on this subject, and please let me know if I can be of any further assistance.

Sincerely,

Harrison Schmitt
United States Senator

HS:jwj
Enclosure

October 24, 1978
Watson Letter to Schmitt

October 24, 1978

COPY

The Honorable Harrison Schmitt
1251 Dirksen Senate Office Building
Washington, D. C. 20510

Dear Senator Schmitt:

Thank you so much for sending me the Senate report on Recombinant DNA. Your minority statement I found very important, for without such strong counter arguments the passage of restrictive legislation was bound to remain a very real possibility. Now I gather the chance of any legislation this year is zero, and I'm sure that in 1979 the perceived need for such will recede at a quickening pace. I note that you have been invited to speak at the Wye Kent Symposium next April, and I hope we might again meet there.

I enclose a little piece of autobiography that arose out of a speech I gave last January before my fellow biochemists and molecular biologists. You might find some of the "biohazard history" revealing.

Thank you again for your major help to our scientific world.

Yours sincerely,

J. D. Watson
Director

JDW/mh
enclosure

November 8, 1978
Schmitt Letter to Watson

HARRISON SCHMITT
NEW MEXICO

United States Senate
WASHINGTON, D.C. 20510

November 8, 1978

Dr. James D. Watson
Director, Cold Spring Harbor Laboratory
Cold Spring Harbor
Long Island, New York 11724

Dear Dr. Watson:

You will perhaps recall that on September 25th I communicated with you and enclosed a copy of an Oversight Report by the Senate Subcommittee on Science, Technology and Space on "Recombinant DNA Research and Its Applications." In minority views in that report, I stated my opposition to "unwarranted and excessive regulation of a field of scientific research where the risks are purely theoretical and inconsistent with other knowledge."

Senator Stevenson, Chairman of the Subcommittee on Science, Technology, and Space, has now stated that he will be introducing legislation early in the next session of Congress that will be directed toward regulation of recombinant DNA research. Enclosed is a copy of his statement from the Congressional Record of October 14, 1978, stating his views on this issue and his intent to submit a bill in January.

I would be most interested in your views and those of your colleagues on Senator Stevenson's position on this issue. This would be of great help when this issue is debated on the Senate floor.

Sincerely,

Jack

Harrison Schmitt
United States Senator

HS:jwj
Enclosure

STANFORD UNIVERSITY MEDICAL CENTER
STANFORD, CALIFORNIA 94305

DEPARTMENT OF BIOCHEMISTRY

PAUL BERG
Willson Professor of Biochemistry

January 5, 1979

Senator Harrison Schmitt
United States Senate
Washington, D.C. 20510

Dear Senator Schmitt,

Please forgive me for the delay in answering your letter of November 8 asking my views on Senator Stevenson's intention to introduce legislation next year to regulate recombinant DNA research. I left for a nearly month-long trip to Japan shortly after your letter arrived and the accumulated chores and obligations that greeted my return kept me from responding sooner.

Lest there be any doubt about my views let me be quite direct: I am unalterably and unequivocally opposed to legislation, whether federal, state or local, that regulates research using recombinant DNA techniques. Such drastic and far-reaching action (or overreaction) is, in my view, unnecessary and, therefore, unwarranted. But, more importantly, I am concerned that such legislation and the inevitable bureaucracy it would engender, could cripple important biologic and medical research that have become reliant upon the recombinant DNA break-through. In attempting to guard against what were never more than conjectural and are now highly improbable and unrealistic fears, such legislation would require costly expenditures and procedures in a vain attempt to obtain the impossible: assurances of zero risk. For me and many others that price is too great, the cost-benefit ratio unacceptably high. As I see it the Nation would gain little from such legislation but the losses would be excessive. Instead of seeking new ways of regulating this field of science, I believe we should be exploiting, as vigorously as we can, the tremendous opportunities afforded by the recombinant DNA methodology. In doing so we should also be alert and ready to respond to any surprises that signal danger. But certainly we should not be erecting political and administrative barriers or bureaucracies that serve no useful purpose. Unfortunately, Secretary Califano is determined to do otherwise (See Nature's comments - enclosure 1 and the Federal Register December 22, 1978).

As you well know creativity, intuition, knowledge and persevering effort are the hallmarks of first-rate science; the administrative edifice that has already been constructed to govern recombinant DNA research frustrates the first two and impedes the last. Having a cumbersome, inefficient and unresponsive bureaucracy, centered in Washington-Bethesda, controlling the design and performance of experiments is incompatible with effective research.

Interminable meetings of local committees that have too little understanding of the experimental details, opportunities or putative risks they must judge, consume scandalous amounts of time. Before experiments can be initiated to determine the significance or usefulness of promising leads, breakthroughs or hypotheses there are frequently months of review, discussion and paper pushing. And, if in the progress of the research some alteration in the protocol becomes necessary or advisable, further time-consuming consultations and deliberations are required. Is it any surprise that an increasing number of leading American scientists, have had enough of burdensome restrictions, and stultifying administrative demands; anticipating still worse times,some are turning to foreign laboratories for relief (see news clipping, enclosure 2). However one feels about the propriety of their decisions and actions, we can not escape the signal and its implications: Important science, research that society wants done and is paying for, will leave the United States to be done in a more hospitable setting. But what are the prospects for those scientists who do not have this option? Very likely, morale and enthusiasm will wane, the overall research effort will be eroded and the momentum generated by the recombinant DNA breakthrough could be dissipated. This may please the "less-knowledge is better" and environmentalist forces but for most it is a calamity we can ill afford.

The recombinant DNA controversy has been and continues to puzzle me. Those of us who signed the 1974 "Moratorium" letter did so not because we were convinced of imminent danger or disaster (as it has often been misrepresented), but because we believed the issue needed to be raised, examined carefully and critically before, not after, proceeding full-speed ahead. But what began and should have remained principally a scientific issue, was captured and exploited by sensationalist journalists (see enclosure 1 for a recent example), by environmentalist and single-issue public interest groups who fail to consider the wider consequences of their own narrow-mindedness and by demagogic scientists and politicians who prefer to manipulate rather than understand the issues. For me and many of my colleagues this exercise in science and public policy has become nightmarish and disastrous.

January 5, 1979
Page 3

Rather than legislative or administrative actions to curb recombinant DNA research we need to look seriously at whether the unlikely risks that remain, warrant huge expenditures of money to support the present machinery that regulates recombinant DNA research. My own early concerns that some recombinant DNA experiments might carry risks have long been dissipated. The data, discussions and experience of the last four years have convinced me that our earlier concerns are no longer warranted.- I now believe that Society has more to fear from the intrusions of government in the conduct of scientific research than from recombinant DNA research itself. Initially, I was full of hope, even optimistic, that we could act wisely and decisively to develop procedures that would promote research using this promising new tool and minimize or eliminate whatever risks that pertained. But from any vantage point, I see only a victory for the politicians of science rather than scientists. I and most of my colleagues are growing increasingly apprehensive of the outcome of such victories. To counter individuals and organizations promoting "Science for the People" we need "People for Science".

I greatly appreciate your invitation to state my views on this matter. I am sorry I could not do it sooner. I also wanted, particularly, to commend you on your science letter opposing oppressive legislation to control scientific research; it is reassuring to know that there are some in our government councils who share, through experience and understanding, our love for science. I often recall your statement that if you were reentering science today, biology would be your first choice. I'd be happy to fill you in and keep you abreast of developments in the recombinant DNA field should you decide its really not too late. Perhaps you'd be interested in some of the scientific happenings in our lab; A recent New York Times article summarizes some current work that will appear in Nature during mid January.

Respectfully,

Paul Berg

Paul Berg

PB:vs

Enclosures

Document 12.15

October 26, 1978
Beckwith Letter to Gartland

DEPARTMENT OF MICROBIOLOGY AND MOLECULAR GENETICS
HARVARD MEDICAL SCHOOL
25 SHATTUCK STREET
BOSTON, MASSACHUSETTS 02115

October 26, 1978

Dr. William Gartland
Recombinant DNA Office
National Institute of Allergy
and Infectious Disease
NIH, Bldg. 31, Rm. 7A04
Bethesda, Md 20014

Dear Dr. Gartland,

I wish to express a concern about a particular class of recombinant DNA experiments which may be hazardous and which may not have been considered in the drawing up of NIH guidelines. I have come to consider this problem as a result of experiments being done by a collaborator of ours who is using certain of our strains in recombinant DNA experiments.

The general problem is as follows: several laboratories (Boyer, UCSF; Gilbert, Harvard; Hofnung, Institut Pasteur, Paris) have used recombinant DNA techniques to construct strains which produce hybrid proteins between the peptide hormones insulin and/or somatostatin and bacterial proteins. In the case of the San Francisco group, the hybrid protein was cytoplasmic. In the Harvard studies, the hybrid is secreted into the bacterial periplasm. I understand that this group is now attempting to isolate mutants of these strains which will secrete the hybrid protein into the culture medium. In the case of the Paris group, they are using strains we constructed in which β-galactosidase, ordinarily a cytoplasmic protein, is exported to the exterior of the bacterial cell surface. They have fused the gene for somatostatin to this exported β-galactosidase in order to put somatostatin on the cell surface.

In all these cases one has a situation in which, in essence, a peptide hormone (human or closely related) is produced in modified form by the bacteria. The modification is the attachment of the hormone to a bacterial protein. This modified form of the hormone may also result in a modified antigenic activity of the hormone portion of the hybrid protein. I have talked with a number of immunologists in the Boston area (Drs. Schur, Unanue and Dessein), with infectious diseases experts (Drs. Gorbach and Fields) and others with knowledge in the recombinant DNA area (Drs. Baltimore and Benjamin). In all cases, these individuals agreed that there was a real possibility that these modified peptide hormones could break tolerance/induce an auto-immune response in humans to their own hormones if they reached the appropriate sites in the body. This, in turn, could clearly have severe effects on the health of the affected individuals. I believe that the individuals I have consulted with are not, in general, alarmist about recombinant DNA matters and their concerns should be taken seriously.

Dr. William Gartland 2. October 26, 1978

It may be that of the experiments I have described, there is a hierarchy of potential dangers. I was drawn to think of this in the first place by the use of our strains by Hofnung and colleagues to put a hybrid somatostatin on the cell surface of E. coli. In this case, the modified hormone would be exposed to the "outside world" and would be in a position to elicit an immune response. Similarly, any strains which secrete individual hormones into the media would raise this possibility. Strains producing modified hormones in the cytoplasm or periplasm might create less of a problem.

It is my understanding that this potential problem is not covered by the guidelines. I would like for you and the Recombinant DNA Advisory Committee to consider this problem and let me know whether it is considered that these concerns are valid, and, if so, what steps will be taken. For instance, it seems reasonable to me to carry out animal tests with these bacterial strains to see whether they can induce the breaking of tolerance. Also, I would like to know the containment which would be necessary for such experiments.

I realize that there is a general impression that E. coli K12 does not persist long enough in the body to cause problems. But my own feeling is that not enough experimentation has been done to scientifically substantiate that impression. It seems to me that particularly when there is a concrete possibility of health hazard, that a more detailed examination of the problem should be carried out.

Since in one of the cases I have described, strains constructed in this laboratory have been used, I personally feel responsible for any consequences of the work. I am eager to hear the response of yourself and the RAC. I am sending a copy of this letter to Dr. Susan Gottesman who worked as a graduate student in this laboratory and, I believe, is a member of the Advisory Committee.

Sincerely,

Jon Beckwith

Jon Beckwith

JB/am

cc: Susan Gottesman
Walter Gilbert
Maurice Hofnung
Malcolm Casadaban
Herbert Boyer

March 22, 1979
Rowe Letter to Dixon

DEPARTMENT OF HEALTH, EDUCATION, AND WELFARE
PUBLIC HEALTH SERVICE
NATIONAL INSTITUTES OF HEALTH
BETHESDA, MARYLAND 20014 20205

March 22, 1979

Dr. Frank Dixon
Director of Research
Scripps Clinic and Research
Foundation
LaJolla, California 92037

Dear Frank:

In the course of discussions of possible biohazards of recombinant DNA research with E. coli host-vector systems, concern has been expressed, as for example in the enclosed materials from Drs. King and Beckwith, about possible immunological diseases resulting from the production of eukaryotic proteins by an enteric bacterium. These concerns are of two general types - production of an autoimmune disease as a result of altered antigenicity or recognition of a host protein, or disturbance of hormone function as a result of an antibody response to the hormone, such as anti-insulin.

As a member of the NIH Recombinant DNA Advisory Committee, I am asking several persons (Richard Asofsky, Frank Austen, Baruj Benacerraf, William Paul, Norman Talal, Philip Paterson, and you) who are acknowledged experts in the field of immunology and/or autoimmune disease to give a thorough, objective evaluation of the evidence for or against the possibility of such disease mechanisms occurring, as well as any other relevant considerations. I would like very much to have a rather detailed analysis, with literature citations. I propose to circulate these analyses to other immunologists for comment, and then to the Recombinant DNA Advisory Committee prior to their next meeting on May 21, 1979.

It is by means of this type of expert input and thoughtful analysis that we will be able to finally come to terms with what is real and what is not real about risk scenarios of recombinant DNA research.

Your help in this could be of immense importance to NIH and to all of biomedical research.

With best personal regards, Sincerely yours,

/s/

Wallace P. Rowe, M.D.
Chief, Laboratory of Viral Diseases
National Institute of Allergy and
Infectious Diseases

2 Encls

1. Jonathan King: Recombinant DNA and Autoimmune Disease. J. Infect. Dis., 137: 663-667, 1978. (From Falmouth Meeting Report.)

2. Letter from Jon Beckwith to William Gartland, dated October 26, 1978.

Document 12.16

January 11, 1979
Nature (**277**:83)

Meanwhile, back in Washington . . .

DNA critics appointed to advisory committee

Two active participants in recent debates over the adequacy of guidelines for research using recombinant DNA techniques have been appointed to the committee responsible for advising the director of the National Institutes of Health on a number of important aspects of such research.

The two—Dr Richard Goldstein, assistant professor of microbiology and molecular genetics at Harvard Medical School, and Dr Sheldon Krimsky, acting director of the program in urban, social and environmental policy at Tufts University—are among 14 new members appointed to the Department of Health, Education and Welfare's Recombinant DNA Advisory Committee by HEW Secretary Joseph A. Califano.

The appointments follow Mr Califano's announcement last month of a number of changes to existing arrangements for regulating recombinant DNA research. The advisory committee is responsible for giving advice on new types of bacteria for use in such research, on whether certain presently prohibited experiments should be conducted, whether additional categories of research should be exempted from the guidelines, and on possible future changes in the guidelines.

Mr Califano also announced that the size of the advisory committee was to be increased to 25, and that the scope of its membership was to be extended to give greater weight to non-scientific representation. The new members of the committee include a professor of education, a professor of law, a prominent environmentalist and a laboratory technician.

In addition to Dr Goldstein and Dr Krimsky, the new members include: Dr Karim Ahmed, senior staff scientist with the Natural Resources Defense Council; Zelma Cason, chief of cytotechnology in the department of cytology at the University of Mississippi Medical Center; Patricia King, professor of law at Georgetown University Law Center in Washington; Dr Samuel Proctor, professor of education at Rutgers University; and Ray Thornton, retiring member of the US House of Representatives and chairman of the House subcommittee on science, research and technology in the last Congress.

The new scientific and medical members are: Dr David Baltimore, (MIT); Dr Francis Broadbent (University of California, Davis); Dr Richard Novick (New York Public Health Research Institute); Dr David Parkinson, (University of Pittsburgh); Dr Damon Pinon, (University of California); Dr Luther Williams (Purdue University); and Dr Frank Young (University of Rochester).

There is thought to have been considerable controversy over some of the other names of "public interest" representatives suggested for possible membership of the committee, particularly since some scientists involved in the research felt that too large a non-technical component might impede the committee's functioning. □

Document 12.17

1979

Recombinant DNA and Genetic Experimentation (p. 151)

RECOMBINANT DNA
AND
GENETIC EXPERIMENTATION

Proceedings of a Conference on Recombinant DNA, jointly organised by the Committee on Genetic Experimentation (COGENE) and The Royal Society of London, held at Wye College, Kent, UK, 1-4 April, 1979

Editors
JOAN MORGAN
and
W. J. WHELAN

A HISTORY OF THE RECOMBINANT DNA GUIDELINES IN THE UNITED STATES

D. S. Fredrickson

National Institutes of Health, U. S. Department of Health, Education and Welfare, Bethesda, Maryland 20205, U.S.A.

On December 16, 1978, a telegram purporting to be from the Vatican was hand-delivered to the office of Joseph A. Califano, Jr., Secretary of Health, Education, and Welfare. "*Habemus regimen recombinatum,*" it proclaimed, in celebration of the end of a long struggle to revise the NIH Guidelines for Research Involving Recombinant DNA Molecules. It was not the first telegram the Secretary had received on this subject. From Peking the preceding June, I had responded to his cabled instructions concerning proposed revisions of the Guidelines which I had taken to China. The U.S. liaison officers found my reply inappropriate for transmission, but I delivered it on my return, imprinted on a rice paper temple-rubbing.

These sophomoric tricks were moments of comic relief in a three-year period of coping with the scientific, political, and legal problems created by the advent of the "new biology." The following pages summarize my impressions of this turbulent experience.

My personal history of the DNA Guidelines in the United States recognizes three phases to date. Phase I is the period between the early concern about possible hazards of recombinant DNA technology and the delivery to NIH of proposed rules for conducting research. Phase II covers the promulgation of the NIH Guidelines in 1976 and the thirty months before their official revision. Phase III began on January 2, 1979, with a new set of rules painfully formulated during this unprecedented curtailment of experimentation in biology.

THE END OF THE BEGINNING

In this a collection are reminiscences of the first apprehensions (1973), the decision to develop guidelines (1974), the Asilomar agreements (1975), and the exhausting constructions of the NIH Recombinant DNA Advisory Committee (RAC). Three versions of guidelines had emerged from RAC meetings after Asilomar. In La Jolla, California, on December 5, 1975, the Committee, with the "variorum edition before it, finally succeeded in scaling conjectural hazards by parliamentary procedure. Chairman Hans Stetten went to the telephone to inform the NIH Director in Bethesda that the nation had acquired rules for recombinant DNA research. Much later I was told how he had returned to the conferees, shoulders drooping, success drained from his face. "He wants to have a public hearing on them", he mumbled.

PUBLIC AIRING BEGINS

From the beginning the decision to "go public" was variously understood and was resented by many. After all, the Director, NIH, has long had authority to promulgate guidelines for investigators the agency supports. There is no requirement for hearings or public comment. I became aware of new responsibilities heading my way sometime in the autumn of 1975, when I had been Director for only two or three months. At that time, I had barely heard of restriction enzymes and could not even have explained the crucial distinctions between Federal *guidelines* and *regulations*. From the first I was inclined—and after a little study and consultation, quite determined—to air in an open and public manner the scientific and social issues. This was the only way to decompress rising tensions and to prepare to defend whatever actions would be taken against certain criticism.

The Director's Advisory Committee (DAC) was convened in February 1976 for public discussion of the Guidelines. The transcript, like all the other relevant documents on the subject, is available in the "public record" published by NIH.[1] The hearing demonstrated the difficulties of holding a town meeting on molecular biology and exposed the full range of opinions on the risks of the new technology. It was apparent that our decisions would have to run a gamut of adversarial reactions and that some, in the end, might well be tested in the courts. After the hearing, the voice of Judge Bazelon lingered longest in my mind: ". . . the healthiest thing that can happen is to let it all hang out, warts and all, because if the public doesn't accept it, it just isn't worth a God-damn."

We made some changes in the proposed Guidelines after the DAC meeting, mainly adding administrative structure. We then set out to acquaint key people and agencies with the details, for NIH supported most but by no means all of the affected research. The widening circle included the National Science Foundation, the Department of Agriculture, other Federal agencies whose authorities were crossed by the NIH Guidelines, the staffs of Congressional committees with jurisdiction over biomedical research, and representatives of industry doing what private research of this sort there was at the time.

ISSUANCE OF THE GUIDELINES

The NIH Guidelines were issued on June 23, 1976. It was front page news, but the reactions were muted. We also established the Office of Recombinant DNA Activites (ORDA), under the direction of William Gartland.

The NIH Guidelines were just that—guidelines, not regulations, which have more of the force of law. The verbs tended to be "shoulds," though some "shalls" had been substituted after the February hearing. It was stated that the Guidelines would be frequently revised, but no special procedures for doing so were laid out. They were expected to evolve as understanding of the subject grew. As it turned out, it was not the subject of the Guidelines, but "due process" for changing and adminstering them, which became the focus for opposition to the research. For the next two and a half years, the Guidelines were to be practically frozen, while the science expanded impatiently within.

EXTENSION OF THE GUIDELINES BEYOND NIH

NIH had no illusions that it was creating guidelines for all the recombinant research in the world. Scientists are citizens of different nations whose laws can supersede intellectual accord. Even within the United States, extension of the same rules to all laboratories could not be achieved by any simple move.

Two different kinds of protest about this incompleteness were brewing in 1976-78. One encouraged extension of the jurisdiction of state and local communities to regulation of laboratory research, a legal area hitherto unexploited. The other sought to persuade DHEW, its Food and Drug Administration (FDA), and other regulatory agencies to use certain narrow authorities to force compliance with common rules. If that failed, the Department was to seek a Federal law to that end. Some fervent advocates of legislation fought for preempting local jurisdictions from enacting more stringent standards if they wished. Others just as vehemently opposed Federal preemption. A Balkanization of recombinant DNA research was one of the most serious and extraordinary threats of this period.

In May 1976 we informed our Department superiors about our intention to issue guidelines, and urged then-Secretary David Mathews to ask the President to direct all relevant Federal agencies to coordinate recombinant DNA activities through an interagency committee. Mathews agreed, but no words emanated from the White House. In July, Senators Kennedy and Javits addressed a letter to President Ford advocating the extension of NIH Guidelines to all Federal and private research. Local hearings in Cambridge, Mass.; Ann Arbor, Mich.; San Diego, Calif.; and New York added to a sense of urgency.

The President's letters were finally dispatched in September, and the Federal Interagency Committee on Recombinant DNA Research was promptly convened in Bethesda. The research agencies readily agreed to use the NIH Guidelines for the research they supported or conducted. The committee then undertook to examine the regulatory authorities of each of the member agencies and to develop recommendations for possible new legislation. Later the committee would examine patent policy and the international aspects of regulating DNA research.

NEPA AND THE FRIENDS OF THE EARTH

A full discussion of the National Environmental Policy Act (NEPA) with reference to laboratory research in general and to how the NIH Guidelines for recombinant DNA research became involved would fill a volume. NEPA, a law passed in 1969, requires the Federal agencies to determine whether contemplated actions will significantly affect the environment. If so, the action must be heralded by an Environmental Impact Statement (EIS). In the spring of 1976 we were made aware that if we released the Guidelines before issuing an EIS, we could be charged with violating NEPA.

Although an EIS had become common in proposals to level mountains or build dams, the adaptation of NEPA to conjectural hazards of laboratory research was a

[1]Office of the Director, NIH (1976-78). Recombinant DNA research, Vols. 1-4 (4,015 pp. in all); for sale by Superintendent of Documents, U.S. Govt. Printing Office, Washington, D.C. 20402, and available in about 600 public libraries of the GPO depository system. (GPO stock no. for Vol. 1, 017-040-00398-6; Vol. 2, 017-040-00422-2; Vol. 3, 017-040-00429-0, and appendix, 017-040-00430-3; and Vol. 4, 017-040-00443-5, and appendix, 017-040-00442-7.) The environmental impact statement, cited in footnote 2, was published as a supplement to Vol. 2 (not included in above page-count.)

nightmare. The situation was aggravated by ambiguous and arbitrary procedures for implementing NEPA within DHEW. The tortoise-like march from *draft* to *final* EIS could take years. But delaying the issuance of the Guidelines pending completion of the EIS process was never an alternative. The voluntary agreements made at Asilomar were losing their hold on the scientists, confusion was mounting, and dissidents in various communities threatened to obtain either local regulation or prohibition of the research if Federal standards were not quickly forthcoming. It was obvious that the public interest would be better served—and the opportunity of scientists to continue experiments, better protected—with guidelines than without them, even if an EIS were not published until after their issuance.

We therefore released the Guidelines in June with an announcement that an EIS was to follow. The draft EIS was filed in September 1976, the final one in October 1977.[2] In May 1977 two suits against NIH were launched in separate Federal courts. One, brought by an organization called The Friends of the Earth, sought to enjoin all recombinant research. The other sought to block the Rowe-Martin risk assessment study. The final EIS was entered as part of the Government's defense in the latter suit (Mack v. Califano). In finding for the Government in March 1978, the Court concluded that NIH, in its EIS, had indeed "taken a hard look" at the consequences of experiments with recombinant DNA. The plaintiff was denied an injunction.

THRUSTS TOWARD LEGISLATION

A bill was introduced in the Senate (S. 621) in February 1977 by Senator Bumpers (D., Ark.), with a companion bill in the House by Rep. Ottinger (D., N.Y.). They were the first of 12 bills to regulate DNA research submitted to the 95th Congress. On February 23, representative scientific leaders were invited to NIH to read selected passages of the proposed new legislation, including heavy penalty provisions. Two weeks later, the last traces of their indifference were dispelled by the acrimonious tone of a forum on "genetic engineering" at the National Academy of Sciences.

The following May the Interagency Committee conveyed to the new HEW Secretary, Joseph A. Califano, Jr., its conclusions that a Federal statute would be required if the Guidelines were to be extended to all recombinant research in the country. It also offered the elements of what it considered an "ideal" law—elements that were quickly converted to an Administration bill introduced by Senator Kennedy (S. 1217) on April 1, 1977. Kennedy then revised the bill radically. In an intensive reaction to this and other proposed laws, scientists and their organizations soon made strong appeals to the Congress. The ardor of the legislators for statutory regulation cooled progressively during 1977-78.

REVISION IS NEEDED

Within six months of their appearance, the NIH Guidelines clearly needed revision. The molecular biologists who had constructed them, if given that chance again, would surely have engaged other disciplines on the route from Asilomar to Bethesda. Especially lacking had been the counsel of experts on infections, who had a better perspective of the improbability that *E. coli* K-12 could be converted into an epidemic pathogen. And more thought should have been given to the containment levels for dealing with viral DNA, to the prohibitions, and to the coverage of organisms known to exchange DNA in Nature.

It is also notable that the guidelines constructed at about the same time in the U.S. and the U.K. were quite different in form. The Americans wrote an extensive codification and the British opted for common-law evolution of minimum rules. Both sets, however, were meant to be interpreted and administered centrally, by a GMAG or an ORDA. The U.S. scientists did not want local committees to second-guess their experimental protocols. More comfortable with central decision-making by study sections in Bethesda, they preferred ORDA's interpretations and administration.

But delays in administrative actions were inevitable. The requirements for prior NIH approval of all changes in ongoing probjects particularly irked investigators. In rejecting at the start nearly all suggestion of control by their institutions, the scientists had made it difficult to regain a proper balance between local application and national standards.

As a broker between the molecular biologists and the various public interests, NIH also failed to perfect the Guidelines before issuing them. We did not incorporate mechanisms for revision. Discretionary powers to make interpretative judgments and minor changes, essential in so complex and fast-moving a subject, were lacking. There were reasons, however, for avoiding imitation of the formal and formidable procedures of the regulatory agencies. .The more we embedded the Guidelines in inflexible administrative molds, the less chance there would be for timely accommodation to the tide of new information that was already rising.

[2]Office of the Director, NIH (1977). National Institutes of Health Environmental Impact Statement on NIH Guidelines for Research Involving Recombinant DNA Molecules Part One (147 pp.) and Part Two (appendices, 438 pp.); for sale by Superintendent of Documents, U.S. Govt. Printing Office, Washington, D.C. 20402, and available in about 600 public libraries of the GPO Depository system. (GPO stock no. 017-040-001413-3.)

REVISION BEGINS

In January 1977 the RAC commenced to prepare a revised set of Guidelines. A workshop was held in June at Falmouth, Mass., to synthesize old and new information about bacterial host-vector systems. In September the proposed revision—a complete rewriting of the text—was formally presented to me and published in the *Federal Register*. A two-day hearing was held in December 1977 at which the RAC members defended their proposed changes. Most critics raised questions of process; but some containment levels were severely challenged, and additional meetings of experts on viruses and plant pathogens led to further alterations.

In Departmental clearance, the revised Guidelines encountered more difficulties than the original. Recombinant DNA research had emerged as a scientific issue with immense appeal to laymen, and Secretary Califano's staff took a strong and sophisticated interest in how all relevant law and administrative practices pertained to the new draft. By now, some dissidents and a militant fraction of the environmental movement had also launched a concerted campaign to exact, if science wished to proceed, a more generous tax in procedure.

On July 28, 1978, the proposed revision, accompanied by our environmental impact assessment and a Director's decision paper, was published in the *Federal Register.* An introductory memorandum from the Secretary invited the public to comment and announced that, after a 60-day period, there would be another hearing chaired by Peter Libassi, the HEW General Counsel.

REVISION COMPLETED

The Libassi hearing took place on September 15 at the HEW headquarters in Washington. NIH staff and I—the "Kitchen RAC"— then dissected the comments received in testimony and 170 letters, and joined in numerous discussions with Mr. Libassi and his committee. Special meetings were also held with a group of environmentalists who wished to reinforce some of their earlier demands. We also met with representatives of pharmaceutical firms, other Federal research agencies, and members of institutional biosafety committees to discuss their problems with the proposed revisions. A culmination of the Libassi hearings was the reconstitution of the RAC to broaden its public (nonscientific) membership and to combine the technical and policy reviews, usually carried out at NIH in a two-tiered process. The appointment of RAC members was shifted from the NIH Director to the HEW Secretary. The revised Guidelines were released late in December, to become effective on January 2, 1979.

The detailed analysis led by the HEW General Counsel had added a few weeks to the long period of revision. One result was a modest additional burden of procedural safeguards, but this was offset by the removal of any grounds for complaint from the most fervent dissident that the public had not been exhaustively consulted.

There were important achievements in the revision. The new Guidelines contain provisions for continuous and orderly evolution of the rules—even to their eventual elimination when the need passes. Many experiments now judged to be harmless are exempt, and containment for other kinds of experiments has been reduced. Also, the discretion and responsibility for observing the rules are beginning to return to the research institutions, where I believe they belong.

Attempts to enact statutory regulation of recombinant DNA experimentation in the United States need not be revived soon. One hopes they will not, for some of the medieval features of the first bills tended to reappear as later ones passed through the committees. A problem remains, however, in the limits on NIH's ability to protect proprietary data submitted to the RAC. Actions taken by Secretary Califano upon release of the Guidelines, to have regulatory agencies (the Food and Drug Administration, the Environmental Protection Agency, etc.) use their existing authorities to extend the Guidelines over research in the private sector, have been helpful in exploring an alternative to a new law.

MORAL

It is possible that the "recombinant DNA affair" will someday be regarded as a social aberration, with the Guidelines preserved under glass. Even so, we can say the beginnings were honorable. Faced with real questions of theoretical risks, the scientists paused and then decided to proceed with caution. That decision gave rise to dangerous overreaction and exploitation, which gravely obstructed the subsequent course. Uncertainty of risk, however, is a compelling reason for caution. It will occur again in some areas of scientific research, and the initial response must be the same. After that, the lessons learned here should help us through the turbulence that is sure to come.

(This paper was given by W.J. Gartland, Director, Office of Recombinant DNA Activities, NIGMS, Bethesda, Maryland, USA.)

DIEU ET
MON DROIT

FURTHER DEVELOPMENTS IN BRITAIN

On August 1, 1978, it became a legal obligation in Britain to report to both the GMAG and the Health and Safety Executive all recombinant DNA experiments. On September 11, Janet Parker, a photographic technician at Birmingham University Medical School, died of smallpox; five days previously Henry Bedson, Professor of Virology, had committed suicide. Bedson's laboratory had been doing conventional virological research, not recombinant DNA research with smallpox virus. The staff of the smallpox laboratory had not scrupulously followed the advice of the Dangerous Pathogens Advisory Group (DPAG), a body analogous to the GMAG but responsible for work with real pathogens. Unlike the GMAG, the DPAG was composed entirely of medical specialists and had no public-interest or trade-union members. The DPAG's work had not been subject to the public scrutiny that had surrounded the GMAG and recombinant DNA. In Britain (and no doubt elsewhere) an ironic situation—which in Birmingham had tragic consequences—had been allowed to develop. Recombinant DNA experiments with only conjectural hazards had come to be more closely regulated than conventional research with known lethal pathogens. The Birmingham accident cast grave doubts on the ability of a profession to regulate the safety and conduct of its own members, as well as on the ability of society to maintain a sense of perspective over such issues as recombinant DNA research.

In Britain the trade union ASTMS, which had two members on the GMAG and also of which Janet Parker was a member, made no secret of its desire to be more deeply involved in science policy and the control of research and saw the GMAG as an avenue of influence in this area. At a one-day symposium organized by the trade union on October 27, 1978, the General Secretary of the ASTMS clearly stated the union's aim (Doc. 13.1). At this symposium Sydney Brenner (MRC Cambridge) and Jonathan King also explained their contrasting views about the need for regulation of recombinant DNA research (Doc. 13.2 and Doc. 13.3).

A few days after the ASTMS symposium, the GMAG began to discuss a new approach to the assessment of the conjectural hazards of recombinant DNA experiments, which was largely the work of Sydney Brenner (Doc.

13.4). An editorial in *Nature* greeted the proposals with an enthusiastic headline—"Now Reason Can Prevail"—but a blue-ribbon committee of the Royal Society was less enamoured of it (Doc. 13.5), and *Nature*'s editorial incensed Waclaw Szybalski (Doc. 13.5). At a public meeting British molecular biologists were particularly critical of the fact that so-called self-cloning experiments, in which recombinant DNA methods are simply used to mimic *in vitro* genetic events that occur in nature, were still not excluded from the regulations (Doc. 13.6). In addition, the prospect of large sums of public money being spent by the British government on risk-assessment experiments—"tens of millions of pounds" was the figure mentioned by Secretary of State Shirley Williams on a television program—caused J. D. Watson and Robert Pritchard (Leicester) to protest (Doc. 13.7 and Doc. 13.8). Why waste so much time and money attempting to assess risks without there being any evidence of a risk to assess?

January 1979 was a difficult time for the GMAG. It had a new chairman, it was being expanded (but a bid by the Trade Union Congress to increase trade-union representation met with no success), and it had quickly to reach a decision about the Brenner scheme for assessing containment conditions. In fact, in this matter it had no real options. In the United States the revised NIH guidelines had come into force, and if Britain's molecular biologists were to stand a chance of competing, the GMAG would somehow have to reduce the high levels of containment it required for experiments. In a letter to *Nature* John Maddox, a public-interest member of the GMAG, made no bones about that and saw the Brenner scheme as a way of bringing it about (Doc. 13.9). In a reply to Maddox, Alan Williamson was less sanguine (Doc. 13.9). In March 1979 the GMAG, still cautious, decided to use the new risk-assessment procedures for a trial period alongside the old Williams report guidelines; at the same time it announced the modest concession that at least some self-cloning experiments were to be exempted. To complete the story, by the end of 1979 the GMAG, having found that the Brenner procedure allowed greater flexibility and for most experiments a relaxation of containment and alignment with U.S. practice, decided to use it instead of the Williams guidelines (Doc. 13.10). Since then the decisions of the GMAG have led progressively to very significant reductions in the containment requirements for most experiments.

A report by a joint working party of two government advisory boards and the Royal Society was published in March 1980 (*Biotechnology*: Report of a Joint Working Party, H.M.S.O., London, 1980) and leaves no doubt as to how the GMAG is expected to proceed:

> We are concerned that controls on recombinant DNA research in the United Kingdom should not be more severe or demanding than in other countries. As new research defines more precisely the nature of the hazards, we recommend that the Government, in consultation with others abroad, should modify the regulatory regime appropriately. We recommend that GMAG and the HSE should continue as rapidly as possible to reduce constraints upon genetic manipulation experiments while maintaining an adequate degree of safety. We support the special attention that is being given to procedures to encourage industrial application of the technology of recombinant DNA and the use of its products on a large scale. We also support the continuing study by GMAG of biological containment with the objective of reducing the requirements for physical containment. We recommend also that GMAG attends even more urgently than hitherto to the possible prejudicial consequences to British industry if constraints in the United Kingdom on genetic manipulation are excessive compared with those in other countries.

Document 13.1

November 2, 1978

New Scientist (**80**:339)

Union seeks bigger role in British science

Roger Lewin

The Association of Scientific, Technical and Managerial Staffs (ASTMS) aims to achieve greater influence over scientific research in the UK, not just in ensuring the safety of its members in laboratories, but also in controlling the direction of science through government bodies. This was the clear message that emerged from a one-day symposium the union organised last Friday in London on the social and safety implications of genetic engineering.

Clive Jenkins, general secretary of ASTMS, said "I want to see the trade union members have a greater input into science policy, not just to protect them, but also to enrich them and their country". This is clearly not an empty ambitiion because Shirley Williams, Secretary of State at the Department of Education and Science, commended the union's contribution to control of genetic engineering through the Genetic Manipulation Advisory Group (GMAG) and suggested that other areas may be appropriate treatment: 'The GMAG experience could be a guide to the control of science," she said.

Although representation on research council committees may be ASTMS's ultimate goal, it recognises very strongly its more immediate responsibility in handling hazards in the workplace. Talking of genetic manipulation, Jenkins pointed to a gross anomaly in the way the two aspects of the science are regulated: genetic engineering, which involves chopping up genes with enzymes and then splicing the bits to other pieces of genetic material, is subject to the relatively tight surveillance of GMAG and local safety committees; classic genetic selection, which may involve work with highly dangerous viruses such as those for smallpox and lassa fever, comes under much less scrutiny. (Research with these hazardous organisms is overseen by the Dangerous Pathogens Advisory Group, DPAG.)

So, on one hand there is the new science of genetic engineering which, though yet to be demonstrated to be dangerous, is closely controlled; and on the other is manipulation of known hazardous organisms which, though it must be done in a licensed laboratory, is not under the continuous monitoring employed in genetic engineering. How should the anomaly be resolved?

Jenkins made his position clear: referring to the recent Birmingham smallpox case (*New Scientist*, 19 October, p 155) he said that "If there had been a local safety committee and adequate consultation at Birmingham (as would be necessary under a GMAG-type system) the disaster at Birmingham would not have occurred. The DPAG structure must be re-examined," he continued, "and we will want union representation there."

Of GMAG's 19 member committee only eight are experts in the field (there are four Trades Union Congress representatives, two of whom are from ASTMS). Jim Coombes, research director of Hoechst UK, believes that "It is inconsistent to have a minority of experts on a committee that is supposed to advise scientists on procedure". Although Coombes is clearly happier with DPAG, whose 15 members are experts in either microbiology or epidemiology, he suggests it would be best to scrap both bodies and set up a biological hazards advisory group to cover both natural and artificial genetic manipulation. Such a group would have much less power than GMAG currently exerts.

Sydney Brenner, soon to be chairman of the Medical Research Council's Laboratory of Molecular Biology in Cambridge, is also unhappy with the anomaly. "Why pick on us?" he asked, "when non-genetic engineers in neighbouring laboratories can juggle with dangerous viruses almost without hindrance." Brenner must represent most of the scientific community in wanting less bureaucratic intervention, not more.

Last Friday's meeting made scientific history in that it was a gathering organised by a trade union on a specifically scientific topic. Donna Haber, an ASTMS official and a member of GMAG, said "We wish to help encourage scientific research in this and other areas, but only if our concerns are adequately dealt with. If they are not, we are just as ready to confront employers over health and safety." Haber also looks to the reconstitution of DPAG along the lines of GMAG, and she puts particular emphasis on the establishment of local safety committees.

ASTMS's determination to have a bigger say in the affairs of science is undoubtedly solid, and it is therefore not surprising that the recent smallpox incident and the debate over genetic engineering are being used as wedges to be driven into the once-intact fabric of scientific elitism. "People no longer trust experts—and for good reason," said Shirley Williams. ASTMS would like to write a new contract between science and society out of which a new kind of trust can grow. □

Document 13.2
November 2, 1978
Nature (**276**:2)

Six months in category four

By Sydney Brenner

LAST WEEK, the British trade union, the Association of Scientific, Technical, and Managerial Staffs (ASTMS), called the first public debate in Britain on the potential hazards of genetic manipulation. In the end there was little debate, which may indicate the level of public understanding of the question, but there were stimulating contributions from the platform. Here we reproduce shortened versions of the talks of Sydney Brenner, director-designate of the Medical Research Council's unit of molecular biology at Cambridge, and Jonathan King, Professor of Biology at the Massachusetts Institute of Technology. On a following page Eleanor Lawrence disentangles the threads of the biology involved, for the meeting blurred the distinction between genetic manipulation proper (using restriction enzymes *in vitro*) and other work (such as the investigation of dangerous pathogens, or recombination in natural systems). Finally, an ASTMS officer explains her union's plans to radicalise university research laboratories in Britain.
TODAY Britain's Genetic Manipulation Advisory Group (GMAG) is considering whether to adopt a new set of principles for assessing the hazards of individual experiments, much more precise than the 'rule of thumb' phylogenetic system so far adopted in the UK, the US, and elsewhere. Sydney Brenner was its inventor.
NEXT WEEK, if GMAG adopts the new guideline, *Nature* will publish a condensed version of them.

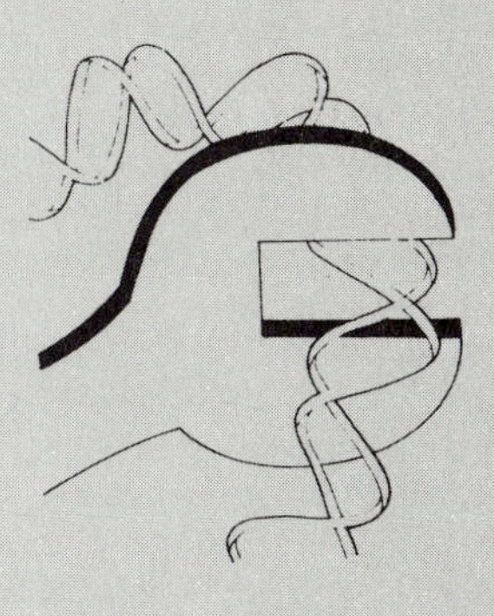

'I've been associated with this debate almost from its inception; and I can say that I have given a considerable amount of my time—perhaps four years of my life—to it, and certainly something like ten shelves in my office to the paper that has flowed out of the discussions. These discussions have been very extensive. It will be impossible here to go into all of the details of the subject. But one must look at the historical constraints under which the subject was approached.

Interest in the subject of 'genetic engineering' or 'genetic manipulation' was first aroused by a letter written by some Americans—Paul Berg and others—requesting a moratorium on this research. I must tell you that there was an immediate response in this country. I think that whereas the moratorium was enforced in this country, it was only treated as a voluntary matter in other countries. As an employee of the Medical Research Council (MRC) I received a letter at the time, from the then head of the organisation, instructing me to obey the moratorium. Of course it was quite easy to obey because we weren't doing experiments.

The Ashby Committee in the UK, followed by meetings at Asilomar, which was international, followed by the Williams working party, resulted in the institution of the Genetic Manipulation Advisory Group (GMAG). And I think that it is a unique institution in the world, in that it involves formally the participation of representatives of organisations who are not scientists. And insofar as the trades unions are part of the public, one has good public input.

Now why have we had this very elaborate organisation, GMAG? I think it is very important to understand why this particular area has been singled out for very special attention. It has been singled out for special attention because of the statement made very early that there were particular dangers associated with genetic manipulation—that is, risks that were so particular, and potentially so great, that people had better start to take special action about them.

That created the historical background for why we have a GMAG for genetic manipulation. But there are lots of other things you can do in laboratories over which there is no control at all. Among we 'genetic manipulators' (if such a profession exists) there is a questioning of why this area has been singled out. It is the perception of people in the field that right next door there are people who do much worse things—you know, with the fags hanging out of the mouth, and a cup of tea, doing it any old way they like.

Now I think this is quite important, because one should desperately avoid the situation that one branch of science is singled out for the delivery of social punishment. Now I don't want to say this has happened, but I do want to emphasise that this is a kind of connection between science and society that we must avoid: that if scientists have been bad boys, as many people think—science *is* being questioned—then the genetic manipulators can so to speak carry the can and get six months in category four. That is a psychological situation that must be avoided.

We must begin to ask new questions about this whole area of research. One has heard the statement made this morning that a researcher can put antibiotic resistance into *Shigella sonnei* with no control; and you have to ask why isn't anybody controlling him? No one's controlling him because he is using natural mechanisms. That is, he's using the mechanisms used in nature to generate new strains—mutation, recombination, even perhaps genes that jump around from one piece of DNA to another; all he is applying to these mechanisms is special selection. He is fishing out of nature events that may be extremely rare and enriching them in a laboratory. His technology is that of enrichment. He creates by his techniques a local high concentration of such elements, of which there are enormous numbers, that lie all over the place.

What a genetic manipulator does is to do things *in vitro*. In theory, if you could do genetic manipulation by avoiding the use of test tubes then in fact you would be doing something that was a natural mechanism, and so in theory you might argue that you could escape from the GMAG regulations. I'm not offering that as a possibility, but I just want you to realise that the difference between genetic manipulation and other biology as understood by GMAG is that it is a sort of 'confined area' of genetic construction. And it is clear that you can make genetic constructions in organisms with natural mechanisms.

And so work on recombinant influenza viruses, work on recombinant bacteria, the transfer of promiscuous plasmids from one bacterium to another, that goes on, but doing the same work—indeed the identical experiment could be formulated—with a restriction enzyme would put you under the GMAG regulations. That is a paradoxical situation that we have to look into.

Why does one have to look into this? Because of the question of what constitutes a risk. There are now moves in the United States to change the so-called 'guidelines' (which may be looked on by some as attempts to amend the Ten Commandments); but the question is what are the grounds for these amendments? Now of course there is the important ground that we have new knowledge, but I think there is another ground which should be honestly stated: namely, that the risk analysis was not properly done. (This is my personal opinion). Now why wasn't it properly done?

"We 'genetic manipulators' question why this area has been singled out."

First, there are things we can actually say are dangerous. It is a remarkable thing there are natural objects we can say are risky. Smallpox virus is risky. A lot of bacteria are risky. The people who work with them know they are risky, and they will therefore take precautions. You don't play games with smallpox. No one can say it is a fundamental belief of his that smallpox isn't dangerous. One can convince him very quickly it's dangerous. So there is immediate feedback to the workers involved.

So with these things there is no problem in declaring what is dangerous. We have—if you like—knowledge about that. We know from historical, clinical, and epidemiological evidence that certain things will knock you off with a high probability. And there seems to me to be no question that those should be declared as such, and if people are fools enough not to accept control on such work, then we will see events such as we have seen in the past—not only in this country but in other countries as well.

There are 5,000 recorded cases of laboratory infection in the United States, and some of them—with *rickettsia*—had fatal consequences. There's no doubt about this. I think it's on this sort of evidence that we have to rely to judge the effectiveness of physical containment. The experiments have been done for us in the past—alas they should not have been done as experiments, but they are there, and we have to rely on that information.

But how are we to assess the potential and certainly the conjectural risks? Who will say what is dangerous? The difficulty here in doing a risk assessment is that we are in a field where we are dealing not with nuclear reactors, not with toxic substances, but with self-replicating biological entities. We are far away from questions of linear dosage. In most of the other cases, chemical and other substances, one can make calculations quite clearly which are based on linear dosage relations. You know what the toxic dose is, or one can do experiments to establish it or estimate it quite accurately, and therefore you can make conditions to prevent that toxic dose from reaching the workers and the public at large.

The trouble with the creation—or if I could call it the enhancement—of organisms is that they could depend very largely on their selective amplification. But that is the key thing. It is selective amplification that counts. Now let me make this clear what I mean. Years ago when I first started to travel in aeroplanes, with a very crude knowledge of physics and aerodynamics, I used to sit in those DC3s, and I had that marvellous vision with which I could look right into the aircraft engine; and I could see in all detail all the parts going round and round, the spinning of crankshafts and pistons and so on; and then I could actually see those hairline cracks developing. It used to worry me. I think many people look at genetic manipulation with that kind of internal vision. They see lying on the Petri dish the one horror bacterium, the one horror colony, the colony that is going to escape off that Petri dish and create global disaster.

I think that that feeling forms one of the most difficult hurdles to cross in trying to do objective risk assessment.

And I believe that to be objective is very important. In the past in this field, we have had no more than the balance of example and counter-example. I have sat in on hundreds of arguments which have got into details that even mediaeval theologians could not have reached as to the total number of nucleotide base pairs that should be allowed under section B subsection A paragraph 1.2. And the situation has been, in my opinion, that people have been arguing about whether the leaf points up, or down, without actually realising on what branch they are standing. And it is a very important thing in this risk assessment—if we are to do it properly and responsibly—that we come to realise that this is a tree, that some branches are high, and some branches are low, and that it doesn't matter if the leaf points up or down if you are going to fall 130 feet. Of course it matters quite a lot if you are only a couple of inches from the ground because there won't be a leaf, but I'm just trying to say that the business of doing it by example and counter-example—that is, to offer an example scenario that you could in this way create a thing, without having any true estimate of the abundance of these examples and counter-examples—has seemed to me to have be-devilled this field right from the beginning.

And so I had a strong belief that what is necessary in companion with the control of the activities and the categorisation and so on is that it is necessary for all of us—I mean scientists and public participants—to get together and analyse the risks. The existing guidelines are wrong.

Now the guidelines are wrong in an interesting way. I think they are wrong because they have not been scaled with respect to each other. That's the first point. I think they are wrong because —and this I find particularly bad, as a biologist—because they seem to me to perpetrate biological myths—that is myths about the world—which don't exist any more. In large measure they resemble the view of biology which in fact underpins the guidelines in anti-vivisection: the concept that if you are cold blooded animals you don't feel pain, because the idea is that warm blood and emotion and pain all go together. (You can take frogs and maul them around as much as you like—but not cats and not dogs and not horses. That's an interesting biological classification.)

There is a strong scientific onus that we do *not* enshrine in legislation myths of biology. That seems to me to be an intellectual responsibility that we should share. But I think that there is another responsibility, which is the social one, and that is that we should not be imposing on a subsection of our scientific community and technicians and practitioners controls that appear to them—and objectively—completely extreme compared to what is going on in other fields.

Indeed, if one analyses this, a strong case can be made out that genetic manipulation is actually a method of attenuating dangerous things, and that it is not in itself intrinsically dangerous. In fact it can be argued very strongly that it is a way of containing things, of moving them away from organisms that are their targets, and locking them up in other organisms where they can do nothing. For myself, if there were some national emergency overnight that scientists had to get to work quickly on lassa fever, I would say that the first thing that we should do is clone it in bacteria. Let's clone it in bacteria, let's lock it up, because there it is rendered non-infectious. Then we can work with lassa fever virus sequences safely in bacteria. (Let me just say that I think lassa fever virus cloning in bacteria is a big no-no experiment, both here and in the United States, and it would take the direst national emergency to make me work with lassa fever virus, and I'd be very, very careful myself!)

But this is a paradox of the whole field—I put it to you—it sounds awful, but I think I must put it as strongly as that—I would work with the lassa fever virus in *E. coli* any day. I'd consider that to be my best guarantee of safety.

I think you must ponder these things because at first sight they appear totally paradoxical. That is because we have not thought out the hazards directly. I think we have not solved all the problems, and I think there is an enormous amount of work that still remains to be done.❞

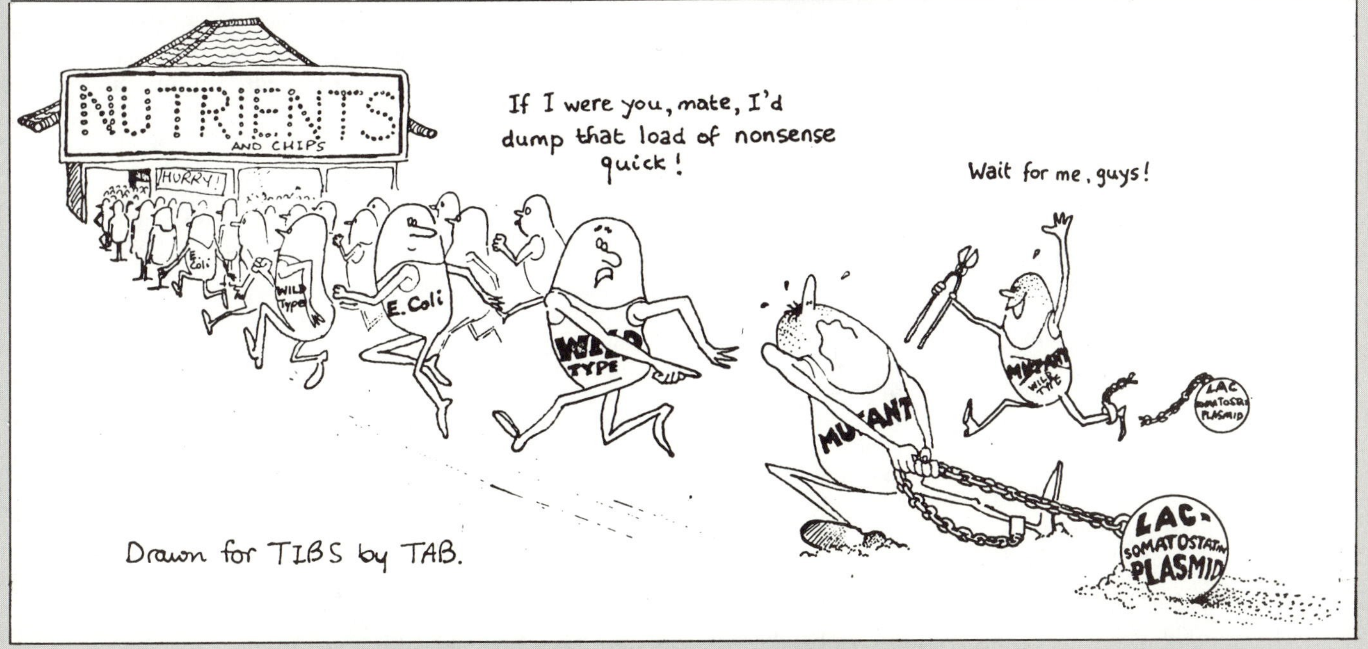

Document 13.3

November 2, 1978
Nature (**276**:4)

New diseases in new niches

By Jonathan King

"Recombinant DNA certainly represents a major breakthrough in our ability to study the organisation of the genetic material of higher organisms; and I should say that from the research point of view recombinant DNA technology can be employed safely. Although in order to employ it safely you have to assess very carefully what's dangerous about it. But perhaps more relevant here is the new production technology, technology that will be used to manufacture commodities for sale. The transformation from research tools to production technology has proceeded far more rapidly than many scientists envisioned. Within a year or two in the United States Eli Lilly corporation expects to be producing human insulin through the growth of thousands of gallons of *Escherichia coli* containing human DNA sequences spliced into a bacterial plasmid.

Now the deployment of new production technologies has more often than not been associated with the generation of unfortunate side effects on the health and welfare of the human population most notably those employed at the point of production. I have here a few historical examples. For example the mechanisation of cotton textile manufacture resulted in a drastic increase in damage to the respiratory tract of the operatives (byssinosis or brown lung). Developments in the German chemical industry—such as the synthesis of the aniline dyes which were used to colour the textiles—entailed the production of potent bladder carcinigens—4-aminobiphenyl and β-naphthylamine.

Most of us can be reasonably assured that most of those chemical carcinogens that are already out there through previous mishandling will not reproduce and increase themselves in the environment. The risk is finite. In the case of bacteria we do have to worry that these organisms—or at least the genes that are linked to the plasmids within these bacteria—will move through the ecosystem, transfer for example from the debilitated strains to wild strains of bacteria, and get into strains which perhaps are well-adapted in a particular niche out there. And then we won't be able to clean them up, for you can't remove, for example, *E. coli* from the ecosystem. It is an intimate part of mammalian life.

Now at this point I wish to clarify and punctuate a very crucial aspect of risk assessment. In trying to assess hazard we must consider what would be the properties, for example, of a wild strain of *E. coli*—even an epidemic strain of *E. coli*—expressing, for example, the human gene for insulin. Now some in the audience will heatedly reply—or they would if I were at home—"but they'll never get into wild strains, they're in debilitated strains, you've got nothing to worry about, you're raising a false spectre". This is of course putting the cart before the horse. The only way that hybrid DNA will be contained, is if people understand that if it is *not* contained there may be problems.

Thus in assessing the risk of cotton dust, we do not examine the effect of cotton fibres on human skin; we examine the effect of cotton fibres on human lungs. Of course, manufacturers in the industry often say "no that's wrong, because the cotton fibres will never get into the workers' lungs". But we know that that it is only if people understand acutely what will happen if those fibres *do* get inside the lung, that action is taken. And knowledge in the past has not been sufficient, it's taken much more action than knowledge.

I must ask: where have infectious diseases come from? After all if I'm going to make a feasible case that there is something to worry about in generating a new human disease, I'm behoven to explore the question of the generation of old ones.

The virulent form of *cholera Vibrio* infection, with characteristic rice-water stools, was first reported in the highly densely populated and unsanitary city of Calcutta in 1817. From there it was carried by the British navy to the newly emerging industrial areas in the north-west of England. Manchester in this period was rapidly converting from cottage production of cotton textiles to full-scale factory production. The workers needed to man the mills were either forced off agricultural holdings or one way or another brought into the city, and essentially forced to live in housing not of their own design. The nature of this housing is quite well documented—miserably crowded, very little light or ventilation, no sanitation, no proper water supply, no means of disposing of waste, thus garbage and excrement polluting the waters used for drinking and washing and food preparation.

Cholera thrived in these industrial districts because these organisms multiply in the intestine, where they elaborate a toxin; and this toxin binds and penetrates the cells of the intestine, and inactivates a protein of the intestinal cells which is needed in protein synthesis. The organism however goes out in the faeces, and if you live in an area with a contaminated water supply and you drink that stuff—boom you get cholera.

The growth of textile manufacture in the north of England, and also in the midlands, gave rise to many other niches. One of them was damaged lungs from cotton fibres, particularly among operatives of the carding and combing room, where the cotton is taken from the boll to the fibre. These individuals were unusually susceptible to tuberculosis and pneumonia infections, since with the primary barrier of the lung and respiratory tract broken, these organisms move down the respiratory tree and eventually get down to the alveoli of the lung, when you have profound tuberculosis and pneumonia.

Given the conditions that they lived in at home, where there was very very close person-to-person contact, contaminated food, no pasteurisation, etc. there were once again created special conditions for these organisms to thrive. Remember also that in this period also children were employed in the mills working 12 to 14 hours a day, malnourished, lacking sunlight, they were hypersusceptible to tuberculosis and related infections. The extraordinary mortality rates among children at that time are due in large part to that.

Thus this is a situation in which technological changes created conditions, both in the factory and at home, in which particular strains of pneumococcus, tuberculosis, and cholera could thrive. Fortunately these conditions do not exist any longer. Over the last 150 years through a kind of alliance between progressive scientists and public health people, and—very importantly—the labour movement, which played a primary role in the fight for decent sanitation and public health, we have reached conditions where water supplies and proper sewers are considered social necessities rather than individual privileges, decent food is available to most people.

I have never thought that the dangers of recombinant DNA research had to do with the generation of a spreading epidemic—some organism that would spread throughout the world. Those conditions don't exist now. It's not the *nature* of mortality from infectious disease. In fact over the past 30 years, the development and the production of antibiotics has created a possibility where even the rare emergence of some of these organisms can be controlled by antibiotics.

On the other hand, recently we've witnessed the ironic side-effects of the mis-use of the antibiotic technology. The Swann committee dealt with the problem here in the UK. In the United States, though we may have had a Swan committee, it came down on the other side, and we still introduce massive amounts of antibiotics routinely into the feed of chickens, cattle, and hogs, and thus into the general environment.

As a result there is intense selection throughout most industralised countries for organisms bearing resistance to antibiotics. And this very often is coded by plasmids, the very plasmids that are the parents of those plasmids that are used in the recombinant DNA technology. The other problem of course is hospitals and private medical practice where antibiotics are tremendously overused, once again creating niches. For example a little pool of water on a sink that's not cleaned up as often as it should be, is a niche where bacteria can thrive only if they contain plasmids conferring resistance to antibiotics.

Now let's switch to the current period of time and ask, in industrialised nations, what kind of problems do we have from infectious diseases? (And again by 'infectious' I don't mean person-to-person spread—I mean parasitic organisms.) I just look briefly at North Carolina and New Jersey in the United States. In North Carolina there are still a great many cotton mills, and byssinosis is still present at a very high rate. (That's because in the United States until very recently brown lung was not recognised as a disease. At least no doctor was willing to testify in court that a cotton worker had lost lung function from that disease.) Also the mechanisation of the cotton industry, which involves grinding up the bracts around the boll into a very fine dust—and it is the bracts that are the primary causative agents of lung disease—increases the problems of the cardroom workers with byssinosis.

So these people occasionally go to the hospital, when they get older, and they are diagnosed as having emphysema or some obstructive problem in the bronchi. There they receive some surgical procedure or some unconventional procedure. What happens then, and so much so that it's a major health problem in the United States (and I believe also in England), is they pick up a hospital infection, a hospital pneumonia, because the primary physical barrier has been breached. A very high percentage of these infections are due to *E. coli*, a strain that in the old days was never ever thought of as a pathogen. *E. coli* is a laboratory bug, it's the medical students' bacterium, it's something we don't have to worry about.

In Essex county in New Jersey workers still get bladder cancer at ten times the national average. (New Jersey also has the highest concentration of the dye industry.) These people occasionally go to hospitals (if they have health insurance—in the United States that's only a quarter of the population), and maybe they get chemotherapy for their bladder cancer. And they are immunosuppressed. And as a result of this they pick up hospital acquired infections, also very often *E. coli*.

Now these infections are not of the spreading epidemic type. It's a different thing; it's a point source; it's more like the bladder carcinogens coming out of the aniline dye industry. There's a place where the organisms are surviving—there's some reservoir—and then somebody comes in who's debilitated or weakened and they pick up the infection.

I don't mean to scare you—I'm now talking about ordinary antibiotic resistant *E. coli* infections—not recombinant DNA—but just to give you a sense of the extent of this I calculate that at present in the United States each year 90,000 individuals suffer surgical wound infections with *E. coli* with 2,700 associated deaths; 40,000 individuals contract *E. coli* pneumonia infections, with 10,0000 associated deaths; and 17,000 people contract *E. coli* bloodstream infections with 4,000 associated deaths. Of course the primary infection is urinary tract infection—about 300,000 cases in the United States, primarily in women.

Now these people are weakened—but that's what happens when you go to the hospital, you're in an automobile accident, or maybe a little baby and your immune system hasn't fully developed; we have a right to be weakened, there's nothing wrong with being weakened. In this period of history the people who get bacterial infections are a different population from those who got bacterial infections 200 years ago.

"The only way that hybrid DNA will be contained is if people understand that if it is not contained there may be problems"

I want to emphasize that one of the problems with these infections is that they carry plasmids specifying antibiotic resistance. And so when the person gets infected (1) the person's already weak; (2) if it's a bloodstream infection it's hard to treat; and (3) the plasmids are specifying antibiotic resistance. The organisms don't necessarily spread through the environment but the plasmids certainly spread from strain to strain. Thus you don't have to pick up a new bug—you pick up a new plasmid.

I mentioned child mortality in the 1800s in Manchester. We don't have that problem now; child mortality has a different character, it's not typhus or typhoid or dysentery; but we do have premature babies who are kept alive in conditions where in the old days they would have died. In the United States these premature infants get a very high rate of meningitis infections—and this is meningitis from *E. coli*. It's a nasty infection, 40 to 80% mortality, and the children who survive have neurological and behavioural abnormalities. The strain of bacterium which causes this —*E. coli* with a K1 antigen—is not a particularly virulent strain of *E. coli*, it's in the mothers, it's out there in the normal population; but if it gets into the meninges, into the spinal cord tissue of a premature infant, then you get a very nasty infection. Again it's special conditions.

Now, let's go back to the older view and say what kind of conditions make bacteria nasty with respect to human disease. Well, we have a few cases: synthesis of a toxin, would be one case, synthesis of a protein increasing the colonising capacity—the ability to stick somewhere—and resistance to antibiotic therapy. The last one I've covered. Cholera, let me remind you, makes a protein that binds to human cells. Diphtheria toxin—the classic case—comes from infection by a phage, not a plasmid, which specifies a protein that's exported outside the cell: that's the thing that makes you sick.

Now I'd like to note two general features of pathogenesis. First, many of these proteins that are associated with the pathogenic character of historically studied bacteria are coded by plasmids; and the transmission from cell to cell of these plasmids is a major medical problem. (It was in the past, and is presently.) Secondly, the feature of many of these toxins is that they have the property of interacting with mammalian cells. You know most proteins in bacteria don't interact with me. They interact with each other. If you want to get a protein that will interact tightly with me, you go to my cells. Go to my kidney cells, and you'll find many many proteins that interact with kidney cells for example, those proteins on the surface of one kidney cell responsible for sticking to the other kidney cell so that I have a kidney.

A number of people in trying to counter the question of the hazards of recombinant DNA research have said human DNA has been getting into bacteria time and time again all through human history. And it has even been proposed explicitly by the very notable American cell biologist Lewis Thomas, though he was making the opposite argument, that some of the diseases I've talked about, like cholera toxin, that the origin of that protein may have been a rare recombinational event, at some time in history, in which a mammalian gene for protein got into bacteria, and that's the way you got a protein that interacts with mammalian cells. So the model here is that when that happens it can be a very unfortunate event.

Now let me move to the question of whether there is really a serious hazard, or how many hairline cracks are there in Sydney Brenner's aircraft engine, in the introduction of mammalian genes into enteric bacteria? I'm sorry I'm forced into the thing that Sydney describes as example and counter-example, but the risk-assessment research hasn't been done—in the United States it has been blocked, and in my estimation it's been blocked because some risk is going to emerge. It's very unfortunate that one's forced into somewhat extreme examples, because the primary data, which might refute some of these, is not available.

We have genes which increase the pathogenicity of the host organism in the niches *already* occupied—that's one possibility; the second is a gene that makes a protein that allows the bug to grow in a place it didn't before—colonise a new place, stick to an organ it didn't before—for example for *E. coli* to colonise the upper respiratory tract, which it doesn't right now; and the third is a more subtle effect, something I've been very concerned with : interfering with the immune response of the host organism.

What about insulin? I don't know what would happen if a newborn infant picked up an *E. coli* meningitis infection and that bacterium was spewing out human insulin; but it seems to me there's every reason to be concerned. And I can't tell you what would happen if the genes for somatostatin, another potent hormone which has been cloned for commercial manufacture, were to be put into a wild strain of *E. coli*; again there's every reason to be concerned. And any endocrinologist not in the pay of the company producing it would have to give pause to the uncontained DNA.

In terms of a new niche, well certainly one of the classes of proteins that people want to study—I'd like to study it myself—is the class of proteins in human cells that allows one cell to stick to another. It's very important in the area of cancer research : how does a cell know when to stop dividing? One kind of DNA that *will* be cloned is the DNA that makes proteins of the cell surface. So people will be isolating bacteria that express in their surface a protein that involves binding to mammalian cells. There's every reason to think that that might, (though it may not), increase the pathogenicity of the wrong strain of bacteria.

Dealing with the immune system—that's a little more complicated. I have published a paper on that in the *Journal of Infectious Disease* called "Recombinant DNA and autoimmune disease" which deals with a model of rheumatic fever: you get a streptococcal infection, a sore throat, from bacteria showing a protein that looks like your heart protein; your body makes antibody against the bacteria, you get over the bacterial infection, and the circulating antibody attacks your own heart protein, leading to rheumatic complications. Well imagine for example an *E. coli* infection with a protein of the synovial membrane of the elbow. You are circulating antibody against the joint. That is one of the established models of arthritis : antibodies attacking your own joints.

Now, in conclusion, the hazards of introducing an extraordinarily immense pool of genetic information into wild strains of microorganisms are serious; every effort must be made to see that the genes stay within the debilitated strains; and that these debilitated strains stay within the laboratory. Now this is going to require absolutely the fullest participation of laboratory workers and production workers. And you are not going to be able to do it without strong union participation.

In the area of recombinant DNA technology, the way you know that these organisms are staying in the safe strains, staying in the strains in which they can be handled usefully from a research point of view and perhaps from a production point of view, is to make sure that laboratory workers are protected. If you can ensure that laboratory workers don't pick up these strains whether in their nasopharyngeal passages, or in a cut, or in a urinary tract infection, then you know they are not getting out into the population. The only way we know that can happen is if that sector of the population is fully involved in the safety standards.

Let me close with the fact that I think that just as in the area of occupational carcinogenesis, of black lung, of brown lung, the protection of people from disease involves a tight alliance between essentially progressive scientists and public health people—people who put that as their primary goal—and the trade union movement. That same alliance is the one that's needed with this new technology, and with many of the other new technologies—and more so now than in the past.

Furthermore that has to happen at the international level; because a small sector of the scientific population which is disturbed by this control, which is disturbed by having technicians having a say in the decisions, flies around to international conferences to figure out how to get out of the guidelines, how to weaken the guidelines, how to avoid having trade union participation, and it can be very important that people like ASTMS make contact, for example, with groups in the United States, the oil, chemical, and atomic workers, the French unions, the German unions, the Belgian unions, who represent the same sector of the population, not only to protect themselves but to make sure that in the long run that we're all protected and that the benefits of this new technology—which do exist, and I do believe in them—are realised.

My last point is: don't trust the NIH. Sydney Brenner was right that a better precedent has been established here than has been established in the United States. A fortune is going to be made from the cloning of insulin in bacteria. Four million doses are sold three times a week in the United States. They are not going to sell that insulin cheap; they are going to sell it expensive, because it's human insulin. They are going to *produce* it cheap. That is a very very powerful force behind the scenes, but it penetrates the scientific deliberations. So that in the United States we have scientists who in their public statements say "I'm only interested in the increase of human knowledge", while at the same time they have engaged lawyers to dissociate themselves from NIH funding, and get private funding, so that they can take out the patents.

Now we don't have a law in the United States that keeps anybody from introducing any gene into an epidemic strain of *E. coli* or *Salmonella* or *Shigella*; and there are forces in that direction because of patents. If somebody else has a patent on making insulin in *E. coli* and you recognise that market what do you do? Take two of the venture capital corporations in the United States, Genentech and Cetus Corporation, both in California; the one has backing from International Nickel and the other from Standard Oil of Indiana. Their front men may be research scientists who say "we are only interested in the expansion of human knowledge", but the background there is very different. ❜

Photo by Paul Conklin for the Academy Forum.

Jonathan King

Document 13.4
November 9, 1978
Nature (**276:**104)

Genetic manipulation: new guidelines for UK

● Britain's Genetic Manipulation Advisory Group (GMAG) is preparing to adopt a new approach to assessing the risks of 'genetic engineering' experiments. GMAG's statement —approved in principle at a meeting on 2 November—appears below. The scheme is based on an attempt to assign numerical risks to individual experiments as a product of three factors: the 'access factor', the 'expression factor', and the 'damage factor'. But the numbers are used only to rank the risks so that the experiments can be assigned consistently to the containment categories I to IV. Qualitative matters are also taken into account, for biology has not yet been reduced entirely to numbers.

The scheme will be used when there are enough data to employ it; otherwise assessment will revert to the old Williams 'phylogenetic' guidelines. Its great virtue appears to lie in its objective scientific basis and its consequent encouragement to research on risks.

GMAG's Sub-Committee on the Validation of Safe Vectors* and, more recently, an *ad hoc* Working Party on Guideline Criteria†, have been considering on GMAG's behalf the merits of an alternative approach to the categorisation of genetic manipulation experiments first outlined by Dr Sydney Brenner (MRC Laboratory of Molecular Biology, Cambridge). GMAG, having considered the enthusiastic advice it has received, is persuaded of the general soundness of the new system, and subject to the consideration of comment on this discussion document and consultation with the Department of Education and Science and the Health and Safety Executive, hopes to introduce the new principles—to be used for the time being in parallel with the existing Williams Guidelines[1]—early in 1979. Before refining the system, GMAG wishes to take into account the opinions of the scientific community, and it is therefore hoped that publication of the following will allow all concerned to make their views known to the Group‡. In the very near future, GMAG will be arranging a meeting with users and with members of local Biological Safety Committees.

While seeing justification and opportunities for change, GMAG considers that there is no case for changing the general arrangements governing this area of research in the UK: the definition of genetic manipulation remains unchanged; the regulations[2] to notify work being done must apply; the roles of GMAG and HSE have been established; facilities for physical containment (categories I-IV) have been defined and substantial investment has gone into their construction.[3] Proposals are put forward in this paper, however, which if implemented would lead to change in the scientific considerations upon which the categorisations of individual experiments are decided by GMAG.

At present, GMAG broadly follows the Report of the Williams Working Party in allotting experimental protocols to the four categories of physical containment. The chief conclusion of the arguments which follow is that those experiments for which it is possible to carry out an approximate risk assessment should where appropriate be given categorisations distinguished by the labels I*, II*, III* and IV*. (It must be emphasised that no change is implied in the specification of physical containment facilities, but only in the *categorisation* of certain experiments.) It is suggested that the two systems of categorisation should persist in parallel for as long as may be necessary. At this stage relatively few experiments can be given the new starred categorisations, but a steady transfer of experiments and classes of experiments from one system to the other should be possible.

Elements of risk assessment

Assessment of the risk of experiments in genetic manipulation must follow a formal and comprehensive risk analysis, which is possible only when the biology of the host-vector system and of its interaction with the cells of organisms at risk is reasonably well known. At present the relevant biological data are not generally available, and many systems of potential value in genetic manipulation will therefore be excluded from the categorisation procedure now proposed. With the passage of time, however, it will be possible to include within the new basis for categorisation an increasing proportion of experimental proposals.

In general, risk analysis entails a consideration of the following steps.

(a) *Escape from containment*

The escape of organisms carrying foreign genetic elements from the containment specified for them is always a possibility, either because the facilities are not designed in such a way as to provide absolute assurance that organisms will not escape or because of some malfunction or accident. Thus it is proper to associate with each category of containment a specific probability (C_i, where i runs from I to IV) that organisms will escape. The quantities C_i have the units "probability per unit time per organism contained".

The ratios of these probabilities will be measures of the relative efficacy of the four different levels of containment, and there are at present only rough estimates of the ratio C_{IV}/C_I, the relative safety of the most stringent and the least stringent of the containment categories now in use. The ratio is likely to be less than 10^{-6} (corresponding to a factor of 100 for the interval between each category of containment) and may be less than 10^{-9}. It is important that the true range of safety offered by the four categories of containment should urgently be defined more precisely. On present indications, the range of efficacy spanning the least and the most stringent categories of containment is comparable with the degree to which the viability of some host organisms is reduced relative to the wild-type strains; this points to questions for the

Professor J. H. Subak-Sharpe (Chairman), Dr S. Brenner, Professor D. C. Ellwood, Dr M. Fried, Dr M. Moss (Secretary), Professor K. Murray, Dr N. Murray, Professor M. H. Richmond, Dr P. W. J. Rigby, Dr R. Weiss, Dr N. Wilkie, Professor R. Williamson.

†*Mr J. Maddox (Chairman), Professor K. R. Dumbell, Professor M. H. Richmond, Dr R. Weiss, Dr J. H. Morris (Secretary).*

‡*Secretary, GMAG, at the Medical Research Council, 20 Park Crescent, London W1N 4AL.*

future about the relative protection from conjectured hazard of physical and biological containment.

(b) *Persistence in the environment*
The most direct concern in the assessment of the risks of genetic manipulation is that organisms carrying foreign genetic elements will gain access to the tissues of laboratory workers, but it is important that those concerned with the safety of genetic manipulation should consider the posibility that, as an intermediate step, organisms may lodge and even replicate in the laboratory into which they have escaped (as, for example, in culture dishes). In this and other parts of the analysis, care must be taken to identify all the possible circumstances in which organisms carying particular foreign genetic elements may be relatively at an advantage, so that the genes concerned are amplified and not attenuated.

(c) *Access to organisms at risk*
Organisms which escape as aerosols may be inhaled by laboratory workers or ingested, thus gaining access directly to the nasal tracts, the lungs and the gut. It is important that unfamiliar routes of access should not be overlooked—for example, the effects on human lungs of the inhalation of *E. coli* organisms carrying genes capable of producing, say, insulin are at present unknown.

(d) *Replication*
The risk analysis of experiments in genetic manipulation differs from that of, say, the operation of nuclear power plants in that agents which may potentially cause damage may in some circumstances be capable of replication. In the risk analysis of experiments in genetic manipulation, great care should therefore be taken to identify pathways that permit the replication of organisms carrying foreign genetic elements or which permit the transfer of the foreign genetic elements to other organisms which are capable of replication. One factor in a quantitative measure of a particular hazard would no doubt be provided by the duration and the intensity of the exposure of susceptible cels to organisms carrying potentially damaging genetic elements, and it is theoretically possible to calculate quantities such as these, given assumptions about the viability of manipulated organisms in the environments in which they lodge. Such formal calculations are however unlikely to be meaningful in present circumstances, so that more qualitative comparisons must at present suffice.

Great care must be taken to ensure that full allowance is made for the unavoidable incompleteness of present knowledge of the behaviour of foreign genetic elements in their hosts. Thus allowance must be made for the possibility that a "disabled" strain will revert (by genetic mutation) to the wild type, or that a foreign gene will be transferred (by the mobilisation of the plasmid) to wild-type bacteria. As things are, sufficient numerical data do not exist bearing on the chance that such events will occur, so that, for some time to come, researchers will have to rely on order-of-magnitude estimates of some of these quantities.

The conjectured hazards
Concern that genetic manipulation may entail hazards for living things stems from the possibility that foreign genetic elements may affect the normal functioning of particular groups of cells or the coordinated functioning of organisms at risk. Another of the impediments to formal and quantitative risk assessment that is likely to persist for some time is that of comparing the possible (but still hypothetical) biological consequences of foreign genetic elements with the damage done by naturally occurring disease. Even so, a tentative listing of some of the possible interactions between foreign genetic elements is suggestive of some of the points to which researchers should direct attention.

(*i*) *Cell death*
Cells may be killed when they are infected by viruses or when they are exposed to toxins produced by bacteria. Specific microorganisms and viruses often affect specific groups of cells in the body of a higher organism—poliomyelitis kills people because the poliomyelitis virus kills cells in the central nervous system when it has access to them, for example. In the analysis of the risks of genetic manipulation, the questions arise whether bacteria or viruses carrying foreign genetic elements are more virulent than the originals, whether the range of their specificity has been extended and whether bacteria carrying the intact genomes of pathogenic viruses make it possible for cells at risk to be exposed to potentially damaging viruses by unnatural routes.

(*ii*) *Changes of genetic constitution*
The introduction of foreign genetic elements into human somatic cells could have a number of harmful consequences. The state of differentiation of the cell might be affected. Cryptic viral elements already included in the DNA of the cell might be mobilised, or the foreign genetic elements themselves might be incorporated in the genome of the intact cell, increasing the range of cryptic viral function embodied in the cell. The chief causes of anxiety on such grounds have so far been in connexion with the occurrence of cancer, but understanding is far from complete. Experiments in which animal viruses multiply in human cells (in tissue cultures) without killing the cells may deserve special attention.

(*iii*) *Immunological consequences*
There are several hypothetical mechanisms whereby untoward immunological reactions might be provoked by foreign genetic elements. Thus foreign genetic elements might be incorporated in the genome of a human cell in such a way that the product of that gene (possibly a protein) was expressed on the cell surface, thus exposing the cells affected to attack by the body's immunological defences; this may correspond to the mechanism of some naturally occurring autoimmune diseases. On the other hand, the introduction of foreign genetic elements might be accomplished in such a way that their presence simulates that of the normal components of body tissues, with the result that the range of the sensitivity of the immune system would be narrowed. In principle, it is also possible that some of the cells of the immune system are especially vulnerable to organisms carrying particular genetic elements.

(*iv*) *Disturbances of endocrine function*
Many hormones are believed to exert their physiological effects by interacting with specific biochemical elements either on cell surfaces or in the nuclear DNA. Foreign genetic elements could in principle interfere with either of these reactions, but it is also the case that infection by bacteria carrying genes which direct the synthesis of human hormones *and* which are expressed and then secreted into the body could be damaging to the individuals concerned.

(*v*) *Disturbances of cell metabolism*
Foreign genetic elements inadvertently introduced into somatic cells could in principle disturb the metabolism of these cells, changing their metabolic rate, the rate of cell division or even the process (still unknown) by which somatic cells age.

(*vi*) *Consequences for microorganisms*
Experiments in genetic manipulation in which bacteria might artificially be endowed with resistance to antibiotics of medical or veterinary importance are potentially hazardous. Another early suggestion of a damage mechanism is the possibility that apparently innocuous organisms (such as the laboratory strains of *E. coli* K12) could be converted by genetic manipulation into infectious pathogens; relevant in this regard is the conclusion of a conference held at Falmouth (New Hampshire, USA) in 1977 that *E. coli* K12 at least cannot be converted to an epidemic pathogen. There are three kinds of reasons for this conclusion: the gentic basis of pathogenicity is complex and does not reside in a single gene, the wild-type strains of *E. coli* which have evolved are probably in some sense optimal for the ecological niches they occupy and the viability of laboratory strains of *E. coli* is so much less than that of wild-type bacteria that such organisms carrying pathogenic elements would probably be eliminated rapidly. This conclusion does not imply that individuals could not be damaged by *E. coli* carrying pathogenic elements such as the gene for cholera toxin, but merely that the production of an *infectious* pathogenic agent is improbable. There is a need for a consideration along the same lines of other organisms capable of infecting human beings.

(*vii*) *Ecological consequences*
Similar considerations apply to the interaction of foreign genetic elements and species other than human beings. To the extent that species were differently affected and that foreign genetic elements were widespread, ecological changes are in principle possible. Changes in the ecological balance of microorganisms are perhaps especially important.

For those assessing recombinant DNA experiments, this list should make it possible to single out those kinds of interactions between foreign genetic elements and cells which are likely to be widely regarded as potentially serious and damaging. Possible immunological and oncogenic consequences point to the need for a relatively high category of physical containment.

Application in the immediate future
In an ideal world, it would be possible to make numerical estimates of the risk of experiments in genetic manipulation by multiplying together the probabili-

Examples of the new categories

● The cloning of pure mammalian globin genes in an approved disabled *E. coli* host-vector system in circumstances in which the globin genes are not expressed would be I*. For the time being, the use of non-disabled strains of *E. coli* (including K12) or the expression of the globin genes in a disabled host-vector system would place the experiment in category II*. The cloning of expressible globin genes in a non-disabled host-vector system would, for the time being, be a category III* experiment. The allocations to these higher categories are justified by the present lack of certainty that the secretion of mammalian globin in the body tracts (including lungs and the genitourinary tract) to which *E. coli* has access would be without risk. As this uncertainty is removed, lower categorisation may be considered appropriate.

Category I* is justified for the cloning of non-expressible globin genes in a disabled host-vector system on the following grounds—all three of the factors A, E and D are small (notionally less than unity). If it emerges that D is small even when globin genes are expressed, the experiments provisionally allocated to II* and III* might be reduced in categorisation.

This experiment may therefore be considered typical of those destined for Category I*. It should be noted that two of the three factors afford a measure of biological protection, and that in spite of the apparent innocuousness of globin, GMAG would probably be unwilling to agree that an experiment of this kind should be undertaken with wild-type *E. coli.*

● The cloning of genetic elements from the genome of the mouse with the use of defective polyoma virus as a vector, and with cultured mouse cells as hosts (with appropriate helper virus to facilitate the replication of the virus and the expression of functionable genes) would also be a category I* experiment. Formally, such experiments have much in common with those in which fragments of the *E. coli* genome are redistributed within the same organism with the help of *lambda* bacteriophage. In reality, however, the possibility that polyoma virus may replicate in cells other than those of the mouse cannot be entirely discounted (replication in rat cells has been demonstrated, while the possibility that polyoma might replicate in human cells has not been vigorously explored). Accordingly, the cloning of *pure* mouse genes in polyoma and cultured mouse cells is suggested as a category I* experiment, with mouse-genome shotgun experiments in category II* for the time being.

● Category I* will also include those experiments for which at present there is no substantial evidence of risk. The chief candidates of this kind are the experiments in which *E. coli* genes are cloned in *E. coli* organisms.

● If these illustrations are taken as a paradigm of experiments in category I*, then it would follow that disabled *E. coli* carrying but capable of expressing the gene for, say, wheat germ gluten would be categorised II* to allow for the possibility that at least some members of the population might react immunologically against the production of gluten in such a way. Other experiments that might be given the same categorisation on further consideration by the technical panel might include the incorporation of non-expressed interferon genes in *E. coli* K12, or expressed interferon genes in a more disabled strain of *E. coli* K12 (eg χ1776). Shotgun experiments with mouse DNA cloned in polyoma virus would also be included.

● Similarly, and again on further examination by the technical panel, category III* might include disabled *E. coli* strains carrying expressed human growth hormone genes; the expression of cholera toxin genes in disabled strains of *E. coli* would remain in category IV*.

ties representing the various events (*a*) to (*d*) (see page 105), summing over all possible pathways leading from the escape of a genetically altered organism from its containment to the causation of a particular consequence for cells exposed and then multiplying by factors which take account of the costs of the damage that results. This is not now possible, however. Some may even argue that it will never be possible or necessary.

Nevertheless it is possible to use the framework of this formal risk assessment to rank the relative risks of some experiments in genetic manipulation, albeit in a somewhat qualitative manner. To that extent, the arguments first put forward by Dr Sydney Brenner in the GMAG Safe Vectors Sub-Committee are a valuable means of moving towards a more objective system of categorisation.

What is now proposed is that experiments with which researchers are familiar should be allocated to the four containment categories on the basis of a qualitative and relative assessment of their risks; and that the potential risks of novel experiments should then be assessed in relation to these familiar paradigm experiments.

For this purpose, it is convenient (as Brenner has argued) to define a group of three factors which may be taken to represent the chance that events following the escape of genetically altered organisms into the environment may lead to particular conjectured hazards. If data were available, it would then be possible to think of calculating the risk of an experiment by multiplying the numerical values of the three factors together. For the time being, however, such a procedure would provide a spurious sense of precision and it is proposed that there should be a comparison (which necessarily entails the exercise of people's judgment) of the values of the three factors relative to those which obtain for the paradigm experiments.

Analysis might therefore proceed as follows. First, it is supposed that there is an escape of organisms carrying foreign genetic elements. The overriding element in the assessment of the risk of a particular kind of disturbance is the likelihood that the organisms or the genetic elements they carry will have access to susceptible cells of the target organisms. Thus there is defined an

- **Access factor** (*A*). This is plainly a composite of factors representing the chance that escaped organisms will actually enter the human body, that they will survive there and that they will penetrate whatever membranes must be penetrated so as to reach the tissues containing susceptible cells. For many experiments, biological containment is dominant—for example the disabled strain of *E. coli* K12 known as χ1776 may be less viable than wild-type strains of *E. coli* by a factor of 10^8

In some experiments in genetic manipulation, the potential risk depends crucially on whether the foreign genetic element is able to express its normal function. This would, for example, be the case in an experiment in which the gene specifying the cholera toxin was incorporated in a plasmid of *E. coli*. Thus it is convenient to define an

- **Expression factor** (*E*), which would be large (or notionally have the numerical value 1) if the foreign gene were efficiently translated into the protein product of the gene and if the products were then secreted from the altered organism, which would have a similar value in those circumstances (as for example in the carriage of a genetic element which can be integrated into the genome of target cells without being translated) but which would be small in those experiments in which steps had been taken deliberately to ensure that the gene would not be expressed.

Shotgun experiments require further consideration. The fear that "unknown" genetic elements might cause unknown and damaging changes in organisms at risk has been widespread for the past four years. The Williams Report suggested that experiments involving random lengths of DNA should be categorised somewhere in between the categorisation appropriate for genetic elements known to be damaging (eg cholera toxin) and those thought to be without risk. This conclusion remains valid, but knowledge of the genetic constitution of organisms is growing quickly, and for some it is now reasonably certain that a genetic element drawn at random is likely to be less of a relative hazard than one deliberately selected. This, for example, may be the case with many animal viruses, whose genetic structures appear to be so economical of DNA that it is hard to believe that a piece drawn at random would have physiological significance. Thus it is convenient to define a

- **Damage factor** (*D*), which expresses the chance that a genetic element will cause damage. *D* will be large (notionally, unity) if, for example, the host organism were known to contain a gene specifying a bacterial toxin; will be less if it is a genetic element drawn at random from the same bacterial genome; and will be very small if the product of the gene is known to be without physiological effect on the cells of the target organism (as may be the case where the gene product is, for example, human globin).

It is proposed that, in the months ahead, GMAG should take steps to allow researchers (if they wish) to submit their proposals to GMAG with evidence which will allow GMAG to make an assessment of the three factors and advise on the new basis. Given the present incompleteness of biological understanding and the lack of data bearing on the numerical determination of these factors, it is clear that, for the time being, decisions will continue to depend on judgment rather than calculation. It is also clear that experiments using familiar systems will be the most susceptible to this alternative treatment. One advantage of following such a course of action will nevertheless be to encourage the accumulation of data bearing on the ultimately more objective assessment of risks.

Acknowledgments
The credit for urging on GMAG the risk assessment approach to genetic manipulation and for drawing it first to the attention of GMAG's Safe Vectors Sub-Committee goes to Dr Sydney Brenner, whose cooperation in developing the system into its present form is also warmly acknowledged. GMAG also wishes to thank the Guideline Criteria Working Party for their work in drafting the present paper, and innumerable other experts who have generously given their advice. ☐

References

1 *Report of the Working Party on the Practice of Genetics Manipulation. HMSO. Cmnd* 6600, 1976.
2 *Health and Safety (Genetic Manipulation) Regulations* 1978. *SI* 1978, *No.* 752.
3 *First Report of the Genetic Manipulation Advisory Group. HMSO., Cmnd* 7215, 1978.

Additional copies of this document are available at a price (including postage) of 25p each, or 20p each for orders of five or more. Send cheque, payable to *Nature*, to Promotion Manager, Nature, 4 Little Essex Street, London WC2.

Document 13.5
February 15, 1979
Nature (**277**:509)

Flaws in GMAG's guidelines

'We urge that time be allowed . . . before any firm decisions about major changes in the approach to categorisation are made' — **Royal Society**

• A statement issued by the Council of the Royal Society and prepared by an ad hoc group chaired by Professor W. F. Bodmer. Other members were: Sir David Evans, Professor D. C. Phillips, Dr M. G. P. Stoker, Sir Frederick Warner, and Sir Robert Williams.

WE welcome the attempt of the Genetic Manipulation Advisory Group's paper (*Nature*, 9 November, page 104) at a more rational approach to the categorisation of the experiments which come under its jurisdiction, using the framework of risk assessment. The chemical and nuclear industries have considerable experience of quantitative risk assessment, based on fault-tree analysis, and hazard and operability studies, backed by extensive data on reliability. In the longer term we believe that advantage should be taken of this experience and so we are requesting the Royal Society's recently established Risk Study Group on the Assessment and Perception of Risk, under the chairmanship of Sir Frederick Warner, to include the GMAG paper and the application of risk assessment techniques to genetic manipulation in its considerations.

Even in areas such as the chemical and nuclear industries, where extensive background data are available, there is often disagreement among experts. We believe that the present knowledge of the quantities needed for a quantitative risk assessment in the area of genetic manipulation is quite inadequate for this now to form a basis for categorisation of experiments. The absolute risk associated with any experiment involves the product of many probabilities (all, of course, less than one) some of which are known and known to be low, and so this risk may in many cases be estimated to be very low, even given our present incomplete knowledge. Categorisation of experiments, however, requires the estimation of relative risks, or ratios between probabilities which in most cases will be small, and, at the same time, quite imprecisely defined, and so we do not see how these ratios can yet be calculated with sufficient accuracy to be useful for categorisation.

Particular areas of ignorance are the probabilities of human error and the real risks associated with the ingestion of the experimental product, namely the host carrying a recombinant vector. In the case, for example, of vaccine development such as for polio, which probably involved much greater potential risks than most work with recombinant DNA, extensive data were collected on the *in vivo* effects of potential vaccines in animals and humans, on the basis of which the magnitude of the risks could be confidently assessed.

There is an urgent need for more comparable data on the recombinant DNA products and on their associated hosts and vectors, such as *E. coli* K12 and its mutants, phage lambda and other plasmid vectors. Until such data, and other relevant information required for quantitative risk assessment, are available we believe the approach suggested in the GMAG paper, and appropriate developments of it, should be used for studies of the existing categorisation, rather than for categorisation itself. If, for example, further testing fully supported the present evidence that *E. coli* K12 could not be converted into an epidemic pathogen by recombinant DNA techniques, then it might be possible to assign nearly all recombinant DNA experiments using normal *E. coli* K12 host-vector systems to category I, or at most category II.

The real risks, if any, associated with most experimental products of genetic manipulation are, as already pointed out, the most difficult to assess since they are all, so far, conjectural. While it is appropriate to encourage research in areas which might be thought relevant to the assessment of these conjectured risks, as in the GMAG paper, we believe that the discussion of such conjectured risks should be adequately documented by reference to the scientific literature and subject to comparable standards of scientific scrutiny.

The discussion in the GMAG paper falls far short of these standards. For example in describing risks (the section headed "The Conjectured Hazards") there is a progression from "conjectured" through "likely" and "potentially severe" to "possible" with no justification whatsoever. The mere existence of a list of conjectured risks introduces the notion of their probability, when some may even be impossible. In other contexts, risks of this sort have been classified, by consensus of those knowledgeable in the field, into (1) negligible, (2) possible but perhaps doubtful, and (3) definite though possibly small risks. It might be instructive to do the same with some of the conjectured risks that have been attributed to work with recombinant DNA.

The difficulty in applying the suggested new approach to categorisation is illustrated by some of the chosen examples in the GMAG paper. "Self-cloning", for example of *E. coli* K12 genes into *E. coli* K12 host-vector systems, is assigned to category I and made the subject of extensive notification and detailed reporting to GMAG, whereas the Health and Safety Executive's definition of genetic manipulation, accepted as unchanged by GMAG, specifically excludes such experiments from its remit (through the use of the phrase ". . . incorporation into a host organism *in which they do not naturally occur* . . ."—our italics).

We deplore any extension of GMAG's remit, implicit or explicit, especially into areas where there is evidence that the risks are negligible and which form a significant part of experimental work in modern biology. Any such extension should be subject to very careful further consideration and justification and should not arise automatically from the adoption of a new procedure.

It is also not at all clear why experiments involving the expression of cholera toxin genes in *E. coli* K12 need to be performed under conditions which are considerably more stringent than those required for work with *Vibrio cholera itself*. We urge that any recommendation for a higher level of containment for recombinant DNA

March 1, 1979
Nature (278:10)

work with a pathogen, than for work with the pathogen itself, at equivalent doses, be carefully justified. In many cases involving, for example, work with certain animal viruses it is likely, as has often been pointed out before, that recombinant DNA experiments with parts of the viral genome will entail much less risk than work with the whole virus.

We believe it to be very important that GMAG consider its procedures carefully in relation to the new US National Institutes of Health (NIH) Guidelines which came into effect on 2 January 1979. These new guidelines were prepared after extensive open, documented and subsequently published consultation with many groups in the United States and abroad, including in particular Europe and the United Kingdom, especially through the European Molecular Biology Organisation. In the earlier considerations of the problems posed by genetic manipulation, following the "Ashby" (Cmnd 5880) and "Williams" (Cmnd 6600) Reports, the approach followed independently by the United Kingdom was quite close to that followed in the United States and indeed at times in the forefront.

The new NIH Guidelines, which are being adopted in some form by most other countries that are advanced in this field in Europe and elsewhere differ, however, quite significantly not only from the new GMAG proposals, but also from some of its current practice. They differ, for example, with respect to self-cloning experiments and in the use, in general, of set categories for given types of experiments, rather than accumulated case law. A set list of categorisations obviates the need, once an institution is appropriately registered, for repeated referral back to a central authority before starting any particular experiment. Now that a significant amount of case law has accumulated, we urge GMAG to consider following this approach for an increasing proportion of experiments under its remit, and so simplify its activities and those of the people it serves, while in no way jeopardising recommended standards of safety.

We believe that in cases where GMAG's procedures lead to substantial differences in categorisation from those recommended by the NIH Guidelines, the difference must be carefully justified. A lack of comparability between procedures in the United Kingdom and elsewhere, especially in the United States, could seriously jeopardise this country's research and development efforts in this important area of work and in particular, threatens our contribution to extensive international cooperation in research in this field.

In this connection we would draw attention to the coordinating activities of COGENE, the Committee on genetic experimentation set up by the International Council of Scientific Unions. It should be noted that COGENE jointly with the Royal Society is organising an important meeting in April 1979 to bring together those concerned from many countries to consider these problems.

We have emphasised the need to obtain much more quantitative data to form the basis for a quantitative risk assessment. The pooling of data on risks with the US and other countries, as is done, for example, in the nuclear industry, would greatly extend the available data base needed for a proper quantitative risk assessment.

The timing of requests for responses to the GMAG paper, just before Christmas, and its earlier working paper in the summer, has been unfortunate and the period allowed for a response very short. It is not clear why such a major change in approach should be treated with such urgency. We urge that time be allowed to collect more risk data, to assess the risk assessment approach in relation to current categorisation, and to assess the impact of the new NIH Guidelines, before any firm decisions about major changes in the approach to categorisation are made. □

Risks of recombinant DNA regulations

SIR,—A more appropriate title for your editorial (November 9, page 103) would have been 'The absence of reason continues' or 'Reason ought to prevail', rather than 'Now reason can prevail'. Let me explain why I consider the regulations of the recombinant DNA technique and the new GMAG proposal as unreasonable.

- Since the risks of the recombinant DNA technique are principally in the realm of hypothetical scenarios and science fiction, it is totally unreasonable to propose any complex regulations.
- GMAG is proposing to estimate the imaginary risks of various types of experiments, but totally ignores the true risks of regulations themselves. As I have discussed before (*Trends in Biochemical Sciences* **3**, NZ43; 1978), the regulations lead to many real dangers to science and society while their benefits are restricted to supporting the livelihood of the bureaucrats who administer the regulatory machinery. It is totally unreasonable to institute regulations without first evaluating the benefit-to-risk ratio of the proposed regulations themselves, and this ratio in the case of recombinat DNA regulations approaches zero.
- As wisely argued by J. D. Watson (*The New Republic* **180**, 12; 1979) it is folly to control any human endeavour which may benefit society, just because it is impossible to provide a perfect proof that no risks exist or can be imagined. There is a high probability that such bureaucratic controls will be detrimental to the best interests or society, rather than having any benefits.

The GMAG proposals on the complex numerical estimations of the purely imaginary risks bear a resemblance to the Hans Christian Andersen fable on the 'Emperor's New Clothes' and appear to be a rather pathetic example of where the recombinant DNA folly could lead us. Let's hope that at the end reason will prevail and all regulations will be converted to a one-sentence statement proposed by D. Stetten, the first chairman of the NIH Recombinant DNA Advisory Committee: "The conditions of containment appropriate for any recombinant DNA experiment are those which are dictated by the most virulent or dangerous organism entering into that experiment".

Yours faithfully,
W. SZYBALSKI

McArdle Laboratory for Cancer Research, University of Wisconsin, USA

Document 13.6

January 4, 1979
Nature (**277**:3)

Bacteriologists lobby GMAG's first public meeting

A LOBBY for the exemption of experiments in which recombinant DNA techniques are used simply to rearrange the genes within *E. coli* and other bacteria made a strong protest against continuing control at GMAG's first public meeting last week.

GMAG, Britain's Genetic Manipulation Advisory Group, was fulfilling its promise to hold open meetings on its proposed new system for the risk assessment of recombinant DNA experiments (*Nature* **276**, 104; 1978).

The new chairman of GMAG, Sir William Henderson (above right), opened the meeting optimistically, saying that those who feel that GMAG's hand is too heavy "do not sufficiently allow for the depth of public interest, and how the composition of GMAG takes this into account". But all on GMAG, said Sir William "wish to improve the procedures and consult".

Those at the meeting, held on the last working day before Christmas (22 December), were mostly scientists. The consultation amounted largely to two protests from the floor: that the new scheme was difficult to operate in practice because of lack of sufficient knowledge; and that the re-arrangement of genes within a bacterium such as *E. coli* using recombinant DNA techniques (but not otherwise) must still legally be notified to the Health and safety Executive and then to GMAG, and must be carried out under Category I conditions.

These conditions, although the least stringent of GMAG's four categories of physical containment, are still more rigorous than normal laboratory practice and can mean that the laboratory has to be modified. This type of experiment has now been exempted from control under the proposed new NIH guidelines in the US, but hopes that the new proposals in the UK would automatically exempt these experiments have been disappointed.

Question after question from the floor drew an acknowledgment from Professor Mark Richmond of GMAG that the group would have to make a decision on this problem soon. John Maddox, one of GMAG's public interest representatives, defended the categorisation and notification of such experiments on political grounds. While accepting that the procedure for notification would be made simpler and easier, he took the view that public concern would be lessened if GMAG were seen to be controlling all experiments using recombinant DNA techniques, even though there might be scientific grounds for exemptions in certain cases.

From the floor Professor R. Pritchard warned that this type of apparent illogicality undermined the credibility of the whole notification procedure in the eyes of researchers and their local safety committees.

Nevertheless, the feeling of the meeting was that the new risk assessment scheme was more rational than previous guidelines. But its precise application worried people.

The most immediate problem is posed by the 'gaps' in biological knowledge which become apparent when a consistent attempt is made to calculate the risk of an experiment. For example, the causes of pathogenicity are not well understood, so one cannot estimate all the possible biological effects of a modified bacterium on the human body.

In cases like this Dr Sydney Brenner—who invented the scheme and was present at the meeting—falls back on averages. Of all known bacteria, what fraction are pathogenic? This fraction could be applied as the probability that any new bacterium were pathogenic.

Dr Peter Rigby of Imperial College, who according to Dr Brenner "has worked out a magnificent set of examples", raised the final issue: that all the calculation is in vain if the precise containment effectiveness of the four containment categories is unknown. Some claim the containment factor between categories I and IV to be 10^{12}; some 10^{8}. "It must be measured" said Dr Rigby, extracting a cautious commitment that GMAG act.

Eleanor Lawrence

"To recombinant DNA!"

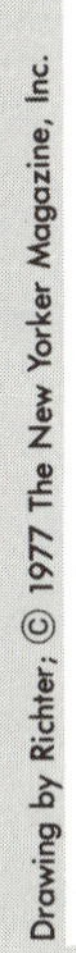

Document 13.7

March 8, 1979

Nature (**278**:113)

Let us stop regulating DNA research

By J. D. Watson

The Royal Society critique (*Nature*, 15 February 1979, page 510) of the proposed new approach by Britain's Genetic Manipulation Advisory Group (*Nature*, 9 November 1978, page 102) to recombinant DNA regulation is correct on at least one point. We do not have the facts on hand to make the quantitative risk assessments to decide whether a proposed experiment should be done under maximal, minimal, or no precautions.

Assigning numbers in the manner suggested by Dr Sydney Brenner implies that we have some real measure of prospective risk. However, to my knowledge there is no evidence that any prospective recombinant DNA experiment poses any realistic threat to any scientist who uses the tools of his trade, much less to society itself. I thus do not see any logical way to arrive at such numerical risks, and the whole GMAG exercise strikes me as having no more validity than quantitative religion.

To be sure some experiments like putting *E. coli* genes back into *E. coli* are thought safe by virtually all informed individuals. In contrast, the insertion of the gene for the botulinus toxin into *E. coli* might strike some competent scientists as risky. But how much additional peril will be generated beyond that involved with working with botulinus organism itself is not obvious.

The *a priori* guesses that different knowledgeable minds will give may at best tell us something about their respective neuroses or lack of them. Each of us has different thresholds of fear, be it rational or irrational. For my part, I already see so many unambiguous hazards, like dioxin, that I remain incapable of adding still one more unless I see some real numbers. Recombinant-DNA-induced diseases to me fall in the category of UFOs or witches. Others may take them seriously, but they should not expect me to join in. I would not spend a penny trying to see if they exist and would regard as pure folly the decision to spend sizable sums of money on such projects.

I am thus appalled by the recent proposals of the Honourable Joseph Califano and Shirley Williams that their respective governmental departments (HEW and the Department of Education and Science) initiate extensive experiments to probe the presumptive safety of recombinant DNA experiments. The variety of prospective experiments is so legion that we have no possibility of checking even a tiny fraction for their potential danger.

Moreover, the answers we could obtain would have no real carry-over to other even closely related situations. So we would be spending vast sums of money (Shirley Williams has proposed tens of million pounds) merely to give the facade of responsibility. In fact, we would be irresponsible by diverting money away from projects which unambiguously might advance science, if not society itself.

We must thus be careful not to undertake risk assessment experiments merely because they can be done. Here I part company with the Royal Society's position that we should see, for example, whether nonpathogenic strains of *E. coli* can be converted into pathogens through random insertion of foreign DNA. Such data are very unlikely to affect the way we work with recombinant DNA. Unless we directly take DNA from pathogenic bacteria, the chances of a positive answer are low. But even if we were to generate a new pathogen, it is not obvious that it would pose any special risk.

Almost totally lacking in the past six years of seemingly never-ending conjectures about the dangers from recombinant DNA experimentation is recognition of the fact that the world is already filled with large varieties of pathogenic microorganisms of many types. Moreover, they are constantly mutating to give us new forms that we have not seen before. Some are constantly making us ill, and sometimes even doing us in. For the most part, however, we can fight them at least to a draw, and, of necessity, most of us do not spend much time worrying whether the next time they will get us. Otherwise we would be completely paralysed from any constructive behaviour.

Instead of continuing to waste masses of paper and the time of countless individuals who have real jobs to carry out, I believe we should quickly and resolutely abandon any form of recombinant DNA regulation. Concurrently our national leaders should announce that they will help push DNA research as fast as our national and corporate treasuries can permit.

At the same time we should, of course, remain alert to the rare possibility that one or more recombinant DNA research workers will come down with a disease we have not seen before. If that happens we naturally would have a story that deserves serious journalism, and for the first time there would be real facts for the risk assessor. But until that situation arises, and I'm doubtful that it ever will, we must make clear to the public that there is no more reason to fear recombinant DNA than there is to panic about the Loch Ness Monster. □

Document 13.8
March 8, 1979
Nature (**278**:113)

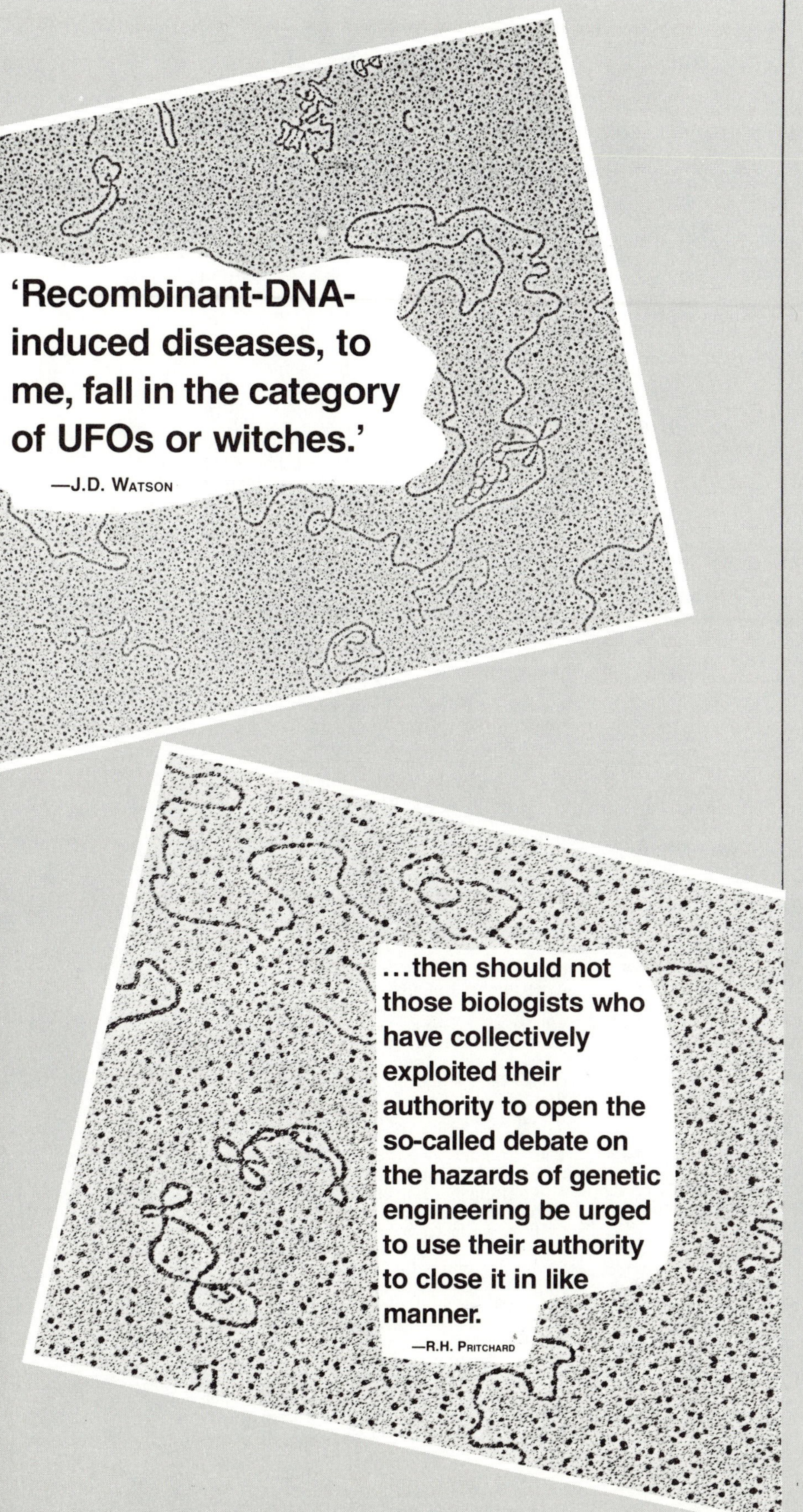
'Recombinant-DNA-induced diseases, to me, fall in the category of UFOs or witches.'
—J.D. WATSON
...then should not those biologists who have collectively exploited their authority to open the so-called debate on the hazards of genetic engineering be urged to use their authority to close it in like manner.
—R.H. PRITCHARD

Photos courtesy Tom Broker, Cold Spring Harbor Laboratory.

Reassess hazards of recombinant DNA

SIR,—The comments (15 February, page 509) by a committee of the Royal Society on the recent published statement of the Genetic Manipulation Advisory Group will be widely welcomed and supported. However, I should like to comment on one of the suggestions they make.

It is argued that there is an urgent need for more research to evaluate the risks associated with the ingestion of organisms containing recombinant DNA. Before devoting yet more resources to the general hypothesis that recombinant DNA synthesised *in vitro* poses novel hazards to man, is there not a prior need to reconsider in which areas of recombinant DNA research there remain sound arguments indicating that the hypothesis should be taken seriously? The committee has properly criticised the Genetic Manipulation Advisory Group for publishing what are often called hazard scenarios without adequate argument or evidence supported by citation of relevant literature. But could not a similar criticism be made with equal validity against the hazards hypothesis in its entirety? The report of the working party chaired by Lord Ashby which was set up to evaluate this hypothesis made a number of assertions in favour of it but none of them was documented in the way the Royal Society committee advocates. Admittedly this report was written when responses to the possibilities suggested in the Berg letter were more emotional than rational. All the more reason, now that four years have passed, for a cool reassessment of the questions before we embark on experimentation.

The most urgent need in 1979 surely is for a re-examination of the hazards hypothesis to determine which, if any, aspects of it remain credible enough to deserve further attention on scientific grounds. This reassessment can only begin if those who remain convinced of hazards are prepared to state their case in published form with the rigour urged by the Royal Society committee, and in a manner that permits rebuttal like any other scientific argument.

If such a reassessment suggests that areas of significant risk remain then let us devote resources to quantifying the risks in these areas. If, on the other hand, this reassessment suggests that the hypothesis is unfounded or if no one is prepared to argue the case to the contrary, then should not those biologists who have collectively exploited their authority to open the so-called debate on the hazards of genetic engineering be urged to use their authority to close it in like manner. The danger in what the Royal Society proposes is that it could lend further weight to a hypothesis that may not deserve it.

Yours faithfully,
R. H. PRITCHARD

Department of Genetics,
University of Leicester, UK.

Document 13.9

March 1, 1979
Nature (**278**:118)

April 12, 1979
Nature (**278**:597)

GMAG: NIH guidelines are not the way

SIR,—Professor Alan Williamson is well within his rights to complain (1 February, page 346) that there was "no real debate" about the proposed basis of GMAG's new way of working at the public meeting on 21 December, although it is by no means clear at whom the complaint should be directed. By inference, GMAG is somehow to blame. Professor Williamson could just as well have blamed those among his colleagues who chose to use up what time there was in echoing (and amplifying) what the platform had already acknowledged—that there must be special arrangements for some (but not all) homogenic experiments. He might also blame himself for not having brought up the issues to which he now draws attention.

Of these, perhaps the most important is Professor Williamson's assertion that the UK could with advantage adopt the NIH guidelines and be done with further introspection. It is therefore worth saying that the NIH guidelines are in several respects defective, at least if one supposes that some of the conjectured hazards of genetic manipulation may (or might be) real. First, their classification of animal virus experiments is irrational, and takes no account of conjectured immunological hazards in particular. Second, the potential importance of expression compared with mere replication has been entirely overlooked in the "broadly based public debate" that Professor Williamson admires. Third, even the new NIH guidelines are not constructed in such a way that what must be presumed to be a growing class of experiments without risk can be progressively identified and excluded from rigorous restraint.

I am one of those who believe that if GMAG does not change its procedures soon, we shall find that events force on us a decision between the NIH guidelines and doing the research elsewhere. It is of course ironical that we should be faced with such a choice when the hazards of genetic manipulation remain conjectural. In my opinion, however, it is disingenuous of Professor Williamson and other would-be genetic manipulators to complain at the constraints which they invited a mere four years ago. Making a bonfire of the regulations now will cause only trouble. Surely it is in the interests of the scientific community that they should be dismantled by rational means.

GMAG's proposed way of working offers precisely such a possibility, provided that genetic manipulators will play their part in gathering the necessary data. My own belief is that if the proposed system were adopted and willingly accepted, we should find that before the end of 1980 all but a handful of experiments were no more restricted than by the requirements of "good microbiological practice". Is not that a chance worth taking?

Yours faithfully,

JOHN MADDOX
Nuffield Foundation, London, UK

Dismantling DNA regulations

SIR,—I am delighted that John Maddox has taken up the debate on the problems related to regulations governing genetic manipulation (1 March, page 10). His analysis of the situation is perceptive when he states that the choice, if GMAG's procedures don't change soon, will be only whether to work with NIH guidelines here or elsewhere. In this regard John Maddox and I probably disagree only on time scale. He clearly feels that all may be resolved by the end of 1980 and asks if this is not enough, is it "a chance worth taking"?

If we are going to gamble we must know the odds and the stakes. At stake is the future of most genetic manipulation work in this country. The odds are heavily against anyone trying to continue in this highly competitive field if work has to be conducted under regulations so much more restrictive than elsewhere.

It is particularly bitter irony to find myself grouped with those scientists who invited constraints four years ago. That group was dominated by US scientists who are now working under relaxed NIH guidelines, while their colleagues here languish in the aftermath of their public outburst. From the outset, I have thought that those who invited constraints were doing both science and the public a huge disservice; I see no evidence to change my mind. A bureaucracy has been established and will be hard, if not impossible, to dismantle rationally.

In attacking the NIH guidelines, John Maddox subscribes to currently fashionable scare stories, those of "conjectured immunological hazards" and other problems "of expression". In fact when we talk of known proteins we can estimate possible dangers because we handle the proteins in their regular environment. We can therefore assess how to take precautions in handling the same products in genetic manipulation experiments. Such decisions, at least at the research laboratory level, are a matter of day-to-day care and not a matter for committee rulings and elaborate bureaucratic procedures. If there were risks which could be quantified and if there were a need for regulations which which could be formulated rationally, then we might start to build from scratch. Unfortunately, we are not in a position to build, but rather in the position of dismantling excessively restrictive, empirically conceived, regulations.

If we were to accept the idea of dismantling the regulations by rational means we could fall into an expensive trap. Any conjectured risk would need to be quantified and the limit to conjecture and to expense involved in collecting the necessary data cannot be foreseen. What I asked for previously was an acceptance of the spirit of the NIH guidelines. Let us recognise the need for care when there is a known hazard, but let us not get lost in a morass of conjectured risks.

Yours faithfully,

ALAN R. WILLIAMSON
Institute of Biochemistry, University of Glasgow, UK.

at MEDICAL RESEARCH COUNCIL
20 Park Crescent London W1N 4A
Telephone 01-636 5422

GMAG NOTE NO 11

THE GMAG CATEGORISATION SCHEME - PROVISIONAL EFFECTIVE DATE: 1 JANUARY 1980.

In this Note "genetic manipulation" and "work" are defined in the Regulations on Genetic Manipulation (S1 1978 No. 752) - copies of which are available from the Health and Safety Executive, EMAS, 25 Chapel Street, London NW1 5DT.

1. Introduction

Following the publication of a categorisation scheme based on risk assessment (GMAG Note 8 and Nature 276 104-108, 1978) the Group has now considered a large number of proposals submitted to them under this risk assessment procedure in parallel with the categorisation scheme outlined in the Williams Report (Cmnd. 6600). The Group has recently reviewed the categorisation scheme based on risk assessment and agreed that it had worked successfully. It is the Group's view that with the aid of the guidance notes detailed below the scheme can be extended to all work involving genetic manipulation.

The GMAG has advised the Health and Safety Executive that it intends from 1 January 1980 that all proposals to do work involving genetic manipulation will be considered by the Group under the risk assessment scheme detailed in this Note and the *suffix sed during "the experimental period" will cease to be used. The categorisation scheme outlined in the Williams Report would not therefore, be applicable to this work after 1 January 1980.

Although it is not expected that local safety committees and workers will have difficulty with the risk assessment scheme the Group will provide advice where there are cases of uncertainty.

The Group expect that workers and local safety committees will provide reasons for their suggested categorisation of each proposal and, if necessary, backed by experimental evidence.

Submission of proposals

Following discussions with local safety committees ll work designated as Category I should be notified to HSE and GMAG, as at present, with work proceeding under the appropriate conditions. Work designated by the local safety committee as Category II in which expression is not being sought nor likely to occur may also, following notification to GMAG, proceed under the appropriate conditions.

For work designated Category II in which expression is sought or otherwise likely to occur, except that with well characterised polypeptides with no evidence of adverse biological activity (eg globin), and for all work designated III and IV the advice of GMAG must be obtained before the work is undertaken.

Any work involving genetic manipulation or genetically manipulated organisms, including that with excepted organisms, which involves culture volumes of 10 litres and above must not proceed until the GMAG's advice has been obtained.

I Guidance Notes on the GMAG categorisation scheme

The following Guidance Notes on the risk assessment scheme have been prepared by GMAG's Technical Panel. The Panel considered the possible risks associated with a number of experiments under the headings of Access, Expression and Damage, which are described in GMAG Note 8. The consequences of a wide range of procedures which are being currently used were considered by the Panel. The following scheme represents the present assessment of the categories for 'shot-gun' experiments involving primarily E.coli host/vector systems, but it also includes certain viral vectors.

At present there are four different categories of containment in many different laboratory environments and operated by personnel with varying skills. The series of examples outlined in these Notes are intended as guidance to scientists and local safety committees in assessing the appropriate containment facility in the light of local circumstances. The Guidance Notes define levels of risk which enable an experiment to be up-or down-graded by one category of containment. In order that the separate risks for each heading may be combined, numbers differing by 10^3 are given for each level. These levels are crude since they apply to ranges of experiments and it is expected that more precise figures under the headings of expression, access and damage may be obtainable by experiment in individual instances. The values presented are all probabilities per unit bacterium: a value of 1 means that all bacteria are expected to have access, express a polypeptide or cause some biological damage, 10^{-3} means a chance of this occurring is 1 in 10^3 bacteria.

ACCESS

The host/vector system* can be defined before the experiment and quantitative evidence is now becoming available about survival and mobilisation for many of the commonly used systems. Table 1 illustrates an acceptable set of figures.

* Details of the Host/vector systems approved by the Group are given in GMAG Note 9.

Table 1

Hosts systems such as wild type _E.coli_	1
E.coli K12 or similar	10^{-3}
Approved disabled host/vector systems including certain viral vectors in tissue culture lines (see discussion below)	10^{-6}

10^{-6} is a figure that has been argued on the basis of Curtiss' experiments with X1776. It is considered that some of the newer EK2 equivalent certified systems might well be considered safer. For example, MRC1 on the basis of animal and human experiments might be 10^{-9}.

It is assumed that plasmids cannot be transferred to other vectors in approved disabled host/vector systems and that this is not an available means of access for these. Table 1 also takes into account survival in the gut which would be the most probable pathway of access for a bacterium secreting a polypeptide. The infection of tissues by _E.coli_ was considered but the Panel were assured that healthy individuals with normal immunological status are most unlikely to be colonised by _E.coli_ in minor wounds. In the important instance of bladder infection, this derives by infection from the gut flora, and the frequency correlates with urinary tract defects. Colonisation other than by the gut route implied by Table 1 would therefore be, by some factor, less likely.

EXPRESSION

Knowledge about the expression of polypeptides in recombinant DNA is rapidly increasing and it is probable that a number of sites in plasmids may express. It is also necessary to take into account the fact that pseudo-coli promoters occur in eukaryotic DNA and that adventitious expression of gene sequences, such as those derived from cDNA, may result. We therefore consider that recombinants in plasmids from which expression is not selected for should be placed in an intermediate level of risk until non-expression of the relevant mRNA has been demonstrated at the 0.1% level. This reflects the difficulty of obtaining expression in the absence of selection. Genomic DNA in phage or in demonstrated non-expressing sites elsewhere would still carry the lowest risk as there is evidence that most genomic coding sequences contain 'introns' which cannot be processed from transcripts to give functional mRNA in _E.coli_ or yeast.

Table 2

Source of DNA	
Any DNA from which expression as a polypeptide is sought	1
cDNA from mRNA in any plasmid site from which expression is not sought	10^{-3}
Genomic DNA in a known plasmid expressing site	10^{-3}
Any DNA in a recombinant where non-expression has been demonstrated; depending on sensitivity of detection procedure	10^{-6}
Genomic DNA in lambda or other demonstrated non-expressing site	10^{-6}

DAMAGE

The Technical Panel was convinced that damage was only likely through the expression of a polypeptide. There does not seem to be any likelihood of single-stranded RNA surviving in the gut and no evidence was presented that double-stranded RNA was a product of any of the systems under consideration. The 10^{-9} level in the following Table reflects this. We have had some difficulty in defining polypeptide products at each level of risk, but believe that the investigator should show that bacteria making the required product have no adverse biological activity in animals before it is assigned to the 10^{-6} category.

Table 3

Damage	
Expression of toxins, hormones and similar biologically active molecules or parts thereof or of a product which enhances the pathogenicity of the host organism.	1
Uncharacterised polypeptides of unknown biological activity	10^{-3}
Well characterised polypeptides with no evidence of adverse biological activity, eg. globin	10^{-6}
No polypeptide expressed	10^{-9}

As a general rule, and overriding other considerations, when attempts are made to gain expression of toxins, hormones or similar products, these procedures should be carried out only in non-mobilisable vectors regarded as disabled, thus preventing transfer to a wild type host.

The tables refer to the 'shot-gun' type of experiments. We consider that all DNA other than that obtained by a previous cloning procedure has a substantial chance of including other sequences and should therefore remain in the same category despite partial purification by biochemical procedures. Recombinants made from defined cloned sequences may also merit a lower category of containment if they can be shown not to express a polypeptide or, as the result of an experiment, be shown not to cause damage. Previous considerations of sequence purity were based on the possible presence of unknown eukaryotic sequences which could increase the host range or pathogenicity of bacteria, or contain, say, a cancer enhancing gene for man. Damage caused by such sequences is not now held to be likely because of the demonstrated multi-factorial nature of the relevant disease processes.

The Williams guidelines laid considerable stress on phylogenetic relatedness. In the new scheme this now appears properly under Damage. While, in general, biologically active polypeptides from organisms more distantly related to man are less likely to affect him, this is not, of course, universally true since there are many potent toxins in lower vertebrates and invertebrates. Individual judgement will therefore be necessary.

Table 4 codifies results from combining the suggested risk assignments from the previous tables, for most commonly used systems.

Table 4

Overall risk assessment for making recombinants from total or partially purified DNA in most commonly used systems

1. cDNA or synthetic gene

DAMAGE	ACCESS	EXPRESSION RISK*	CALCULATED	NEW CATEGORY
a) Toxins, hormones and similar biologically active molecules or parts thereof or a product which enhances the pathogenicity of the host organisms.	Disabled Host	+	10^{-6}	III Ø
		±	10^{-9}	II
		-	10^{-12}	I
b) Uncharacterised polypeptides of unknown biological activity	E.coli K12 or equivalent	+	10^{-6}	III
		±	10^{-9}	II
		-	10^{-12}	I
	Disabled Host	+	10^{-9}	II
		±	10^{-12}	I
		-	10^{-12}	I

*Based on a site of incorporation
ØThere may be certain circumstances in which the GMAC or local safety committee may consider category IV containment would be desirable.

DAMAGE	ACCESS	EXPRESSION RISK*	CALCULATED	NEW CATEGORY
c) Well characterised polypeptides with no adverse biological effect	*E.coli* K12 or equivalent	+	10^{-9}	II
		±	10^{-12}	I
	Disabled Host	all	10^{-12}	I
2. Genomic DNA				
a) Including sequences coding for toxins, hormones and similar biologically active molecules or parts thereof or a product which enhances the pathogenicity of the host organism	*E.Coli* K12 or equivalent	Expressing	10^{-6}	IIIØ
		Non-expressing	10^{-9}	II
	Disabled Host	Expressing	10^{-9}	II
		Non-expressing	10^{-12}	I
b) Including sequences coding for uncharacterised polypeptides of unknown biological activity	*E.coli* K12 or equivalent	Expressing	10^{-9}	II
		Non-expressing	10^{-12}	I
	Disabled host	all		I
c) All other including spacers, satellites, etc.	*E.coli* K12 or equivalent	all		I

*Based on a site of incorporation
ØThere may be certain circumstances in which the GMAG or local safety committee may consider category IV containment would be desirable.

Notes

1. ± signifies no selection for expression, but non-expression not demonstrated.

2. Non-expressing sites include lambda sites and sites on plasmids which have been shown not to express the designated class of polypeptide.

3. It is assumed that nonsense polypeptides made by reading frame shifts present no hazard, but that part-protein may.

4. Advice on the containment of genetic manipulation experiments involving the use of dangerous pathogens should also be sought from the Dangerous Pathogens Advisory Group.

Viral Vectors

It was assumed that expression was probable when viral vectors were propagated in eukaryotic cells therefore a figure of 1 is appropriate in table 2. Thus the access and damage factors are the variable parameters. Established cell lines are known to be exceedingly difficult to transmit to humans therefore a figure 10^{-6} seems appropriate in table 1. It is recognised however, that primary cells derived from an individual could after, for example, transformation easily recolonise that individual. The status of such experiments will therefore depend very considerably on individual circumstances. The damage factor may be treated in the same manner as prokaryotic host/vector systems.

Notification of Projects and the Authority to Proceed

Under the earlier "Williams" guidelines, projects considered by the local Safety Committee to be in Category I or II could be undertaken immediately after notification of GMAG and HSE. It is now intended that all experiments considered to be in Category I and II should be on the same basis, with one important exception.

GMAG considers that the Access and Expression factor can now be adequately defined, but that some doubt remains about the possibility of damage due to the expression of biologically active polypeptides. Therefore GMAG have decided that (1) All category I projects can be undertaken immediately after notification, and (2) those Category II procedures where the decision between II and some higher category depends on an assessment of the damage likely from the expression of a polypeptide should be delayed until after GMAG has given its advice. This means that where the figure 1 is appropriate in Table 2 all projects must await advice from GMAG except those in which expression is sought for "well-characterised polypeptides with no evidence of adverse biological effect" (see table 4, 1c).

It is anticipated that results from direct tests of damage by the expression (by bacteria) of various polypeptides will be available within a reasonable time. This will allow a clearer definition of which procedure, if any, should be undertaken in the higher categories of containment.

Summary

This scheme is relatively simple to understand and it corresponds to views about the relative hazards as understood at present. The order of magnitude differences are so large that they should not give the false impression that we know exactly what any given probability is, but at the same time, they both allow the calculation of a self-consistent overall risk and provide a yardstick against which different figures provided for specific experiments may be compared.

Photo by Countrywide Photographic (Ashford Ltd.).

Participants at the COGENE Conference, Wye College, Kent, England, April 1979.

IS THE END IN SIGHT?

Under the revised NIH guidelines that came into force on January 2, 1979, very few experiments indeed required the highest level of physical containment (P4). Using biologically disabled host-vector systems, most of the experiments people wished to do required between P1 and P3 containment laboratories. Unlike P4 laboratories, these were within the resources of most academic institutions. But the red tape involved and the high cost of installation and maintenance made even P2 and P3 laboratories very expensive.

In his introduction to the revised NIH guidelines, Donald Fredrickson had written in December 1978, "No new facts or unconsidered older ones have emerged to support the fears of harmful effects, and one prominent early proponent* of guidelines has repudiated his support for them. At the least, there is growing sentiment that the burden of proof is shifting towards those who would restrict recombinant DNA research." This shift in the burden of proof was one of the major achievements of the 1977–1978 campaign to revise the guidelines. Molecular biologists had fought their way out of the corner into which they had been hassled between 1975 and 1977, when they were told they must prove that their experiments carried no biohazard even though there had never been any experimental or logical evidence of significant risk. At last the traditional notion that one is innocent until proven guilty, rather than the reverse, had been reestablished.

In his review in *Nature* of one of several books that had appeared and chronicled the recombinant DNA debate, Robert Pritchard reached the conclusion "that in matters of policy we should base our decisions on the quality of the arguments and not upon the status of those who expound them." (Doc. 14.1.) Most molecular biologists, thoroughly sobered and not a little disillusioned by the events between Asilomar and early 1979, had reached the same conclusion.

The newly revised guidelines could of course be lived with, especially by those doing experiments with the standard host-vector systems; those proposing more novel experiments could anticipate greater amounts of paperwork and longer delays, but eventually they, too, would be granted permission to proceed. But if there were no real scientific grounds for guidelines

*J. D. Watson.

and regulations, then why should they be retained? Perhaps for too long molecular biologists had trodden the road of appeasement. At an international conference convened by COGENE at Wye, Kent, England, on April 1–4, 1979, the majority conclusion was that the guidelines should go (Doc. 14.2). Predictably, the Coalition for Responsible Genetic Research in New York was disturbed by reports of statements made at the COGENE Conference and by the way the RAC was handling its affairs (Doc. 14.3).

If scientific considerations led to the conclusion that there was no justification for guidelines, other considerations still caused many to believe that immediate and wholesale abolition of regulations would be politically unacceptable. Taking the offensive with a piecemeal approach might, however, succeed. At the May 21–23 meeting of the RAC, two members, Allan Campbell (Stanford) and Wallace Rowe (NIH), proposed that all recombinant DNA experiments involving *E. coli* K12 host-vector systems be simply exempted from the guidelines (Doc. 14.4). The suggestion was not new; Szybalski, in fact, had stated it explicitly at the December 1977 hearing at the NIH (see Chapter 9), where he said:

> Let me make one major suggestion, which would render the Guidelines scientifically more sound, but still quite conservative. I suggest that with only a few exceptions, one should exclude from the Guidelines all experiments employing EK1 or EK2 host-vector combinations which carry novel recombinant DNA. If you think it would be helpful, you should require only a very simple registration of these EK1 or EK2 experiments. *E. coli* K12 host-vector combinations carrying DNA coding for known polypeptide toxins could be among the experiments which would still remain under the Guidelines.

December 1977 was not, however, May 1979, and Szybalski's proposal had gone unheeded.

The news that the RAC had received and discussed a proposal to exempt *E. coli* K12 experiments, and had set up an *ad hoc* subcommittee to consider the issue further, set in motion a write-in campaign that would swell the NIH's mail bags for the rest of the year as the RAC and NIH moved toward a decision. There were few surprises about the correspondents' positions on this issue (Doc. 14.5). At its meeting on September 5–6, the RAC (which had been increased to 25 members) voted in favor of a slightly modified version of the proposed *E. coli* K12 exemption (10 for, 4 against, 1 abstention; 10 members absent). This version, published in the *Federal Register* (November 30, 1979), became known as the *E. coli* K12/P1 recommendation:

> Those recombinant DNA molecules that are propagated in *E. coli* K12 hosts not containing conjugation-proficient plasmids or generalized transducing phages, when lambda or lambdoid bacteriophages or non-conjugative plasmids are used as vectors, are exempted from the Guidelines, subject to the prohibitions of I-D-1 through I-D-6. Prior to initiation of the experiments, investigators wishing to carry out such experiments must submit a registration document that contains (a) a description of the source(s) of DNA, (b) nature of the inserted DNA sequences and (c) the hosts and vectors to be used. This registration document must be dated and signed by the investigator and filed only with the local IBC (Institutional Biosafety Committee) with no requirement for review by the IBC prior to initiation of experiments in these categories.

As in the past, several environmentalist organizations immediately demanded an environmental impact statement (Doc. 14.5). Members of the RAC itself, some encouraged by allegations in the press of improper procedures at the September meeting, disputed what constituted a proper quorum and majority for such a decision (Doc. 14.5). The Coalition for Responsible Genetic Research certainly felt the result of the vote was inadequate, while *Nature*'s correspondent felt it was unlikely that the NIH would accept the RAC's proposal (Doc. 14.6).

In the *Federal Register* on November 30, 1979, the Director of the NIH published his analysis of the pros and cons of the RAC's *E. coli* K12/P1 recommendation and referred to a 312-page book compiled by his staff entitled *Background Documents on E. coli K12/P1 Recommendations.* Donald Fredrickson announced his intention to follow the RAC's advice concerning P1 containment for experiments with *E. coli* K12 host-vector systems, but contrary to the RAC's proposal he decided not to exempt such experiments from the guidelines. He further proposed that Institutional Biosafety Committees be notified of *E. coli* K12 experiments either before or after their commencement, depending on whether or not a deliberate attempt was being made to achieve expression in *E. coli* K12 of the foreign genes. Although this was a compromise, for research workers it constituted a very significant reduction in the bureaucratic impediments surrounding the vast majority of currently envisaged experiments. Publication of the revised guidelines in the *Federal Register* on January 29, 1980 (excerpted in Doc. 14.7) represented a considerable but by no means complete victory for those intent upon removing unnecessary regulations of recombinant DNA research.

During 1980 the RAC recommended further simplification of the guidelines, and by November it and the NIH felt confident enough to leave review of experiments explicitly covered in the guidelines to local institutional biosafety committees. In the U.K. the GMAG made, if anything, even more far-reaching relaxations. The RAC, the GMAG, and other comparable national committees are fast evolving into vestigial organs of the bureaucracy.

Document 14.1
March 29, 1979
Nature (**278**:481)

Angels and devils of science

R. H. Pritchard

Recombinant DNA: The Untold Story. By J. Lear. Pp. 280. (Crown: New York, 1978.) $8.95.

THIS book contains a brief account of the discoveries and discoverers that made DNA splicing exploitable as a method of amplifying genes and creating new hybrids. It also contains a blow by blow account of the rise and fall of the conviction, and the public debate it led to, that the synthesis of hybrids from organisms whose DNAs do not naturally encounter one another is hazardous because it could lead inadvertently to the creation of new pathogens not controllable by established methods.

This hypothesis in its general form was always a highly speculative one unsupported by any evidence. Its validity was determined not by reasoned argument but by shows of hands at mass meetings and *ex cathedra* pronouncements from groups of eminent scientists. The scientific community was led into a headlong rush to impose upon itself the most stringent safety precautions against totally hypothetical hazards. It caused intelligent scientists to engage in acts of public self-immolation and to public announcements of the destruction of valuable but completely harmless strains of microorganisms. Later there was cheating to get around the regulations and sneaking to get round the cheaters. It has cost millions of dollars and millions in other currencies so far to finance containment laboratories and containment bureaucracies. It has been a damaging episode in the relationship between the scientific community and the public, and there must be important lessons to be learned from it. Readers of Lear's book are in danger of learning the wrong lesson. He is not a dispassionate reporter. He has a particular point of view which leads him to a particular interpretation of the events he records. On the other hand, he has researched his material to a considerable depth. His account of events rings true. He writes entertainingly and often draws amusing and recognisable pen portraits of the central characters. His account of the science itself is admirably lucid with few howlers. It is a book well worth reading.

John Lear doesn't question the validity of the Frankenstein-monster hypothesis. Thus the story he tells is a story of the triumph of good over evil, with the devil threatening a comeback towards the end and the final outcome even now undecided. Good is accountability of scientists to society. The devil, on the other hand, believes in unbridled licence for anyone to do whatever experiment he chooses, accountable to no-one but himself. It is a fairy story, of course, but it seems to be one that Lear believes in.

According to Lear a telephone call from one Robert Pollack to a Professor Berg in 1971 sparked off the great debate. Pollack was worried that Berg proposed to propagate the genome of the monkey virus SV40 in *Escherichia coli* by splicing it into a bacterial virus called λ. Berg subsequently abandoned the experiment. Both men later organised a conference on Biohazards in Biological Research which seems to have been devoted primarily to the hazards of handling tumour virus. The conference did not get widespread publicity inside or outside the scientific community. Lear argues that it was because the press was not alerted. Later in discussing events which really did start the ball rolling he argues that the Pollack–Berg message needed time to sink in. There is another possibility. Pollack's concern was based on a reasoned (although not necessarily valid) extrapolation from well-founded observations. SV40 produces tumours in some animals and causes transformation of human cell lines to forms with cancer-like properties. It also forms natural hybrids with adenoviruses which cause respiratory infections. Anyone familiar with these facts, with the frequency with which animal viruses form hybrids naturally, and with the uncertainty surrounding their pathogenicity, would not lightly synthesise large quantities of SV40 DNA in *E. coli* and it would not be unnatural for a virologist to express his worries to a biochemist whose background might not have alerted him to the potential hazards. But these commendable initiatives could not possibly engender an impassioned debate between the angels and devils of the scientific world. Who would act the devil? Scientists are no more mischievous, no less intelligent and no more suicidal than the population at large.

A conference two years later provided a more favourable breeding ground for devils. A new and trivially simple method of gene splicing was announced. Anybody could do it. The particular and rational concern (about the lack of care in handling the DNA of suspected pathogens like animal viruses) began to merge with a general and irrational one (that any new combination of genetic material will produce organisms with unknown, unpredictable, and hence potentially hazardous properties).

Devils would soon point out that to be a hazard to the population at large the novel hybrid would not only need to be pathogenic but have a positive adaptive value. To them the probability that, for example, a random insertion of fruitfly genes into *E. coli* could produce such an organism seemed, to put it simply, rather less than the probability that a random splicing of bits of the computer tape of an automatic knitting machine programmed to produce pullovers into the tape of a machine programmed to produce sox would not only lead to the generation of a wearable garment (the pullox) but one which would capture a significant share of the market. Naive angels (who seemed to believe that a cholera epidemic could be caused by *E. coli* containing a cholera toxin gene, that diarrhoea could be caused by *E. coli* carrying a cellulase gene, and tooth decay by *E. coli* with a tooth decay gene kit from *Streptococcus mutans*) thought the probability was too high to be ignored. Smart angels agreed that the probability was negligible but asserted that it could not be claimed to be zero. And if it wasn't zero, to ignore it proved one to be a devil. Lovable angels argued that a little nonsense was a small price to pay for a safe code of practice for animal viruses. There were also naive lovable angels who were properly concerned about the spread of plasmid-borne drug resistance in bacteria. They wanted to ban the synthesis of bugs resistant to new combinations of antibiotics by gene splicing. When devils pointed out that to solve a problem the first step was to properly diagnose it, naive lovable angels comforted themselves with the thought that a ban could do no harm, and it might even help just a bit.

Such a confusion of issues, such a mixture of motives, plus a widespread lack of awareness of population biology and ecology among the protagonists, were necessary raw materials for a controversy. But to raise the emotional temperature and politicise the issue required another factor. Pollack's spark needed a torch and it has been suggested to me by many people that the torch was the deep concern amounting to guilt of many United States scientists with the misuse of scientific knowledge—a concern that was born out of Los Alamos and came to a head for biologists when defoliants were used in an unpopular war. Lear's book offers support for this view at many points and a Roy Curtiss III, who emerges as its hero, is a good exemplar of this factor. He circulates one thousand copies of a sixteen-page signed testimony of self-criticism, admitting his innocent guilt and the frightening consequences of his work, pledging himself to be good in the future, and demanding much more severe constraints than were being mooted at the time. His stature in Lear's eyes is further enhanced by the revelation that he went to the library some years later and discovered that his work wasn't as frightening as he had thought. Thus he not only earns his role as a repository of the conscience of the scientific community but demonstrates his honesty and humility as well.

The bandwagon of public accountability came to a euphoric climax at Asilomar. A panel of assorted angels had been working hard on a code of practice which was accepted, after being modified a bit, because no one else had a better one. For the first time the scale of the hazards of genetic engineering were divined and graded and the revelations confirmed by a show of hands (allowing lesser mortals to publicly affirm their concern). For the first time the public could see that scientists believed that the theoretical hazards were more dangerous than real ones like, say, cholera or typhoid fever.

Before Asilomar public displays of concern could be indulged in at no cost. After Asilomar the bill was presented in the form of a blanket of controls out of all proportion to any reasonable assessment of hazards. Scientists handling totally harmless organisms found themselves embraced by a code of practice which assumed that they were dangerous pathogens. With minds thus concentrated there was growing pressure against the enshrinement of the code in legislation. Some people just went ahead and ignored it. To Lear this lobbying and cheating proves his thesis that there are enemies of the people who will use any means fair or foul in the persuit of unbridled scientific licence even to the point of danger to others. He has found his devils and in his enthusiasm to expose them his objectivity fails him. He even discovers a convention among scientists, which is flouted by his devils of course, but also by himself when it suits his purpose. This is the convention, new to me, that it is improper to cite unpublished work.

I should like to be able to record that those scientists who were collectively most influential in setting in motion the so-called debate on the hazards of genetic manipulation had, after due reflection, collectively used their influence to counterbalance the effect of their earlier statements. They have not done so and consequently Lear's thesis is bound to retain some credibility.

This book occasionally alludes to events abroad. The British Parliament reacted with alacrity to the 1974 petition and the Working Party it established completed its work with similar speed. We get a pat on the back for this but I fear that what it demonstrates is the action of reflexes trained not to be caught with pants down in an unusually conservative society. Unfortunately, the same reflexes make any relaxation of the code of practice here much more difficult than in the more open and adventurous United States. Not yet for us the return to reason represented by the recent statement of the Director of the US National Institutes of Health that the burden of proof is shifting towards those who would restrict recombinant DNA research.

This episode should help to educate the public about science and scientists. The realisation that eminence does not necessarily equal wisdom, that expertise in one area can equal ignorance in another, that scientists are no more objective than anyone else, is healthy enough. The conclusion I draw from this is that on matters of policy we should base our decisions on the quality of arguments and not upon the status of those who expound them, as we have learned to do outside science. If this precept had been followed, Lear would have no story to tell. □

R. H. Pritchard is Professor and Head of the Department of Genetics at the University of Leicester, UK.

April 12, 1979
Nature (**278**:590)

Guidelines should go, DNA meeting concludes

Eleanor Lawrence reports on a meeting where biologists scourged themselves for going public on conjectural risks

LAST week in the village of Wye, deep in the Kent countryside, an audience predominantly composed of molecular biologists overwhelmingly reiterated the now widely-held view that recombinant DNA research poses no special risks.

The meeting was convened by the Royal Society and COGENE, the Committee on Genetic Experimentation of the International Council of Scientific Unions, to discuss the status of recombinant DNA work and the guidelines controlling it.

Essentially most scientists working in the field now believe that the original fears were based on bad scientific judgment, and that recombinant DNA experiments at the very worst can pose no more hazard than that of working with the most dangerous organism involved in the experiment. Therefore, they argue, regulations for recombinant DNA research are unwarranted and should be abolished. The inconsistency inevitable in guidelines designed to guard against conjectural hazards and the bureaucracy involved in their implementation pose a threat to the freedom of scientific enquiry.

The scientific basis for the change of heart appears to rest first on advice from experts in infectious diseases that it is virtually impossible to convert the laboratory strain of the common gut organism *E. coli* into an epidemic pathogen by the random insertion of a block of foreign genes.

The fear that the insertion of animal virus genes into *E. coli* would result in a new route for the virus to bypass normal host defences is now also held to be groundless. The current conventional wisdom is that work with cloned viral DNA poses, if anything, even less risk than work with the virus itself, and that cloned viral DNA fragments offer the safest way to study the molecular biology of the most lethal viruses such as Lassa or smallpox.

Joe Sambrook of Cold Spring Harbor Laboratory observed that had Professor Bedson been working in Birmingham with cloned smallpox sequences rather than the complete virus, both he and Janet Parker would be alive today. (The World Health Organisation is indeed considering the possibility that cloned fragments of smallpox DNA might be the safest way of conserving the smallpox genome for posterity). Although little direct evidence addressing this question was available in 1977—when the Ascot Workshop's recommendation of such views to the NIH was influential in gaining considerable relaxation of guidelines for work with animal viruses—experimental support in the case of certain DNA viruses, at least, has recently been obtained from the NIH 'worst case' polyoma virus experiment (see box).

Participants in the debate also marshal evolutionary arguments, such as the growing appreciation that genetic exchange occurs across wide species barriers in microorganisms, as general ammunition. But it is not clear that these arguments necessarily address the particular point that still worries those who see the need for guidelines. They ask to be assured that the specific product of any given recombinant DNA experiment is not going to be hazardous to those who may be exposed to it, either in the laboratory, or in the general environment.

Among those who call for the end of regulation a more subjective attitude is that expressed most forcibly by J. D. Watson of Cold Spring Harbor. Watson now attributes the call for the moratorium as mixture of fears over research with tumour viruses themselves and an attack of mild liberal guilt, and considers that he and this fellow signatories displayed a complete lack of scientific judgment. "We were jackasses" he told the conference. "It was a decision I regret; one that I am intellectually ashamed of". Watson adopted this position soon after the original 'Berg letter' was written.

Another signatory, Stanley Cohen of Stanford University, also felt the group's original action was irresponsible on scientific grounds, as well as politically naive. It was based simply on a "lack of certainty there was no risk" and was therefore an "irresponsible scientific argument".

The third signatory present, Norton Zinder, holds a somewhat different view. Although he now thinks their original fears to be groundless, in the circumstances as they saw them at the time there was no other action that could be taken.

Given that they now largely have the support of their scientific colleagues on guideline committees, the problem now facing those who wish to get rid of public guidelines was neatly summed up by Mark Richmond, a member of the UK Genetic Manipulation Advisory Group. Publication of the Berg letter, he pointed out, was in effect a political act even though there had been no political intention. But through this political act the subject has lost its innocence, and it is now going to be hard for the scientists involved to regain a position in which their views can be regarded as objective.

"How are you going to reassure the public that you're not arguing from self-interest", said Richmond, "but that you are now arguing objectively—particularly when, looked at from the outside, you seem to make enormous quantum jumps between what you require for conditions at one time and what you required for the same experiments a couple of years ago"

R. Pritchard, Professor of Biochemistry at Leicester and a well-known opponent of the UK GMAG reiterated his opinion that the effect of the Berg letter was largely due to the eminent names appended to it, and that its message could only be countermanded by the authors themselves. They should state publicly and unequivocally what they now believe, he said, to the loudest round of applause of the meeting.

Not all the participants, however, judged the display of public recantation before the Inquisition to be warranted. In the perception of a historian of science, Charles Weiner of MIT, the recombinant DNA debate has not been the wasteful and illborn episode it seems to most of those concerned.

Weiner has followed the tortuous course of the debate from the beginning and has collected in his *Recombinant DNA Archive* at MIT much of the formal documentation, supplemented by the informal accounts of those involved. Although molecular biologists found the experience of coming face to face with the apprehensions of society a traumatic one, they have, in Weiner's opinion, come through with their science and honour relately unscathed.

Despite the trauma and despite the bureaucracy said Weiner, recombinant DNA research flourishes. Has the whole debate and its attendant publicity, he wondered, stimulated rather than repressed the research? But he was concerned at the attitude he now sees amongst younger scientists, to whom the message of the recombinant DNA debate now seems to be "shut up or be shut down". □

Document 14.3

May 6, 1979

Simring Letter to Gartland

72 Jane Street
New York N Y 10014
(212) 675-7173

Coalition for Responsible Genetic Research

May 6, 1979

Dr. William Gartland
Recombinant DNA Advisory Committee
NIH, Bldg. 31, Room 4A 52
Bethesda, MD 20014

Dear Dr. Gartland:

On behalf of the Coalition for Responsible Genetic Research, I wish to make some general comments on the recent treatment of the recombinant DNA issue by the NIH and some proposals for consideration at the next meeting of the RAC, May 21-23, 1979. I also enclose comments on the proposed actions under the guidelines and on the proposed risk assessment program published in the Federal Register April 13 and April 2.

I. General comments on recent treatment of the recombinant DNA issue by the NIH leadership

We are disturbed by the cynicism and general lack of caution and restraint in dealing with the recombinant DNA issue that is emanating from the NIH leadership. We consider Dr. Setlow's comment, that the "RAC will not be allowed to die as a committee until it has vomitted enough," and Dr. Fredrickson's use of slides of critics of NIH policy (shown in his absence by Dr. Gartland) at the recent COGENE meeting in England to be not only in poor taste but also to create an atmosphere in which responsible and rational consideration of the issues will be difficult to achieve.

There are distinct parallels between the treatment of the recombinant DNA issue by the NIH and the treatment of the nuclear issue by those involved in promoting nuclear power. For the latter group, it appears that nothing less than an absolute disaster will convince them that they are dealing with serious risks to the health and well-being of the general public. With regret, we observe that the "Three Mile Island" syndrome also characterizes the handling of the genetic manipulation problem by the NIH. Safeguards are treated with cynicism; decisions are rushed through because of special interests in the promotion of the technology; data that is at best ambiguous and that could be taken to indicate problems are ignored or treated selectively.

We feel that the cynicism and impatience on the part of the NIH leadershiptowards those who advocate caution has already had a damaging effect on the handling of issues by the RAC. At the first meeting of the expanded committee in February, a large number of decisions were rushed through. The time allowed for consideration of the issues involved was totally inadequate. Moves from some of the new members to slow down the pace of decision and to improve procedures were often strenuously resisted or voted down (see attached article from Environment, May, 1979).

The proposals below are made in the hope that procedures for handling future agendas of the RAC can be upgraded. Otherwise, it seems inevitable that at some point, serious errors are going to be made.

II. Proposals for consideration at the meeting of the RAC, May 21-23, 1979

1. Changes in RAC Procedures*

a) A quorum for the committee should be defined.

b) Meetings should not run over the previously announced time of adjournment. At the last meeting, at least two important decisions were reached after the official termination time: i) a decision to approve the proposal of Dr. David Baltimore to use defective mouse leukemia virus to insert genes into mouse cells in tissue culture; ii) a decision on interim procedures to deal with proposals for large-scale recombinant DNA processes. We propose that these decisions be reconsidered by the full committee.

c) Votes should be recorded by name. Since the committee is responsible for making public policy for the recombinant DNA field, we feel the general public has a right to know in detail how decisions were made. As Ms. Patricia King stated at the last meeting, this record will be critical in the future for assessing the value of the RAC as a device for making public policy for technical issues.

d) A glossary should be provided with each technical document so that the issues involved are intelligible to all members of the committee. (Scientists making proposals to ORDA might be asked to provide such as document.)

e) Materials for committee meetings should be circulated at least two weeks ahead of the meeting time. To allow a reasonable turnaround time, this means that the period for responses to proposed actions published in the Federal Register should end at least three weeks before an RAC meeting. The present arrangements do not give adequate time for RAC members to absorb the material for committee meetings. Therefore we propose that the informal decisions reached at the last meeting, to mail materials one week ahead, be reconsidered.

2. General principles for containment of risks from novel uses of recombinant DNA techniques

The RAC should address directly the question of how to assess the risks from novel uses of recombinant DNA techniques and arrive at consensus on the general principles involved. We observe that some members of the committee tend to assign low containment levels unless some specific scenario for harm comes to mind. We feel this approach courts disaster. The techniques of nuclear fission are well understood. Even so, the formation of the hydrogen bubble at Three Mile Island was a totally unforeseen event.

We propose that for any novel use of recombinant DNA techniques for which no risk assessment studies have been carried out, initial physical containment

---

*Some of these proposals were made by members of the RAC at the last meeting but not acted upon.

should be set at no less than P3. Containment levels should be changed only after comprehensive risk assessment studies have been carried out and agreement reached as to the significance of the empiricical data.

3. Decisions on large-scale processes

a) All decisions on large-scale processes should be made at the formal meetings of the RAC.

b) A full risk assessment of the process in question should take place before consideration of an increase in scale.This should include i) assessment of the safety of the organisms in use; ii) consideration of possible contamination of the large volumes in question; iii) re-examination of the effectiveness of physical containment levels developed for small-scale experiments when applied to large volumes; iv) consideration of possible escape routes provided by a large volume.

c) All facilities involved in large-scale processes should be required to obtain a license from some authority.* The licensing authority should have responsibility for establishing mandatory training and medical surveillance programs, and for site visits.

III. Comments on proposed actions under the guidelines published in the Federal Register, April 13, 1979

1. Proposed exemption under I-E-5 for experiments involving EK1 and EK2 host-vector systems (item #6)

We are strongly opposed to this irresponsible and short-sighted proposal. It would make a mockery of the attempt to develop rational containment policies for recombinant DNA work. This proposal also seriously undermines the NIH's own proposed program of risk assessment (Federal Register, April 2, 1979).If the NIH is seriously considering pursuing further risk assessment experiments on E.coli K-12 systems, a decision to dismantle precautions for those systems before the results are in is unjustified and irrational.

2. Proposed exemption under I-E-5 for cloning in tissue culture cells (#7)

We are strongly opposed to this proposal. Recent findings have shown clearly that many cells contain genetic sequences relating to the transforming genes of tumor viruses which may be latent or cryptic in the cell. In addition, the critical sequences in transforming viruses may represent a very small fraction of the genome. Therefore, the introduction of foreign DNAs linked to viral sequences (even if only one-fourth of the genome of the virus) into tissue culture cells may still provide the possibility of generating within that cell a source of genetic variation for endogenous or exogenous viruses.

3. Criteria for characterized clones (#9)

9a: Absence of potentially hazardous genes

The statement that "in E.coli the risk of induced autoimmunity from exposure to clones that produce proteins that are either human hormones or other biologically

*E.g. EPA, OSHA, or CDC.

active molecules is considered insignificant" is scientifically inaccurate and irresponsible. The most accurate statement that can be made at present regarding the risk of inducing autoimmunity is that this risk is unknown. Discussion of this risk at the February meeting of the RAC demonstrated wide variation in scientific opinion in the absence of data and clearly showed the need for empirical studies to resolve the uncertainties involved. This need is also reflected in the proposed risk assessment plan (Federal Register, April 2, 1979) which calls for "animal studies of hormone-producing strains of E.coli generated by recombinant DNA technology" and studies of "the possible occurrence of autoantibodies or autoreactive cells due to the production of cells due to the production of eukaryotic polypeptides by bacteria that colonize higher organisms."

IV. Comments on Proposed Plan for a Program to Assess the Risks of Recombinant DNA Research (Federal Register, April 2, 1979)

1. Significance of the Rowe-Martin experiment on the infectivity of polyoma DNA cloned in E.coli K-12

The statement that this experiment produced "no evidence that the inserted DNA produced any special hazard" is misleading. In fact, this experiment produced some positive results whose significance is still being investigated. The results of this experiment should be reviewed in depth by the RAC at its May meeting.Additional comments by members of the Coalition will be submitted to ORDA.

2. Part III. Scientific Aspects of Recombinant DNA Risk Assessment Plan: Higher Eukaryotes (Federal Register, p.19303, col.2)

There is little evidence to support the statement that the "possibility of (1) creating novel nondefective viruses as a result of the insertion of a new DNA fragment or (2) altering the host range of the viral vector" is "unlikely." Specific risk assessment studies on these possibilities should be planned and carried out.

Sincerely yours,

Francine Simring

Francine Robinson Simring
Executive Director

Document 14.4

May 31, 1979 July 31, 1979
Nature (**279**:360) *Federal Register*, Excerpt

Scientists debate safety of research on *E coli* strain

A STRONG bid is being made to exempt a large body of recombinant DNA experiments using a disabled strain of the bacterium *Escherichia coli* from the research guidelines laid down by the National Institutes of Health.

Exemption has been proposed to the NIH's Recombinant DNA Advisory Committee (RAC) by Dr Wallace Rowe, head of the Laboratory of Viral Diseases at the National Institute of Allergy and Infectious Diseases, and Dr Allan Campbell, Professor of Biology at Stanford University.

The two scientists base their recommendation on new information which has recently emerged on the potential risks of recombinant DNA research. This evidence, they claim, makes it unreasonable to apply restrictions on most experiments involving the cloning of plasmids inside the 'disabled' *E. coli* K-12 strain, over and above what would normally be considered as safe laboratory practice.

Other members of the committee meeting in Washington last week, however, expressed concern at the implications of such a move. They suggested that, given the many uncertainties still involved, a total exemption from current controls would be premature. However, the committee agreed to set up a working party to study the available data on the safety of experiments using the K-12 strain and suggested that it should consider a proposal that all such experiments are carried out under the minimal P1 physical containment conditions.

In a statement to the committee, Dr Rowe said that information pointing to the safety of K-12 host vector systems included extensive analysis of the biology of the organism and molecular segments cloned inside it, the negative results of monitoring laboratory personnel for acquisition of the bacterium and its plasmids and the results of risk experiments involving the cloning of polyoma virus DNA which he and other research workers had carried out at Fort Detrick last year.

"The basic message from every one of these experiments is that there is no cause for concern about recombinant DNA research with K-12. I do not know of a single piece of new data that has indicated that K-12 recombinant DNA research could generate a biohazard", Dr Rowe said.

For a recombinant organism to become a health hazard, it had to escape from a laboratory, survive outside the laboratory environment, establish itself in some ecological niche, and finally have some detrimental effect on a higher organism. "Since the original guidelines, data and information have accumulated that indicate that each of these steps is so highly improbable with the K-12 cloning system that each step alone gives sufficient assurance of safety, much less the combination of all four."

Dr Rowe's interpretation of the safety of the K-12 system, however, is not shared by all scientists. In a letter to the RAC commenting on the proposal to exempt most of the experiments using this strain from the guidelines, Dr Roy Curtiss, Professor of Microbiology at the University of Alabama in Birmingham, who has done much work on the development of enfeebled organisms, said he felt such a move would be premature.

In particular, said Dr Curtiss, although it was generally agreed at the Falmouth conference of 1977 that one could not convert K-12 into an epidemic pathogen, a number of studies had since indicated that "the overall probability for transmission of recombinant DNA from *E. coli* K-12 hosts and vectors are higher than I or others believed."

Some of these observations included the excretion of EK1 hosts by human subjects several weeks after receiving a dose, the survival of the disabled λ1776 strain in the human gut for up to four days, and the selective survival of EK1 hosts in individual mice and rats.

"I surmise that if the participants at the Falmouth conference had been aware of these data, more consideration would have been given to possible consequences of transmission of recombinant DNA to indigenous microorganisms of various natural environments," he said.

In addition to the scientific arguments about the safety of such experiments—other scientists, in particular Dr Jonathan King of MIT and Dr Jon Beckwith of Harward Medical School have raised the possibility of biologically active recombinant organisms inducing an auto-immune response from the human body—the committee's discussions also brought in wider considerations.

Dr Rowe has made no secret of his feelings about the guidelines, calling them "wastful, expensive, inefficient, inflexible and inhibitory". The perception or risk need not be zero before restrictions were lifted, he said. "We have to do a cost-benefit analysis. If the guidelines are harming research, which I believe they are, then this has to be entered into the equation". Dr Campbell expressed similar sentiments in support of the proposed exemption. "In my judgement the NIH guidelines as presently written and administered do more harm than good. We should now be examining the guidelines section by section for the hazards that exist."

However, other members of the committee expressed the view that total exemption for a large group of experiments was inadvisable, since risk assessment experiments had not gone far enough to demonstrate that dangers did not exist, while laboratories could not be relied upon to adopt safe procedures (such as a ban on mouth pipetting). "Society is facing a mounting load of hazards, and the RAC is in the vanguard of responsible research in general. You cannot legislate morality, but you can provide a framework for responsibility", said Dr Richard Novick, chairman of plasmid biology at the Public Health Research Institute in New York.

After considerable debate, it was decided to set up a working group to synthesise the data both supporting and not supporting the proposal made by Dr Rowe and Dr Campbell. It was also agreed that the proposal should be modified so that, rather than considering a total exemption for this class of experiments, the committee should consider reducing them to requiring P1 containment levels, as well as registration with local institutional biohazard committees. □

Federal Register / Vol. 44, No. 148 / Tuesday, July 31, 1979 / Notices

SUPPLEMENTARY INFORMATION: The National Institutes of Health will consider the following changes and amendments under the Guidelines for Research Involving Recombinant DNA Molecules (43 FR 60108), as well as actions under these Guidelines.

1. *Proposed Exemption for E. coli K–12 Host-Vector Systems.* The RCA Working Group on *E. coli* K–12 host-vector systems will report to the full committee for discussion/action documentation for a proposed exemption under I–E–5 of the Guidelines for experiments involving EK1 and EK2 host-vector systems. This proposed action would exempt certain categories of recombinant DNA molecules in addition to those already stated in Sections 1–E–1 to –4. The proposed exemption is as follows:

Those recombinant DNA molecules that are propagated in *E. coli* K–12 hosts not containing conjugation-proficient plasmids or generalized transducing phages, when lambda or lambdoid bacteriophages or non-conjugative plasmids are used as vectors, can be handled at P1 and are exempted from the Guidelines.

Document 14.5

September 19, 1979
Cavalieri Letter to Fredrickson

SLOAN-KETTERING INSTITUTE *for* CANCER RESEARCH

DONALD S. WALKER LABORATORY, 145 BOSTON POST RD., RYE, N.Y. 10580 OWENS 8-1100

September 19, 1979

Dr. Donald S. Fredrickson
Director
National Institutes of Health
Bethesda, Maryland 20205

Dear Dr. Fredrickson:

I am appalled by the recent vote (Sept. 6-7) of the RAC regarding the NIH guidelines. What might have been a laudible effort by this committee has become dirty with common political maneuvering. Whether or not the vote of 10-4 (with one abstension) is "legal" is irrelevant. It is clear that the spirit of this committee will have vanished should you accept this decision to eliminate virtually all safeguards for the general community. This is after all a public issue; the convenience of scientists is hardly reason to replace sane procedures with irresponsible ones.

The May meeting of the RAC, which I attended, bordered on a sham; the September meeting, with heavy input from industrial interests was stacked for achieving purposes not concerned with public interests.

Yours truly,

Liebe F. Cavalieri, Ph.D.
Member, Sloan-Kettering Institute
Professor of Biochemistry, Graduate
School of Medical Sciences
Cornell Medical College

LFC/nc

RESEARCH UNIT OF MEMORIAL SLOAN-KETTERING CANCER CENTER

November 26, 1979
Chamot Letter to Harris

Department for Professional Employees, AFL-CIO
815 16th Street, N.W., Washington, D.C. 20006 Phone 202/638-0320

November 26, 1979

Honorable Patricia Harris
Secretary
Department of Health, Education & Welfare
200 Independence Avenue, SW, Room 615F
Washington, DC 20201

Dear Mrs. Harris:

We are very concerned that the National Institutes of Health is in the process of lowering safety standards applying to genetic engineering research, even though major questions of safety remain unanswered (especially in the case of organisms other than weakened E. coli strains). We are distressed that this weakening of the NIH Guidelines is coming just when major commercial applications of the new technology are approaching reality, and no other specific regulations or legislation currently exist upon which to base technical decisions regarding threats to worker and community safety from these new organisms.

This AFL-CIO Department, comprising 26 national unions, represents over one and one-half million professional and technical workers, including laboratory staffs and production workers in chemical and health related industries. Attached is a copy of a resolution on recombinant DNA research which was passed by the delegates to our biennial convention on November 13, 1979.

We ask that you use the authority of your office to prevent the premature weakening of the NIH Guidelines for Recombinant DNA Research. We also urge that you take the responsibility for overseeing the performance of this research out of the hands of those who have a prime responsibility for funding and encouraging it, the NIH.

Sincerely,

Dennis Chamot

Dennis Chamot, Ph.D.
Assistant Director

DC/mw
Encl.

University of Wisconsin–Madison
LABORATORY OF GENETICS
Genetics Building
445 Henry Mall
Madison, Wisconsin 53706

July 10, 1979

Dr. Donald Frederickson, Director
National Institute of Health
9000 Rockville Peak
Bethesda, MD 20205

Dear Dr. Frederickson,

Enclosed is a letter, drafted at the Nucleic Acids Research Conference, supporting exemption of K12 experiments from the NIH guidelines. It was signed by 183 scientists.

This is in response to the publication in the Federal Register of the agenda of the Recombinant DNA Advisory Committee.

Yours truly,

Frederick R. Blattner
Professor of Genetics

August 15, 1979

THE JOHNS HOPKINS UNIVERSITY
SCHOOL OF MEDICINE

...excluding from such exemption presently prohibited experiments. In agreement with Rowe and Campbell, I believe that we have had sufficient experience with E. coli K-12 host-vector systems to conclude that routine use of such systems does not require regulation to protect the safety of laboratory personnel or the public. In my opinion simple, standard microbiological practice is quite adequate for this purpose, as it is for demonstrably hazardous enteric pathogens.

Sincerely,

Daniel Nathans, M.D.
Professor and Director
Department of Microbiology

PRO·BONO·HUMANI·GENERIS · The Rockefeller University · 1901

THE ROCKEFELLER UNIVERSITY
1230 YORK AVENUE · NEW YORK, NEW YORK 10021

December 14, 1979

... Therefore, I want to let you know that I am strongly in favor of the relaxation of the NIH Guidelines on Recombinant DNA as proposed by the Director's decision described in the Federal Register of Nov. 30. I hope that the support and trust which you carry from so many of us will enable you to facilitate the implementation of this decision.

With best wishes for the Holiday Season,

Sincerely yours,

Rollin D. Hotchkiss
Professor of Genetics

STANFORD UNIVERSITY MEDICAL CENTER
STANFORD, CALIFORNIA 94305

DEPARTMENT OF BIOCHEMISTRY

August 20, 1979

...issue in the manner we did. I am persuaded by the evidence I have seen that molecular cloning of any DNA segments in E. coli K-12 using the array of present day cloning vectors is no longer of any real concern, certainly not enough to warrant the continued inclusion of such experiments within the purview of The Guidelines and the IBC's. Perhaps it would still be useful to maintain a log of recombin[illegible] DNA experiments in this category so that these experiments can contribute t[illegible] ongoing assessment of the safety of such research activities.

With best personal rega[illegible]
Sincerely,

Paul

Paul Berg

CARNEGIE INSTITUTION OF WASHINGTON
DEPARTMENT OF EMBRYOLOGY
115 WEST UNIVERSITY PARKWAY
BALTIMORE, MARYLAND 21210
TELEPHONE 467-1414

August 14, 1979

...It has reached a point where further testing is essentially a waste of time. Critics of the research will continue to ask for more tests after each negative report. An enlightened approach must now accept the weight of evidence and liberalize these unecessary restrictions.

I hope that you will pass this letter on to the committee that is evaluating this change in the Recombinant DNA Guidelines.

Sincerely yours,

Donald D. Brown,
Director

McARDLE LABORA[illegible]
FOR CANCER RESEAR[illegible]
UNIVERSITY OF WISCONSIN · MADISON, WI[illegible]

MEDICAL CENTER

December 14, 1979

Dear Don:

I have read the proposed guidelines for research on recombinant DNA published in the Federal Register of November 30, 1979. I am glad to see this simplification in the regulations governing research with recombinant DNA. It is clear from all of the research that has been carried out using recombinant DNA technology that the only "surprises" have been to make unplanned expression even less likely than thought. It is clear that the earlier fears of effects of eukaryotic DNA in E. coli were exaggerated.

Sincerely yours,

Howard M. Temin

October 4, 1979
Curtiss Letter to Fredrickson

the University of Alabama in Birmingham / UNIVERSITY STATION / BIRMINGHAM, ALABAMA 35294

the Medical Center / DEPARTMENT OF MICROBIOLOGY / October 4, 1979

Dr. Donald Fredrickson
Director
National Institutes of Health
Bethesda, Maryland 20014

Dear Don:

I am opposed to the change in the NIH Guidelines for Recombinant DNA Research recently recommended to you by the Recombinant DNA Advisory Committee to exempt all recombinant DNA experiments using Escherichia coli K-12 hosts (not containing conjugative plasmids or lysogenic for generalized transducing phages) with lambda or lambdoid bacteriophage and nonconjugative plasmid vectors. My reasons are as follows:

First, much of the data on the safety of the E. coli K-12 systems pertains to the genetically disabled EK2 systems and such data, whether on survival or transmissibility, are irrelevant in justifying the exemption proposed since wild-type E. coli K-12 hosts with wild-type nonconjugative plasmid and wild-type lambda phage vectors would be permissible to use.

Second, data on the safety of the various E. coli K-12 host-vector systems as were presented and discussed at the Falmouth meeting in 1977 were already evaluated by the membership of a previous Recombinant DNA Advisory Committee and were used to justify lowering the levels of containment required for many experiments as permitted by the revised NIH Guidelines for Recombinant DNA Research as issued on January 2, 1979. One can thus ask what is the basis for the current committee's recommendation to use these same data to justify an additional lowering of containment?

Third, essentially all data obtained since 1977 by NIH contractors conducting risk assessment experiments to evaluate the safety of the E. coli K-12 systems have indicated that host strains and bacteriophage lambda vectors survive better in many environments than previously believed (see Recombinant DNA Technical Bulletin, Volume 2, No. 2, July, 1979). As a consequence, transmission of recombinant DNA from K-12 to other microbes is a more probable event than was previously believed (see pages 3-4 from my letter of May 11, 1979 to William Gartland).

Fourth, although the polyoma risk assessment experiments provided data to indicate that viral genetic information was more safely studied when in E. coli than when in the intact virus, they also indicated that a new pathogenic agent,

Dr. Donald Fredrickson
October 4, 1979
Page Two

even though of very low infectivity, could be created. It is thus apparent that initial concerns about the inadvertant endowment of new ecological niches for determinants of pathogenicity were justified.

Fifth, although I know of no reason other than the above to suspect any hazard associated with recombinant DNA activities using E. coli K-12 host-vector systems, there must remain uncertainty since none of us is totally clairvoyant. Indeed, initial users of DDT, thalidomide and even the internal combustion engine were totally unable to predict the ultimate health hazards associated with their use. Also, most of the risk assessment experiments discussed and proposed by the participants at the Falmouth Conference (J. Infect. Dis. 137:704) have yet to be conducted and only recently have some of the more important studies been included in NIH's proposed risk assessment program (Federal Register, April 2, 1979). It would thus appear that the current Recombinant DNA Advisory Committee no longer considers the advice of the Falmouth participants or the data from the experiments they proposed to be relevant in assessing the risks of using the E. coli K-12 host-vector systems.

Sixth, I should note that our laboratory uses recombinant DNA techniques for experiments requiring P1, P2 and P3 containment using donor DNA from eukaryotes and both pathogenic and non-pathogenic prokaryotes. Adherence to the NIH Guidelines has not hindered our work.

I therefore consider the exemption of all experiments using the E. coli K-12 host-vector systems to be premature and not justified on the basis of objective review of available scientific evidence. I also believe that the Recombinant DNA Advisory Committee's recommendation to you was based more on the politics of science than on the data of science.

Sincerely yours,

Roy Curtiss III

RCIII/pp
cc: Dr. William Gartland

October 30, 1979
Cohen Letter to Fredrickson

STANFORD UNIVERSITY MEDICAL CENTER

DEPARTMENT OF GENETICS

October 30, 1979

Dr. Donald Frederickson
Director
National Institutes of Health
Bethesda, Maryland 20014

Dear Don:

I have had the opportunity to read some of the comments transmitted to you recently by Roy Curtiss and others expressing opposition to the change in the NIH Guidelines for Recombinant DNA Research recommended by the Recombinant DNA Advisory Committee. Specifically, these comments concerned the proposal to exempt all recombinant DNA experiments using Escherichia coli K12 hosts (not containing conjugative plasmids or lysogenic for generalized transducing phages) with lambda or lambdoid bacteriophage and nonconjugative plasmid vectors.

In contrast to Roy, I strongly support the proposed change in the Guidelines. Roy's opposition to the change seems based on the premise that the proposed exemption for cloning in E. coli K12 is equivalent to the abolition of all safety measures in such work. In fact, what is being proposed by the Advisory Committee is reliance on standard operating procedures rather than on a cumbersome and costly administrative apparatus to ensure biosafety for work in this area. Certainly there is no indication that cloning of DNA in E. coli K12 poses any hazard that warrants extraordinary precaution, and in fact, experience thus far indicates that recombinant DNA methods provide increased safety for work with genetic material that does encode hazardous products. Since there is no evidence of special hazard in the experiments proposed for exemption, there would seem to be no need for a special administrative process to promote safety; standard microbiological practices appear to be adequate for work with a variety of organisms that are capable of causing serious disease or environmental damage; if special regulations or enforced guidelines are not necessary to protect the public health and the environment from organisms that are known to be hazardous, they surely should not be necessary to deal with organisms that give no indication of being hazardous at all. The continuation of special procedures for work involving cloning of DNA in E. coli K12 would be justified only if there were some valid basis for believing there are special hazards.

I strongly urge you to approve the change recommended by the Recombinant DNA Advisory Committee. This change would place the biohazard issues in appropriate perspective for a major segment of recombinant DNA work by relying on standard microbiological practices, rather than on special regulations, to promote biosafety.

Sincerely yours,

Stan

Stanley N. Cohen
Professor

SNC:ps

DEPARTMENT OF GENETICS, STANFORD UNIVERSITY SCHOOL OF MEDICINE, STANFORD, CALIFORNIA 94305 • (415) 497-5052

September 21, 1979
Hartzman Letter to Fredrickson

FRIENDS OF THE EARTH 72 JANE STREET • NEW YORK, NEW YORK 10014 • (212) 675-5911

September 21, 1979

Dr. Donald S. Fredrickson
Director
National Institutes of Health
Bethesda, Maryland 20014

Dear Dr. Fredrickson:

We are writing to request that an environmental impact statement be prepared for the proposed exemptions from the recombinant DNA guidelines which were considered at the September 5-6 meeting of the Recombinant Advisory Committee. The proposed exemptions are of such breadth and importance as to require full compliance with the National Environmental Policy Act (NEPA). The proposed exemptions are certainly controversial, as evidenced by the 10 to 4 vote, with one abstention, by the Recombinant Advisory Committee. Furthermore, recently obtained risk assessment data is subject to differing interpretations, leading us to believe that further assessment is necessary before such action is taken. The proposed exemptions plainly constitute a major federal action which may significantly affect the environment.

As you know, we have been dissatisfied with previous actions in the recombinant DNA program, both with respect to NEPA compliance and the rational consideration of risks in the conduct of recombinant research. The question of appropriate safeguards should not be decided as a result of pressure from the scientific community, because of letters signed by x hundred scientists, or because of paperwork involved; but rather on the basis of substantive considerations- the possibility of adverse environmental impacts and a careful exploration of alternatives.

Our concern is two-fold: that hazards be properly taken into account in pursuing recombinant DNA research, and that the law be obeyed in conducting this federal program. NEPA applies to <u>all</u> federal agencies, and the NEPA regulations promulgated by the Council on Environmental Quality are mandatory for <u>all</u> federal agencies. As a matter of law an environmental impact statement should have been prepared prior to consideration of the proposed exemptions by the Recombinant Advisory Committee.

As part of our request for an environmental impact statement, we are asking that the Recombinant Advisory Committee reconsider the proposal in light of the environmental impact statement. We do not see anything unusual in this, as well as our earlier requests. All other federal agencies, and many state agencies because of state NEPA's, comply with the required procedures. There is no reason why the NIH or the scientific community should be exempted from these legal requirements. Our polity is based on the rule of law, not of men, and the least we can expect of NIH actions is conformity to the law.

Sincerely yours,
Richard M. Hartzman
Richard M. Hartzman, Esq.

cc: F. Peter Libassi, Esq.
Robert Nicholas, Esq., CEQ

October 10, 1979

Dr. Donald Fredrickson
Director
National Institutes of Health
Bethesda, MD 20014

Dear Dr. Fredrickson:

Richard Hartzman of Friends of the Earth has forwarded me a copy of his September 21, 1979, letter to you regarding the need for preparation of an Environmental Impact Statement supporting the Recombinant DNA Advisory Committee's decision to exempt experiments from the guidelines. Mr. Hartzman raises some significant questions concerning NIH's compliance with the requirements of the National Environmental Policy Act.

Because of the sweeping nature of the approved exemptions and the seriousness of Mr. Hartzman's requests, we would appreciate receiving a copy of your reply to Mr. Hartzman's letter. We would expect this reply to provide your legal rationale for your decision not to prepare an impact statement prior to your potential acceptance of the Recombinant DNA Advisory Committee's action. We intend to follow this situation closely.

Sincerely,
Leslie Dach

Leslie Dach
Science Associate

LD:cs

Environmental Defense Fund, 1525 18th Street NW, Washington, DC 20036 (202) 833-1484
OFFICES IN: NEW YORK, NY (NATIONAL HEADQUARTERS); WASHINGTON, DC; BERKELEY, CA; DENVER, CO

September 28, 1979
Simring Letter to Fredrickson

72 Jane Street
New York N Y 10014
(212) 675-7173

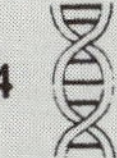

Coalition for Responsible Genetic Research

September 28, 1979

Dr. Donald Fredrickson, Director
National Institutes of Health
Bethesda, Md. 20014

Dear Dr. Frederickson:

The Coalition for Responsible Genetic Research requests that the National Institutes of Health prepare environmental impact statements on the recent proposals of the Recombinant DNA Advisory Committee (September 6-7, 1979) to exempt some 85% of recombinant DNA research from government guidelines, to permit exemptions from the 10-liter limit on culture volume, etc. as required by the National Environmental Policy Act.

The need for environmental impact statements is especially compelling since recent studies under sponsorship of the NIH indicate that

1. naked polyoma DNA can cause infection in mice
2. gene-splice products can cause tumors in mice
3. recombinant DNA organisms can survive (for four days) in the human gut and in sewage.

We cannot afford a biological Three Mile Island accident. Yet the ingredients are present:

- premature expansion of the tecnnology
- non-compliance with the law
- inadequate assessment of the risks
- lack of objective information to the public
- lack of effective public participation in decision-making

It is essential to consider with great caution the legal, scientific and social ramifications of the RAC proposals (voted for by less than a majority of the committee members). The opportunity to view the situation with the benefit of foresight is given to us by past experience with the numerous public health and safety problems of other technologies.

We strongly urge that no action be taken on the aforementioned RAC proposals until appropriate environmental impact statements are prepared and the proposals reconsidered by the RAC in light of those documents.

Yours very truly,

Francine Robinson Simring
Executive Director

FRS:fh

October 10, 1979
Ozonoff Letter to Fredrickson

Boston University Medical Center

School of Medicine
80 East Concord Street
Boston, Massachusetts 02118

Department of Socio-Medical Sciences
and Community Medicine

October 10, 1979

Dr. Donald Fredrickson
Director
National Institutes of Health
Bethesda, MD 20014

Dear Dr. Fredrickson:

The recent actions of RAC concerning the relaxation of rules under which NIH sponsored recombinant DNA research may be conducted are of great concern to many of us who have been following this issue. The fact that such a major action was taken on the vote of less than a majority of RAC members, together with the magnitude of the change argues for a reconsideration, and most certainly, the preparation of an environmental impact statement as required by NEPA. While it is true that the action was supported by many scientists (principally those who have a special interest in that type of work) there are many more hundreds of us who had no opportunity to comment on the proposed changes and who have grave misgivings about the wisdom of this action. Moreover, it comes at a most peculiar juncture in the risk assessment process, since papers printed in ORDA's own Recombinant DNA Technical Bulletin and the published Rowe-Martin experiments present very worrisome findings.

At the very least, it is not too much to ask that the time be taken to prepare an EIS, as required by federal law. I would expect nothing less.

Sincerely yours,

David Ozonoff, M.D.

David Ozonoff, M.D., M.P.H.
Chief, Section on Environmental Health
Boston University School of
Public Health

cc: Robert Nicholas, Esq.
Peter Libassi, Esq.

September 14, 1979
Goldstein Letter to Fredrickson

DEPARTMENT OF MICROBIOLOGY AND MOLECULAR GENETICS
HARVARD MEDICAL SCHOOL
25 SHATTUCK STREET
BOSTON, MASSACHUSETTS 02115

14 September 1979

Dr. Donald S. Fredrickson
Director
National Institutes of Health
Bethesda, Maryland 20205

Dear Dr. Fredrickson:

I write in reference to recent actions of the Recombinant DNA Advisory Committee during the meeting held September 6-7.

I believe it important that you are made aware of the fact that a minority of the total RAC members (i.e. 10 of 25) were allowed to pass on the major action to exempt *E. coli* K12 host recombinant DNA studies from the present NIH guidelines. That *less than* a simple majority of authorized RAC members can exempt *80-85% of* current recombinant DNA studies from NIH regulation makes me seriously question the legitimacy of the RAC to carry out its advisory role in a democratic fashion.

Procedural questions are not my only concern with respect to the above decision by the RAC. This important decision was also premature in terms of the present risk assessment studies being carried out by NIH contractors. What is most ironic here is that some of these ongoing studies address the most pertinent question involved with *E. coli* K-12 recombinant DNA work -- the transfer of a recombinant segment to other hosts. Rather than focusing on this question, which was at the heart of the discussions at the Falmouth Risk Assessement Meeting, the chairperson allowed the presentation of unpublished and unverified studies by industrial representatives in favor of the *E. coli* K12 exemption and the over 10 liter exemption.

I do hope you will consider what I have mentioned above before you accept or reject the exemption on the *E. coli* K12 work. The situation at the recent RAC meeting has given me reason to consider resignation from the RAC. I know my feelings are shared by several other RAC members.

I look forward to your reply.

Sincerely,

Richard Goldstein
Associate Professor

September 24, 1979
Krimsky Letter to Fredrickson

TUFTS UNIVERSITY

Graduate Program in Urban Social and Environmental Policy

September 24, 1979

Dr. Donald S. Fredrickson
Director
National Institutes of Health
Bethesda, Maryland 20205

Dear Dr. Fredrickson:

As you know, at the last meeting of the Recombinant DNA Advisory Committee, a proposal to exempt a large class of recombinant DNA experiments from the NIH guidelines was approved by a 10 to 4 vote with one abstention.

I was one of the RAC members who voted against the proposal. Since I know you will be giving serious attention to this major action I would like to summarize the reasons for my vote which fall into four categories: (i) one area of potential risk has not been investigated experimentally; (ii) there are conflicting interpretations of the mouse-polyoma experiment; (iii) survival experiments with some debilitated strains of E.coli showed that survival was higher than previously expected; (iv) plasmid transfer experiments with debilitated and ordinary strains of E.coli have not been completed.

The RAC subcommittee on risk assessment has proposed that a conference be held "as expeditiously as possible," to evaluate certain conditions that have a direct bearing on the safety of E.coli K12 host-vector systems. These include:

(1) studies of hormone-producing strains of E.coli, to evaluate direct adverse effects.
(2) studies to determine the possible occurence of autoantibodies or autoreactive cells due to the production of eukaryotic polypeptides (including hormones) by bacteria that colonize higher organisms.

I believe these areas of concern are in themselves of sufficient importance to warrant holding off the blanket exemption of E.coli K12 (with plasmid and phage restrictions) from the guidelines. RAC has been advised that experiments to answer these questions are not difficult to accomplish. Consequently, I feel strongly that NIH should fund the relevant studies to resolve this area of uncertainty.

The mouse-polyoma experiments are a critical part of the risk assessment information. However, there are conflicting interpretations of the significance of these experiments. Professors Ethan Signer and Jonathan King (both of MIT)emphasize that an artificial recombinant was created in those experiments (bacteriophage lambda containing two copies of polyoma DNA) which proved to be infectious and tumorigenic to mice. This, they assert, shows that a new laboratory hybrid was created that was not found in nature and that could be a new source of infection and tumors.

On several occasions I raised this interpretation to my scientific colleagues on the RAC. I have not been satisfied with the responses of those who shrug off the Signer-King interpretation. The argument that these newly created hybrids are less infectious or less tumorigenic than the free polyoma does not speak to the fact that the polyoma-phage-E.coli system may bring the polyoma to new niches. So we have less infectious or tumorigenic polyoma in new places.

(If I were a mouse,with intellect,I would surely be concerned about the possibility of spreading polyoma DNA in E.coli.)

There is yet another aspect to the polyoma experiment that has been troublesome to me. I have been unable to grasp its generalizability. Can we expect to have the same result (lower infective capacity of intact virus in E.coli than free virus) if other viruses were used in the risk assessment experiment? Does the experiment set an upper bound on risk for this class of experiments? The clearest argument that was addressed to this point was by my RAC colleague David Baltimore at the last meeting. But I found his logic at the least confusing and at the most flawed. He maintained that the mouse polyoma experiment was among the most sensitive that could be designed for determining whether a DNA fragment could be released from its carrier molecule while in E.coli. If the polyoma experiment is indeed the most sensitive one that can be done for that purpose, then the probability of detecting the event is clearly greater than other similar experiments (with other DNA viruses and other animals). But the probability of detecting the event should be independent of the occurence of the event itself (except for some unusual cases in quantum mechanics).

Since the outcome of the polyoma experiment has been used as one of the principal reasons for the exemption of E.coli K12 from the guidelines, I would like to see some reviewed scientific papers on these questions that meet the tests of time and criticism. And because I found the justifications for generalizing the polyoma experiment unsatisfactory, I gave the results much less value in the total risk assessment than other RAC members.

Dr. Stuart Levy of Tufts University's School of Medicine is under contract to study the plasmid transfer and survival of certain strains of E.coli. Dr. Levy has reported that the survival of one strain was higher than previously expected. He emphasized the impartance of testing the survival of Chi 1776 with different plasmids since he finds that "plasmids may influence survival of a host bacterium in yet undefined ways." (Recombinant DNA Bulletin, July 1979).

It is for the above reasons that I feel it is premature to exempt the vast majority of recombinant DNA experiments with E.coli K12. I hope that you defer the exemptions until all the risk assessment work is completed and until there has been appropriate discussion within the scientific community of the significance of the mouse-polyoma experimental results.

Sincerely yours,

Sheldon Krimsky, Ph.D.
Acting Director, Program in Urban
Social and Environmental Policy

Medford, Massachusetts 02155
617 628-5000

November 1, 1979
Walters Letter to Fredrickson

The Joseph and Rose Kennedy Institute of Ethics
Center for Bioethics
Georgetown University
Washington, D.C. 20057
202/625-2371

November 1, 1979

Donald S. Fredrickson, M.D.
Director
National Institutes of Health
Building One, Room 137
Bethesda, MD 20014

Dear Dr. Fredrickson:

I am writing in response to published charges that improper procedures were followed by the Recombinant DNA Advisory Committee at its September meeting. The allegations concern the Committee's approval of reduced physical containment levels and revised oversight procedures for most types of research involving E. coli K-12 host-vector systems.

Improprieties alleged, in particular, by a New Scientist article ("Dirty Tricks in Genetic Engineering Committee?" New Scientist, 4 October 1979, p. 3) include the following:

1. The Recombinant Advisory Committee (RAC) made its decision hastily.
2. The agenda of the meeting was changed, and the vote on this issue was moved ahead.
3. Scientists and industry officials teamed up to force the proposed change in guidelines through the Committee.
4. Non-scientists on the Committee did not understand the issues at stake.
5. Only 10 of the 25 RAC members voted in favor of the proposed changes in guidelines.

Of these allegations, the first, third, and fourth are demonstrably false. The second is true but reflects only a desire by the Chairperson to accomodate a Working Group of the Committee. And the fifth is true but entirely appropriate according to well-established procedures for Federal advisory committees. In the following paragraphs I will comment on each of the allegations in turn.

1. Few decisions by the RAC have been made so deliberately and carefully as the recommendation to reduce physical containment levels and revise oversight procedures for research involving E. coli K-12 hosts and vectors. The original Rowe-Martin proposal to exempt most E. coli K-12 research was published in the Federal Register on April 13, 1979, discussed at length at the May RAC meeting and studied by a Working Group containing persons holding various views on the issue. A revised proposal was published in the Federal Register on July 31, 1979, commented on by numerous correspondents, discussed and further amended at the September RAC meeting, and approved by a vote of 10 for, 4 against, and 1 abstention.

2. The only change in the agenda made prior to the discussion of the E. coli K-12 host-vector issue was the deferral of a report by the Working Group on RAC Procedures. This report was delayed to accomodate one of the members of the Working Group who was unavoidably delayed in his arrival on the first day of the two-day meeting. Thus, far from indicating any malicious intention on anyone's part, the alteration in the agenda demonstrated the Chairperson's interest in providing a fair hearing for each Committee member's point of view.

fifteen members present at the time this important revision was discussed voted according to their best assessments of the evidence. A two-thirds majority of those present and voting favored the proposal.

3. No one forced any action on the RAC. Numerous scientists commented on the proposed reduction of containment levels and revision of oversight procedures for research with E. coli K-12 hosts and vectors. By my count, thirteen individuals and two groups commented on the proposed revision. All but one communication expressed support for the revision; the remaining communication did not oppose the revision but recommended that the views of Dr. Sydney Brenner be taken into account in the evaluation of the proposal.

It is true that industrial scientists reported unpublished data on the autoimmunity question during the September RAC meeting. In my view, this report was simply discounted by RAC members because no data were available for examination.

There is no reliable way, short of interviewing each RAC member, to determine the precise grounds for his or her vote on this proposed revision. Speaking for myself as one nonscientist on the Committee, my vote in favor of the proposed revision was based on the data from the Falmouth meeting, the data presented in the reports of the Rowe-Martin experiments, the report of the Working Group charged with studying this proposal, and the solicited opinions of several eminent immunologists on the autoimmunity issue.

4. It is difficult to respond dispassionately to a charge which seems to question the competence of at least some nonscientist members of the Committee. My observation as one of those nonscientists is that all members of the Committee, both scientists and nonscientists, have made a conscientious effort to study the materials provided in advance of each meeting and to do whatever additional background reading is needed to prepare themselves for discussion of Committee matters. The fact that nonscientists voted both for and against the proposed revision at least indicates that nonscientists were not stampeded into adopting one position or the other.

5. It is true that only fifteen members of RAC voted on this important proposed revision. The members who were present and voting regretted that the remaining members were prevented by other commitments or by unavoidable delays from taking part in the vote. However, thirteen members do constitute a quorum of the Committee, and it simply has not been possible to have all members of the Committee present for the whole of every RAC meeting. Thus, the

If any question had been raised before the vote concerning the number of Committee members present, or if a deferral of the vote until the second day of the meeting had been requested, I have not the slightest doubt that the question would have been seriously discussed or the deferral granted. If I remember correctly, not one word was spoken about this procedural matter in Committee discussions, either before or after the vote.

In closing, if the New Scientist report had merely questioned the prudence of the Committee's recommendation concerning research with E. coli K-12 hosts and vectors, I would not have written this lengthy response. Reasonable people differ in their judgments about the acceptability of potential risks. However, the New Scientist's charges about procedural improprieties in the Committee's decisionmaking on this issue are either false or misleading. I hope that this letter can make a small contribution toward setting the public record straight.

Sincerely yours,

LeRoy Walters

LeRoy Walters, Ph.D.
Director, Center for Bioethics
Kennedy Institute

LW:cwn

cc: William J. Gartland

November 14, 1979
Simring Letter to Harris

72 Jane Street
New York N Y 10014
(212) 675-7173

Coalition for Responsible Genetic Research

November 14, 1979

Dr. Patricia Harris
Secretary HEW
Washington, D.C.

RE: National Institutes of Health regulation of genetic engineering

Dear Dr. Harris:

Despite mounting evidence that there is greater hazard than previously thought by proponents of genetic engineering, the NIH has elected to lower safety standards by exempting some 80-85% of the research from regulation. Dr. Roy Curtiss, of the University of Alabama Medical Center, who developed the host organism used in the technology, opposes the exemptions as "premature".

Further, only 14 members* of the 25 members of the NIH advisory committee on genetic engineering (Recombinant DNA Advisory Committee) were present to vote on the exemptions; only 10 members voted affirmatively, 4 negatively (Dr. Susan Gottesman, Dr. Karim Ahmed, Dr. Sheldon Krimsky, Dr. Richard Goldstein), 1 abstaining.

Whereas this action is of such broad scope and significance as to have impact throughout the country, with repercussions abroad; and

Whereas new data is being developed through current risk assessment experimentation sponsored by the NIH,

We respectfully urge that:

1. NIH Director Donald Fredrickson's decisions be published forthwith in the Federal Register for public comment.
2. NIH submit an Environmental Impact Statement, concerning exemptions and other major federal actions to be assumed, in order to comply with the National Environmental Policy Act.
3. HEW meet with Dr. Roy Curtiss and those RAC members who voted against the exemptions.

Only by proceeding cautiously can we hope to avoid for this controversial technology the shoals of unrealistic optimism and technological inadequacy upon which the nuclear energy industry has foundered.

Please call upon us for any assistance which you feel that we might render to you and your department. With all good wishes that your stay in office will help fulfill the function of the NIH to promote the health of the American people,

* and Chairperson

Yours very truly,

Francine Robinson Simring
Executive Director

FRS:fh

cc: Stevenson, Waxman,
Cranston

December 20, 1979

Dr. Donald S. Fredrickson
Director
National Institutes of Health
Bethesda, Maryland 20205

Dear Dr. Fredrickson,

As you know, I am a member of the Recombinant Advisory Committee who voted against the proposal to lower required containment for *E. coli* K12 recombinant DNA experiments. During this additional comment period, questions have been raised about the procedures used by our committee when this proposal was approved. I would like to register my opinion on this matter.

In my opinion, the RAC discussion was extensive and relevant. There was ample opportunity for expression of a variety of opinions, both at the September meeting and during the previous discussion at the May meeting. The voting procedures used by the committee were appropriate and in accordance with HEW regulations. The vote as registered almost certainly reflects the judgement of the committee as a whole even though several members were absent. In the end, what we really differed about were the conclusions that we drew from the data; I suspect that no amount of future data, discussion or experience would make the committee unanimous on this issue.

I voted against the proposal because I felt that the data on possible consequences of expression of a harmful product by an *E. coli* containing recombinant DNA had not been fully explored, and that therefore P1-EK1 containment might not be assumed to be adequate in all cases. Members of our committee, each with access to the same data, came to different conclusions. The problem is not what the data shows but what the overall risk of *E. coli* experiments might be and how much risk is acceptable. The split vote of the RAC at least partially reflects this range of possible conclusions from the same data, and expresses the complexity of the problem more than anything else. This is as it should be: we are making our best guesses about unknown and speculative hazards, and we need the differences of opinion to insure that we think about everything. I do not think that such a split vote should be overinterpreted; the committee will become either totally unable to act at all or forced to a false unanimity if decisions are held up simply on the basis of split votes.

I hope these comments will be helpful.

Sincerely,

Susan Gottesman

Susan Gottesman

Document 14.6

November 1, 1979

Nature (**282**:3)

NIH director unlikely to grant exemption from controls for DNA experiments

The director of the US National Institutes of Health, Dr Donald Fredrickson, seems unlikely to accept in full the recommendation of the Recombinant DNA Advisory Committee (RAC) that a large body of recombinant DNA experiments be exempted from the guidelines established by the NIH, following widespread criticism of the implications of such a move.

Sources within the NIH say that Dr Fredrickson will probably approve a significant reduction in the safety precautions needed to carry out most experiments using as host the disabled K-12 strain of the bacterium *Escherichia coli*; these are said to account for between 80 and 85% of all experiments using recombinant DNA techniques.

But whereas the RAC, in a split 10 to 4 vote, decided at its last meeting in September to recommend *exemption* for the guidelines for such experiments — with merely the ruling that they should be registered with local biohazard committees, and should preferably be carried out at the minimal Pl physical containment level — Dr Fredrickson is likely to insist that the Pl conditions be enforced. This will include strict adherence to technician training requirements, and a ban on practices such as mouth-pipetting.

The NIH has had a flood of comments since the proposal to exempt the experiments was first put forward by RAC members Dr Allen Campbell and Dr Wallace Rowe at the committee's meeting in May. Many scientists wrote supporting the proposal, including in particular a petition signed by 183 scientists attending the Gordon conferences on nucleic acids and on biological regulatory mechanisms.

Others, however, have expressed concern about the implications of such a drastic move. Professor Roy Curtiss, of the department of microbiology at the University of Alabama in Birmingham, wrote to Dr Fredrickson criticising the proposed exemptions as premature, given the uncertainties that still exist over the hazards of recombinant organisms. "I believe that the RAC's recommendation to you was based more on the politics of science than on its data", he wrote.

In addition, the Natural Resources Defense Council in New York has demanded that an environmental impact statement be carried out for the proposed safeguard reductions, arguing that "the proposed exemptions are of such breadth and importance as to require full compliance with the National Environmental Policy Act."

In view of the controversy surrounding the committee's recommendation and its implications, Dr Frederickson has been carrying out a detailed review both of the arguments used during the RAC meeting, and of the data used to support them.

The results of this review will not be known for several weeks. However it is thought that Dr Fredrickson is reluctant to face the criticism that total exemption would involve.

For example, two weeks ago the citizens of Amherst, which has introduced a local ordinance requiring recombinant DNA research at the University of Amherst to be conducted under the NIH guidelines, agreed to require adherence to the 1978 revision of the guidelines, rather than their stricter 1976 original version.

However according to Dr Bruce Levin, a research scientist at the university who has been closely involved in the local debates, there would probably be strong resistance from the local community if a large proportion of the experiments were to be exempt from regulation.

Dr Fredrickson, however, is said to be prepared to accept most of the arguments in favour of a substantial reduction in required containment levels for many experiments involving the K-12 strain; one of the few areas still to be resolved is whether local committees would be required to give prior approval to experiments using biologically active materials, such as active polypeptides or active proteins.

The NIH is paying particular attention in its review to uncertainties that have arisen as a result of various risk assessment experiments, such as those which indicate that bacteria into which plasmids have been introduced can survive considerably longer than expected in the human gut.

Meanwhile staff members for Senator Adlai Stevenson's science and technology subcommittee are preparing legislation that would require all non-federally supported research involving recombinant DNA techniques — in particular that carried out by private companies — to register their experiments with the NIH. At present companies can register; but the arrangement is voluntary. □

Document 14.7

January 29, 1980

Federal Register, Excerpt

6732 Federal Register / Vol. 45, No. 20 / Tuesday, January 29, 1980 / Notices

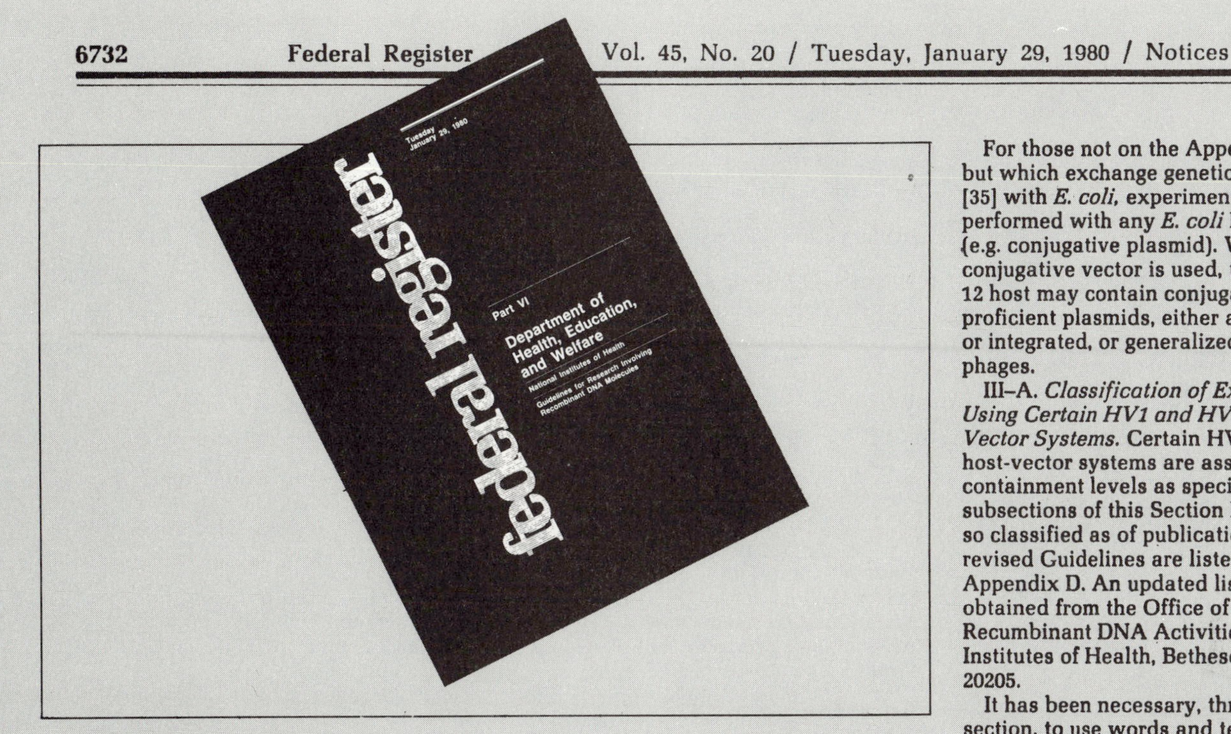

III. Containment Guidelines for Covered Experiments

Part III discusses experiments covered by the Guidelines. The reader must first consult Part I, where listings are given of prohibited and exempt experiments.

Containment guidelines for permissible experiments are given in Part III. Changes in these levels for specific experiments (or the assignment of levels to experiments not explicitly considered here) may not be instituted without the express approval of the Director, NIH. (See Sections IV–E–1–b–(1)–(a), IV–E–1–b–(1)–(b), IV–E–1–b–(2)–(b), IV–E–1–b–(2)–(c), and IV–E–1–b–(3)–(b).)

In the following classification of containment criteria for different kinds of recombinant DNAs, the stated levels of physical and biological containment are minimal for the experiments designated. The use of higher levels of biological containment (HV3>HV2>HV1) is encouraged if they are available and equally appropriate for the purposes of the experiment.

III–O. *Classification of Experiments Using the E. coli K-12 Host-Vector Systems.* Most recombinant DNA experiments currently being done employ *E. coli* K–12 host-vector systems. These are the systems for which we have the most experience and knowledge.

Some experiments using *E. coli* K–12 host-vector systems are prohibited (see Section I–D).

Some experiments using *E. coli* K–12 host-vector systems are exempt from the Guidelines (see Section I–E).

Other experiments using *E. coli* K–12 shall use P1 physical containment and, except as specified in the last paragraph of this section, an EK1 host-vector system (i.e. (a) the host shall not contain conjugation-proficient plasmids or generalized transducing phages, and (b) lambda or lambdoid bacteriophages or non-conjugative plasmids shall be used as vectors). For these experiments no Memorandum of Understanding and Agreement (MUA) as described in Section IV–D–1–c need be submitted, nor is any registration with NIH necessary. However, for these experiments, prior to their initiation, investigators must submit to their Institutional Biosafety Committee (IBC) a registration document that contains a description of (a) the source(s) of DNA, (b) the nature of the inserted DNA sequences, and (c) the hosts and vectors to be used. This registration document must be dated and signed by the investigator and filed only with the local IBC. The IBC shall review all such proposals but such review is not required prior to initiation of experiments. An exception, however, which does require prior review and approval by the IBC is any experiment in which there is a deliberate attempt to have the *E. coli* K–12 efficiently express any gene coding for a eukaryotic protein.

Experiments involving the insertion into *E. coli* K–12 of DNA from prokaryotes that exchange genetic information with *E. coli* by known physiological processes will be exempted from these Guidelines if they appear on the "list of exchangers" set forth in Appendix A (see Section I–E–4).

For those not on the Appendix A list but which exchange genetic information [35] with *E. coli*, experiments may be performed with any *E. coli* K–12 vector (e.g. conjugative plasmid). When a non-conjugative vector is used, the *E. coli* K–12 host may contain conjugation-proficient plasmids, either autonomous or integrated, or generalized transducing phages.

III–A. *Classification of Experiments Using Certain HV1 and HV2 Host-Vector Systems.* Certain HV1 and HV2 host-vector systems are assigned containment levels as specified in the subsections of this Section III–A. Those so classified as of publication of these revised Guidelines are listed in Appendix D. An updated list may be obtained from the Office of Recombinant DNA Activities, National Institutes of Health, Bethesda, Maryland 20205.

It has been necessary, throughout this section, to use words and terms marked with footnote reference numbers. The footnotes (Part V) define more fully what the terms denote.

III–A–1. *Shotgun Experiments.* These experiments involve the production of recombinant DNAs between the vector and portions of the specified cellular source, preferably a partially purified fraction. Care should be taken either to preclude or eliminate contaminating microorganisms before isolating the DNA.

III–A–1–a. *Eukaryotic DNA Recombinants.*

III–A–1–a–(1). *Primates.* P2 physical containment + an HV2 host-vector or P3+HV1.

III–A–1–a–(2). *Other Mammals.* P2 physical containment + an HV2 host-vector or P3 + HV1.

III–A–1–a–(3). *Birds.* P2 physical containment + an HV2 host-vector, or P3 + HV1.

III–A–1–a–(4). *Cold-Blooded Vertebrates.* P2 physical containment + an HV1 host-vector or P1 + HV2. If the eukaryote is known to produce a potent polypeptide toxin,(*34*) the containment shall be increased to P3 + HV2.

III–A–1–a–(5). *Other Cold-Blooded Animals and Lower Eukaryotes.* This large class of eukaryotes is divided into two groups:

III–A–1–a–(5)–(a). Species that are known to produce a potent polypeptide toxin(*34*) that acts in vertebrates, or are known pathogens listed in Class 2,(*1*) or are known to carry such pathogens must use P3 physical containment + an HV2 host-vector. When the potent toxin is not a polypeptide and is likely not to be the product of closely linked eukaryote genes, containment may be reduced to P3 + HV1 or P2 + HV2. Species that produce potent toxins that affect invertebrates or plants but not vertebrates require P2 + HV2 or P3 + HV1. Any species that has a demonstrated capacity for carrying particular pathogenic microorganisms is included in this group, unless the organisms used as the source of DNA have been shown not to contain those agents, in which case they may be placed in the following group.(*2A*)

III–A–1–a–(5)–(b). The remainder of the species in this class including plant pathogenic or symbiotic fungi that do not produce potent toxins: P2 + HV1 or P1 + HV2. However, any insect in this group must be either (i) grown under laboratory conditions for at least 10 generations prior to its use as a source of DNA, or (ii) if caught in the wild, must be shown to be free of disease-causing microorganisms or must belong to a species that does not carry microorganisms causing disease in veterbrates or plants.(*2A*) If these conditions cannot be met, experiments must be done under P3 + HV1 or P2 + HV2 containment.

III–A–1–a–(6). *Plants.* P2 physical containment + an HV1 host-vector, or P1 + HV2. If the plant source makes a potent polypeptide toxin,(*34*) the containment must be raised to P3 physical containment + an HV2 host-vector. When the potent toxin is not a polypeptide and is likely not to be the product of closely linked plant genes, containment may be reduced to P3 + HV1 or P2 + HV2.(*2A*)

III–A–1–b. *Prokaryotic DNA Recombinants.* P2 + HV1 or P1 + HV2 for experiments with phages, plasmids and DNA from nonpathogenic prokaryotes which do not produce polypeptide toxins(*34*). P3 + HV2 for experiments with phages, plasmids and DNA from Class 2 agents(*1*).

III–A–2–a. *Viruses of Eukaryotes* (summary given in Table III; see also exception given at asterisk at end of Appendix D).

III–A–2–a–(1)–(a). *Nontransforming viruses.*

III–A–2–a–(1)–(a)–(*1*). *Adeno-Associated Viruses, Minute Virus of Mice, Mouse Adenovirus (Strain FL), and Plant Viruses.* P1 physical containment + and HV1 host-vector shall be used for DNA recombinants produced with (i) the whole viral genome, (ii) subgenomic DNA segments, or (iii) purified cDNA copies of viral mRNA.(*37*)

III–A–2–a–(1)–(a)–(*2*). *Hepatitis B.*

III–A–2–a–(1)–(a)–(*2*)–(*a*). P1 physical containment + an HV1 host-vector shall be used for purified subgenomic DNA segments.(*38*)

III–A–2–a–(1)–(a)–(*2*)–(*b*). P2 physical containment + an HV2 host-vector, or P3 + HV1, shall be used for DNA recombinants produced with the whole viral genome or with subgenomic segments that have not been purified to the extent required in footnote 38.

III–A–2–a–(1)–(a)–(*2*)–(*c*). P2 physical containment + an HV1 host and a vector certified for use in an HV2 system, or P3 + HV1, shall be used for DNA recombinants derived from purified cDNA copies of viral mRNA.(*37*)

III–A–2–a–(1)–(a)–(*3*). *Other Nontransforming Members of Presently Classified Viral Families.(36)*

III–A–2–a–(1)–(a)–(*3*)–(*a*). P1 physical containment + an HV1 host- subgenomic DNA(*38*) segments or (ii) purified cDNA copies of viral mRNA.(*37*)

III–A–2–a–(1)–(a)–(*3*)–(*b*). P1 physical containment + an HV1 host and a vector certified for use in an HV2 system shall be used for DNA recombinants produced with the whole viral genome or with subgenomic segments that have not been purified to the extent required in footnote 38.

III–A–2–a–(1)–(a)–(b). *Transforming Viruses.(37A)*

III–A–2–a–(1)–(b)–(*1*). *Herpes Saimiri, Herpes Ateles, and Epstein Barr Virus.(39)*

III–A–2–a–(1)–(b)–(*1*)–(*a*). P1 physical containment + an HV1 host-vector shall be used for DNA recombinants produced with purified nontransforming subgenomic DNA segments.(*38*)

III–A–2–a–(1)–(b)–(*1*)–(*b*). P2 physical containment + an HV1 host and a vector certified for use in an HV2 system, or P3 + HV1, shall be used for (i) DNA recombinants produced with purified subgenomic DNA segments containing an entire transforming gene(*38*) or (ii) purified cDNA copies of viral mRNA.(*37*)

III–A–2–a–(1)–(b)–(*1*)–(*c*). P3 physical containment + an HV1 host-vector, or P2 + HV2, shall be used for DNA recombinants produced with the whole viral genome or with subgenomic segments that have not been purified to the extent required in footnote 38.

III–A–2–a–(1)–(b)–(*2*). *Other Transforming Members of Presently Classified Viral Families.(36)*

III–A–2–a–(1)–(b)–(*2*)–(*a*). P1 physical containment + an HV1 host-vector shall be used for DNA recombinants produced with purified nontransforming subgenomic DNA segments.(*38*)

III–A–2–a–(1)–(b)–(*2*)–(*b*). P2 physical containment + an HV1 host and a vector certified for use in an HV2 system, or P3 + HV1, shall be used for (i) DNA recombinants produced with the whole viral genome, (ii) subgenomic DNA segments containing an entire transforming gene, (iii) purified cDNA copies of viral mRNA, (*37*) or (iv) subgenomic segments that have not been purified to the extent required in footnote 38.

III–A–2–a–(2). *DNA Transcripts of RNA Viruses.*

III–A–2–a–(2)–(a). *Retroviruses.*

III–A–2–a–(2)–(a)–(*1*). *Gibbon Ape, Woolly Monkey, Feline Leukemia and Feline Sarcoma Viruses.(39).*

III–A–2–a–(2)–(a)–(*1*)–(*a*). P1 physical containment + an HV1 host-vector shall be used for DNA recombinants produced with purified nontransforming subgenomic DNA segments.(*38*).

III–A–2–a–(2)–)a)–(*1*)–(*b*). P2 physical containment + an HV1 host and a vector certified for use in an HV2 system, or P3 + HV1, shall be used for DNA recombinants produced with purified subgenomic DNA segments (*38*) containing an entire transforming gene.

III–A–2–a–(2)–(a)–(*1*)–(*c*). P2 physical containment + an HV2 host-vector, or P3 + HV1, shall be used for DNA recombinants produced with (i) the whole viral genome, (ii) purified cDNA copies of viral mRNA,(*37*) or (iii) subgenomic segments that have not been purified to the extent required in footnote 38.

III–A–2–a–(2)–(a)–(*2*). *Other Members of the Family Retroviridiae.(36)*

III–A–2–a–(2)–(a)–(*2*)–(*a*). P1 physical containment + an HV1 host-vector shall be used for DNA recombinants produced with purified nontransforming subgenomic DNA segments.(*38*).

III–A–2–a–(2)–(a)–(*2*)–(*b*). P2 physical containment + an HV1 host and a vector certified for use in an HV2 system, or P3 + HV1, shall be used for DNA recombinants produced with (i) subgenomic DNA segments containing an entire transforming gene, (ii) the whole viral genome, or (iii) purified cDNA copies of viral mRNA,(*37*) or (iv) subgenomic segments that have not been purified to the extent required in footnote 38.

III–A–2–a–(2)–(b). *Negative Strand RNA Viruses.* P1 physical containment + an HV1 host-vector shall be used for DNA recombinants produced with (i) cDNA copies of the whole genome, (ii) subgenomic cDNA segments, or (iii) purified cDNA copies of viral mRNA.(*37*)

III–A–2–a–(2)–(c). *Plus-Strand RNA Viruses.*

III–A–2–a–(2)–(c)–(*1*). *Types 1 and 2 Sabin Poliovirus Vaccine Strains and Strain 17D (Theiler) of Yellow Fever Virus.* P1 physical containment + an HV1 host-vector shall be used for DNA recombinants produced with (i) cDNA copies of the whole viral genome, (ii) subgenomic cDNA segments, or (iii) purified cDNA copies of viral mRNA.(*37*)

III–A–2–a–(2)–(c)–(*2*). *Other Plus-Strand RNA Viruses Belonging to Presently Classified Viral Families.(36)*

VI. Voluntary Compliance

VI–A. *Basic Policy.* Individuals, corporations, and institutions not otherwise covered by the Guidelines are encouraged to do so by following the standards and procedures set forth in Parts I–IV of the Guidelines. In order to simplify discussion, reference hereafter to "institutions" are intended to encompass corporations, and individuals who have no organizational affiliation. For purposes of complying with the Guidelines, an individual intending to carry out research involving recombinant DNA is encouraged to affiliate with an institution that has an Institutional Biosafety Committee approved under the Guidelines.

Since commerical organizations have special concerns, such as protection of proprietary data, some modifications and explanations of the procedures in Parts I–IV are provided below, in order to address these concerns.

VI–B. *IBC Approval.* The NIH Office of Recombinant DNA Activities (ORDA) will review the membership of an institution's Institutional Biosafety Committee (IBC) and, were it finds the IBC meets the requirements set forth in Section IV–C–2, will give its approval to the IBC membership.

It should be emphasized that employment of an IBC member solely for purposes of membership on the IBC does not itself make the member an institutionally affiliated member of purposes of Section IV–D–2–a.

Except for the unaffiliated members, a member of an IBC for an institution not otherwise covered by the Guidelines may participate in the review and approval of a project in which the member has direct financial interest, so long as the member has not been and does not expect to be engaged in the project. Section IV–2–d is modified to that extent for purposes of these institutions.

VI–C. *Registration.* Upon approval of a recombinant DNA research project by the IBC, an institution may register the project by submitting to ORDA the information required in the Administrative Practices Supplement.

VI–D. *Certification of Host-Vector Systems.* A host-vector system may be proposed for certification by the Director, NIH, in accordance with the procedures set forth in Section II–D–2–a.

Institutions not otherwise covered by the Guidelines will not be subject to Section II–D–3 by complying with these procedures.

In order to ensure protection for proprietary data, any public notice regarding a host-vector system which is designated by the institution as proprietary under Section VI–F–1 will be issued only after consultation with the institution as to the content of the notice.

VI–E. *Requests for Exceptions, Exemptions, Approvals.* Requests for exceptions from prohibitions, exemptions, or other approvals required by the Guidelines should be requested by following the procedures set forth in the appropriate sections in Parts I–IV of the Guidelines.

In order to ensure protection for proprietary data, any public notice regarding a request for an exception, exemption, or other approval which is designated by the institution as proprietary under Section VI–F–1 will be issued only after consultation with the institution as to the content of the notice.

VI–F. *Protection of Proprietary Data.* In general, the Freedom of Information Act requires Federal agencies to make their records available to the public upon request. However, this requirement does not apply to, among other things, "trade secrets and commercial and financial information obtained from a person and privileged or confidential." 18 U.S.C. 1905, in turn makes it a crime for an officer or employee of the United States or any Federal department or agency to publish, divulge, disclose, or make known "in any manner or to any extent not authorized by law any information coming to him in the course of his employment or official duties or by reason of any examination or investigation made by, or return, report or record made to or filed with, such department or agency or officer or employee thereof, which information concerns or relates to the trade secrets, [or] processes . . . of any person, firm, partnership, corporation, or association." This provision applies to all employees of the Federal Government, including special Government employees. Members of the Recombinant DNA Advisory Committee are "special Government employees."

VI–F–1. In submitting information to NIH for purposes of complying voluntarily with the Guidelines, an institution may designate those items of information which the institution believes constitute trade secrets or privileged or confidential commercial or financial information.

VI–F–2. If NIH receives a request under the Freedom of Information Act for information so designated, NIH will promptly contact the institution to secure its views as to whether the information (or some portion) should be released.

VI–F–3. If the NIH decides to release this information (or some portion) in response to a Freedom of Information request or otherwise, the institution will be advised; and the actual release will not be made until the expiration of 15 days after the institution is so advised, except to the extent that earlier release, in the judgment of the Director, NIH, is necessary to protect against an imminent hazard to the public or the environment.

VI–F–4. Projects should be registered in accordance with procedures specified in the *Administrative Practices Supplement.* The following information will usually be considered publicly available information, consistent with the need to protect proprietary data:

a. The names of the institution and principal investigator.

b. The location where the experiments will be performed.

c. The host-vector system.

d. The source of the DNA.

e. The level of physical containment.

VI–F–5–a. Any institution not otherwise covered by the Guidelines, which is considering submission of data or information voluntarily to NIH, may request presubmission review of the records involved to determine whether, if the records are submitted, NIH will or will not make part of all of the records available upon request under the Freedom of Information Act.

VI–F–5–b. A request for presubmission review should be submitted to ORDA, along with the records involved. These records must be clearly marked as being the property of the institution, on loan to NIH solely for the purpose of making a determination under the Freedom of Information Act. ORDA will then seek a determination from the HEW Freedom of Information Officer, the responsible official under HEW regulations (45 CFR Part 5), as to whether the records involved (or some portion) are or are not available to members of the public under the Freedom of Information Act. Pending such a determination, the records will be kept separate from ORDA files, will be considered records of the institution and not ORDA, and will not be received as part of ORDA files. No copies will be made of the records.

VI–F–5–c. ORDA will inform the institution of the HEW Freedom of Information Officer's determination and follow the institution's instructions as to whether some or all of the records involved are to be returned to the institution or to become a part of ORDA files. If the institution instructs ORDA to return the records, no copies or summaries of the records will be made or retained by HEW, NIH, or ORDA.

VI–F–5–d. The HEW Freedom of Information Officer's determination will represent that official's judgement, as of the time of the determination, as to whether the records involved (or some portion) would be exempt from disclosure under the Freedom of Information Act, if at the time of the determination the records were in ORDA files and a request were received from them under the Act.

THE COMPLIANCE OF INDUSTRY

The campaign to exempt *E. coli* K12 experiments from the guidelines was basically conducted by and for the benefit of academic researchers. During 1979 the NIH, the RAC, and other government agencies were also occupied with another major longstanding issue, the compliance of industrial laboratories with the NIH guidelines. As the excerpts from the guidelines published in the *Federal Register* of January 29, 1980, show, this issue had also been resolved at least temporarily by 1980. (See Doc. 14.7.)

Since 1976 the central question concerning industry and the guidelines was this: Could industry be relied upon to comply, when such compliance necessitated a limited disclosure of possibly patentable information, or were statutory powers necessary? The attempts and failures in Congress during 1977 to legislate for recombinant DNA have been documented in Chapter 6. It is worth noting that during 1977 the academic world appeared to have mounted a far more active lobby than the industries, at which the legislation was allegedly primarily aimed.

The reported comment of Gilbert Omenn was symptomatic of the executive branch's lack of interest. (See Doc. 12.5.) In virtual isolation within the legislature, Senator Adlai Stevenson persevered with the idea of a law (Doc. 15.1) and in January 1980 introduced a bill to Senator Kennedy's Health Subcommittee (Doc. 15.2). It failed, however, to gain any significant support from the Committee members.

On the other hand, if compliance was to remain voluntary, both industry and the NIH had an interest in devising a code of practice that among other things would include procedures to protect proprietary rights. Voluntary versus mandatory compliance had been discussed repeatedly by the Interagency Committee on Recombinant DNA since its formation in November 1976. This Committee had quickly reached the conclusion that existing legislation was inadequate (see Chapter 6). However, despite that conclusion, Secretary of HEW Califano requested the FDA and the EPA to consider applying their authority to new regulations. A notice from the FDA appeared in the *Federal Register* of December 22, 1978, immediately following the revised NIH guidelines (see Chapter 12). It announced the intent to propose under existing legislation new regulations requiring that any submis-

Photo courtesy Genentech.

sion to the agency of activities involving recombinant DNA should contain evidence of full compliance with the NIH guidelines (Doc. 15.3).

In June 1979 the Environmental Defense Fund expressed to Secretary Califano its concern about the unregulated industrial use of recombinant DNA techniques (Doc. 15.4). An excerpt of the minutes of the Tenth meeting of the Interagency Committee on July 17, 1979, however, reveals the fate of the FDA's empty threat and the committee's consensus of support for the voluntary approach (Doc. 15.5). Shortly after this Interagency Committee meeting, the Director of the NIH published in the *Federal Register* (August 3, 1979) a proposed supplement to the guidelines, outlining a system of procedures for a purely voluntary compliance. This move elicited protests from the Environmental Defense Fund and the AFL-CIO (Doc. 15.6). It is noteworthy that the U.S. trade-union movement played a much lesser role than its counterparts in some European countries throughout the recombinant DNA debate.

The RAC was discussing the issue at the same time that these developments were taking place. At its May 21–23, 1979, meeting, after what must have been a confused discussion and series of votes, the RAC finally voted in favor of mandatory compliance. It also voted on how it should deal with requests to handle large volumes of recombinant DNA materials—another matter close to industry's heart (see Doc. 15.7). Before the RAC met again, the Interagency Committee had reached its consensus in favor of voluntary compliance (see Doc. 15.5). The RAC had been upstaged, and at its next meeting in September it deferred discussions of industrial compliance until December. At the September meeting, however, the notion of voluntary compliance for a restricted "experimental period" was first introduced (Doc. 15.7). Three months later, in December, the RAC resolved after a protracted discussion that the voluntary compliance should be reviewed in June 1980 (Doc. 15.7).

Within the RAC there are enthusiasts for mandatory compliance and therefore new legislation. But given the executive's and legislature's disinterest, there is more than a touch of unreality to the RAC's discussion of this issue. Like a safety valve, however, the RAC provides an outlet for letting off heads of steam.

On January 29, 1980, a code of practice for the voluntary compliance of industry was incorporated into the guidelines. The scheme appeared to function satisfactorily, except that industry apparently found the RAC too slow in reaching decisions about alternative host-vector systems, more commercially attractive than *E. coli* K12, and about permission to use cultures far in excess of the 10-liter limit of the guidelines (Doc. 15.8). Of late these complaints have been heard less frequently.

Document 15.1

November 21, 1979

Stevenson Letter to Harris

United States Senate

COMMITTEE ON COMMERCE, SCIENCE
AND TRANSPORTATION
WASHINGTON, D.C. 20510

November 21, 1979

The Honorable Patricia Roberts Harris
Secretary
Department of Health, Education, and Welfare
Washington, D.C. 20201

Dear Secretary Harris:

The Senate Subcommittee on Science, Technology, and Space has been deeply interested in the progress of recombinant DNA research since our first oversight hearings two years ago. We continue to follow developments in the field closely, both to help ensure the protection of public health and the environment and to facilitate the orderly development of this extremely promising technology.

I understand that the Director of the National Institutes of Health, Dr. Fredrickson, is nearing decisions on two important changes in the NIH Recombinant DNA Research Guidelines. The first involves the recommendation of the Department's advisory committee to remove all but minimal controls on experiments using the E. coli K-12 host-vector system. The second proposal is to institute a system of voluntary registration of experiments by researchers and institutions not now subject to the Guidelines. I am writing to urge your careful review of these proposals.

I am aware that the so-called exemption of E. coli K-12 experiments is controversial even within the scientific community. Some researchers and others have questioned whether experiments that have been conducted to ascertain the risks of K-12 containing recombinant DNA have been either conclusive or exhaustive. I do not presume to question the scientific judgment of the committee or the Director, but I do wish to call your attention to three important procedural issues which deserve consideration before a decision is made. First, is it wise to proceed with so significant a relaxation of the Guidelines before completion of the risk assessment studies, many of them relating specifically to K-12, outlined by the NIH in its Federal Register notice of September 13, 1979? Secondly, does the brief experience under the revised Guidelines justify confidence in the willingness and ability of local institutional review committees to maintain the laboratory standards and practices which the RAC believes should still be followed? Finally, if the Director's recommendation differs from the RAC's proposal, should there be another opportunity for public comment and departmental review?

The Honorable Patricia Roberts Harris
November 21, 1979
Page Two

The issue of industry compliance with the Guidelines is not, of course, a scientific one; and here, I believe, the proposed amendment to the Guidelines stating the terms under which research may be registered voluntarily with NIH is seriously flawed. Most firms have stated their intention to observe the containment requirements and prohibitions of the Guidelines; but without NIH approval of scale-up experiments and use of new host-vector systems, they may not be able to proceed with important applications of recombinant DNA techniques. Voluntary registration provides no assurance, however, that all firms will register their research nor that any single firm will register all of its work. Thus, NIH may well be in the position of approving activities on the basis of partial information. By the same token, companies that register in good faith can have no confidence that discovery of unregistered activities will not cast doubt upon all NIH actions under the voluntary registration system.

I intend to introduce legislation requiring notification to the Department of all recombinant DNA work by institutions not presently subject to the Guidelines, It would authorize penalties for failure to register activities. It would provide statutory protection against disclosure of proprietary information. But it would not create a full-fledged regulatory scheme. I hope that I will have your support of this small but essential step to provide a complete public record of recombinant DNA activities and permit development of their commercial applications.

With every good wish.

Sincerely,

ADLAI E. STEVENSON, Chairman
Subcommittee on Science, Technology, and Space

Crystals of insulin produced by recombinant DNA methods.

Document 15.2

February 15, 1980
Science (**207**:745)

Control of Commercial Gene Splicing

A bill requiring companies to notify the government of their gene splicing activities was introduced last month by Senator Adlai Stevenson.

Congress's failure to pass legislation on recombinant DNA research has left commercial gene splicers under no formal obligation to abide by the National Institutes of Health's safety rules, although probably all companies are at present doing so. The NIH has tried to cover this lacuna by persuading companies to register their projects with the NIH recombinant DNA committee on a voluntary basis; committee members swear to protect the trade secrets revealed to them.

"I remain convinced," Stevenson said in proposing his bill, "that the voluntary registration scheme . . . serves neither the public's nor industry's interest." If only a few companies fail to register, the whole system could be discredited, Stevenson observes. The fact that a company can withdraw information from the NIH if there is disagreement over its proprietary nature could in the senator's view put the government in the false position of "condoning certain activities and assuring the public of their safety on the basis of incomplete data."

At present companies can submit data to the NIH and withdraw it if the Freedom of Information officer decides it is not proprietary. The Pharmaceutical Manufacturers Association is happy with the NIH's voluntary registration system and opposes Stevenson's bill.

The bill has been referred to Senator Kennedy's health subcommittee, which alone has the power to do anything with it, but Stevenson plans to have his science and space subcommittee hold oversight hearings on the matter.

All gene splicing experiments with *E. coli* K12 may now be conducted in minimal (P1) containment. The NIH committee's proposal to this effect (*Science*, 21 September 1979) has been accepted; the new guidelines are published in the *Federal Register* of 29 January, marking a further stage in their vexed but preplanned evolution.

Nicholas Wade

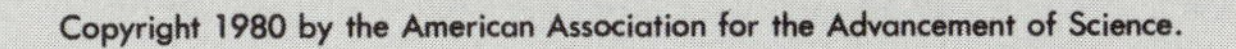

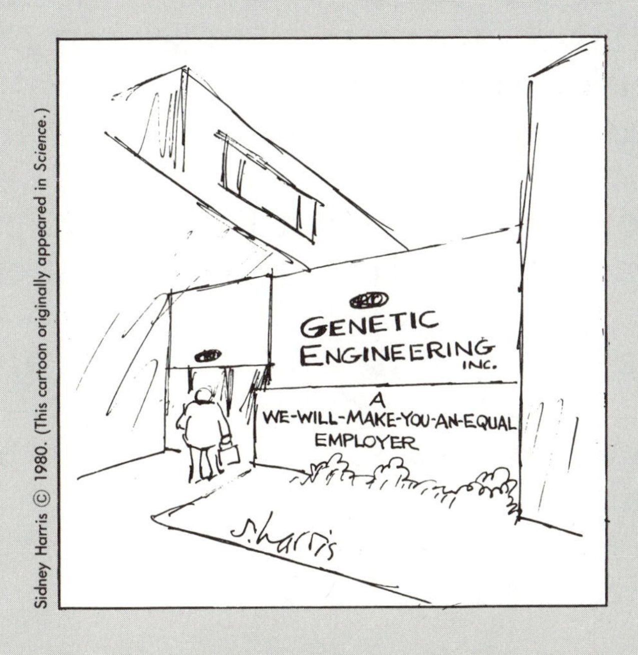

Sidney Harris © 1980. (This cartoon originally appeared in *Science*.)

Document 15.3

December 22, 1978

Federal Register, Excerpt

[4110-03-M]

DEPARTMENT OF HEALTH, EDUCATION, AND WELFARE

Food and Drug Administration

[21 CFR Part 59]

[Docket No. 78N-0012]

RECOMBINANT DNA

Intent to Propose Regulations

AGENCY: Food and Drug Administration.

ACTION: Notice of Intent to Propose Regulations.

SUMMARY: The Food and Drug Administration (FDA) intends to propose regulations to require assurances in future submissions to the agency that any recombinant deoxyribonucleic acid (DNA) work that has been or will be performed in connection with these submissions fully complies with the National Institutes of Health (NIH) Guidelines for Recombinant DNA Research. This notice is being issued because the Commissioner of Food and Drugs believes it would be helpful to invite public comment on whether such action would be appropriate and desirable.

From *Federal Register*, Vol. 43, Part VIII, p. 60134.

Document 15.4

June 5, 1979
Dach Letter to Califano

1525 18th Street, NW, Washington, D.C. 20036 • 202/833-1484

June 5, 1979

Hon. Joseph Califano, Secretary
Department of Health,
Education and Welfare
Room 615F
200 Independence Avenue, SW
Washington, DC 20201

Dear Secretary Califano:

As you are aware, the Environmental Defense Fund has actively monitored DHEW's and Congress' activities in regulation of recombinant DNA research. We are writing now to bring to your attention a number of issues which we feel are of central importance. We have discussed these issues in some detail with Peter Libassi and HEW and NIH staff during a meeting held on May 25, 1979.

Primarily, we are concerned about the public health hazards posed by unregulated industrial use of recombinant DNA techniques. Evieence is mounting that a number of industries are now or will in the near future be engaging heavily in recombinant DNA techniques. There is little guarantee that such work will be carried out in accordance with the current NIH guidelines. Indeed, one company has already indicated that it will proceed contrary to the guidelines. EDF has carefully reviewed NIH's proposed additions to the guidelines aimed at providing a voluntary registration and regulatory program for industry research. We feel, categorically, that such a voluntary approach is doomed to failure. Industry will not comply but will nevertheless use the voluntary program as a smokescreen to avoid meaningful regulation. In addition, NIH has neither the regulatory experience nor statutory authority to police industrial research. EDF therefore urges you not to promulgate the proposed additions to section IV of the guidelines. Instead, we urge you to publicly support the need for Congressional legislation to deal with this growing problem and to urge your fellow Cabinet members to use their existing authority to promulgate appropriate regulations.

Second, we were greatly disturbed by the conduct of the May Recombinant DNA Advisory Committee (RAC) meeting. The meeting was a procedural nightmare. The RAC process currently does not deserve the support or confidence of the American people. The number of basic procedural flaws was outstanding, including the failure to define a quorum, to require record votes, or to prohibit members with financial conflicts of interest from voting, and the creation of an agenda so full that thoughtful consideration of the issues was impossible. These procedural failings must be remedied before RAC meets again. Moreover, EDF requests that HEW convene a meeting of consumer, environmental, and labor representatives to discuss and review the proposed procedural changes before they are implemented.

Third, the staff of ORDA must be increased. The ORDA 1978 budget was only $154,000 -- about one quarter of one percent of the amount of grants awarded by NIH for the conduct of recombinant DNA research. The RAC meeting was punctuated by requests from committee members for staff functions that ORDA said it was unable, because of resource constraints, to provide. An increased ORDA budget should pay for better NIH monitoring of IBC activity and ongoing recombinant DNA research and the development of background papers on proposed RAC actions to aid public and RAC member analysis of upcoming agenda items. (ORDA should also prepare an "economic impact assessment" for the DNA guidelines. The public hears regularly from the scientific establishment that the current NIH guidelines are seriously delaying research. Yet no hard figures on the increased cost or time of compliance are available. I strongly feel that such figures would belie much of the claims made by the opponents of regulation.)

Fourth, the number of public members on RAC needs to be substantially increased. The scientific members of RAC have, for the most part, voted their constituencies -- the academic scientific establishment. To counter that bias, more public members are needed. An increased ORDA staff would provide the needed technical support for these members. Finally, HEW should make available grant monies to the public to support public monitoring of RAC and NIH decisions concerning recombinant DNA. Other federal agencies, for example EPA, currently provide grant monies for citizen workshops and citizen monitoring of agency activities. Such funding is needed so that the public can meaningfully oversee the NIH's implementation of the DNA guidelines.

Although we may not have been always satisfied with the final decisions, EDF has found your and Mr. Libassi's staff open and responsive to our concerns. We appreciate that relationship and hope that Mr. Libassi's departure will not deter continued dialogue on this issue.

At the same time, we are concerned that the procedural problems with the current RAC process be remedied immediately. If such remedies are not forthcoming before the next RAC meeting, EDF will consider legal and other strategies for achieving these needed changes.

Sincerely,

Leslie Dach
Science Associate

LD:cs
cc: Hon. Edward M. Kennedy
Hon. Henry A. Waxman
Hon. Adlai Stevenson
Dr. Joseph G. Perpich, Assoc. Director
Program Planning and Evaluation, NIH

Document 15.5
July 17, 1979
Excerpt

INTERAGENCY COMMITTEE ON RECOMBINANT DNA RESEARCH

Minutes of Tenth Meeting
July 17, 1979
8:30–10:00 a.m.
National Institutes of Health
Bethesda, Maryland 20205

The meeting, chaired by Dr. Donald S. Fredrickson, Director, National Institutes of Health, was devoted to (1) a review of the implementation of the revised NIH Guidelines, (2) an NIH proposal for extending the Guidelines to private-sector research, and (3) agency discussion on this proposed approach and alternatives. (See list of attendees, Attachment A.)

Application of the Revised Guidelines to Private-Sector Research

Dr. Fredrickson opened the discussion by citing the memorandum that Secretary Joseph A. Califano, upon the release of the revised Guidelines in December, had sent to Donald Kennedy, Commissioner of Food and Drugs, directing him to propose that all recombinant DNA research submitted to the Food and Drug Administration comply with the NIH Guidelines in order to meet regulatory requirements. In response, the FDA published a notice of intent to propose regulations for this purpose. The Secretary also sent letters to Administrator Douglas M. Costle, Environmental Protection Agency, in December 1978, to Secretary of Agriculture Bob Bergland in February 1979, and to Secretary of Labor Ray Marshall in July 1979, asking them to consider how their regulatory authorities might be used to require that recombinant DNA research conducted in the private sector comply with the NIH Guidelines. (See letter, Attachment D.)

Dr. Fredrickson asked agency representatives to comment on the letters. Dr. Rosa Gryder of FDA reported that the agency had received 13 letters commenting on the notice. In general, the Pharmaceutical Manufacturers Association and other private-sector respondents challenged FDA's authority to propose such regulations. Dr. Hugo Graumann of Agriculture, citing Secretary Bergland's reply to Secretary Califano (Attachment D), reported that Agriculture has a regulatory authority under certain circumstances but sees no present need to invoke it. Dr. Murray from the Environmental Protection Agency noted that EPA has an advisory committee looking at this entire matter and that Administrator Costle would soon have their views for a reply to the Secretary. Dr. Logan of the Occupational Safety and Health Administration (OSHA), Department of Labor, said that a response to the Califano letter would go out as soon as possible.

Dr. Fredrickson reported that he and Dr. Donald Kennedy, Commissioner of Food and Drugs, in light of the comments FDA had received, had jointly drafted a supplement to the NIH Guidelines to provide industry with a voluntary approach to compliance. Back in May, Peter Libassi, then General Counsel for the Department of Health, Education, and Welfare, had held meetings with representatives from the Pharmaceutical Manufacturers Association and public interest groups to review the draft supplement. (In the fall of 1978, as chairman of the HEW committee responsible for the final Guidelines, Mr. Libassi had reviewed the proposed revision with these groups.) The PMA representatives now endorsed the voluntary approach in principle but wished to make further recommendations on the protection of proprietary information. The environmental and public representatives, on the other hand, objected to the voluntary approach and urged that alternatives of legislation or regulation be considered. They specifically requested that the Secretary direct OSHA to exercise its authority.

Dr. Fredrickson noted that in light of the Libassi meeting he had spoken to Secretary Califano, who had requested the advice of the Interagency Committee on the voluntary approach and the NIH draft supplement. Specifically, the Secretary had sought the Committee's views on the following questions:

- Are the measures described in the supplement the best way to ensure compliance with safety standards for recombinant DNA research in the private sector?
- Is NIH the best agency to provide a mechanism for voluntary compliance?
- Are the NIH Guidelines as amended in the supplement suited to the industrial setting, and are all of the procedural requirements appropriate? If not, how should they be revised?

Dr. Fredrickson then called for discussion. He noted that recombinant DNA research has proceeded under guidelines internationally and that only in Britain have regulations been invoked. Committee discussion then turned to the draft supplement (Attachment E). A number of representatives asked why industry would wish to comply. What were the incentives? Dr. Fredrickson noted that industry representatives have repeatedly stated that they would prefer to comply with the NIH Guidelines and that, clearly, if they did not, the threat of regulation or legislation would be ever present.

Dr. Gryder of FDA proposed that NIH try to obtain information on all recombinant DNA projects in the private sector. Dr. Fredrickson noted that an attempt had been made in 1977, at Senator Kennedy's request, to develop a survey under which industrial organizations, mainly the PMA, would query their members; but NIH does not intend at present to revive the effort. The supplement, however, would be sent to as many industry contacts as possible for their information.

Dr. Fredrickson then asked for the Committee's views on NIH's acting as a lead agency to handle voluntary compliance. Dr. Duda of the Department of Energy questioned whether NIH has the necessary experience to deal with the industrial sector, with all of the attendant problems concerning proprietary data. He noted that the NIH experience has been mainly with academia rather than industrial organizations and laboratories.

On the other hand, a number of Committee members, including Dr. Lewis of the National Science Foundation and Dr. Graumann of Agriculture, strongly believed that NIH should provide for voluntary compliance. They noted the key importance of one Federal locus to handle these activities and clearly nominated NIH in view of its necessary expertise. Mr. Murray of the EPA concurred.

Mr. Walsh of the Department of State agreed that NIH should be the locus, at least in the absence of legislation or the invoking of regulations, but believed that protection of proprietary data might be troublesome for NIH. Dr. Gryder of FDA acknowledged the burden of dealing with proprietary data, but held that mandatory registration of projects should be considered in addition to the voluntary approach. Dr. Fredrickson pointed out, however, that mandatory registration would require a statute, and there was general agreement that legislation was not warranted at the present time to accomplish this end.

Agency Views on the Draft Supplement

Dr. Fredrickson then requested each agency's views on the draft supplement. All of the *research* agencies represented on the Committee approved of the NIH approach to try to achieve voluntary compliance. These agencies include:

- Department of Agriculture
- National Science Foundation
- Department of Defense
- Department of Energy
- Veterans Administration.

The *regulatory* agencies endorsed the NIH approach as well. These include:

- Food and Drug Administration
- Environmental Protection Agency
- Occupational Safety and Health Administration
- Nuclear Regulatory Commission
- Center for Disease Control.

Mr. Murray of EPA said that his agency has the issue under study and that the NIH approach seems adequate as a first step. Dr. Logan of OSHA supported the voluntary guidelines, but said that OSHA would consider additional guidance in research laboratories to protect workers. The following agencies also concurred in the NIH approach:

- Department of Commerce
- Department of State
- Health, Education, and Welfare
- Office of the Assistant Secretary for Health, HEW.

Mr. David Cannon of the Department of Justice stressed the importance of making sure that all persons at NIH and all relevant committees dealing with proprietary information be fully apprised of the criminal sanctions for disclosure.

Summary and Conclusions

Dr. Fredrickson reported that NIH has received a request from the Eli Lilly Company to permit scale-up of recombinant DNA cultures to 150 liters for production of insulin. RAC members are now reviewing that request, and it will be a key item on the agenda for their next meeting. Questions were asked about the company Genentech and whether it was in compliance with the Guidelines. Dr. Gartland noted that Genentech had proceeded to scale up under the old Guidelines and has not yet applied for the prior approval from NIH that the revised Guidelines require. Dr. Fredrickson pointed out that the supplement to the NIH Guidelines for the private sector would encourage industry to register activities and comply with the Guideline standards. He also observed, in response to Dr. Gryder's comments, that perhaps Commerce could be helpful in leading an industry survey to determine whether all private laboratories that conduct this research are complying with the Guidelines under the special supplement.

Dr. Fredrickson thanked the Committee for its advice and consensus support for the voluntary approach. He promised to keep the members informed about the Department's review of this matter. Dr. Gartland noted earlier that the RAC had voted 9 to 6, with 6 abstentions, in favor of mandatory compliance in the private sector. Dr. Fredrickson said that all views would be taken into account in drafting the preamble to the supplement for publication in the *Federal Register*, with subsequent RAC review, debate, and approval for action by the NIH Director.

Respectfully submitted,

Joseph G. Perpich, M.D., J.D.
Associate Director for
Program Planning and Evaluation
National Institutes of Health
July 26, 1979

Document 15.6

August 16, 1979

Dach Letter to Fredrickson

Environmental Defense Fund, 1525 18th Street NW, Washington, DC 20036 (202) 833-1484
OFFICES IN: NEW YORK, NY (NATIONAL HEADQUARTERS); WASHINGTON, DC; BERKELEY, CA; DENVER, CO

August 16, 1979

Dr. Donald Fredrickson
Director
National Institutes of Health
Bethesda MD 20205

RE: Proposed Supplement to the NIH Guidelines for Recombinant DNA Research; 44 Federal Register 45868-69.

Dear Dr. Fredrickson:

While EDF agrees with NIH's desire to maximize voluntary compliance with the NIH recombinant DNA guidelines by experiments not now covered, EDF strongly opposes any procedure which results in NIH approval of such experiments. EDF does support RAC establishment of containment guidelines for experiments voluntarily brought to its attention and RAC certification of host-vector systems.

EDF is concerned that, in the absence of statutory authority enabling NIH inspection and enforcement of industrial experiments and providing for stiff penalties for violations, approval is a meaningless exercise. Indeed, its only benefit is to the industrial sponsor, who is then free to proclaim that its experiments are safe and above public concern. From a political standpoint, we fear that a voluntary program, no matter how insufficient, will provide a public relations weapon to industry and an argument against mandatory controls.

NIH is totally without authority to insure that approved experiments are actually carried out in compliance with the guidelines. When violations are uncovered, RAC members will be prohibited from speaking out due to the agencies' interpretation of 18 U.S.C. §1905. NIH has no recourse to administrative or judicial compliance orders, no matter how significant or imminent the potential risk to health and the environment. Unfortunately, publicity and other informal means will not insure compliance. NIH is unable to assure the public that all recombinant DNA research and development projects undertaken by industry will be reported to NIH. Indeed, it is likely that a company will gladly report those experiments it expects to easily meet likely containment requirements and keep totally secret those it fears are more controversial. In areas of toxic chemical regulation, industry has systematically failed to report significant health risks to the appropriate regulatory authority. This failure has prompted introduction of legislation establishing criminal penalties for failing to report potential significant health hazards.

On the basis of its experience in toxic chemical regulation, EDF is firmly convinced that voluntary compliance programs have never worked and never will. The burden of proof is clearly on the proponents of such programs to demonstrate how this case will be any different from the many that have gone before. To ask that industry first be given time to implement a voluntary program before mandatory controls are deemed necessary is to be blind to the regulatory history of the past decade.

EDF does not believe that NIH refusal to approve industrial experiments will in any way delay the development of beneficial recombinant DNA products. If industry feels it needs NIH approval in order to proceed, let it forcefully make that case to Congress.

For these reasons EDF opposes NIH approval of industrial recombinant DNA research and development projects. If NIH feels that industry should comply with the guidelines, we urge NIH to make these feelings known to the press and to Congress and to use every means at its disposal to press for a legislative broadening of its authority. As you are aware, RAC did vote at its last meeting to recommend mandatory industry compliance.

EDF also objects to any consideration by RAC of exceptions to the large volume prohibition on a case-by-case basis before adequate procedural mechanisms for such review are in place. RAC is currently debating such provisions, and no individual requests for exceptions should be approved outside of whatever review mechanism is established.

EDF does support NIH guidelines for membership on IBC's not presently covered by the guidelines. Again we do not see the need for NIH approval. Interested citizens and companies, armed with the NIH guidelines, can make these determinations themselves. We do suggest that the guidelines for industry IBC membership be changed from those proposed in 44 Federal Register 45868. At the very least, two worker representatives should be mandated. It is the workers who are potentially at the highest risk for recombinant DNA related health effects. They should have a voice in reviewing such proposals. We further suggest that the number of non-affiliated members be raised to three. We feel that this increase is required to overcome the greater insularity and secrecy of industrial versus academic institutions.

As previously indicated, EDF can support NIH's objective of certifying host-vector systems proposed by industry. Certification is more of a categorization than an approval of an operating plan and as such does not depend on subsequent monitoring to insure compliance. EDF can also support RAC establishment of containment guidelines for experiments proposed by industry. RAC would then lend its expertise to facilitate the proper determination of safety levels but not give its stamp of approval to a process it cannot control.

We appreciate the opportunity to submit these comments and await RAC's debate on this issue.

Sincerely,

Leslie Dach
Science Associate

LD/jas

August 23, 1979
Taylor Letter to Fredrickson

AMERICAN FEDERATION OF LABOR AND CONGRESS OF INDUSTRIAL ORGANIZATIONS

815 SIXTEENTH STREET, N.W.
WASHINGTON, D.C. 20006

(202) 637-5000

August 23, 1979

Dr. Donald S. Fredrickson
Director
National Institutes of Health
Bethesda, Md. 20205

Dear Dr. Fredrickson:

A proposed supplement to the NIH Guidelines for Recombinant DNA Research appeared in the August 3, 1979, Federal Register. These additions provide a mechanism for voluntary compliance with the Guidelines by corporations otherwise not covered by them.

I would like to express most strongly our view that these new sections not be added to the Guidelines. The history of attempts to provide adequate health and safety protections to American workers is replete with tragic examples of employer callousness and employee suffering. Voluntarism is not sufficient.

Currently, with commercial interest in this technology increasing rapidly, a major problem is lack of complete knowledge even of who is doing experiments with recombinant DNA techniques. Those with something to hide will not be revealed by any voluntary system.

We firmly believe that regulatory authority over health and safety of workers must be the responsibility of the Occupational Safety and Health Administration. Promulgation of your inadequate and impotent Guideline additions can only hamper efforts to provide proper oversight by OSHA.

Therefore, these Guideline sections should not be adopted. NIH has no business intruding upon the affairs of non-grantees or other government agencies.

Sincerely,

George H. R. Taylor
Director
Department of Occupational Safety and Health

GT/bw
opeiu #2, afl-cio

Document 15.7

May 21–23; September 6–7; December 6–7, 1979
RAC Meetings Minutes, Excerpts

DEPARTMENT OF HEALTH, EDUCATION, AND WELFARE
PUBLIC HEALTH SERVICE
NATIONAL INSTITUTES OF HEALTH

Recombinant DNA Advisory Committee

MINUTES OF MEETING

MAY 21–23, 1979

The Recombinant DNA Advisory Committee (RAC) was convened for its fifteenth meeting at 9 a.m. on May 21, 1979, in the Terrace Room, Linden Hill Hotel, 5400 Pooks Hill Road, Bethesda, Maryland. Dr. Jane K. Setlow, (Chairman) Biologist, Brookhaven National Laboratory presided. In accordance with Public Law 92-463 the meeting was open to the public.

Committee members present were:

Dr. Abdul K. Ahmed; Dr. David Baltimore; Dr. Francis E. Broadbent; Dr. Allan M. Campbell; Mrs. Zelma Cason; Dr. Peter R. Day; Dr. Richard Goldstein; Dr. Susan K. Gottesman; Dr. Richard B. Hornick; Ms. Patricia A. King; Dr. Sheldon Krimsky; Dr. Elizabeth M. Kutter; Dr. Richard P. Novick; Dr. David K. Parkinson; Dr. Ramon Pinon; Dr. Samuel D. Proctor; Dr. Emmette S. Redford; Dr. Wallace P. Rowe; Dr. John Spizizen; Mr. Ray H. Thornton; Dr. LeRoy Walters; Dr. Luther S. Williams; Dr. Frank E. Young; Dr. Milton Zaitlin; and Dr. William J. Gartland, Jr., Executive Secretary.

A Committee roster is attached. (Attachment I)

The following ad hoc consultants to the Committee were present:

Dr. Mary-Dell Chilton, University of Washington
Dr. Julian Davies, University of Wisconsin

The following non-voting members and liaison representatives were present:

Dr. George Duda, Department of Energy; Dr. Louis C. LaMotte, Center for Disease Control; Dr. Herman Lewis, National Science Foundation; Dr. Sue Tolin, Department of Agriculture; and Dr. William J. Walsh, Department of State.

Dr. Krimsky made a motion that the RAC recommends supporting mandatory compliance by non-NIH funded institutions (with the NIH recombinant DNA Guidelines). Dr. Krimsky stated that he cannot see the justification for treating industry differently from academia. Dr. Parkinson said that he is reluctant to propose additional regulations. He added that it is likely that FDA and OSHA have authority, and he requested a discussion of occupational health with representatives of OSHA and NIOSH at the next meeting of the RAC. Dr. Walters said that he prefers trying a voluntary scheme with incentives for industry to participate. Dr. Talbot said that the Secretary of HEW is on record as opposing the use of Section 361 of the PHS act to regulate recombinant DNA research. Dr. Walters stated that everyone agrees that private industry should be in compliance with the Guidelines and should register projects. There is a difference of opinion with regard to strategy, i.e., whether there should be a voluntary or mandatory system.

DEPARTMENT OF HEALTH, EDUCATION, AND WELFARE
PUBLIC HEALTH SERVICE
NATIONAL INSTITUTES OF HEALTH

Recombinant DNA Advisory Committee

MINUTES OF MEETING

SEPTEMBER 6–7, 1979

The Recombinant DNA Advisory Committee (RAC) was covened for its sixteenth meeting at 9 a.m. on September 6, 1979, in Conference Room 10, Building 31, National Institutes of Health, 9000 Rockville Pike, Bethesda, Maryland. Dr. Jane K. Setlow, (Chairman) Biologist, Brookhaven National Laboratory presided. In accordance with Public Law 92-463 the meeting was open to the public, except for the review of proposals involving proprietary information as the last item of business on September 7, 1979.

Committee members present for all or part of the meeting were:

Dr. Abdul Karim Ahmed; Dr. David Baltimore; Dr. Winston Brill; Dr. Francis Broadbent; Dr. Allan Campbell; Mrs. Zelma Cason; Dr. Richard Goldstein; Dr. Susan Gottesman; Dr. Sheldon Krimsky; Dr. Werner Maas; Dr. James Mason; Dr. Elena Nightingale; Dr. Richard Novick; Dr. David Parkinson; Dr. Samuel Proctor; Mr. Ray Thornton; Dr. LeRoy Walters; Dr. Luther Williams; Dr. Frank Young; Dr. Milton Zaitlin; and Dr. William J. Gartland, Jr., Executive Secretary.

A Committee roster is attached. (Attachment I)

III. PROCEDURES FOR PROTECTION OF PROPRIETARY INFORMATION

The RAC reviewed a document (696) prepared by NIH

staff, entitled "Guide for the Control of Privileged Recombinant DNA Information." This document establishes minimum requirements to control and protect documents that contain privileged recombinant DNA information. Dr. Gottesman said that the document is very precise with respect to ORDA, but less precise with respect to responsibilities of RAC members. In response to a statement by Dr. Walters, Mr. Riseberg said that "personal supervision" means supervision by the RAC member and not by an assistant or secretary of the RAC member. Dr. Gottesman said that this would probably require use of a unique locked file cabinet. Mr. Thornton, who has had experience with handling confidential information in Congress and in the Navy, said that it is the responsibility of the authorized individual to maintain confidentiality. He said that caution must be exercised, but that in most cases confidential material can be mingled with personal files. Mr. Riseberg said that the problem is control and not the mixing of confidential and non-confidential materials. He noted that FDA considers a number of items, such as locked files, to be sufficient to ensure adequate control. Dr. Walters expressed concern that if the precedent is a requirement for locked files it will involve buying new equipment to protect confidential information. Dr. Setlow said that the alternative would be to look at confidential material only during closed portions of the RAC meeting. In response to a question by Dr. Walters, Mr. Riseberg said that a safe deposit box at a bank would be a reasonable way of safekeeping documents. The RAC requested that a copy of Section 1905, Title 18 U.S.C. be attached to the document being discussed. Dr. Gartland said that RAC members would be asked to sign the "Commitment to Protect Privileged Recombinant DNA Information" (Attachment A to Document 696) before the meeting goes into closed session to review proprietary information.

DEPARTMENT OF HEALTH, EDUCATION, AND WELFARE
PUBLIC HEALTH SERVICE
NATIONAL INSTITUTES OF HEALTH

Recombinant DNA Advisory Committee

MINUTES OF MEETING

DECEMBER 6–7, 1979

The Recombinant DNA Advisory Committee (RAC) was convened for its seventeenth meeting at 9 a.m. on December 6, 1979, in Conference Room 10, Building 31, National Institutes of Health, 9000 Rockville Pike, Bethesda, Maryland. Dr. Jane K. Setlow, (Chairman) Biologist, Brookhaven National Laboratory presided. In accordance with Public Law 92-463 the meeting was open to the public, except for the review of proposals involving proprietary information as the last item of business on December 7, 1979.

Committee members present for all or part of the meeting were:

Dr. Adbul Karim Ahmed; Dr. David Baltimore; Dr. Winston Brill; Dr. Francis Broadbent; Dr. Allan Campbell; Mrs. Zelma Cason; Dr. Richard Goldstein; Dr. Susan Gottesman; Dr. Jean Harris; Ms. Patricia King; Dr. Sheldon Krimsky; Dr. Werner Maas; Dr. James Mason; Dr. Elena Nightingale; Dr. Richard Novick; Dr. Samuel Proctor; Mr. Ray Thornton; Dr. LeRoy Walters; Dr. Luther Williams; Dr. Frank Young; Dr. Milton Zaitlin; and Dr. William J. Gartland, Jr., Executive Secretary.

Dr. Goldstein suggested that the following wording be substituted for (3) in Dr. Krimsky's motion:

> "That the RAC recommends to the Secretary of HEW that Federal legislation rather than voluntary compliance is required for the regulation of non-Federally funded research in the area of recombinant DNA research."

Dr. Krimsky accepted the proposed amendment. Dr. Young proposed an amendment that would delete the preamble which reads as follows:

> "Whereas a compliance program based exclusively on the good faith of such institutions, involving no sanctions, and no accountability for breaches in compliance, is untenable in concept."

Dr. Krimsky accepted the proposed amendment. The Krimsky motion as amended failed by a vote of five in favor, nine opposed, with four abstentions. Dr. Goldstein requested that his vote in favor be recorded.

Dr. Novick moved 774/2 with the addition of item 3 of the previous motion as amended by Dr. Goldstein. Ms. King said the proposal is now inconsistent. Paragraph 1 of 774/2 states that it is desirable to establish a uniform standard of conduct for the performance of experiments involving recombinant DNA techniques while the new added paragraph specifies the type of uniformity that is required. This motion was defeated by a vote of seven in favor, ten opposed, with two abstentions. Dr. Goldstein asked that his vote in favor of the motion be recorded.

Ms. King moved the proposal as it had appeared in the *Federal Register* (774/2). Dr. Gottesman proposed to amend the last paragraph of the proposal to read as follows:

"At the same time, the committee regards the concept of voluntary compliance as experimental; in order to ensure further consideration after an initial trial period, the Committee agrees to conduct a review of the voluntary compliance program at its June 1980 meeting."

Dr. Zaitlin favored this amendment which would eliminate the reference to a June 1, 1980, termination date. Dr. Novick said he preferred to retain a termination date in the proposal. The RAC accepted Dr. Gottesman's amendment by a vote of ten in favor, seven opposed, with one abstention. A vote on 774/2 as amended by Dr. Gottesman, was accepted on this motion by a vote of four in favor, three opposed, with one abstention.

The text of the motion as adopted by the RAC is as follows:

"Whereas it is desirable to establish a uniform standard of conduct for the performance of experiments involving recombinant DNA techniques,

And whereas the RAC has recommended mandatory compliance with the NIH Guidelines for non-federally funded institutions,

And whereas there is currently no extant legal framework within which this can be effected,

The RAC congratulates the Pharmaceutical Manufacturers Association and its member companies for the cooperative spirit that they have shown in agreeing to comply with the NIH Guidelines voluntarily under provisions of the supplement to the Guidelines adopted by the RAC at its meeting of September 6-7, 1979.

At the same time, the committee regards the concept of voluntary compliance as experimental, in order to ensure future consideration after an initial trial period, the Committee agrees to conduct a review of the voluntary compliance program at its June 1980 meeting."

Document 15.8

May 29, 1980
Nature (**285**:279)

Recombinant DNA

Guidelines for scale-up

Washington

Like a man with a new pair of shoes, the US National Institutes of Health are experiencing some discomfort in administering industry's voluntary compliance with the safety guidelines covering recombinant DNA research.

Pressures on the NIH are coming from two directions. On the one hand, increasing competition from European and Japanese industry is leading US companies to complain that lengthy procedures for approving new host/vector systems and large-scale fermentation experiments are becoming a commercial handicap.

On the other, the members of the NIH's Recombinant DNA Advisory Commission are feeling their way uncertainly into the engineering and safety aspects of large-scale fermentation techniques of which few possess either detailed knowledge or expertise.

The system of voluntary compliance was proposed last year by NIH director Dr. Donald Fredrickson, largely as a way of avoiding the need for new legislation extending the guidelines to the private sector. Since the system was introduced in January, eleven companies have registered their Institutional Biohazard Committees (IBCS) with the NIH, and seven large-scale experiments have been approved.

Last year the containment guidelines were reduced for most experiments to a level requiring minimal physical safegaurds, with the result that few companies now feel them to be excessively restrictive. Indeed some take comfort from the fact that Japan has recently introduced the NIH guidelines in their original, more stringent, form.

There is greater concern, however, about the time taken up by NIH's certification procedures, in particular that required to review new host/vector systems—a responsibility which many RAC members now consider to be one of the committee's most important functions.

At present, for example, the disabled bacterium strain *Escherichia coli* K-12 is the only approved host/vector system for experiments at the P1 minimum containment level (although the use of *Saccharomyces cerevisiae* strain of yeast is expected to be recommended for approval at RAC's next meeting).

The result, according to Dr. Peter Farley, President of the Berkeley-based Cetus Corporation, is that European companies who can locate their operations in countries with minimal certification requirements already have a major commercial advantage through being able to exploit other host/vector systems, such as *Bacillus subtilis*.

Time is also being taken up by the need for special approval by the Director of NIH for each experiment carried out using more than 10 litres of culture—a relatively arbitrary limit set in the early days of the guidelines and apparently based on conventional laboratory practices.

Amendments to the guidelines which would accelerate the approval process for commercial developments have now been recommended to NIH by Dr. Irving Johnson, Vice-President of Research for Eli Lilly and Company, which has already received permission for several scaled-up experiments in the production of human insulin.

In particular, Dr. Johnson is proposing that once approval has been given for a particular experiment to be carried out above the ten-liter limit, further volume increases using the same biological materials at the same containment levels would be required merely to be approved by the local IBC.

More controversial is Dr. Johnson's suggestion that, in order to make up for RAC members' lack of experience with large-scale applications, industry representatives should be added to the committee with expertise in areas such as fermentation technology and engineering.

And he is also proposing that a permanent subcommittee be set up to advise the Director of NIH on actions related to large-scale applications, with authority to recommend approval of preliminary plans for large-scale operational facilities.

The various proposals will be discussed by the RAC when it meets next week. Their reception is uncertain.

Committee members are aware that they lack expertise in either worker safety or fermentation engineering. "Many have not been entirely happy with their role" says committee member Dr. David Baltimore of the Massachusetts Institute of Technology.

Yet there is disagreement over whether this lack should be made up primarily by expanding the scope of the committee, through increased use of consultants or the addition of new members. Or, alternatively, whether greater involvement should be encouraged from agencies which already have statutory responsibility for such matters, in particular the Department of Labor's Occupational Safety and Health Administration.

Either way, it seems increasingly unlikely that there will be new legislation imposing the guidelines on the private sector. A bill seeking to do so, and requiring the registration of all industrial activities using recombinant DNA techniques, has been introduced by Senator Adlai Stevenson Jr.; but it has so far failed to gain any significant support, either inside Congress or without.

David Dickson

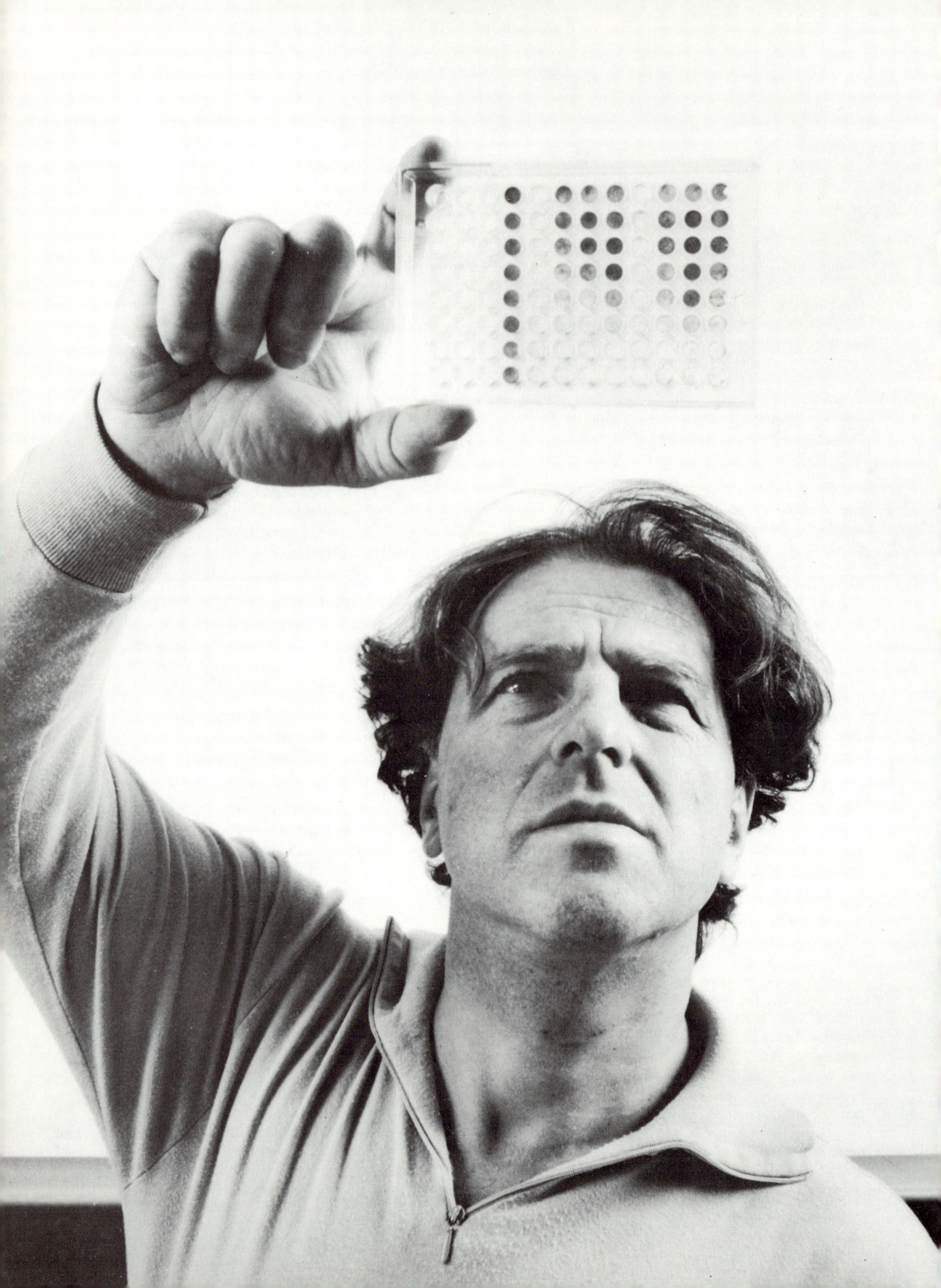

NEW INDUSTRY, NEW PATENTS

Ever since 1973, those in favor of the unimpeded development and exploitation of recombinant DNA technology had been comparing the potential benefits that could confidently be foreseen with the conjectural hazards that were being imagined. (See, for example, Doc. 3.2.) Prominent among the benefits was the possibility of converting bacteria and other microorganisms into fermenters of valuable human and other proteins. In the early days (at Asilomar, for example) no one was quite sure how long it would take to develop the genetic manipulations necessary to achieve the full expression of human genes in bacteria. The potential was, however, indisputable and clearly realizable, given the time and freedom to perform the appropriate experiments. This book describes the extent to which that freedom has been curtailed. Despite those severe impediments, the great expectations have been brought close to realization more quickly than most would have dared hope or predict. Such rapid development is, however, not without its growing pains.

Before the advent of the recombinant DNA technique, molecular biology was very much an academic pursuit. Unlike their colleagues in branches of physics and chemistry, very few molecular biologists had or even considered links of any sort with industry; they were more accustomed to defending themselves against the criticism that there were precious few practical benefits indeed to show for the spectacular progress in determining the molecular basis of life. Recombinant DNA revolutionized all that; it has spawned a multimillion-dollar industry and has become the focus of speculative investment by companies with venture capital to spare (Doc. 16.1, Doc. 16.2, and Doc. 16.3).

"Biotechnology is one of the biggest industrial opportunities of the late twentieth century"—says the London *Economist. "Where genetic engineering will change industry"*—proclaims *Business Week. "One way ahead for British Biotechnology?"*—asks *Nature*. When an academic community suddenly finds itself to be the subject of hundreds of such headlines (Doc. 16.4), as well as the hot topic and tip of investment analysts, a white hope of governments trying to salvage ailing economies and the target of undreamed of blandishments from private industry, strains, conflicts, jealousies, and confusion be-

Photo of Charles Weissmann by Ralph Crane, *Life Magazine*, © 1980, Time Inc.

tween old and new loyalties inevitably result. And the inevitable has been reinforced by the particular way in which the commercialization of gene cloning has developed. Its outstanding practitioners have neither resigned from universities to take jobs with large established chemical and pharmaceutical companies nor have they become consultants. Rather, they have been recruited with offers of stock options into new, small-venture capital companies backed by large corporations such as Inco, Monsanto, Standard Oil, Social, and Schering-Plough. These recruits, retaining their full-time academic jobs and enjoying the relatively free exchange of information in academic life and the services of the cream of young scientists and postdoctorals, have in their publicly funded university laboratories pursued research directly related to their company's projects and profit. To be enjoying so blatantly the best of both worlds creates envy.

Critics—in addition to believing that the molecular biologist-businessmen are behaving both unethically, in the way they exploit public resources, and distastefully, in the way they first announce their results at press conferences for the good of the companies' shares—raise further objections. They are concerned that preoccupation with patents and the patentable will destroy the academic tradition of free exchange and publication of information and cause history to be rewritten to suit the patent file (see Doc. 16.5, and below). They believe that some of the best minds in the field are being diverted to development and technology rather than to the solution of more basic problems, and that this is playing into the hands of governments which, especially in periods of economic difficulty, prefer to support "mission-oriented" research with direct practical goals at the expense of basic research.

Such criticisms are not empty, but there is another side. First, in a capitalist society routing public money to private enterprise and therefore to individuals is hardly novel, neither is the desire to make money. [Exactly how much money has been made is unclear, but according to Nicholas Wade (*Science*, Vol. 206, p. 663) the paper value of the four most prominent small gene-splicing companies—Cetus, Genentech, Genex, and Biogen—was about $225 million in November 1979.] A desire to see one's research put to practical use is as genuine as it is understandable. It is also clear that the speed with which the potential benefits of recombinant DNA are being brought close to realization owes something, perhaps in fact a great deal, to this entrepreneurial system. Furthermore, as the small companies gain confidence, evolve, and open new laboratories of their own, not only are new and welcome jobs for molecular biologists created but the initial need to do company work in academic laboratories, if it was to be done at all, is diminished. We feel confident that just as branches of chemistry and physics have evolved an acceptable association with industry, so will molecular biology, without rending itself apart.

Schooled to success in academic molecular biology, nothing if not competitive, any molecular biologist-businessman should have no problems keeping his latest work secret until he is ready to reveal it to competitors. Familiarity with patent law is another matter, however, and the "who did or thought of the experiment first" disputes, not new to the field, acquire added significance in the patent courts (Doc. 16.5).

On June 16, 1980, the U.S. Supreme Court handed down a decision on a fundamental question of great consequence to genetic engineers, when it ruled, five to four, that forms of life carrying a manmade genetically engineered component can be patented. The test case was Ananda Chakrabarty's petition to patent a strain of *Pseudomonas* that he had engineered, not by the recombinant DNA methods but by conventional genetic manipulations (Doc. 16.6). The historic decision will no doubt greatly encourage genetic engineers, and we include the full text here (Doc. 16.7).

Although it was welcome news to genetic engineers and their financial backers, the Supreme Court's ruling has stimulated some American religious groups to call for a review of U.S. patent laws. The National Council of Churches of Christ would also apparently wish to involve the United Nations in an attempt to evolve international guidelines (Doc. 16.8). Pope John Paul II has also expressed his fears of the consequences of genetic manipulations to a U.N. agency, UNESCO in Paris (Doc. 16.9). What might be said in debate in the U.N. General Assembly on genetic engineering defies our imaginations.

More recently, in November 1980 a patent (Doc. 16.10) that covers very many of the basic techniques of recombinant DNA work was awarded to Herbert Boyer (University of California, San Francisco) and Stanley Cohen (Stanford). Stanford is to be the beneficiary, and early in 1981 it held a seminar to inform industrial companies on how to obtain a license. If the second patent applied for by Boyer and Cohen to cover any organism produced by methods under the first patent is also granted, Stanford's revenues should be quite substantial.

The last document in this collection (Doc. 16.11) is the share prospectus of Genentech, the recombinant DNA company of which Herbert Boyer is a co-founder. When Genentech went public in September 1980, its shares rose from the starting price of $35.00 to $89.00 in minutes. Fortunes will be made and perhaps some lost, but the fact that the speculation today is over recombinant DNA stocks rather than its imaginary biohazards is the best evidence that an unparalleled episode in the history of science has ended.

May 16, 1980
Science (**208**:688)

Cloning Gold Rush Turns Basic Biology into Big Business

Cloning a gene can help raise $50 million for your company. Will the laboratory suffer?

The date on which molecular biology became big business was 16 January 1980. Reporters had been notified by telegram that a "major announcement" in molecular biology would be made by the company Biogen and two members of its scientific advisory board, Charles Weissman of the University of Zurich and Walter Gilbert of Harvard. The news delivered at the Boston Park Plaza Hotel was that Weissman had cloned and got expression of the human leucocyte interferon gene in biologically active form.

"should not be interpreted to mean that it has any significant advantage in either technology or patent protection. Achieving expression of the [interferon] gene is only the first of many steps required to demonstrate a commercial process."

By all accounts, neither Weissman nor Gilbert exaggerated the significance of the news to the assembled reporters. Nonetheless, the mere context of the occasion, which linked the recombinant DNA technique with the possibility of manufacturing a promising anti-cancer

pany's paper value at $50 million. A few months before the January press conference, Biogen had decided to raise more capital on the basis of a self-assessed paper value of more than $100 million. Biogen president Robert Cawthorn says that the purpose of the press conference was not primarily to help raise more capital for the company. "The intent was to draw attention to Biogen. The day may come when we want to go public. So it is better that the public knows something about Biogen." But Cawthorn confirms

Biotechnology Survey

"Biotechnology is one of the biggest industrial opportunities of the late twentieth century," the London *Economist* opined last month. Corporate investors have become so enamored of recombinant DNA that the paper value of the four small enterprises that specialize in gene splicing has more than doubled in the last 6 months alone.

The commercialization of molecular biology is a fundamentally healthy process. But the recombinant DNA technique, promising as it is, seems on occasion in danger of being oversold. And some problems lie ahead for academic biologists as they work out the ground rules for assisting in the transfer of biotechnology to industry while preserving the university's role as a source of independent advice. The following articles consider the present stage in the evolution of the gene splicing industry, as well as the less publicized but probably more immediately significant technology of monoclonal antibodies.

A major announcement in molecular biology this was not. For one thing, the cloning of human fibroblast interferon had already been achieved and published in a Japanese journal. Even the commercial significance of the news was far from clear. Biogen is only one of several competitors in the race to produce human interferon, a substance of possible though still unproven use in cancer therapy. As one Wall Street analyst, Scott King of F. Eberstadt and Co., advised his clients, Biogen's announcement

drug, sufficed to produce a major impact on the public imagination and on Wall Street. The stock of Schering-Plough, which owns 16 percent of Biogen and rights to its interferon process, rose by 8 points, temporarily adding some $425 million to the paper value of the company's shares.

The purpose of the press conference, however, was not to help Schering-Plough but to help Biogen. Schering-Plough's purchase last year of 16 percent of Biogen for $8 million had set the com-

that the company is looking for new investors, preferably those who are minded to stake at least $10 million apiece.

Molecular biology has come of age.

The sudden eagerness of investors to put money into genetic engineering has begun to reach the frenzy of a gold rush. "Every venture capitalist has looked at this—there are so many people out there it is like a bloody battlefield," notes one Wall Street analyst. The clearest manifestation of the cloning gold rush is that the paper value of the four most pub-

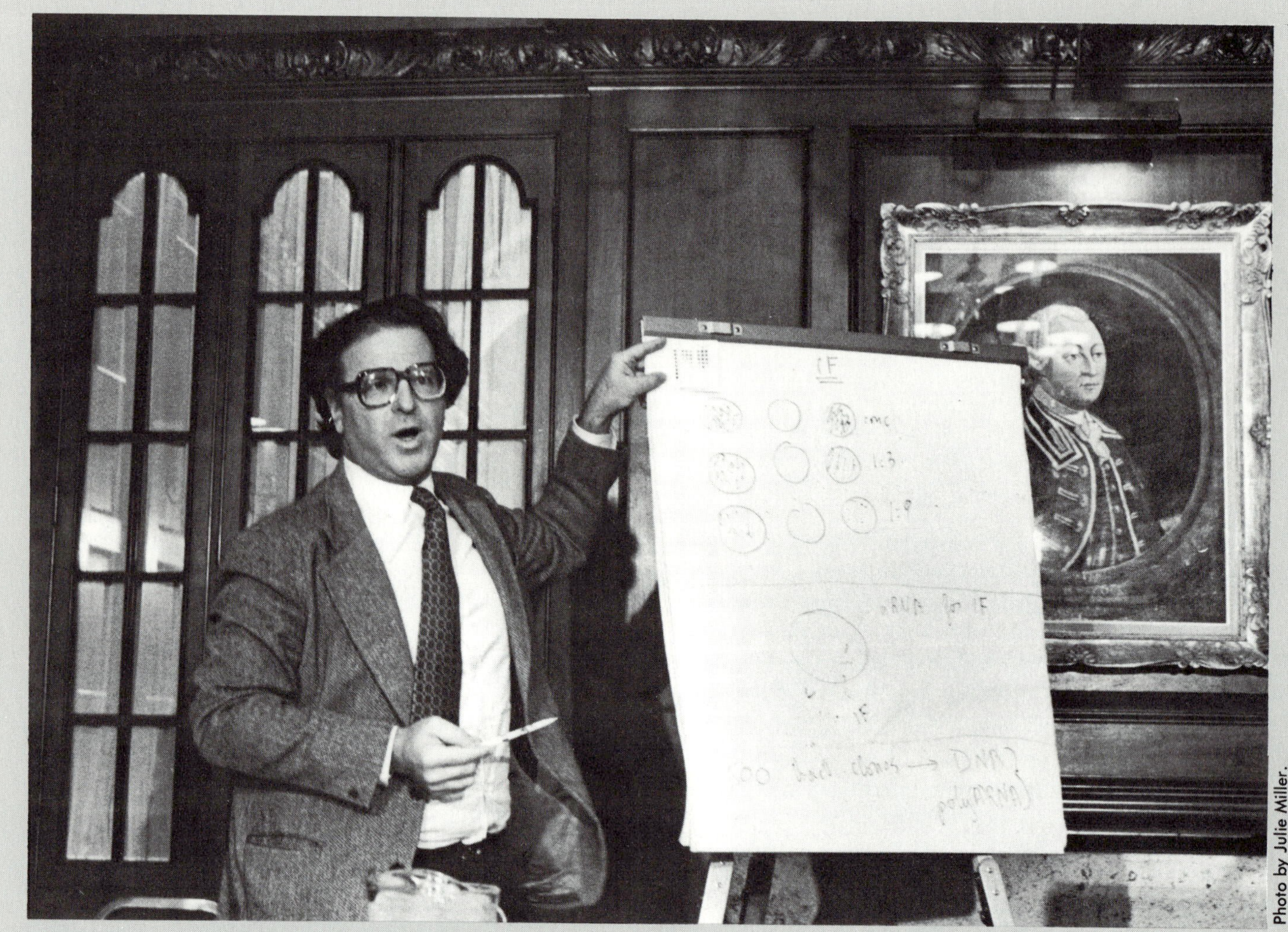

Photo by Julie Miller.

Charles Weissmann at Biogen's $50 million press conference.

licized gene splicing enterprises (Biogen, Cetus, Genentech, and Genex) has more than doubled in the last 6 months, to a total worth of over $500 million. Biogen is only following the general pattern. Yet none of the companies has so far brought any gene spliced product to market and it could be several years before any succeeds in doing so.

Every study of innovation upholds the transfer of knowledge from academy to industry as a socially laudable objective. The investment hoopla surrounding molecular biology is a sign of this process going with a swing. Nor is it immediately obvious why the transfer of this particular technology should present any special problems. But the possibility of certain growing pains in a basically healthy process is already evident.

One is the turbulence being caused in academe as large sums of money become available to a community of researchers that has not previously had to decide how to deal with the profit motive. Some welcome the development, others feel strongly that the business ethic is different from the scientific ethic and that basic research into molecular biology may be compromised by the subject's commercial exploitation. A danger of another sort lies in the possibly excessive expectations which now surround the field's commercial future. Even some of the gene splicing entrepreneurs agree that there has been a lot of hyperbole written about their activities, although all deny seeking publicity in order to raise money.

"There are millions of dollars floating around. If you claim you've done something fancy you can raise a lot of money," says a leading molecular biologist. "The question is, what will this do to the academic atmosphere?"

One fear is that it will reduce the ability of academics to provide independent advice on controversial issues, a function of a university which may be no less important than its role in the transfer of knowledge. Stanley Cohen and Herbert Boyer, who hold the basic patent application on the recombinant DNA technique, have renounced all royalties they may receive if the patent is issued. One of his reasons, Cohen says, is that he was involved in the public debate about control of recombinant DNA research, and "the fact that I had already turned over all royalties to Stanford enabled me to speak out in ways which would not have been possible if my motives were being questioned."

Those within the community of molecular biologists are unlikely to question a colleague's position on an issue merely because he has some financial interest in it, but the wider public is more skeptical. In recent arguments before the Supreme Court on the patentability of genetically altered bacteria, Genentech cited a prominent molecular biologist in arguing that worries about the dangers of genetic engineering had all but disappeared. But the U.S. solicitor general in his reply brief contended that because of this researcher's involvement with a gene splicing company, "He is thus hardly an impartial observer in the debate over the biohazards associated with genetic engineering developments."

Another concern is that the direct involvement of academics in the commercialization of recombinant DNA will allow commercial interests to influence the goals and nature of academic research. "Just as war-related academic research compromised a generation of scientists, we must anticipate a similar demise in scientific integrity when corporate funds have an undue influence over academic research," Sheldon Krimsky of Tufts University has contended. Krimsky, a member of the NIH Recombinant DNA Advisory Committee, considers that for a researcher to be a principal in a company as well as an academic poses too great a conflict of interest in terms of dividing time and resources between university and company research.

Gilbert, who along with Herbert Boyer of the University of California, is one of the academic pioneers in the commercialization of gene splicing, sees in it an opportunity for rewriting the usual relationship between university and industry. In gene splicing, it is the academics who will be in control, if the Biogen model is successful. "Industry likes academics to be consultants. What we are seeing here is an attempt by academics to control the industrial development."

Gilbert and the eight other members of Biogen's scientific advisory board, of which Gilbert is chairman, hold 15 percent of the company's stock with the option to increase their holding to 30 percent. The stock holding is worth only so much paper until the company goes public, unless a private market should develop, but the fact that the Biogen board members are presumably millionaires, even if only on paper, has excited an array of emotions, envy no doubt included, among those who feel equally well qualified. The issue is a matter of some sensitivity to Biogen's scientific principals. Gilbert, who was erroneously introduced to the NBC public by John Chancellor as a "recent millionaire," dismisses the paper value as a "crazy reflection of the way the company structure is controlled." The higher the paper value can be set, "the less control we have to give up to people buying into the company."

Biogen at present employs 16 scientists in an 8000-square foot laboratory in Geneva. Gilbert believes that the company's activities will eventually provide employment for hundreds, perhaps thousands, of scientists. The existence of the companies, the job market they create for postdoctorates, "is already changing people's attitudes," Gilbert observes. The friction being created in universities by the commercial exploitation of recombinant DNA is in his view a temporary phase: "In the short term there are stresses. We will go through these growing pains for 5 years or so, after which the pure and applied aspects of the field will be further apart."

Stanley Cohen is another who perceives a change in biological researchers' attitudes toward industry. "Five years ago, questions of patents and involvement in industry were often viewed in a negative light. But the whole view of the biological community towards industry has evolved over the past few years," Cohen says.

Whatever the reason for the new attitude, whether a conservative trend in the country at large, or biologists' sudden perception of the commercial opportunities opening up for their profession, the ground rules for researchers' involvement still remain to be defined in several areas. One is the release of information to the public. Both Genentech and Biogen have been criticized for announcing results at press conferences prior to their publication in the scientific literature. "Competition and the increasing involvement of academic scientists in the field of commercial application may be part of the problem. Free inquiry and the pressures of competition associated with the application of technology are not necessarily compatible," Stanford University science writer Spyros Andreopoulos complained in a recent article in the *New England Journal of Medicine*.

In practice, the traditional embargo on prepublication release is often honored in the breach. Genentech's cloning of somatostatin was announced before publication by the president of the National Academy of Sciences at a Senate committee hearing, perhaps to counter the committee's interest in an incident involving an infraction of the NIH gene splicing rules. Why shouldn't the companies also announce their results to serve their own purposes? Andreopoulos's concern is that exaggerated claims made or implied at such press conferences cannot be checked out by reporters if the results are not generally available. Though this may be true, the press conferences given by the companies are only one among several ingredients that have contributed to the high expectations surrounding commercial gene splicing.

The business world was perhaps surprisingly late in awaking to the wonders of gene splicing, but has made up for its tardiness with enthusiasm. There are still remarkably few investment analysts who follow the field. Best known among them is Nelson Schneider of E. F. Hutton in Washington, D.C. A seminar arranged by Schneider in New York last September gave the financial world its first serious exposure to gene splicing. Investors' interest has been whetted by a series of articles in which Schneider's upbeat commentaries on the commercial future of gene splicing have been echoed, naturally enough, by the principals of the four best known gene splicing companies.

"Some people feel there has been a nationwide promotion campaign to benefit selected individuals and companies, that this is all a contrivance that will adulterate science. But this just isn't so," notes Fred Middleton, vice president of Genentech. Genuine public interest is the reason that so much has been written on the subject; as for Genentech, "We are trying to avoid overexposure at this point," Middleton states.

"I am the first one to say that the whole field has been subject to too much hype, particularly in relation to the business projects," observes Ronald Cape, chairman of Cetus. Yet the hyperbole, or enthusiasm, is part of the climate in which investment money is pouring into the four gene splicing companies. As one raises more capital, the others are forced to follow suit. "Now that everyone is spending money hand over fist, if we don't expand as fast as the other people, we will fall out of our position, and if we didn't have the big stockholders we have, we would have to say this is the end of the game," explains Cape.

Cetus, worth $100 million in November last year, is trying to raise another $55 million based on a paper value of $250 million. Its 1979 income, almost entirely from contract research, is said to be about $7 million. Genentech, worth $65 million 6 months ago, has not raised any capital since then, but claims its corporate worth now to exceed $100 million. Biogen has upped its valuation from $50 million to $100 million, and Genex has ascended from $9 million or so to a vaunted $75 million.

The first gene spliced products to reach the market are research enzymes, produced by New England BioLabs and Bethesda Research Laboratories (see p. 690). None of the four new companies yet has a marketable product and only Genentech has sought NIH permission to proceed to large-scale culture with a recombinant DNA project. Last month the NIH approved five such projects, concerning human growth hormone, somatostatin, the A and B chains of human insulin, human proinsulin, and thymosinalpha-1. By this criterion, Genentech is five necks ahead, but in a race that has a long way yet to go.

There is, of course, a solid basis for confidence in the commercial future of gene splicing, but some observers wonder if the present levels of expectation haven't gotten out of hand. The established pharmaceutical firm of Upjohn is worth about $1 billion yet Cetus is claiming to be worth $250 million, notes Charles Newhall of New Enterprise Associates in Baltimore: "That means Cetus is valued at one-quarter of Upjohn. What stream of products will come from Cetus that could justify that investment?" Newhall wonders.

Boosters of gene splicing often liken its golden future to that of the microelectronics industry. But in the world of bugs and chips there's many a slip twixt cup and lip. Analysts such as Schneider of E. F. Hutton and Daniel Adams, former head of the venture capital division of Inco, take the view that the corporations who are investing in the little gene splice companies are big enough to know what they are doing. "A company is worth whatever people are willing to pay for its stock, and more and more people are believing in the economic potential of the area," says Schneider. According to Adams, who had a major role in founding Biogen, "None of these companies is worth anything by conventional criteria because you can't put price/earning multiples on them, but a value of $50 million to $100 million is not excessive in that I think the value will ultimately be realized. The people who are paying these prices are not dumb. It is not a question of widows and orphans being taken to the cleaners."

Others are not so convinced that the corporations are that much less gullible than the widows. Says one Wall Street analyst, "People really don't understand anything at all about this field; all they want to know is how to invest in it." "These business people who come to my laboratory are not interested in the complications, they are interested only in the bottom line of what might happen. They want to be in on something that could be big, and it is worth it to them to risk $10 million," says a well-known molecular biologist.

Even the corporate players in the gene splicing game have doubts about some of the investments that other corporations are making. "Many corporations are looking into this area with some concern, because of fear that they are being passed by. There are some who are willing to pay almost anything to get into it and I think that is totally wrong," comments Donald Murfin, president of Lubrizol Enterprises, the venture capital division of Lubrizol. Lubrizol Enterprises now owns 25 percent of Genentech; it paid $10 million for 15 percent of the company last September, and recently bought out Inco's holding for $15 million.

Another interesting line of skepticism comes from the principals of several small companies which, with an established position in other areas of biotechnology, are now gearing up to join the gene splicing derby. Their criticism, though it is easy to see a motive behind it, is not without interest. The present funding levels of the four best publicized companies, says Orin Friedman, president of Collaborative Research of Boston, "have lost touch with reality." Friedman has recently set up a new company, Collaborative Genetics, to specialize in gene splicing applications, but in his view "The enormous publicity given to the commercial potential of recombinant DNA may be counterproductive because it is creating unreal expectations. I think there are potentialities in the technology, but based on my experiences with other technologies, the gap between a laboratory process and reduction to commercial reality is going to take much longer than the impression created in the numerous articles about the subject," Friedman observes.

"Without the PR, there is no question that the high flying money perceives you as being of little value. But these paper valuations have a way of folding," notes Stephen Turner, president of Bethesda Research Laboratories. Turner raises the analogy of a chain letter, with the man in the street being the ultimate recipient when the gene splicing companies go public.

"That is 100 percent absolutely and completely ridiculous," says E. F. Hutton's Schneider. "None of these companies has any thought of going public and if they were we would tell them not to because there is no guarantee as yet that this technology will produce anything."

There is no guarantee either that the companies now developing particular gene splicing technologies will be able to hold onto their advantage. What if the academic research community should develop a general method for cloning and amply expressing the product of any known gene? With the basic technology available to all, the advantage might move away from the little companies, whose major asset is access to leading molecular biologists, and toward enterprises that either have large sales forces, as do the pharmaceutical companies, or possess advanced expertise in fermentation technology, as do the Japanese.

The cloning gold rush has entered an interesting but unpredictable phase. There is certainly gold to be found, but no one can be quite sure just how soon, or how easy it will be to protect whatever is struck.—Nicholas Wade

Document 16.2

May 16, 1980
Science (**208**:690)

Three New Entrants in Gene Splicing Derby

Three small companies which are already established in the biotechnology business have recently entered the gene splicing race alongside the better known contestants. They are Bethesda Research Laboratories, New England BioLabs, and Collaborative Genetics.

Bethesda Research Laboratories announced in March that it had cloned one of the genes involved in the biosynthesis of proline, a commercially significant amino acid. BRL was founded in January 1976 by Stephen Turner, formerly with Becton Dickinson, a medical supply firm. When restriction enzymes came into prominence, Turner wanted Becton Dickinson to produce them but, meeting with indifference, he decided to start his own company.

Photo by Witwicki

New England BioLabs with president Donald Comb (fourth from right, front row)

With $30,000 of his own money, Turner hired a technician, rented 1000 square feet in Rockville, Maryland, and went round the nearby campuses of NIH and Johns Hopkins selling restriction enzymes from an ice bucket.

By the end of 1976, he had sold $100,000 worth of restriction enzymes and had hired two more technicians. From that point the company has expanded with great rapidity. It now employs 150 people, including 35 Ph.D.'s, and has sales of more than $2.5 million.

Almost half of BRL's sales are still in restriction enzymes, but Turner is diversifying as fast as possible into other fields, chiefly the recombinant DNA and hybridoma technologies. Recombinant DNA projects include production of amino acids and cellulase. Peter Kretchmer, formerly of NIH, is joining BRL to head its new department of genetics. BRL has a scientific board that includes Jack G. Chirikjian of Georgetown University and others.

Turner's goal is to make his company the Sears Roebuck of molecular biology: "We are part of the flow of information and materials. Our mission is to supply the tools and techniques of molecular biology, wherever it may lead," he says. Turner considers that by making new techniques available as soon as they are developed, BRL can help shrink the lead time and break down the "feudalistic kind of structure" which makes it hard for those outside the elite institutions to get immediate access to what the front-runners are doing.

The current BRL catalog offers several hundred different products of relevance to molecular biology, including almost everything a cloner might desire, from restriction enzymes to plasmids and molecular linkers.

Almost 90 percent of BRL is owned by Turner and other principals of the company. The rest is owned by a small venture capital firm, New Enterprise Associates of Baltimore. Only recently has the company raised venture capital. "Having gone 4 years without outside capital means that you can bring it in on your own terms," Turner says.

New England BioLabs, of Beverly, Massachusetts, is also a producer of restriction enzymes. The company was founded in 1974 by Donald G. Comb, then at the Harvard Medical School, and is wholly owned by Comb and his wife. Sales last year were $3 million, Comb says.

New England BioLabs and Bethesda Research Laboratories dominate the restriction enzyme market. Each believes it has the larger share. On the basis of literature citations, New England BioLabs considers it commands 75 percent of the market, with BRL taking the rest. BRL, on the other hand, believes from questionnaires that it has 65 percent and its rival 25 percent. Whatever the exact shares, the two little companies seem to serve their market well enough to have kept bigger competitors out.

Comb says he is interested in keeping his company small—he has only 22 people, compared with BRL's 150—but has started to recruit cloners for a project to do with malaria.

New England BioLabs apparently has the distinction of being the first company to bring a recombinant DNA-made product to market, even if to a somewhat specialized group of consumers. Since 1975 the company has been selling DNA ligase from *Escherichia coli* produced from a gene cloned by Robert Lehman of Stanford. Three other recombinant DNA-made enzymes are featured in the company's catalog.

A third small company entering the gene splicing field is Collaborative Research of Waltham, Massachusetts. Founded in 1962, the company now employs 85 people and has sales of under $5 million, tissue culture being a major part of its business. In November last year, Collaborative Research president Orin Friedman set up a majority-owned subsidiary, Collaborative Genetics, which is focused on the manipulation of yeast, by gene splicing and other means, for energy transformation and industrial application.

Collaborative Genetic's advisory board includes David Baltimore and David Botstein of MIT, Ronald Davis of Stanford, and Gerald Fink of Cornell. "We have in fact cornered the market in yeast genetics," claims Friedman. Despite which, Friedman is highly guarded in his expectations of producing ethanol more efficiently from genetically engineered yeast. Much effort has been devoted to seeing how gene splicing might help to improve the venerable art of fermentation. But, says Friedman, "There are no obvious or simple ways in which the recombinant DNA technology can be applied to make major improvements in alcohol production by yeast."—NICHOLAS WADE

Document 16.3

January— June 1980
Newspaper Article and Headlines

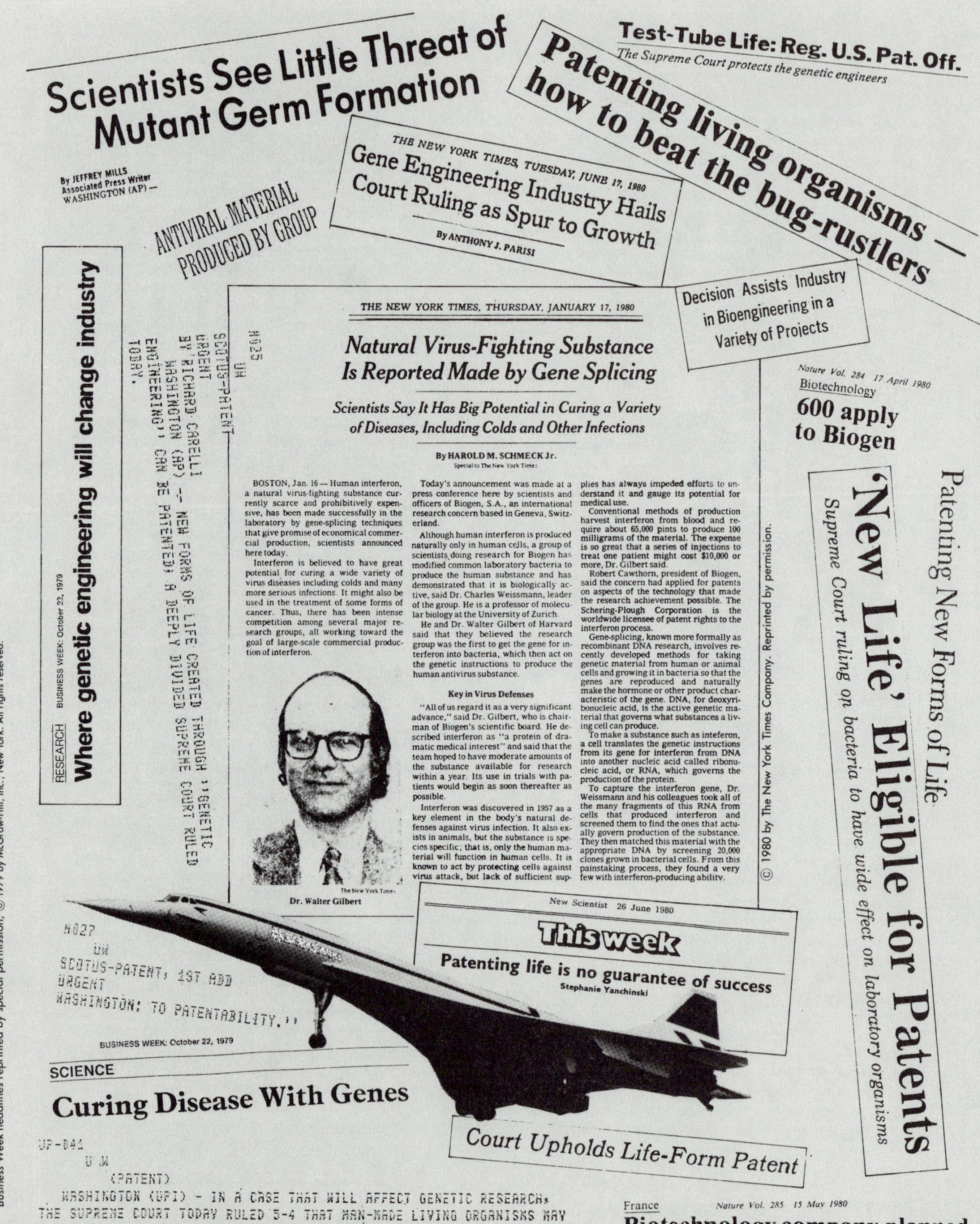

Scientists See Little Threat of Mutant Germ Formation

By JEFFREY MILLS
Associated Press Writer
WASHINGTON (AP) —

Test-Tube Life: Reg. U.S. Pat. Off.

The Supreme Court protects the genetic engineers

Patenting living organisms — how to beat the bug-rustlers

THE NEW YORK TIMES, TUESDAY, JUNE 17, 1980

Gene Engineering Industry Hails Court Ruling as Spur to Growth

By ANTHONY J. PARISI

ANTIVIRAL MATERIAL PRODUCED BY GROUP

Decision Assists Industry in Bioengineering in a Variety of Projects

RESEARCH BUSINESS WEEK: October 22, 1979

Where genetic engineering will change industry

H025
UH
SCOTUS-PATENT
URGENT
BY RICHARD CARELLI
WASHINGTON (AP) -- NEW FORMS OF LIFE CREATED THROUGH ''GENETIC ENGINEERING'' CAN BE PATENTED, A DEEPLY DIVIDED SUPREME COURT RULED TODAY.

THE NEW YORK TIMES, THURSDAY, JANUARY 17, 1980

Natural Virus-Fighting Substance Is Reported Made by Gene Splicing

Scientists Say It Has Big Potential in Curing a Variety of Diseases, Including Colds and Other Infections

By HAROLD M. SCHMECK Jr.
Special to The New York Times

BOSTON, Jan. 16 — Human interferon, a natural virus-fighting substance currently scarce and prohibitively expensive, has been made successfully in the laboratory by gene-splicing techniques that give promise of economical commercial production, scientists announced here today.

Interferon is believed to have great potential for curing a wide variety of virus diseases including colds and many more serious infections. It might also be used in the treatment of some forms of cancer. Thus, there has been intense competition among several major research groups, all working toward the goal of large-scale commercial production of interferon.

Today's announcement was made at a press conference here by scientists and officers of Biogen, S.A., an international research concern based in Geneva, Switzerland.

Although human interferon is produced naturally only in human cells, a group of scientists doing research for Biogen has modified common laboratory bacteria to produce the human substance and has demonstrated that it is biologically active, said Dr. Charles Weissmann, leader of the group. He is a professor of molecular biology at the University of Zurich.

He and Dr. Walter Gilbert of Harvard said that they believed the research group was the first to get the gene for interferon into bacteria, which then act on the genetic instructions to produce the human antivirus substance.

Key in Virus Defenses

"All of us regard it as a very significant advance," said Dr. Gilbert, who is chairman of Biogen's scientific board. He described interferon as "a protein of dramatic medical interest" and said that the team hoped to have moderate amounts of the substance available for research within a year. Its use in trials with patients would begin as soon thereafter as possible.

Interferon was discovered in 1957 as a key element in the body's natural defenses against virus infection. It also exists in animals, but the substance is species specific; that is, only the human material will function in human cells. It is known to act by protecting cells against virus attack, but lack of sufficient supplies has always impeded efforts to understand it and gauge its potential for medical use.

Conventional methods of production harvest interferon from blood and require about 65,000 pints to produce 100 milligrams of the material. The expense is so great that a series of injections to treat one patient might cost $10,000 or more, Dr. Gilbert said.

Robert Cawthorn, president of Biogen, said the concern had applied for patents on aspects of the technology that made the research achievement possible. The Schering-Plough Corporation is the worldwide licensee of patent rights to the interferon process.

Gene-splicing, known more formally as recombinant DNA research, involves recently developed methods for taking genetic material from human or animal cells and growing it in bacteria so that the genes are reproduced and naturally make the hormone or other product characteristic of the gene. DNA, for deoxyribonucleic acid, is the active genetic material that governs what substances a living cell can produce.

To make a substance such as inteferon, a cell translates the genetic instructions from its gene for interferon from DNA into another nucleic acid called ribonucleic acid, or RNA, which governs the production of the protein.

To capture the interferon gene, Dr. Weissmann and his colleagues took all of the many fragments of this RNA from cells that produced interferon and screened them to find the ones that actually govern production of the substance. They then matched this material with the appropriate DNA by screening 20,000 clones grown in bacterial cells. From this painstaking process, they found a very few with interferon-producing abilitv.

The New York Times

Dr. Walter Gilbert

Nature Vol. 284 17 April 1980
Biotechnology

600 apply to Biogen

Patenting New Forms of Life

'New Life' Eligible for Patents

Supreme Court ruling on bacteria to have wide effect on laboratory organisms

New Scientist 26 June 1980

This week

Patenting life is no guarantee of success

Stephanie Yanchinski

H027
UH
SCOTUS-PATENT, 1ST ADD
URGENT
WASHINGTON: TO PATENTABILITY.''

BUSINESS WEEK: October 22, 1979

SCIENCE

Curing Disease With Genes

Court Upholds Life-Form Patent

UP-041
U W
(PATENT)
WASHINGTON (UPI) - IN A CASE THAT WILL AFFECT GENETIC RESEARCH, THE SUPREME COURT TODAY RULED 5-4 THAT MAN-MADE LIVING ORGANISMS MAY BE PATENTED.

France Nature Vol. 285 15 May 1980

Biotechnology company planned

Document 16.4

Applied Genetic News Subscription Flyer, Excerpt

A NEW NEWSLETTER FOR A NEW INDUSTRY — SUBSCRIBE NOW!

SUBSCRIPTION FORM

Please enter my name as a Subscriber to **APPLIED GENETICS NEWS.** Introductory subscription rate is $125.00 for the year for 12 issues (published monthly).

☐ 1 year ($125.00) ☐ 2 years ($235.00) ☐ 3 years ($338.00)

☐ Check enclosed ☐ Please bill me

(For non-U.S. subscriptions, please add $12/year for Airmail).

Send to: **Business Communications Co., Inc.,** P.O. Box 2070C, Dept. AG1, Stamford, CT 06906 U.S.A.

Published by: BUSINESS COMMUNICATIONS CO., INC.

APPLIED GENETICS NEWS
A Monthly Newsletter — A BCC Inc. Publication — P. O. Box 2070C, Stamford, CT 06906 — 203/325-2208
Edited by Charles Joslin

Creating life or sustaining it —
No matter, individual and industry alike will feel its impact!

The Growth Business of the Future, BIOTECHNOLOGY has already attracted $150 million in venture capital, much of it by big companies such as Du Pont, Standard Oil of California, Hoffman-LaRoche, G. D. Searle, Schering-Plough, Lubrizol, Innoven (Monsanto and Emerson Electric), National Distillers, Koppers, General Electric, and Inco, Ltd. — most of them working with companies such as Biogen, Cetus, Genentech, and Genex. It's a big business on the move.

Alert personnel in a wide range of industries see markets worth over $130 billion a year where bacteria and enzymes modified by new evolving genetic engineering techniques can revolutionize industrial production technology. And this is only the beginning.

- The pharmaceutical industry, for one, is on the brink of a big roller coaster ride. Biological engineering techniques are expected to lead to more efficient manufacturing techniques and to cures of dozens, perhaps hundreds, of diseases. In the process of altering human life, the industry itself will become richer and, undoubtedly, extensively restructured.
- Other industries — chemicals, plastics, mining, energy, forest products, pollution control, food — could also be turned upside down by who gets what new biological technology first. **Applied Genetics** will cut across traditional industry boundry lines like nothing else has since money.
- Whoever you are, whatever you are doing, there is no doubt that you will be affected. New attitudes, developments, and contributions in biotechnology have already become essential for realistic short-range planning.

APPLIED GENETICS NEWS reveals the story as it's happening. It gives the news and puts it into context, tells what it means **now,** and what it could mean later. What's happening, who's doing what, and what it costs. Basic research, genetic production technology, patents, legal developments, funding and finance, key people, laboratories and companies, industrial applications, who is ahead in the medical race. And we tell where the cross-links are.

If you read any of the major or specialized magazines, from **Newsweek** to **American Metal Markets** (even your local newspaper or biology texts), you've seen stories on the new biological technology — occasional catch-up stories marking major breakthroughs.

If you read **APPLIED GENETICS NEWS,** you will know what is happening every month — see the breakthroughs as they appear and where they are headed.

Document 16.5

April 3, 1980

Nature (**284**:388)

United States

Inventorship dispute stalls DNA patent application

Is scientific history being rewritten to make research results more patentable? **David Dickson** reports

A PATENT application covering some of the basic techniques of genetic engineering is being held up by the US Patent Office in a dispute over whether the applicants can legally claim to be the sole inventors. The patent request has been filed jointly by Stanford University and the University of California, San Francisco, (UCSF) on behalf of Dr Stanley Cohen and Dr Herbert Boyer. It covers techniques developed at the two institutions in the early 1970s, which demonstrated for the first time the feasibility of replicating biologically-functional foreign genes introduced into a living organism.

Central to the patent application, first filed in 1974, is a paper published in the *Proceedings of the National Academy of Sciences* (**70**, 3240, 1973) in the preceding year, which describes the successful construction and replication of a plasmid capable of transferring antibiotic resistance into *Escherichia coli.*

In addition to Cohen and Boyer, the paper names two other authors Dr Annie Chang of Stanford and Dr Robert Helling, Associate Professor of Botany at the University of Michigan. According to the rules of the Patent Office, joint authorship means the two are considered co-inventors of the processes described.

Stanford and UCSF argue that although the research described in the paper and the patent application was built on techniques developed elsewhere, it was Cohen and Boyer who provided the creative input necessary to demonstrate their potential. They did this by conceiving and demonstrating how the techniques could be combined to create functional recombinant plasmids.

But Helling, who is not named in the patent application, disputes that his role was marginal. He has refused to sign a disclaimer, required by the Patent Office before it approves the application, agreeing that he was not an inventor of the processes described.

"I felt that we were all equal in this, and do not want to sign a letter saying that I was just another laboratory worker", he told *Nature* last week. "I was part and parcel of the whole thing; I don't feel that I should sign something that I do not believe is true."

Helling is not the only scientist who has refused to sign (although his refusal has been the most disruptive). Dr John Morrow, of Johns Hopkins University, the first author on a 1974 *PNAS* paper cited by Cohen and Boyer (to show how a bacterial plasmid could be used to reproduce *Xenopus* DNA), complains in particular about the secrecy which has covered the application. "I am not prepared to sign a disclaimer to a patent application that I have not seen," he says.

And there may be further disputes in store. In addition to the process patent, Stanford and UCSF are also asking protection for the plasmid developed, arguing that it was the development of this particular pSC101 plasmid that made the success of the techniques possible.

Meanwhile, patents are being granted on the processes used. Dr Roy Curtiss, of the University of Alabama, has recently been granted a patent for the techniques used to create the disabled χ1776 strain of *E. coli* (for which both process and product patents have already been issued in the UK).

The issue has raised at least two separate arguments. The first is on the morality of granting private licensing rights to the results of research carried out on public funds. As this is currently accepted practice, most scientists feel that Stanford and UCSF are justified in seeking to benefit from research which the two institutions pioneered — particularly since the two universities stress that the patents would remain freely available to anyone who wanted to use the techniques for research purposes. And both Cohen and Boyer have agreed that royalty proceeds should go to support research at the universities, and that neither will benefit directly.

A further difficulty arises from the clash between scientific conventions of co-authorship, and the legal implications of these conventions when it comes to claims of inventorship. The latter is a legal definition which courts have ruled is enshrined in the Constitution, but is seldom considered by scientists in determining whose name should go on a research paper. Patents can be granted for techniques whose details have been published less than a year previously, but where such prior publication has taken place, all those listed as co-authors are treated as *prima facie* co-inventors.

In the Stanford/UCSF case, all agree that the development of recombinant DNA methodologies relied heavily on discoveries made at several institutions, for example on work on restriction enzymes at Johns Hopkins University and by Paul Berg's group at Stanford, or on the functioning of ligases at the National Institutes of Health.

As Cohen puts it: "Scientific advances such as the one we have been involved in are in fact the result of multiple discoveries carried out by many individuals over a long period of time". The difficulty lies in evaluating the significance of any particular contribution.

Patent attorneys for the two universities argue that Boyer and Cohen provided the "major inspiration and direction" for the experiments described in the *PNAS* paper, and they have told Helling that, unless he signs the disclaimer — or they can convince the patent examiner by other means that he has no legal claim as an inventor — then they will seek a court order instructing the issuing of a patent, and 'deposing' him as an inventor. But a full-scale legal dispute is not a prospect that anyone fancies.

Legally-defined constraints imposed by commercial considerations are proving a problem for scientists concerned to retain the integrity of the research process.

"It is difficult to demand high standards of your students when you see some of the things that are going on", says Dr Mark Ptashne of Harvard University, who claims that colleagues are beginning to write papers and cite references. "In such a way that their patent applications are valid".

Dr Zsolt Harsanyi, who is currently heading a major study of the implications of recombinant DNA technology for the office of Technology Assessment, confirms that "in the past couple of months" scientists have begun to talk about the potential for rewriting scientific history as a "serious problem".

Some are now using this as an argument against permitting the patenting of research results at all. Others suggest that fields such as electronics and chemistry have faced such conflicts for years, and have learned to live with the consequences.

In the case of molecular biology, the learning process does not look like being a comfortable one. □

June 18, 1976
Rosenzweig Letter to Fredrickson

OFFICE OF THE VICE PRESIDENT
FOR PUBLIC AFFAIRS

STANFORD UNIVERSITY
STANFORD, CALIFORNIA 94305

June 18, 1976

Dr. Donald S. Frederickson, Director
National Institute of Health—Bethesda, Maryland

Dear Dr. Frederickson:

From Paul Berg and others I know that you are aware of the discussions taking place at Stanford now over the wisdom of proceeding(on behalf of Stanford and the University of California) with an application for patent protection for discoveries in the area of Recombinant DNA. As you know, we began to move in this direction with the knowledge and consent of NIH and NSF; as you also know, the whole matter of patent protection is now the subject of lively debate here. The purpose of this letter is to solicit your views.

As further background to what you already have, you might be interested in the enclosed memorandum in which I have attempted to summarize some of the major questions and address them in a way that makes sense to me, at least. When I wrote this, I was speaking for myself only, and I was not trying to articulate University policy. From reactions I have received, I would guess that most of the University's senior officers would agree with my conclusions (though there is dissent), as do many, though not all, faculty. One point on which there is substantial unanimity among the officers of the University is that, if this line of work is to be developed in a way that provides income to the holder of a patent, there is no institution or group that has a stronger claim to that income than Stanford and the University of California -- using as the standard for that judgment, the value of the money earned to the future progress of scientific research and education.

Let me emphasize that we do not yet have conclusions. We are proceeding with the necessary steps in the patent application process and we have had discussions with a prospective licensee. We have taken no irrevocable steps, but we are rapidly approaching the stage at which binding decisions will need to be taken. By that I mean not days, but perhaps a small number of months. Your contribution to our deliberations would be extremely valuable. I would especially welcome your views on the suggestions at the end of my memorandum, but I do not want to limit you to those matters, alone.

If you think it would be useful to discuss these questions at greater length in person, I would be happy to go East, or if you prefer, arrange a session here at Stanford or in San Francisco. The issues we are dealing with are complex, interesting and important, and the way they are resolved is likely to have a lasting effect on science and education. I think that is not too strong a statement. I hope you agree and that we will have the benefit of your views.

Encl.

Sincerely,

Robert M. Rosenzweig
Vice President for Public Affairs

cc: R.W. Lyman
W.F. Massy
C.Rich
P.Berg
S.Cohen

Document 16.6

June 27, 1980
Science (**208:**1445)

Court Says Lab-Made Life Can Be Patented

In a decision made by five votes to four, the U.S. Supreme Court ruled on 16 June that "a live, human-made micro-organism is patentable subject matter"—in other words, that forms of life can be patented if there is a man-made element to them.

The decision seems certain to open the way for the Patent and Trademark Office to begin processing the heap of patent applications based on recombinant DNA technology. The Patent Office has been delaying action on these applications—now numbering more than a hundred—while arguing in the courts that it did not have the power under existing law to grant patents on living things.

The Supreme Court's decision to the contrary is likely to provide a helpful boost to the fledgling genetic engineering industry. The boost is mostly psychological, however, because most companies will still rely on secrecy to protect their proprietary information. With bacteria being improved so rapidly, it is process patents, already available, that are important at present. Patenting microorganisms themselves may be more important in a decade or so when large investments are made in designing a particular bug.

The Supreme Court's ruling is significant because it resolves the long-standing issue of whether forms of life can be patented. The test case which forms the pretext for the decision is a patent application filed by General Electric as long ago as 7 June 1972. The application concerned a form of *Pseudomonas* bacterium genetically engineered to digest oil slicks, by Ananda Chakrabarty, now at the University of Illinois.

The bottom line of the Supreme Court's decision is that it does not matter whether something is living or not: Its patentability depends on whether it is a product of nature or man-made. Chakrabarty's bacterium, in the Supreme Court's view, "is the result of human ingenuity and research," and therefore is patentable. The Supreme Court's thinking therefore follows the reasoning of a lower court, the Court of Customs and Patents Appeals, which concluded after its review of the case: "In short, we think the fact that micro-organisms, as distinguished from chemical compounds, are alive, is a distinction without legal significance."

In the view of some observers, the Supreme Court has acted somewhat erratically on the issue of patenting life, without adding very much intellectual substance to the debate. The most interesting part of the fight has been that between the Patent Office and the Court of Customs and Patents Appeals, which has twice directed the Patent Office to grant patents on microorganisms (*Science*, 9 November 1979). The Supreme Court, when asked to arbitrate, at first ordered the Court of Customs and Patent appeals to reconsider its position in light of another Supreme Court patent ruling, known as Parker *v.* Flook, which concerned the patentability of methods of calculation.

The salient feature of the Flook decision was the Court's admonition to "proceed cautiously when we are asked to extend patent rights into areas wholly unforeseen by Congress." Since that was exactly the position of the Patent Office on the issue of patenting life, the Supreme Court seemed to be directing the lower court to reverse itself and decide in the Patent Office's favor.

The Court of Customs and Patent appeals, however, had other ideas, one of which may have been the recollection that only since 1966 have its decisions been subject to review by the Supreme Court, and even then through a train of somewhat accidental circumstances. "To conclude on the light Flook sheds on these cases," the appeals court dryly replied, "very simply . . . we find none."

With this judicial snub, the case returned to the Supreme Court, which this time had to bite the bullet. But instead of overruling the appeals court, as might have been expected from its earlier decision, the Supreme Court has rolled over and adopted both the appeals court's position and its reasoning. What has prompted this apparent change of mind?

Ananda Chakrabarty

One observer suggests that the court's ideas "evolved" as it became more familiar with the case. Another, more cynically, cites the turnover of law clerks since the previous term. The majority opinion, written by Chief Justice Burger, omits to explain why the court found Flook relevant in 1978 but not in 1980.

The point is taken up, however, in a terse minority opinion by Justice Brennan who, if *The Brethren* is to be believed, does not always hold his brother Burger's reasoning in high esteem. Noting the caution in Flook about extending patent rights beyond what Congress had foreseen, and the fact that Congress in 1930 and 1970 passed acts extending patent rights to plants but specifically excluding bacteria, Brennan observes in a withering footnote that "I should think the necessity for caution is that much greater when we are asked to extend patent rights into areas Congress has foreseen and considered but has not resolved." The majority's decision, in his view, "extends the patent system to cover living material even though Congress plainly has legislated in the belief that [present law] does not encompass living organisms. It is the role of Congress, not this Court, to broaden or narrow the reach of the patent laws."

Congress is expected to hold hearings on the subject and could always pass a new law.—NICHOLAS WADE

TEXT OF SUPREME COURT's JUNE 16 MICROORGANISM PATENT DECISION

NOTICE: This opinion is subject to formal revision before publication in the preliminary print of the United States Reports. Readers are requested to notify the Reporter of Decisions, Supreme Court of the United States, Washington, D.C. 20543, of any typographical or other formal errors, in order that corrections may be made before the preliminary print goes to press.

No. 79-136

Sidney A. Diamond, Commissioner of Patents and Trademarks, Petitioner, *v.* Ananda M. Chakrabarty et al. } On Writ of Certiorari to the United States Court of Customs and Patent Appeals.

[June 16, 1980]

MR. CHIEF JUSTICE BURGER delivered the opinion of the Court.

We granted certiorari to determine whether a live, human-made micro-organism is patentable subject matter under 35 U. S. C. § 101.

I

In 1972, respondent Chakrabarty, a microbiologist, filed a patent application, assigned to the General Electric Company. The application asserted 36 claims related to Chakrabarty's invention of "a bacterium from the genus *Pseudomonas* containing therein at least two stable energy-generating plasmids, each of said plasmids providing a separate hydrocarbon degradative pathway." [1] This human-made, genetically engi-

[1] Plasmids are hereditary units physically separate from the chromosomes of the cell. In prior research, Chakrabarty and an associate discovered that plasmids control the oil degradation abilities of certain bacteria. In particular, the two researchers discovered plasmids capable of degrading camphor and octane, two components of crude oil. In the work represented by the patent application at issue here, Chakrabarty discovered a process by which four different plasmids, capable of degrading four different oil components, could be transferred to and maintained stably in a single *Pseudomonas* bacteria, which itself has no capacity for degrading oil.

neered bacterium is capable of breaking down multiple components of crude oil. Because of this property, which is possessed by no naturally occurring bacteria, Chakrabarty's invention is believed to have significant value for the treatment of oil spills.[2]

Chakrabarty's patent claims were of three types: first, process claims for the method of producing the bacteria; second, claims for an inoculum comprised of a carrier material floating on water, such as straw, and the new bacteria; and third, claims to the bacteria themselves. The patent examiner allowed the claims falling into the first two categories, but rejected claims for the bacteria. His decision rested on two grounds: (1) that micro-organisms are "products of nature," and (2) that as living things they are not patentable subject matter under 35 U. S. C. § 101.

Chakrabarty appealed the rejection of these claims to the Patent Office Board of Appeals, and the Board affirmed the Examiner on the second ground.[3] Relying on the legislative history of the 1930 Plant Patent Act, in which Congress extended patent protection to certain asexually reproduced plants, the Board concluded that § 101 was not intended to cover living things such as these laboratory created micro-organisms.

The Court of Customs and Patent Appeals, by a divided vote, reversed on the authority of its prior decision in *In re Bergy*, 563 F. 2d 1031 (1978), which held that "the fact that

[2] At present, biological control of oil spills requires the use of a mixture of naturally occurring bacteria, each capable of degrading one component of the oil complex. In this way, oil is decomposed into simpler substances which can serve as food for aquatic life. However, for various reasons, only a portion of any such mixed culture survives to attack the oil spill. By breaking down multiple components of oil, Chakrabarty's micro-organism promises more efficient and rapid oil-spill control.

[3] The Board concluded that the new bacteria were not "products of nature," because *Pseudomonas* bacteria containing two or more different energy-generating plasmids are not naturally occurring.

micro-organisms . . . are alive . . . [is] without legal significance" for purposes of the patent law.[4] Subsequently, we granted the Government's petition for certiorari in *Bergy*, vacated the judgment, and remanded the case "for further consideration in light of *Parker* v. *Flook*, 437 U. S. 584." 438 U. S. 902 (1978). The Court of Customs and Patent Appeals then vacated its judgment in *Chakrabarty* and consolidated the case with *Bergy* for reconsideration. After re-examining both cases in the light of our holding in *Flook*, that court, with one dissent, reaffirmed its earlier judgments. — F. 2d — (1979).

The Government again sought certiorari, and we granted the writ as to both *Bergy* and *Chakrabarty*. — U. S. — (1979). Since then, *Bergy* has been dismissed as moot, — U. S. — (1980), leaving only *Chakrabarty* for decision.

II

The Constitution grants Congress broad power to legislate to "promote the Progress of Science and the useful Arts, by securing for limited times to authors and inventors the exclusive right to their respective writings and discoveries." Art. I, § 8. The patent laws promote this progress by offering inventors exclusive rights for a limited period as an incentive for their inventiveness and research efforts. *Kewanee Oil Co.* v. *Bicron Corp.*, 416 U. S. 470, 480–481 (1974); *Universal Oil Co.* v. *Globe Co.*, 322 U. S. 471, 484 (1944). The authority of Congress is exercised in the hope that "[t]he productive effort thereby fostered will have a positive effect on society through the introduction of new products and processes of manufacture into the economy, and the emanations by way of increased employment and better lives for our citizens.", *Kewanee*, *supra*, at 480.

The question before us in this case is a narrow one of statu-

[4] *Bergy* involved a patent application for a pure culture of the micro-organism *Streptomyces vellosus* found to be useful in the production of lincomycin, an antibiotic.

tory interpretation requiring us to construe 35 U. S. C. § 101, which provides:

> "Whoever invents or discovers any new and useful process, machine, manufacture, or composition of matter, or any new and useful improvement thereof, may obtain a patent therefor, subject to the conditions and requirements of this title."

Specifically, we must determine whether respondent's micro-organism constitutes a "manufacture" or "composition of matter" within the meaning of the statute.[5]

III

In cases of statutory construction we begin, of course, with the language of the statute. *Southeastern Community College* v. *Davis*, 442 U. S. 397, 405 (1979). And "unless otherwise defined, words will be interpreted as taking their ordinary, contemporary, common meaning." *Perrin* v. *United States*, — U. S. —, — (1979). We have also cautioned that courts "should not read into the patent laws limitations and conditions which the legislature has not expressed." *United States* v. *Dubilier Condenser Corp.*, 289 U. S. 178, 199 (1933).

Guided by these canons of construction, this Court has read the term "manufacture" in § 101 in accordance with its dictionary definition to mean "the production of articles for use from raw materials prepared by giving to these materials new forms, qualities, properties, or combinations whether by hand labor or by machinery." *American Fruit Growers, Inc.* v. *Brogdex Co.*, 283 U. S. 1, 11 (1931). Similarly, "composition of matter" has been construed consistent with its common usage to include "all compositions of two or more substances and . . . all composite articles, whether they be the results of

[5] This case does not involve the other "conditions and requirements" of the patent laws, such as novelty and nonobviousness. 35 U. S. C. §§ 102, 103.

chemical union, or of mechanical mixture, or whether they be gases, fluids, powders, or solids." *Shell Dev. Co.* v. *Watson,* 149 F. Supp. 279, 280 (DC 1957) (citing 1 A. Deller, Walker on Patents § 14, p. 55 (1st ed. 1937)). In choosing such expansive terms as "manufacture" and "composition of matter," modified by the comprehensive "any," Congress plainly contemplated that the patent laws would be given wide scope.

The relevant legislative history also supports a broad construction. The Patent Act of 1793, authored by Thomas Jefferson, defined statutory subject matter as "any new and useful art, machine, manufacture, or composition of matter, or any new or useful improvement [thereof]." Act of Feb. 21, 1793, ch. 11, § 1, 1 Stat. 318. The Act embodied Jefferson's philosophy that "ingenuity should receive a liberal encouragement." V Writings of Thomas Jefferson, at 75–76. See *Graham* v. *John Deere Co.,* 383 U. S. 1, 7–10 (1966). Subsequent patent statutes in 1836, 1870, and 1874 employed this same broad language. In 1952, when the patent laws were recodified, Congress replaced the word "art" with "process," but otherwise left Jefferson's language intact. The Committee Reports accompanying the 1952 act inform us that Congress intended statutory subject matter to "include anything under the sun that is made by man." S. Rep. No. 1979, 82d Cong., 2d Sess., 5 (1952); H. R. Rep. No. 1923, 82d Cong., 2d Sess., 6 (1952).[6]

This is not to suggest that § 101 has no limits or that it embraces every discovery. The laws of nature, physical phenomena, and abstract ideas have been held not patentable. See *Parker* v. *Flook,* 437 U. S. 584 (1978); *Gottschalk* v. *Benson,* 409 U. S. 63, 67 (1973); *Funk Seed Co.* v. *Kalo Co.,* 333

[6] This same language was employed by P. J. Federico, a principal draftsman of the 1952 recodification, in his testimony regarding that legislation: "[U]nder section 101 a person may have invented a machine or manufacture, which may include anything under the sun that is made by man. . . ." Hearings on H. R. 3760 before Subcommittee No. 3 of the House Committee on the Judiciary, 82d Cong., 1st Sess., 37 (1951).

U. S. 127, 130 (1948); *O'Reilly* v. *Morse,* 15 How. 61, 112–121 (1853); *Le Roy* v. *Tatham,* 14 How. 155, 175 (1852). Thus, a new mineral discovered in the earth or a new plant found in the wild is not patentable subject matter. Likewise, Einstein could not patent his celebrated law that $E=mc^2$; nor could Newton have patented the law of gravity. Such discoveries are "manifestations of . . . nature, free to all men and reserved exclusively to none." *Funk, supra,* at 130.

Judged in this light, respondent's micro-organism plainly qualifies as patentable subject matter. His claim is not to a hitherto unknown natural phenomenon, but to a nonnaturally occurring manufacture or composition of matter—a product of human ingenuity "having a distinctive name, character [and] use." *Hartranft* v. *Wiegmann,* 121 U. S. 609, 615 (1887). The point is underscored dramatically by comparison of the invention here with that in *Funk.* There, the patentee had discovered that there existed in nature certain species of root-nodule bacteria which did not exert a mutually inhibitive effect on each other. He used that discovery to produce a mixed culture capable of inoculating the seeds of leguminous plants. Concluding that the patentee had discovered "only some of the handiwork of nature," the Court ruled the product nonpatentable:

> "Each of the species of root-nodule bacteria contained in the package infects the same group of leguminous plants which it always infected. No species acquires a different use. The combination of the six species produces no new bacteria, no change in the six bacteria, and no enlargement of the range of their utility. Each species has the same effect it always had. The bacteria perform in their natural way. Their use in combination does not improve in any way their natural functioning. They serve the same ends nature originally provided and act quite independently of any effort by the patentee." 333 U. S., at 127.

Here, by contrast, the patentee has produced a new bacterium with markedly different characteristics from any found in nature and one having the potential for significant utility. His discovery is not nature's handiwork, but his own; accordingly it is patentable subject matter under § 101.

IV

Two contrary arguments are advanced, neither of which we find persuasive.

(A)

The Government's first argument rests on the enactment of the 1930 Plant Patent Act, which afforded patent protection to certain asexually reproduced plants, and the 1970 Plant Variety Protection Act, which authorized patents for certain sexually reproduced plants but excluded bacteria from its protection.[7] In the Government's view, the passage of these Acts evidences congressional understanding that the terms "manufacture" or "composition of matter" do not include living things; if they did, the Government argues, neither Act would have been necessary.

We reject this argument. Prior to 1930, two factors were thought to remove plants from patent protection. The first was the belief that plants, even those artificially bred, were products of nature for purposes of the patent law. This posi-

[7] The Plant Patent Act of 1930, 35 U. S. C. § 161, provides in relevant part:

"Whoever invents or discovers and asexually reproduces any distinct and new variety of plant, including cultivated sports, mutants, hybrids, and newly found seedlings, other than a tuber propogated plant or a plant found in an uncultivated state, may obtain a patent therefor. . . ."

The Plant Variety Protection Act of 1970, provides in relevant part:

"The breeder of any novel variety of sexually reproduced plant (other than fungi, bacteria, or first generation hybrids) who has so reproduced the variety, or his successor in interest, shall be entitled to plant variety protection therefor. . . ." 7 U. S. C. § 2402 (a).

See generally, 3 A. Deller, Walker on Patents, Chapter IX (2d ed. 1964); R. Allyn The First Plant Patents (1934)

tion appears to have derived from the decision of the Patent Office in *Ex parte Latimer*, 1889 C. D. 123, in which a patent claim for fiber found in the needle of the *Pinus australis* was rejected. The Commissioner reasoned that a contrary result would permit "patents [to] be obtained upon the trees of the forests and the plants of the earth, which of course would be unreasonable and impossible." *Id.*, at 126. The *Latimer* case, it seems, came to "set[] forth the general stand taken in these matters" that plants were natural products not subject to patent protection. H. Thorne, Relation of Patent Law to Natural Products, 6 J. Pat. Off. Soc. 23, 24 (1923).[8] The second obstacle to patent protection for plants was the fact that plants were thought not amenable to the "written description" requirement of the patent law. See 35 U. S. C. § 112. Because new plants may differ from old only in color or perfume, differentiation by written description was often impossible. See Hearings on H. R. 11372 before the House Committee on Patents, 71 Cong., 2d Sess., 4 (1930), p. 7 (memorandum of Patent Commissioner Robertson).

In enacting the Plant Patent Act, Congress addressed both of these concerns. It explained at length its belief that the work of the plant breeder "in aid of nature" was patentable invention. S. Rep. No. 315, 71st Cong., 2d Sess., 6–8 (1930); H. R. Rep. No. 1129, 71st Cong., 2d Sess., 7–9 (1930). And it relaxed the written description requirement in favor of "a description . . . as complete as is reasonably possible." 35 U. S. C. § 162. No Committee or Member of Congress, however, expressed the broader view, now urged by the Government, that the terms "manufacture" or "composition of mat-

[8] Writing three years after the passage of the 1930 Act, R. Cook, Editor of the Journal of Heredity, commented: "It is a little hard for plant men to understand why [Article I § 8] of the Constitution should not have been earlier construed to include the promotion of the art of plant breeding. The reason for this is probably to be found in the principle that natural products are not patentable." Florists Exchange and Horticultural Trade World, July 15, 1933, at 9.

ter" exclude living things. The sole support for that position in the legislative history of the 1930 Act is found in the conclusory statement of Secretary of Agriculture Hyde, in a letter to the Chairmen of the House and Senate committees considering the 1930 Act, that "the patent laws . . . at the present time are understood to cover only inventions or discoveries in the field of inanimate nature." See S. Rep. No. 315, *supra,* at Appendix A; H. R. Rep. No. 1129, *supra,* at Appendix A. Secretary Hyde's opinion, however, is not entitled to controlling weight. His views were solicited on the administration of the new law and not on the scope of patentable subject matter—an area beyond his competence. Moreover, there is language in the House and Senate Committee reports suggesting that to the extent Congress considered the matter it found the Secretary's dichotomy unpersuasive. The reports observe:

> "There is a clear and logical distinction *between the discovery of a new variety of plant and of certain inanimate things,* such, for example, as a new and useful natural mineral. The mineral is created wholly by nature unassisted by man. . . . On the other hand, a plant discovery resulting from cultivation is unique, isolated, and is not repeated by nature, nor can it be reproduced by nature unaided by man. . . ." S. Rep. No. 315, *supra,* at 6; H. R. Rep. No. 1129, *supra,* at 7 (emphasis added).

Congress thus recognized that the relevant distinction was not between living and inanimate things, but between products of nature, whether living or not, and human-made inventions. Here, respondent's micro-organism is the result of human ingenuity and research. Hence, the passage of the Plant Patent Act affords the Government no support.

Nor does the passage of the 1970 Plant Variety Protection Act support the Government's position. As the Government acknowledges, sexually reproduced plants were not included under the 1930 Act because new varieties could not be repro-

duced true-to-type through seedlings. Brief for United States 27, n. 31. By 1970, however, it was generally recognized that true-to-type reproduction was possible and that plant patent protection was therefore appropriate. The 1970 Act extended that protection. There is nothing in its language or history to suggest that it was enacted because § 101 did not include living things.

In particular, we find nothing in the exclusion of bacteria from plant variety protection to support the Government's position. See *supra,* at n. 7. The legislative history gives no reason for this exclusion. As the Court of Customs and Patent Appeals suggested, it may simply reflect congressional agreement with the result reached by that court in deciding *In re Arzberger,* 112 F. 2d 834 (1940), which held that bacteria were not plants for the purposes of the 1930 Act. Or it may reflect the fact that prior to 1970 the Patent Office had issued patents for bacteria under § 101.[9] In any event, absent some clear indication that Congress "focused on [the] issues . . . directly related to the one presently before the Court," *SEC* v. *Sloan,* 436 U. S. 103, 120–121 (1978), there is no basis for reading into its actions an intent to modify the plain meaning of the words found in § 101. See *TVA* v. *Hill,* 437 U. S. 153, 189–193 (1978); *United States* v. *Price,* 361 U. S. 304, 313 (1960).

(B)

The Government's second argument is that micro-organisms cannot qualify as patentable subject matter until Congress expressly authorizes such protection. Its position rests on the fact that genetic technology was unforeseen when Congress

[9] In 1873, the Patent Office granted Louis Pasteur a patent on "yeast, free from organic germs of disease, as an article of manufacture." And in 1967 and 1968, immediately prior to the passage of the Plant Variety Protection Act, that office granted two patents which, as the Government concedes, state claims for living micro-organisms. See Reply Brief of United States, at 3, and n. 2.

enacted § 101. From this it is argued that resolution of the patentability of inventions such as respondent's should be left to Congress. The legislative process, the Government argues, is best equipped to weigh the competing economic, social, and scientific considerations involved, and to determine whether living organisms produced by genetic engineering should receive patent protection. In support of this position, the Government relies on our recent holding in *Parker* v. *Flook*, 437 U. S. 584 (1978), and the statement that the judiciary "must proceed cautiously when . . . asked to extend patent rights into areas wholly unforeseen by Congress." *Id.*, at 596.

It is, of course, correct that Congress, not the courts, must define the limits of patentability; but it is equally true that once Congress has spoken it is "the province and duty of the judicial department to say what the law is." *Marbury* v. *Madison*, 1 Cranch 137, 177 (1803). Congress has performed its constitutional role in defining patentable subject matter in § 101; we perform ours in construing the language Congress has employed. In so doing, our obligation is to take statutes as we find them, guided, if ambiguity appears, by the legislative history and statutory purpose. Here, we perceive no ambiguity. The subject matter provisions of the patent law have been cast in broad terms to fulfill the constitutional and statutory goal of promoting "the Progress of Science and the useful Arts" with all that means for the social and economic benefits envisioned by Jefferson. Broad general language is not necessarily ambiguous when congressional objectives require broad terms.

Nothing in *Flook* is to the contrary. That case applied our prior precedents to determine that a "claim for an improved method of calculation, even when tied to a specific end use, is unpatentable subject matter under § 101." 437 U. S., at 595, n. 18. The Court carefully scrutinized the claim at issue to determine whether it was precluded from patent protection under "the principles underlying the prohibition against pat-

ents for 'ideas' or phenomena of nature." *Id.*, at 593. We have done that here. *Flook* did not announce a new principle that inventions in areas not contemplated by Congress when the patent laws were enacted are unpatentable *per se*.

To read that concept into *Flook* would frustrate the purposes of the patent law. This Court frequently has observed that a statute is not to be confined to the "particular application[s] . . . contemplated by the legislators." *Barr* v. *United States*, 324 U. S. 83, 90 (1945). Accord, *Browder* v. *United States*, 312 U. S. 335, 339 (1941); *Puerto Rico* v. *Shell Co.*, 302 U. S. 253, 257 (1937). This is especially true in the field of patent law. A rule that unanticipated inventions are without protection would conflict with the core concept of the patent law that anticipation undermines patentability. See *Graham* v. *John Deere Co.*, 383 U. S., at 12–17. Mr. Justice Douglas reminded that the inventions most benefiting mankind are those that "push back the frontiers of chemistry, physics, and the like." *A. & P. Tea Co.* v. *Supermarket Corp.*, 340 U. S. 147, 154 (1950) (concurring opinion). Congress employed broad general language in drafting § 101 precisely because such inventions are often unforeseeable.[10]

To buttress its argument, the Government, with the support of *amicus*, points to grave risks that may be generated by research endeavors such as respondent's. The briefs present a gruesome parade of horribles. Scientists, among them Nobel laureates, are quoted suggesting that genetic research may pose a serious threat to the human race, or, at the very least, that the dangers are far too substantial to permit such research to proceed apace at this time. We are told that genetic research and related technological developments may spread

[10] Even an abbreviated list of patented inventions underscores the point: telegraph (Morse, No. 1647); telephone (Bell, No. 174,465); electric lamp (Edison, No. 223,898); airplane (the Wrights; No. 821,393); transistor (Bardeen & Brattain, No. 2,524,035); neutronic reactor (Fermi & Szilard, No. 2,708,656); laser (Schawlow & Townes, No. 2,929,922). See generally Revolutionary Ideas, Patents & Progress in America, Office of Patents (1976).

pollution and disease, that it may result in a loss of genetic diversity, and that its practice may tend to depreciate the value of human life. These arguments are forcefully, even passionately presented; they remind us that, at times, human ingenuity seems unable to control fully the forces it creates—that, with Hamlet, it is sometimes better "to bear those ills we have than fly to others that we know not of."

It is argued that this Court should weigh these potential hazards in considering whether respondent's invention is patentable subject matter under § 101. We disagree. The grant or denial of patents on micro-organisms is not likely to put an end to genetic research or to its attendant risks. The large amount of research that has already occurred when no researcher had sure knowledge that patent protection would be available suggests that legislative or judicial fiat as to patentability will not deter the scientific mind from probing into the unknown any more than Canute could command the tides. Whether respondent's claims are patentable may determine whether research efforts are accelerated by the hope of reward or slowed by want of incentives, but that is all.

What is more important is that we are without competence to entertain these arguments—either to brush them aside as fantasies generated by fear of the unknown, or to act on them. The choice we are urged to make is a matter of high policy for resolution within the legislative process after the kind of investigation, examination, and study that legislative bodies can provide and courts cannot. That process involves the balancing of competing values and interests, which in our democratic system is the business of elected representatives. Whatever their validity, the contentions now pressed on us should be addressed to the political branches of the government, the Congress and the Executive, and not to the courts.[11]

[11] We are not to be understood as suggesting that the political branches have been laggard in the consideration of the problems related to genetic research and technology. They have already taken action. In 1976, for example, the National Institutes of Health released guidelines for NIH-

We have emphasized in the recent past that "[o]ur individual appraisal of the wisdom or unwisdom of a particular [legislative] course . . . is to be put aside in the process of interpreting a statute." *TVA* v. *Hill*, 437 U. S. 153, 194 (1978). Our task, rather, is the narrow one of determining what Congress meant by the words it used in the statute; once that is done our powers are exhausted. Congress is free to amend § 101 so as to exclude from patent protection organisms produced by genetic engineering. Compare 42 U. S. C. § 2181, exempting from patent protection inventions "useful solely in the utilization of special nuclear material or atomic energy in an atomic weapon." Or it may choose to craft a statute specifically designed for such living things. But, until Congress takes such action, this Court must construe the language of § 101 as it is. The language of that section fairly embraces respondent's invention.

Accordingly, the judgment of the Court of Customs and Patent Appeals is affirmed.

Affirmed.

sponsored genetic research which established conditions under which such research could be performed. 41 Fed. Reg. 27902. In 1978 those guidelines were revised and relaxed. 43 Fed. Reg. 60080, 60108, 60134. And committees of the Congress have held extensive hearings on these matters. See, *e. g.*, Hearings on genetic engineering before the Subcommittee on Health of the Senate Committee on Labor and Public Welfare, 94th Cong., 1st Sess. (1975); Hearings before the Subcommittee on Science, Technology, and Space of the Senate Committee on Commerce, Science, and Transportation, 95th Cong., 1st Sess. (1978); Hearings before the Subcommittee on Heatlh and the Environment of the House Committee on Interstate and Foreign Commerce, 95th Cong., 1st Sess. (1977).

SUPREME COURT OF THE UNITED STATES

No. 79–136

Sidney A. Diamond, Commissioner of Patents and Trademarks, Petitioner,
v.
Ananda M. Chakrabarty et al.

On Writ of Certiorari to the United States Court of Customs and Patent Appeals.

[June 16, 1980]

MR. JUSTICE BRENNAN, with whom MR. JUSTICE WHITE, MR. JUSTICE MARSHALL, and MR. JUSTICE POWELL join, dissenting.

I agree with the Court that the question before us is a narrow one. Neither the future of scientific research, nor even the ability of respondent Chakrabarty to reap some monopoly profits from his pioneering work, is at stake. Patents on the processes by which he has produced and employed the new living organism are not contested. The only question we need decide is whether Congress, exercising its authority under Art. I, § 8, of the Constitution, intended that he be able to secure a monopoly on the living organism itself, no matter how produced or how used. Because I believe the Court has misread the applicable legislation, I dissent.

The patent laws attempt to reconcile this Nation's deep-seated antipathy to monopolies with the need to encourage progress. *Deepsouth Packing Co.* v. *Laitram Corp.*, 406 U. S. 518, 530–531 (1972); *Graham* v. *John Deere Co.*, 383 U. S. 1, 7–10 (1966). Given the complexity and legislative nature of this delicate task, we must be careful to extend patent protection no further than Congress has provided. In particular, were there an absence of legislative direction, the courts should leave to Congress the decisions whether and how far to extend the patent privilege into areas where the common understand-

ing has been that patents are not available.[1] Cf. *Deepsouth Packing Co.* v. *Laitram Corp.*, *supra.*

In this case, however, we do not confront a complete legislative vacuum. The sweeping language of the Patent Act of 1793, as re-enacted in 1952, is not the last pronouncement Congress has made in this area. In 1930 Congress enacted the Plant Patent Act affording patent protection to developers of certain asexually reproduced plants. In 1970 Congress enacted the Plant Variety Protection Act to extend protection to certain new plant varieties capable of sexual reproduction. Thus, we are not dealing—as the Court would have it—with the routine problem of "unanticipated inventions." *Ante*, at 12. In these two Acts Congress has addressed the general problem of patenting animate inventions and has chosen carefully limited language granting protection to some kinds of discoveries, but specifically excluding others. These Acts strongly evidence a congressional limitation that excludes bacteria from patentability.[2]

First, the Acts evidence Congress' understanding, at least since 1930, that § 101 does not include living organisms. If newly developed living organisms not naturally occurring had been patentable under § 101, the plants included in the scope

[1] I read the Court to admit that the popular conception, even among advocates of agricultural patents, was that living organisms were unpatentable. See *ante*, at 7–8, and n. 8.

[2] But even if I agreed with the Court that the 1930 and 1970 Acts were not dispositive, I would dissent. This case presents even more cogent reasons than *Deepsouth Packing Co.* not to extend the patent monopoly in the face of uncertainty. At the very least, these Acts are signs of legislative attention to the problems of patenting living organisms, but they give no affirmative indication of congressional intent that bacteria be patentable. The caveat of *Parker* v. *Flook*, 437 U. S. 584, 596 (1978), an admonition to "proceed cautiously when we are asked to extend patent rights into areas wholly unforeseen by Congress," therefore becomes pertinent. I should think the necessity for caution is that much greater when we are asked to extend patent rights into areas Congress has foreseen and considered but has not resolved.

of the 1930 and 1970 Acts could have been patented without new legislation. Those plants, like the bacteria involved in this case, were new varieties not naturally occurring.[3] Although the Court, *ante*, at 7, rejects this line of argument, it does not explain why the Acts were necessary unless to correct a pre-existing situation.[4] I cannot share the Court's implicit assumption that Congress was engaged in either idle exercises or mere correction of the public record when it enacted the 1930 and 1970 Acts. And Congress certainly thought it was doing something significant. The committee reports contain expansive prose about the previously unavailable benefits to be derived from extending patent protection to plants.[5] H. R.

[3] The Court refers to the logic employed by Congress in choosing not to perpetuate the "dichotomy" suggested by Secretary Hyde. *Ante*, at 9. But by this logic the bacteria at issue here are distinguishable from a "mineral . . . created wholly by nature" in exactly the same way as were the new varieties of plants. If a new act was needed to provide patent protection for the plants, it was equally necessary for bacteria. Yet Congress provided for patents on plants but not on these bacteria. In short, Congress decided to make only a subset of animate "human-made inventions," *ibid.*, patentable.

[4] If the 1930 Act's only purpose were to solve the technical problem of description referred to by the Court, *ante*, at 8, most of the Act, and in particular its limitation to asexually reproduced plants, would have been totally unnecessary.

[5] Secretary Hyde's letter was not the only explicit indication in the legislative history of these Acts that Congress was acting on the assumption that legislation was necessary to make living organisms patentable. The Senate Judiciary Committee Report on the 1970 Act states the Committee's understanding that patent protection extended no further than the explicit provisions of these Acts:

"Under the patent law, patent protection is limited to those varieties of plants which reproduce asexually, that is, by such methods as grafting or budding. No protection is available to those varieties of plants which reproduce sexually, that is, by seeds. S. Rep. No. 91-1246, 91st Cong., 2d Sess., 3 (1970).

Similarly, Representative Poage, speaking for the 1970 Act, after noting the protection accorded asexually developed plants, stated that "for plants

Rep. No. 91-1605, 91st Cong., 2d Sess., 1-3 (1970); S. Rep. No. 315, 71st Cong., 2d Sess., 1-3 (1930). Because Congress thought it had to legislate in order to make agricultural "human-made inventions" patentable and because the legislation Congress enacted is limited, it follows that Congress never meant to make patentable items outside the scope of the legislation.

Second, the 1970 Act clearly indicates that Congress has included bacteria within the focus of its legislative concern, but not within the scope of patent protection. Congress specifically excluded bacteria from the coverage of the 1970 Act. 7 U. S. C. § 2402 (a). The Court's attempts to supply explanations for this explicit exclusion ring hollow. It is true that there is no mention in the legislative history of the exclusion, but that does not give us license to invent reasons. The fact is that Congress, assuming that animate objects as to which it had not specifically legislated could not be patented, excluded bacteria from the set of patentable organisms.

The Court protests that its holding today is dictated by the broad language of § 101, which "cannot be confined to the 'particular application[s] . . . contemplated by the legislators.'" *Ante*, at 12, quoting *Barr* v. *United States*, 324 U. S. 83, 90 (1945). But as I have shown, the Court's decision does not follow the unavoidable implications of the statute. Rather, it extends the patent system to cover living material even though Congress plainly has legislated in the belief that § 101 does not encompass living organisms. It is the role of Congress, not this Court, to broaden or narrow the reach of the patent laws. This is especially true where, as here, the composition sought to be patented uniquely implicates matters of public concern.

produced from seed, there has been no such protection." 122 Cong. Rec. 40295 (1970).

Document 16.8

July 16, 1980
The Blue Sheet (**23:**1)

WALLACE WERBLE, SR., FOUNDER

The Blue Sheet®

DRUG RESEARCH REPORTS

TRADEMARKS REG. U.S. PAT. OFFICE

Founded 1957 – $230.00 a year

A WEEKLY SPECIALIZED PUBLICATION PROVIDING IN-DEPTH, INTERPRETATIVE COVERAGE OF GOVT. POLICIES, FUNDING & ACTIVITIES IN DRUG & MEDICAL RESEARCH (NIH/FDA), HEALTH MANPOWER & HEALTH CARE DELIVERY SYSTEMS & PLANS

ONE NATIONAL PRESS BUILDING, WASHINGTON, D.C. 20045 – PHONE (202) 624-7600

Vol. 23, No. 29 THE NEWS THIS WEEK July 16, 1980

GENETIC ENGINEERING MAY BE ADDRESSED BY ETHICS PANEL The President's Commission for the Study of Ethical Problems in Medicine and Biomedical and Behavioral Research instructed its staff July 12 to conduct a preliminary survey of ethical issues involved in the applications of genetic engineering. Results of the survey will be presented to the commission Sept. 15-16, which will then decide whether to formally consider the issue.

Commission interest in genetic engineering was prompted by a letter from the Natl. Council of the Churches of Christ, which urged a reexamination of patent laws in response to the Supreme Court's July 16 ruling that man-made organisms are patentable ("The Blue Sheet" June 25, p. 13-14 and June 18, p. 4).

The church group did not specifically ask the commission to tackle the ethical issues of genetic engineering. Instead, "we intend to request that President Carter provide a way for representatives of a broad spectrum of our society to consider these matters and advise the govt. on its necessary role," the letter said. "In the long-term interest of all humanity, our govt. must launch a thorough examination of the entire spectrum of issues involved in genetic engineering to determine before it is too late what oversight and controls are necessary."

Congress should "begin immediately" to review and revise patent laws "to deal with the new questions related to patenting life forms," and the UN should "evolve internatl. guidelines related to genetic engineering," the letter advised. "We pledge our own efforts to examine the religious and ethical issues involved in genetic engineering."

Questions the council want answered include: "Who shall determine how human good is best served when new life forms are being engineered? Who shall control genetic experimentation and its results which could have untold implications for human survival? Who will benefit and who will bear any adverse consequences, directly or indirectly?"

Neither the staff nor the commission were very enthusiastic about tackling the issue. Staff director Alexander Capron, a *Penn.* Law professor, said the staff "isn't itching to take on another issue" because "we have enough responsibilities already."

Commissioner Arno Motulsky, a *U. Wash.* geneticist, questioned whether "ethical problems exist in the human application" of genetic engineering. "My first impression is negative to conducting a study," he said. Motulsky claimed that the public and press have been "needlessly worried" over the potential of genetic engineering to create monstrosities. "Changing germinal cells would be a problem, but I presently don't see any way of doing that now," he said.

However, Capron emphasized that "this is a field which is moving with incredible rapidity." Current genetic engineering research advances were unthinkable years ago, and future research that might alter human growth cells would pose immediate ethical problems, he said. "Changing the gene that produces diabetes isn't the same as giving someone insulin," Capron added.

Document 16.9

June 2, 1980

Pope John Paul II UNESCO Speech, Excerpt

The Pope's UNESCO Address

The World As An Environment For Humanity

We are well aware, ladies and gentlemen, that the future of man and mankind is threatened, radically threatened, in spite of very noble intentions, by men of science. And it is menaced because the tremendous results of their research work and their discoveries, especially regarding natural science, have been and continue to be exploited — to the prejudice of ethical imperatives — to ends which having nothing to do with the prerequisites of science, but with the ends of destruction and death to a degree never before attained, causing unimaginable ravages. Although science is called to be a service to men's lives, only too often do we see that it is enslaved by these destructive goals, destructive of the true dignity of man and human life. Such is the case when scientific research is itself directed toward these goals or when its results are applied to ends contradictory to those of humanity. This can be verified as well in the realm of genetic manipulations and biological experiments as well as in those of chemical, bacteriological or nuclear armaments.

Document 16.10

December 2, 1980
Boyer— Cohen Patent

United States Patent [19] [11] **4,237,224**
Cohen et al. [45] **Dec. 2, 1980**

[54] **PROCESS FOR PRODUCING BIOLOGICALLY FUNCTIONAL MOLECULAR CHIMERAS**

[75] Inventors: **Stanley N. Cohen**, Portola Valley; **Herbert W. Boyer**, Mill Valley, both of Calif.

[73] Assignee: **Board of Trustees of the Leland Stanford Jr. University**, Stanford, Calif.

[21] Appl. No.: **1,021**

[22] Filed: **Jan. 4, 1979**

Related U.S. Application Data

[63] Continuation-in-part of Ser. No. 959,288, Nov. 9, 1978, which is a continuation-in-part of Ser. No. 687,430, May 17, 1976, abandoned, which is a continuation-in-part of Ser. No. 520,691, Nov. 4, 1974.

[51] **Int. Cl.**[3] **C12P 21/00**
[52] **U.S. Cl.** .... 435/68; 435/172; 435/231; 435/183; 435/317; 435/849; 435/820; 435/91; 435/207; 260/112.5 S; 260/27R; 435/212
[58] **Field of Search** .... 195/1, 28 N, 28 R, 112, 195/78, 79; 435/68, 172, 231, 183

[56] **References Cited**

U.S. PATENT DOCUMENTS

3,813,316 5/1974 Chakrabarty .... 195/28 R

OTHER PUBLICATIONS

Morrow et al., Proc. Nat. Acad. Sci. USA, vol. 69, pp. 3365–3369, Nov. 1972.
Morrow et al., Proc. Nat. Acad. Sci. USA, vol. 71, pp. 1743–1747, May 1974.
Hershfield et al., Proc. Nat. Acad. Sci. USA, vol. 71, pp. 3455 et seq. (1974).
Jackson et al., Proc. Nat. Acad. Sci. USA, vol. 69, pp. 2904–2909, Oct. 1972.
Mertz et al., Proc. Nat. Acad. Sci. USA, vol. 69, pp. 3370–3374, Nov. 1972.
Cohen, et al., Proc. Nat. Acad. Sci. USA, vol. 70, pp. 1293–1297, May 1973.
Cohen et al., Proc. Nat. Acad. Sci. USA, vol. 70, pp. 3240–3244, Nov. 1973.
Chang et al., Proc. Nat. Acad. Sci, USA, vol. 71, pp. 1030–1034, Apr. 1974.
Ullrich et al., Science vol. 196, pp. 1313–1319, Jun. 1977.
Singer et al., Science vol. 181, p. 1114 (1973).
Itakura et al., Science vol. 198, pp. 1056–1063 Dec. 1977.
Komaroff et al., Proc. Nat. Acad. Sci. USA, vol. 75, pp. 3727–3731, Aug. 1978.
Chemical and Engineering News, p. 4, May 30, 1977.
Chemical and Engineering News, p. 6, Sep. 11, 1978.

Primary Examiner—Alvin E. Tanenholtz
Attorney, Agent, or Firm—Bertram I. Rowland

[57] **ABSTRACT**

Method and compositions are provided for replication and expression of exogenous genes in microorganisms. Plasmids or virus DNA are cleaved to provide linear DNA having ligatable termini to which is inserted a gene having complementary termini, to provide a biologically functional replicon with a desired phenotypical property. The replicon is inserted into a microorganism cell by transformation. Isolation of the transformants provides cells for replication and expression of the DNA molecules present in the modified plasmid. The method provides a convenient and efficient way to introduce genetic capability into microorganisms for the production of nucleic acids and proteins, such as medically or commercially useful enzymes, which may have direct usefulness, or may find expression in the production of drugs, such as hormones, antibiotics, or the like, fixation of nitrogen, fermentation, utilization of specific feedstocks, or the like.

14 Claims, No Drawings

1

PROCESS FOR PRODUCING BIOLOGICALLY FUNCTIONAL MOLECULAR CHIMERAS

The invention was supported by generous grants of NIH, NSF and the American Cancer Society.

CROSS-REFERENCE TO RELATED APPLICATIONS

This application is a continuatin-in-part of applicatin Ser. No. 959,288, filed Nov. 9, 1978, which is a continuation of application Ser. No. 687,430 filed May 17, 1976, now abandoned, which was a continuation-in-part of application Ser. No. 520,691, filed Nov. 4, 1974, now abandoned.

BACKGROUND OF THE INVENTION

1. Field of the Invention

Although transfer of plasmids among strains of *E. coli* and other Enterobacteriaceae has long been accomplished by conjugation and/or transduction, it has not been previously possible to selectively introduce particular species of plasmid DNA into these bacterial hosts or other microorganisms. Since microorganisms that have been transformed with plasmid DNA contain autonomously replicating extrachromosomal DNA species having the genetic and molecular characteristics of the parent plasmid, transformation has enabled the selective cloning and amplification of particular plasmid genes.

The ability of genes derived from totally different biological classes to replicate and be expressed in a particular microorganism permits the attainment of interspecies genetic recombination. Thus, it becomes practical to introduce into a particular microorganism, genes specifying such metabolic or synthetic functions as nitrogen fixation, photosynthesis, antibiotic production, hormone synthesis, protein synthesis, e.g. enzymes or antibodies, or the like—functions which are indigenous to other classes of organisms—by linking the foreign genes to a particular plasmid or viral replicon.

BRIEF DESCRIPTION OF THE PRIOR ART

References which relate to the subject invention are Cohen, et al., Proc. Nat. Acad, Sci., USA, 69, 2110 (1972); ibid, 70, 1293 (1973); ibid, 70, 3240 (1973); ibid, 71, 1030 (1974); Morrow, et al., Proc. Nat. Acad. Sci., 71, 1743 (1974); Novick, Bacteriological Rev., 33, 210 (1969); and Hershfeld, et al., Proc. Nat. Acad. Sci., in press; Jackson, et al., ibid, 69, 2904 (1972);

SUMMARY OF THE INVENTION

Methods and compositions are provided for genetically transforming microorganisms, particularly bacteria, to provide diverse genotypical capability and producing recombinant plasmids. A plasmid or viral DNA is modified to form a linear segment having ligatable termini which is joined to DNA having at least one intact gene and complementary ligatable termini. The termini are then bound together to form a "hybrid" plasmid molecule which is used to transform susceptible and compatible microorganisms. After transformation, the cells are grown and the transformants harvested. The newly functionalized microorganisms may then be employed to carry out their new function; for example, by producing proteins which are the desired end product, or metabolities of enzymic conversion, or be lysed and the desired nucleic acids or proteins recovered.

2

DESCRIPTION OF THE SPECIFIC EMBODIMENTS

The process of this invention employs novel plasmids, which are formed by inserting DNAhaving one or more intact genes into a plasmid in such a location as to permit retention of an intact replicator locus and system (replicon) to provide a recombinant plasmid molecule. The recombinant plasmid molecule will be referred to as a "hybrid" plasmid or plasmid "chimera." The plasmid chimera contains genes that are capable of expressing at least one phenotypical property. The plasmmid chimera is used to transform a susceptible and competent microorganism under conditions where transformation occurs. The microorganism is then grown under conditions which allow for separation and harvesting of transformants that contain the plasmid chimera.

The process of this invention will be divided into the following stages:

I. preparation of the recombinant plasmid or plasmid chimera;

II. transformation or preparation of transformants; and

III. replication and transcription of the recombinant plasmid in transformed bacteria.

Preparation of Plasmid Chimera

In order to prepare the plasmid chimera, it is necessary to have a DNA vector, such as a plasmid or phage, which can be cleaved to provide an intact replicator locus and system (replicon), where the linear segment has ligatable termini or is capable of being modified to introduce ligatable termini. Of particular interest are those plasmids which have a phenotypical property, which allow for ready separation of transformants from the parent microorganism. The plasmid will be capable of replicating in a microorganism, particularly a bacterium which is susceptible to transformation. Various unicellular microorganisms can be transformed, such as bacteria, fungii and algae. That is, those unicellular organisms which are capable of being grown in cultures of fermentation. Since bacteria are for the most part the most convenient organisms to work with, bacteria will be hereinafter referred to as exemplary of the other unicellular organisms. Bacteria, which are susceptible to transformation, include members of the Enterobacteriaceae, such as strains of *Escherichia coli;* Salmonella; Bacillaceae, such as *Bacillus subtilis;* Pneumococcus; Streptococcus, and *Haemophilus influenzae.*

A wide variety of plasmids may be employed of greatly varying molecular weight. Normally, the plasmids employed will have molecular weights in the range of about 1×10^6 to 50×10^6d, more usually from about 1 to 20×10^6d, and preferably, from about 1 to 10×10^6d. The desirable plasmid size is determined by a number of factors. First, the plasmid must be able to accommodate a replicator locus and one or more genes that are capable of allowing replication of the plasmid. Secondly, the plasmid should be of a size which provides for a reasonable probability of recircularization with the foreign gene(s) to form the recombinant plasmid chimera. Desirably, a restriction enzyme should be available, which will cleave the plasmid without inactivating the replicator locus and system associated with the replicator locus. Also, means must be provided for providing ligatable termini for the plasmid, which are complementary to the termini of the foreign gene(s) to allow fusion of the two DNA segments.

Another consideration for the recombinant plasmid is that it be compatible with the bacterium to be transformed. Therefore, the original plasmid will usually be derived from a member of the family to which the bacterium belongs.

The original plasmid should desirably have a phenotypical property which allows for the separation of transformant bacteria from parent bacteria. Particularly useful is a gene, which provides for survival selection. Survival selection can be achieved by providing resistance to a growth inhibiting substance or providing a growth factor capability to a bacterium deficient in such capability.

Conveniently, genes are available, which provide for antibiotic or heavy metal resistance or polypeptide resistance, e.g. colicin. Therefore, by growing the bacteria on a medium containing a bacteriostatic or bacteriocidal substance, such as an antibiotic, only the transformants having the antibiotic resistance will survive. Illustrative antibiotics include tetracycline, streptomycin, sulfa drugs, such as sulfonamide, kanamycin, neomycin, penicillin, chloramphenicol, or the like.

Growth factors include the synthesis of amino acids, the isomerization of substrates to forms which can be metabolized or the like. By growing the bacteria on a medium which lacks the appropriate growth factor, only the bacteria which have been transformed and have the growth factor capability will clone.

One plasmid of interest derived from *E. coli* is referred to as pSC101 and is described in Cohen, et al., Proc. Nat. Acad. Sci., USA, 70, 1293 (1972), (referred to in that article as Tc6-5). Further description of this particular plasmid and its use is found in the other articles previously referred to.

The plasmid pSC101 has a molecular weight of about 5.8×10^6d and provides tetracycline resistance.

Another plasmid of interest is colicinogenic factor EI (ColE1), which has a molecular weight of 4.2×10^6d, and is also derived from *E. coli.* The plasmid has a single EcoRI substrate site and provides immunity to colicin E1.

In preparing the plasmid for joining with the exogenous gene, a wide variety of techniques can be provided, including the formation of or introduction of cohesive termini. Flush ends can be joined. Alternatively, the plasmid and gene may be cleaved in such a manner that the two chains are cleaved at different sites to leave extensions at each end which serve as cohesive termini. Cohesive termini may also be introduced by removing nucleic acids from the opposite ends of the two chains or alternatively, introducing nucleic acids at opposite ends of the two chains.

To illustrate, a plasmid can be cleaved with a restriction endonuclease or other DNA cleaving enzyme. The restriction enzyme can provide square ends, which are then modified to provide cohesive termini or can cleave in a staggered manner at different, but adjacent, sites on the two strands, so as to provide cohesive termini directly.

Where square ends are formed such as, for example, by HIN (Haemophilus influenzae RII) or pancreatic DNAse, one can ligate the square ends or alternatively one can modify the square ends by chewing back, adding particular nucleic acids, or a combination of the two. For example, one can employ appropriate transferases to add a nucleic acid to the 5' and 3' ends of the DNA. Alternatively, one can chew back with an enzyme, such as a λ-exonuclease, and it is found that there is a high probability that cohesive termini will be achieved in this manner.

An alternative way to achieve a linear segment of the plasmid with cohesive termini is to employ an endonuclease such as EcoRI. The endonuclease cleaves the two strands at different adjacent sites providing cohesive termini directly.

With flush ended molecules, a T_4 ligase may be employed for linking the termini. See, for example, Scaramella and Khorana, J. Mol. Biol. 72: 427–444 (1972) and Scaramella, DNAS 69: 3389 (1972), whose disclosure is incorporated herein by reference.

Another way to provide ligatable termini is to leave employing DNAse and Mn^{++} as reported by Lai and Nathans, J. Mol. Biol, 89: 179 (1975).

The plasmid, which has the replicator locus, and serves as the vehicle for introduction of a foreign gene into the bacterial cell, will hereafter be referred to as "the plasmid vehicle."

It is not necessary to use plasmid, but any molecule capable of replication in bacteria can be employed. Therefore, instead of plasmid, viruses may be employed, which will be treated in substantially the same manner as the plasmid, to provide the ligatable termini for joining to the foreign gene.

If production of cohesive termini is by restriction endonuclease cleavage, the DNA containing the foreign gene(s) to be bound to the plasmid vehicle will be cleaved in the same manner as the plasmid vehicle. If the cohesive termini are produced by a different technique, an analogous technique will normally be employed with the foreign gene. (By foreign gene is intended a gene derived from a source other than the transformant strain.) In this way, the foreign gene(s) will have ligatable termini, so as to be able to covalently bonded to the termini of the plasmid vehicle. One can carry out the cleavage or digest of the plasmids together in the same medium or separately, combine the plasmids and recircularize the plasmids to form the plasmid chimera in the absence of active restriction enzyme capable of cleaving the plasmids.

Descriptions of methods of cleavage with restriction enzymes may be found in the following articles: Greene, et al., *Methods in Molecular Biology,* Vol. 9, ed. Wickner, R. B., (Marcel Dekker, Inc., New York), "DNA Replication and Biosynthesis"; Mertz and Davis, 69, Proc. Nat. Acad. Sci., USA, 69, 3370 (1972);

The cleavage and non-covalent joining of the plasmid vehicle and the foreign DNA can be readily carried out with a restriction endonuclease, with the plasmid vehicle and foreign DNA in the same or different vessels. Depending on the number of fragments, which are obtained from the DNA endonuclease digestion, as well as the genetic properties of the various fragments, digestion of the foreign DNA may be carried out separately and the fragments separated by centrifugation in an appropriate gradient. Where the desired DNA fragment has a phenotypical property, which allows for the ready isolation of its transformant, a separation step can usually be avoided.

Endonuclease digestion will normally be carried out at moderate temperatures, normally in the range of 10° to 40° C. in an appropriately buffered aqueous medium, usually at a pH of about 6.5 to 8.5. Weight percent of total DNA in the reaction mixture will generally be about 1 to 20 weight percent. Time for the reaction will vary, generaly being from 0.1 to 2 hours. The amount of endonuclease employed is normally in excess of that required, normally being from about 1 to 5 units per 10 μg of DNA.

Where cleavage into a plurality of DNA fragments results, the course of the reaction can be readily followed by electrophoresis. Once the digestion has gone to the desired degree, the endonuclease is inactivated by heating above about 60° C. for five minutes. The digestion mixture may be worked up by dialysis, gradient separation, or the like, or used directly.

After preparation of the two double stranded DNA sequences, the foreign gene and vector are combined for annealing and/or ligation to provide for a functional recombinant DNA structure. With plasmids, the annealing involves the hydrogen bonding together of the cohesive ends of the vector and the foreign gene to form a circular plasmid which has cleavage sites. The cleavage sites are then normally ligated to form the complete closed and circularized plasmid.

The annealing, and as appropriate, recircularization can be performed in whole or in part in vitro or in vivo. Preferably, the annealing is performed in vitro. The annealing requires an appropriate buffered medium containing the DNA fragments. The temperature employed initially for annealing will be about 40° to 70° C., followed by a period at lower temperature, generaly from about 10° to 30° C. The molar ratio of the two segments will generally be in the range of about 1–5:-5–1. The particular temperature for annealing will depend upon the binding strength of the cohesive termi. While 0.5 hr to 2 or more days may be employed for annealing, it is believed that a period of 0.5 to 6 hrs may be sufficient. The time employed for the annealing will vary with the temperature employed, the nature of the salt solution, as well as the nature of the cohesive termini.

The ligation, when in vitro, can be achieved in conventional ways employing DNA ligase. Ligation is conveniently carried out in an aqueous solution (pH 6–8) at temperatures in the range of about 5° to 40° C. The concentration of the DNA will generally be from about 10 to 100 g/ml. A sufficient amount of the DNA ligase or other ligating agent e.g. T_4 ligase, is employed to provide a convenient rate of reaction, generally ranging from about 5 to 50 U/ml. A small amount of a protein e.g. albumin, may be added at concentrations of about 10 to 200 g/ml. The ligation with DNA ligase is carried out in the presence of magnesium at about 1–10 mM.

At the completion of the annealing or ligation, the solution may be chilled and is ready for use in transformation.

It is not necessary to ligate the recircularized plasmid prior to transformation, since it is found that this function can be performed by the bacterial host. However, in some situations ligation prior to transformation may be desirable.

The foreign DNA can be derived from a wide variety of sources. The DNA may be derived from eukaryotic or prokaryotic cells, viruses, and bacteriophage. The fragments employed will generally have molecular weights in the range of about 0.5 to 20×10^6d, usually in the range of 1 to 10×10^6d. The DNA fragment may include one or more genes or one or more operons.

Desirably, if the plasmid vehicle does not have a phenotypical property which allows for isolation of the transformants, the foreign DNA fragment should have such property. Also, an intact promoter and base sequences coding for initiation and termination sites should be present for gene expression.

In accordance with the subject invention, plasmids may be prepared which have replicons and genes which could be present in bacteria as a result of normal mating of bacteria. However, the subject invention provides a technique, whereby a replicon and gene can coexist in a plasmid, which is capable of being introduced into a unicellular organism, which could not exist in nature. The first type of plasmid which cannot exist in nature is a plasmid which derives its replicon from one organism and the exogenous gene from another organism, where the two organisms do not exchange genetic information. In this situation, the two organisms will either be eukaryotic or prokaryotic. Those organisms which are able to exchange genetic information by mating are well known. Thus, prior to this invention, plasmids having a replicon and one or more genes from two sources which do not exchange genetic information would not have existed in nature. This is true, even in the event of mutations, and induced combinations of genes from different strains of the same species. For the natural formation of plasmids formed from a replicon and genes from different microorganisms it is necessary that the microorganisms be capable of mating and exchanging genetic information.

In the situation, where the replicon comes from a eukaryotic or prokaryotic cell, and at least one gene comes from the other type of cell, this plasmid heretofore could not have existed in nature. Thus, the subject invention provides new plasmids which cannot naturally occur and can be used for transformation of unicellular organisms to introduce genes from other unicellular organisms, where the replicon and gene could not previously naturally coexist in a plasmid.

Besides naturally occurring genes, it is feasible to provide synthetic genes, where fragments of DNA may be joined by various techniques known in the art. Thus, the exogenous gene may be obtained from natural sources or from synthetic sources.

The plasmid chimera contains a replicon which is compatible with a bacterium susceptible of transformation and at least one foreign gene which is directly or indirectly bonded through deoxynucleotides to the replicon to form the circularized plasmid structure. As indicated previously, the foreign gene normally provides a phenotypical property, which is absent in the parent bacterium. The foreign gene may come from another bacterial strain, species or family, or from a plant or animal cell. The original plasmid chimera will have been formed by in vitro covalent bonding between the replicon and foreign gene. Once the originally formed plasmid chimera has been used to prepare transformants, the plasmid chimera will be replicated by the bacterial cell and cloned in vivo by growing the bacteria in an appropriate growth medium. The bacterial cells may be lysed and the DNA isolated by conventional means or the bacteria continually reproduced and allowed to express the genotypical property of the foreign DNA.

Once a bacterium has been transformed, it is no longer necessary to repeat the in vitro preparation of the plasmid chimera or isolate the plasmid chimera from the transformant progeny. Bacterial cells can be repeatedly multiplied which will express the genotypical property of the foreign gene.

One method of distinguishing between a plasmid which originates in vivo from a plasmid chimera which originates in vitro is the formation of homoduplexes between an in vitro prepared plasmid chimera and the plasmid formed in vivo. It will be an extremely rare event where a plasmid which originates in vivo will be the same as a plasmid chimera and will form homoduplexes with plasmid chimeras. For a discussion of homoduplexes, see Sharp, Cohen and Davidson, J. Mol. Biol., 75, 235 (1973), and Sharp, et al, ibid, 71, 471 (1972).

The plasmid derived from molecular cloning need not homoduplex with the in vitro plasmid originally employed for transformation of the bacterium. The bacterium may carry out modification processes, which will not affect the portion of the replicon introduced which is necessary for replication nor the portion of the exogenous DNA which contains the gene providing the genotypical trait. Thus, nucleotides may be introduced or excised and, in accordance with naturally occurring mating and transduction, additional genes may be introduced. In addition, for one or more reasons, the plasmids may be modified in vitro by techniques which are known in the art. However, the plasmids obtained by molecular cloning will homoduplex as to those parts which relate to the original replicon and the exogenous gene.

II. Transformation

After the recombinant plasmid or plasmid chimera has been prepared, it may then be used for the transformation of bacteria. It should be noted that the annealing and ligation process not only results in the formation of the recombinant plasmid, but also in the recircularization of the plasmid vehicle. Therefore, a mixture is obtained of the original plasmid, the recombinant plasmid, and the foreign DNA. Only the original plasmid and the DNA chimera consisting of the plasmid vehicle and linked foreign DNA will normally be capable of replication. When the mixture is employed for transformation of the bacteria, replication of both the plasmid vehicle genotype and the foreign genotype will occur with both genotypes being replicated in those cells having the recombinant plasmid.

Various techniques exist for transformation of a bacterial cell with plasmid DNA. A technique, which is particularly useful with *Escherichia coli*, is described in Cohen, et al., ibid, 69, 2110 (1972). The bacterial cells are grown in an appropriate medium to a predetermined optical density. For example, with *E. coli strain C600*, the optical density was 0.85 at 590 nm. The cells are concentrated by chilling, sedimentation and washing with a dilute salt solution. After centrifugation, the cells are resuspended in a calcium chloride solution at reduced temperatures (approx. 5°–15° C.), sedimented, resuspended in a smaller volume of a calcium chloride solution and the cells combined with the DNA in an appropriately buffered calcium chloride solution and incubated at reduced temperatures. The concentration of Ca^{++} will generally be about 0.01 to 0.1 M. After a sufficient incubation period, generally from about 0.5–3.0 hours, the bacteria are subjected to a heat pulse generally in the range of 35° to 45° C. for a short period of time; namely from about 0.5 to 5 minutes. The transformed cells are then chilled and may be transferred to a growth medium, whereby the transformed cells having the foreign genotype may be isolated.

An alternative transformation technique may be found in Lederberg and Cohen, I. Bacteriol., 119, 1072 (1974), whose disclosure is incorporated herein by reference.

III. Replication and Transcription of the Plasmid

The bacterial cells, which are employed, will be of such species as to allow replication of the plasmid vehicle. A number of different bacteria which can be employed, have been indicated previously. Strains which lack indigenous modification and restriction enzymes are particularly desirable for the cloning of DNA derived from foreign sources.

The transformation of the bacterial cells will result in a mixture of bacterial cells, the dominant proportion of which will not be transformed. Of the fraction of cells which are transformed, some significant proportion, but normally a minor proportion, will have been transformed by recombinant plasmid. Therefore, only a very small fraction of the total number of cells which are present will have the desired phenotypical characteristics.

In order to enhance the ability to separate the desired bacterial clones, the bacterial cells, which have beeen subjected to transformation, will first be grown in a solution medium, so as to amplify the absolute number of the desired cells. The bacterial cells may then be harvested and streaked on an appropriate agar medium. Where the recombinant plasmid has a phenotype, which allows for ready separation of the transformed cells from the parent cells, this will aid in the ready separation of the two types of cells. As previously indicated, where the genotype provides resistance to a growth inhibiting material, such as an antibiotic or heavy metal, the cells can be grown on an agar medium containing the growth inhibiting substance. Only available cells having the resistant genotype will survive. If the foreign gene does not provide a phenotypical property, which allows for distinction between the cells transformed by the plasmid vehicle and the cells transformed by the plasmid chimera, a further step is necessary to isolate the replicated plasmid chimera from the replicated plasmid vehicle. The steps include lysing of the cells and isolation and separation of the DNA by conventional means or random selection of transformed bacteria and characterization of DNA from such transformants to determine which cells contain molecular chimeras. This is accomplished by physically characterizing the DNA by electrophoresis, gradient centrifugation or electron microscopy.

Cells from various clones may be harvested and the plasmid DNA isolated from these transformants. The plasmid DNA may then be analyzed in a variety of ways. One way is to treat the plasmid with an appropriate restriction enzyme and analyze the resulting fragments for the presence of the foreign gene. Other techniques have been indicated above.

Once the recombinant plasmid has been replicated in a cell and isolated, the cells may be grown and multiplied and the recombinant plasmid employed for transformation of the same or different bacterial strain.

The subject process provides a technique for introducing into a bacterial strain a foreign capability which is genetically mediated. A wide variety of genes may be employed as the foreign genes from a wide variety of sources. Any intact gene may be employed which can be bonded to the plasmid vehicle. The source of the gene can be other bacterial cells, mammalian cells, plant cells, etc. The process is generally applicable to bacterial cells capable of transformation. A plasmid must be available, which can be cleaved to provide a linear segment having ligatable termini, and an interact replicator locus and system, preferably a system including a gene which provides a phenotypical property which allows for easy separation of the transformants. The linear segment may then be annealed with a linear segment of DNA having one or more genes and the resulting recombinant plasmid employed for transformation of the bacteria.

By introducing one or more exogeneous genes into a unicellular organism, the organism will be able to produce polypeptides and proteins ("poly(amino acids)") which the organism could not previously produce. In some instances the poly(amino acids) will have utility in themselves, while in other situations, particularly with enzymes, the enzymatic product(s) will either be useful in itself or useful to produce a desirable product.

One group of poly(amino acids) which are directly useful are hormones. Illustrative hormones include parathyroid hormone, growth hormone, gonadotropins (FSH, luteinizing hormone, chorionogonadatropin, and glycoproteins), insulin, ACTH, somatostatin, prolactin, placental lactogen, melanocyte stimulating hormone, thyrotropin, parathyroid hormone, calcitonin, enkephalin, and angiotensin.

Other poly(amino acids) of interest include serum proteins, fibrinogin, prothrombin, thromboplastin, globulin e.g. gamma-globulins or antibodies, heparin, antihemophilia protein, oxytocin, albumins, actin, myosin, hemoglobin, ferritin, cytochrome, myoglobin, lactoglobulin, histones, avidin, thyroglobulin, interferin, kinins and transcortin.

Where the genes or genes produce one or more enzymes, the enzymes may be used for fulfilling a wide variety of functions. Included in these functions are nitrogen fixation, production of amino acids, e.g. polyiodothyronine, particularly thyroxine, vitamins, both water and fat soluble vitamins, antimicrobial drugs, chemotheropeutic agents e.g. antitumor drugs, polypeptides and proteins e.g. enzymes from apoenzymes and hormones from prohormones, diagnostic reagents, energy producing combinations e.g. photosynthesis and hydrogen production, prostaglandins, steroids, cardiac glycosides, coenzymes, and the like.

The enzymes may be individually useful as agents separate from the cell for commercial applications, e.g. in detergents, synthetic transformations, diagnostic agents and the like. Enzymes are classified by the I.U.B. under the classifications as I. Oxidoreductases; II. Transferases; III. Hydrolases; IV. Lyases; V. Isomerases; and VI. Ligases.

EXPERIMENTAL

In order to demonstrate the subject invention, the following experiments were carried out with a variety of foreign genes.

(All temperatures not otherwise indicated are Centrigrade. All percents not otherwise indicated are percents by weight.)

EXAMPLE A

A. Preparation of pSC101 Plasmid

Covalently closed R6-5 DNA was sheared with a Virtis stainless steel microshaft in a one milliliter cup. The R6-5 DNA was sheared at 2,000 r.p.m. for 30 minutes in TEN buffer solution (0.02 M Tris-HCl (pH 8.0)-1 mM EDTA (pH 8.0)-0.02 M NaCl), while chilled at 0°-4°.

The sheared DNA sample was subjected to sucrose gradient sedimentation at 39,500 r.p.m. in a Spinco SW 50.1 rotor at 20°. A 0.12 mil fraction was collected on a 2.3 cm diameter circle of Whatman No. 3 filter paper, dried for 20 minutes and precipitated by immersion of the disc in cold 5% trichloroacetic acid, containing 100 μg/ml thymidine. The precipitate was filtered and then washed once with 5% trichloroacetic acid, twice with 99% ethanol and dried. pSC101 was the 27S species having a calculated molecular weight of 5.8×10^6 d.

B. Generalized Transformation Procedure

E. coli strain C600 was grown at 37° in H1 medium to an optical density of 0.85 at 590 nm. At this point the cells were chilled quickly, sedimented and washed once in 0.5 volume 10 nM NaCl. After centrifugation, the bacteria was resuspended in half the original volume of chilled 0.03 M calcium chloride, kept at 0° for 20 minutes, sedimented, and then resuspended in 0.1 of the original volume of 0.03 M of calcium chloride solution. Chilled DNA samples in TEN buffer were supplemented with 0.1 M calcium chloride to a final concentration of 0.03 M.

0.2 ml of competent cells treated with calcium chloride was added to 0.1 ml of DNA solution with chilled pipets and an additional incubation was done for 60 minutes at 0°. The bacteria were then subjected to a heat pulse at 42° for two minutes, chilled, and then either placed directly onto nutrient agar containing appropriate antibiotics or, where indicated, diluted 10 times in L-broth and incubated at 37° before plating. The cell survival is greater than 50% after calcium chloride treatment and heat pulse. Drug resistance was assayed on nutrient agar plates with the antibiotics indicated in specific experiments.

EXAMPLE I: Construction of Biologically Functional Bacterial Plasmids in vitro

A. Covalently closed R6-5 plasmid DNA was cleaved by incubation at 37° for 15 minutes in a 0.2 ml reaction mixture containing DNA (40 μg/ml, 100 mM Tris.HCl (pH 7.4)), 5 mM $MgCl_2$, 50 mM NaCl, and excess (2 U) EcoRI endonuclease in 1 μl volume. An additional incubation at 60° for 5 minutes was employed to inactivate the endonuclease.

The resulting mixture of plasmid fragments was employed for transformation of *E. coli* strain C600 in accordance with the procedure previously described. A single clone was examined further which was selected for resistance to kanamycin and was also found to carry resistance to neomycin and sulfonamide, but not to tetracycline, chloramphenicol, or streptomycin after transformation of *E. coli* by EcoRI generated DNA fragments of R6-5. Closed circular DNA obtakined from this isolate (plasmid designation pSC102) by CsCl-ethidium bromide gradient centrifugation had an S value of 39.5 in neutral surcrose gradients.

Treatment of pSC102 plasmid DNA with EcoRI resistriction endonuclease in accordance with the above-described procedure resulted in the formation of 3 fragments that were separable by electrophoresis in agarose gels. Intact pSC102 plasmid DNA and pSC101 plasmid DNA, which had been separately purified by dye-buoyant density centrifugation, were treated with EcoRI endonuclease followed by annealing at 0°-2° for about six hours. The mixture was then subjected to ligation with pSC101 and pSC102 in a ratio of 1:1 respectively, by ligating for 6 hours at 14° in 0.2 ml reaction mixtures containing 5 mM $MgCl_2$, 0.1 mM NAD, 100 μg/ml of bovine-serum albumin (BSA), 10 mM ammonium sulphate (pH 7.0), and 18 U/ml of DNA ligase. (J. Mertz and Davis, Proc. Nat. Acad. Sci., USA, 69, 3370 (1972); and Modrich, et al., J. Biol. Chem., 248, 7495 (1973). Ligated mixtures were incubated at 37° for 5 minutes and then chilled in ice water. Aliquots containing 3.3–6.5 μg/ml of total DNA were used directly for transformation.

Transformation of *E. coli* strain C600 was carried out as previously described. For comparison purposes, transformation was also carried out with a mixture of pSC101 and pSC102 plasmid DNA, which had been subjected to EcoRI endonuclease, but not DNA ligase. The antibiotics used for selection were tetracycline (10 μg/ml) and kanamycin (25 μg/ml). The results are reported as transformants per microgram of DNA. The following table indicates the results.

TABLE I

Transformation of *E. coli* C600 by a mixture of pSC101 and pSC102 DNA

Treatment of DNA	Transformation frequency for antibiotic resistence markers: Tetracycline	Kanamycin	Tetracycline + kanamycin
None	2×10^5	1×10^5	2×10^2
EcoRI	1×10^4	1.1×10^3	7×10^1
EcoRI + DNA ligase	1.2×10^4	1.3×10^3	5.7×10^2

Kanamycin resistance in the R65 plasmid is a result of the presence of the enzyme kanamycin monophosphotransferase. The enzyme can be isolated from the bacteria by known procedures and employed in an assay for kanamycin in accordance with the procedure described in Smith, et al., New England J. Medicine, 286, 583 (1972).

In the preparation for the enzyme extracts, the E. coli are grown in ML-broth and harvested in a late logarithm phase of growth. The cells are osmotically shocked (see Nossal, et al., J. Biol. Chem. 241, 3055 (1966), washed twice at room temperature with 10 ml 0.01 M Tris and 0.03 M NaCl, pH 7.3, and the pellet suspended in 10 ml 20% sucrose, 3×10^3 M EDTA and 0.033 M Tris (pH 7.5), stirred for 10 minues at room temperature and centrifuged at 16,000 g for 5 minutes. The pellet is then suspended in 2 ml of cold 5×10^{-4} M MCl_2, stirred for 10 minutes at 2° and centrifuged at 26,000 g for 10 minutes to yield a supernatant fluid referred to as the osmotic shockate. The solution should be stored at −20° or lower. (See Benveneste, et al., FEBS Leters, 14 293 (1971).

The osmotic shockate may then be used in accordance with the procedure of Smith, et al., supra.

EXAMPLE II: Genome Construction between Bacterial Species in vitro: Replication and Expression of Staphylococcus Plasmid Genes in *E. coli*

S. aureus strain 8325 contains the plasmid pI258, which expresses resistance to penicillin, erythromycin, cadmium and mercury. (Lindberg, et al., J. Bacteriol., 115, 139 (1973)). Covalently closed circular pSC101 and pI258 plasmid DNA were separately cleaved by incubation at 37° for 15 minutes in 0.2 ml reaction mixtures by EcoRI endonuclease in accordance with the procedure described previously. Aliquots of the two cleaved species were mixed in a ratio of 3 μg of pI258:1 μg of pSC101 and annealed at 2°–4° for 48 hours. Subsequent ligation was carried out for six hours at 14° as described previously and aliquots containing 3.3–6.5 μg/ml of total DNA were used directly in the transformation as described previously.

Other transformations were carried out employing the two plasmids independently and a mixture of the two plasmids. Selection of transformants was carried out at antibiotic concentrations for tetracycline (Tc, 25 μg/ml) or pencillin (Pc, 25 OU/ml). The transformation was carried out with *E. coli* strain C600 $r_K^- m_K^-$. The following table indicates the results.

TABLE III

Transformation of C600 $r_K^- m_K^-$ by pSC101 and pI258 Plasmid DNA

DNA	Transformants/μg DNA: Tc	Pc
PSC101 closed circular	1×10^6	<3
pI258 closed circular	<3.6	<3.6
pSC101 + pI258 untreated	9.1×10^5	<5
pSC101 + pI258 EcoRI-treated	4.7×10^3	10

The above table demonstrates that bacteria can be formed which have both tetracycline resistance and penicillin resistance. Thus, one can provide the phenotypical property penicillin resistance in bacteria from DNA, which is indigenous to another biological organism. One can thus use *E. coli* for the production of the enzyme, which imparts penicillin resistance to bacteria, and assay for penicillin in a manner similar to that employed for kanamycin. Penicillinase is used for destroying penicillin in blood serum of patients treated with penicillin in order to determine whether pathogenic organisms whose growth is inhibited by penicillin may be present.

EXAMPLE III: Replication and Transcription of Eukaryotic DNA in *E. coli*

The amplified ribosomal DNA (rDNA) codeing for 18S and 28S ribsomal RNA of the South African toad, *Xenopus laevis* was used as a source of eukaryotic DNA for these experiments. Dawid, et al., J. Mol. Biol., 51, 341 (1970). E. coli-X. laevis recombinant plasmids were constructed in vitro as follows:

The reaction mixture (60 μl) contained 100 mM Tris.HCl (pH 7.5) 50 mM NaCl, 5 mM $MgCl_2$, 1.0 μg of pSC101 plasmid DNA and 2.5 μg of *X. laevis* rDNA, and excess EcoRI restriction endonuclease (1 μl, 2 U). After a 15 minute incubation at 37°, the reaction mixture was placed at 63° for 5 minutes to inactivate EcoRI endonuclease. The product was then refrigerated at 0.5° for 24 hours, to allow association of the short cohesive termini.

The reaction mixture for ligation of phosphodiester bonds was adjusted to a total volume of 100 μl and contained in addition to the components of the endonuclease reaction, 30 mM Tris.HCl (pH 8.1), 1 mM sodium EDTA, 5 mM $MgCl_2$, 3.2 nM NAD, 10 mM ammonium sulphate, 5 μg BSA, and 9 U of *E. coli* DNA ligase. All components were chilled to 5° before their addition to the reaction mixture. The ligase reaction mixture was incubated at 14° for 45 minutes, and then at 0.5° for 48 hours. Additional NAD and ligase were added and the mixture incubated at 15° for 30 minutes and then for 15 minutes at 37°. The ligated DNA was used directly in the plasmid transformation procedure previously described. The DNA was used to transform *E. coli* strain C600 $r_K^- m_K^-$ and tetracycline resistant transformants ($3.3 \times 10^3/\mu g$ of pSC101 DNA) were selected and numbered consecutively CD1, CD2, etc. Plasmid DNA was isolated from a number of the transformants.

^{32}P-labeled 18 S and 28 S *X. laevis* rRNA were hybridized with DNA obtained from the plasmids CD4, CD18, CD30, and CD42. CD4 DNA annealed almost equally with both the 18 S and 28 S rRNA species. CD18 plasmid DNA hybridized principally with 28 S *X. laevis* rRNA, while the DNA of plasmids CD30 and CD42 annealed primarily with 18 S rRNA. These data indicate that portions of the *X. laevis* rDNA were, in fact, incorporated into a plasmid recombinant with pSC101, which was capable of transforming *E. coli*, so as to be capable of replicating X. laevis rDNA.

Transcription of *X. laevis* DNA was also carried out in *E. coli* minicells. The minicell producing *E. coli* strain P678-54 was transformed with plasmid DNA isolated from *E. coli* strain C600 $r_K^- m_K^-$ containing CD4, CD18, or CD42. Many cells containing the plasmids were isolated and incubated with [^3H] uridine; RNA purified from such minicells was hybridized with *X. laevis* rDNA immobilized on nitrocellulose membranes in order to determine whether the *X. laevis* rDNA linked to the pSC101 replicon is transcribed in *E. coli*. The results in the following table show that RNA species capable of annealing with purified *X. laevis* rDNA are synthesized in *E. coli* minicells carrying the recombinant plasmids, CD4, CD18, and CD42, but not by minicells carrying the pSC101 plasmid alone.

Minicells containing plasmids were isolated as described by Cohen, et al., Nature New Biol., 231, 249 (1971). They were incubated with [^3H] uridine (50 μCi/ml, 30 Ci/mol) as described by Roozen, et al., J. Bacteriol., 107, 21 (1971) for 10 minutes at 37°. Minicells collected by centrifugation were resuspended in Tris.HCl (20 mM, pH 7.5)-5 mM $MgCl_2$-1 mM EDTA pH 8.0 and rapidly frozen and thawed 3 times. RNA was extracted as described in Cohen, et al., J. Mol. Biol., 37, 387 (1968). Hybridization assays were carried out in nitrocellulose membranes as described in Cohen, et al., ibid, at saturating levels of pSC101 DNA. Hybridizations involving *X. laevis* DNA were not performed at DNA excess. Counts bound to blank filters (5-10 c.p.m.) were substracted from experimentally determined values. ^3H count eluted from filters containing *X. laevis* DNA were rendered acid soluble by ribonuclease A 20 μg/ml, 0.30 M NaCl-0.030 M sodium citrate, 1 hour, 37°. The following table indicates the results.

TABLE III

Plasmid carried by minicells	[^3H] RNA synthesized by *E. coli* minicells: Input cpm	[^3H] RNA counts hybridized to *X. laevis* rDNA: 0.2μg	[^3H] RNA counts hybridized to *X. laevis* rDNA: 0.4μg	[^3H] RNA counts hybridized to pSC101 DNA 18μg
CD42	4810	905 (19%)	1436 (30%)	961 (20%)
CD18	3780	389 (10%)	—	1277 (34%)
CD4	5220	789 (15%)	—	1015 (19%)
pSC101	4170	0 (0%)	—	1500 (36%)

EXAMPLE IV: Plasmid ColEl as a Molecular Vehicle for Cloning and Amplification of Trp Operon

In a volume of 200 μl (100 mM Tris.HCl (pH 7.5)-5 mM $MgCl_2$-50 mM NaCl), 5.7 μg of ColEl (*E. coli* JC411Thy$^-$/ColEl) (Clewell, et al., Proc. Nat. Acad. Sci., USA, 62, 1159 (1969) and 6.0 μg DNA from bacteriophage ϕ80pt190 (Deeb, et al., Virology, 31, 289 (1967) were digested to completion with homogeneously purified EcoRI endonuclease, monitoring the digestion by electrophoresis of the fragments in an agarose gel. The endonuclease was inactivated by heating at 65° for 5 minutes, the digest dialyzed overnight against 5 mM Tris.HCl, pH 7.5, and the sample concentrated to 50 μl. The fragments were ligated as described in Dugaiczyk, et al., Biochemistry, 13, 503 (1974) at a concentration of 75 pmoles/ml of fragments.

Transformation was carried out as previously described except that the cells were grown to $A_{590} = 0.600$ and following exposure to DNA were incubated in L-broth for 90 minutes. The cells were collected and resuspended in 10 mM NaCl before plating. Cells employed as recipients for the transformations were *E. coli* strains C600 trpR', ΔtrpE5(MV1), C600 trpR$^-$ trpE 10220 recA(MV2), C600 ΔtrpE5(MV10) and C600 ΔtrpE5 recA(MV12). (trpR$^-$ is the structural gene for the trp repressor and ΔtrpE5 is a trp operon deletion entirely within trpE and removing most of the gene.) Approximately 2 μg of the DNA was used to transform the cells.

Cultures were plated on Vogel-Bonner agar supplemented with 50 μg/ml of the non-selective amino acids, 0.2% glucose and 5 μg/ml of required vitamins. Transformants to colicin immunity were initially selected on a lawn of a culture of a mutant strain carrying ColE1. Clones were then selected for their ability to grow in the absence of tryptophan. Cells capable of producing tryptophan were isolated, which could be used for the production of exogenous tryptophan. The subject example demonstrates the introduction of a complete operon from foreign DNA to provide a transformant capable of replicating the operon and transcribing and translating to produce enzymes capable of producing an aromatic amino acid.

EX. V: Cloning of Synthetic Somatostatin Gene

The deoxyribonucleotide sequence for the somatostatin gene was prepared in accordance with conventional procedures. (Itakura et al, Science, 198 1056 (1977)). To prepare the recombinant plasmid, plasmid pBR 322 was digested with Eco RI. The reaction was terminated by extraction with a mixture of phenol and chloroform, the DNA precipitated with ethanol and resuspended in 50 μl of T_4 DNA polymerase buffer. The reaction was started by the addition of 2 units of T_4 DNA polymerase. After 30 min at 37°, the mixture was extracted with phenol and chloroform and the DNA precipitated with ethanol. The λplac5 DNA (3 μg) was digested with the endonuclease Hae III and the digested pBR 322 DNA blunt end ligated with the Hae III-digested λplac5 DNA in a final volume of 30 μl with T_4 DNA ligase (hydroxylopatite fraction) in 20 mM tris-HCl (pH 7.6), 10 mM $MgCl_2$, 10 mM dithiothreitol and 0.5 mM ATP for 12 hrs at 12°. The ligated DNA mixture was dialyzed against 10 mM tris-HCl (pH 7.6) and used to transform *E. coli* strain RR1. Transformants were selected for tetracycline resistance and ampicillin resistance on antibiotic (20 μg/ml) X-gal (40 μg/ml) medium. Colonies constitutive for the synthesis of β-galactosidase were identified by their blue color and of 45 colonies so identified, 3 of them were found to contain plasmids with 2 Eco R1 sites separated by ~200 base pairs.

The plasmid so obtained pBH10 was modified to eliminate the Eco R1 site distal to the lac operator and plasmid pBH20 was obtained.

Plasmid pBH20 (10 μg) was digested with endonucleases Eco R1 and Bam HI and treated with bacterial alkaline phosphatase (0.1 unit of BAPF, Worthington) and incubation was continued for 10 min at 65°. After extract with a phenol-chloroform mixture, the DNA was precipiated with ethanol. Somatostatin DNA (50 μl containing 4 μg/ml) was ligated with the Bam HI-Eco R1 alkaline phosphatase=treated pBH20 DNA in a total volume of 50 μl with 4 units of T_4 DNA ligase for 2 hrs at 22° and the recombinant plasmid used to transform *E. coli* RR1. Of the Tc^r transformants isolated (10), four plasmids has Eco R1 and Bam HI sites. Base sequence analysis indicated that the plasmid pSOM1 had the desired somatostatin DNA fragment inserted. Because of the failure to detect somatostatin activity from cultures carrying plasmid pSOM1, a plasmid was constructed in which the somatostatin gene could be located at the COOH-terminus of the β-galactosidase gene, keeping the translation in phase. For the construction of such a plasmid, pSOMI (50 μg) was digested with restriction enzymes Eco RI and Pst I. A preparative 5 percent polyacrylamide gel was used to separate the large Pst I-Eco RI fragment that carries the somatostatin gene from the small fragment carrying the lac control elements (12). In a similar way plasmid pBR322 DNA (50 μg) was digested with Pst I and Eco RI restriction endonucleases, and the two resulting DNA fragments were purified by preparative electrophoresis on a 5 percent polyacrylamide gel. The small Pst I-Eco RI fragment from pBR322 (1 μg) was ligated with the large PstI-Eco RI DNA fragment (5 μg) from pSOM1. The ligated mixture was used to transform *E. coli RR*1, and transformants were selected for Ap^r on X-gal medium. Almost all the Ap^r transformants (95 percent) gave white colonies (no lac operator) on X-gal indicator plates. The resulting plasmid, pSOM11, was used in the construction of plasmid pSOM11-3. A mixture of 5 μg of pSOM11 DNA and 5 μg of λplac5 DNA was digested with Eco RI. The DNA was extracted with a mixture of phenol and chloroform; the extract was precipitated by ethanol, and the precipitate was resuspended in T4 DNA ligase buffer (50 μl) in the presence of T4 DNA ligase (1 unit). The ligated mixture was used to transform *E. coli* strain RR1. Transformants were selected for Ap^r on X-gal plates containing ampicillin aidn screened for constitutiveβ-galactosidase production. Approximately 2 percent of the colonies were blue (such as pSOM11-1 and 11-2). Restriction enzyme analysis of plasmid DNA obtained from these clones revealed that all the plasmids carried a new Eco RI fragment of approximately 4.4 megadaltons, which carries the lac operon control sites and most of the β-galactosidase gene (13, 14). Two orientations of the Eco RI fragment are possible, and the asymmetric location of a Hind III restriction in this fragment can indicate which plasmids had transcription proceeding into the somatostatin gene. The clones carrying plasmids SOM11-3, pSOM11-5, pSOM11-6, and pSOM11-7 contained the Eco RI fragment in this orientation.

It is evident from the above results, that both DNA from a eukaryotic source and RNA transcribed from the eukaryotic DNA can be formed in a bacterial cell and isolated. Thus, the subject process provides a simple technique for producing large amounts of eukaryotic DNA and/or RNA without requiring the production and maintenance of the eukaryotic organism or cells. The employment of DNA for production of ribosomal RNA is merely illustrative of using a genome from a eukaryotic cell for formation of a recombinant plasmid for replication in a bacteria. Genomes from a eukaryotic cell for formation of genotypical properties, such as the production of enzymes, could have equivalently been used. As evidenced by the transformation with DNA from a bacteriophage, and entire operon can be introduced into a bacterial cell and the cell becomes capable of its transcription, translation, and production of a functional gene product. Thus, a wide variety of auxotrophic properties can be introduced into a bacterial cell.

In accordance with the subject invention, DNA vehicles are provided, which are covalently closed circular extrachromosomal replicons or genetic elements, including plasmids and viral DNA. The vehicles generally will have molecular weights in the range of about 1 to 20×10^6 and are characterized by having an intact replicon, which includes a replicator locus and gene. The vehicle is capable of clevage by a restriction enzyme to provide a linear segment having an intact replicon and cohesive termini, which may be directly obtained by the cleavage or by subsequent modification of the termini of the linear segment. The vehicle will be capable of transforming a bacterial cell and to that extent is compatible with the cell which will provide replication and translation. Preferably, the vehicle will have a phenotypical property which will allow for segregation of the transformant cells. Phenotypical properties include resistance to growth inhibiting materials, such as antibiotics, peptides and heavy metals, morphological properties, color, or the like, and production of growth factors, e.g. amino acids.

The vehicle is combined with DNA indigenous to a biological organism other than the cell which provides replication and provides a genotypical or phenotypical property which is alien to the cell. The source of the DNA can be prokaryotic or eukaryotic, thus including bacteria, fungi, vertebrates, e.g. mammals, and the like.

The plasmid vehicle and the alien DNA having complementary cohesive termini can be annealed together and covalently linked to provide a recombinant plasmid, which is capable of transforming a bacterial cell, so as to be capable of replication, transcription, and translation. As a result, a wide variety of unique capabilities can be readily introduced into bacteria, so as to provide convenient ways to obtain nucleic acids and to study nucleic acids from a foreign host. Thus, the method provides the ability to obtain large amounts of a foreign nucleic acid from bacteria in order to be able to study the function and nature of the nucleic acid. In addition, the subject method provides means for preparing enzymes and enzymic products from bacteria where the natural host is not as convenient or efficient a source of such product. Particularly, bacteria may allow for more ready isolation of particular enzymes, uncontaminated by undersirable contaminants, which are present in the original host. In addition, the products of the enzymic reactions may be more readily isolated and more efficiently produced by a transformant than by the original host. Besides enzymes, other proteins can be produced such as antibodies, antigens, albumins, globulins, glycoproteins, and the like.

Although the foregoing invention has been described in some detail by way of illustration and example for purposes of clarity of understanding, it will be obvious that certain changes and modifications may be practiced within the scope of the appended claims.

We claim:

1. A method for replicating a biologically functional DNA, which comprises:

transforming under transforming conditions compatible unicellular organisms with biologically functional DNA to form transformants; said biologically functional DNA prepared in vitro by the method of:

(a) cleaving a viral or circular plasmid DNA compatible with said unicellular organism to provide a first linear segment having an intact replicon and termini of a predetermined character;

(b) combining said first linear segment with a second linear DNA segment, having at least one intact gene and foreign to said unicellular organism and having termini ligatable to said termini of said first linear segment, wherein at least one of said first and second linear DNA segments has a gene for a phenotypical trait, under joining conditions where the termini of said first and second segments join to provide a functional DNA capable of replication and transcription in said unicellular organism;

growing said unicellular organisms under appropriate nutrient conditions; and

isolating said transformants from parent unicellular organisms by means of said phenotypical trait imparted by said biologically functional DNA.

2. A method according to claim 1, wherein said unicellular organisms are bacteria.

3. A method according to claim 2, wherein said transformation is carried out in the presence of calcium chloride.

4. A method according to claim 3, wherein said phenotypical trait is resistance to growth inhibiting substance, and said growth is carried out in the presence of a sufficient amount of said growth inhibiting substance to inhibit the growth of parent unicellular organisms, but insufficient to inhibit the growth of transformants.

5. A method according to claim 1, wherein said unicellular organism is *E. coli*.

6. A method according to claim 1, wherein said predetermined termini are staggered and cohesive.

7. A method according to claim 6, wherein said joining conditions includes enzymatic ligation.

8. A method according to claim 6, wherein said cohesive ends are formed by staggered cleavage of said viral or circular plasmid DNA and a source of said second segment with a restriction enzyme.

9. A method acording to claim 6 wherein said cohesive termini are formed by addition of nucleotides.

10. A method according to claim 1, wherein said predetermined termini are blunt end and said joining conditions include enzymatic ligation.

11. A method for replicating a biologically functional DNA comprising a replicon compatible with a host unicellular organism joined to a gene derived from a source which does not exchange genetic information with said host organism, said method comprising:

isolating said biologically functional DNA from transformants prepared in accordance with claim 1;

transforming unicellular microorganisms with which said replicon is compatible with said isolated DNA to provide second transformants; and

growing said second transformants under appropriate nutrient conditions to replicate said biologically functional DNA.

12. A method for producing a protein foreign to a unicellular organism by means of expression of a gene by said unicellular organism, wherein said gene is derived from a source which does not exchange genetic information with said organism, said method comprising:

growing transformants prepared in accordance with any of claims 1 and 11 under appropriate nutrient conditions, whereby said organism expresses said foreign gene and produces said protein.

13. A method according to claim 12, wherein said protein is an enzyme.

14. A method according to claim 11, wherein said method is repeated substituting said biologically functional DNA from transformants prepared in accordance with claim 1 with second or subsequent transformants to produce additional transformants.

* * * * *

D.N. Leff/NSF-Mosaic

Stanley Cohen.

MARCH 9, 1981

$1.50

TIME

Prince Charles Picks a Bride

Shaping Life in the Lab

The Boom In Genetic Engineering

Genentech's Herbert Boyer

MAGEE

Document 16.11

August 19, 1980
Genentech Stock Prospectus, Excerpt

PRELIMINARY PROSPECTUS DATED AUGUST 19, 1980

A registration statement relating to these securities has been filed with the Securities and Exchange Commission but has not yet become effective. Information contained herein is subject to completion or amendment. These securities may not be sold nor may offers to buy be accepted prior to the time the registration statement becomes effective. This prospectus shall not constitute an offer to sell or the solicitation of an offer to buy nor shall there be any sale of these securities in any State in which such offer, solicitation or sale would be unlawful prior to registration or qualification under the securities laws of any such State.

Genentech, Inc.
Genentech, Inc.
Genentech, Inc.
Genentech, Inc.
Genentech, Inc.

1,000,000 SHARES
COMMON STOCK

All the shares of Common Stock offered hereby are being sold by Genentech, Inc. (the "Company" or "Genentech").

Prior to this offering, there has been no public market for the Common Stock of the Company. It is currently estimated that the initial public offering price per share will be between $25 and $30. See "Underwriting" for information relating to the method of determining the initial public offering price.

The Common Stock offered hereby involves a HIGH DEGREE OF RISK. See "Introductory Statement—Certain Factors to be Considered" for information with respect to the Company's short operating history, uncertainty of financial results and capital needs and other risk factors.

THESE SECURITIES HAVE NOT BEEN APPROVED OR DISAPPROVED BY THE SECURITIES AND EXCHANGE COMMISSION NOR HAS THE COMMISSION PASSED UPON THE ACCURACY OR ADEQUACY OF THIS PROSPECTUS. ANY REPRESENTATION TO THE CONTRARY IS A CRIMINAL OFFENSE.

	Price to Public	Underwriting Discounts(1)	Proceeds to Company(2)
Per Share			
Total Minimum			
Total Maximum(3)			

(1) See "Underwriting" for information relating to indemnification of the Underwriters and other matters.

(2) Before deducting expenses payable by the Company estimated at $

(3) Assuming full exercise of the 30-day option granted by the Company to the Underwriters to purchase, on the same terms, up to an additional 100,000 shares to cover any over-allotments. See "Underwriting."

The shares are offered by the several Underwriters when, as and if issued by the Company and accepted by the Underwriters and subject to their right to reject orders in whole or in part. It is expected that delivery of the shares will be made on or about September , 1980.

BLYTH EASTMAN PAINE WEBBER
INCORPORATED

HAMBRECHT & QUIST

The date of this Prospectus is September , 1980

PROSPECTUS SUMMARY

The information below should be read in conjunction with the detailed information and financial statements appearing elsewhere in this Prospectus.

The Company

Genentech, Inc. develops and produces products with commercial potential using genetically engineered microorganisms created by means of recombinant DNA technology or "gene splicing." Since inception in 1976, Genentech has applied genetic engineering techniques developed by it and others to produce a variety of important health care products, including human insulin, human growth hormone and human interferon. Other applications of the Company's technology under development include the industrial and agricultural fields.

The Offering

Common Stock offered by the Company	1,000,000 Shares(1)
Common Stock outstanding after the offering ..	7,472,102 Shares(1)
Use of Proceeds	For capital expenditures and to increase working capital for product research and development and clinical programs.
Proposed NASDAQ symbol	GENE

(1) Does not include up to 100,000 shares which may be sold to the Underwriters to cover over-allotments.

IN CONNECTION WITH THIS OFFERING, THE UNDERWRITERS MAY OVER-ALLOT OR EFFECT TRANSACTIONS WHICH STABILIZE OR MAINTAIN THE MARKET PRICE OF THE COMMON STOCK OF THE COMPANY AT A LEVEL ABOVE THAT WHICH MIGHT OTHERWISE PREVAIL IN THE OPEN MARKET. SUCH STABILIZING, IF COMMENCED, MAY BE DISCONTINUED AT ANY TIME.

POSITIONS OPEN | POSITIONS OPEN | POSITIONS OPEN

Recombinant DNA Scientists And Technicians

Southern Medical and Pharmaceutical Corporation, a Florida based corporation, currently engaged in the production of human interferons and in research and development of recombinant DNA techniques, is currently increasing its staffing for its recombinant DNA division.

The corporation is soliciting curriculum vitae from qualified scientists at the Ph.D. and Master's Degree level with backgrounds in genetic engineering and recombinant DNA technology for immediate employment and/or for future placement of Ph.Ds and post doctoral candidates after completing their current educational or research endeavors. Southern Medical is in particular need of Ph.D. level scientists with strong backgrounds in DNA sequencing techniques, molecular biology and biochemistry.

The company is located on the beautiful west coast of Florida, an area with superb recreational, cultural and educational opportunities. Salaries are negotiable depending on the qualifications of the applicant. Please forward your curriculum vitae to 128 South Moon Avenue, Brandon, Florida 33511.

Southern Medical And Pharmaceutical Corporation
128 South Moon Avenue
Brandon, Florida 33511

...ual Opportunity Employer

MOLECULAR BIOLOGIST

Excellent opportunity for PhD Microbiologist (or Biochemist with experience in microbiology) to pursue basic and applied research in bioconversion of biomass, xenobiotic biodegradation, and pathway evolution. Applicants should have experience in recom-...

ICI

ICI CORPORATE LABORATORY. BIOSCIENCE GROUP

MOLECULAR BIOLOGIST/MICROBIOLOGIST

A vacancy has arisen for a molecular biologist/microbiologist to work on the cloning and expression of genes from Gram positive organisms in both Gram positive and Gram negative host systems. We expect that suitable candidates will have a good honours degree and/or suitable laboratory experience. Interested applicants should send a detailed curriculum vitae with the names of two referees to

Miss A. McLear, Personnel Officer, Corporate Laboratory, PO Box 11, The Heath, Runcorn.

(7267)A

BIOGEN, INC.

A Leader in Biotechnology

Biogen believes that the most important technological ...developments in the future will be the fruit of significant ...pplications of recombinant DNA science.

As a pioneering leader in developing and applying this ...chnology to a spectrum of major industries — food, ...rmaceutical, chemical, agriculture, mining. Biogen is ...ning a world of diversified opportunities for a Director ...esearch and Scientists at PhD levels in the following ...:

...ecombinant DNA
...ochemistry
...netics
...robiology
...nic Chemistry

...SITIONS IN OUR NEW ... AREA FACILITIES

...ve salary — comprehensive benefit package. ...ticulars and prompt consideration, write ...in confidence to:

GENETICA

We are a new comp... research. Backed by the largest Fren... (involved in industrial biotechnology sinc... deliberately created an independant organizati... interrelationships within our research team.

Highly qualified personnel in our private co... receive attractive salaries and all the general advantages of the employment sy... France (knowledge of French is not necessary).

SPECIALISTS IN RECOMBINANT D...

...YOU WANT A CHALLENGING CA...

...SITION IN PARIS

At SYNTEX on the SAN FRANCISCO PENINSULA, our emphasis is on products related to the Life Sciences ... to health and growth and an improved standard of living. This premise will give an added dimension to your career at SYNTEX and offer you a superb scientific environment.

MOLECULAR BIOLOGISTS

...ider these key challeng... ...xpanding staff o... ...m.

CELLTECH

...ew British company, ...oth Government and City ...itutions, which is dedicated ...oitation of discoveries in the ...enetics and Biotechnology. ...ept of Celltech as a specialist ...acting as a bridge between ...n science and large-scale ...y is novel for the UK.

Chemists

We need people to carry out the chemical synthesis of oligonucleotides and direct experience of this field would be a great advantage; however, we are also willing to train suitably qualified people and experience of phosphorus chemistry, solid-phase synthesis or heterocyclic chemistry would be considered a suitable qualification. Experience in using computers would be an advantage.

Molecular Biologists &

Recombinant DNA Research

Frito-Lay, Inc., is a recognized leader in the growing $12 billion snack food industry. During the past 5 years, our revenues have doubled to almost $2 billion, and 1980 will be our 12th consecutive year of record earnings. As you would expect, state-of-the-art technology is a key ingredient in our success.

Frito-Lays's centralized Corporate Research and Development Department is expanding. Currently, positions are available in Recombinant DNA Research. The positions involve designing and conducting a research program in recombinant DNA with the objective of genetic alteration of plants.

To qualify, you must have a Ph.D. in biological science or biochemistry with thorough experience in the area of molecular biology and recombinant DNA technology. Experience with plant tissue would be helpful.

rito-Lay, Inc., offers an excellent starting salary and a comprehensive enefits package as well as the chance to grow with a leader. Send replies, cluding salary history, is strictest confidence to:

ITO-LAY, INC.
NAGER, TECHNICAL PLACEMEN
. BOX 35034
LLAS, TEXAS 75
l Opportuni

Molecular Geneticists

armaceutical company has several challenging positions available r M.S. level scientists in its Molecular Biology Department. Can- ould have proven experience and demonstrated proficiency in re- DNA methodologies and a strong background in procarytic molecu- cs.

xcellent salaries and comprehensive benefits which include reloca- ROFIT SHARING.

Recombinant DNA Scientists and Technicia

Southern Medical and Pharmaceutical Corporati Florida based corporation, currently engaged in the duction of human interferons and in research development of recombinant DNA techniques ntly increasing its staffing for its recombinant D

oration is soliciting curriculum vitae fr scientists at the PhD and Master's Degree le ckgrounds in genetic engineering and recor DNA technology for immediate employme for future placement of PhDs level scientists wit backgrounds in DNA sequencing techniques cular biology and biochemistry.

company is located on the beautiful west coast of rida, an area with superb recreational, cultural and ucational opportunities. Excellent salaries, working onditions, fringe benefits and attractive stock options re available to qualified scientists.

Southern Medical and Pharmaceutical Corporation,
128 South Moon Avenue,
Brandon, Florida 33511

An equal opportunity employer

(NW320)A

UNIVERSITY OF DUBLIN
Trinity College

Postdoctoral Fellow in Recombinant DNA Research

A Postdoctoral Fellow is required to participate in an applied recombinant DNA research project with Dr David McConnell, funded by industry. Experience in recombinant DNA technology, especially in cDNA cloning and/or cloning in yeast is desirable. Salary negotiable, within the range IR£6,000 — £7,500 p.a.

Further information can be obtained from the Staff Office.

Curriculum vitae, relevant material and the names of two referees should be sent to:

The Staff Office,
West Theatre,
Trinity College,
Dublin 2.
Tel: 772941 ext. 1775,
by Friday 27th February 1981. (7630)A

MOLECULAR GENETICIST—PhD

• Southern Location

Ethyl Corporation, a Fortune 500 company located in Baton Rouge, Louisiana, is seeking an innovative individual to join our expanding Biological Processes Group. This group has corporate responsibility for investigation and development of novel biological processes for production of industrial products.

The position requires a PhD degree and 1-5 years postdoct industrial experience, including training in the techniques plications of recombinant DNA. Responsibilities will include tive research directed toward novel biological processes f existing products, development of bench-scale processe strate feasibility and interaction with engineering staff up and commercialization. This is an excellent o creative individual to contribute to all phases of dev biological process.

Baton Route is located 80 miles northwest of New Orleans o Mississippi River, where the climate is extremely favorable and cost of living among the very lowest. In addition to an excellent opportunity for professional recognition and growth, Ethyl offers a comprehensive salary/benefits package and an attractive working environment. For confidential consideration, please send detailed resume with salary requirements to:

MR. A. J. ROBICHAUX, *Recruiting Supervisor*

ETHYL
CORPORATION
P.O. Box 341
aton Rouge, LA 70821
lover

Ethyl

Internation

Management to exploit adva

Biotechnolog

titute, is a "not-for
of California. It c
gement consultan
a fee income of £
l operations. Londo
t and Africa.
cology and Physic
esearch Laborator
ation. Those alreac
estry and Minerals
new London based
tivities to Industry
appointed will hav

Research Scientists

Molecular/Cellular Genetics–Agriculture

AGRIGENETICS CORPORATION is a vigorous medium-sized seed and genetic research company enjoying rapid growth, with excellent base-line positions in most major crops and in bacterial inoculants. The company has broadened its existing plant breeding and research activities to include programs in cell biology, microbiology, and the molecular genetics of higher plants. Accordingly, it is inviting applications for newly

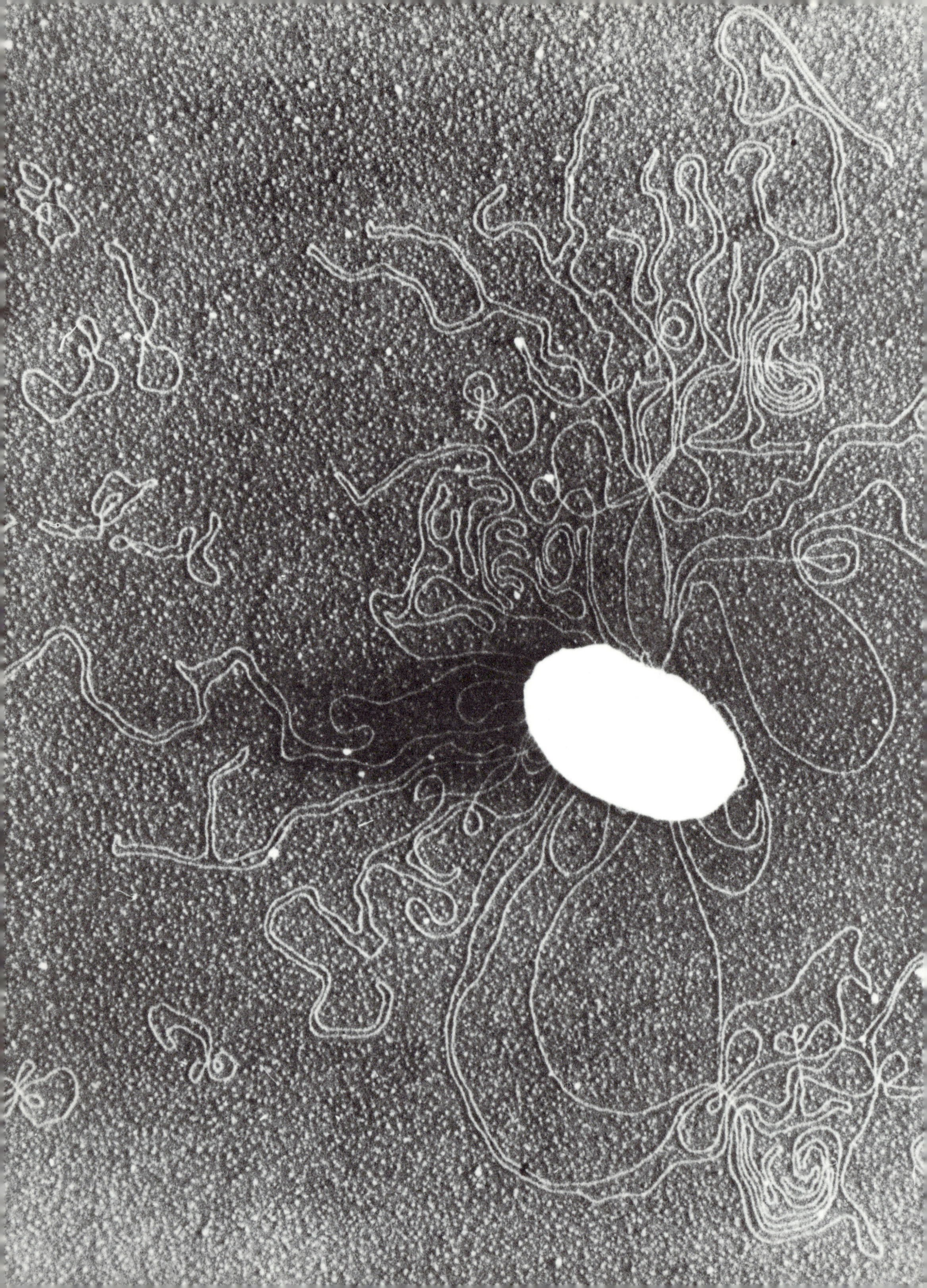

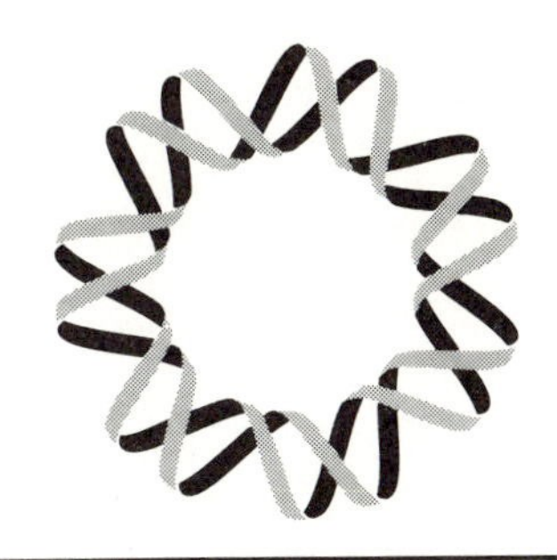

Scientific Background

Over and over during the public debate about recombinant DNA, those who opposed its speedy introduction by the academic research community asked, Why so much haste? Wouldn't prudence be a better course, with the initial experiments done under maximal security conditions in spots remote from major population centers? Shouldn't we keep up such tough precautions until molecular geneticists could reassure the population that no one was at risk? Such proposals, however, increasingly fell on deaf ears among potential users of recombinant DNA. They, almost to a man, felt growing frustration, if not anger, as month after month passed under the initial NIH guidelines that many of us believed were travesties against common sense.

The depth of this feeling, which initially seemed so stridently unnecessary to many nonscientists, is easy to explain. Recombinant DNA was no ordinary scientific advance, and once it was available it became virtually impossible to go on acting as if it didn't exist. Wide new classes of experiments were suddenly made feasible, possibly allowing research to be accomplished over the next decade that might not otherwise be accomplished over the next century. To make this clear, we shall now recapitulate the basics of molecular genetics and show why this field was approaching a road block until the prospect of recombinant DNA appeared on the horizon.

Electron micrograph of a giant bacterial chromosome and several small plasmids spilling out of an E. coli cell. (Courtesy of Huntington Potter and David Dressler.)

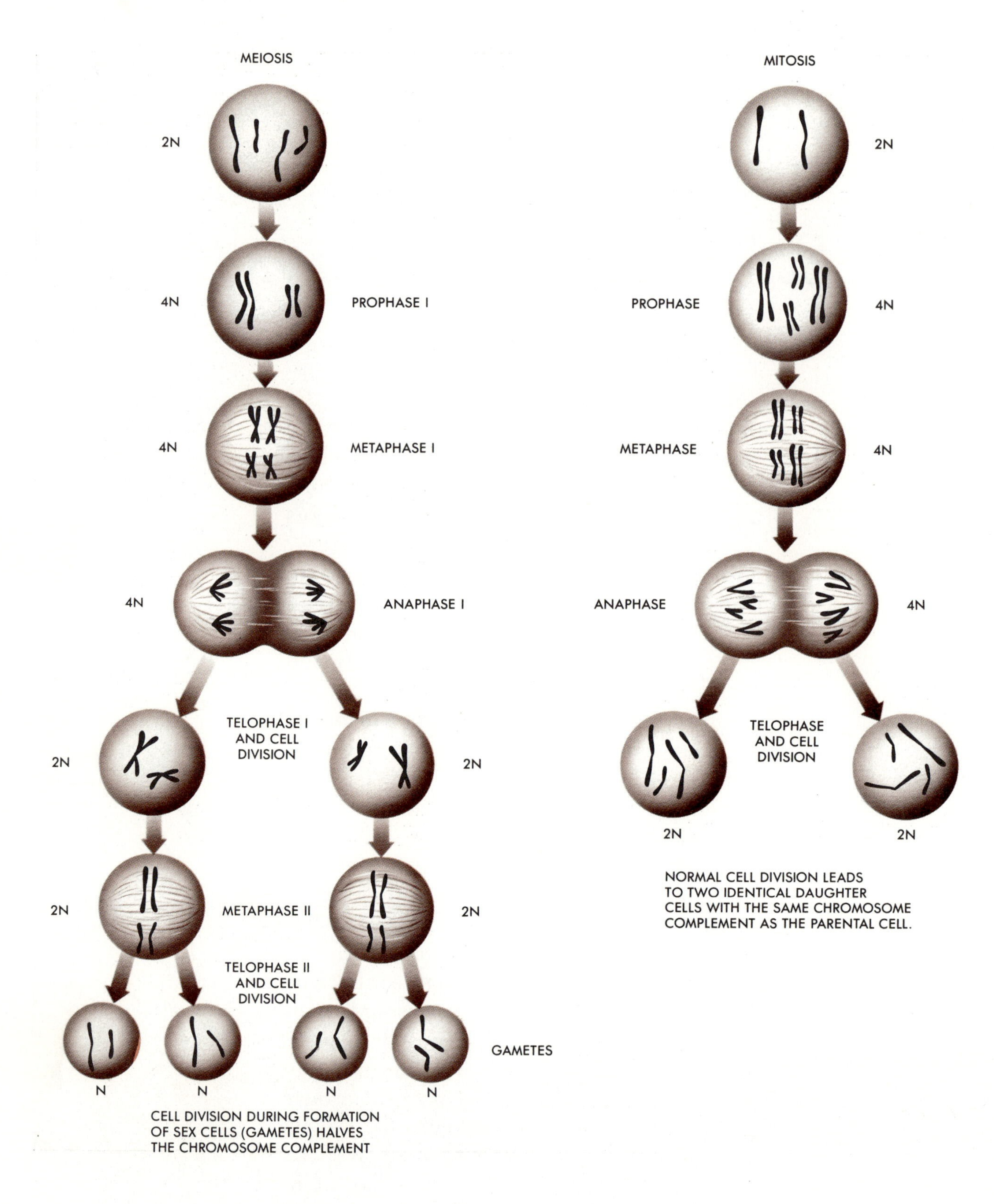

CELL DIVISION DURING FORMATION OF SEX CELLS (GAMETES) HALVES THE CHROMOSOME COMPLEMENT

NORMAL CELL DIVISION LEADS TO TWO IDENTICAL DAUGHTER CELLS WITH THE SAME CHROMOSOME COMPLEMENT AS THE PARENTAL CELL.

DNA Is the Primary Genetic Material

Chromosomes—Bearers of Heredity

Modern biology, as we know it, had its beginnings in the early nineteenth century, when in Germany the development of both powerful light microscopes and techniques for fixing and staining living tissues led to the generalization that all living material, whether of plant or animal origin, was made up of tiny boxlike entities called cells. The cell is the fundamental unit of life; every cell arises as the result of the growth and subsequent splitting of a parental cell into two daughter cells. Each contains an inner membrane-bounded body called the nucleus. The region between the nucleus and the outer membrane surrounding all cells is the cytoplasm. By the 1860s rodlike bodies within the nucleus were identified; these bodies stained strongly with certain basic dyes, and so they were called chromosomes (chromo comes from the Greek word for color). The individual chromosomes in a cell can be distinguished by their size and shape, and for any species the number per cell is constant and also even. The number is even because each distinct chromosome is present in two copies known as an homologous pair. The total even number of chromosomes is called the diploid number (2N), while the number of distinct chromosomes is called the haploid number (N). There are two copies of each distinct chromosome in each cell. Prior to cell division the diploid set reproduces itself to yield a cell with 4N chromosomes. This allows each of the two daughter cells that result from cell division to inherit the parental diploid (2N) constitution.

The partitioning of chromosomes during ordinary cell division (mitosis) is exact; each daughter cell receives one copy of each chromosome present in the parental cell. In contrast, in meiosis, during the formation of the sex cells (sperm and egg), the chromosome number is reduced to one-half (N represents haploid), with each sex cell receiving one partner of each homologous chromosome pair. Following union of the sperm (N) and egg (N) to produce a fertilized egg, the chromosome number is restored to 2N.

As the movement of chromosomes during mitosis and meiosis was being recognized in the late 1880s, it became clear that their behavior mimicked the expected behavior of the cellular elements then being postulated to pass identical traits from parental to daughter cells, as well as to bear the hereditary contributions of the male and female parents. Proof for the idea that chromosomes are carriers of heredity could only come, however, after the rediscovery, in 1901, of the genetic experiments by the monk Gregor Mendel. In 1865 at Brno, now in Czechoslovakia, he had crossed peas and proved that traits like color and shape are controlled by hereditary factors that we now call genes. Each diploid cell contains two copies of each gene: one derived from the male parent, the other from the female. During crosses between individuals with different genetic compositions, their genes become reassorted, thereby leading to the generation of numerous genetically related but distinct offspring from a single pair of parents.

The first trait to be assigned a chromosome location was sex itself. The work was done at Columbia University, where in 1905 Nettie Stevens and Edmund Wilson discovered the existence of the so-called sex chromosomes. One chromosome, the "X," while present in two copies (2X) in the female, is present in only one copy in the male, which also contains a morphologically distinct Y chromosome. During the halving of chromosome numbers as the sex cells are formed, all eggs will necessarily receive single X chromosomes while sperm will contain either an X or a Y chromosome. Fertilization by the sperm containing X chromosomes will generate female (XX) progeny, while fertilization by those carrying a Y chromosome will yield male (XY) progeny. The proposal that sex traits are loaded on a single pair of chromosomes neatly explained the 1:1

CHROMOSOMES OF MALE AND FEMALE FRUITFLY DROSOPHILA

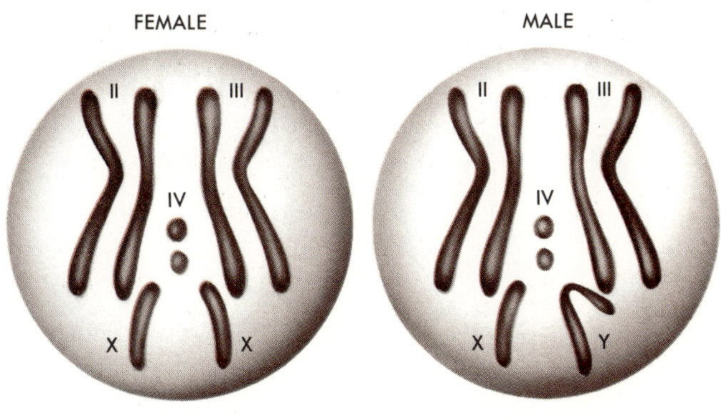

1:1 RATIO OF MALES AND FEMALES

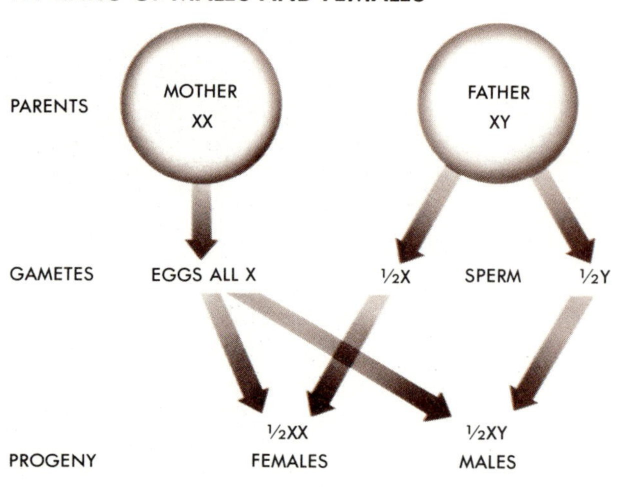

ratio of males and females. It was natural to speculate that all traits, not just those determining sex, might be chromosomally located. The first confirmation came very soon from genetic crosses with the red-eyed fruitfly *Drosophila* in T. H. Morgan's laboratory at Columbia around 1910–1915. In the first such experiments a changed, mutated gene leading to white eye was also located, or mapped, on the X chromosome. Over the next several years many more mutants were mapped to either the sex chromosome or one of the nonsex chromosomes (autosomes). The coexistence of two mutants on the same chromosome was shown by their tendency to reassort together (or to be linked) in genetic crosses. But this linkage was never complete; it was a consequence of physical exchanges (called crossing over) occurring when homologous chromosomes come together in pairs before forming haploid sex cells. Most importantly, crossing over provides a way to order genes along chromosomes because those genes that are far apart on a chromosome will have a better chance to cross over than will closely spaced ones. The arrangements found were always linear, and by 1912 the hypothesis that chromosomes are the bearers of heredity was virtually unassailable.

Early breeding experiments utilized genetically stable variants that appeared spontaneously at low frequency. These stable variants, called mutants, arise by mutation. We now know that many spontaneous variants result from changes in single genes. Such genes are mutant genes, as opposed to normal, or "wild-type," genes. In 1926 Hermann Müller and Lewis Stadler independently discovered that X-rays induce mutations, thereby providing geneticists with a much, much larger number of mutant genes than available when the so-called spontaneous mutations were the only source of variability.

As genetic experiments grew in momentum, it became apparent that many genes were located in a strictly linear order on each chromosome; estimates for the number of genes present on all the chromosomes in a given sperm (egg) soon exceeded 500. By the late 1930s it seemed probable that at least 1500 genes were contained within the four chromosomes of *Drosophila*.

POPPET-BEAD MODEL:
CHROMOSOME 2 OF THE FRUITFLY DROSOPHILA

ANTENNAE (SHORT)
WING SHAPE (DUMPY)
LEGS (SHORT)
WING SHAPE (CURVED)
EYE COLOR (BROWN)

REASSORTMENT OF GENES BY CROSSING OVER

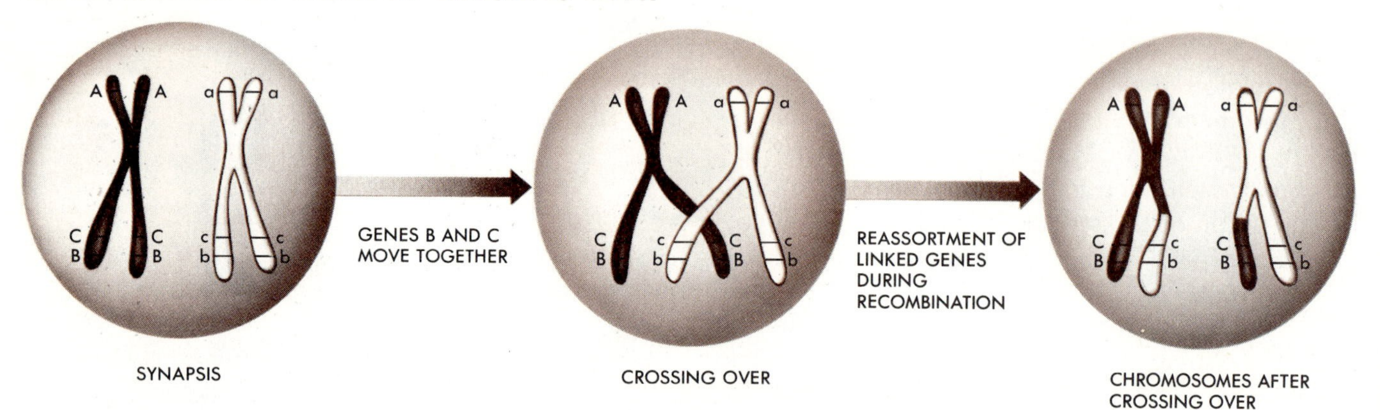

Initially it was taken for granted that crossing over occurred only between genes because chromosomes were conceived as being constructed like poppet beads; that is, genes were held together by connectors that easily broke and rejoined. By the late 1930s, though, when many thousands of progeny from single *Drosophila* crosses were examined, scientists discovered rare examples of apparent crossing over within a gene that determines eye color. But at that time there was no way to know that the two mutants being examined contained mutations in the same gene. Conceivably they were in adjacent genes that were functionally related and therefore, when mutated, produced seemingly identical changes (such as white eyes). Could crossing over cause rearrangements in the genes themselves? The answer had to await the development of systems employing many, many mutations in the same gene, and where millions—not mere thousands—of progeny from the same genetic cross could be examined.

The One-Gene-One-Protein Hypothesis

When chromosomes were first described in the 1860s, virtually nothing was known about the chemical reactions of cells. Even in 1901, at the time of the rediscovery of Mendel's law, the subject of biochemistry as we now know it was still in its infancy. Yet there was steadily increasing evidence that life depended on the presence in all cells of catalytic agents that vastly accelerated the rate at which chemical bonds are made and broken. These catalytic agents were intially called ferments because they were first observed to convert glucose to ethyl alcohol in the absence of oxygen (the process of anaerobic fermentation). Now enzymes is the common term for these agents; any given chemical reaction in a cell has a specific enzyme as a catalytic agent. Most cells contain perhaps more than 1000 different enzymes. Although initially much controversy surrounded the chemical identity of enzymes, as early as the late 1880s there were hints that at least some enzymes were proteins. But it was not until the early 1930s that the protein nature of *all* enzymes was universally accepted.

By the early 1900s the monumental work of the German chemist Emil Fischer had established that the primary chemical feature marking a protein is the peptide bond. This bond links numerous different amino acids together in long chainlike molecules called polypeptides. A protein molecule contains one to several polypeptide chains. The number of different amino acids within proteins remained unclear until the early 1940s, when the number began to converge toward 20. Insulin was the first protein whose exact sequence of amino acids was worked out; this research was accomplished by Frederick Sanger in the early 1950s in Cambridge, England. Sanger established the order of 52 amino acids on the two polypeptide chains that make up the active molecule. Since then, proteins consisting of almost 1000 amino acids have been sequenced.

Long before the essence of proteins was found to reside in their amino-acid sequences, scientists speculated that genes somehow controlled the synthesis of specific enzymes. This concept was first clearly proposed by the English physician Archibald

AMINO-ACID BUILDING BLOCKS OF A PROTEIN

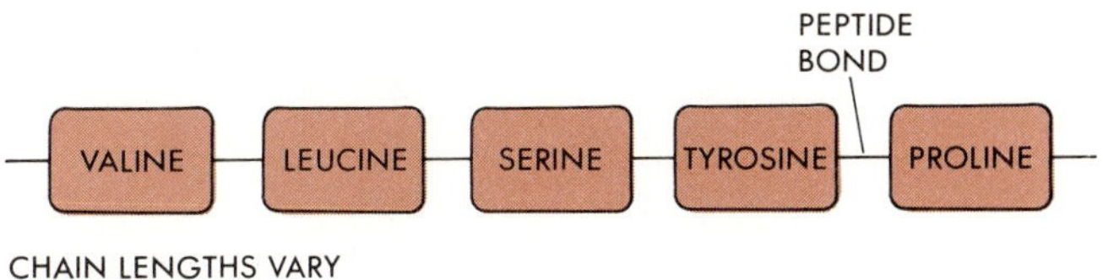

CHAIN LENGTHS VARY FROM 50 to >2000

THE 20 LETTERS OF THE PROTEIN ALPHABET*

Glycine	GLY	Lysine	LYS
Alanine	ALA	Arginine	ARG
Valine	VAL	Asparagine	ASN
Isoleucine	ISO	Glutamine	GLN
Leucine	LEU	Cysteine	CYS
Serine	SER	Methionine	MET
Threonine	THR	Tryptophan	TRP
Proline	PRO	Phenylalanine	PHE
Aspartic Acid	ASP	Tyrosine	TYR
Glutamic Acid	GLU	Histidine	HIS

* Proteins are made up of different combinations of these 20 building blocks, called amino acids.

THE SUBUNIT COMPOSITION OF INSULIN

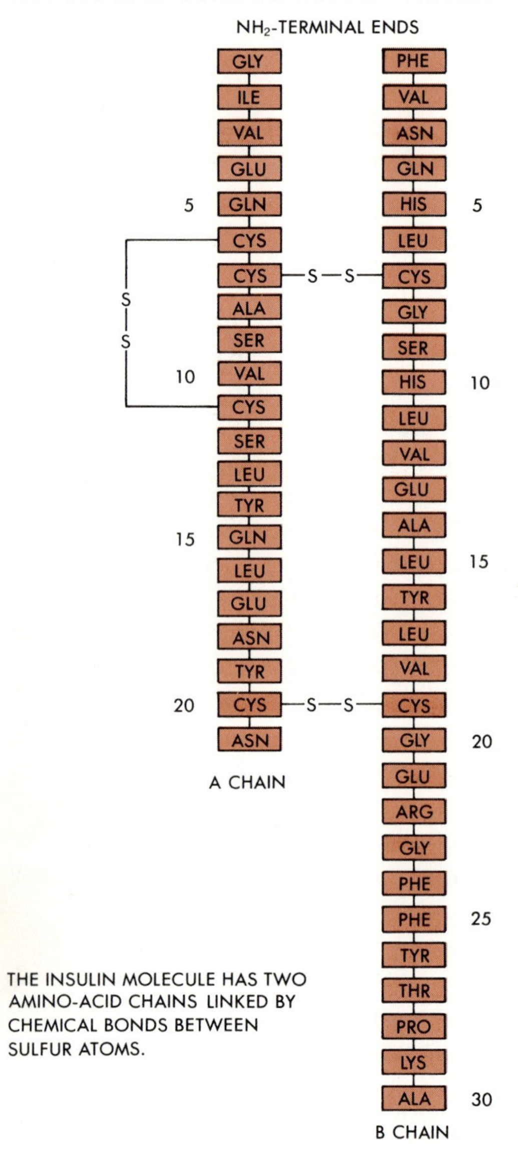

THE INSULIN MOLECULE HAS TWO AMINO-ACID CHAINS LINKED BY CHEMICAL BONDS BETWEEN SULFUR ATOMS.

Garrod, who studied what he called "inborn errors" in metabolism. In the genetic disease phenylketonuria, the amino acid phenylalanine cannot be converted to the related amino acid tyrosine. This "error" leads to a buildup in the blood of the toxic intermediary metabolite phenylpyruvate. Garrod correctly hypothesized that this biochemical failure arose because the respective mutant genes cannot provide for the presence of the enzyme that converts phenylpyruvate into tyrosine. Further evidence favoring the "one-gene-one-enzyme" hypothesis only slowly accumulated until the early 1940s, when at Stanford University the geneticist George Beadle and the biochemist Edward Tatum used X-rays to induce mutations in the mold *Neurospora*. As they had hoped, some of their mutant *Neurospora* strains had lost their ability to make necessary molecules such as the amino acid arginine and the vitamin biotin. In each case the cause of the metabolic block was the failure to synthesize a specific enzyme.

Because enzymes are always proteins, the one-gene-one-enzyme concept was likely to be broadened to one-gene-one-protein. Important support for this formulation came in the early 1950s at Caltech when the chemist Linus Pauling studied the genetic disease sickle-cell anemia. Pauling showed that the mutant "sickle" genes lead to altered hemoglobin molecules, which have a reduced ability to bind with oxygen. Important as it was, though, the one-gene-one-protein idea could provide no clue to the molecular mechanisms involved, so long as the nature of the gene itself remained a total mystery.

Chromosome Composition: Protein and DNA

For many years it was hoped that as microscopes improved, we might eventually see genes sitting side by side along chromosomes. But even when the first electron microscopes (EM) appeared in the early 1940s with a potential resolution over 100 times greater than the light microscope, there were disappointments. The first EM pictures of chromosomes showed no repeating pattern at the molecular level, thus suggesting a highly irregular gene structure that would not be simple to interpret. Attempts to purify chromosomes away from other cellular constituents were much more informative given the caveat that it was impossible to obtain really pure chromosomes.

Two main chromosomal components were almost invariably found: (1) deoxyribonucleic acid (DNA), and (2) a class of small positively charged proteins known as the histones, which, being basic, neutralized the acidity of DNA. DNA had been known to be a major constituent of the nucleus (hence the name "nucleic" acid) ever since its discovery in 1869 by the Swiss scientist Frederick Miescher. In the 1920s, with the DNA-specific purple dye developed by the German chemist Robert Feulgen, DNA was found to be sited exclusively on the chromosomes. DNA therefore had the location we expected for a genetic material. In contrast, the histones could apparently be ruled out as genetic

BASES OF NUCLEIC ACIDS

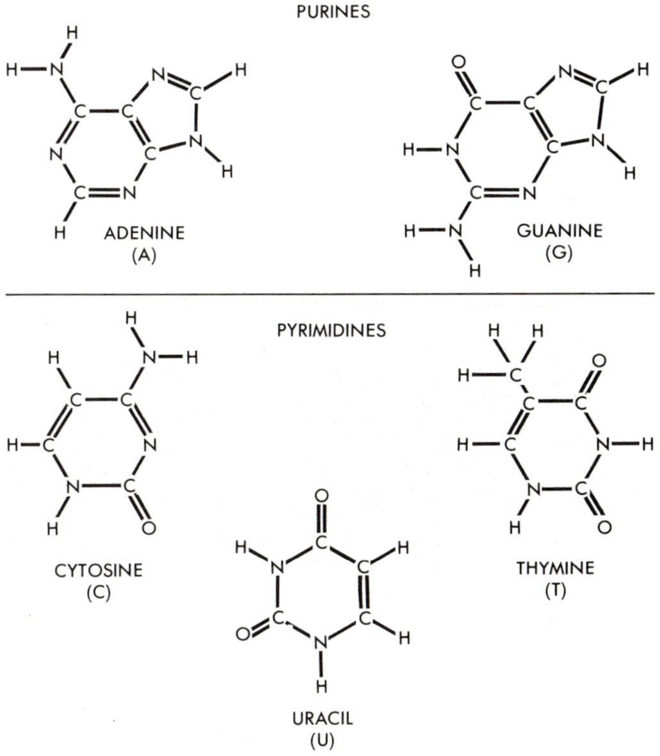

PURINES

ADENINE (A)

GUANINE (G)

PYRIMIDINES

CYTOSINE (C)

URACIL (U)

THYMINE (T)

components because they were absent from many sperm and replaced by even smaller basic proteins, the protamines. But most biochemists were not inclined to focus attention on DNA. They thought it would not be nearly as specific as the proteins, of which an unlimited number can be constructed by linking together the 20 amino acids in different orders. So perhaps some minor and not yet well-characterized protein component of the chromosomes would be found to be the true genetic material.

Cells Contain RNA as Well as DNA

Already late in the nineteenth century it had been discovered that cells have a second class of nucleic acid, now called ribonucleic acid (RNA). Yet the manner in which DNA and RNA differed was not exactly known until over 50 years later. Both classes resemble proteins in being constructed from end-to-end linkages of many smaller building blocks. The building blocks of nucleic acid, however, are more complex than any amino acid and are called nucleotides. Each nucleotide contains a phosphate group, a sugar moiety, and either a purine or a pyrimidine base (flat ringed-shaped molecules containing carbon and nitrogen). When nucleotides are linked together in large numbers, they are polynucleotides. Early on, the sugar component of RNA was known to be different from that of DNA. Yet it was not until the 1920s that the work of Phillip Levine of the Rockefeller Institute revealed that the sugar of DNA is deoxyribose (hence the name deoxyribonucleic acid) and the sugar of RNA is the closely related ribose (hence the name ribonucleic acid). Two purines and two pyrimidines are found in both DNA and RNA. The two purines, adenine and guanine, are used in both DNA and RNA. The pyrimidine cytosine is likewise found in both DNA and RNA. In contrast the pyrimidine thymine is found only in DNA, while the structurally similar pyrimidine uracil is used in RNA.

Unlike DNA, which appeared to locate exclusively in the nucleus, RNA is in the cytoplasm as well as the nucleus. Within the nucleus RNA is concentrated in several chromosomally attached dense granules called nucleoli, which are absent from the vast length of most chromosomes.

A NUCLEOTIDE OF THE DNA ALPHABET

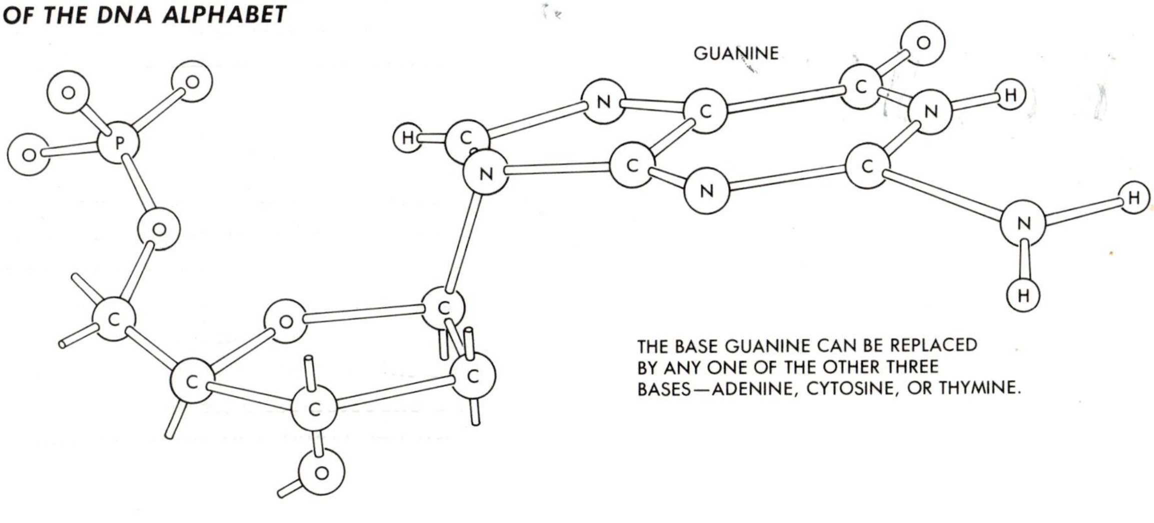

GUANINE

THE BASE GUANINE CAN BE REPLACED BY ANY ONE OF THE OTHER THREE BASES—ADENINE, CYTOSINE, OR THYMINE.

Avery's Discovery of DNA as a Genetic Molecule

Deciding whether the gene is made of DNA, proteins, or even of RNA was impossible so long as there was no direct assay for a genetic substance. The best of all possible proofs would be to show specific genetic changes in a cell after its exposure to a certain chemical compound. The first hint that this could be done came in 1928 from the accidental observation by the Englishman Frederick Griffith that strains of nonvirulent pneumococcus bacteria (having no outer capsule) could be converted into capsulated virulent strains by exposing them to virulent cells that had been killed by high temperature. This discovery was confirmed quickly and extended in Oswald T. Avery's Rockefeller Institute laboratory by the finding that virulent bacteria could be broken open to yield an active fraction still capable of "transforming" nonvirulent cells into their virulent equivalents. At this stage Avery, along with Maclyn McCarty and Colin MacLeod, took on the task of chemically identifying the active ingredient. To their surprise the active transforming fraction did not contain the sugarlike components that comprised the outer capsule. Proteins likewise seemed not to be involved because the ability to transform was totally resistant to enzymes that break peptide bonds. Instead their most potent transforming fractions were essentially pure DNA. Equally important, these fractions lost all transforming activity when exposed to DNase, an enzyme that breaks the bonds linking deoxynucleotides together. In contrast, no loss of transforming activity was observed when they added RNase, the RNA-degrading enzyme.

These findings, first announced in 1944, greatly surprised the vast majority of geneticists and biochemists, who were long convinced that anything as specific as the gene must be a protein. Many skeptics preferred to believe that Avery and his colleagues had somehow missed seeing the "genetic protein," and that DNA was required for activity in their assay only because it functioned as an unspecific scaffold to which the real protein genes were fixed. Upon reflection, though, the pinpointing of DNA should not have been that unexpected. There were by then strong hints that it was a very large molecule containing hundreds of nucleotides. If the sequence of the four main nucleotides is irregular, then the number of potentially different DNA sequences is the astronomically large 4^n (n = number of nucleotides in a chain).

Viruses as Packaged Genetic Elements Moving from Cell to Cell

By then there was increasing interest in DNA as a result of its recent discovery in several highly purified viruses. The nature of these tiny disease-causing particles, which multiply only in living cells, was long disputed. Some scientists considered them a form of naked genes; other preferred to think of them as the smallest form of life. Only when it became possible to purify them away from cellular debris and look at them in the electron microscope did their nature begin to be revealed. They were clearly *not* minute cells; rather, they lost their identity as discrete particles when they multiplied within cells. The best guess, therefore, was that they were parasites at the genetic level, and that by studying how they multiplied, we might develop the very best systems for analyzing gene structure and replication.

At this point most attention turned toward analyzing the growth cycle of those viruses that multiply in bacteria—the bacteriophages (the word "phages" is from the Greek word for eating). Most favored for study were a group of phages that multiply within the common intestinal bacterium *Escherichia coli* (*E. coli*). Given names such as T1, T2, T4, and λ, a single parental phage particle can multiply to several hundred progeny particles within roughly 20 minutes. Analysis of their genetic properties started when mutants arose during the phage multiplication cycle. After several independently arising mutant phages infected a single bacterium, some of the progeny phages that were produced appeared normal. Viruses were thus also capable of genetic recombination. Subsequent experiments employing many different mutants suggested that each virus particle contained several different genes linearly arranged along the viral chromosome.

By purely genetic experiments, however, scientists could not decide whether it was the DNA or one of the protein components that carried the genetic specificity. This point was not settled until 1952, when in Cold Spring Harbor, Long Island, Alfred Hershey and Martha Chase showed that only the DNA of phages entered the host bacteria. Their surrounding protein coats remained outside and thus could be ruled out as potential genetic material.

PHAGE GENETIC MATERIAL IS DNA, NOT PROTEIN

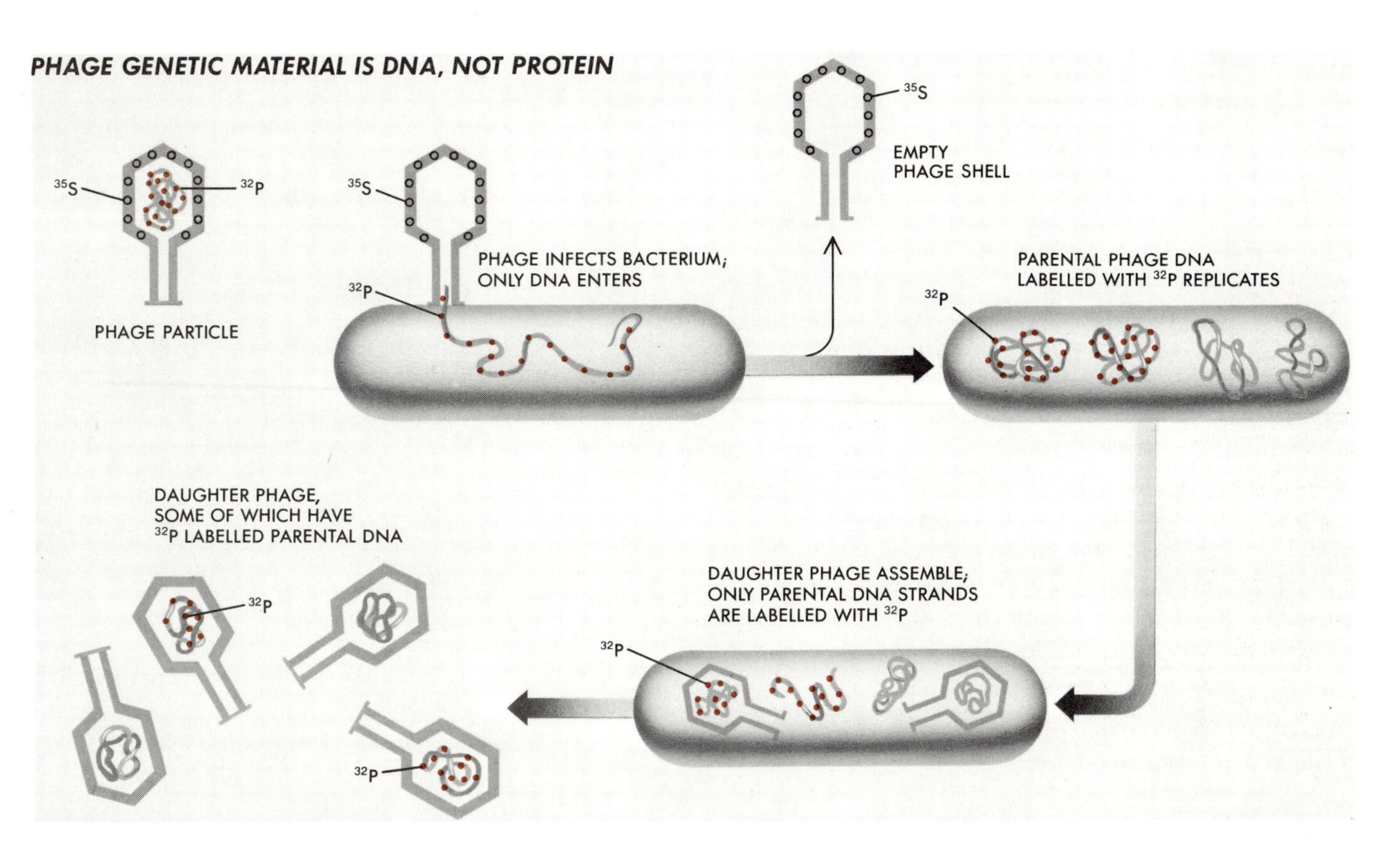

Linkage of DNA and RNA Molecules by 5′-3′ Phosphodiester Bonds

Organic chemists thus had a strong incentive to treat nucleic acids as potentially more significant than proteins. Particularly important was the decision of Alexander Todd, a leading British chemist, to focus on nucleotide chemistry when university basic research resumed at the end of World War II. Todd's long-term goal was to clarify the precise chemical bonds that held together the component nucleotides in DNA and RNA. By 1953 his laboratory showed that in DNA the phosphate groups always joined a specific carbon atom (5′) of one deoxyribose molecule to the 3′ carbon atom of the adjacent deoxyribose to form a phosphodiester bond. The alternating . . . ^{5}sugar^{3}-phosphate-^{5}sugar^{3}-phosphate . . . groups are called the polynucleotide backbones, to which the purine and pyrimidine bases are attached as effective side groups.

A longer time was taken to settle the linkages within RNA; these were not discovered until two years later in 1955. RNA was also found to have a highly regular backbone, employing again only 5′-3′ phosphodiester links to hold together its component nucleotides.

PHOSPHODIESTER BONDS LINK THE BASES TO FORM THE DNA BACKBONE

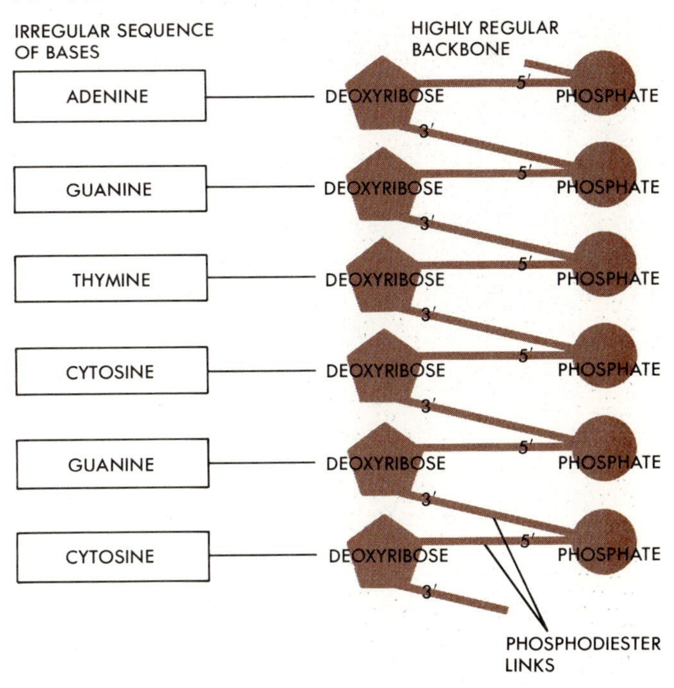

Establishing the Diameter of DNA

Although in the late 1930s Swedish physical chemists already had evidence from its behavior in solution that DNA was asymmetrically shaped, direct measurements of its size became possible only when the electron microscope came into general use within several years after the end of World War II. All carefully prepared samples showed extremely elongated molecules, many thousands of angstroms (Å = 10^{-10} meter) in length, and approximately 20 Å thick. All the molecules were unbranched, which confirmed the highly regular backbone structure proposed by organic chemists. From the length and the fact that each nucleotide base is just over 3 Å thick, it was clear that most molecules were composed of many thousands of nucleotides and very possibly were much larger than any other natural chainlike (polymeric) molecules.

Base Composition Between Different Species

While work in England centered on the backbone, several American laboratories, particularly that of the Austrian biochemist Erwin Chargaff, then working at Columbia University College of Physicians and Surgeons, were using newly developed chromatographic techniques to separate quantitatively the purine and pyrimidine components from DNA of quite disparate plants and animals. By 1951 it became clear that the four bases were not generally present in equal amounts, thus finally disproving the once popular idea that DNA was made of repeating identical tetranucleotide units in which each base was present once. Such a structure would not permit DNA to be a genetic substance, and its firm disproval created much more general acceptance for the idea that genes were made of DNA.

Furthermore, the number of purines was almost equal to the number of pyrimidines, with the amount of the purine adenine (A) closely approximating that of the pyrimidine thymine (T). Correspondingly, the amount of the purine guanine (G) approximated that of the pyrimidine cytosine (C). At first the quantitative separation methods were not sufficiently accurate to prove true equality; but as they improved, the measured ratios of A/T and G/C grew closer to 1:1. No obvious significance was initially attached to these ratios (now known as Chargaff's ratios), but in retrospect this was not surprising since the fundamental DNA unit was considered to be the single polynucleotide chain. The fact that the truth might be otherwise could become clear only after DNA was taken up as an object of serious study by scientists whose primary concern was with its three-dimensional shape.

Discovery of the Double Helix

Because rotation around many chemical bonds in the backbone of polynucleotides is possible, there might be many different three-dimensional arrangements of DNA. If so, we might never know what genes look like. To test this unsettling possibility, the techniques of X-ray crystallographers were needed. Crystallographers study how X-rays are diffracted from crystallinelike compounds in order to determine the spatial arrangement of the atoms within the compounds.

X-rays were used to study structures as early as 1914, starting with the pioneering work of the English physicists W. H. Bragg and W. L. Bragg (father and son). The first molecules were simple salts such as sodium chloride, and even by 1945 the largest molecules yet worked out had molecular weights of less than 500. Yet probing work had started on the much, much larger proteins, which generally have molecular weights greater than 15,000. Preliminary work on DNA was commenced in Leeds by W. T. Astbury, just before the start of World War II. He used DNA sent over from Sweden by the physical chemist E. Hammersten, who was studying its size and shape.

Dry DNA has the appearance of irregular white fluffs of cotton, but it becomes highly tacky when it takes on water and can be drawn out into thin fibers, in which thin polynucleotide chains line up in parallel arrays, often semicrystalline in nature. With such fibers Astbury obtained his first DNA X-ray diagram. The fact that DNA gave a reproducible specific X-ray picture by itself was very important. Its backbone did not randomly twist about; instead, there was a clearly preferred shape that, if identified, might reveal how DNA functioned as the gene.

High-level crystallographic analysis of DNA began seriously in 1949 at the physics department of King's College, London. Leading the work was the physicist Maurice Wilkins, who, with his co-worker Rosalind Franklin, soon obtained remarkably detailed X-ray pictures by using carefully purified

DNA prepared in Bern by Rudolf Signer. The same basic diffraction pattern was obtained from DNAs with quite different ratios of AT/GC. The underlying structure was thus not only very regular but in some way—as yet to be determined—independent of the exact order of the purine and pyrimidine bases along a chain.

The X-ray diagrams furthermore revealed that DNA chains have a helical arrangement, in which the flat 3.4 Å thick purine and pyrimidine bases are stacked on top of one another in the center of the molecule, with the sugar–phosphate backbone on the outside twisting around the central axis and making one turn every 34 Å. Ten nucleotides along each chain are thus used to make one helical turn. Even more significantly, the measured 20 Å diameter was much larger than expected from a molecule containing only one polynucleotide chain. The question then became one of how to construct a highly regular DNA molecule in which two or perhaps three chains were twisted around each other.

In the spring of 1953, this matter was resolved by the proposal of James Watson and Francis Crick that DNA is a double helix in which two polynucleotide chains running in opposite directions are held

HYDROGEN BONDING BETWEEN BASE PAIRS

THYMINE
HYDROGEN BOND
ADENINE
DEOXYRIBOSE
DEOXYRIBOSE

CYTOSINE
GUANINE
DEOXYRIBOSE
DEOXYRIBOSE

BASE PAIRING OF TWO DNA CHAINS

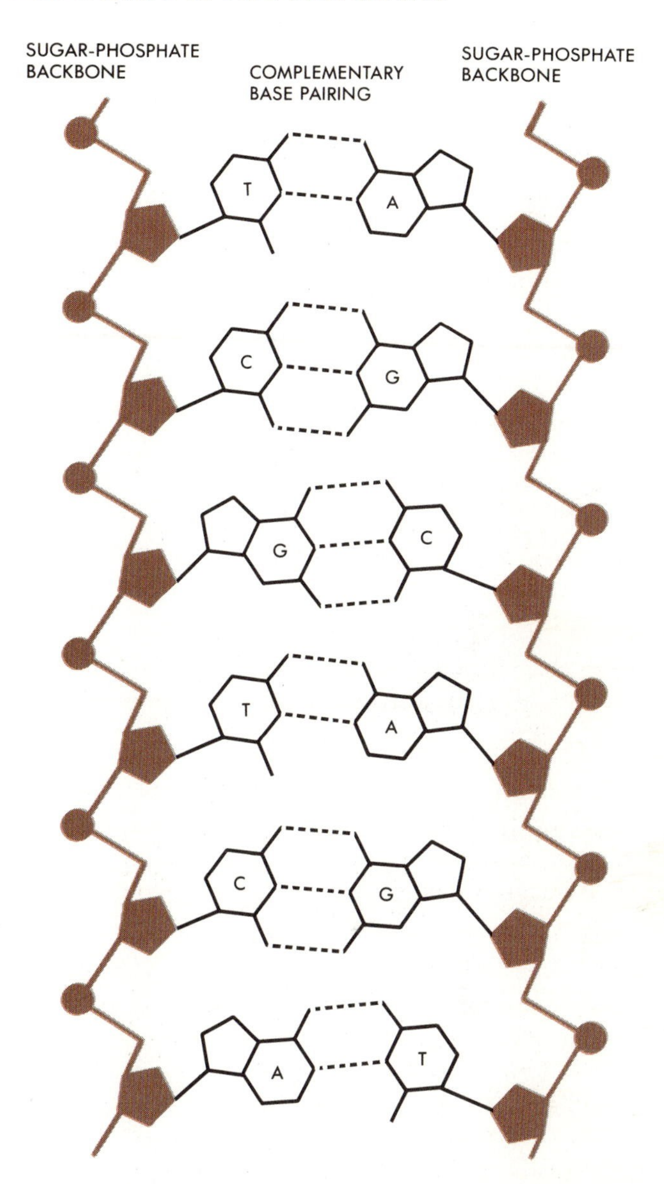

together by hydrogen bonds (a weak form of chemical bond) between pairs of centrally located bases. In the double helix the purine adenine (A) is always hydrogen bonded to the pyrimidine thymine (T), while the purine guanine (G) always hydrogen bonds to the pyrimidine cytosine (C). At any point along a chain, any of the four bases can be inserted to yield AT, TA, GC, and CG base pairs. The double helix is therefore compatible with all possible DNA sequences.

Because of this specific base pairing, if we know the sequence of one chain (e.g., TCGCAT), we also know that of its partner (AGCGTA). We refer to the opposing sequences as complementary and the corresponding polynucleotide partners as the complementary chains. Despite the relative weakness of the hydrogen bonds holding the base pairs together, each DNA molecule contains so many base pairs that the complementary chains never spontaneously separate under physiological conditions. If, however, DNA is exposed to near-boiling temperature, so many base pairs fall apart that the double helix separates into its two complementary chains (this process is called denaturation). The existence of the double helix provides a structural chemical explanation for Chargaff's rules: A = T; G = C. Only with these specific pairs can all the backbone sugar–phosphate groups have identical orientations and permit DNA to have the same structure with any sequence of bases.

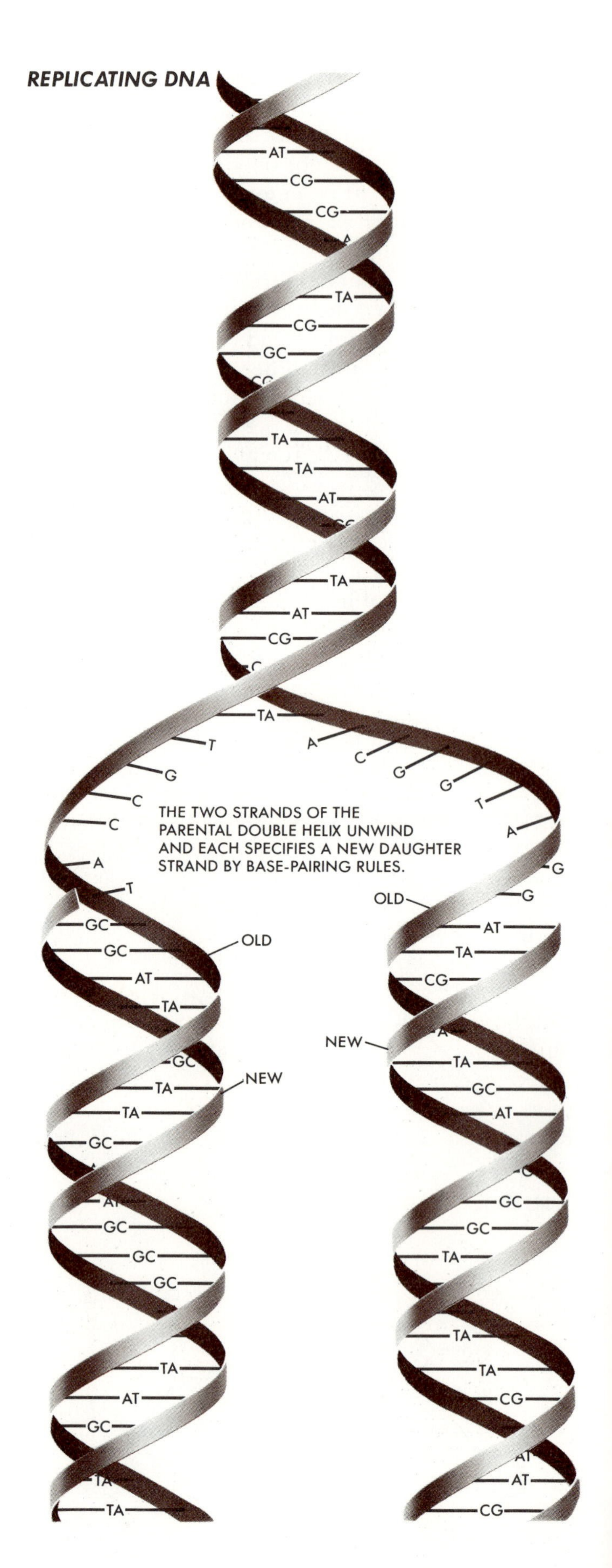
REPLICATING DNA
THE TWO STRANDS OF THE PARENTAL DOUBLE HELIX UNWIND AND EACH SPECIFIES A NEW DAUGHTER STRAND BY BASE-PAIRING RULES.
OLD
NEW
OLD
NEW

The Complementary Nature of DNA Is at the Heart of Its Capacity for Self-Replication

Before we knew what genes looked like, it was almost impossible to speculate wisely about how they could be exactly duplicated prior to cell division. Any proposal had to be general, and the best (most simple) proposal was that of Linus Pauling and Max Delbrück, his physicist colleague at Caltech. In 1940 they suggested that the surface of the gene somehow acted as a positive mold, or template, for the formation of a molecule of complementary (negative) shape, in much the same way that a plaster-of-paris cast can make an identical copy of a piece of sculpture. The complementary-shaped negative could subsequently serve as the template for the formation of its own complement, thereby producing an identical copy of the original mold.

Thus when it was realized that the two chains of DNA had complementary shapes, it was impossible not to get very excited and promptly propose that the two strands of the double helix should be regarded as a pair of positive and negative templates; each specifies its complement and thereby generates two daughter DNA molecules of identical sequence to that of the parental double helix. If this was indeed the way DNA duplicates, then we should find that the parental strands separate before duplicating and each daughter molecule contains one of the parental chains.

Proof of Strand Separation During DNA Replication

It took some five years before there was firm evidence of strand separation. Proof came from the experiments of Matthew Meselson and Franklin Stahl at Caltech. They had the clever idea of using high-density differences to separate parental from daughter DNA molecules. They first grew cultures of the bacterium *E. coli* in a medium highly enriched in the heavy isotopes ^{13}C and ^{15}N. By virtue of its ^{13}C and ^{15}N isotopic content, the DNA in these bacteria was much heavier than the normal light DNA coming from cells grown in the presence of the much more abundant natural isotopes ^{12}C and ^{14}N. Because of its greater density, heavier DNA can be clearly separated from light DNA by spinning in an extremely high-speed centrifuge.

When heavy DNA-containing cells are transferred to a normal "light" medium and allowed to multiply for one generation, all the heavy DNA is replaced by DNA half way in density between heavy and light. The disappearance of the heavy DNA indicates that DNA replication is not a conservative process in which the complementary strands of the double helix stay together. Instead, its replacement by the hybrid-density DNA indicates a semiconservative replication process, in which the two parental strands (heavy) separate to serve as templates for their complementary strands (light) and each daughter molecule has one heavy and one light chain.

Whether the complementary strands completely separated before replication started was not immediately known. Now abundant electron-microscope evidence of Y-shaped replication forks indicates that strand separation and replication go hand in hand. As soon as a section of the double helix begins to separate for replication, the resulting single-stranded regions are quickly used as templates and become converted to new double-helical regions.

Genetic Messages Within DNA Are Written in a Four-Letter Alphabet

Once the double helix was identified, it became possible to speculate more precisely on the one-gene-one-protein relationship. First of all, the genetic information in DNA must be conveyed solely by the linear sequences of its four bases, each of which represents one of the four letters (A, T, G, C) in the DNA alphabet. Gene mutations must therefore represent changes in the sequence of bases either by substituting one base pair for another or by adding, or deleting, one or many base pairs. Mutant proteins in turn would represent changes in amino-acid sequence, the simplest mutants being proteins in which one amino acid was replaced by another.

The first experiments showing mutant proteins bearing single amino-acid replacements involved the sickle hemoglobin molecules produced in humans with the genetic disease sickle-cell anemia. Working in Cambridge, England, Vernon Ingram demonstrated that several independently arising sickle mutations resulted in mutant hemoglobin chains that differed from wild-type normal hemoglobin chains through specific amino-acid substitutions at unique sites along the hemoglobin chain. This discovery hinted that many, if not most, mutations represented single base-pair changes as opposed to more drastic alterations in the base sequence. Probing still deeper into the gene–protein relationship was not then possible with this system because there was no conceivable way to map genetically the several known "sickle" mutants along the

THREE MECHANISMS OF MUTATION

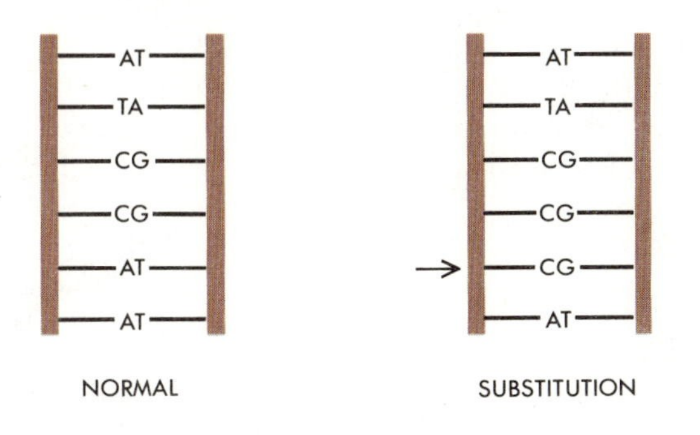

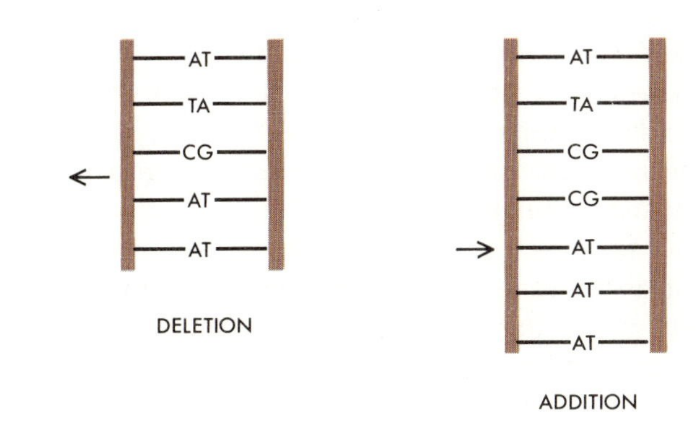

respective human chromosome. Nor could any way be seen to isolate specifically the DNA coding for the respective hemoglobin gene.

Development of Fine Structure Genetics

The great speed at which the basic facts of molecular genetics emerged following the discovery of the double helix was possible only because of the prior decision taken collectively in the mid-1940s to focus, whenever possible, on genetic experiments with the bacterium *Escherichia coli* (*E. coli*) and its various phages. Though some *E. coli* strains cause disease, for a long time totally nonpathogenic strains (e.g., B and K12) have been adapted for laboratory use and grow in culture extremely well (dividing as frequently as once every 20 minutes). They thus provide almost perfect systems for studying the organization of genes in the simplest known cells, as well as serving as ideal vehicles for observing more closely the nature of viruses.

The first rigorous genetic experiments with *E. coli* and its phages were started by Max Delbrück and Salvador E. Luria. In the early 1940s they observed that some *E. coli* cells mutated to resistance to specific phages, and they made accurate measurements of the spontaneous mutation rates. Soon afterwards the first phage mutants were isolated by Alfred D. Hershey; he showed that they recombine genetically. Yet it was not until the structure of DNA was known that the genetic structure of a single phage was thoroughly explored. The decisive experiments were done at Purdue University by Seymour Benzer, who isolated many hundreds of mutations within the *r*II gene of T4 to test whether genetic recombination (crossing over) occurs within the genes themselves. Quickly he found that to be the case, and in fact that most crossing over takes place within rather than between genes.

The same conclusion was soon afterwards reached from genetic studies of *E. coli* itself. Successful crosses between mutant bacteria were first done at Yale University in 1946 by Joshua Lederberg and Edward Tatum. After they mixed together pairs of different *E. coli* mutants, each bearing several different mutational deficiencies, progeny bacteria appeared that lacked the nutritional requirements of either of their respective parents. They soon established a tentative genetic map and proposed that *E. coli* has a chromosome capable of crossing over. The stage was thus set for a growing number of other geneticists to join Lederberg in exploiting the fact that very large numbers of progeny bacteria could quickly be examined from a single genetic cross. Genetics could thus be studied much faster on bacteria, which can multiply every 20 minutes, than on any higher organism. Though most research went first to ordering the various genes along the single *E. coli* chromosome and establishing the existence of two different sexes, attention later turned to the structure of single bacterial genes. The situation proved to be similar to that in the *r*II gene of bacteriophage T4; the mutations in certain bacterial genes occurred in a strictly linear order. This was expected if one assumed that the mutable sites are the successive base pairs of the corresponding DNA molecules.

Colinearity of the Gene and Its Polypeptide Products

The gene could now be precisely defined as the collection of adjacent nucleotides that specify the amino-acid sequences of the cellular polypeptide chains. Simplicity argued that the corresponding nucleotide and amino-acid sequences would be colinear, and this hypothesis was soon confirmed by correlating the relative locations of mutations in a gene with changes in its polypeptide product. The best early data were obtained at Stanford University by Charles Yanofsky, who studied mutations in a gene coding for an enzyme needed to make the amino acid tryptophan. He demonstrated very convincingly that the relative order of each amino-acid replacement was the same as that of its respective mutation along the genetic map. The molecular processes underlying colinearity, however, were not at all obvious because the 20 different amino acids far exceeded the number of different nucleotides in DNA. A one-to-one correspondence between nucleotides and amino acids could not exist. Instead, groups of nucleotides must somehow specify (code for) each amino acid.

RNA Carries Information of DNA to Cytoplasmic Sites of Protein Synthesis

A direct template role for DNA in the ordering of amino acids in proteins is not possible because DNA is located exclusively on the chromosomes in

ORDER OF GENE MUTATIONS IS THE SAME AS THAT OF AMINO-ACID CHANGES

TRYPTOPHAN SYNTHETASE: GENE AND PROTEIN

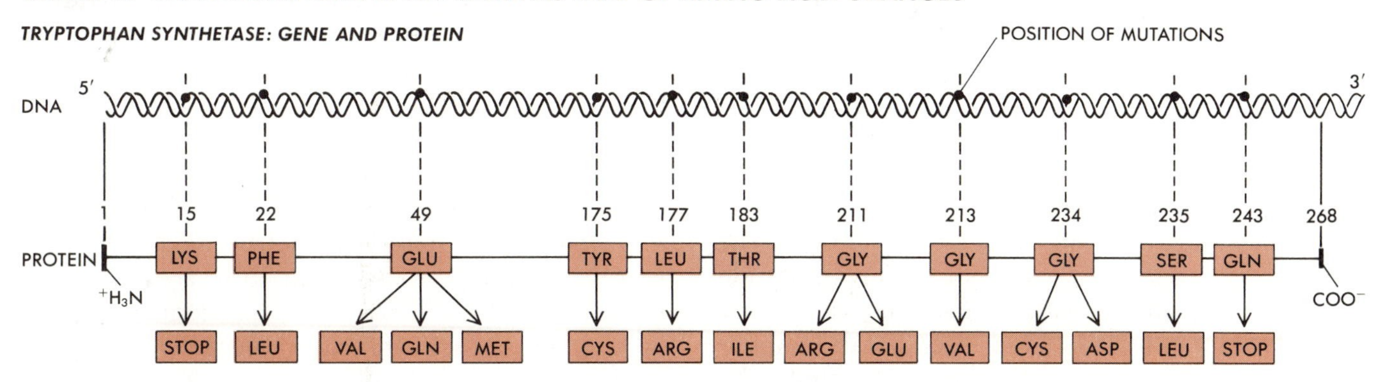

ALTERED AMINO-ACID RESIDUES

the nucleus, whereas most, if not all, cell protein synthesis occurs in the cytoplasm. The genetic information of DNA (the nucleotide sequence) must thus be transferred to an intermediate molecule, which then moves into the cytoplasm, where it orders the amino acids. Speculation that this intermediate molecule was RNA began seriously as soon as the double helix was discovered. Primarily, the cytoplasm of cells making large numbers of proteins always contained large amounts of RNA. Even more importantly, the sugar–phosphate backbones of DNA and RNA are quite similar, and it was easy to imagine the synthesis of single RNA chains upon single-stranded DNA templates to yield unstable hybrid molecules in which one strand was DNA and the other strand RNA. Here it is important to note that the unique base of RNA, uracil, is chemically very similar to thymine in that it specifically base-pairs to adenine. The relationship between DNA, RNA, and protein, as conceived in 1953, was thus

REPLICATION ↻ DNA ——→ RNA ——→ PROTEIN,
TRANSCRIPTION TRANSLATION

where single DNA chains serve as templates for either complementary DNA (the process of DNA replication) or complementary RNA molecules (the process of transcription). In turn, the RNA molecules serve as the templates that order the amino acids within the polypeptide chains of proteins during the process of translation, so named because the nucleotide language of nucleic acids is translated into the amino-acid language of proteins.

Imagining How Amino Acids Line Up on RNA Templates

The groups of nucleotides that code for an amino acid are called codons. From the beginning it seemed likely that most, if not all, codons comprised sets of three adjacent nucleotides. Groups of two can be arranged in only 16 different permutations ($4 \times 4 = 16$), four too few to code for the 20 different amino acids. Groups of three (AAA, AAC, AAU . . .), however, result in many more independent permutations ($4 \times 4 \times 4 = 64$) than are logically needed to specify all the amino acids. So we speculated whether the codons might be overlapping such that given bases would help to specify more than one amino acid. If that were true, we would expect to find restrictions on which amino acids could be linked together. The first known amino-acid sequences were thus eagerly scanned by the physicist George Gamow, to see whether some amino acids never occurred next to each other. By 1957 it was clear that no such restrictions of sequence existed

SINGLE DNA CHAIN SERVES AS TEMPLATE FOR RNA

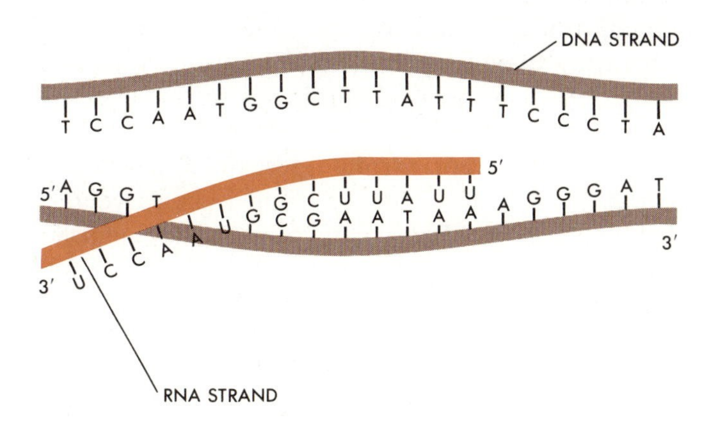

and that the successive codons along an RNA chain did not overlap.

Though initially we wondered whether the amino acids might be linked by fitting into specific cavities along the surfaces of RNA molecules, we could not ignore for long the fact that there was no obvious complementarity between the specific portions (the side groups) of many amino acids and the purine and pyrimidine bases of RNA. The side groups of the amino acids leucine and valine, for example, cannot form any hydrogen bonds, and unless somehow modified would not be expected to be attracted to any RNA template. This dilemma led Francis Crick to propose that many, if not all, amino acids had to be first attached to some form of adaptor molecules before they could chemically bind to an RNA template. Testing the adaptor suggestion, however, was impossible until techniques were developed for dissecting biochemically the exact steps of amino acids becoming incorporated into growing polypeptides.

Roles of Enzymes and Templates in the Synthesis of Nucleic Acids and Proteins

The initial reaction of the theoretically inclined geneticists to the double helix was the pure delight that came from seeing how it could function as a template. In a real sense their 50-year-old quest was over. In contrast the biochemists, who were then just working out how enzymes participated in the synthesis of the nucleotides and amino acids, saw their role as really now beginning. They realized that making phosphodiester and peptide bonds would also require specific enzymes. The discovery of such enzymes would demand finding conditions in which, say, DNA, RNA, or protein is made in extracts of disrupted cells. The first such experiments were difficult because not only do most cells contain the enzymes to make proteins and nucleic acid, but they frequently also possess active enzymes that can break them down. In the first successful experiments very little DNA, RNA, or protein was made. For example, test-tube-made DNA could be detected only by using radioactively labelled precursors (e.g., ^{14}C thymine) to distinguish it from the much larger amounts of unlabelled DNA that preexisted in the cells used to make the "cell-free extracts." The possession of an active extract, in which, say, DNA is made, allows further experimentation with extracts that are fractionated to discover both the exact precursors and the nature of the enzyme(s) needed for the assembly process. Such experiments should, furthermore, reveal the chemical identity of any templates that might be required.

By 1960 both DNA and RNA had been successfully synthesized in highly purified cell-free extracts, and the nature of their immediate precursors was firmly established. In both cases they were nucleotides containing three adjacent phosphate groups and hence are called nucleoside triphosphates. Two of these phosphate groups are split off when adjacent nucleotides are linked together, with the energy present in the broken bonds used to make the phosphodiester links of the sugar–phosphate backbone. The enzymes that directly make the phosphodiester bonds are called polymerases (because of the polymeric nature of the nucleic acids); the enzymes that make DNA are called DNA polymerases, and those that make RNA are known as RNA polymerases.

DNA polymerases catalyze the formation of DNA only in the presence of preexisting DNA templates, and the newly made DNA chains contain sequences complementary to their templates. For RNA polymerase to make RNA, DNA must be present; its template role was likewise shown by finding complementarity between DNA template and RNA product.

Three Forms of RNA Involved in Protein Synthesis

Fractionation of active cell-free extracts that had incorporated radioactively labelled amino acids into

RNA MOLECULES IN *E. COLI*

Type	*Relative amount (%)*	*Mass (kilodaltons)*
Ribosomal RNA (rRNA)	80	1.2×10^3 0.55×10^3 3.6×10^1
Transfer RNA (tRNA)	15	2.5×10^1
Messenger RNA (mRNA)	5	Heterogeneous

polypeptides revealed the temporary attachment of the newly made protein to the semispherical 200 Å-diameter cytoplasmic particles that contain RNA and are called ribosomes. Until 1960 it was generally believed that the major RNA components of ribosomes (ribosomal RNA) were the templates that ordered the amino acids. All ribosomes were found to be built from two unequally sized subunits (molecular weights, or MWs, of one and two million), each of which contained roughly equal amounts of protein and RNA. The existence of the ribosomal subunits had no simple explanation, nor could we propose any plausible hypothesis for why their apparent RNA templates fell into two fixed sizes (MWs 0.6 and 1.2 million), when polypeptides showed so much variation in size; some chains contained as few as 50 amino acids, others contained up to 2000.

Then to everyone's great surprise it was discovered that neither of the major ribosome RNA components had a template role. Instead, the true templates were a minor fraction of RNA representing less than 2 percent of all cellular RNA. Because it carries the specificity of the genes to the cytoplasm, it is called messenger RNA (mRNA). In the cytoplasm it moves over the surface of the ribosome, bringing successive codons into position for ordering their respective amino acids. Ribosomes are thus unspecific factories that by themselves have no specificity and to which any messenger RNA molecule can attach.

Equally important was the discovery that prior to their incorporation into protein, the amino acids are chemically linked to small RNA molecules called transfer RNA (tRNA). The amino-acid–tRNA complexes line up next to the codons of mRNA, with the actual recognition and binding being mediated by tRNA components. No contacts exist between the individual amino acids and the mRNA codons. Molecules of tRNA are thus the adaptors whose existence had been predicted several years before by Francis Crick.

For each of the 20 different amino acids, a specific enzyme catalyzes its linkage to its specific tRNA molecules. The amino acids are always linked to one specific end of their respective tRNA molecules; the binding of tRNA to mRNA is mediated by sets of internal nucleotides that have sequences complementary to their respective codons and thus are called anticodons.

STRUCTURE OF TRANSFER RNA

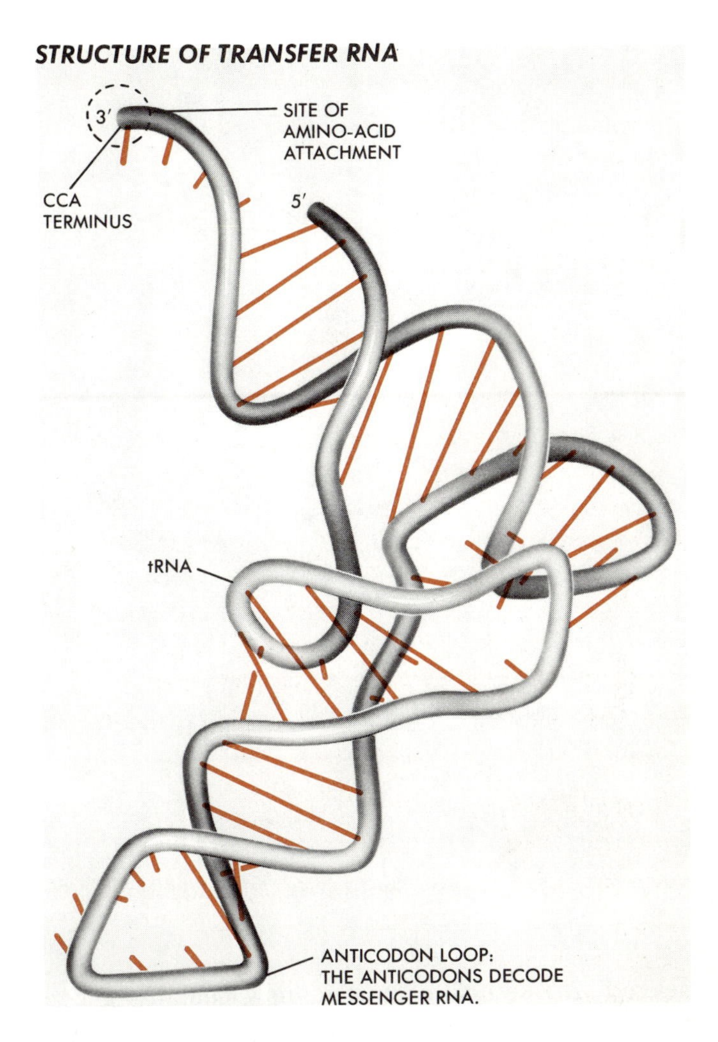

Genetic Evidence That Codons Contain Three Successive Bases

An exhaustive genetic study of a large number of phage T4 mutants that contained additions (or deletions) of single base pairs led Sydney Brenner and Francis Crick to the major statement that each codon contains three bases. In Cambridge, England, they made genetic crosses to show that the addition or subtraction of either one or two base pairs invariably led to highly abnormal nonfunctional proteins. In contrast, if three base pairs were either added or subtracted, the resulting proteins frequently were totally active. They concluded, as we now know most correctly, that the genetic code is read in stepwise groups of three base pairs. If one (or two) bases are added (or deleted), the resulting reading frame is upset, leading to the use of a completely new collection of codons that invariably code for amino-acid sequences that make no functional sense. In contrast, when groups of three base pairs are inserted or

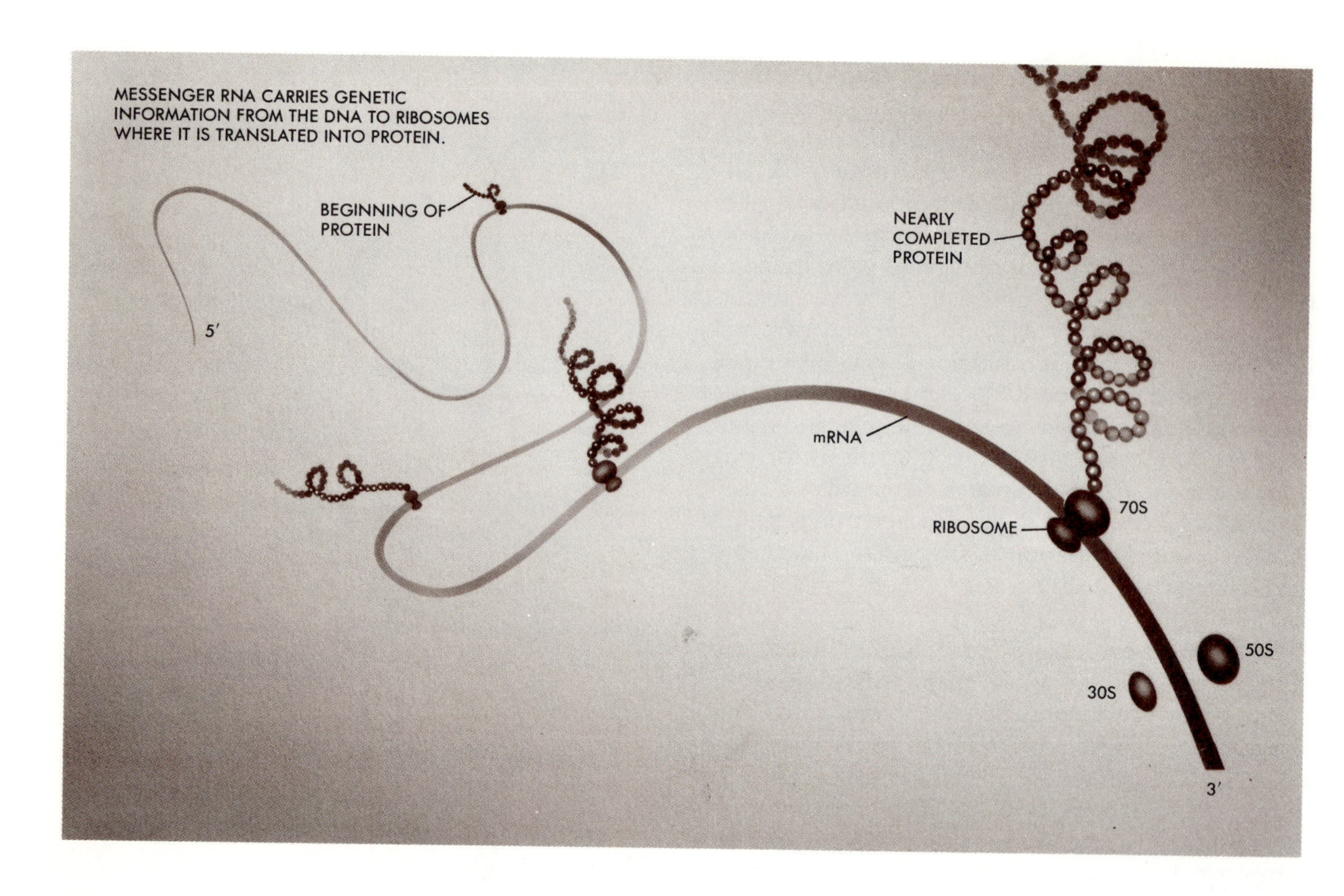
MESSENGER RNA CARRIES GENETIC INFORMATION FROM THE DNA TO RIBOSOMES WHERE IT IS TRANSLATED INTO PROTEIN.
BEGINNING OF PROTEIN
NEARLY COMPLETED PROTEIN
5′
mRNA
70S
RIBOSOME
50S
30S
3′

deleted, the resulting protein, now containing one more (or less) amino acid, remains otherwise unchanged and thus often retains full biological activity.

Synthetic mRNA to Make the Codon Assignments

The realization that ribosomes are by themselves unspecific and only become programmed to make specific RNA by binding mRNA molecules led Marshall Nirenberg and Heinrich Matthaei in 1961 to do their historic experiment in which they used as mRNA an enzymatically made regular polynucleotide poly-U (UUUUUU . . .). When poly-U was added to cell extracts containing ribosome molecules depleted of normal mRNA, they observed that only polyphenylalanine was synthesized. UUU thus coded for the amino acid phenylalanine. Soon poly-A (AAAAAA . . .) was found to code for strings of lysine residues, while poly-C (CCCCCC . . .) yielded polypeptides containing only proline. Over the next several years synthetic polynucleotides containing random mixtures of two or more nucleotides were used to work out tentatively many other codons.

The Genetic Code Fully Deciphered by June 1966

Most of the remaining still unidentified codons could be established when H. Gobind Khorana found ways to make codons by using repeating copolymers (e.g., GUGUGU . . . , AAGAAG . . . , GUUGUU . . .). By 1966 the search for the genetic code was over, and we unambiguously knew that (1) all codons contain three successive nucleotides, (2) many amino acids are specified by more than one codon (the so-called degeneracy of the code), and (3) 61 of the 64 possible combinations of the three bases are used to code for specific amino acids. The three combinations that do not specify any amino acid (UAA, UAG, UGA) were all found to code for stop signals that indicate chain termination.

The finding of stop codons at first created the expectation that specific start codons might also exist, especially since it was becoming more and more certain that all proteins begin with the amino

THE GENETIC CODE

First position (5′ end)	*Second position* U	C	A	G	*Third position (3′ end)*
U	Phe	Ser	Tyr	Cys	U
	Phe	Ser	Tyr	Cys	C
	Leu	Ser	Stop	Stop	A
	Leu	Ser	Stop	Trp	G
C	Leu	Pro	His	Arg	U
	Leu	Pro	His	Arg	C
	Leu	Pro	Gln	Arg	A
	Leu	Pro	Gln	Arg	G
A	Ile	Thr	Asn	Ser	U
	Ile	Thr	Asn	Ser	C
	Ile	Thr	Lys	Arg	A
	Met	Thr	Lys	Arg	G
G	Val	Ala	Asp	Gly	U
	Val	Ala	Asp	Gly	C
	Val	Ala	Glu	Gly	A
	Val	Ala	Glu	Gly	G

Note: Given the position of the bases in a codon, it is possible to find the corresponding amino acid. For example, the codon 5′ AUG 3′ on mRNA specifies methionine, whereas CAU specifies histidine. UAA, UAG, and UGA are termination signals. AUG is part of the initiation signal, in addition to coding for internal methionines.

acid methionine. But there is only one methionine codon (AUG), and it codes for internally located methionine as well as initiator methionine. AUGs that are used to start polypeptides are all closely preceded by a purine-rich sequence (e.g., AGGA) that may help to position the starting AUG opposite the ribosomal cavity containing the initiating amino-acid–tRNA complex.

Average-Sized Genes Contain at Least 1200 Base Pairs

Because all codons were found to contain three base pairs, the number of base pairs in a gene must be at least three times the number of amino acids in its respective polypeptide. An average-sized protein of 400 amino acids was thus thought to require a section of DNA comprising some 1200 nucleotide pairs. This number was much smaller than the number of base pairs in even the smallest DNA molecule. So most DNA molecules were thought to harbor many, many genes. That this is indeed the case became clear when the exact relationship be-

TRANSLATION OF MESSENGER RNA

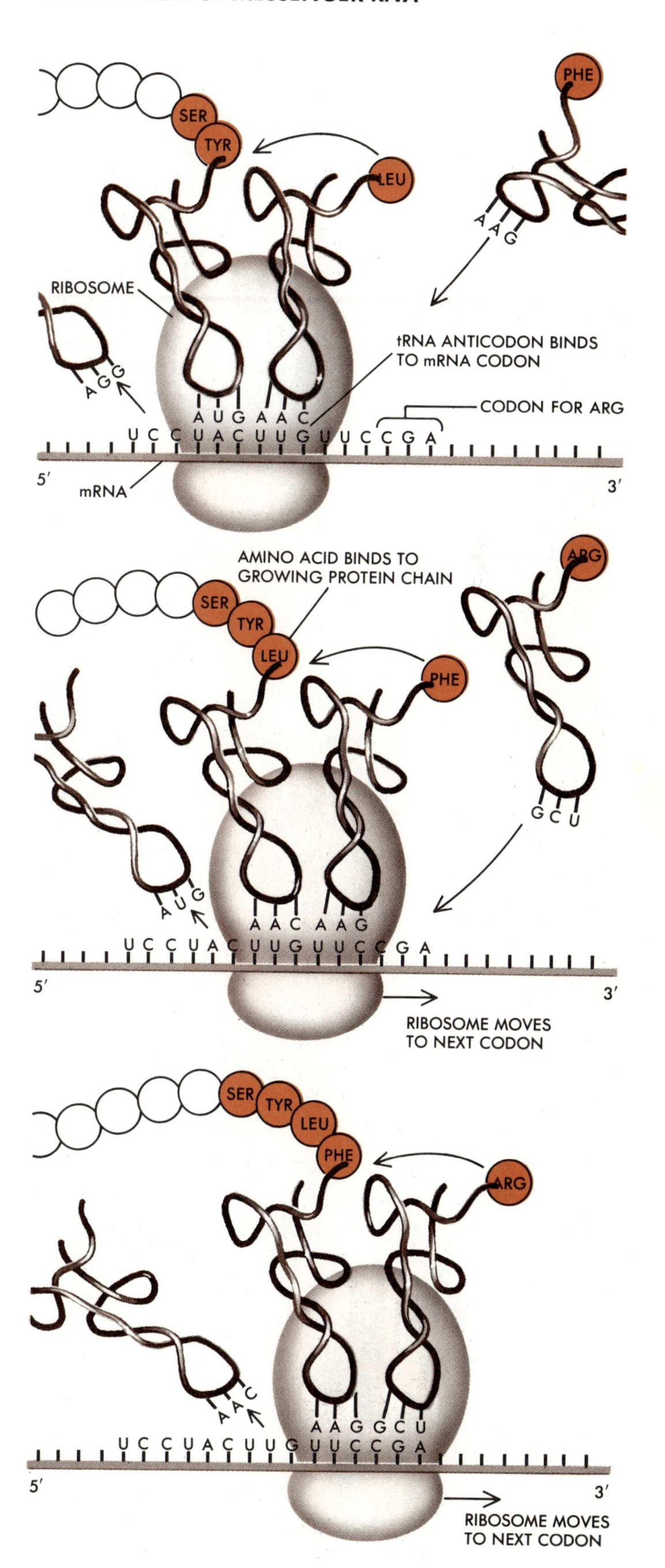
PHE
SER
TYR
LEU
RIBOSOME
tRNA ANTICODON BINDS TO mRNA CODON
CODON FOR ARG
5′
mRNA
3′
AMINO ACID BINDS TO GROWING PROTEIN CHAIN
ARG
RIBOSOME MOVES TO NEXT CODON
RIBOSOME MOVES TO NEXT CODON

tween DNA molecules and chromosomes was more precisely defined. When the double helix was found, it was believed that many distinct DNA molecules were used to construct all but the smallest chromosome. This picture suddenly changed with the realization that long DNA molecules are inherently fragile and easily break into much smaller fragments. When much more care was taken to prevent shearing, some DNA molecules containing as many as 200,000 base pairs were quickly seen.

Now our best guess is that the chromosome of the *E. coli* contains a single DNA molecule made up of more than 4 million base pairs. Only one DNA molecule can be found within the even larger chromosomes of higher plants and animals that on the average contain some twenty times more DNA than the *E. coli* chromosome. A chromosome is thus more properly regarded as a single genetically specific DNA molecule to which is attached a large number of positively charged protective(?) structural proteins (e.g., histones), as well as other proteins whose functions have yet to be determined.

Viruses as Sources of Homogenous DNA Molecules

Until recently, the only DNA molecules seriously studied had been isolated from DNA viruses, each of which contains a single DNA molecule. They may be either linear or circular in shape, with replication generally starting at a unique internal site and moving away bidirectionally until the duplication process is completed. The various DNA phages, particularly those that multiply on *E. coli,* were the favored early source of DNA to study because they are easily grown in large amounts and many have relatively small DNA molecules that do not easily break in solution. The well-studied linear DNA of phage T7, for example, consists of approximately 35,000 base pairs, along which some 30 genes have been mapped and the number of amino acids in each of their polypeptide products has been determined. Over 90 percent of its base pairs are used to specify these products, and so the individual genes must be quite close to one another.

Abnormal Transducing Phages Provide Unique Segments of Bacterial Chromosomes

The chromosomes of *E. coli*—if not of all bacteria—are circular, with bidirectional DNA replication always initiated at a specific site. They are much too long to visualize in their entirety in the electron microscope, and without genetic tricks there would be no method to select any specific section to study. Even the most careful isolation procedures necessarily shear bacterial chromosomes into tens of pieces with no two fragments having the same ends. Luckily, some high-powered bacterial genetics changed this bleak picture. Careful genetic examination of certain phages revealed that a small percentage genetically recombined their DNA with that of their host bacterial cells to yield abnormal phages in which functional fragments of bacterial DNA were inserted in the phage chromosomes. The phages carrying these hybrid chromosomes are called transducing phages and can program still other strains of bacteria to manufacture proteins they normally cannot make. With genetic tricks, specific transducing phages carrying one to several desired bacterial genes can be isolated, and, as we shall relate further on, most incisive use has been made of a phage that carries the *E. coli* gene involved in the breakdown of the sugar lactose to the simpler sugars glucose and galactose.

Plasmids as Autonomously Replicating Minichromosomes

In addition to the main circular chromosomes (4×10^6 base pairs) that carry almost all the genes, many bacteria possess large numbers of tiny circular chromosomes, which may contain only several thousand base pairs. These minichromosomes, called plasmids, were first noticed as genetic elements, unlinked to the main chromosome and carrying genes that conveyed resistance to antibiotics such as the tetracyclines or kanamycin. That these genes were found on plasmids as opposed to main-chromosomal DNA was not a matter of chance. Antibiotic resistance requires relatively large amounts of the enzymes that chemically neutralize the antibiotics. Perhaps the quickest way to achieve this is to place their respective genes on multicopy plasmids.

Replication of plasmid DNA is carried out by the same set of enzymes used to duplicate the main chromosomal DNA. When male and female cells mate prior to genetic recombination, a copy of the male chromosome is transferred into the female cells. In contrast, many plasmids cannot be transferred during mating, and once a gene is on such a

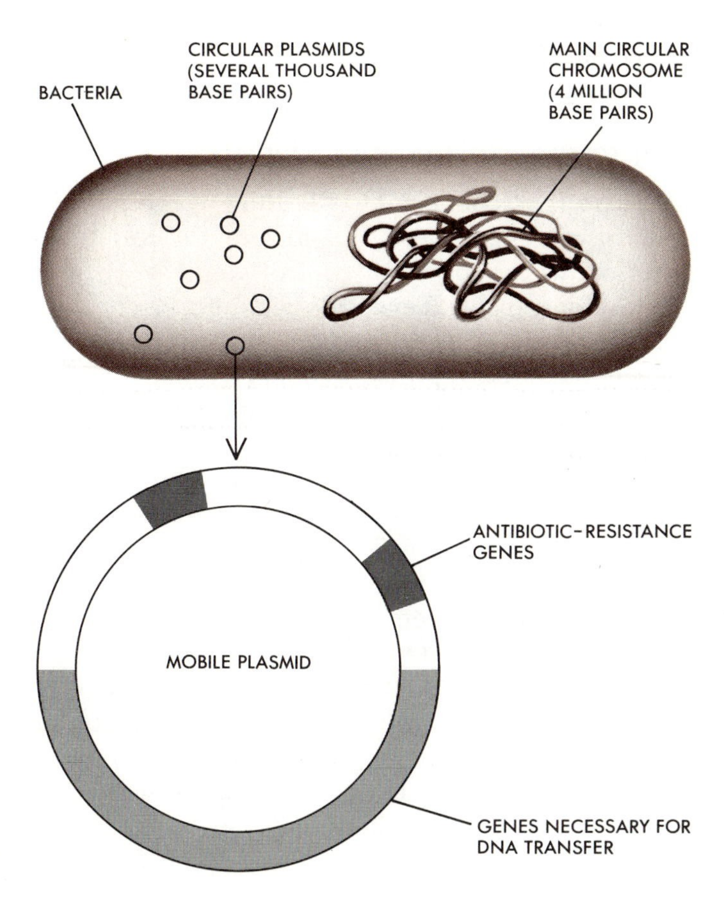

nonmobilizable plasmid it cannot be easily moved either onto another plasmid or onto a main chromosome.

Because plasmid DNA is so much smaller than even highly fragmented main-chromosomal DNA, it is easily separable, and highly purified plasmid DNA is readily obtained. When added to plasmid-free bacteria, it is taken up in a functional form to yield bacteria that will soon contain many copies of the plasmid. In general, for reasons as yet unclear, a given bacterial cell can only harbor one form of plasmid; two different plasmids cannot coexist with each other.

Specific Repressor Molecules Regulate the Rate of mRNA Synthesis

Bacteria generally exist in environments that change rapidly, and some of the enzymes they may need at one moment may be unwanted the next hour. Correspondingly, they may quickly need to use another enzyme whose presence was previously unnecessary. One way of meeting these challenges would be to carry on simultaneously the synthesis of all such enzymes independently of whether their substrates are present. This would be inherently wasteful, however, and in fact bacteria do not operate this way. Instead, bacterial genes are constructed so that many function at highly variable rates, with the respective mRNA molecules made at appreciable rates only when their genes receive signals from outside to go into action.

The compounds that transmit these signals are called inducers. For example, the enzyme β-galactosidase is normally made only at high rates when *E. coli* cells are growing in the presence of its inducer, lactose. This sugar can only function as a food source when it is cleaved into the simpler sugars glucose and galactose, and β-galactosidase is the enzyme that catalyzes this splitting. The presence of lactose greatly increases the rate at which the enzyme RNA polymerase can bind to the beginning of the gene that codes for β-galactosidase and begin the synthesis of the respective mRNA. In turn, the higher amounts of β-gal mRNA lead to correspondingly more β-galactosidase.

Genetic analysis proved crucial to the working out of the molecular details of this adaptive phenomenon, with the essential clues emerging from the study of mutants that were unable to vary the amount of β-galactosidase. The key findings were that independent of whether lactose was present or absent, some mutants made maximum amounts of enzyme while other mutants produced only traces. Such results led Jacques Monod and François Jacob of the Institut Pasteur in Paris to postulate two things: (1) the existence of a specific repressor molecule that binds near the beginning of the β-galactosidase gene at a specific site called the operator; the repressor, by binding to the operator site on the DNA, sterically prevents RNA polymerase from commencing synthesis of β-gal mRNA; and (2) lactose acts as an inducer which, by binding to the repressor, prevents the repressor from binding to the operator. In the presence of lactose the repressor is inactivated and the mRNA is made. Upon removal of lactose, the repressor regains its ability to bind to the operator DNA and switch off the genes.

Bacterial Genes with Related Functions Organized into Operons

The *E. coli* gene for β-galactosidase is located adjacent to two additional genes involved in lactose metabolism. One gene codes for lactose permease, a protein that facilitates the specific entry of lactose into bacteria, while the second codes for thiogalactosidase transacetylase, an enzyme which may help to remove lactoselike compounds that β-galactosidase cannot split into useful metabolites. The same mRNA that codes for β-galactosidase codes for the permease and the acetylase; thus when lactose is added to *E. coli* cells, the relative amounts of all three proteins rise coordinately. The collections of adjacent genes that are transcribed into single mRNA molecules are called operons. Some operons are large, with, for example, the eleven proteins involved in the synthesis of the amino acid histidine all translated off one extremely large mRNA molecule containing over 10,000 nucleotides.

The site where RNA polymerase binds to the beginning of a gene is called the promoter. The operators to which repressors bind are always close to their respective promoters. In this way the binding of a repressor necessarily blocks the simultaneous binding of RNA polymerase. The control of a promoter by a repressor is thus an example of negative control. Promoters can also be under control of positive effector molecules. By binding upstream from the promoter (away from the start of mRNA synthesis), they increase the rate at which mRNA chains are made. Positive control elements most likely act by helping to open up the two chains of the double helix at the promoter site, thereby facilitating the binding of RNA polymerase.

Constitutive Synthesis of Repressor Molecules

Each repressor is coded for by a specific gene. The gene for the lactose repressor lies immediately in front of the operon. In other cases, however, the repressor genes may occur widely separated from the operon's genes. The rate at which repressors are made is normally invariant (constitutive synthesis), with the exact rate being a function of the structure of its specific promoter. Normally the promoters of repressor genes function at very low rates, leading to the presence of only a few repressor mRNA molecules in the average cell. Promoter mutants exist, however, which allow much higher rates of repressor mRNA synthesis and, correspondingly, much higher (say, tenfold) numbers of repressor molecules per cell. Even in the presence of high levels of inducer (e.g., lactose), such mutant cells make smaller than usual amounts of the respective induced proteins (e.g., β-galactosidase).

Repressors Isolated and Identified

Because the genetic studies leading to the postulation of repressors were so complete, the role of repressors as key bacterial control elements seemed almost inescapable. Final proof, though, had to await the development of biochemical procedures by which individual repressors could be isolated, chemically identified, and shown specifically to bind to their respective operators. These steps depended on the development of genetic techniques to increase the number of repressor molecules per cell beyond the few copies normally present. As long as repressor amounts were that low, there was no effective way to tackle them.

Two main tricks were used in conquering the lactose repressor. The first was the genetic manipulation of the *E. coli* genome. The gene coding for the lac repressor and the beginning of the lac operon was attached to the phage λ chromosome. When such phages multiply in *E. coli,* several hundred copies of the λ-lac chromosome are produced, as well as corresponding large numbers of lac repressor mRNA molecules. In this way the amounts of repressor per cell were amplified some tenfold. Still further enrichment came from constructing the λ-lac strains with mutant promoters, which overproduce repressor mRNA by another factor of ten. The resulting amounts of repressor became sufficient in 1966 to allow Walter Gilbert and Benno Müller-Hill to demonstrate at Harvard University that the lactose repressor was a protein of MW 38,600, which has two specific binding sites—one for lactoselike compounds, the other for DNA containing the lactose operator. Virtually simultaneously Mark Ptashne, also at Harvard, isolated the repressor coded by phage λ that controls the rate at which several classes of λ-specific mRNA is coded. This repressor is a 26,000-MW polypeptide chain and likewise binds only to its specific operator.

More recently, several positive regulators of RNA polymerase binding and hence gene expression have been isolated from *E. coli* and also shown to be proteins. The best-understood positive regulator protein signals to the appropriate genes that glucose is not available as a food source. When glucose is absent, there is a buildup in the amounts of the intracellular regulator cyclic-AMP (cAMP). This

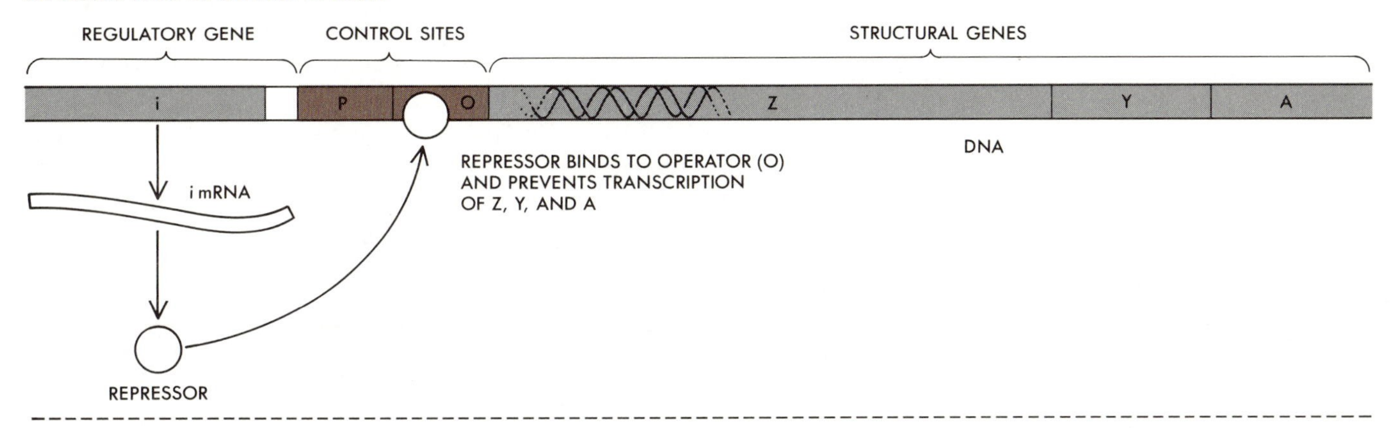

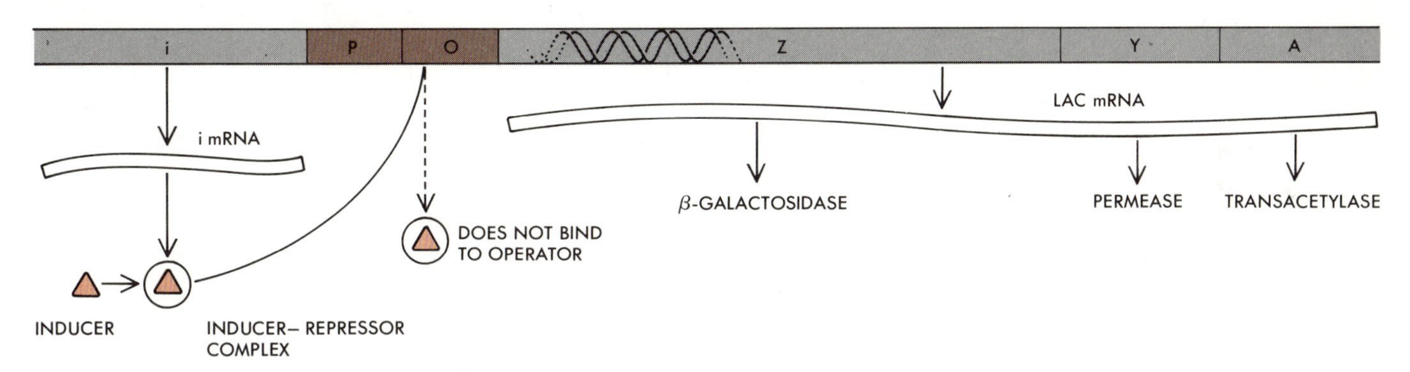

cAMP then in turn binds to the DNA-binding protein known as CAP. The resulting cAMP–CAP complexes, by binding to the respective promoters, help to activate operons whose respective enzymes can break down alternative sugars such as lactose or galactose.

In bacteria the regulation of DNA functioning is thus controlled in part, if not exclusively, by the binding of specific regulatory proteins to control sequences situated at the beginning of their various genes (operons).

Difficulties in Probing Gene Regulation in Higher Plants and Animals

By the late 1960s it was commonplace to ask whether the genes of higher cells were also regulated by specific DNA-binding proteins. In particular, would such proteins be the key to understanding the mysteries of embryology, through which fertilized eggs divide and differentiate, eventually giving rise to the highly specific cell types that make up our issues and organs? Here the study of single animal or plant cells growing in culture, as opposed to the study of whole organisms, was the way to proceed, and virtually all serious embryologists were hoping soon to develop model cell-culture systems in which the functions of single genes could be followed during differentiation. Merely watching proteins come and go during differentiation, however, would not by itself be a major step forward. We wanted to discover the signals that would turn a given gene on or off, and to know how so many proteins can have their expression so exquisitely coordinated.

Here the only signals we already knew of in higher organisms were certain steroid hormones that had recently been shown to bind to specific receptor molecules in the cytoplasm. These hormone–receptor complexes then moved to the nucleus, where they clearly bound to chromosomes and somehow turned on (off) the synthesis of many specific proteins. All attempts to pursue this phenomenon at the DNA level, however, came to naught. We had no way of finding the particular sections of DNA to which the hormone–receptor complexes were binding specifically and of distinguishing specific binding from nonspecific binding to any section of DNA. Moreover, though mutant cells could be found in which hormones did not function normally, there was no simple way to map genetically the relevant genes. No way then, or now, exists for doing conventional genetic crosses between higher cells growing in culture. As a result, even if operators and repressors exist in higher cells, the probability of genetically demonstrating them was indeed slim, if not nonexistent. And as long as it remained impossible to isolate from the large chromosomes of higher organisms specific DNA fragments that coded for particular proteins, direct tests for the presence of specific DNA-binding proteins were also out of the question.

Animal Viruses: Model Systems for Gene Expression in Higher Cells

Although the chromosomal DNA of higher cells seemed virtually hopeless to study, strong arguments could be made that some real answers might come through studying how DNA viruses multiply in higher animal cells. Many of these viruses grew well in cultured cells and could easily be labelled with radioactive isotopes. Moreover, some—like the monkey virus SV40 and the mouse virus polyoma—had remarkably small circular DNA molecules containing maybe as few as 5000 base pairs and most likely coding for less than ten proteins. They thus resembled the simple bacterial viruses from which we had learned so much. We were also drawn to these viruses because several were being found to transform those cells in which they cannot multiply well into cancerous cells capable of forming tumors in appropriate animals. Each of these viruses thus might contain one or more genes specifically coding for proteins capable of transforming a normal cell into its cancerous equivalent. By studying these viruses, we might learn some of the basic principles of gene expression in higher cells, as well as basic facts about the origin and nature of cancer cells.

Deep understanding of the DNA tumor viruses thus became the goal of an increasing number of scientists who in the late 1960s had decided to switch their research from bacteria and their viruses to cells of higher animals. By 1970, SV40 and polyoma viruses were both shown to have a life cycle neatly divisible into early and late phases. During the early phase, mRNA is made that codes for a viral protein that accumulates in the cell nucleus, where it plays a necessary role in the replication of SV40 DNA that occurs during the late phase of the viral life cycle. A strong hint existed that this protein might be the key to the cancer-causing ability of SV40, when we discovered that it was the only SV40-coded product consistently present in cells made cancerous by the virus. Since its discovery, therefore, this protein has been known as the T (tumor) protein. The late stage of the SV40 life cycle is marked by the synthesis of SV40 DNA as well as the synthesis of mRNA, which codes the structural proteis of the virus particle's outer shell.

How to follow up these facts at the molecular level, however, was not obvious because there was as yet no way to relate the SV40 or polyoma virus mRNAs and the corresponding proteins to specific sections of the SV40 and polyoma DNAs. The circular forms of the viral DNAs at first seemed to preclude the finding of any specific reference point (e.g., an end) from which, with the electron microscope, the binding sites of the early and late mRNA could be mapped. The realization that animal cells and their viruses would never be as simple to work with as bacteria or their phages was something we would always have to live with. There is no economically feasible way to grow animal cells on the same large scale as bacteria. The molecular understanding of how any DNA animal virus gives rise to cancer was therefore far from attainable because there were no procedures then to handle the small amounts of their key nucleic-acid and protein components.

Methodologies for Creating Recombinant DNA Molecules

Development of Ways to Sequence Nucleic Acids

For deeper insights about the organization of DNA, methods had to be developed to reveal exact nucleotide sequences—first of selected regions of the gene, then of an entire gene, and finally of an entire chromosome. The first nucleic-acid sequences to be established were not of DNA but of the relatively small tRNA molecules that are built up from only 75–80 nucleotides. Already in 1964, the structure of the yeast tRNA molecule to which the amino acid alanine attaches was worked out. To do this, Robert Holley and his colleagues at Cornell University had to find specific enzymes that broke the tRNA chains reproducibly into smaller and smaller discrete fragments, until they were a size that could be sequenced directly by simple stepwise degradation procedures.

With each year these methodologies greatly improved, and by 1975 the complete sequence of the RNA chromosome* of the single-stranded RNA phage MS2 was worked out in Walter Fiers' laboratory in Ghent. For the first time the precise way in which a simple chromosome was put together could be visualized. For the first time we knew the exact codons that specify the amino acids of the three proteins coded by the three genes of phage MS2, as well as the stop codons that signal chain termination. Few nucleotides separated the three genes, but unexpectedly long untranslated regions (129 and 174 bases) existed at the two ends. Here, as on all other messenger-RNA-like molecules, the two physical ends never act as start or stop signals.

Direct sequencing of any DNA molecule was not then possible because there was no way to cut DNA at specific points to produce discrete reproducible fragments having unique sequences. The then available deoxyribonucleases (DNases) all cut DNA into hopelessly heterogeneous collections of small fragments whose order within the original DNA could never be deciphered.

Restriction Enzymes Make Sequence-Specific Cuts in DNA

All the enzymes first found to break the phosphodiester bonds of nucleic acids, the so-called nucleases, showed very little sequence dependency; the most specific was the T1 RNase, isolated from a mold, which only cuts next to guanine residues. Highly preferred sites of cleavage on certain RNAs were found, but these reflected the way single-stranded RNA molecules fold into complex three-dimensional arrangements rather than the enzymes' tendency to cut within specific base sequences. The prevailing opinion was that highly specific nucleases would never be found, and therefore the isolation of discrete DNA fragments, even from viral DNA, would not be possible. The only grounds to think differently were observations, beginning as early as 1953, that when DNA molecules from one strain of *E. coli* are introduced into a different *E. coli* strain (for example, *E. coli* strain B vs. *E. coli* strain C), they rarely functioned genetically. Instead the foreign DNAs are almost always quickly fragmented into smaller pieces. Quite infrequently the infecting DNA molecule would not be broken down because it had somehow become modified so that it and all its descendants could now multiply on the new bacterial strain. In 1966 chemical analysis of a small viral DNA modified in such a way that it could survive in a different strain of *E. coli* revealed the presence of one to several methylated bases not present in the unmodified DNA. Methylated bases are not inserted as such into growing DNA chains but arise through the enzymatically catalyzed addition of methyl groups (CH_3) to preexisting, usually newly synthesized, DNA chains.

The stage was thus set in the late 1960s for Stewart Linn and Werner Arber, working in Geneva, to find in extracts of cells of *E. coli* strain B both a specific modification enzyme that methylated unmethylated DNA and a "restriction" nuclease that broke down unmethylated DNA. Over the next several years, the discovery of restriction nucleases and their companion modification methylases in two other *E. coli* strains opened up the possibility that many site-specific nucleases might exist. None of these early *E. coli* restriction enzymes lived up to their finders' first hopes, however, because although they recognized specific unmethylated sites, they cleaved the DNA at random locations far removed from their recognition sites.

Specific restriction nucleases that did cleave their recognition sites were soon found. The first was discovered in 1970 by Hamilton Smith of Johns Hopkins University, who followed up his accidental finding that the bacterium *Haemophilus influenzae* rapidly broke down foreign phage DNA. This degradative activity was subsequently found in cell-free extracts and shown to be due to a true restriction nuclease because it easily broke down *E. coli* DNA, while it failed to cut up the DNA of the *Haemophilus* cells from which it had been extracted. When highly purified, *Hin*dII, as this enzyme is

*RNA replaces DNA as the genetic material in many viruses (e.g., tobacco mosaic, influenza, polio, certain RNA phages). Replication follows the same pattern used for DNA, with single RNA chains serving as templates to make chains of complementary sequence. The specific replication enzymes are coded on the viral RNA chromosomes and called replicases. Equally important, one of the complementary partners can also function as mRNA by combining with ribosomes and coding directly for the amino acids of the viral proteins.

SOME RESTRICTION ENZYMES AND THEIR CLEAVAGE SEQUENCES

Microorganism	Abbreviation	Sequence (5′ → 3′ / 3′ → 5′)
Bacillus amyloliquefaciens H	*Bam*HI	G↓G A T C C C C T A G↑G
Brevibacterium albidum	*Bal*I	T G G↓C C A A C C↑G G T
Escherichia coli RY13	*Eco*RI	G↓A A T T C C T T A A↑G
Haemophilus aegyptius	*Hae*II	Pu G C G C↓Py Py↑C G C G Pu
Haemophilus aegyptius	*Hae*III	G G↓C C C C↑G G
Haemophilus haemolyticus	*Hha*I	G C G↓C C↑G C G
Haemophilus influenzae R_d	*Hind*II	G T Py↓Pu A C C A Pu↑Py T G
Haemophilus influenzae R_d	*Hind*III	A↓A G C T T T T C G A↑A
Haemophilus parainfluenzae	*Hpa*I	G T T↓A A C C A A↑T T G
Haemophilus parainfluenzae	*Hpa*II	C↓C G G G G C↑C
Providencia stuartii 164	*Pst*I	C T G C A↓G G↑A C G T C
Streptomyces albus G	*Sal*I	G↓T C G A C C A G C T↑G
Xanthomonas oryzae	*Xor*II	C G A T C↓G G↑C T A G C

called, was found to bind to this set of sequences, in which the arrows indicate the exact cleavage sites, and Py and Pu represent any pyrimidine or purine residue:

5′GTPy↓PuAC3′
3′CAPu↑PyTG5′

Since then restriction enzymes that cut specific sequences have been isolated from some 230 bacterial strains, and over 70 different specific cleavage sites have been found. Some of these enzymes recognize specific groups of four bases, whereas many more recognize groups of six. The ones that bind to only four bases cut many more bonds in a given DNA molecule than those restriction enzymes that have to recognize a specific group of six. A six-base restriction sequence by chance may not exist even once in a given viral DNA molecule. For example, the GAATTC/CTTAAG recognition sequence of the *E. coli* RI enzyme is not present in phage T7 DNA, which is 35,000 base pairs long.

Restriction Maps Are Highly Specific

The various fragments generated when a specific viral DNA is cut by a restriction enzyme can be easily separated by using electric fields to move the negatively charged DNA pieces through flat porous agarose gels (a process called electrophoresis). The rate at which the fragments move is a function of their lengths, with small fragments moving much faster than large fragments. Depending on the concentration of agarose, the larger fragments may hardly be able to move into the gel. Restriction fragments move unharmed through such agarose gels and can be eluted as biologically intact double helices. Staining of such gels with dyes that bind to DNA generates a series of bands (a restriction map), each corresponding to a restriction fragment whose molecular weight can be established by calibration with DNA molecules of known weights. Different restriction enzymes necessarily give different restriction maps for the same viral DNA molecule. In general, the most useful enzymes are the ones that recognize rare recognition sequences and therefore produce small numbers of fragments that can easily be separated from one another on the agarose gels.

The first restriction map was obtained in 1971 by Daniel Nathans, a colleague of Hamilton Smith's

RESTRICTION FRAGMENTS OF DNA SEPARATED BY ELECTROPHORESIS

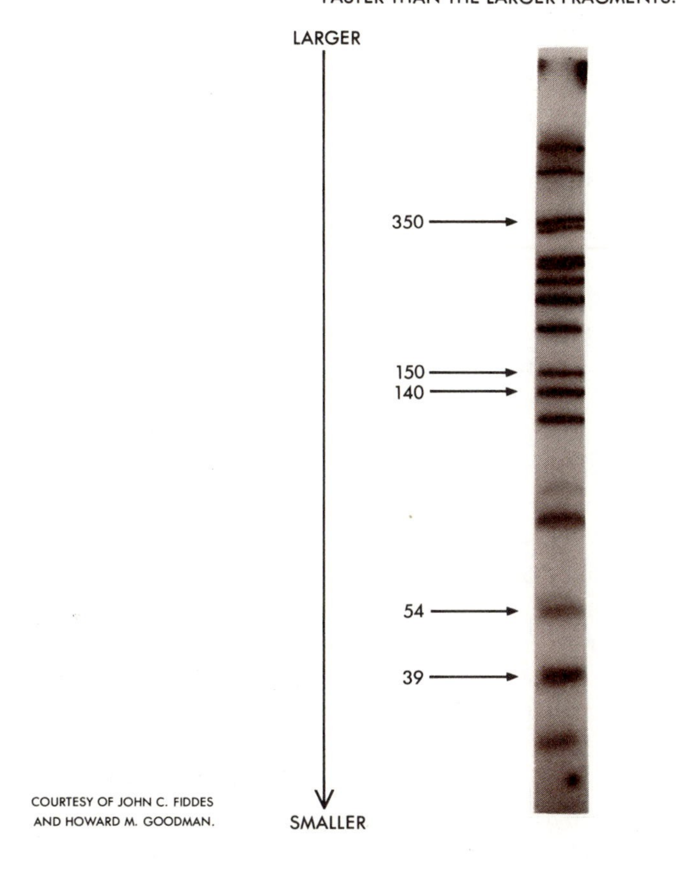

RESTRICTION MAP OF SV40 PRODUCED WITH ENZYMES HindII AND HindIII

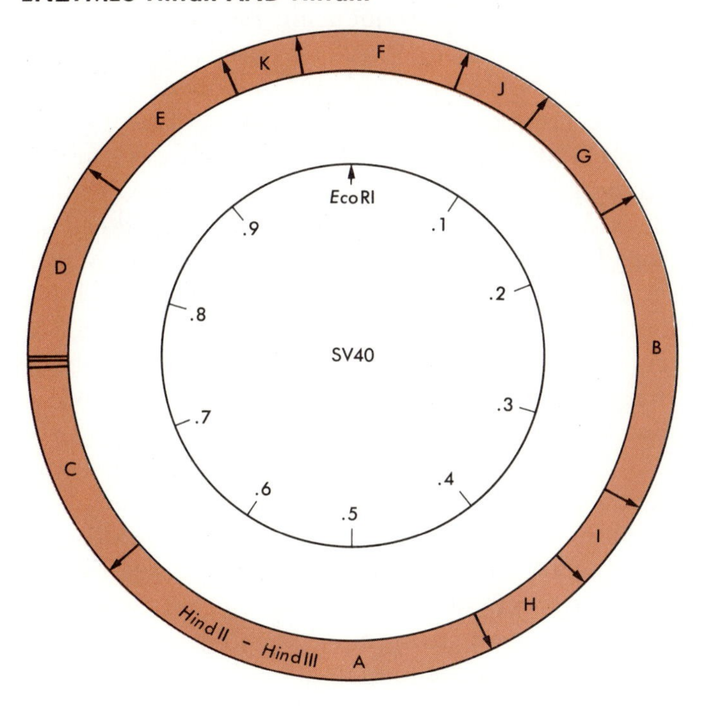

at Johns Hopkins. He used the *Hin*dII enzyme to cut the circular DNA of the monkey virus SV40 into eleven specific fragments. The order in which these eleven fragments occur in the SV40 DNA could be deduced by studying the patterns of fragments produced as the digestion proceeded to completion. The first cut breaks the circular molecule into a linear structure that is then cut into progressively smaller fragments. By following the pattern of production of, first, the overlapping intermediate-sized fragments and from them the fragments of the complete digest, Nathans produced a restriction map that locates the sites on the circular viral DNA that are attacked by the restriction enzyme. By repeating the experiment with other enzymes, we can build a more detailed map with many different restriction sites.

With this information it became possible to determine the position on the circular viral DNA of regions of biological importance. For example, by briefly radioactively labelling replicating viral DNA and then digesting it with *Hin*dII, Nathans proved that the replication of SV40 DNA always begins in one specific *Hin*dII fragment and proceeds bidirectionally around the circular DNA molecule. Subsequently, by using other enzymes, each of which generates a different set of fragments, including the enzyme *Eco*RI which cuts SV40 DNA only once, experimenters could precisely locate the site of the initiation of DNA replication at some 1700 base pairs away from the *Eco*RI site.

The restriction maps and restriction fragments were then used to identify the regions of the viral DNA that specify the mRNAs of the viral proteins at different stages during viral replication. To do this, radioactively labelled mRNA from infected cells was isolated at early and late times after infection. Pure restriction fragments of the viral DNA were prepared and denatured so the two DNA chains of the double helix separated. The mRNA was then mixed with the separated DNA strands in conditions that allow the RNA to form RNA–DNA double helices with DNA strands that have a complementary base sequence. Such RNA–DNA hybridization experiments revealed that both early and late viral RNA are coded by continuous DNA regions, each spanning about half of the total SV40 DNA. The promoters of both the early and the late mRNAs are near the origin of DNA replication, but synthesis of early mRNA proceeds counterclock-

GILBERT AND MAXAM SEQUENCING PROCEDURE

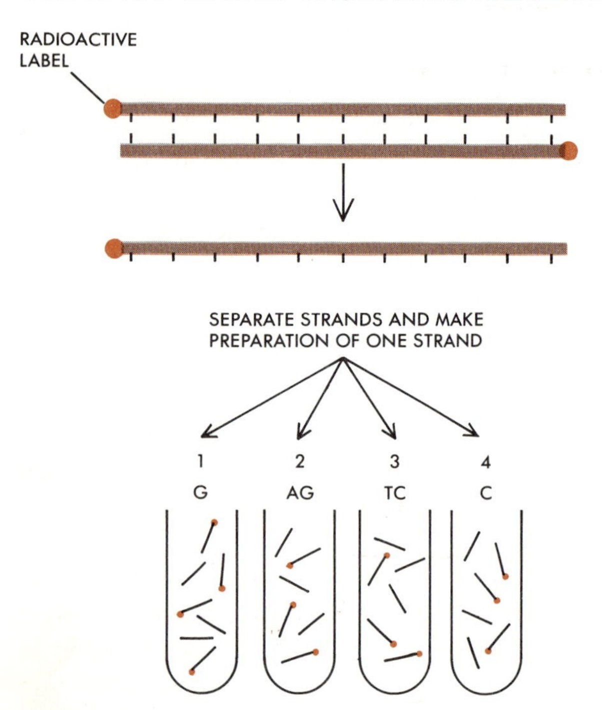

CHEMICAL AGENT IN TUBES DESTROYS ONE OR TWO OF THE FOUR BASES AND SO CLEAVES STRANDS AT THOSE SITES. THE REACTION IS CONTROLLED SO THAT ONLY SOME STRANDS ARE CLEAVED AT EACH SITE, GENERATING A SET OF FRAGMENTS OF DIFFERENT SIZES.

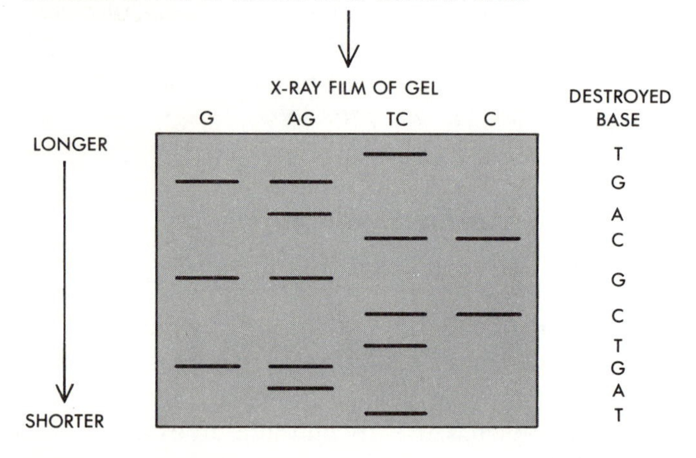

FRAGMENTS FROM TUBES ARE SEPARATED ACCORDING TO SIZE BY GEL ELECTROPHORESIS AND THE RADIOACTIVE FRAGMENTS PRODUCE IMAGES ON AN X-RAY FILM.

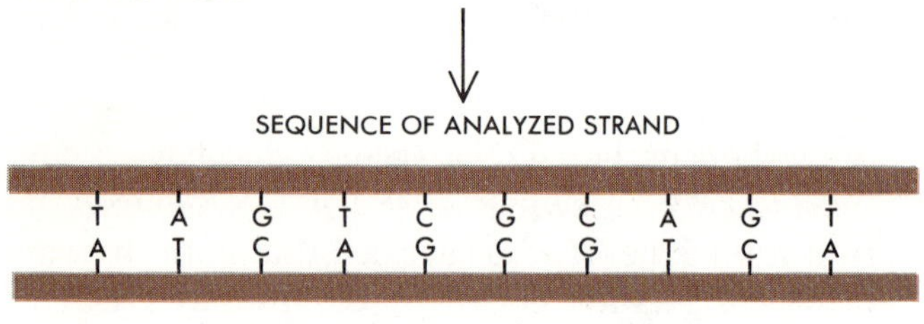

IMAGES ON THE X-RAY FILM DETERMINE WHICH BASE WAS DESTROYED TO PRODUCE EACH RADIOACTIVE FRAGMENT. THIS INFORMATION YIELDS THE BASE SEQUENCE OF THE ANALYZED STRAND.

wise and that of late mRNA clockwise. These techniques, which allow us to identify the genetically significant regions of DNA and then physically dissect them, are important in themselves. But they also paved the way to extremely powerful new methods of DNA sequencing and recombinant DNA techniques.

Restriction Fragments Lead to Powerful New Methods for Sequencing DNA

When the first restriction fragments became available, there was no good method to sequence them directly. The only realistic way to proceed was to use RNA polymerase to synthesize their complementary RNA chains, on which the elegant new RNA-sequencing procedures of Fred Sanger could be employed. In the mid 1960s Sanger had stopped sequencing proteins and turned his attention to the need for fast simple procedures for working out the sequences of long stretches of RNA. By employing Sanger's procedures, Sherman Weissman at Yale University and Walter Fiers in Ghent established by the end of 1976 the sequence of more than half of the over 5200 base pairs of the DNA of SV40 virus.

A breakthrough came when methods were developed allowing the sequences of 100–500 base-pair fragments of DNA to be read directly off parallel series of long gels that each separate hundreds of fragments successively differing in length by a single nucleotide. Sanger devised the first of these direct DNA-sequencing methods, the "plus minus" method, in 1975. It is based on the elongation of DNA chains with DNA polymerase. With this technique the 5386 base-pair sequence of the small DNA phage ϕ X176 was quickly determined. An equally powerful method based on the chemical degradation of DNA chains was developed at Harvard University by Walter Gilbert and Allan Maxam in 1977. All the 5226 base pairs of SV40 DNA became quickly known, as did that of the small recombinant plasmid pBR377, whose 4362 bases were determined in less than a year by Greg Sutcliffe in Gilbert's laboratory.

A few months later Sanger devised a second technique based on chain elongation. Specific inhibitors of DNA chain elongation are used to produce DNA chains that always end at a specific base (G, C, A, or T), thereby allowing the exact sequence along any given chain to be unambiguously worked out. With this technique, the 5577 base-pair se-

***Hin*dII CUTS DNA AND MAKES BLUNT ENDS**

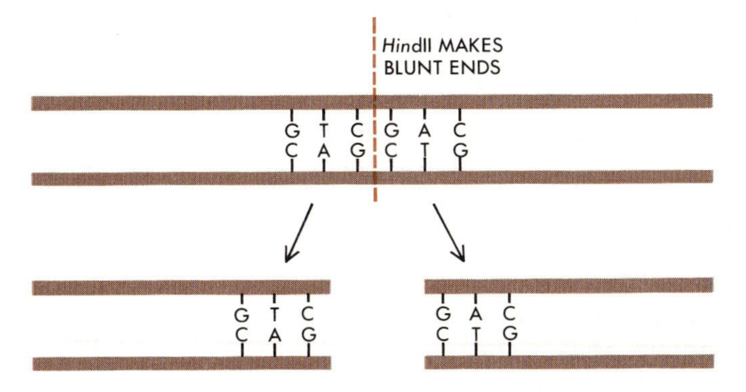

quence of the phage G4, a relative of ϕ X174, was determined quickly.

The exact sequence of any reasonably-sized piece of DNA (up to 10,000 base pairs) is now a feasible project for any well-trained genetically inclined biochemist, and already the base sequences are known for the operators and promoters used in the regulation of several bacterial operons (lactose and galactose). From the sequence of a gene it is a simple matter to deduce the amino-acid sequence of the protein it specifies. Nowadays it is often faster to determine the sequence of a protein by this indirect route rather than by directly sequencing the protein. Many potential months—if not years—of protein sequencing can be replaced often by just weeks of DNA sequencing.

*Eco*RI Produces Fragments Containing Sticky (Cohesive) Ends

Restriction enzymes like *Hin*dII break double helices at the center of their recognition sites to produce blunt-ended fragments that are base-paired to their ends and that have no tendency to stick together. In contrast, the *Eco*RI enzyme cuts near the outside of the recognition site, making staggered cuts that create short four-base single-stranded tails on the ends of each fragment. Many other enzymes in addition to *Eco*RI make staggered cuts, and many specific tail sequences are known. Complementary single-stranded tails tend to associate by base pairing and thus are often called cohesive, or sticky, ends. For example, the linear rods that *Eco*RI generates by cutting circular SV40 DNA often temporarily recyclize by base pairing between their tails. Fragments held together by such base pairing can be permanently rejoined by adding the enzyme DNA ligase to catalyze the formation of new phosphodiester bonds at the gap sites.

Because base pairing occurs only between complementary base sequences, the cohesive AATT ends produced by *Eco*RI will not, for example, pair with the AGCT ends produced by *Hin*dIII. But,

***Eco*RI CUTS DNA AND MAKES STICKY ENDS**

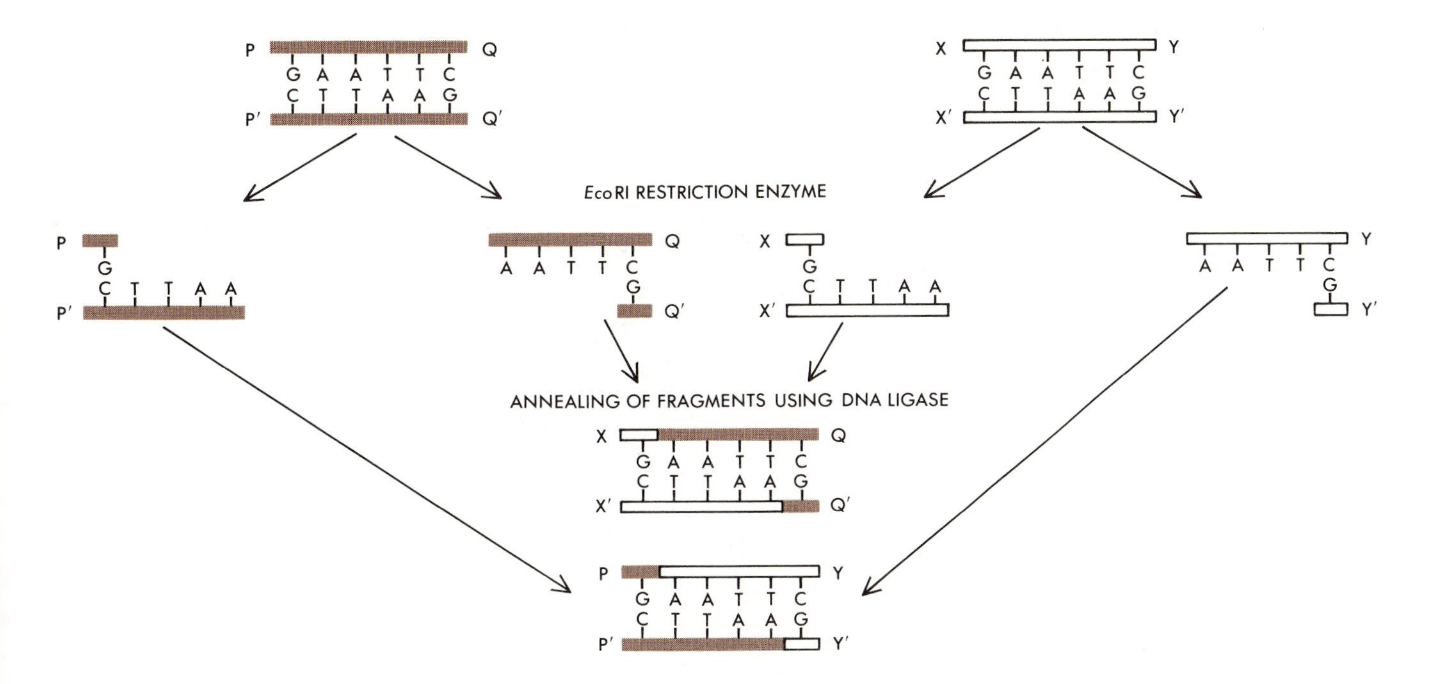

any two fragments (regardless of their origin) produced by the same enzyme can stick together and later be joined together permanently by the action of the enzyme DNA ligase. Such experiments were first done at Stanford in 1972 by Janet Mertz and Ron Davis, who realized that *Eco*RI in conjunction with DNA ligase should provide a general way to achieve *in vitro* (test-tube) site-specific genetic recombination.

Enzymology of DNA Replication

The enzyme DNA ligase that seals together the restriction fragments is but one of many enzymes that we now know are involved in DNA replication. The first enzyme to make DNA chains in the test-tube, DNA polymerase I, was for almost ten years believed to be the main—if not the only—polymerase needed to link up deoxynucleotides into DNA chains. In 1967, however, an *E. coli* mutant was found that had almost no DNA polymerase I yet made DNA at normal rates. Within a year, two new DNA polymerases were found, of which DNA polymerase III is now considered the DNA polymerase involved in most chain elongation. Much of the inherent complexity of DNA replication arises because the two chains of the double helix run in opposite directions (5′ → 3′ and 3′ → 5′), and their daughter strands must likewise run in opposite directions. Yet the elongation of all individual daughter chains occurs in the 5′ → 3′ direction. This apparent paradox was resolved by the realization that one daughter strand grows continuously longer in one direction while the other daughter strand is made discontinuously from smaller pieces that are individually elongated in the opposite direction. This feature necessitates both a special polymerase to fill in the gaps (most likely DNA polymerase I) and the joining enzyme DNA ligase first described in 1967.

In addition, other enzymes edit the DNA to remove erroneously incorporated bases or repair chains damaged by agents like ultraviolet light or X-rays. Still more proteins are needed to separate the parental strands at the replication fork, as well as to bind temporarily to single-stranded regions prior to their conversion to double helices. Also needed are several specific proteins involved in initiating DNA chains.

Most of these proteins were found over the last twenty years in the laboratory of Arthur Kornberg at Stanford University or by his former collaborators working elsewhere. Stanford University has thus been the site where the enzymology of DNA has been practiced at its best, and it is not surprising that the first test-tube-made recombinant DNA molecules were made there.

ENZYMES INVOLVED IN DNA REPLICATION

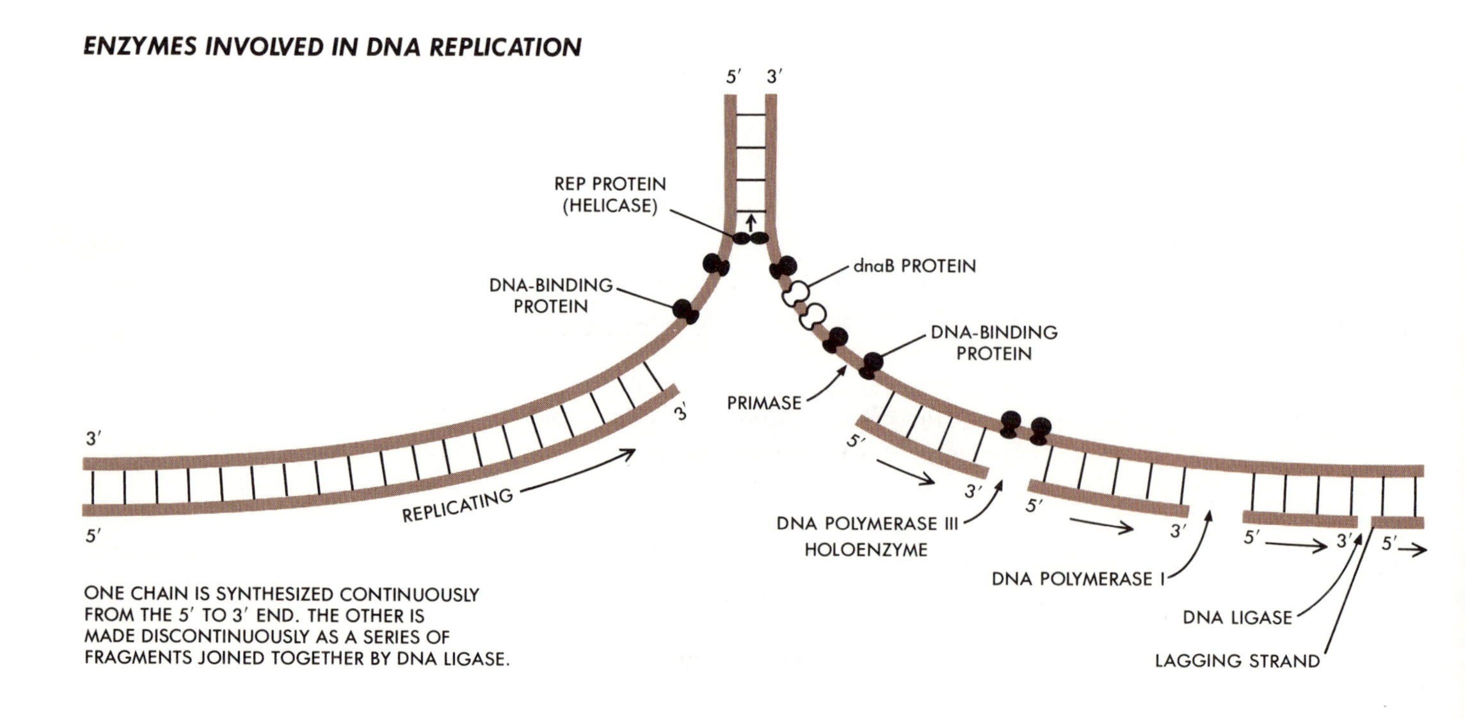

ONE CHAIN IS SYNTHESIZED CONTINUOUSLY FROM THE 5′ TO 3′ END. THE OTHER IS MADE DISCONTINUOUSLY AS A SERIES OF FRAGMENTS JOINED TOGETHER BY DNA LIGASE.

Enzymatic Addition of Sticky Ends to Blunt-Ended DNA Molecules

The *E. coli* enzyme terminal transferase that adds nucleotides to the 3′ ends of DNA chains provides a general method for creating cohesive ends on blunt-ended DNA fragments. For example, if polydeoxy A (AAAA . . .) is added to the two 3′ ends of one double-stranded fragment and polydeoxy T (TTTT . . .) to the 3′ ends of another fragment, the two fragments, when mixed together,

ADDING POLYDEOXY-A AND POLYDEOXY-T TAILS

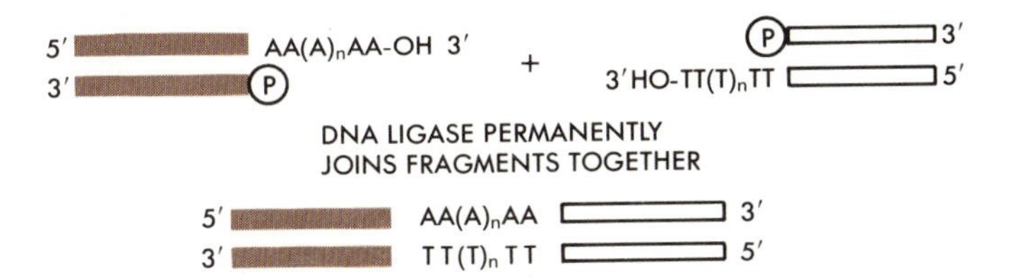

can form base pairs between their complementary tails. Appropriate enzymes can be added to fill in any single-stranded gaps, and finally DNA ligase can be used to join permanently the two fragments together. These procedures, also developed at Stanford in 1971–1972 by Peter Lobban and Dale Kaiser, and by David Jackson and Paul Berg, provide a second general method for creating recombinant DNA molecules. They do, however, introduce regions of $\frac{\text{AAAA}\ldots}{\text{TTTT}\ldots}$ base pairs at the junctions between the fused fragments. Such additional base sequences could affect the function of the joined molecules, and whenever possible, cohesive ends generated by restriction enzyme cuts are used to create recombinant DNA molecules.

Small Plasmids as Vectors for the Amplification (Cloning) of Foreign Genes

The realization that *Eco*RI generates specific cohesive ends that can later be sealed up by DNA ligase was followed within a year by the development of the first practical method for systematically cloning specific DNA fragments, regardless of their origin. The essential trick was the random insertion of the *Eco*RI fragments of a DNA molecule into circular plasmid DNA that also had been cut with *Eco*RI. This procedure led to hybrid plasmids, which can be used to infect plasmid-free bacteria. Each bacterial cell acquires a recombinant plasmid carrying a specific foreign DNA fragment. In the first such experiments, carried out in early 1973 by Herbert Boyer and Stanley Cohen and their collaborators at Stanford and University of California, San Francisco, the small *E. coli* plasmid pSC101 was used because it contained only a single *Eco*RI recognition site and was converted by *Eco*RI into linear rods. When such rods were mixed with foreign DNA fragments also possessing cohesive ends generated by *Eco*RI, and DNA ligase was added, new hybrid plasmids were created. Each contained one or more pieces of foreign DNA inserted into the RI site of the plasmid.

ELECTRON MICROGRAPH OF PLASMID SC101

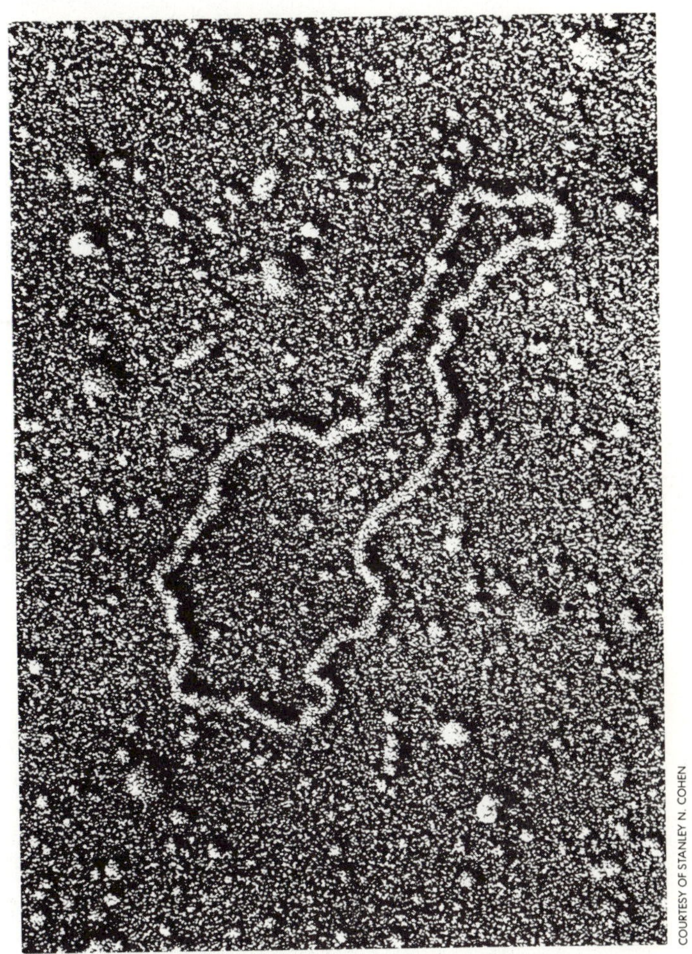

COURTESY OF STANLEY N. COHEN

In the original Boyer–Cohen use of plasmid pSC101, the foreign DNA was that of another plasmid, and the recombinant was a new plasmid containing two origins of replication. The possibility then existed of soon doing further experiments in which all sorts of foreign DNA, both from microbes and from higher plants and animals, could be inserted into these plasmids. For example, the *E. coli* chromosome (4×10^6 base pairs) contains about 500

CLONING DNA IN A PLASMID

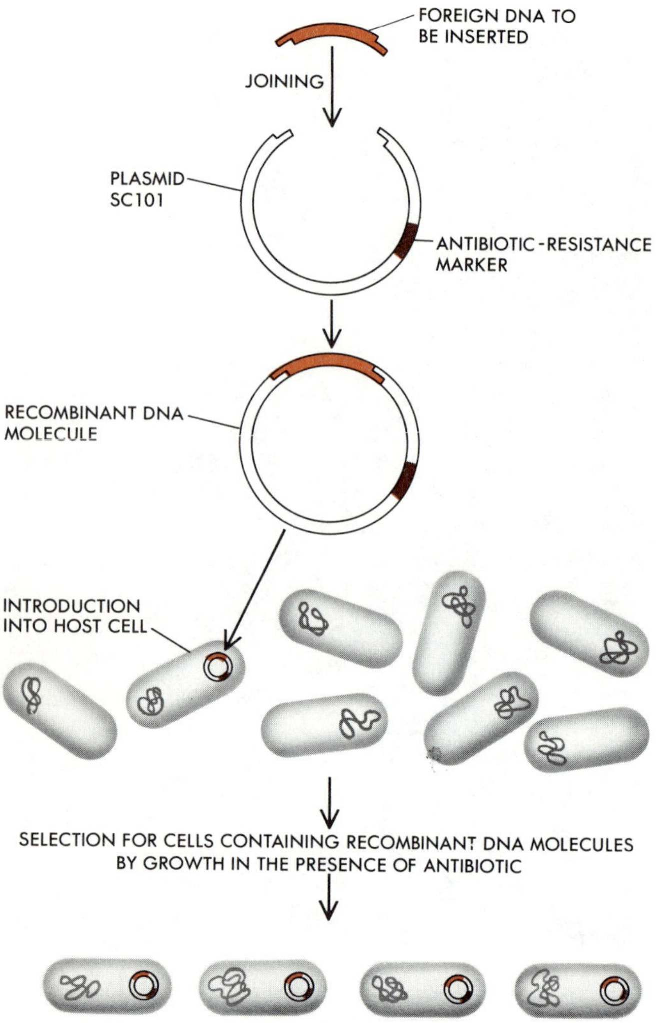

FOREIGN DNA TO BE INSERTED
JOINING
PLASMID SC101
ANTIBIOTIC-RESISTANCE MARKER
RECOMBINANT DNA MOLECULE
INTRODUCTION INTO HOST CELL
SELECTION FOR CELLS CONTAINING RECOMBINANT DNA MOLECULES BY GROWTH IN THE PRESENCE OF ANTIBIOTIC

different recognition sites for *Eco*RI. By random insertion of the *Eco*RI fragments into pSC101, it would be possible to clone all the *E. coli* genes in the form of fragments easily isolatable for subsequent genetic as well as biochemical manipulation.

Chromosomes of Higher Organisms Become Open to Molecular Analysis

Most important were the possibilities that recombinant DNA opened up for the analysis of DNA from plants and animals. Although transducing phages carrying specific parts of bacterial chromosomes had long been available, we had not been able to construct transducing animal viruses carrying specific chromosomal genes such as the ones that code for hemoglobin or the muscle proteins actin and myosin. Now by randomly inserting large numbers of different restriction fragments of human DNA into the proper bacterial plasmids ("shotgunning" DNA), we had a high likelihood of subsequently generating one or more bacterial clones containing the recombinant plasmid carrying the specific human gene we wanted. The moment we could devise an appropriate selection technique for a specific mRNA species (e.g., human hemoglobin mRNA), finding the right clone would be no problem. Furthermore, given the right clones, direct searches could be made for DNA-binding proteins with possible control roles. Moreover, fine structure analysis of a given gene should soon be virtually commonplace, with the genes that code the genetically most mysterious antibodies being prime candidates for the first vertebrate DNAs to be thoroughly analyzed.

Tumor Virology Now on a Solid Molecular Basis

With the arrival of new recombinant DNA procedures, we now had the perfect method to grow desired DNA restriction fragments, if not the intact genomes of DNA tumor viruses. Virtually effortlessly we should soon have large amounts of all tumor virus DNAs. Previously when growing small batches of DNA tumor viruses by conventional cell culture methods, we were never absolutely sure that we were not subjecting ourselves to some risk of cancer. If such a risk existed, we thought it to be small because the viruses we used (like SV40 or polyoma) were quite similar to common human viruses that did not apparently cause tumors. So we did not believe that by working on tumor viruses we had been exposing ourselves to any major new risk to which we were not already exposed. Nonetheless, it would be most satisfying to be able to stop growing intact tumor viruses in large batches, and it was more than obvious that recombinant DNA procedures should be taken up by tumor virologists at maximum possible speed.

Scientists' Early Concerns About the Implications of Unrestricted Gene Cloning

Unfortunately an earlier talk by Janet Mertz, then a student in Paul Berg's laboratory at Stanford, about the possibility of cloning SV40 in *E. coli* and its phages had already started a counterreaction. Particularly concerned was the cell biologist Robert Pollack, then working on SV40-mediated transformation of mouse cells at Cold Spring Harbor. Given his doubts about the total safety of SV40 itself, should he not be even more concerned with the less clearly assayable prospect that SV40-carrying bacteria could conceivably act as a vector to transmit human cancer? After hearing Janet Mertz talk in the summer of 1971 at Cold Spring Harbor about planning to use the currently studied terminal transferase tails (e.g., AAAA and TTTT) to genetically engineer SV40, Pollack telephoned Berg to ask him if he had worried about the potential biohazards of recombinant DNA. Although Berg indicated he saw no reason to panic, he said he would consider postponing such work, and in fact later made the decision to refrain from cloning tumor virus genomes.

These events, which took place two years before the antibiotic-resistant plasmids were employed as cloning vectors, no doubt reflected an undercurrent of concern among molecular biologists that they knew too little about tumor viruses to treat them casually. Also to be considered was the intense excitement created by the new restriction enzymes like *Eco*RI. Their mere existence was sufficient to keep Berg's laboratory, as well as that of other leading tumor virologists, very busy over the next several years without the need to employ recombinant DNA technologies immediately.

A further concern surfaced over the following year as the opinion spread that the DNA of mice—and possibly all higher cells—harbored latent RNA tumor virus DNA. Many of the recombinant plasmids in which we hoped to find human antibody genes might instead contain genes capable of causing cancer. If so, should we regard virtually all recombinant DNA experiments using vertebrate DNA as potentially dangerous? The answers were mixed. Those hoping soon to do a particular experiment, say with human DNA, had no worries about either his or her own safety or that of others. Yet often this same person would openly question the advisability of medically oriented microbiologists making recombinant plasmids containing mixtures of many antibiotic resistance markers, arguing that maybe they would spread out of control through human populations. And virtually everybody was united in arguing against the unrestricted freedom to do experiments that might have military consequences. These experiments instinctively sounded nasty, even though no one concerned knew any of the technical aspects of biological warfare that had long remained tightly classified in military secrecy.

Here it is important to note that objections against unrestricted use of recombinant DNA started even before the Boyer–Cohen introduction of plasmids as cloning vectors. The first debates as to whether we should use recombinant DNA thus preceded our actual ability to move full speed ahead. Very early such conversations aroused no deep emotions among either those who said they were concerned or within those who thought the whole matter was academic silliness created by naive souls who had never worked with real pathogens. Almost everyone was deep into his or her experiments of the moment, and concern over experiments that would not come to pass for at least a year or two seemed of little consequence.

But after the first *Eco*RI–pSC101 cloning experiments were announced, it became clear that it would soon be impossible to go on doing molecular genetics as if recombinant DNA did not exist. The only question was whether to move ahead as fast as possible or to try to invent methods that would reassure the worriers without straight-jacketing most future explorations of recombinant DNA. As the documents we have reproduced in this book show, although the decision on which way to proceed could initially have gone either way, the cautious approach prevailed. The exploitation of recombinant DNA research was subjected to regulations and massive bureaucracy that only now are being lifted. At a cost of what must have been many millions of dollars, the recombinant DNA debate was pursued worldwide. Delays and frustrations were forced on those wanting to get on with what were bound to be exciting experiments. Just how exciting those experiments have been will now be unfolded.

The Isolation of Cloned Genes

In the mid-1970s our ability to exploit recombinant DNA methods to full potential faced, in addition to bureaucratic paperwork, several obstacles. One was artificial—the development of disabled hosts and vectors that would have no significant probability of surviving outside the laboratory and that would satisfy the criteria of biological containment established in the guidelines. Another, and more fundamental, was the development of methods for identifying bacteria that carry the cloned genes in which we were interested.

The Need for Formal Certification of "Safe" Bacteria and Plasmid Vectors

The release of the original NIH guidelines in June 1976 did not let recombinant DNA research immediately take off. In the first place, none of the so-called "safe" bacterial hosts and plasmid and phage vectors had yet been developed and certified as meeting the safety criteria specified by the guidelines. At Asilomar in February 1975, there was talk that only a few weeks' work would be necessary to make "safe" bacteria or phage, but it was not until late 1976 that the first of the "safe" bacteria (EK2 category) became available. Certification of the plasmid (phage) vectors, into which DNA fragments were to be inserted, consumed more time. Even after the Recombinant DNA Advisory Committee (RAC) gave its approval to the now widely used pMB9 and pBR322 plasmids, several frustrating months passed before they were formally certified by the NIH directorate in the late spring of 1977. Not surprisingly, the RAC's approval was interpreted as a green light by several European laboratories, which saw no reason to be held back by time-consuming procedural restraints demanded by HEW Secretary Joseph Califano.

Development of "Safe" Bacteria and Plasmid Vectors

The first "safe" *E. coli* K12 strain was developed in 1976 by Roy Curtiss III (University of Alabama), who named it *chi 1776* after the Bicentennial. Among the many defects that should prevent the escape of *chi 1776* to the outside world was a metabolic requirement for diaminopimelic acid, an intermediate in the biosynthesis of lysine that is not present in human intestines. *Chi 1776* also possesses a fragile cell wall, which bursts open in low salt concentrations or in the presence of even a trace of detergent. Unfortunately, those who began using *chi 1776* found it difficult to work with. It grew to much lower cell densities than ordinary *E. coli* K12 strains, and even when enough cells had been grown, it proved more difficult than usual to introduce recombinant DNA molecules into *chi 1776* cells. Much effort was therefore spent developing other "safe" derivatives of *E. coli* K12 that eventually were officially certified as approved EK2 hosts.

The guidelines also made it necessary to use only plasmids that had been modified by mutational events so they could not move, say, within the human intestine, by a sexual process from a "safe" to an "unsafe" strain of bacteria. Fortunately, it is a simple matter to construct new plasmids lacking all the genes that control the movement, or mobilization, of plasmids from cell to cell and thereby render them nonmobilizable. But each time a plasmid was

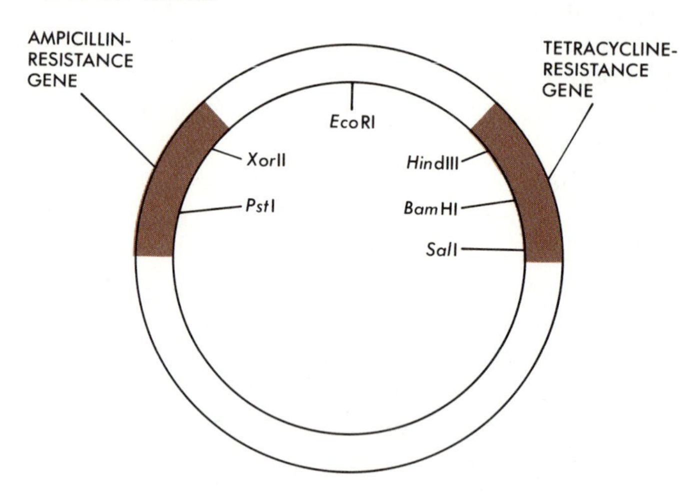

in any way modified, it required new certification by the RAC—a procedure involving much paperwork and delay. As a result, those plasmid vectors that were initially certified as "safe" have been used over and over, rather than risk further delays from the certification process.

Why Use Drug-Resistance Plasmids?

One of the most obvious concerns about the biosafety of recombinant DNA research was that it might contribute to the further spread of bacterial drug resistance. The extensive use and abuse of antibiotics in medicine and animal husbandry have caused many natural strains of bacteria to become resistant to the most common antibiotics. Usually this resistance depends on acquisition by the bacterial cell of a plasmid that specifies enzymes that can break down the drugs.

As vectors for recombinant DNA, plasmids with genes for antibiotic resistance offer a great advantage. When they are used together with host bacterial cells that are plasmid-free and therefore antibiotic-sensitive, the entry of the resistance plasmid carrying the recombinant DNA can be easily detected. Those cells that acquire the plasmid become resistant to the antibiotic. The plasmid BR322, for example, carries genes for resistance against two antibiotics, ampicillin and tetracycline. Moreover, sites recognizable by restriction enzymes are within these antibiotic-resistance genes. So when a piece of foreign DNA is recombined into one or the other resistance gene, that gene is inactivated. This means that the successful insertion of a piece of foreign DNA into one of the antibiotic-resistance genes is easily detected. The genetic potential for that resistance is eliminated. The next step—the successful introduction of the plasmid (now known to carry foreign DNA) into a host bacterium—is also easy to test because it results in the bacterium acquiring the resistance specified by the second and still intact resistance gene.

By using plasmids carrying antibiotic-resistance genes already widespread in nature, by mutating the plasmids so they cannot spontaneously move from cell to cell, and by using "safe" strains of bacteria, experiments with drug-resistance plasmids can have all the advantages without significant risk of contributing to the spread of antibiotic resistance. As some medical microbiologists pointed out during the recombinant DNA debate, even if these precautions were not taken, the contribution of recombinant DNA research to the spread of antibiotic resistance would have been trivial compared to that resulting from the excessive daily use of antibiotics in medicine and agriculture.

Probes for Cloned Genes

If we take the total DNA of a human cell, cut it into fragments with a suitable restriction enzyme, join the fragments to plasmid vectors, and introduce the recombinant DNAs into a population of bacteria, we have a so-called human gene "library." It is, however, a library without a proper index, and we are faced with the problem of separating the bacterium carrying the human gene we want from the millions of others. To overcome this "needle-in-the-haystack" dilemma, we now use nucleic-acid probes, which exploit the complementarity of nucleic-acid sequences. To obtain such probes, we first isolate—in as pure a form as possible—radioactively labelled mRNA specified by the gene we want to isolate. Given the possession of such mRNA, we can test our gene library for those bacteria that have DNA with chains complementary to our mRNA probe. Here hybridization screening procedures are used, which specifically detect DNA–RNA double helices.

There is a snag, however. It is generally impossible to obtain totally pure mRNA probes even for proteins that are made in large amounts in certain differentiated cells. For example, in immature red blood cells, roughly only half the mRNA is for hemoglobin, the red protein which in the mature red blood cells accounts for well over 90 percent of the total protein. If we use an impure preparation of mRNA as a probe for a cloned gene, we run the risk of many false positives, and the less pure the mRNA, the greater that risk. Fortunately, by a circuitous route involving what amounts to cloning the mRNA itself, we can obtain pure probes.

INSERTION INTO DRUG-RESISTANCE GENES

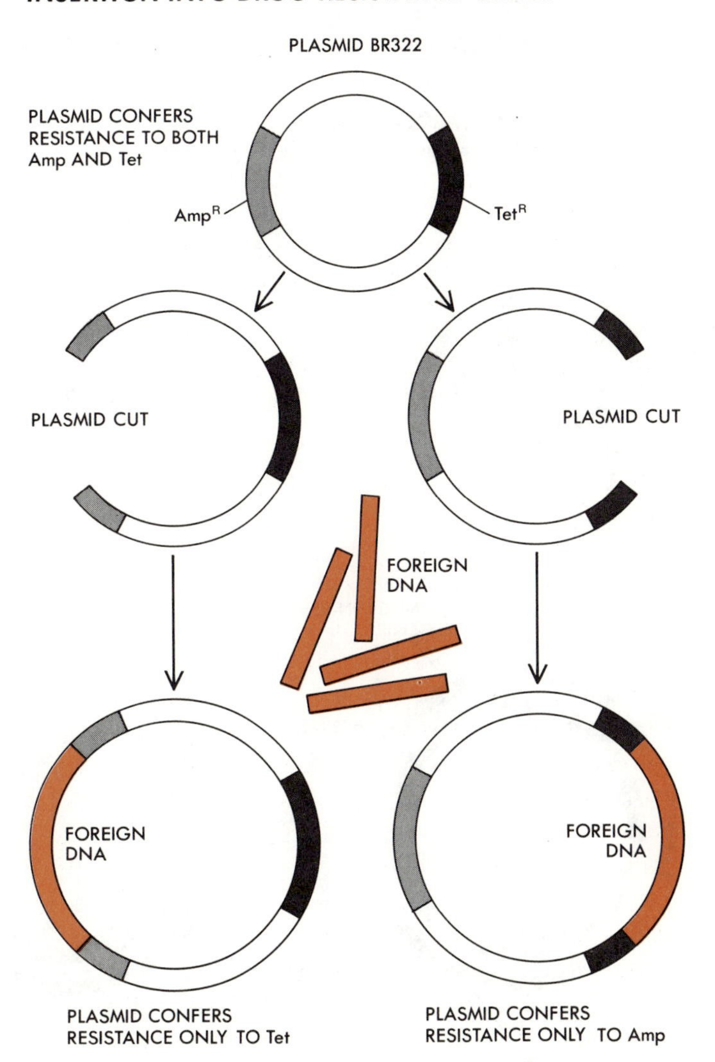

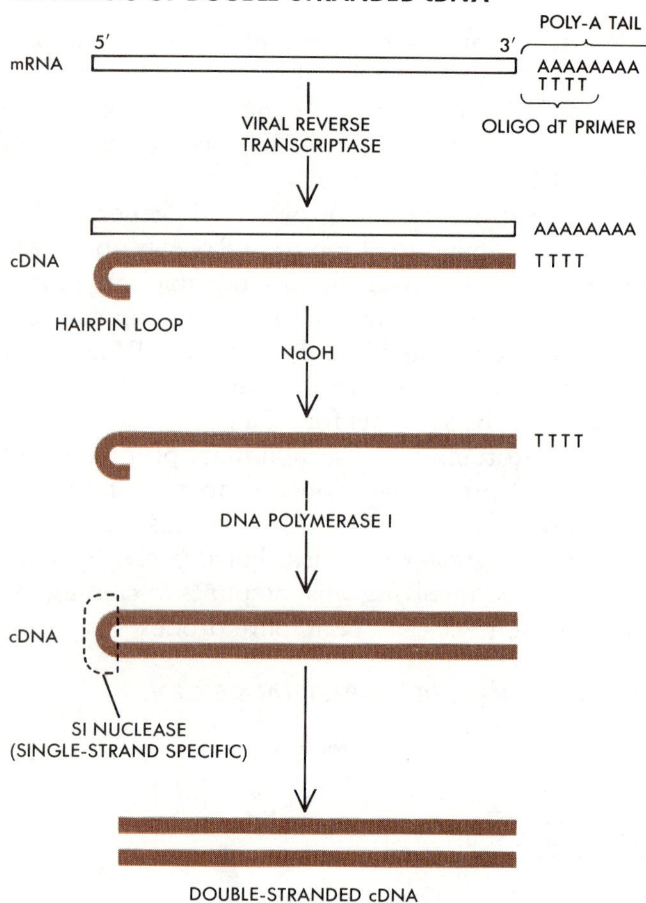

Synthesis and Cloning of cDNA

Virtually every mRNA molecule has at its 3′ end a run of adenine nucleotide residues called a poly-A tail. When short chains of poly-T are mixed with the mRNA, they hybridize with the poly-A tails to give an mRNA molecule with a double-stranded poly-A–poly-T tail. An enzyme called reverse transcriptase, obtained from certain RNA tumor viruses, can use the poly-T chain as a primer and the mRNA chain as a template, to synthesize a DNA strand complementary to the mRNA sequence. This enzyme reverses the first step of gene expression, hence its name.

The result of this reaction is an mRNA–DNA hybrid molecule from which the DNA and mRNA strands can be separated. Then the DNA strand carrying the exact specificity of its mRNA parent can, in turn, be used as a template by reverse transcriptase to synthesize the complementary DNA strand. The end result is a double-stranded DNA molecule complementary to an mRNA, which is therefore referred to as a cDNA. Then the cDNA molecule can be inserted into a plasmid vector by adding artificial sticky single-stranded ends to the DNA and can be propagated as a recombinant DNA in bacteria. But here we should note that even these

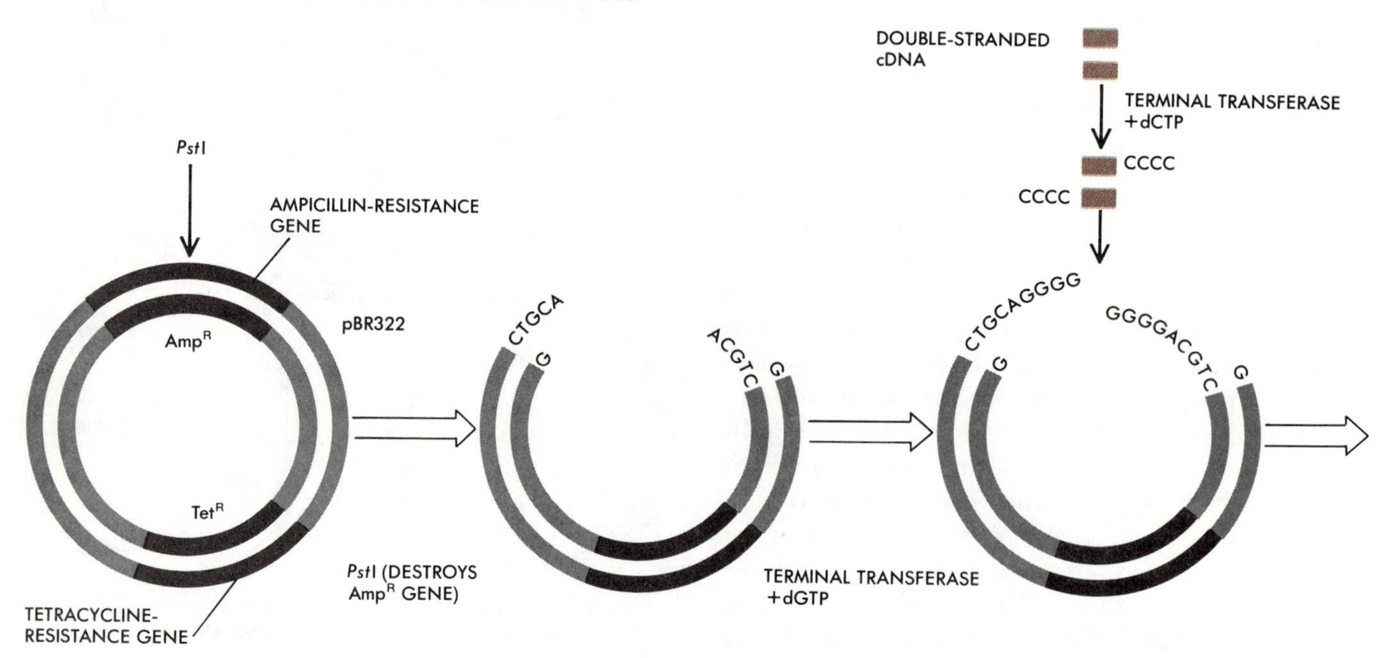

first-step procedures were identified as potentially dangerous at the Asilomar Conference, and so they were effectively proscribed until the RAC came forth with its first set of guidelines.

Identifying Specific cDNA Probes

If the set of experiments just described is done with a mixed population of mRNAs—which in practice is always the case because we cannot obtain pure species of mRNA—the cDNA will also be mixed. But the system can be arranged so each bacterial cell receives only one cDNA recombinant molecule. Each cell and its descendant will carry only one sort of cDNA. The problem now is to discover which sort of cDNA is present in each cell.

In order to do this, single cells are picked and grown into large homogeneous cultures. The cDNA-containing plasmids from this genetically pure, clonal population of cells are then extracted, and their DNA is denatured so the strands separate. The separated DNA strands can then be bound to nitrocellulose filters. When mixtures of mRNAs are passed through these filters, those and only those mRNA molecules complementary to the pure species of cDNA on the filters are retained. Later the bound mRNA can be washed off the filters,

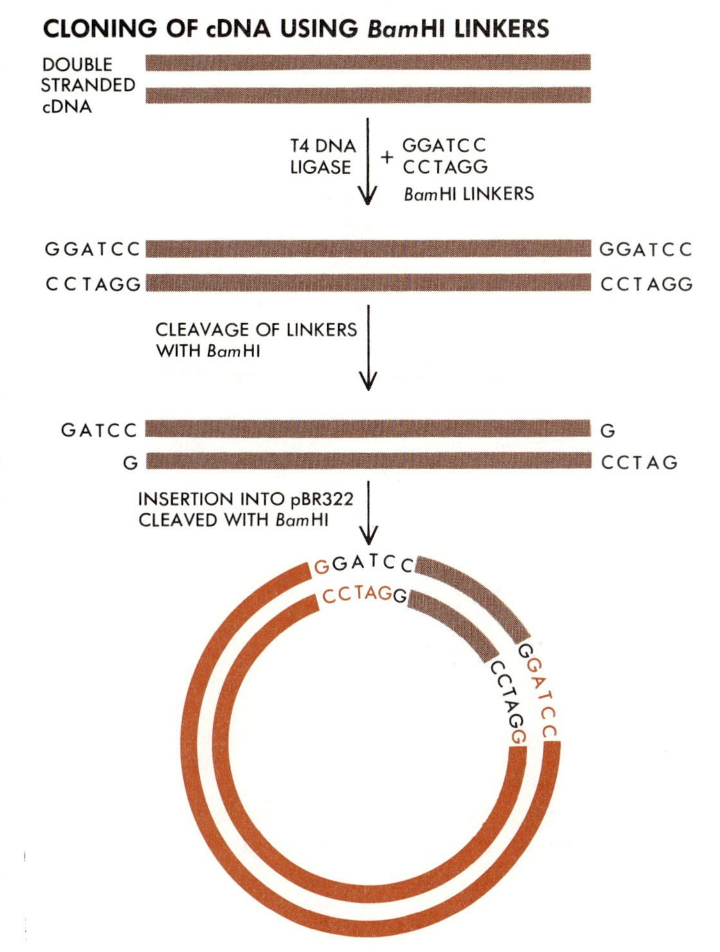

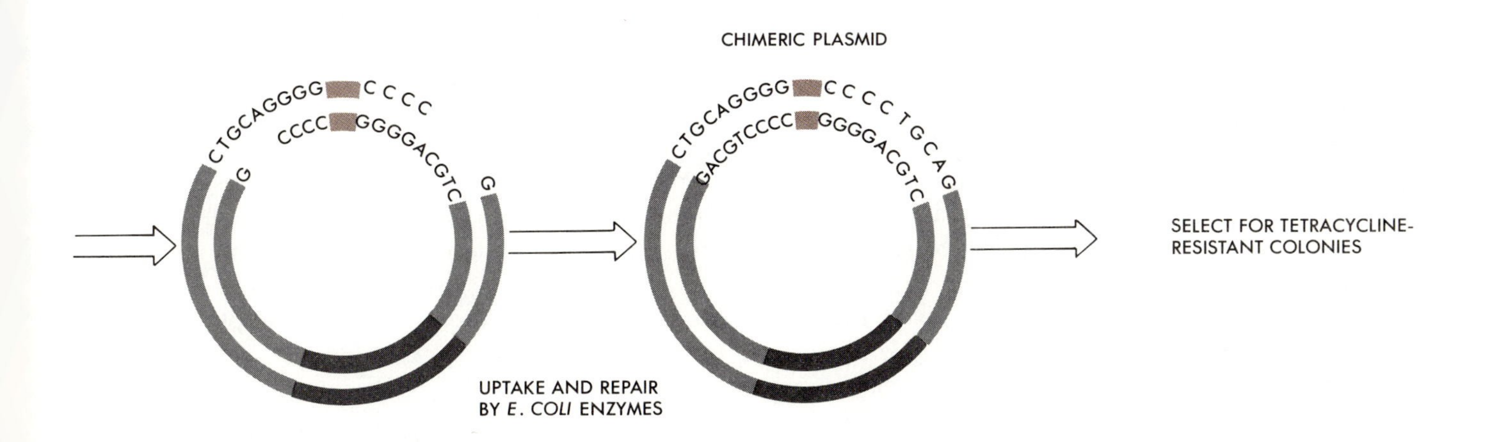

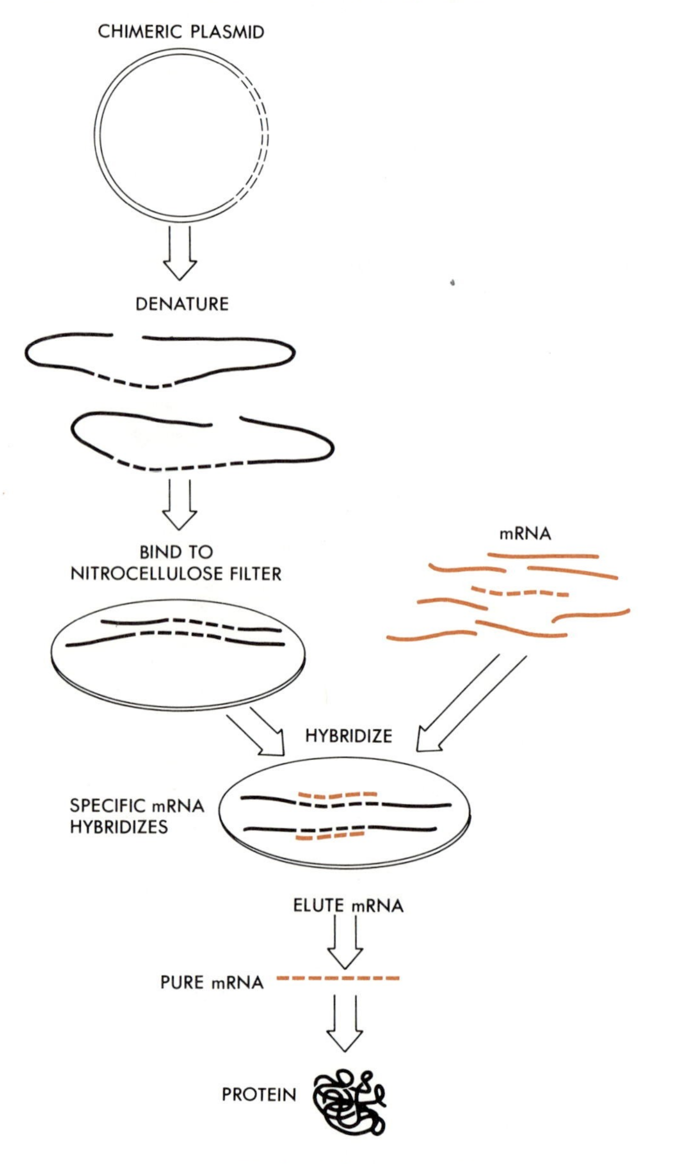

concentrated, and added to a cell-free system that translates mRNA into protein. The protein specified by the cDNA insert often can be identified. When this is the case, we have a pure cDNA probe for a single gene and its corresponding mRNA.

Clearly, the more abundant a given mRNA species is within an mRNA preparation, the easier it is to obtain its respective cDNA probe. So, naturally, recombinant DNA methods were first used to make probes for abundant proteins such as hemoglobin and the antibodies made in myelomas, cancer cells that produce antibody. Early on, cDNA probes were also made for ovalbumin, which comprises more than half the protein made in hormonally stimulated cells in the chicken oviduct.

Once one cDNA probe for a given protein (and gene) has been made, it can be subsequently used with the nitrocellulose filter technique to screen rapidly large numbers of bacterial colonies for other recombinant—or, as they are sometimes called, chimeric—plasmids carrying the same or closely related gene sequences.

In this way many cDNA probes for proteins such as hemoglobin can be isolated and tested to see whether they represent complete rather than partial probes. One procedure measures the sizes of the RNA–DNA heteroduplexes formed after mixing, say, a hemoglobin mRNA preparation with denatured cDNA probes. Those probes that form heteroduplexes with the entire mRNA are then analyzed to see if the nucleotide sequences match those of the respective proteins. As expected, this was precisely the case, with the additional finding that the mRNA molecules of many genes contain more nucleotide sequences at their 5′ ends. These extra sequences specify leader segments of the proteins that are cleaved off soon after synthesis and whose presence is therefore often unsuspected. In general, amino terminal leaders have been found for proteins that attach to cell membranes. The amino terminal leader's function is to burrow through the membrane, thereby initially fixing the nascent polypeptide chain to the membrane.

Genomic Fragments Are Best Cloned in Bacteriophage λ

Once the desired cDNA plasmid probes became available, it was possible to look directly at the structure of the chromosomal genes themselves rather than their mRNA transcripts. Only by examining the potentially much longer fragments obtainable from the chromosomes could the possible regulatory sequences outside the 5′ and 3′ ends of the direct coding sequences of a gene be examined. Many thousands of chimeric plasmids—each bearing a specific fragment of, for example, the total human, mouse, or chicken DNA—are easily prepared, and by screening sufficiently large numbers, the entire human, mouse, or chicken genome can be examined for those fragments carrying, say, hemo-

IDENTIFICATION OF BACTERIAL COLONIES HARBORING SPECIFIC cDNA PLASMIDS

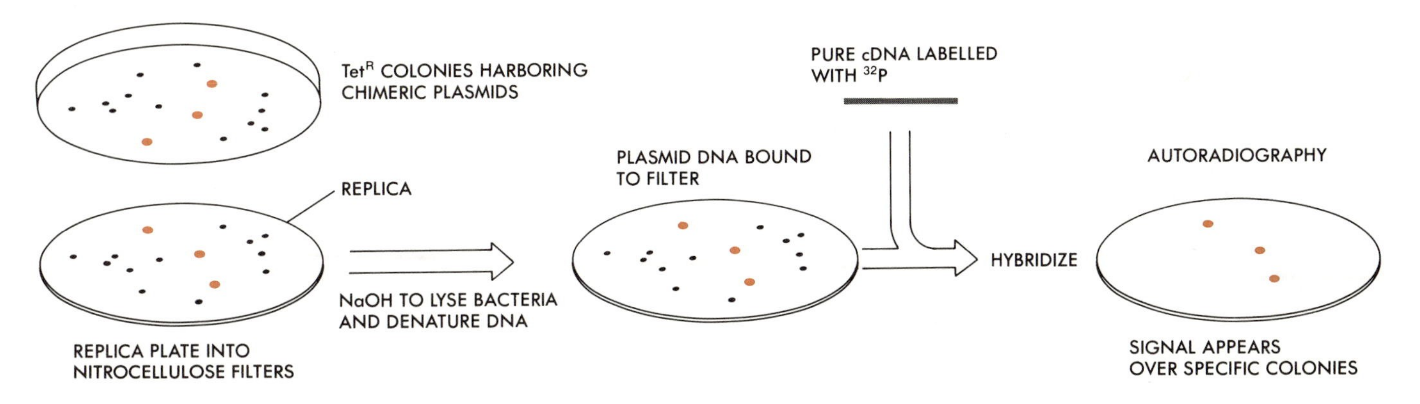

globinlike sequences. Soon, however, it became obvious that plasmids bearing large chromosomal DNA insertions were not stable and tended as they replicated to give rise to smaller plasmids in which increasing amounts of the inserted DNA had been deleted. This is a consequence of the fact that the less DNA a plasmid has, the faster it can multiply. Thus, genetic segments unnecessary for plasmid multiplication invariably tend to be lost. Inserts of cDNA probes are of course subject to the same elimination pressures, but deletions only become commonplace when the inserted fragments are several thousand base pairs larger than most cDNA probes.

In contrast to their instability in plasmids, large chromosomal DNA fragments (about 15,000 base pairs, or 15 kilobases) are essentially stable when inserted into the DNA of specially prepared strains of phage λ. Already at Asilomar it had been suggested that phage λ could be mutated so it would be unable to insert its DNA into that of host *E. coli* cells, and would be at least as safe as, if not safer than, disabled plasmid vectors in *E. coli* strains with properties similar to those of *chi 1776*. Such "safe" λ vectors exploit the fact that the entire central section of phage λ DNA is not necessary for its replication in *E. coli* but only functions to ensure the integration of the phage DNA into the host bacterial chromosome during its lysogenic phase. Special λ strains have been created with recognition sites for the restriction enzyme *Eco*RI located so as to leave intact the left and right end fragments of the viral DNA that are essential for its replication. After *Eco*RI cutting, these end fragments, because of their relatively large sizes, can easily be purified from all the other fragments generated by *Eco*RI and later used to make new λ-like phages containing one left fragment, one right fragment, and one foreign DNA insert in the 15–20 kilobase range. Most fortunately, maturation of phage λ requires that its DNA chromosome be approximately 45 kilobases long; thus the only DNAs constructed *in vitro* that can multiply following such manipulations are ones that are chimeric mixtures of phage ends and foreign DNA of the appropriate length.

Once a "library" of λ phages carrying specific eukaryotic genes has been constructed, it is easily screened with cDNA plasmid probes, again by using the probe's radioactivity to mark out those phage colonies (plaques) bearing complementary DNA sequences. The haploid number of chromosomes of a mammalian cell is a total of about 3×10^9 base pairs of DNA. When DNA is cloned in bacteriophage λ each fragment is on average 1.5×10^4 base pairs. Screening a mere million phage plaques will effectively sample all the DNA of the mammalian cell for a given gene. Thus, for a specific cDNA probe, at worst only a few weeks may be necessary to screen a phage λ "library" for the respective genes.

Developing Procedures for Cloning Genes That Code for Less Abundant Proteins

The most direct way to enrich a crude mRNA preparation for a minor component involves its centrifugation, say, through a sucrose gradient. This procedure separates the mRNA into samples of different sizes, which can then be collected and tested in cell-free translational systems to determine which size class codes for the desired protein. The mRNA

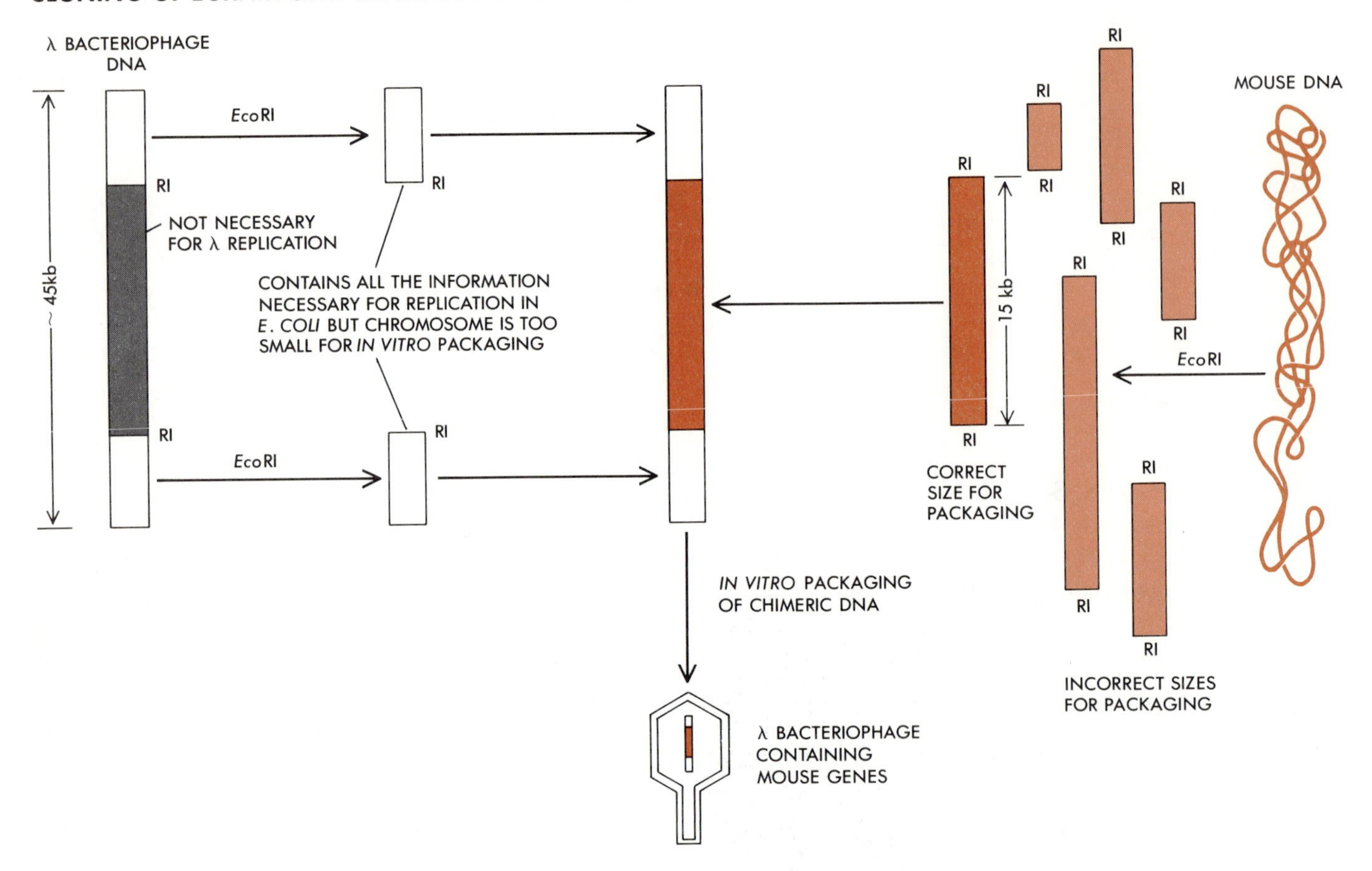

molecules that code for the light chains of antibodies (for example) sediment at approximately 13 sedimentation units (13S), while those that code for the α and β chains of hemoglobin move much more slowly (9S). Further purification can be obtained by subjecting the mRNA samples enriched by centrifugation to electrophoresis in acrylamide gels, a procedure that again separates different sized molecules but with much higher resolution than is possible by simple centrifugation. In this way, mRNA preparations 80–90 percent pure for the antibody light chains have been obtained.

But with less common proteins the desired mRNA remains still a minor component after the foregoing procedures. Fortunately, we now have several new enrichment tricks, one of which is most effective for minor proteins that are cell specific. The rat protein α_{2u}, for example, comprises 1 percent of the protein made in the male rat liver, but it is totally absent in the female rat liver cells. This difference reflects the fact that the synthesis of α_{2u} mRNA is stimulated by the male sex hormone testosterone and strongly inhibited by the female sex hormones, the estrogens. How these sex hormones control mRNA synthesis remains a major unsolved problem, and the α_{2u} gene could provide an excellent system for finding the answer. When the still heterogeneous single-stranded cDNA copied from the appropriately sedimenting fraction of male rat liver mRNA is mixed under hybridizing conditions with much larger amounts of female liver mRNA, the only cDNA molecules that do not form cDNA–mRNA double helices are the ones that code for α_{2u}. Since single-stranded nucleic-acid molecules, unlike their double helical equivalent, do not bind to hydroxyapatite, the single-stranded α_{2u} cDNA becomes much more enriched after just one passage through a hydroxyapatite column. Then it can comprise over half of the total cDNA, and the plasmid probes into which it is inserted can easily be detected.

We can use another trick for selecting a desired cDNA probe when the amino-acid sequence of the desired minor protein is available. This was the case for the β subunit of human chorionic gonadotropin (HCG) whose 145 amino acids had been unambiguously ordered. From the amino-acid sequence, the nucleotide sequence of the corresponding mRNA and cDNA was predicted. This revealed which restriction enzymes will cut the cDNA and how often each will cut it. A mixed population of cloned cDNAs can then be screened with the appropriate restriction enzymes, and those that yield fragments of the expected sizes and in the expected numbers can be further analyzed. With this procedure only six weeks were required to make the cDNA probe of the HCG subunit gene.

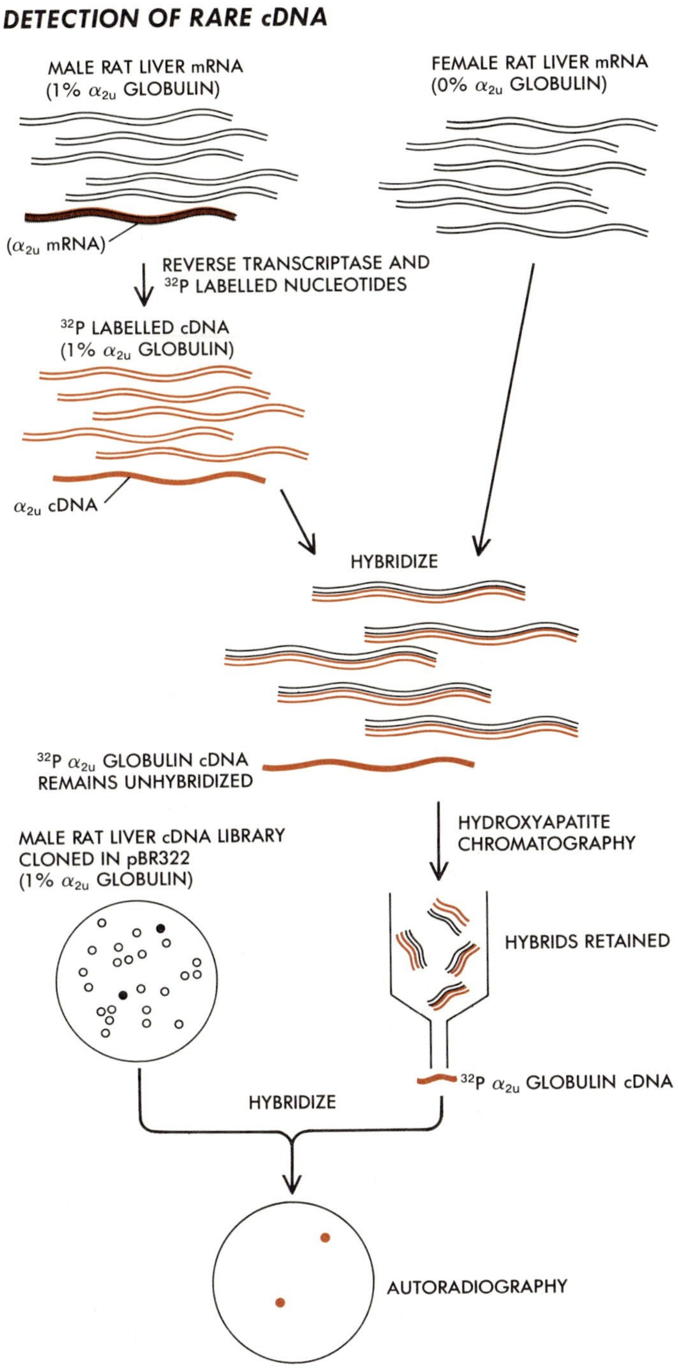

Isolating Eukaryotic Genes Expressed in Bacteria

Specific eukaryotic genes can also be isolated by looking for their expression after their respective cDNA probes have been placed in the appropriately constructed plasmid. Such procedures start with cDNA probes made on enriched (say, by size) mRNA preparations. These cDNAs are then inserted into plasmids within genes that are highly expressed once the plasmid is introduced into a host bacterium. For example, when an insulin cDNA probe is placed in the plasmid BR322 within its gene for β-lactamase, a mixed β-lactamase–insulin polypeptide is produced. After the enzymatic removal of most β-lactamase amino acids, biologically active insulinlike molecules are detectable. Alternatively, the cDNA probes can be placed within an appropriately constructed plasmid that contains the beginning of the *E. coli* lactose operon. The inserted sequences are precisely placed next to the ribosome-binding site of lac mRNA. In this way, the foreign gene is coupled to the high level *E. coli* lac promoter and to the nucleotide sequences optimally suited for binding mRNA molecules to *E. coli* ribosomes. This method was first used to make large amounts of a rare regulatory protein of phage λ. More recently it has been used to make biologically detectable amounts of the antiviral agent interferon. By further manipulating the genetic properties of the host cell and the cloning plasmid, we can increase the expression of the introduced foreign DNA so it makes the system commercially attractive. For example, bacteria that produce human growth hormone to levels of 1 percent of the total bacterial protein have been developed.

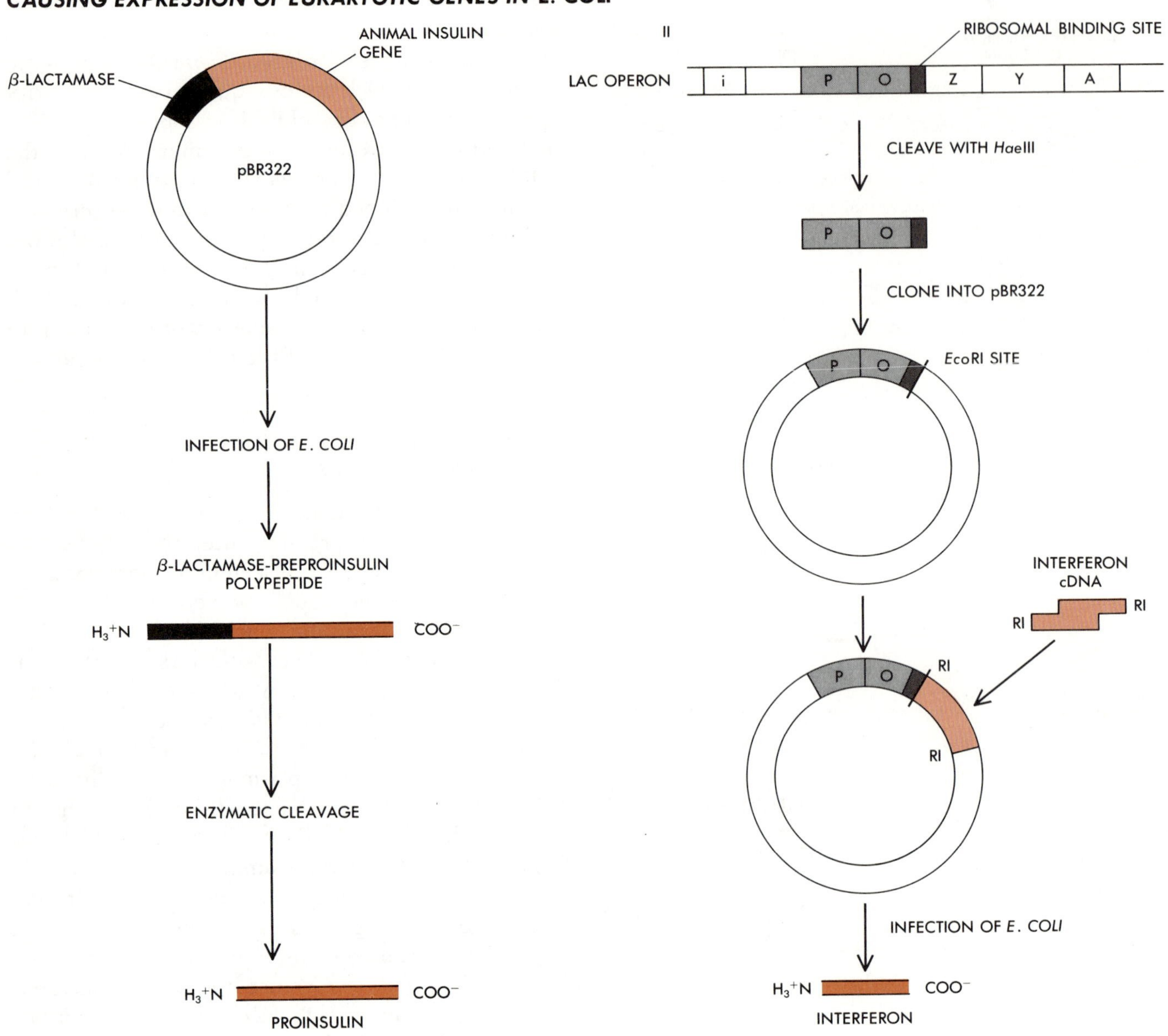
CAUSING EXPRESSION OF EUKARYOTIC GENES IN E. COLI
ANIMAL INSULIN GENE
β-LACTAMASE
pBR322
INFECTION OF E. COLI
β-LACTAMASE-PREPROINSULIN POLYPEPTIDE
H_3^+N
COO^-
ENZYMATIC CLEAVAGE
H_3^+N
COO^-
PROINSULIN
II
RIBOSOMAL BINDING SITE
LAC OPERON
i
P
O
Z
Y
A
CLEAVE WITH HaeIII
P
O
CLONE INTO pBR322
EcoRI SITE
INTERFERON cDNA
RI
RI
RI
RI
INFECTION OF E. COLI
H_3^+N
COO^-
INTERFERON

The Unexpected Complexity of Eukaryotic Genes

Examination of the structure of specific chromosomal DNA fragments effectively began in the spring of 1977, soon after the certification of the first "safe" vectors. Then everyone's attention was focused on the regulatory sequences for mRNA synthesis that were thought likely to lie in front of the 5′ ends of the coding segments themselves. It was taken for granted that the nucleotide sequences within the genes would be identical to the ones in the cDNA probes. Yet the first preliminary results indicated that the restriction fragments obtained from the segments of chromosomal DNA were frequently different from the ones generated from their related cDNA probes. Initially, these observations caused only puzzlement because a priori it seemed impossible for the mRNA (cDNA) sequences to be other than identical to the sequences of the genes from which they were transcribed.

Discovery of Split Genes

Then the first announcements were made during the 1977 Cold Spring Harbor Symposium of an mRNA splicing phemonenon during adenovirus multiplication. The primary viral RNA transcripts within the nucleus of an infected cell are shortened by removing one or more internal sections to produce smaller mRNA molecules; these then move to the cytoplasm where they serve as templates for viral protein synthesis. Quickly the generality of splicing was extended to the SV40-like viruses, and the question immediately arose whether splicing might also be involved in the processing of most chromosomal DNA. For several years it had been known that many eukaryotic mRNAs were first synthesized as large precursors (pre-mRNAs) that were later processed in the nucleus to much smaller products. But until the announcement of adenovirus splicing it had always been assumed that this processing necessarily and exclusively involved removal of long, possibly regulatory sections at the 5′ and 3′ ends of pre-mRNA.

So during the summer of 1977, the patterns from restriction enzyme fragments of chromosomal DNA were reinterpreted to see if they might indicate noncoding sequences interspersed in the coding sequences of genes and whether the primary RNA transcripts are spliced to remove the noncoding sequences as the mRNA is matured. Firm proof of this came from several experimental procedures. Most direct were the electron microscope observations of the DNA–RNA heteroduplexes made between functional mRNA molecules and their corresponding chromosomal genes. Single-stranded DNA sections looping out from such heteroduplexes represent gene sequences that have been removed from the mRNA. The exact sizes and locations of the inserts could be measured more precisely by treating the heteroduplexes with the ribonuclease (S1 endonuclease) that specifically cuts out the unpaired DNA bases and leaves genomic DNA fragments bound to the corresponding sequences that persist in the functional mRNA. Such functional gene sequences are now called exons, while the spliced out sequences are know as introns.

Introns exist in virtually all mammalian and vertebrate genes, and also in the genes of eukaryotic microorganisms (e.g., yeast), though with much lower frequency. Often the noncoding introns of a gene comprise many more nucleotides than its coding exons, which accounts, at least in part, for the previously unexplained large sizes of so many primary pre-mRNA transcripts. The number and size of introns vary widely from one gene to another; genes coding for the long (~900 amino acids) chains of collagen, the connective-tissue protein, possess nearly 40 introns.

Specific Bases of Exon–Intron Boundaries and Involvement of Small Nuclear RNAs in Splicing

By the summer of 1978, just a year after the first split genes were discovered, the sequences of the bases at many exon–intron boundaries had been determined by using the new and rapid DNA-sequencing methods of Gilbert and Maxam. It was hoped that such sequence data would provide evidence in favor of intramolecular base pairing that could bring together the respective upstream and downstream splice sites in an mRNA precursor. So positioned, specific splicing enzymes could carry out the appropriate cutting and joining events. Such hairpin loop structures require a group of upstream bases complementary to a group of downstream bases. But no such complementary groups of sequences were found at 5′ (upstream) and 3′ (downstream) splice points. The upstream and downstream splice sites therefore could not be brought together by self-complementarity. Yet the base sequences at the boundaries between exons and introns were far from random, and after many had been sequenced a pattern emerged. All the sequences at the opposite ends of different introns could be related to two so-called consensus sequences. These consensus sequences suggested an alternative mechanism for bringing the opposite ends of introns together prior to splicing, a mechanism involving small adaptor RNAs that normally are attached to the splicing enzymes.

A precedent for such positioning RNA was already known from the discovery that the site-specific ribonuclease P of *E. coli* contains an essential RNA component. Moreover, large numbers of functionally obscure small RNA molecules (snRNA) had long been known in virtually every form of eukaryotic nucleus. Most excitingly, when the nucleotide sequence of one such snRNA "Ul" from rat cells was examined, a large stretch of bases at its 5′ end was exactly complementary to the consensus sequence at the splice sites. Our best guess now is that "Ul" snRNA forms part of a splicing complex by base pairing with both ends of an intron to bring them into exact register for cutting and splicing.

Despite intense efforts, it has proved extremely difficult to prepare extracts of mammalian cells that will splice pre-mRNA reproducibly in the test-tube. Yet the splicing of precursors of several yeast tRNAs does occur reproducibly *in vitro*, and the first proof that snRNA-like molecules are involved might soon be available from this system.

DIAGRAM OF mRNA–DNA HETERODUPLEXES WHEN:

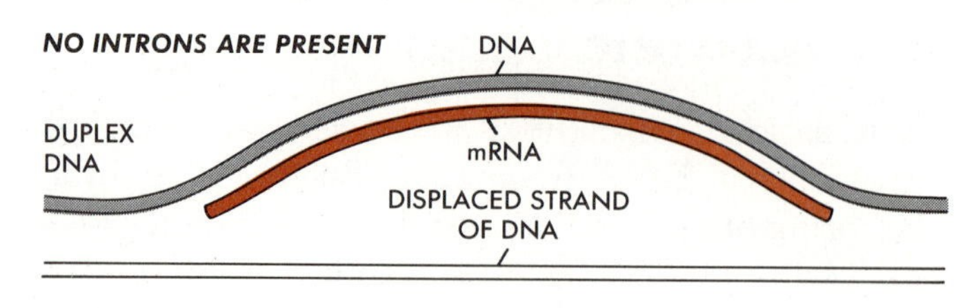

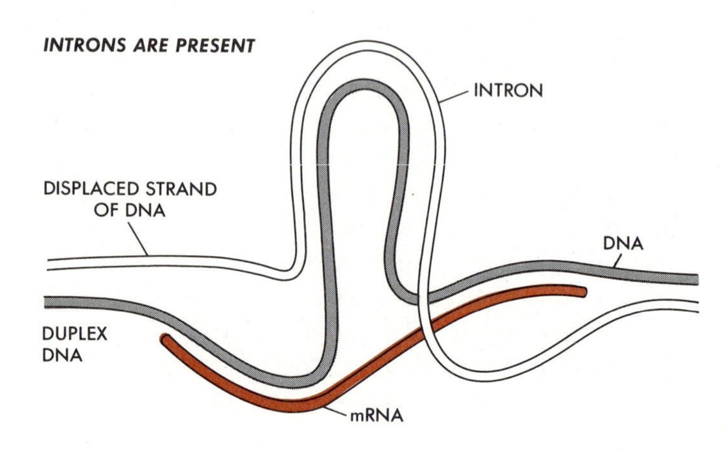

Complete Sequencing of the First Mammalian Genes

Given the speed of the new methods, the sequencing of many complete mammalian genes has become a feasible objective, even though the presence of introns means that usually several thousand nucleotides must be ordered. The chicken ovalbumin gene is more than four times larger than required by the protein's amino-acid sequence, and its primary mRNA transcript comprises some 7700 bases. The majority of the ovalbumin gene sequence was known by January 1979, and the complete sequence was finished late in 1980. By now we also know complete sequences of several insulins, of human growth hormone, of a variety of interferons and histones, of several hemoglobin chains, and of several antibodies.

Though the time necessary for a highly motivated molecular geneticist to sequence an aver-

PATHWAY OF mRNA MATURATION

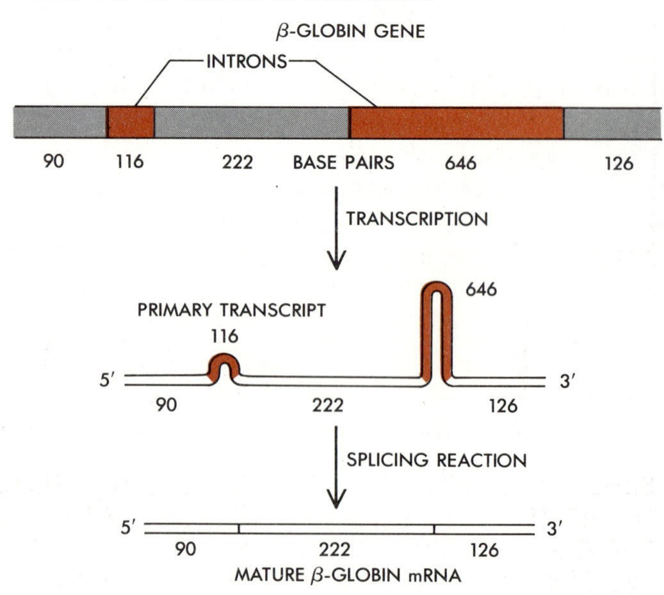

ORGANIZATION OF A SPLIT GENE

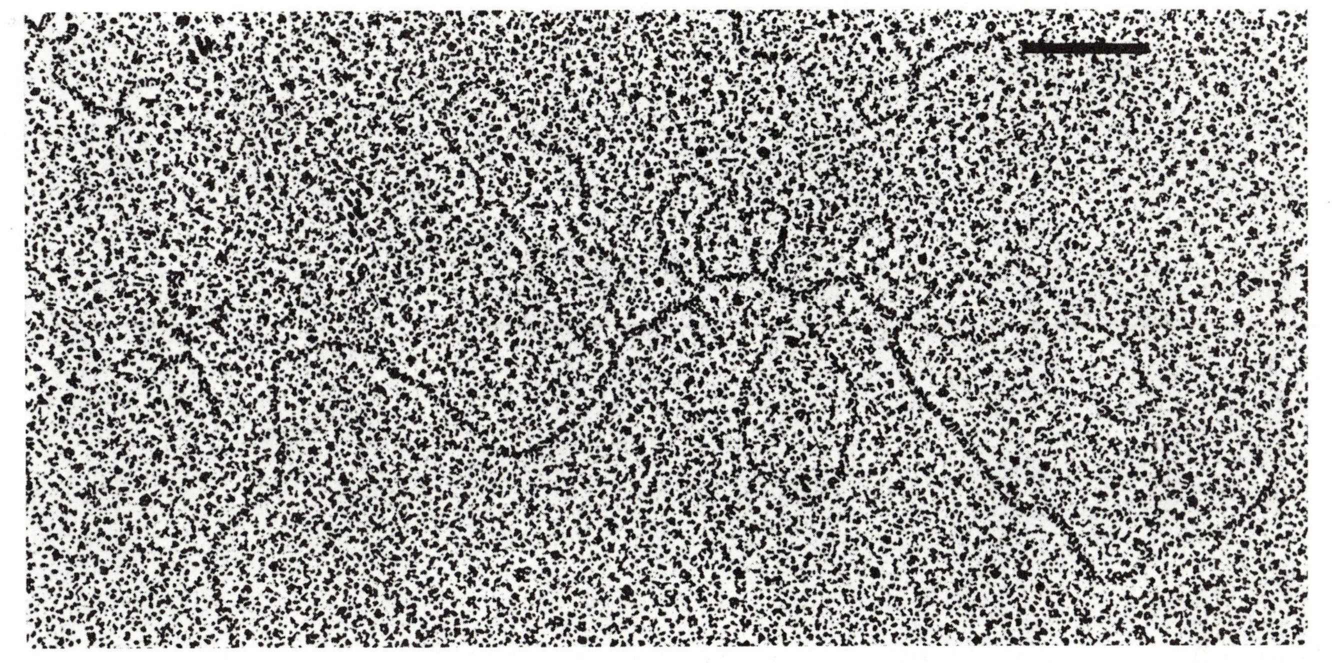

ELECTRON MICROGRAPH COURTESY OF PIERRE CHAMBON.

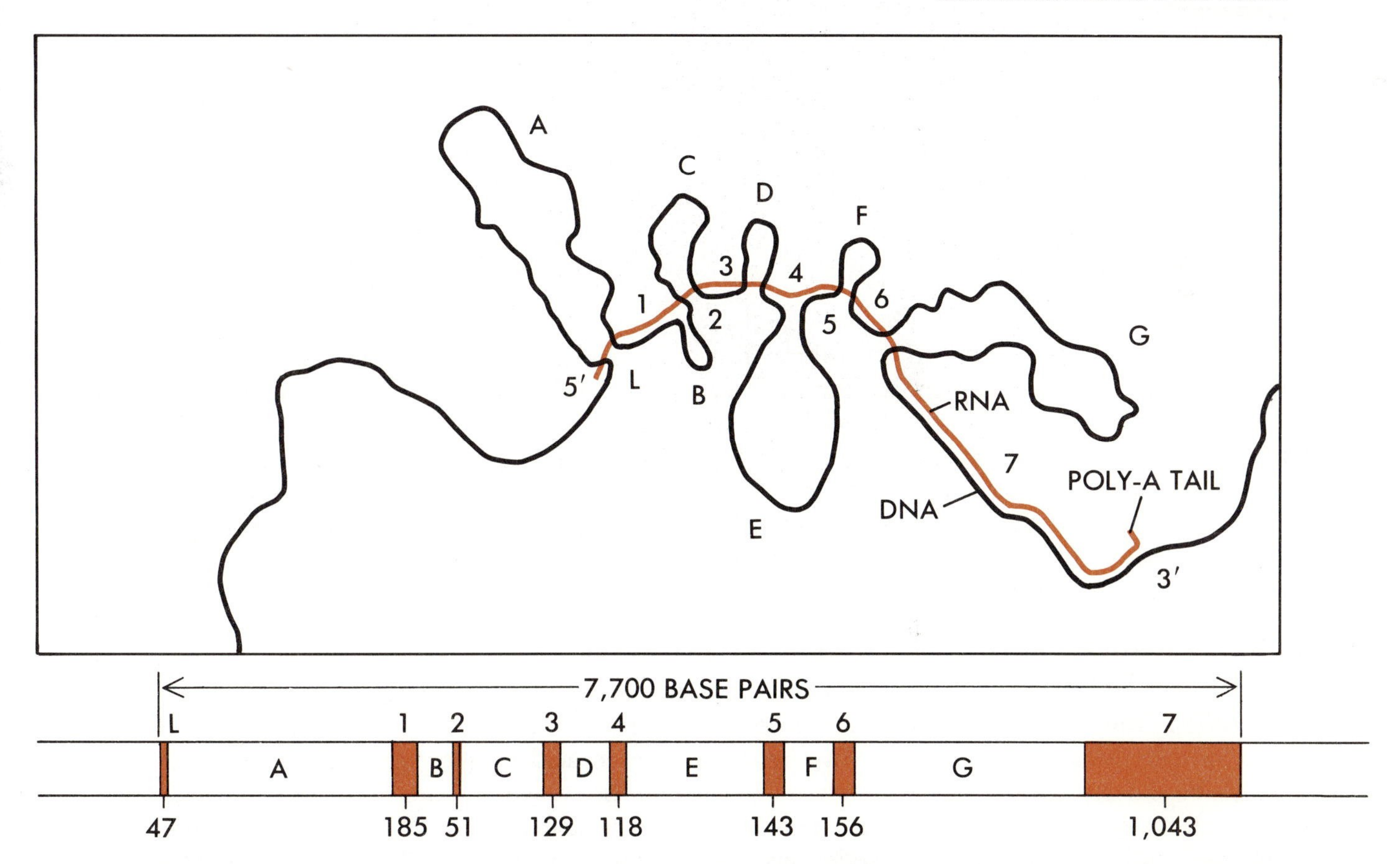

SINGLE-STRANDED DNA CONTAINING THE GENE FOR THE PROTEIN OVALBUMIN WAS ALLOWED TO HYBRIDIZE WITH OVALBUMIN MESSENGER RNA. THE EIGHT EXONS (L, 1-7) OF THE GENE ANNEAL TO THE COMPLEMENTARY REGIONS OF RNA, AND THE SEVEN INTRONS (A-G) LOOP OUT FROM THE HYBRID. THE 5′ AND 3′ ENDS OF THE MESSENGER ARE INDICATED, AS IS THE POLY-A TAIL.

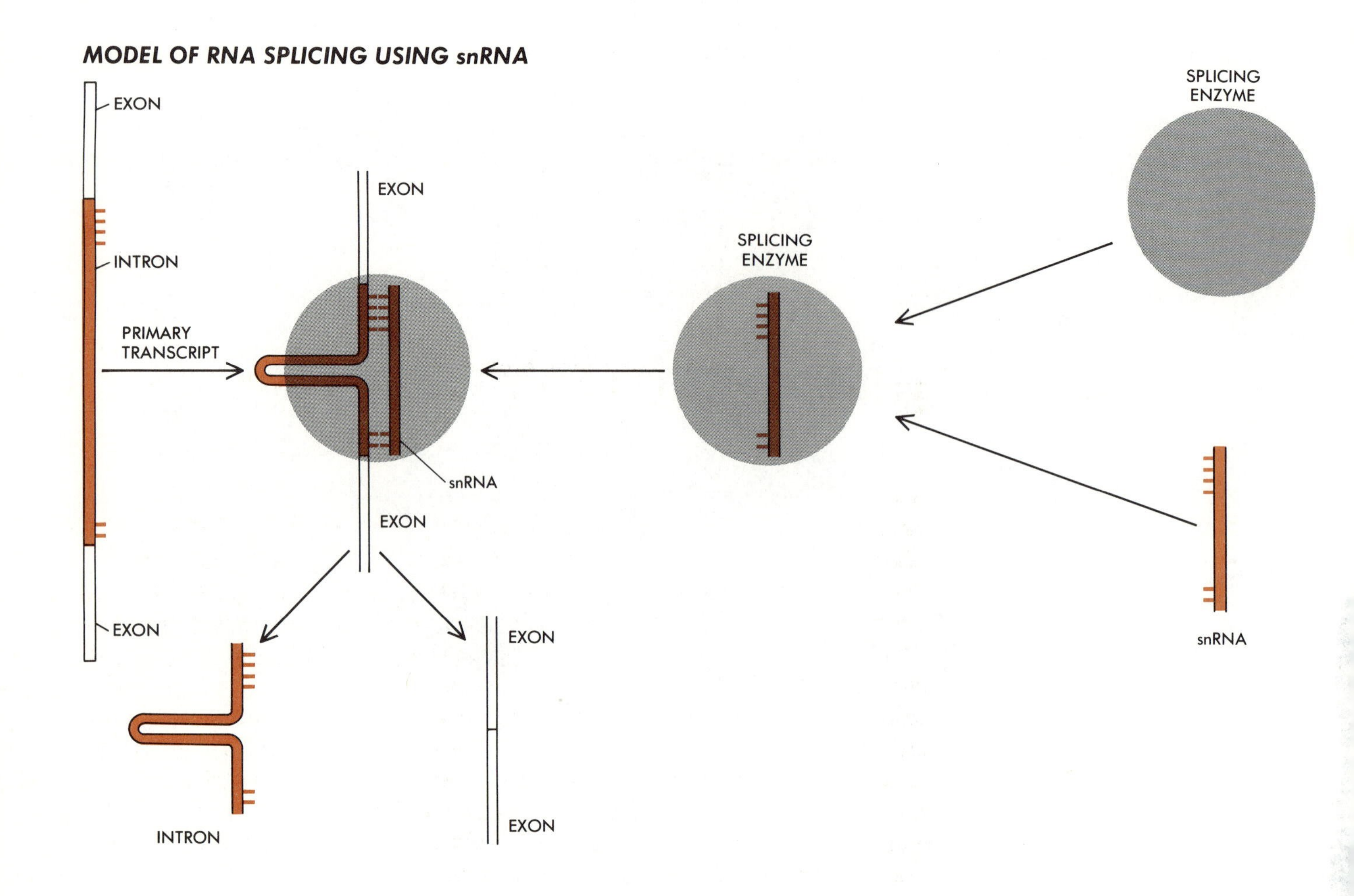

age gene has now been reduced to about a year, the cost of such procedures is not yet trivial. Taking into account time and materials, one could estimate that each base pair in a gene that is sequenced probably costs 5–10 dollars. The need remains for still more automatic and cheaper sequencing procedures, particularly as we move from the sequences of individual genes to those of appreciably larger segments of eukaryotic chromosomes.

Introns Sometimes Mark Functional Protein Domains

At first neither the location nor the number of introns within a given gene made sense. In rats, for example, two closely related genes code for insulin—one has only one intron and the other two. Both the rat insulin I and rat insulin II genes have an intron of almost identical size located immediately downstream from the sequences coding for the insulin leader. The second intron of the rat insulin II gene is located within the so-called C segment of the protein precursor to insulin that is digested away to produce the two-chained structure of mature insulin molecules. Humans have only one insulin gene whose two inserts are located similarly to those of the rat insulin II gene, thus suggesting the descent of rat and human genes from a common ancestor. No obvious functional difference marks the amino acids separated by the second insulin intron, whose location might be accidental. In hemoglobin, though, the amino acids comprising the special functional domain surrounding the heme group are clearly delineated by an intron from the more distal amino acids. As we describe below, introns within antibody genes are precisely located between functional domains. For this reason much protein evolution may have been accomplished by genetic recombination events that bring together domains previously located on separate genes. It is conceivable that the long length of many introns helps to

COMPARISON OF HUMAN AND RAT INSULIN GENES

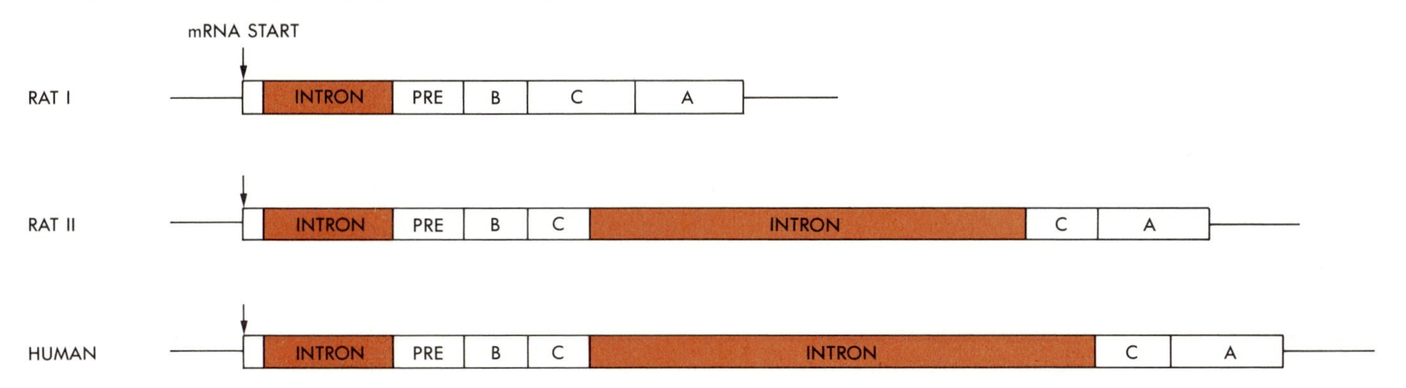

ensure that much genetic crossing over reshuffles intact coding sequences as a package.

Discovery of Clustered Gene Families That May Include Vestigial Evolutionary Relics

With reliable DNA probes for a growing number of vertebrate genes, it has become clear that multiple gene families exist for many of the more abundant proteins. Sometimes multiple copies of a gene have virtually identical sequences, are arranged tandemly, and function simultaneously to synthesize rapidly certain proteins. The histone genes are one such example. In other instances the members of a gene family, though clustered on the chromosome, are not identical. Such genes specify closely related proteins (e.g., embryonic and adult hemoglobin) needed at different stages of differentiation. Also, within the hemoglobin gene clusters of several mammals are hemoglobinlike genes that no longer can function because long deletions are present and essential regulatory signals are absent. These segments of DNA could represent vestigial remnants of formerly functional genes whose physiological roles have, during evolution, been taken over by other members of the hemoglobin gene family.

We are less sure of the meaning of other gene clusters; for example, next to the gene for chicken ovalbumim are two similarly sized genes with identical intron patterns. But we do not know the function of the two corresponding proteins.

Between consecutive, closely related genes in a cluster are often segments of DNA, called spacers, that do not code for protein. These are not to be confused with introns, which occur within genes. The spacers frequently contain more DNA than the split genes themselves. Whether such spacer DNA is totally useless (junk DNA?) is still unclear. In any case, if most spacers have the same sizes as those between the genes in the hemoglobin family, then as little as 10 percent of the nucleotides along a chromosome actually code for specific amino acids. The number of functional genes within a mammalian cell is thus far from the several million estimate initially deduced from a cell's total DNA content. Now we suspect that the true number will be no greater than 100,000, an estimate which may be lowered as we analyze longer stretches of individual vertebrate chromosomes.

HUMAN GLOBIN GENE CLUSTERS

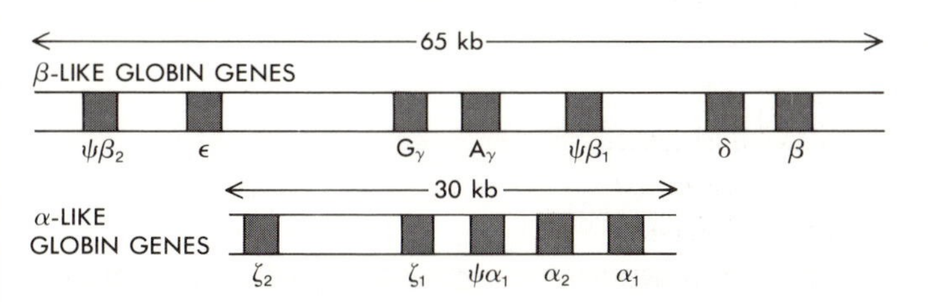

NOTE: ψ GENES ARE THE "PSEUDO" VESTIGIAL GENES.

GLOBIN POLYPEPTIDES	EMBRYO	FETUS	ADULT
α-LIKE	ζ	α	α
β-LIKE	ε	G_γ, A_γ	δ, β

Polypeptide Precursors of Protein Hormones

Some years ago it was discovered that several small pituitary-hormone proteins are derived by cleaving a large precursor protein. Corticotropin (ACTH) and lipotropin (β-LPH) are derived from a precursor polypeptide with a molecular weight of about

30,000. These two hormones are, in turn, broken into even smaller biologically active peptides such as α-melanotropin (α-MSH), the endorphins, and methionine enkephalin.

Further progress toward understanding how the common precursor polypeptide was specifically cleaved to yield the active hormones was prevented because it proved impossible to obtain the precursor molecule pure and in the amounts necessary for determining its amino-acid sequence. All that could be said was that only about one-third to one-half of the precursor molecule ended up in the functional hormones. There was great but frustrated curiosity about the structure and composition of the half of the precursor about which we knew nothing. Perhaps it gave rise to as yet unknown hormones.

Such problems have fallen to recombinant DNA procedures. By early 1979 a cDNA probe of the mRNA for the common precursor of corticotropin and β-lipotropin had been cloned and the sequence of its 1091 base pairs determined. From this base sequence the amino-acid sequence of the precursor polypeptide could be deduced because we know which amino acids are specified by which codons of the genetic code. This information revealed where in the precursor the ACTH and β-LPH sequences occur. Moreover, we discovered a gene sequence for a previously unsuspected third hormone named γ-melanotropin (γ-MSH). Three, not two, related pituitary hormones are produced from this one precursor. In addition, because several nucleotide sequences are repeated in the cDNA, it seems highly likely that during evolution a single ancestral gene underwent triplication to yield a row of three fused genes, each of which has suffered independent mutations to generate three related, but not identical, hormones.

In the future, the most direct way to determine the primary sequences of many proteins will be by cloning and then sequencing their corresponding cDNA probes. This sequence information will simultaneously reveal which codons are used most frequently and tell us more about regulatory sequences at 5′ and 3′ ends of mRNAs. We have, for example, learned that the sequence AAUAA probably plays a role in signalling that gene transcription into RNA should terminate.

If a protein and its mRNA are rare, the preparation of the corresponding cDNA probes will require patience and ingenuity. It takes about a week's work to check by *in vitro* translation analysis which protein a single cDNA probe specifies. To make rapid progress we now need enrichment procedures for the mRNAs of those proteins present in small amounts, which is, of course, the vast majority.

Rearranging Germ-Line DNA Segments to Form Antibody Genes

For several decades we have known that an enormous number (perhaps in the millions) of different antibody (immunoglobulin) molecules exist, each characterized by a unique site that can bind to spe-

MATURATION OF PITUITARY HORMONES

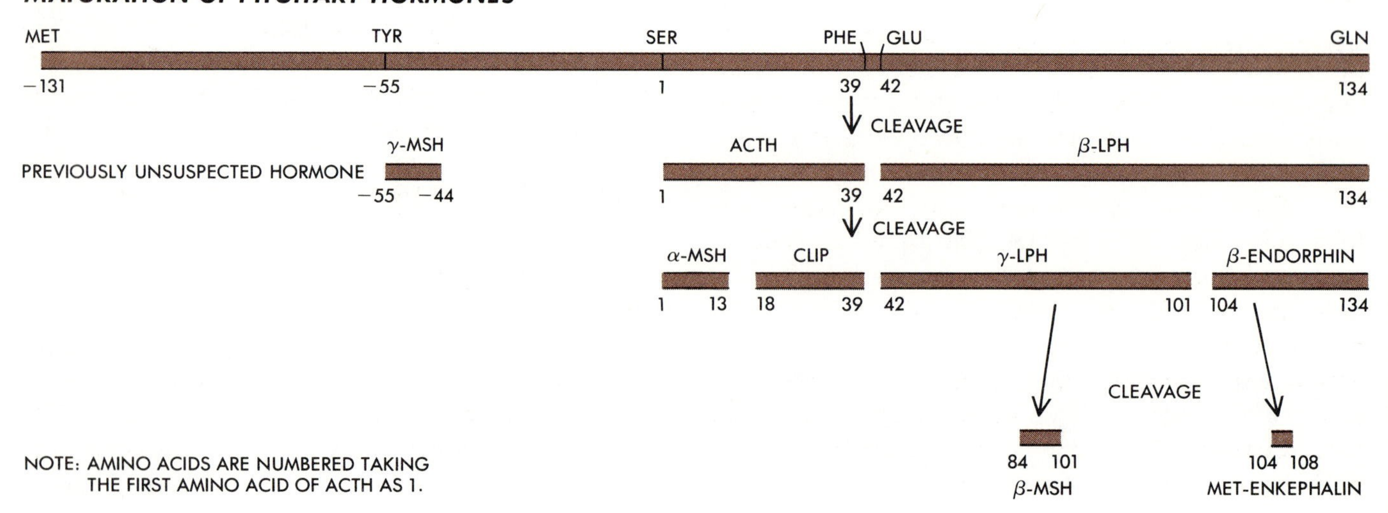

cific molecular determinants (antigens). Many immunologists thought initially that all antibodies were made of the same polypeptide chains and that their uniqueness arose from the way their newly synthesized identical polypeptide chains folded around the respective antigenic determinants. We now know that this theory is wrong. Each antibody has its own amino-acid sequence, and each antibody-producing cell (plasma cell) makes only one antibody. At first, this was a disturbing discovery because it seemed to imply that a separate gene would have to exist for each separate antibody. If so, perhaps a large fraction, if not the majority of the vertebrate DNA, would have to be devoted to coding antibody molecules. But such speculations could not be tested until protein chemists established the basic structure of the antibody molecule.

STRUCTURE OF AN ANTIBODY PROTEIN

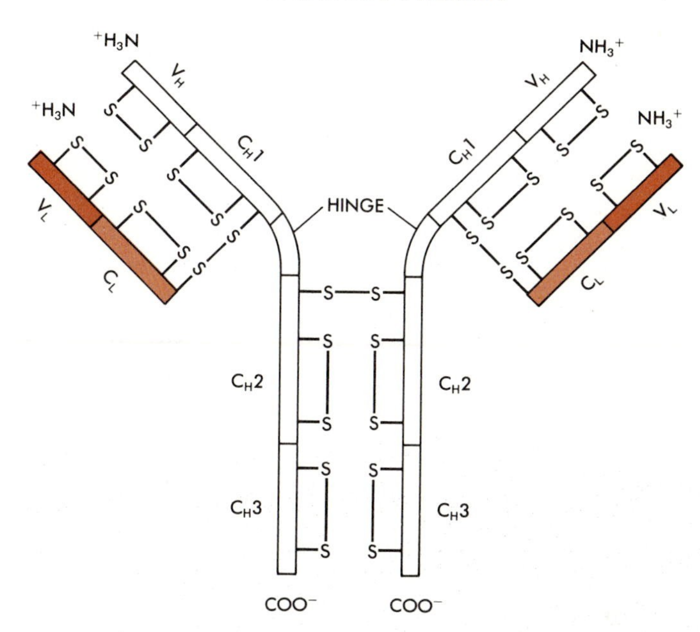

Initial Attempts to Cope with a Potentially Large Number of Antibody Genes

The first insights began to emerge in the early 1960s, when it was realized that the fundamental antibody unit consists of two identical light (L) chains of molecular weight 17,000 and two identical heavy (H) chains of molecular weight 35,000 held together by disulfide bonds. (The terms "light" and "heavy" refer to the differences in molecular weight of the two sorts of chains.) Each such four-chain unit contains two identical binding sites for antigens formed partly by specific amino acids of the light chain and partly by specific heavy-chain amino acids. Once the basic antibody layout had been established, the amino-acid sequences of the component light and heavy chains were determined by using the homogeneous antibodies made by specific myeloma cells. Myelomas are cancerous antibody-producing (or plasma) cells, and in any one animal all the cells of a myeloma tumor are the descendants of one original cancer cell. This explains why all the antibody molecules from any one myeloma have the same amino-acid sequence.

Both light- and heavy-chain sequences vary from one antibody to another, but in a way that no one would have predicted initially. Although each light and each heavy chain has unique sequences, almost all of this specificity is restricted to about 100 amino acids at their amino terminal ends (the variable, or V, region). Half of each light chain and three-quarters of each heavy chain have almost identical sequences (the constant, or C, region).

Separate Genes for V and C Segments

To account for the constant and variable portions of each heavy or light chain, William Dreyer and Claude Bennett, as long ago as 1965, put forward a bold hypothesis. They proposed that the V and C regions are coded by separate genes, and that in the germ line only one C-region gene belongs to each class of heavy (C_H) or light (C_L) chain but many thousands of genes belong to the variable regions V_H and V_L. To form a functional antibody they proposed that in the precursor to the plasma cell a genetic recombination event takes place to bring one of the V genes next to its respective C gene to yield functional V_LC_L and V_HC_H genes. This most perceptive hypothesis won few early converts because it flew in the face of the general belief that the arrangement of DNA within a given chromosome was effectively immutable except at meiosis during the formation of the sex cells.

Using mRNA Probes to Obtain Support for the Joining of V and C Genes

Most people preferred to avoid facing up to the possible need for any molecular recombination between the putative V and C genes until there were direct molecular probes that might identify them. In the first experiments done between 1974 and 1976, mRNA probes that were more than half pure were made from cancerous antibody-producing myeloma cells. These mRNA probes were mixed, under conditions favoring hybridization, with unfractionated homologous myeloma DNA to see if it was possible to count the number of genes for a single antibody type in each myeloma cell. The answer for at least one light-chain mRNA probe was a very low number, perhaps only one.

Such experiments, however, could not distinguish V from C sequences, nor could they indicate differences in the relative locations of V and C sequences in embryonic cells compared with myeloma plasma cells. To do that required a way to cut up the total myeloma and embryonic cell DNAs into reproducible pieces, a procedure that became possible only with the ready availability of restriction enzymes. Using them, Susumu Tonegawa in the spring of 1976 observed that V and C sequences linked together on the same DNA restriction fragment from an antibody-producing mouse myeloma cell were not similarly linked together in embryonic DNA. This classic experiment was done with necessarily impure mRNA probes because of the still-effective prohibitions against making cDNA probes. But as soon as the first suitable cDNA vector was approved by the RAC and a P3 laboratory had been constructed at the Basel Institute of Immunology, the appropriate cDNA probes for V and C regions were made and their specificity directly determined by DNA sequencing. They were then used for selecting genomic segments that carry the respective antibody genes. Possession of such reagents has revolutionized our understanding of the molecular basis of antibody diversity.

JOINING V GENE TO C GENE

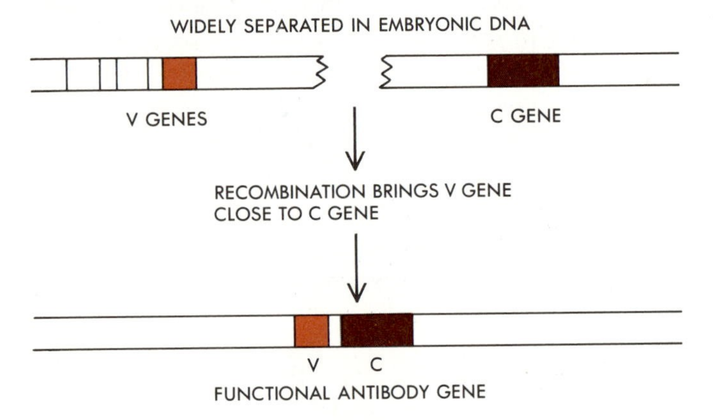

Isolating Functional Antibody Genes from Myeloma Cells

The cDNA probes that establish the nature and number of antibody genes were made from specific myelomas whose antibody products had already been sequenced. Direct comparisons were thus possible between the nucleotide sequences of functional antibody genes and the amino-acid sequences of the antibodies they specify. Many introns were found immediately, and, most importantly, most were located at junctions between functional domains. In the light chain an intron separates almost all of the amino terminal leader sequences from the V segment, while a second intron delineates almost all of the constant sequences. Within heavy-chain genes, introns more extensively separate functionally related domains (in other words, they separate exons coding for domains). Each of the three domains of the C_H protein is clearly delineated by introns, as is the so-called hinge region lying between the second and third constant region domains. All these observations support well the hypothesis that proteins have evolved by the rearrangement of exons.

Embryonic Cells as Sources of Unjoined V and C Genes

The structures of the V and C segments before they are joined to create functional antibody genes were revealed by cloning the appropriate genomic DNA segments from embryonic cells and hybridizing them with probes specific for the V and C regions. C-region probes invariably were very specific, whereas V-region probes often hybridized to many different V genes. Such cross-hybridization reflects the fact that the V regions of different antibodies often differ by only a few amino-acid substitutions. Now we have evidence for at least 200 V_L genes and at least an equal number of V_H genes. Here it is important to note that because the specificity of an antibody is determined by both its V_L and V_H components, the number of potentially different antibodies is at least $V_L \times V_H$ or $200 \times 200 = 4 \times 10^4$. Moreover, this number clearly underestimates the real number of potential antibodies because the V and C genes do not always link together at the same point. A cluster of related, but not identical, J (for joining) segments resides at the end of each C_H and C_L gene, and the linkage to their V_H and V_L partners

SEPARATION OF THE PROTEIN DOMAINS OF IMMUNOGLOBULIN HEAVY CHAINS BY INTRONS

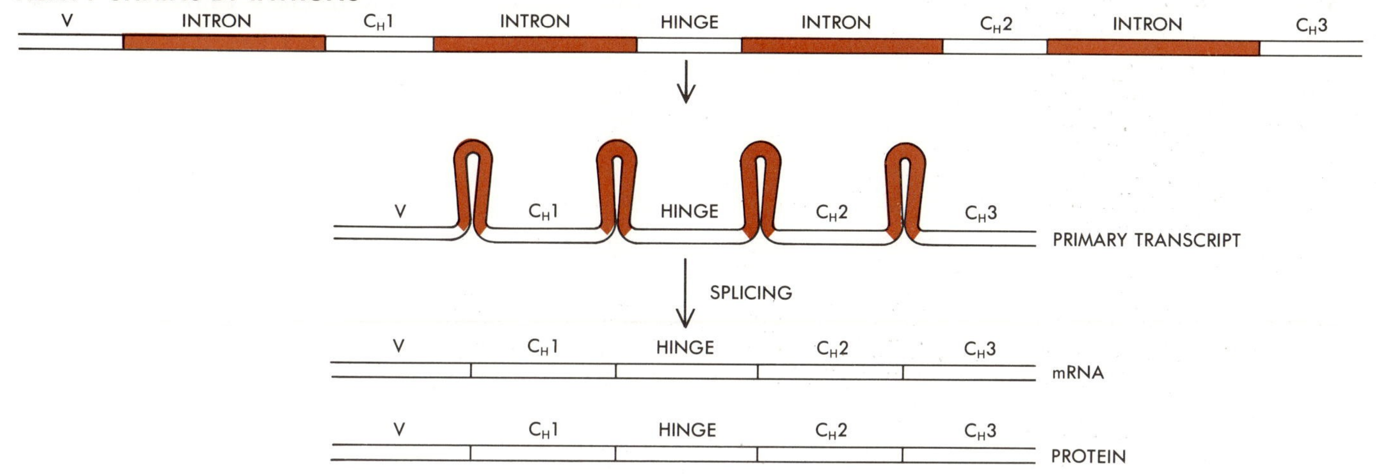

may occur next to any of these J segments. Regardless of which J segment is used, subsequent RNA splicing occurs in such a way that only the bases within the J segment used for V–C joining are retained in the finally processed light- and heavy-chain mRNAs.

How the RNA splicing is regulated so only one J segment is retained remains totally mysterious, as is the nature of events which bring the V and C genes together. One model for DNA joining postulates that the respective V and C segments, initially located far apart on the same DNA molecule, are brought together by a recombination process that eliminates the intervening sequences. The sequences at the ends of the V and JC segments are complementary and could allow formation of hydrogen-bonded hairpin loops that would align the appropriate bases ready for cutting and rejoining. There is some support for this proposal.

A DNA Elimination Event Also Allows a V_H Gene to Be Attached to Two Different C_H Genes

Recombinant DNA procedures have also swiftly solved the puzzle of how a given heavy-chain variable gene (V_H) can be attached first to a constant segment (C_μ) characteristic of immunoglobulin of the M class and then, during the later immune response, transfer its linkage to a constant segment (C_γ) characteristic of the immunoglobulin G class. No change of immunological specificity occurs during this changeover because the heavy-chain types differ only in their "constant" components, all of which are coded by a group of genes clustered together on the same chromosome. How this switch happens became crystal clear as soon as the appropriate C_μ and C_γ genes were cloned and their sequences were compared with those of a functional gene coding for a known γ-type heavy chain (MOPC 141). The key observation was the finding of J-segment bases in the gene coding for the γ heavy chain of MOPC 141 despite the absence of any J segment in the corresponding embryonic C_γ gene. In contrast, several J segments were found at the beginning of the embryonic C_μ gene, one of which exactly corresponded to that observed in the

JOINING V_H GENE TO C_H GENE VIA J SEGMENT

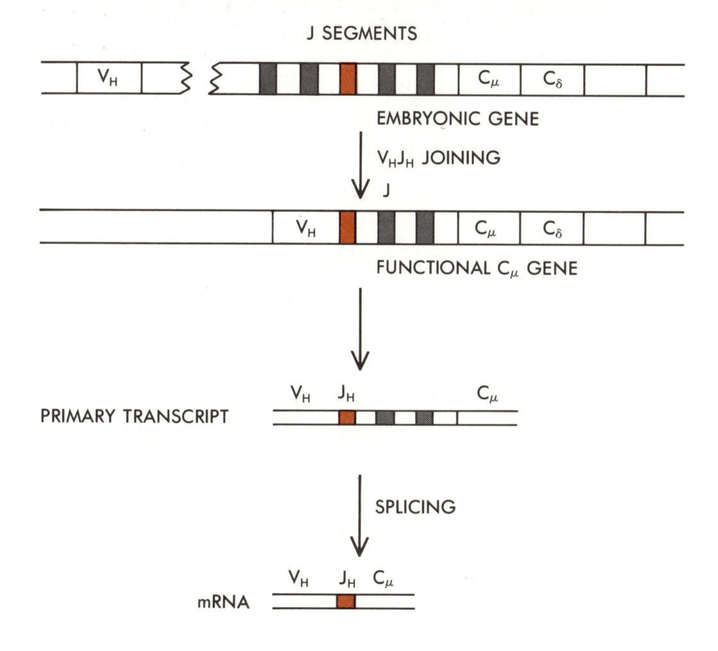

RECOMBINATION SWITCHES FROM C_μ GENE TO C_γ GENE

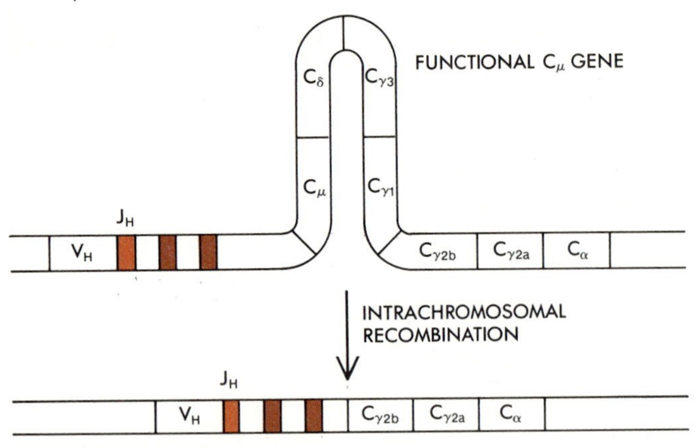

functional γ gene of MOPC 141. Two recombination events are therefore necessary to generate a functional gene for γ-type heavy chains. The first joining event attaches a V_H gene to the C_μ gene at one of its flanking J segments, thereby allowing synthesis of a μ-class heavy chain. This synthesis continues until a second recombination event removes most of the C_μ gene sequences and thereby links the previously joined V_H–J segment to the intron sequences flanking a C_γ gene. This leads to the synthesis of a γ-class heavy chain whose J-coded sequences bear witness to the prior V–JC_μ arrangement.

Alternative Splicing Allows Single Cells to Make Both μ and δ Heavy Chains with Identical V_H Segments

Recombinant DNA procedures have also clarified another previously puzzling observation, namely, that one cell can simultaneously make heavy chains of two types but with the same variable region. The initial explanation proposed for this situation was that one type of heavy chain might be translated from a very stable long-lived mRNA that persisted in the cytoplasm long after the corresponding gene had been eliminated. Analysis of the structure of the genes involved with recombinant DNA techniques, however, suggested an alternative. The variable region (V), a joining region (J), and the two constant regions C_μ and C_δ were found to be contiguous. This organization led to the suggestion that the whole complex might be translated into a large precursor RNA molecule from which, by differential splicing, either one or the other of the constant regions was eliminated. With cDNA probes prepared from the appropriate mRNA this mechanism was confirmed. Differential splicing of a common precursor RNA generates two distinct heavy-chain mRNAs. A dual-splicing potential also determines whether certain classes of antibodies are bound to the plasma membranes of the cells in which they are made or are secreted to the outside. And, of course, alternative splicing of a precursor RNA to yield different types of mRNA occurs during the replication of some viruses—the adenoviruses, for example. Indeed, it was during studies of adenoviruses that the phenomenon was first observed.

Movable Genetic Elements

Recombinant DNA research has also permitted analysis of the interaction between the DNA of tumor viruses and the cells they transform. Most, if not all, cancerous cells transformed by DNA tumor viruses such as SV40 and adenovirus 2 have integrated into their chromosomes viral DNA sequences whose expression is at the heart of the host cell's cancerous properties. Beginning in 1972 molecular biologists began to examine the amount and genomic organization of the tumor viral DNA integrated in the transformed cell by using as probes fragments of DNA obtained by digesting the viral genomes with restriction enzymes. Yet the isolation of the integrated viral DNA along with the flanking cellular sequences became possible only after recombinant DNA procedures had been developed.

Integrated Forms of DNA Tumor Viruses

The first such gene cloning was done in 1978 in France and England, where the required physical containment for recombinant DNA experiments with tumor viral genomes was decided on a case-by-case basis and was less exacting than that required by rigidly categorized and prohibitively stringent American rules. Only in January 1979 did it become possible in the United States to clone tumor virus DNA. As soon as it was allowed, specific restriction enzyme fragments from tumor viral DNAs were cloned to produce specific probes for isolating integrated viral genomes.

From what has emerged it looks as if the integration processes are virtually haphazard, occurring

ALTERNATIVE SPLICING GENERATES DIFFERENT mRNAs

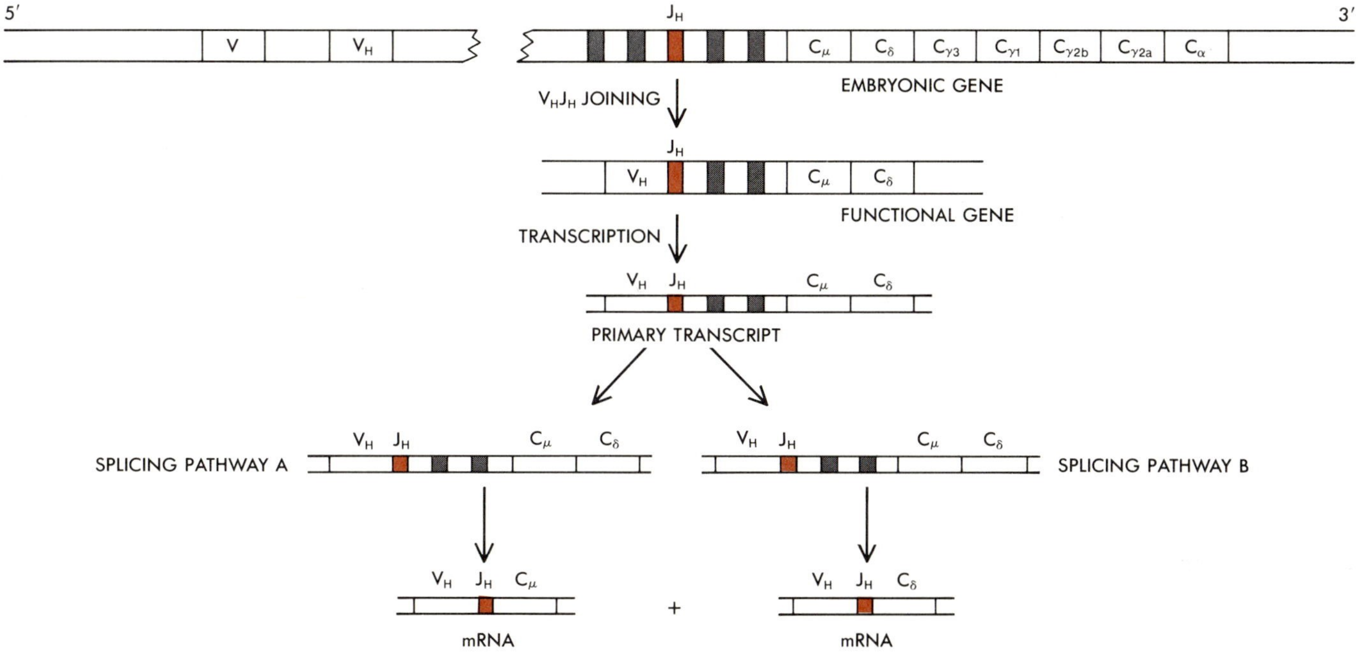

neither at specific sites nor within any specific sequence in either the viral or host chromosomal DNA. Some integrated segments reveal almost bizarre rearrangements of viral sequences, and we suspect that the recombination events that mediate integration involve either replicating DNA or DNA being transcribed into RNA. Both replication and transcription necessarily produce limited regions of single-stranded DNA that could form the temporary base-paired bridges that would be necessary to hold the molecules together as they participate in nonhomologous (illegitimate) recombination events.

Testing the Concept of Movable Genetic Elements

In the early 1950s Barbara McClintock, working at Cold Spring Harbor Laboratory on the corn plant, began to call attention to the novel behavior of genetic elements that she called controlling elements. First noticed because they inhibited the expression of other genes, these control elements do not have fixed chromosomal locations. Instead, they move about the corn genome and inhibit the functions of those genes with which they come into close contact. The controlling elements can be excised as well as inserted; after excision, the function of a previously dormant gene often returns. The genes they associate with are thus inherently unstable and have high mutation rates.

For many years the corn plant provided the only genetic system in which such movable genetic elements were observed. Then preliminary evidence began to accumulate that several of the highly mutable loci in *Drosophila* might be associated with movable control elements. But most geneticists paid little attention to such loci until the discovery in the late 1960s that certain highly pleiotropic (having many effects) mutations in *E. coli* resulted from the insertion of large segments of DNA called insertion sequences (IS). Most importantly, many independent insertion events involved exactly the same DNA sequences. Four main blocks of insertion sequences were found and labelled, IS1, IS2, IS3, and IS4. Multiple copies of the IS1 and IS3 elements are scattered throughout the *E. coli* chromosome. Not only are the IS elements themselves capable of movement, but, when present in closely spaced pairs, they move as a unit, carrying along the genes lying between them. These more complex units are called transposons.

With the arrival of recombinant DNA, it quickly became possible to clone each of the IS elements along with their particular flanking se-

IS1 AND A TRANSPOSON IT GENERATES

INSERTION SEQUENCE 1 (IS1): 768 BASE PAIRS (bp) OF DNA. TWO INVERTED REPEATED SEQUENCES (IR) OF 24 bp SANDWICH THE GENE FOR THE TRANSPOSASE ENZYME INVOLVED IN THE MOVEMENT OF IS1.

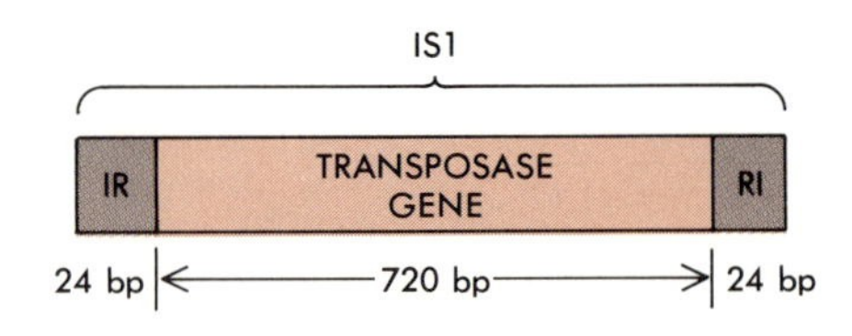

A TRANSPOSON (Tn) COMPRISING TWO IS1 ELEMENTS SANDWICHING THE *E. COLI* GENE FOR HEAT-STABLE TOXIN 1, WHICH CAUSES DIARRHEA.

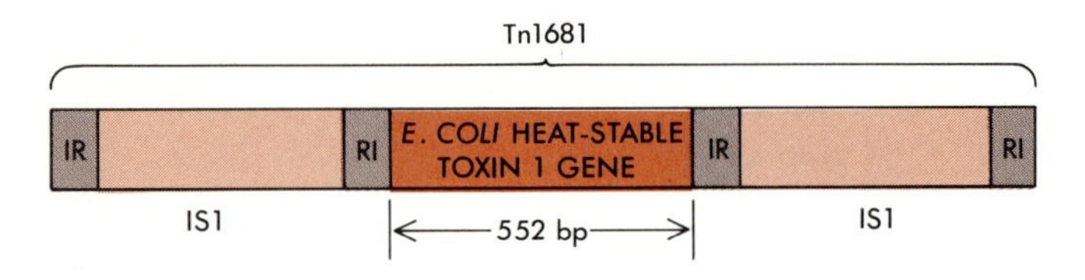

TRANSPOSON INSERTION MODEL

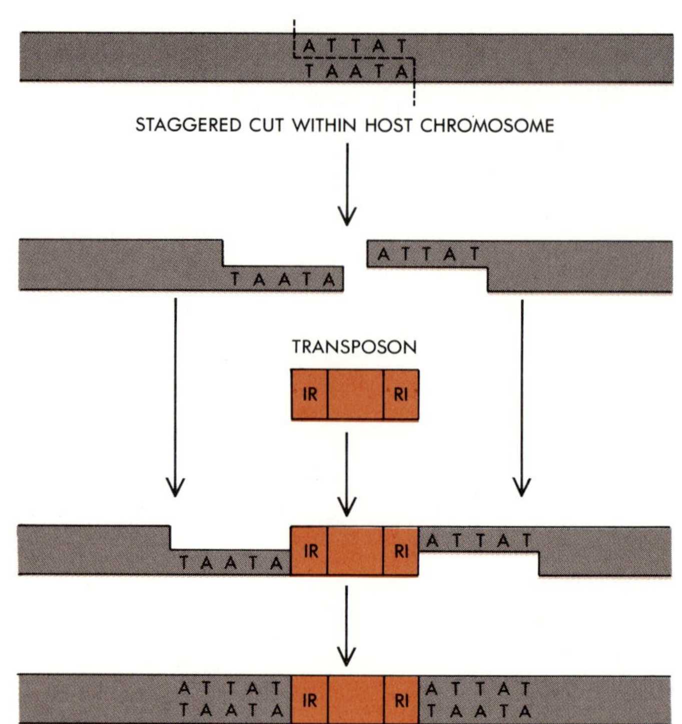

quences. By comparing the sequences at the junctions between IS elements and chromosomal DNA we quickly discovered that these elements can insert themselves into many places in the *E. coli* chromosome. The most important fact is that the immediately adjoining host chromosomal sequence at the left-hand junction point was always identical to the host chromosomal sequence at the right-hand junction. This arrangement indicates that the actual insertion event involves a staggered cut within the host chromosome that creates single-stranded tails into which the IS elements are placed.

It now appears that the "movable elements" may not actually move because their appearance at a new site is not necessarily accomplished by their disappearance from any other previous location. Instead, what apparently happens is that a parental IS segment gives rise to one or more copies of itself which become inserted elsewhere. Whether or not these copies are made by way of an RNA intermediate and resemble the transposonlike retroviruses remains unknown. Transposition is, though, specifically mediated by a protein (transposase) coded by the IS element (transposon) itself. By studying the properties of purified transposases we can learn how they act.

Have RNA Tumor Virus Genomes Evolved from Movable Genetic Elements?

Like the DNA tumor viruses, the RNA tumor viruses (retroviruses) transform normal cells into their cancerous equivalents by integrating themselves into host-cell chromosomes. But they do so in a radically different fashion. After the infecting viral RNA chromosome enters a cell, it is transcribed into a cDNA complement by the enzyme reverse transcriptase, which is coded by the viral nucleic acid. Reverse transcriptase enters the host cell along with its RNA template in the infecting virus particle. (This enzyme is the one that is used to make cDNA in cloning experiments.) The double-helical proviral cDNA then becomes inserted into the host chromosomes. Integration does not change the order of the genes on the infecting RNA chromosome, and so the integrated genome can be transcribed later to give progeny RNA molecules. Some of these are surrounded by viral-specific "coat" proteins to form new RNA tumor virus particles.

Sequence analyses of the ends of several integrated retroviral genomes aroused much excitement because the immediately adjacent cellular DNA sequences on both sides of the viral DNA were the same. The five bases of cellular DNA on the left side were invariably identical to the five host-cell bases on the right side. In addition, at the two ends of integrated proviruses there were identical blocks of viral DNA several hundred base pairs long. These results exactly paralleled those found only a year earlier for transposons in *E. coli*; the results were quickly extended to similar movable genetic elements in yeast and in *Drosophila.* Integrated retroviruses are conceivably movable genetic elements that have acquired the ability to move from cell to cell by possessing genes for proteins which, by aggregating around their RNA transcripts, form infectious virus particles.

Sex Changes in Yeast by Gene Replacement

Haploid yeast cells have two sexual forms, *a* and α, that, when mixed, fuse into diploid cells. The sex of a haploid cell is determined by the mating-type gene, which can be either the *a* or the α form. In some strains of yeast the mating type is a stable property; in others it changes rapidly as if the mating-type gene were converted from one form to the other. Such rapid changes of mating type could not be explained by any conventional genetic models, thus prompting yeast geneticists to explore models where the molecular organization of the mating-type gene could change from one state to another.

Most important were experiments in Japan and Russia that revealed two genes close to the mating-type locus whose presence controlled the specificity of the sex switching. These findings led to the suggestion in 1976 for a cassette model of sex switching. The model postulated that these two additional genes were silent copies of mating-type genes, one an unexpressed *a* type, the other an unexpressed α type. Within this scheme sex switching occurs when a copy of one of the silent genes is inserted into the mating-type locus and the previous tenant is somehow displaced. Each time this occurs the yeast-mating type changes.

Proof of the cassette model was obtained in 1979 when the DNA of the functional mating-type locus was cloned and used to select the putative silent genes. The base sequences of the *a* and α silent genes closely resemble the base sequences of the corresponding functional genes in the mating-type locus.

The mechanisms for moving copies of silent genes and inserting them into the mating-type locus remain obscure, but this movement has clear parallels with the movement of transposons in *E. coli*. In both processes a copy of the parental genetic element is inserted at the new site in the chromosome while the parental element remains intact and in place.

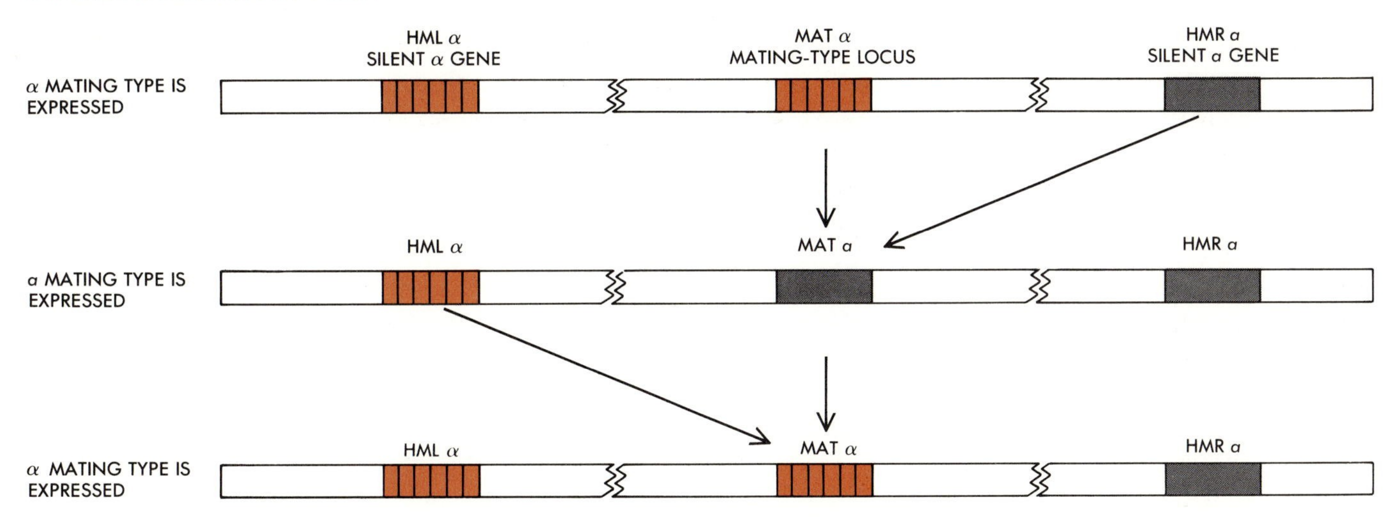

Epilogue

For eight years recombinant DNA research was placed under constraints that were crippling. Only in the last two years have those constraints been relaxed to the present point where for most research with standard host-vector systems they are no longer a serious impediment. That so much new and totally unexpected has been learned since 1973 about the structure, organization, and expression of genes in higher organisms—despite all the obstacles put in the way of recombinant DNA experiments (recall that in the fall of 1976 clones of yeast and fruitfly DNA were ordered to be destroyed)—is the best testimony to the power of these methods. Recombinant DNA is no ordinary technical development; in combination with new ways for sequencing DNA it is a tool of enormous power. It came at a time when the conventional genetic and biochemical techniques that had proven so successful for studying bacteria were becoming inadequate in relation to the complexity of eukaryotic cells and their genomes. Without recombinant DNA methods, further progress in understanding the molecular genetics of higher organisms seemed destined to be painfully slow, disappointingly meagre, and very expensive. Recombinant DNA rescued the field from a future of gradual atrophy. Moreover, it offered the prospect of new industries based on molecular biology, and a rebuttal to the many critics who pointed out that the double helix and all that had contributed virtually nothing of practical benefit to society.

With so much at stake—the future of their subject—it is not surprising that molecular biologists battle doggedly in committees, in lobbies, at public hearings, in the press and other media to have lifted the unnecessary restrictions and censorship that over the past eight years have sometimes threatened to bring a complete halt to this research. Remember that in August 1976, *The New York Times Magazine* devoted several pages to a feature article calling for a worldwide moratorium on all recombinant DNA research. Of course, the persistence with which molecular biologists fought back was also spurred on by a sour reflection. The disastrous way in which the issue had captivated the public's imagination owed more than a little to the strident way the molecular biologists first sounded an alarm about the possible risks of recombinant DNA. That whole story is traced in the documents reproduced in this book.

The first scientific achievements—the discovery of split genes in higher organisms, of gene clusters with their implications for our ideas about gene evolution, of recombination in somatic cells giving rise to functional antibody genes, of movable genetic control elements, of the generality of mRNA splicing—we have briefly sketched in the book's last section. Politics and politicking preoccupied the first years of the recombinant DNA story, but that phase, fortunately, is fast becoming history. This book is our epitaph to that extraordinary episode in the story of modern biology.

Photo courtesy Louise T. Chow, Cold Spring Harbor Laboratory.

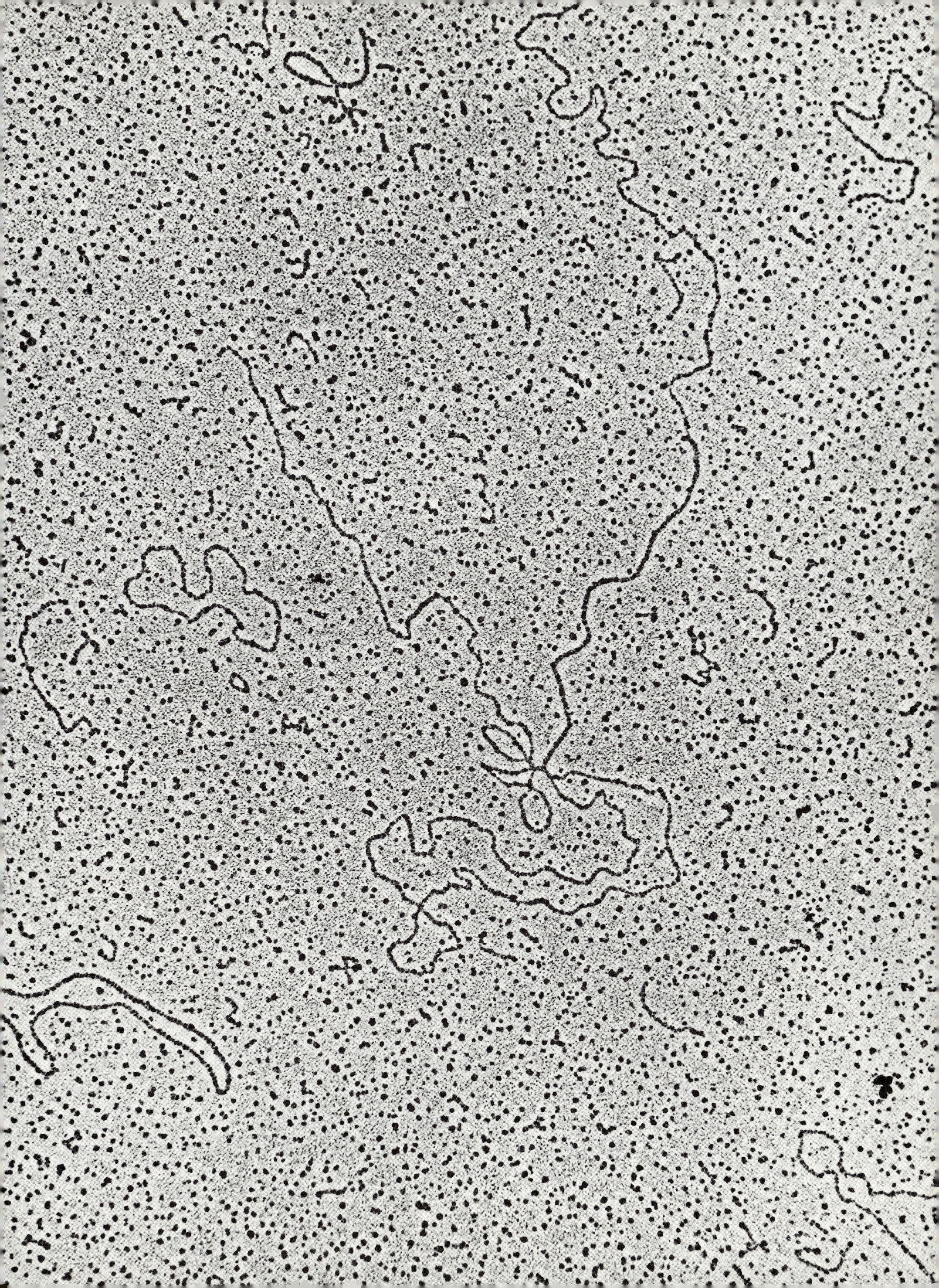

Nobel Laureates Whose Work Led to Genetic Engineering

Werner Arber

Photo by Ross Meurer, Courtesy Cold Spring Harbor Laboratory Library Archives.

Reprinted with permission of *Baltimore Sun*.

The Weather
Cold Ink

The Harvard Crimson

CAMBRIDGE, MASS., WEDNESDAY, OCTOBER 15, 1980

...CLXXI, No. 31

Gilbert Wins Nobel For Genetics Work

By BURTON F. JABLIN

Walter Gilbert '53, American Cancer Society Professor of Molecular Biology, yesterday won the Nobel Prize in chemistry for his work in genetics.

Gilbert will share the $212,000 prize with Frederick Sanger of Cambridge University and Paul Berg of Stanford University, who also were cited for their contributions to genetic research.

Specifically, the Nobel committee lauded Gilbert for his development between 1975 and 1978 of a rapid way to determine the chemical makeup of DNA, the basic material of genes. Sanger won for his development of a different method of determining DNA sequences, and Berg received the prize for discovering a means of transferring genes from one organism to another.

"This prize recognizes the antecedents of an entire field—recombinant DNA research," Gilbert said yesterday, adding, "Everything we do today in recombinant DNA would...

"Their work shows the extent to which molecular biology has made a complete bridge between genetics and the chemical mechanisms thereof," Harrison added.

The efforts of the three Nobel chemistry prize winners has "made possible an explosion in knowledge in the last three or four years," Fotis C. Kafatos, professor of Biology, said yesterday.

"The ability to determine the precise structure of a gene is as fundamental to future progress in biology as the availability of dictionaries has been to the ability to write and understand literature," David Dressler, lecturer on Biochemistry and Molecular Biology, said yesterday.

Gilbert said the techniques developed separately by him and Sanger—which are now used extensively in genetic research laboratories around the world—aLOW

(continued on page 4)

Crimson

...int

...ion

Reprinted with permission of Harvard University.

DAILY EXPRESS

12p • Weather: Showers

THE VOICE OF BRITAIN

PROF'S PRIZE DOUBLE

By CLARE DOVER

...NTIST Professor Sanger pulled off an ...oric double for ...n yesterday.

...was awarded the ...Prize for Chemistry ...e SECOND time.

...the 62-year-old ...t admitted: "I didn't ...ow I had been nominated.

...it is very exciting. It ...eat honour and a great ...gement to have my ...ecognised."

...ssor Sanger, of Cam...e University, was ...l the prize for his ...f how chemical mole... ...arry the message of ...nce and govern the ...of life.

...night he was more ...ed with the bubbles in ...mpagne he bought for ...rch team to celebrate.

...n double is unique in ...year history of the ...undation.

...have been th...

The prize guy! Prof in Nobel double

double laureates in the past, including Marie Curie.

But no one has ever won the chemistry prize twice. Professor Sanger has been at Cambridge since his student days.

He won his first Nobel Prize in 1958 for his work unravelling the structure of insulin.

Then the Prize was worth £14,000. Now it is £89,500.

Professor Sanger said last night: "I haven't ... to the mone... to ...

because the work itself is more important.

"We are trying to achieve something in the scientific world."

His secretary, Miss Peggy Dowding, said Professor Sanger suspected a student hoax when the news came through.

"But of course he thought the same way last time," she added.

"It was only when more calls started to flood in congratulating him that he realised it was true.

Excited

"He was quite overwhelmed. We are all very excited about it because it is the first time it has been awarded in the subject twice."

Professor Sanger's work which won him the prize involved unravelling the 5,375 sub-units which go to make up a particular virus.

He is now ... the...

happens when things go wrong.

As happens frequently, the Nobel Prize was shared this year.

The other winners were American professors Paul Berg, of Stanford University, California, and Walter Gilbert, of Harvard.

In its citation, the Royal Swedish Academy of Sciences said:—

"The investigations of Berg, Gilbert and Sanger have given us a detailed insight into the chemical basis of the genetic machinery in living organisms."

It added that the work would in the long run improve efforts to understand cancer.

Marie Curie shared the 1903 Physics prize with ... band and ... Ch...

Courtesy of Express Newspapers, Ltd.

The Stanford Observer

October 1980

Paul Berg wins Nobel Prize for chemistry

"Stunned is about the best word I can describe it."

These were the words of Paul Berg to the question of how he felt when he learned he had won the 1980 Nobel Prize in chemistry.

But the first word that he and two others had won the $212,000 prize did not come from the Nobel Foundation. It was from a colleague whose son had heard the news on the radio early Tuesday morning.

At first when the telephone rang Berg thought it was bad news, he told a packed press conference at Stanford University School of Medicine. "A telephone call at 5:55 ... normally brings bad news, not ...

after and camped on our doorstep waiting for me to come out."

The 54-year old Willson Professor of Biochemistry was given one-half of the prize "for his fundamental studies of the biochemistry of nucleic acids, with particular regard to recombinant DNA," the Nobel foundation said Tuesday.

The other half of the prize, the Foundation said, will be jointly shared by Walter Gilbert, professor at Harvard University, and Frederick Sanger, director of research at MRC Laboratory of Molecular Biology at the University of Cambridge, England.

Their contributions concerned the determination of base sequences ...ids, the Foundation

"It covers frankly most of the activities that go on in this department (biochemistry).

"Principally, the work that we have done is trying to understand how genes are organized in our chromosomes and how they function both in health and disease and normal development.,"

Berg said the breakthrough that came about five or six years ago was the ability to isolate specific genes or segments of DNA and to be able to study them in molecular detail.

"Our own work has been principally using isolated genes and learning how to reintroduce them into animal cells in order to learn something about them and their function," he said.

"The research we are doing," Berg said, "I hope will lead to a ...found and complete un... ...details

Paul Berg

Reprinted with permission of Stanford University.

...ded by ...rks, ad... Stanford ...of whom ...Kennedy ...980s left ...fellowship

(STO-2)STOCKHOLM.DEC.8(AP)-JOINT WINNERS-The three men who share the 1980 Nobel Prize for Chemistry attending mondays press conference at the Royal Academy of Sciences here.FROM LEFT: Paul Berg,USA.,Stanford University,Calif;Walter Gilbert,USA.,Harvard University,Mass; and Frederick Sanger,British,Cambridge University,England.Sanger previously won a Nobel Chemistry award in 1958. (WIREPHOTO)-344-(PRB21400stf/Tobbe Gustavsson-Swedish Code 777) 1980

Paul Berg Walter Gilbert Frederick Sanger

Photo from Wide World Photos.

Reading List

Textbooks

Ayala, F., and Kiger, J. A., Jr., 1980. *Modern Genetics.* Benjamin/Cummings, Menlo Park, Calif.

Broda, P., 1979. *Plasmids.* W. H. Freeman and Company, San Francisco.

Kornberg, A., 1980. *DNA Replication.* W. H. Freeman and Company, San Francisco.

Lehninger, A., 1975. *Biochemistry: The Molecular Basis of All Structure and Function,* 2nd ed. Worth Publishers, New York.

Stent, G. S., and Calendar, R., 1978. *Molecular Genetics: An Introductory Narrative,* 2nd ed. W. H. Freeman and Company, San Francisco.

Stryer, L., 1981. *Biochemistry,* 2nd ed. W. H. Freeman and Company, San Francisco.

Watson, J. D., 1976. *Molecular Biology of the Gene,* 3rd ed. Benjamin/Cummings, Menlo Park, Calif.

General Books

Cairns, J., Stent, G. S., and Watson, J. D., eds., 1966. *Phage and the Origins of Molecular Biology.* Cold Spring Harbor Laboratory, Cold Spring Harbor, New York.

Hanawalt, P. C., and Haynes, R. H. (Introductions by), 1973. *The Chemical Basis of Life: An Introduction to Molecular and Cell Biology,* Readings from *Scientific American.* W. H. Freeman and Company, San Francisco.

Judson, H. F., 1979. *The Eighth Day of Creation.* Simon & Schuster, New York.

Olby, R., 1974. *The Path to the Double Helix.* University of Washington Press, Seattle.

Portugal, F. H., and Cohen, J. S., 1977. *A Century of DNA: A History of the Discovery of the Structure and Function of the Genetic Substance.* MIT Press, Cambridge, Mass. (Paperback edition, 1980.)

Srb, A. M., Owen, R. D., and Edgar, R. S. (Introductions by), 1970. *Facets of Genetics,* Readings from *Scientific American.* W. H. Freeman and Company, San Francisco.

Watson, J. D., 1968. *The Double Helix.* Atheneum, New York. (Text and paperback editions.) New American Library, New York, 1969. (Paperback edition.)

Watson, J. D., 1980. *The Double Helix: A Norton Critical Edition,* edited by G. S. Stent. W. W. Norton, New York.

Popular Books on Recombinant DNA

Beers, R. F., Jr., and Bassett, E. G., eds., 1977. *Recombinant Molecules: Impact on Science and Society.* Raven Press, New York.

Goodfield, J. G., 1977. *Playing God: Genetic Engineering and the Manipulation of Life.* Random House, New York.

Grobstein, C., 1979. *A Double Image of the Double Helix: The Recombinant-DNA Debate.* W. H. Freeman and Company, San Francisco.

Hutton, R., 1978. *Bio-Revolution: DNA and the Ethics of Man-Made Life.* New American Library, New York.

Jackson, D., and Stich, S., 1979. *The Recombinant DNA Debate.* Prentice-Hall, Englewood Cliffs, N.J.

Lear, J., 1978. *Recombinant DNA: The Untold Story.* Crown Publishers, New York.

Rogers, M., 1979. *Biohazard.* Avon Publishers, New York.

Richards, J., ed., 1978. *Recombinant DNA: Science, Ethics, and Politics.* Academic Press, New York.

Wade, N., 1979. *The Ultimate Experiment,* 2nd ed. Walker & Company, New York.

Biotechnology and the Law: Recombinant DNA and the Control of Scientific Research, 1978. *Southern California Law Review* 51(6).

Milunsky, A., and Annas, G., eds., 1976. *Genetics and the Law.* National Symposium on Genetics and the Law, Boston, 1975. Plenum, New York.

Scientific Books on Recombinant DNA

Abelson, J., and Butz, E., eds., 1980. *Recombinant DNA. Science* 209 (entire issue). American Association for the Advancement of Science, Washington, D.C.

Davis, R. W., Botstein, D., and Roth, J. R., 1980. *A Manual for Genetic Engineering: Advanced Bacterial Genetics.* Cold Spring Harbor Laboratory, Cold Spring Harbor, New York.

Freifelder, D. (Introductions by), 1978. *Recombinant DNA,* Readings from *Scientific American.* W. H. Freeman and Company, San Francisco.

Maniatis, T., Fritsch, E. F., and Sambrook, J., 1981. *A Manual for Genetic Engineering: Molecular Cloning.* Cold Spring Harbor Laboratory, Cold Spring Harbor, New York.

Morgan, J., and Whelan, W. J., eds., 1979. *Recombinant DNA and Genetic Experimentation.* Pergamon Press, Elmsford, N.Y.

Scott, W. A., and Werner, R., eds., 1977. *Molecular Cloning of Recombinant DNA.* Academic Press, New York.

Wu, Ray, ed., 1979. *Recombinant DNA. Methods in Enzymology, Vol. 68.* Academic Press, New York.

DNA Is the Primary Genetic Material

Mendel, G. English translation of Mendel's experiments in plant hybridization, reprinted in:

Peters, J. A., ed., 1959. *Classic Papers in Genetics.* Prentice-Hall, Englewood Cliffs, N.J.

Stern, C., and Sherwood, E. R., eds., 1966. *The Origin of Genetics.* W. H. Freeman and Company, San Francisco.

Stevens, N. M., 1905. Studies in spermatogenesis with especial reference to the "accessory" [sex] chromosome. *Carn. Inst. Wash.*, publ. 36, pp. 1–32, pls i–vii.

Garrod, A. E., 1908. Inborn errors of metabolism. *Lancet* 2:1–7, 73–79, 142–148, 214–220. (Also a book by Oxford University Press, London, 1909.)

Morgan, T. H., 1910. Sex-limited inheritance in *Drosophila. Science* 32:120–122.

Muller, H. J., 1927. Artificial transmutation of the gene. *Science* 46:84–87.

Stadler, L. J., 1928. Mutations in barley induced by X-rays and radium.

Beadle, G. W., and Tatum, E. L., 1941. Genetic control of biochemical reactions in *Neurospora. Proc. Nat. Acad. Sci.* 27:499–506.

Pauling, L., Itano, H. A., Singer, S. J., and Wells, I. C., 1949. Sickle cell anemia: a molecular disease. *Science* 110:543–548.

Sanger, F., and Tuppy, H., 1951. The amino acid sequence in the phenylalanyl chain of insulin. *Biochem. J.* 49:463–490.

Sanger, F., and Thompson, E. O. P., 1953. The amino acid sequence in the glycyl chain of insulin. *Biochem. J.* 53:353–374.

Pauling, L., and Delbrück, M., 1940. The nature of the intermolecular forces operative in biological processes. *Science* 92:77–79.

Avery, O. T., MacLeod, C. M., and McCarty, M., 1944. Studies on the chemical nature of the substance inducing transformation of *Pneumococcal* types. *J. Exp. Med.* 79:137–158.

Chargaff, E., 1951. Structure and function of nucleic acids as cell constituents. *Fed. Proc.* 10:654–659.

Hershey, A. D., and Chase, M., 1952. Independent functions of viral protein and nucleic acid in growth of bacteriophage. *J. Gen. Physiol.* 36:39–56.

Watson, J. D., and Crick, F. H. C., 1953. Molecular structure of nucleic acid. A structure for deoxyribose nucleic acid. *Nature* 171:737–738.

Watson, J. D., and Crick, F. H. C., 1953. The structure of DNA. *Cold Spring Harbor Symp. Quant. Biol.* 18:123–131.

Meselson, M., and Stahl, F. W., 1958. The replication of DNA in *Escherichia coli. Proc. Nat. Acad. Sci.* 44:671–682.

Development of Fine Structure Genetics

Luria, S. E., and Delbrück, M., 1943. Mutations of bacteria from virus sensitivity to virus resistance. *Genetics* 28:491–511.

Hershey, A. D., 1947. Spontaneous mutations in bacterial viruses. *Cold Spring Harbor Symp. Quant. Biol.* 11:67–77.

Lederberg, J., and Tatum, E. L., 1947. Novel genotypes in mixed cultures of biochemical mutants of bacteria. *Cold Spring Harbor Symp. Quant. Biol.* 11:113–114.

Benzer, S., 1955. Fine structure of a genetic region in bacteriophage. *Proc. Nat. Acad. Sci.* 41:344–354.

Yanofsky, C., Carlton, B. C., Guest, J. R., Helinski, D. R., and Henning, U., 1964. On the colinearity of gene structure and protein structure. *Proc. Nat. Acad. Sci.* 51:266–272.

In Vitro Systems for Protein Synthesis

Zamecnik, P. C., and Keller, E. B., 1954. Relationship between phosphate energy donors and incorporation of labeled amino acids into proteins. *J. Biol. Chem.* 209:337–354.

Hoagland, M. B., Keller, E. B., and Zamecnik, P. C., 1956. Enzymatic carboxyl activation of amino acids. *J. Biol. Chem.* 218:345–358.

Hoagland, M. B., Stephenson, M. L., Scott, J. F., Hecht, L. I., and Zamecnik, P. C., 1958. A soluble ribonucleic acid intermediate in protein synthesis. *J. Biol. Chem.* 231:241–257.

Nirenberg, M. W., and Mattaei, J. H., 1961. The dependence of cell-free protein synthesis in *E. coli* upon naturally occurring or synthetic polyribonucleotides. *Proc. Nat. Acad. Sci.* 47:1588–1602.

Discovery of mRNA and Its Role in Codon Establishment

Brenner, S., Jacob, F., and Meselson, M., 1961. An unstable intermediate carrying information from genes to ribosomes for protein synthesis. *Nature* 190:576–581.

Gros, F., Hiatt, H., Gilbert, W., Kurland, C. G., Risebrough, R. W., and Watson, J. D., 1961. Unstable ribonucleic acid revealed by pulse labelling of *Escherichia coli. Nature* 190:581–585.

Crick, F. H. C., Barnett, L., Brenner, S., and Watts-Tobin, R. J., 1961. General nature of the genetic code for proteins. *Nature* 192:1227–1232.

Nirenberg, M. W., 1963. The genetic code: II. *Scientific American* 208(3):80–94. Offprint 153.

Crick, F. H. C., 1966. The genetic code: III. *Scientific American* 215(4): 55–62. Offprint 1052.

Rich, A., and Kim, S. H., 1978. The three-dimensional structure of transfer RNA. *Scientific American* 238(1):52–62.

Establishing Systems to Study Gene Regulation

Jacob, F., and Monod, J., 1961. Genetic regulatory mechanisms in the synthesis of proteins. *J. Mol. Biol.* 3:318–356.

Gilbert, W., and Müller-Hill, B., 1966. Isolation of the lac repressor. *Proc. Nat. Acad. Sci.* 56:1891–1898.

Ptashne, M., 1967. Isolation of the λ phage repressor. *Proc. Nat. Acad. Sci.* 57:306–313.

Ptashne, M., and Hopkins, N., 1968. The operators controlled by the λ phage repressor. *Proc. Nat. Acad. Sci.* 60:1282–1287.

Maniatis, T., and Ptashne, M., 1976. A DNA operator-repressor system. *Scientific American* 234(1):64–76.

Methodologies for Creating Recombinant DNA Molecules

Restriction Enzymes

Linn, S., and Arber, W., 1968. Host specificity of DNA produced by *Escherichia coli,* X. *In vitro* restriction of phage fd replicative form. *Proc. Nat. Acad. Sci.* 59:1300–1306.

Smith, H. O., and Wilcox, K. W., 1970. A restriction enzyme from *Hemophilus influenzae.* I. Purification and general properties. *J. Mol. Biol.* 51:379–391.

Kelly, T. J., Jr., and Smith, H. O., 1970. A restriction enzyme from *Hemophilus influenzae.* II. Base sequence of the recognition site. *J. Mol. Biol.* 51:393–409.

Danna, K., and Nathans, D., 1971. Specific cleavage of simian virus 40 DNA by restriction endonuclease of *Hemophilus influenzae. Proc. Nat. Acad. Sci.* 68:2913–2917.

Roberts, R. J., 1981. Restriction and modification enzymes and their recognition sequences. *Nucleic Acids Research* 9:75–96.

Nucleic-Acid Sequencing

Sanger, F., Brownlee, G. G., and Barrel, B. G., 1965. A two-dimensional fractionation procedure for radioactive nucleotides. *J. Mol. Biol.* 13:373–398.

Holley, R. W., 1966. The nucleotide sequence of a nucleic acid. *Scientific American* 214(2):30–39. Offprint 1033.

Sanger, F., and Coulson, A. R., 1975. A rapid method for determining sequences in DNA by primed synthesis with DNA polymerase. *J. Mol. Biol.* 94:441–448.

Southern, E. M., 1975. Detection of specific sequences among DNA fragments separated by gel electrophoresis. *J. Mol. Biol.* 98:503–517.

Fiers, W., Contreras, R., Duerinck, F., Haegeman, G., Iserentant, D., Merregaert, J., Min Jou, W., Molemans, F., Raeymaekers, A., Berghe, V., Volckaert, G., and Ysebaert, M., 1976. Complete nucleotide sequence of bacteriophage MS2 RNA: primary and secondary structure of replicase gene. *Nature* 260:500–507. Correction in *Nature* 260:810.

Maxam, A. M., and Gilbert, W., 1977. A new method of sequencing DNA. *Proc. Nat. Acad. Sci.* 74:560–564.

Sanger, F., Nicklen, S., and Coulson, A. R., 1977. DNA sequencing with chain-terminating inhibitors. *Proc. Nat. Acad. Sci.* 74:5463–5467.

Sanger, F., Air, G. M., Barrel, B. G., Brown, N. L., Coulson, A. R., Fiddes, J. C., Hutchison III, C. A., Slocombe, P. M., and Smith, M., 1977. Nucleotide sequence of bacteriophage ΦX174 DNA. *Nature* 265:687–695.

Fiers, W., Contreras, F., Haegeman, G., Rogers, R., Vande Voorde, A., Van Heuverswyn, H., Van Herreweghe, J., Volckaert, G., and Ysebaert, M., 1978. Complete nucleotide sequence of SV40 DNA. *Nature* 273:113–120.

Reddy, V. B., Thimmappaya, B., Dhar, R., Subramanian, K. N., Zain, B. S., Pan, J., Ghosh, P. K., Celma, M. L., and Weissman, S. M., 1978. The genome of simian virus 40. *Science* 200:494–502.

Enzymology of DNA Synthesis

Kornberg, A., 1960. Biologic synthesis of deoxyribonucleic acid. 1959 Nobel Prize lecture reprinted in *Science* 131:1503–1508.

Weiss, B., and Richardson, C. C., 1967. Enzymatic breakage and joining of deoxyribonucleic acid, I. Repair of single-strand breaks in DNA by an enzyme system from *Escherichia coli* infected with T4 bacteriophage. *Proc. Nat. Acad. Sci.* 57:1021–1028.

Olivera, B. M., and Lehman, I. R., 1967. Linkage of polynucleotides through phosphodiester bonds by an enzyme from *Escherichia coli*. *Proc. Nat. Acad. Sci.* 57:1426–1433.

Richardson, C. C., 1969. Enzymes in DNA metabolism. In *Annu. Rev. Biochem.* 38:795–840, eds. E. E. Snell, P. D. Boyer, A. Meister, and R. L. Sinsheimer.

Kornberg, A., 1980. Aspects of DNA replication. *Cold Spring Harbor Symp. Quant. Biol.* 43:1–9.

The First Recombinant DNA Molecules

Mertz, J. E., and Davis, R. W., 1972. Cleavage of DNA by RI restriction endonuclease generates cohesive ends. *Proc. Nat. Acad. Sci.* 69:3370–3374.

Jackson, D., Symons, R., and Berg, P., 1972. Biochemical method for inserting new genetic information into DNA of simian virus 40: circular SV40 DNA molecules containing lambda phage genes and the galactose operon of *Escherichia coli*. *Proc. Nat. Acad. Sci.* 69:2904–2909.

Lobban, P., and Kaiser, A. D., 1973. Enzymatic end-to-end joining of DNA molecules. *J. Mol. Biol.* 78:453–471.

Cohen, S., Chang, A., Boyer, H., and Helling, R., 1973. Construction of biologically functional bacterial plasmids *in vitro*. *Proc. Nat. Acad. Sci.* 70:3240–3244.

The Isolation of Cloned Genes

Grunstein, M., and Hogness, D. S., 1975. Colony hybridization: a method for the isolation of cloned DNAs that contain a specific gene. *Proc. Nat. Acad. Sci.* 72:3961–3965.

Benton, W. D., and Davis, R. W., 1977. Screening λgt recombinant clones by hybridization to single plaques *in situ*. *Science* 196:180–182.

Maniatis, T., Kee, S. G., Efstratiadis, A., and Kafatos, F. C., 1976. Amplication and characterization of a β-globin gene synthesized *in vitro*. *Cell* 8:163–182.

Maniatis, T., Hardison, R. C., Lacy E., Lauer, J., O'Connel, C., Quon, D., Sim, G. K., and Efstratiadis, A., 1978. The isolation of structural genes from libraries of eucaryotic DNA. *Cell* 15:687–701.

Nakanishi, S., Inoue, A., Kita, T., Nakamura, M., Chang, A., Cohen, S., and Numa, S., 1979. Nucleotide sequence of cloned cDNA for bovine corticotropin-β-lipotropin precursor. *Nature* 278:423–427.

Fiddes, J. C., and Goodman, H. M., 1980. The cDNA for the β-subunit of human chorionic gonadotropin suggest evolution of a gene by readthrough into the 3′-untranslated region. *Nature* 286:685–687.

Hohn, B., and Murray, K., 1977. Packaging recombinant DNA molecules into bacteriophage particles *in vitro*. *Proc. Nat. Acad. Sci.* 74:3259–3263.

Collins, J., and Hohn, B., 1978. Cosmids: a type of plasmid gene cloning vector that is packageable *in vitro* in bacteriophage heads. *Proc. Nat. Acad. Sci.* 75:4242–4246.

Alt, F. W., Kellems, R. E., Bertino, J. R., and Schimke, R. T., 1978. Selective multiplication of dihydrofolate reductase genes in methotrexate-resistant variants of cultured murine cells. *J. Biol. Chem.* 253:1357–1370.

Kurtz, D. T., and Nicodemus, C. F., 1981. Cloning of α_{2u} globulin cDNA using a high efficiency technique for the cloning of trace messenger RNAs. *Gene* 13:145–152.

Linkers

Heyneker, H. L., Shine, J., Goodman, H. M., Boyer, H., Rosenberg, J., Dickerson, R. E., Narang, S. A., Itakura, K., Lin, S., and Riggs, A. D., 1976. Synthetic *lac* operator DNA is functional *in vivo*. *Nature* 263:748–752.

Scheller, R., Dickerson, R., Boyer, H., Riggs, A., and Itakura, K., 1977. Chemical synthesis of restriction enzyme recognition sites useful for cloning. *Science* 196:177–180.

Seeburg, P. H., Shine, J., Martial, J. A., Baxter, J. D., and Goodman, H. M., 1977. Nucleotide sequence and amplication in bacteria of structural gene for rat growth hormone. *Nature* 270:486–494.

Shine, J., Seeburg, P. H., Martial, J. A., Baxter, J. D., and Goodman, H. M., 1977. Construction and analysis of recombinant DNA of human chorionic somatomammotropin. *Nature* 270:494–499.

Expression of Eukaryotic Genes in Bacteria

Chang, A. C. Y., Nunberg, J. H., Kaufman, R. J., Erlich, H. A., Schimke, R. T., and Cohen, S. N., 1978. Phenotypic expression in *E. coli* of a DNA sequence coding for mouse dihydrofolate reductase. *Nature* 275:617–624.

Villa-Komaroff, L., Efstratiadis, A., Broome, S., Lomedico, P., Tizard, R., Naber, S. P., Chick, W. L., and Gilbert, W., 1978. A bacterial clone synthesizing proinsulin. *Proc. Nat. Acad. Sci.* 75:3727–3731.

Goeddel, D. V., Kleid, D. G., Bolivar, F., Heyneker, H., Yansura, D., Crea, R., Hirose, T., Kraszewski, A., Itakura, K., and Riggs, A., 1979. Expression in *Escherichia coli* of chemically synthesized genes for human insulin. *Proc. Nat. Acad. Sci.* 76:106–110.

Goeddel, D., Heyneker, H., Hozumi, T., Arentzen, R., Itakura, K., Yansura, D., Ross, M., Miozzari, G., Crea, R., and Seeburg, P., 1979. Direct expression in *Escherichia coli* of a DNA sequence coding for human growth hormone. *Nature* 281:544–548.

Guarente, L., Lauer, G., Roberts, T., and Ptashne, M., 1980. Improved methods for maximizing expression of a cloned gene: a bacterium that synthesizes rabbit β-globin. *Cell* 20:543–553.

Nagata, S., Taira, H., Hall, A., Johnsrud, L., Streuli, M., Escödi, J., Boll, W., Cantell, K., and Weissmann, C., 1980. Synthesis in *E. coli* of a polypeptide with human leukocyte interferon activity. *Nature* 284:316–320.

Derynck, R., Remaut, E., Saman, E., Stanssens, P., De Clercq, E., Content, J., and Fiers, W., 1980. Expression of human fibroblast interferon gene in *Escherichia coli*. *Nature* 287:193–197.

Goeddel, D., Yelverton, E., Ullrich, A., Heyneker, H., Miozzari, G., Holmes, W., Seeburg, P., Dull, T., May, L., Stebbing, N., Crea, R., Maeda, S., McCandliss, R., Sloma, A., Tabor, J., Gross, M., Familletti, P., and Pestka, S., 1980. Human leukocyte interferon produced by *E. coli* is biologically active. *Nature* 287:411–416.

Gilbert, W., and Villa-Komaroff, L., 1980. Useful proteins from recombinant bacteria. *Scientific American* 242(4):74–94.

Gene Synthesis

Itakura, K., Hirose, T., Crea, R., Riggs, A., Heyneker, H., Bolivar, F., and Boyer, H., 1977. Expression in *Escherichia coli* of a chemically synthesized gene for the hormone somatostatin. *Science* 198:1056–1063.

Gait, M. J., and Sheppard, R. C., 1977. Rapid synthesis of oligodeoxyribonucleotides: a new solid-phase method. *Nucleic Acids Research* 4:1135–1158.

Khorana, H. G., 1979. Total synthesis of a gene. *Science* 203:614–625.

Chan, S. J., Noyes, B., Agarwal, K., and Steiner, D., 1979. Construction and selection of recombinant plasmids containing full-length complementary DNAs corresponding to rat insulin I and II. *Proc. Nat. Acad. Sci.* 76:5036–5040.

The Unexpected Complexity of Eukaryotic Genes

Discovery of Splicing in Animal Viral Systems

Berget, S. M., Berk, A. J., Harrison, T., and Sharp, P. A., 1978. Spliced segments at the 5′ termini of adenovirus-2 late mRNA: a role for heterogeneous nuclear RNA in mammalian cells. *Cold Spring Harbor Symp. Quant. Biol.* 42:523–529.

Broker, T. R., Chow, L. T., Dunn, A. R., Gelinas, R. E., Hassel, J. A., Klessig, D. F., Lewis, J. B., Roberts, R. J., and Zain, B. S., 1978. Adenovirus-2 messengers—an example of baroque molecular architecture. *Cold Spring Harbor Symp. Quant. Biol.* 42:531–553.

Westphal, H., and Lai, S. P., 1978. Displacement loops in adenovirus DNA–RNA hybrids. *Cold Spring Harbor Symp. Quant. Biol.* 42:555–558.

Split Eukaryotic Genes

Breathnach, R., Mandel, J. L., and Chambon, P., 1977. Ovalbumin gene is split in chicken DNA. *Nature* 270:314–319.

Jeffreys, A. J., and Flavell, R. A., 1977. The rabbit β-globin gene contains a large insert in the coding sequence. *Cell* 12:1097–1108.

Tilghman, S. M., Tiermeier, D. C., Seidman, J. G., Peterlin, B. M., Sullivan, M., Maijel, J. V., and Leder, P., 1978. Intervening sequence of DNA identified in the structural portion of a mouse β-globin gene. *Proc. Nat. Acad. Sci.* 75:725–729.

Tilghman, S. M., Curtis, P. J., Tiemeier, D. C., Leder, P., and Weissmann, C., 1978. The intervening sequence of a mouse β-globin gene is transcribed within the 15S β-globin mRNA precursor. *Proc. Nat. Acad. Sci.* 75:1309–1313.

Konkel, D., Tilghman, S., and Leder, P., 1978. The sequence of the chromosomal mouse β-globin major gene: homologies in capping, splicing and poly (A) sites. *Cell* 15:1125–1132.

Gilbert, W., 1978. Why genes in pieces? *Nature* 271:501.

Crick, F. H. C., 1979. Split genes and RNA splicing in evolution of eukaryotic cells. *Science* 204:264–271.

Fritsch, E., Lawn, R., Maniatis, T., 1980. Molecular cloning and characterization of the human β-like globin gene cluster. *Cell* 19:959–972.

Lauer, J., Shen, C., and Maniatis, T., 1980. The chromosomal arrangement of human α-like globin genes: sequence homology and α-globin gene deletions. *Cell* 20:119–130.

Lomedico, P., Rosenthal, N., Efstratiadis, A., Gilbert, W., Kolodner, R., and Tizard, R., 1979. The structure and evolution of the two nonallelic rat preproinsulin genes. *Cell* 18:545–558.

Bell, G. I., Pictet, R. L., Rutter, W., Cordell, B., Tischer, E., and Goodman, H. M., 1980. Sequence of the human insulin gene. *Nature* 284:26–32.

Yamada, Y., Avvedimento, V. E., Mudryj, M., Ohkubo, H., Vogeli, G., Irani, M., Pastan, I., and de Crombrugghe, B., 1980. The collagen gene: evidence for its evolutionary assembly by amplification of a DNA segment containing an exon of 54 bp. *Cell* 22:887–892.

Wozney, J., Hanahan, D., Morimoto, R., Boedtker, H., and Doty, P., 1981. Fine structural analysis of the chicken pro-α2 collagen gene. *Proc. Nat. Acad. Sci.* 78:712–716.

Chambon, P., 1981. Split genes. *Scientific American* 244(5):60–71.

Mechanism for Gene Splicing

Murray, V., and Holliday, R., 1979. Mechanism for RNA splicing of gene transcripts. *FEBS Letters* 106:5–7.

Rogers, J., and Wall, R., 1980. A mechanism for RNA-splicing. *Proc. Nat. Acad. Sci.* 77:1877–1879.

Lerner, M. R., Boyle, J. A., Mount, S. M., Wolin, S. L., and Steitz, J. A., 1980. Are snRNPS involved in splicing? *Nature* 283:220–224.

Yang, V. W., Lerner, M. R., Steitz, J. A., and Flint, S. J., 1981. A small nuclear ribonucleoprotein is required for splicing of adenoviral early RNA sequences. *Proc. Nat. Acad. Sci.* 78:1371–1375.

Rearranging Germ-Line DNA Segments to Form Antibody Genes

Dreyer, W. J., and Bennet, J. D., 1965. The molecular basis of antibody formation: a paradox. *Proc. Nat. Acad. Sci.* 54:864–869.

Hozumi, N., and Tonegawa, S., 1976. Evidence for somatic rearrangement of immunoglobulin genes coding for variable and constant regions. *Proc. Nat. Acad. Sci.* 73:3628–3632.

Tonegawa, S., Brack, C., Hozumi, N., and Pirrotta, V., 1978. Organization of immunoglobulin genes. *Cold Spring Harbor Symp. Quant. Biol.* 42:921–931.

Sakano, H., Rogers, J. H., Hüppi, K., Brack, C., Traumecker, A., Maki, R., Wall, R., and Tonegawa, S., 1979. Domains and the hinge region of an immunoglobulin heavy chain are encoded in separate DNA segments. *Nature* 277:627–633.

Davis, M. M., Calame, K., Early, P. W., Livant, D. L., Joho, R., Weissman, I. L., and Hood, L., 1980. An immunoglobulin heavy-chain gene is formed by at least two recombinational events. *Nature* 283:733–739.

Kataoka, T., Kawakami, T., Takahashi, N., and Honjo, T., 1980. Rearrangement of immunoglobulin γ1-chain gene and mechanism for heavy-chain class switch. *Proc. Nat. Acad. Sci.* 77:919–923.

Early, P., Rogers, J., Davis, M., Calame, K., Bond, M., Wall, R., and Hood, L., 1980. Two mRNAs can be produced from a single immunoglobulin μ gene by alternative RNA processing pathways. *Cell* 20:313–319.

Movable Genetic Elements

McClintock, B., 1957. Controlling elements and the gene. *Cold Spring Harbor Symp. Quant. Biol.* 21:197–216.

Bukhari, A. I., Shapiro, J. A., and Adya, S. L., eds., 1977. *DNA Insertion Elements, Plasmids, and Episomes.* Cold Spring Harbor Laboratory, Cold Spring Harbor, New York.

Gill, R. E., Heffron, F., and Falkow, S., 1979. Identification of the protein encoded by the transposable element Tn*3* which is required for its transposition. *Nature* 282:797–801.

Calos, M. P., and Miller, J. H., 1980. Transposable elements. *Cell* 20:579–595.

Cohen, S. N., and Shapiro, J. S., 1980. Transposable genetic elements. *Scientific American* 242(2):40–49. Offprint 1460.

Hicks, J., Strathern, J. N., and Klar, A. J. S., 1979. Transposable mating type genes in *Saccharomyces cerevisia. Nature* 282:478–483.

Strathern, J. N., Spatola, E., McGill, C., and Hicks, J. B., 1980. Structure and organization of transposable mating type cassettes in *Saccharomyces* yeasts. *Proc. Nat. Acad. Sci.* 77:2839–2843.

Nasmyth, K., and Tatchell, K., 1980. The structure of transposable yeast mating type loci. *Cell* 19:753–764.

Nasmyth, K. A., Tatchell, K., Hall, B. D., Astell, C., and Smith, M., 1981. A position effect in the control of transcription at yeast mating type loci. *Nature* 289:244–250.

Movable Genetic Elements, 1981. *Cold Spring Harbor Symp. Quant. Biol.* 45 (entire issue). Cold Spring Harbor, New York.

Documents List

Chapter 1 The Road to Asilomar

Chapter 2 The Asilomar Conference

Chapter 13	**Further Developments in Britain**	**Page**

Author Index